T0213176

Basic Modern Algebra with Applications

Mahima Ranjan Adhikari · Avishek Adhikari

Basic Modern Algebra
with Applications

 Springer

Mahima Ranjan Adhikari
Institute for Mathematics, Bioinformatics,
    Information Technology and Computer
    Science (IMBIC)
Kolkata, West Bengal, India

Avishek Adhikari
Department of Pure Mathematics
University of Calcutta
Kolkata, West Bengal, India

ISBN 978-81-322-3498-2   ISBN 978-81-322-1599-8 (eBook)
DOI 10.1007/978-81-322-1599-8
Springer New Delhi Heidelberg New York Dordrecht London

Printed on acid-free paper

Springer is part of Springer Science+Business Media (www.springer.com)

Dedicated
to
NABA KUMAR ADHIKARI (1912–1996),
a great teacher,
on occasion of his birth centenary

Preface

This book is designed to serve as a basic text of modern algebra at the undergraduate level. Modern mathematics facilitates unification of different areas of mathematics. It is characterized by its emphasis on the systematic study of a number of abstract mathematical structures. Modern algebra provides a language for almost all disciplines in contemporary mathematics.

This book introduces the basic language of modern algebra through a study of groups, group actions, rings, fields, vector spaces, modules, algebraic numbers, etc. The term Modern Algebra (or Abstract Algebra) is used to distinguish this area from classical algebra. Classical algebra grew over thousands of years. On the other hand, modern algebra began as late as 1770. Modern algebra is used in many areas of mathematics and other sciences. For example, it is widely used in algebraic topology of which the main objective is to solve topological and geometrical problems by using algebraic objects. Category theory plays an important role in this respect. Grigory Perelman solved the Poincaré conjecture by using the tools of modern algebra and other modern mathematical concepts (beyond the scope of the discussion in this book). A study of algebraic numbers includes studies of various number rings which generalize the set of integers. It also offers a natural inspiration to reinterpret the results of classical algebra and number theory and provides a scope of greater unity and generality. Algebraic number theory provides important tools to solve many problems in mathematics. For example, Andrew Wiles proved Fermat's last theorem by using algebraic number theory along with other theories (not discussed in this book). Moreover, an attempt has been made to integrate the classical materials of elementary number theory with modern algebra with an eye to apply them in different disciplines, specially in cryptography. Some applications unify computer science with mainstream mathematics. It dispels, at least, partly a general feeling that much of abstract algebra is of little practical value. The main purpose of this book is to give an accessible presentation to its readers. The materials discussed here have appeared elsewhere. Our contribution is the selection of the materials and their presentation. The title of the book suggests the scope of the book, which is expanded over 12 chapters and three appendices.

Chapter 1 studies basic concepts of set theory and properties of integers, which are used throughout the book and in many other disciplines. Set theory occupies a very prominent place in modern science. There are two general approaches to set theory. The first one is called "naive set theory" initiated by Georg Cantor around 1870; the second one is called "axiomatic set theory" originated by E. Zermelo in 1908 and modified by A. Fraenkel and T. Skolem. This chapter develops a naive set theory, which is a non-formalized theory by using a natural language to describe sets and their basic properties. For a precise description of many notions of modern algebra and also for mathematical reasoning, the concepts of relations, Zorn's lemma, mappings (functions), cardinality of sets are very important. They form the basics of set theory and are discussed in this chapter. The set of integers plays an important role in the development of science, engineering, technology and human civilization. In this chapter, some basic concepts and properties of integers, such as Peano's axioms leading to the principle of mathematical induction, well-ordering principle, division algorithm, greatest common divisors, prime numbers, fundamental theorem of arithmetic, congruences on integers, etc., are also discussed. Further studies on number theory are given in Chap. 10.

Chapter 2 gives an introduction to the group theory. This concept is used in subsequent chapters. Groups serve as one of the fundamental building blocks for the subject called today modern algebra. The theory of groups began with the work of J.L. Lagrange (1736–1813) and E. Galois (1811–1832). At that time, mathematicians worked with groups of transformations. These were sets of mappings, that, under composition, possessed certain properties. Mathematicians, such as Felix Klein (1849–1925), adopted the idea of groups to unify different areas of geometry. In 1870, L. Kronecker (1823–1891) gave a set of postulates for a group. Earlier definitions of groups were generalized to the present concept of an abstract group in the first decade of the twentieth century, which was defined by a set of axioms. In this chapter, we make an introductory study of groups with geometrical applications along with a discussion of free abelian groups and structure theorem for finitely generated abelian groups. Moreover, semigroups, homology groups, cohomology groups, topological groups, Lie groups, Hopf groups, and fundamental groups are also studied here.

Chapter 3 discusses actions of semigroups, groups, topological groups, and Lie groups. Each element of a group determines a permutation on a set under a group action. For a topological group action on a topological space, this permutation is a homeomorphism and for a Lie group action on a differentiable manifold it is a diffeomorphism. Group actions are used in the proofs of the counting principle, Cayley's Theorem, Cauchy's Theorem and Sylow Theorems for finite groups. Counting principle is used to determine the structure of a group of prime power order. These groups arise in the Sylow Theorems and in the description of finite abelian groups. Orbit spaces obtained by topological group actions, discussed in this chapter, are very important in topology and geometry. For example, n-dimensional real and complex projective spaces are obtained as orbit spaces. Finally, semigroup actions applied to theoretical computer science, yield state machines which unify computer science with mainstream mathematics.

Rings also serve as a fundamental building blocks for modern algebra. Chapter 4 introduces the concept of rings, another fundamental concept in the study of modern algebra. A "group" is endowed with only one binary operation while a "ring" is endowed with two binary operations connected by some interrelations. Fields form a very important class of rings. The concept of rings arose through the attempts to prove Fermat's last theorem and was initiated by Richard Dedekind (1831–1916) around 1880. David Hilbert (1862–1943) coined the term "ring". Emmy Noether (1882–1935) developed the theory of rings under his guidance. A very particular but important type of rings is known as commutative rings that play an important role in algebraic number theory and algebraic geometry. Further, non-commutative rings are used in non-commutative geometry and quantum groups. In this chapter, Wedderburn theorem on finite division rings, and some special rings, such as rings of power series, rings of polynomials, rings of continuous functions, rings of endomorphisms of abelian groups and Boolean rings are also studied.

Chapter 5 continues the study of theory of rings, and introduces the concept of ideals which generalize many important properties of integers. Ideals and homomorphisms of rings are closely related. Like normal subgroups in the theory of groups, ideals play an analogous role in the study of rings. The real significance of ideals in a ring is that they enable us to construct other rings which are associated with the first in a natural way. Commutative rings and their ideals are closely related. Their relations develop ring theory and are applied in many areas of mathematics, such as number theory, algebraic geometry, topology and functional analysis. In this chapter, basic properties of ideals are discussed and explained with interesting examples. Ideals of rings of continuous functions and Chinese remainder theorem for rings with their applications are also studied. Finally, applications of ideals to algebraic geometry with Hilbert's Nullstellensatz theorem, and the Zariski topology are discussed and certain connections among algebra, geometry and topology are given.

Chapter 6 extends the concepts of divisibility, greatest common divisor, least common multiple, division algorithm and fundamental theorem of arithmetic for integers with the help of theory of ideals to the corresponding concepts for rings. The main aim of this chapter is to study the problem of factoring the elements of an integral domain as products of irreducible elements. This chapter also caters to the study of the polynomial rings over a certain class of important rings and proves the Eisenstein irreducibility criterion, Gauss Lemma and related topics. Our study culminates in proving the Gauss Theorem which provides an extensive class of uniquely factorizable domains.

Chapter 7 continues to develop the theory of rings and studies chain conditions for ideals of a ring. The motivation came from an interesting property of the ring of integers $\mathbf{Z}$: that its every ascending chain of ideals terminates. This interesting property of $\mathbf{Z}$ was first recognized by the German mathematician Emmy Noether (1882–1935). This property leads to the concept of Noetherian rings, named after Noether. On the other hand, Emil Artin (1898–1962) showed that there are some rings in which every descending chain of ideals terminates. Such rings are called Artinian rings in honor of Emil Artin. This chapter studies special classes of rings, such as Noetherian rings and Artinian rings and obtains deeper results on ideal theory. This chapter further introduces a Noetherian domain and an Artinian domain and establishes their interesting properties. Hilbert's Basis Theorem, which gives an extensive class of Noetherian rings, is also proved in this chapter. Its application to algebraic geometry is also discussed. This study culminates in rings with descending chain condition for ideals, which determines their ideal structure.

Chapter 8 introduces another algebraic system, called vector spaces (linear spaces) interlinking both internal and external operations. In this chapter, vector spaces and their closely related fundamental concepts, such as linear independence, basis, dimension, linear transformation and its matrix representation, eigenvalue, inner product space, etc., are presented. Such concepts form an integral part of linear algebra. Vector spaces have multi-faceted applications. Such spaces over finite fields play an important role in computer science, coding theory, design of experiments and combinatorics. Vector spaces over the infinite field $\mathbf{Q}$ of rationals are important in number theory and design of experiments while vector spaces over $\mathbf{C}$ are essential for the study of eigenvalues. As the concept of a vector provides a geometric motivation, vector spaces facilitate the study of many areas of mathematics and integrate the abstract algebraic concepts with the geometric ideas.

Chapter 9 initiates module theory, which is one of the most important topics in modern algebra. It is a generalization of an abelian group (which is a module over the ring of integers $\mathbf{Z}$) and also a natural generalization of a vector space (which is a module over a division ring or over a field). Many results of vector spaces are generalized in some special classes of modules, such as free modules and finitely generated modules over principal ideal domains. Modules are closely related to the representation theory of groups. One of the basic concepts which accelerates the study of commutative algebra is module theory, as modules play the central role in commutative algebra. Modules are also widely used in structure theory of finitely generated abelian groups, finite abelian groups and rings, homological algebra, and algebraic topology. In this chapter, we study the basic properties of modules. We also consider modules of special classes, such as free modules, modules over principal ideal domains along with structure theorems, exact sequences of modules and their homomorphisms, Noetherian and Artinian modules, homology and cohomology modules. Our study culminates in a discussion on the topology of the spectrum of modules and rings with special reference to the Zariski topology.

Chapter 10 discusses some more interesting properties of integers, in particular, properties of prime numbers and primality testing by using the tools of modern algebra, which are not studied in Chap. 1. In addition, we study the applications of number theory, particularly those directed towards theoretical computer science. Number theory has been used in many ways to devise algorithms for efficient computer and for computer operations with large integers. Both algebra and number theory play an increasingly significant role in computing and communication, as evidenced by the striking applications of these subjects to the fields of coding theory and cryptography. The motivation of this chapter is to provide an introduction to the algebraic aspects of number theory, mainly the study of development of the theory of prime numbers with an emphasis on algorithms and applications, necessary for studying cryptography to be discussed in Chap. 12. In this chapter, we start with the introduction to prime numbers with a brief history. We provide several different proofs of the celebrated Theorem of Euclid, stating that there exist infinitely many primes. We further discuss Fermat number, Mersenne numbers, Carmichael numbers, quadratic reciprocity, multiplicative functions, such as Euler ϕ-function, number of divisor functions, sum of divisor functions, etc. This chapter ends with a discussion of primality testing both deterministic and probabilistic, such as Solovay–Strassen and Miller–Rabin probabilistic primality tests.

Chapter 11 introduces algebraic number theory which developed through the attempts of mathematicians to prove Fermat's last theorem. An algebraic number is a complex number, which is algebraic over the field $\mathbf{Q}$ of rational numbers. An algebraic number field is a subfield of the field $\mathbf{C}$ of complex numbers, which is a finite field extension of the field $\mathbf{Q}$, and is obtained from $\mathbf{Q}$ by adjoining a finite number of algebraic elements. The concepts of algebraic numbers, algebraic integers, Gaussian integers, algebraic number fields and quadratic fields are introduced in this chapter after a short discussion on general properties of field extension and finite fields. There are several proofs of fundamental theorem of algebra. It is proved in this chapter by using homotopy (discussed in Chap. 2). Moreover, countability of algebraic numbers, existence of transcendental numbers, impossibility of duplication of a general cube and that of trisection of a general angle are shown in this chapter.

Chapter 12 presents applications and initiates a study of cryptography. In the modern busy digital world, the word "cryptography" is well known to many of us. Everyday, knowingly or unknowingly, in many places we use different techniques of cryptography. Starting from the logging on a PC, sending e-mails, withdrawing money from ATM by using a PIN code, operating the locker at a bank with the help of a designated person from the bank, sending message by using a mobile phone, buying things through the internet by using a credit card, transferring money digitally from one account to another over internet, we are applying cryptography everywhere. If we observe carefully, we see that in every case we are required to hide some information to transfer information secretly. So, intuitively, we can guess that cryptography has something to do with security. So, intuitively, we can guess that cryptography and secrecy have a close connection. Naturally, the questions that come to our mind are: What is cryptography? How is it that it is important to our daily life? In this chapter, we introduce cryptography and provide a brief overview of the subject and discuss the basic goals of cryptography and present the subject, both intuitively and mathematically. More precisely, various cryptographic notions starting from the historical ciphers to modern cryptographic notions like public key encryption schemes, signature schemes, secret sharing schemes, oblivious transfer, etc., by using mathematical tools mainly based on modern algebra are explained. Finally, the implementation issues of three public key cryptographic schemes, namely RSA, ElGamal, and Rabin by using the open source software SAGE are discussed.

Appendix A studies some interesting properties of semirings. Semiring theory is a common generalization of the theory of associative rings and theory of distributive lattices. Because of wide applications of results of semiring theory to different branches of computer science, it has now become necessary to study structural results on semirings. This chapter gives some such structural results.

Appendix B discusses category theory initiated by S. Eilenberg and S. Mac Lane during 1942–1945, to provide a technique for unifying certain concepts. In general, the pedagogical methods compartmentalize mathematics into its different branches without emphasizing their interconnections. But the category theory provides a convenient language to tie together several notions and existing results of different areas of mathematics. This language is conveyed through the concepts of categories, functors, natural transformations, which form the basics of category theory and provide tools to shift problems of one branch of mathematics to another branch to have a better chance for solution. For example, the Brouwer fixed-point theorem is proved here with the help of algebraic objects. Moreover, extension problems and classification of topological spaces are discussed in this chapter.

Appendix C gives a brief historical note highlighting contributions of some mathematicians to modern algebra and its closely related topics. The list includes a few names only, such as the names of Leonhard Euler (1707–1783), Joseph Louis Lagrange (1736–1813), Carl Friedrich Gauss (1777–1855), Augustin Louis Cauchy (1789–1857), Niels Henrik Abel (1802–1829), C.G.J. Jacobi (1804–1851), H. Grassmann (1809–1877), Evariste Galois (1811–1832), Arthur Cayley (1821–1895), Leopold Kronecker (1823–1891), Bernhard Riemann (1826–1866), Richard Dedekind (1831–1926), Peter Ludvig Mejdell Sylow (1832–1918), Camille Jordan (1838–1922), Sophus Lie (1842–1899), Georg Cantor (1845–1918), C. Felix Klein (1849–1925), Henri Poincaré (1854–1912), David Hilbert (1862–1943), Amalie Emmy Noether (1882–1935), MecLagan Wedderburn (1882–1948), Emil Artin (1898–1962) and Oscar Zariski (1899–1986).

The authors express their sincere thanks to Springer for publishing this book. The authors are very thankful to many individuals, our postgraduate and Ph.D. students, who have helped in proof reading the book. Our thanks are due to the institute IMBIC, Kolkata, for kind support towards the manuscript development work of this book. Finally, we acknowledge, with heartfelt thanks, the patience and sacrifice of our long-suffering family, specially Minati, Sahibopriya, and little Avipriyo.

Kolkata, India   Mahima Ranjan Adhikari
                                                                              Avishek Adhikari

Contents

Chapter 1
Prerequisites: Basics of Set Theory and Integers

Set theory occupies a very prominent place in modern science. There are two general approaches to set theory. The first is called "naive set theory" and the second is called "axiomatic set theory", also known as "Zermelo–Fraenkel set theory". These two approaches differ in a number of ways. The "naive set theory" was initiated by the German mathematician Georg Cantor (1845–1918) around 1870. According to Cantor, "a set is any collection of definite, distinguishable objects of our intuition or of our intellect to be conceived as a whole". Each object of a set is called an element or member of the set. The phrase "objects of our intuition or of our intellect" offers freedom to choose the objects in forming a set and thus a set is completely determined by its objects. He developed the mathematical theory of sets to study the real numbers and Fourier series. Richard Dedekind (1831–1916) also enriched set theory during 1870s. However, Cantor's set theory was not immediately accepted by his contemporary mathematicians. His definition of a set leads to contradictions and logical paradoxes. The most well known of them is the Russell paradox given in 1918 by Bertrand Russell (1872–1970). Such paradoxes led to axiomatize Cantor's intuitive set theory giving the birth of Zermelo–Fraenkel Set Theory. Some authors prefer to call it Zermelo–Fraenkel–Skolem Set Theory. The reason is that it is the theory of E. Zermelo of 1908 modified by both A. Fraenkel and T. Skolem. In 1910 Hilbert wrote "set theory is that mathematical discipline which today occupies an outstanding role in our science and radiates its powerful influence into all branches of mathematics". In this chapter, we develop a naive set theory, which is a non-formalized theory using a natural language to describe sets and their basic properties. For precise description of many notions of modern algebra and also for mathematical reasoning, the concepts of relations, Zorn's Lemma, mappings (functions), cardinality of sets, are very important. They form the basics of set theory and are discussed in this chapter. Many concrete examples are based on them. The set of integers plays an important role in the development of science, technology, and human civilization. Number theory, in a general sense, is the study of set of integers. It has long been one of the favorite subjects not only for the students but also for many others. In this chapter, some basic concepts and properties of integers, such as Peano's axioms, well-ordering principle, division algorithm, greatest common divi-

M.R. Adhikari, A. Adhikari, *Basic Modern Algebra with Applications*,
DOI 10.1007/978-81-322-1599-8_1, © Springer India 2014

sors, prime numbers, fundamental theorem of arithmetic and modular arithmetic are studied. More study on number theory is in Chap. 10.

1.1 Sets: Introductory Concepts

The concept of 'set' is very important in all branches of mathematics. We come across certain terms or concepts whose meanings need no explanation. Such terms are called undefined terms and are considered as primitive concepts. If one defines the term 'set' as 'a set is a well defined collection of objects', then the meaning of collection is not clear. One may define 'a collection' as 'an aggregate' of objects. What is the meaning of 'aggregate'? As our language is finite, other synonyms, such as 'class', 'family' etc., will exhaust. Mathematicians accept that there are undefined terms and 'set' shall be such an undefined term. But we accept the familiar expressions, such as 'set of all integers', 'set of all natural numbers', 'set of all rational numbers', 'set of all real numbers' etc.

We shall neither attempt to give a formal definition of a set nor try to lay the groundwork for an axiomatic theory of sets. Instead we shall take the operational and intuitive approach to define a set. A *set* is a well defined collection of distinguishable objects.

The term *'well defined'* specifies that it can be determined whether or not certain objects belong to the set in question. In most of our applications we deal with rather specific objects, and the nebulous notion of a set, in these, emerge as something quite recognizable. We usually denote sets by capital letters, such as $A, B, C, \ldots$. The objects of a set are called the *elements* or *members* of the set and are usually denoted by small letters, such as $a, b, c, \ldots$. Given a set A we use the notation throughout '$a \in A$' to indicate that an element a is a member of A and this is read as 'a is an element of A' or 'a belongs to A'; and '$a \notin A$' to indicate that the element a is not a member of A and this is read as 'a is not an element of A' or 'a does not belong to A'. Since a set is uniquely determined by its elements, we may describe a set either by a characterizing property of the elements or by listing the elements. The standard way to describe a set by listing elements is to list elements of the set separated by commas, in braces. Thus a set $A = \{a, b, c\}$ indicates that a, b, c are the only elements of A and nothing else. If B is a set which consists of a, b, c and possibly more, then notationally, $B = \{a, b, c, \ldots\}$. On the other hand, a set consisting of a single element x is sometimes called singleton x, denoted by $\{x\}$. By a *statement*, we mean a sentence about specific objects such that it has a truth value of either true or false but not both. If a set A is described by a characterizing property $P(x)$ of its elements x, the brace notation $\{x : P(x)\}$ or $\{x \mid P(x)\}$ is also often used, and is read as 'the set of all x such that the statement $P(x)$ about x is true.' For example, $A = \{x : x$ is an even positive integer $< 10\}$.

Throughout this book, we use the following notations:

$\mathbf{N}^+$ (or $\mathbf{Z}^+$) is the set of all positive integers (zero is excluded);
$\mathbf{N}$ is the set of all non-negative integers (zero is included);

Z is the set of all integers (positive, negative, and zero);
Q is the set of all rational numbers (i.e., numbers which can be expressed as quotients m/n of integers, where $n \neq 0$);
Q$^+$ is the set of all positive rational numbers;
R is the set of all real numbers;
R$^+$ is the set of all positive real numbers;
C is the set of all complex numbers;
$\exists$ denotes 'there exists', $\forall$ denotes 'for all'.

An *implication* is a statement of the form 'P implies Q' or 'if P, then Q' written as $P \Rightarrow Q$. An implication is false if P is true and Q is false; it is true in all other cases. A logical *equivalence* or *biconditional* is a statement of the form 'P implies Q and Q implies P' and abbreviated to *if and only if* (iff) and written as $P \Leftrightarrow Q$. Thus $P \Leftrightarrow Q$ is exactly true when both P and Q are either true or false.

Two elements a and b of a set S are said to be equal, denoted by $a = b$ iff they are the same elements, otherwise we denote $a \neq b$. Two sets X and Y are said to be equal, denoted by $X = Y$ iff they have the same elements. For example, $\{2, 4, 6, 8\} = \{2x : x = 1, 2, 3, 4\}$.

We introduce a special set which we call the '*empty* (or *vacuous* or *null*) set'; denoted by $\emptyset$, which we think of as 'the set having no elements'. The empty set is only a convention. It is important in the context of logical completeness. As $\emptyset$ is thought of as 'the set with no elements' the convention is that for any element x, the relation $x \in \emptyset$ does not hold. Thus $\{x : x$ is an even integer such that $x^2 = 2\} = \emptyset$. If a set A is such that $A \neq \emptyset$, then A is called a *non-empty* (or *non-vacuous* or *non-null*) set and its null subset $\emptyset_A$ is $\{x \in A : x \neq x\}$. If X and Y are two sets such that every element of X is also an element of Y, then X is called *subset* of Y, denoted by $X \subseteq Y$ (or simply by $X \subset Y$), which reads X is contained in Y (or equivalently, $Y \supseteq X$ or simply $Y \supset X$: which reads Y contains X). For example, $\mathbf{N} \subseteq \mathbf{Z}$ and $\mathbf{Z} \subseteq \mathbf{Q}$. Clearly, $X \subseteq X$ for every set X.

Proposition 1.1.1   (i) $X = Y$ *if and only if* $X \subseteq Y$ *and* $Y \subseteq X$;
(ii) *All null subsets are equal.*

Proof (i) It follows from the definition of equality of sets.

(ii) Let A and B be any two sets and $\emptyset_A$, $\emptyset_B$ be the respective null subsets. If $\emptyset_A \subseteq \emptyset_B$ is false, then there exists at least one element in $\emptyset_A$ which is not in $\emptyset_B$; but this is impossible. Consequently, $\emptyset_A \subseteq \emptyset_B$. Similarly, $\emptyset_B \subseteq \emptyset_A$. Hence by (i) $\emptyset_A = \emptyset_B$ and thus all null subsets are equal. $\square$

Remark There is one and only one null set $\emptyset$, and is contained in every set.

If a set X is a subset of a set Y and there exists at least one element of Y which is not an element of X, then X is called a *proper subset* of Y, denoted by $X \subsetneq Y$. If a set X is not a subset of Y, we write $X \nsubseteq Y$.

Proposition 1.1.2 *A set X of n elements has 2^n subsets.*

Proof The number of the subsets having r ($\leq n$) elements out of n elements of X is the number of combinations of n elements taken r at a time, i.e.,

$$^nC_r = n!/r!(n-r)!.$$

Hence the number of all subsets including X and $\emptyset$ is

$$\sum_{r=0}^{n} {}^nC_r = {}^nC_0 + {}^nC_1 + \cdots + {}^nC_n = (1+1)^n = 2^n.$$

$\square$

To avoid extraneos elements, we assume that all elements under consideration belong to some fixed but arbitrary set, called the *universal set*, denoted by **U**.

Given two sets, we can combine them to form new sets. We can carry out the same procedure for any number of sets, finite or infinite. We do so for two sets first, because it leads to the general construction.

Given two sets A and B, we can form a set that consists of exactly all the elements of A together with all the elements of B. This is called the *union* of A and B.

Definition 1.1.1 The *union* (or join) of two sets A and B, written as $A \cup B$, is the set $A \cup B = \{x : x \in A \text{ or } x \in B\}$ ('or' has been used in the inclusive sense).

Remark In set theory, we say that $x \in A$ or $x \in B$. This means x belongs to at least one of A or B and may belong to both A and B. Clearly, $A \cup A = A$; if B is a subset of A, then $A \cup B = A$; if $A = \{1, 2, 3\}$, $B = \{3, 7, 8, 9\}$, then $A \cup B = \{1, 2, 3, 7, 8, 9\}$; $A \cup \emptyset = A$ for every set A.

Given two sets A and B, we can form a set that consists of exactly all elements common to A and B. This is called the *intersection* of A and B.

Definition 1.1.2 The *meet* (or *intersection*) of two sets A and B, written as $A \cap B$, is the set $A \cap B = \{x : x \in A \text{ and } x \in B\}$.

Clearly, $A \cap A = A$; if B is a subset of A, then $A \cap B = B$; $A \cap \emptyset = \emptyset$; if $A = \{1, 2, 3\}$, $B = \{3, 7, 8, 9\}$, then $A \cap B = \{3\}$.

Theorem 1.1.1 *Each of the operations $\cup$ and $\cap$ is*

(a) *idempotent*: $A \cup A = A = A \cap A$, *for every set A;*
(b) *associative*: $A \cup (B \cup C) = (A \cup B) \cup C$ *and* $A \cap (B \cap C) = (A \cap B) \cap C$ *for any three sets A, B, C;*
(c) *commutative*: $A \cup B = B \cup A$ *and* $A \cap B = B \cap A$ *for any two sets A, B;*
(d) $\cap$ *distributes over* $\cup$ *and* $\cup$ *distributes over* $\cap$:

    (i) $A \cap (B \cup C) = (A \cap B) \cup (A \cap C)$;
    (ii) $A \cup (B \cap C) = (A \cup B) \cap (A \cup C)$ *for any three sets A, B, C.*

Proof Verification of (a)–(c) is trivial. We now give a set theoretic proof of (d)(i). The proof is divided into two parts:

$$A \cap (B \cup C) \subseteq (A \cap B) \cup (A \cap C)$$

and

$$(A \cap B) \cup (A \cap C) \subseteq A \cap (B \cup C).$$

Now, $x \in A \cap (B \cup C) \Rightarrow x \in A$ and $x \in B \cup C \Rightarrow (x \in A$ and $x \in B)$ or $(x \in A$ and $x \in C) \Rightarrow x \in A \cap B$ or $x \in A \cap C \Rightarrow x \in (A \cap B) \cup (A \cap C) \Rightarrow A \cap (B \cup C) \subseteq (A \cap B) \cup (A \cap C)$, as x is an arbitrary element of $A \cap (B \cup C)$. We now prove the reverse inclusion.

Since $B \subseteq B \cup C$, $A \cap B \subseteq A \cap (B \cup C)$.

Similarly, $A \cap C \subseteq A \cap (B \cup C)$.

Consequently, $(A \cap B) \cup (A \cap C) \subseteq A \cap (B \cup C)$.

Hence d(i) follows by Proposition 1.1.1(i).   □

Definition 1.1.3 Two non-null sets A and B are said to be *disjoint* iff $A \cap B = \emptyset$.

For example, if A is the set of all negative integers, then $A \cap \mathbf{N}^+ = \emptyset$.

Definition 1.1.4 The (relative) *complement* (or *difference*) of a set A with respect to a set B, denoted by $B - A$ (or $B \backslash A$) is the set of exactly all elements which belong to B but not to A, i.e., $B - A = \{x \in B : x \notin A\}$.

If $\mathbf{U}$ is the universal set, the complement of A in $\mathbf{U}$ (i.e., with respect to $\mathbf{U}$) is denoted by A^c or A'. This implies $(A')' = A$, $\mathbf{U}' = \emptyset$, $\emptyset' = \mathbf{U}$, $A \cup A' = \mathbf{U}$, $A \cap A' = \emptyset$, $B - A = B \cap A'$, $B - A \neq A - B$ for $A \neq B$.

Definition 1.1.5 The *symmetric difference* of two given sets A and B, denoted by $A \triangle B$, is defined by $A \triangle B = (A - B) \cup (B - A)$.

This implies clearly, $A \triangle B = B \triangle A$, $A \triangle (B \triangle C) = (A \triangle B) \triangle C$, $A \triangle \emptyset = A$, $A \triangle A = \emptyset$, $A \triangle B = (A \cup B) - (A \cap B)$ and $A \triangle C = B \triangle C \Rightarrow A = B$.

It is extremely useful to imagine geometric figures in terms of which we can visualize sets and operations on them. A convenient way to do so is to represent the universal set $\mathbf{U}$ by a rectangular area in a plane and the elements of $\mathbf{U}$ by the points of the area. Sets can be pictured by circles within this rectangle and diagrams can be drawn illustrating operations on sets and relations between them. Such diagrams are known as *Venn diagrams*, named after *John Venn* (1834–1923), a British logician.

Our three set operations are represented by shaded regions in Fig. 1.1.

The cartesian product is one of the most important constructions of set theory. It enables us to express many concepts in terms of sets. The concept of cartesian product owes to the coordinate plane of the analytical geometry. It welds together the sets of a given family into a single new set. With two objects a, b there corresponds

Fig. 1.1 Venn diagrams representing three set operations: union, intersection, and set difference

$$A \cup B \qquad A \cap B \qquad A - B$$

a new object (a, b), called their *ordered pair*. Ordered pairs are subject to one condition $(a, b) = (c, d)$ iff $a = c$ and $b = d$. In particular, $(a, b) = (b, a)$ iff $a = b$; a is called the *first coordinate* and b is called the *second coordinate* of the pair (a, b). The ancestor of this concept is the coordinate plane of analytic geometry.

Definition 1.1.6 Let A and B be two non-null sets (distinct or not). Their *cartesian product* $A \times B$ is the set defined by $A \times B = \{(a, b) : a \in A, b \in B\}$.

Example 1.1.1 (i) If $A = \{1, 2\}$, $B = \{2, 3, 4\}$, then $A \times B = \{(1, 2), (1, 3), (1, 4), (2, 2), (2, 3), (2, 4)\}$.

   (ii) If $A = B = \mathbf{R}^1$ (Euclidean line), then $A \times B$, being the set of all ordered pairs of reals, represents the points in the Euclidean plane.

Proposition 1.1.3 $A \times B \neq \emptyset$ *if and only if* $A \neq \emptyset$ *and* $B \neq \emptyset$.

Proof If $A \times B \neq \emptyset$, then there exists some $(a, b) \in A \times B$ such that $a \in A, b \in B$. This shows that $A \neq \emptyset$. $B \neq \emptyset$. Conversely, if $A \neq \emptyset$ and $B \neq \emptyset$, there exist some $a \in A$ and $b \in B$. As the pair $(a, b) \in A \times B$, $A \times B \neq \emptyset$. □

Proposition 1.1.4 *If* $C \times D \neq \emptyset$, *then* $C \times D \subseteq A \times B$ *iff* $C \subseteq A$ *and* $D \subseteq B$.

Proof Left as an exercise. □

Proposition 1.1.5 *For non-empty sets* A *and* B, $A \times B = B \times A$ *if and only if* $A = B$ (*the operation* $A \times B$ *is therefore not commutative*).

Proof Left as an exercise. □

   If a set A has n elements, then the set $A \times A$ has n^2 elements.
   The set of elements $\{(a, a) : a \in A\}$ in $A \times A$ is called the *diagonal* of $A \times A$.

Theorem 1.1.2 '$\times$' *distributes over* $\cup, \cap$ *and* '$-$':

   (i) $A \times (B \cup C) = A \times B \cup A \times C$;
   (ii) $A \times (B \cap C) = A \times B \cap A \times C$;
   (iii) $A \times (B - C) = A \times B - A \times C$.

Proof Left as an exercise. □

Remark Definition 1.1.6 is extended to the product of n sets for any positive integer $n > 2$. If $A_1, A_2, \ldots, A_n$ are non-empty sets, then their product $A_1 \times A_2 \times \cdots \times A_n$

is the set of all ordered n-tuples $(a_1, a_2, \ldots, a_n)$, where a_i is in A_i for each i. If in particular, $A_1 = A_2 = \cdots = A_n = A$, then their product is denoted by the symbol A^n. These ideas yield well known sets $\mathbf{R}^n$ and $\mathbf{C}^n$.

The operation '$\times$' is not associative: $(A \times B) \times C \neq A \times (B \times C)$ in general [see Ex. 11 of SE-I].

Family of Sets   Let $I \neq \emptyset$ be a set. If for each $i \in I$, there exists a set A_i, then the collection of sets $\{A_i : i \in I\}$ is called a *family of sets*, and I is called an *indexing set* for the family. For some $i \neq j$, A_j may be equal to A_i.

Any non-empty collection Ω of sets can be converted to a family of sets by 'self indexing'. We use Ω itself as an indexing set and assign to each member of Ω the set it represents.

We extend the notions of unions and intersections to family of sets.

Definition 1.1.7   Let A be a given set, and $\{A_i : i \in I$ be a family of subsets of $A\}$.

The union $\bigcup_i A_i$ (or $\bigcup_{i \in I} A_i$ or $\bigcup_{\langle i \in I \rangle} A_i$ or $\bigcup \{A_i : i \in I\}$) of the family is the set $\bigcup_i A_i = \{x \in A : x \in A_i$ for some $i \in I\}$ and the intersection $\bigcap_i A_i$ (or $\bigcap_{i \in I} A_i$ or $\bigcap_{\langle i \in I \rangle} A_i$ or $\bigcap \{A_i : i \in I\}$) of the family is the set $\bigcap_i A_i = \{x \in A : x \in A_i$ for each $i \in I\}$.

Clearly, for any set A, $A = \bigcup\{\{x\} : x \in A\} = \bigcup_{x \in A}\{x\}$. In case of infinite intersections and unions, some non-intuitive situations may occur:

Example 1.1.2   (i) Let $A_p = \{n \in \mathbf{Z} : n \geq p, p = 1, 2, 3, \ldots\}$. Then $A_1 \supset A_2 \supset A_3 \supset \cdots$ and each A_p is non-empty set such that each A_p contains its successor A_{p+1}. Clearly, $\bigcap A_p = \emptyset$.

(ii) Let $A_n = [0, 1 - 2^{-n}]$ and $B_n = [0, 1 - 3^{-n}]$ for each positive integer n. Then each A_n is a proper subset of B_n and $A_n \neq B_r$ for each n and r. Clearly,

$$\bigcup_n A_n = \bigcup_n B_n = [0, 1).$$

Definition 1.1.8   Let A be a non-null set. Its power set $\mathcal{P}(A)$ is the set of all subsets of A.

Clearly, $\emptyset \in \mathcal{P}(A)$ and $A \in \mathcal{P}(A)$ for any set A; $B \subseteq A \Leftrightarrow B \in \mathcal{P}(A)$; $a \in A \Leftrightarrow \{a\} \in \mathcal{P}(A)$.

Proposition 1.1.6 $\bigcap_i \mathcal{P}(A_i) = \mathcal{P}(\bigcap_i A_i)$.

Proof   Left as an exercise.   □

Remark $\bigcup_i \mathcal{P}(A_i) \neq \mathcal{P}(\bigcup_i A_i)$. For example, let $A_1 = \{1\}$, $A_2 = \{2\}$. Then $\bigcup_i \mathcal{P}(A_i)$ has three elements, viz. $\emptyset, \{1\}$ and $\{2\}$, whereas $\mathcal{P}(\bigcup_i A_i)$ has four elements viz. $\emptyset, \{1\}, \{2\}$ and $\{1, 2\}$.

1.2 Relations on Sets

In mathematics, two types of very important relations, such as equivalence relation and ordered relations arise frequently. Sometimes, we need study decompositions of a non-empty set X into disjoint subsets whose union is the entire set X (i.e., X is filled up by these subsets). Equivalence relations on X provide tools to generate such decompositions of X and produce new sets bearing a natural connection with the original set X.

A *binary relation* R on a non-empty set A is a mathematical concept and intuitively is a proposition such that for each ordered pair (a, b) of elements of A, we can determine whether aRb (read as a is in relation R to b) is true or false. We define it formally in terms of the set concept.

Definition 1.2.1 A binary relation R on a non-empty set A is a subset $R \subseteq A \times A$ and a binary relation S from A to B is a subset S of $A \times B$. The pair $(a, b) \in R$ is also denoted as aRb.

A binary relation S from A to B is sometimes written as $S : A \to B$. Instead of writing a binary relation on A, we write only a relation on A, unless there is any confusion.

Example 1.2.1 For any set A, the *diagonal* $\Delta = \{(a, a) : a \in A\} \subseteq A \times A$ is the relation of equality.

In a binary relation R on A, each pair of elements of A need not be related i.e., (a, b) may not belong to R for all pairs $(a, b) \in A \times A$.

For example, if $a \neq b$, then $(a, b) \notin \Delta$ and also $(b, a) \notin \Delta$.

Example 1.2.2 The relation of inclusion on $\mathcal{P}(A)$ is $\{(A, B) \in \mathcal{P}(A) \times \mathcal{P}(A) : A \subseteq B\} \subseteq \mathcal{P}(A) \times \mathcal{P}(A)$.

1.2.1 Equivalence Relation

A fundamental mathematical construction is to start with a non-empty set X and to decompose the set into a family of disjoint subsets of X whose union is the whole set X, called a partition of X and to form a new set by equating each such subset to an element of a new set, called a quotient set of X given by the partition. For this purpose we introduce the concept of an equivalence relation which is logically equivalent to a partition.

Definition 1.2.2 A binary relation R on A is said to be an *equivalence relation* on A iff

(a) R is *reflexive*: $(a, a) \in R$ for all $a \in A$;
(b) R is *symmetric*: if $(a, b) \in R$, then $(b, a) \in R$ for $a, b \in A$;
(c) R is *transitive*: if $(a, b) \in R$ and $(b, c) \in R$, then $(a, c) \in R$ for $a, b, c \in A$.

Instead of speaking about subsets of $A \times A$, we can also define an equivalence relation as below by writing aRb in place of $(a, b) \in R$.

Definition 1.2.3 A binary relation R on A is said to be an *equivalence relation* iff R is

(a') reflexive: aRa for all $a \in A$;
(b') symmetric: aRb implies bRa for $a, b \in A$;
(c') transitive: aRb and bRc imply aRc for $a, b, c \in A$.

Example 1.2.3 Define R on $\mathbf{Z}$ by $aRb \Leftrightarrow a - b$ is divisible by a fixed integer $n > 1$. Then R is an equivalence relation.

Proof Since $a - a = 0$ is divisible by n for all $a \in \mathbf{Z}$, aRa for all $a \in \mathbf{Z}$, hence R is reflexive. If $a - b$ is divisible by n, then $b - a$ is also divisible by n; hence R is symmetric. Finally, if $a - b$ and $b - c$ are both divisible by n, then their sum $a - c$ is also divisible by n; hence R is transitive. Consequently, R is an equivalence relation. (See Example 1.2.9.)   □

Example 1.2.4 Let T be the set of all triangles in the Euclidean plane. Define $\alpha R \beta \Leftrightarrow \alpha$ and β are similar for $\alpha, \beta \in T$. Then R is an equivalence relation.

Definition 1.2.4 A relation R on a set A is said to be *antisymmetric* iff xRy and yRx imply $x = y$ for $x, y \in A$.

Example 1.2.5 Consider the inclusion relation '$\subseteq$' on $\mathcal{P}(A)$ of a given set A. Then $P \subseteq Q$ and $Q \subseteq P$ imply $P = Q$ for two subsets $P, Q \in \mathcal{P}(A)$; hence '$\subseteq$' is antisymmetric.

Remark The relations: reflexive, symmetric (or antisymmetric) and transitive are independent of each other.

Example 1.2.6 A relation which is reflexive, symmetric but not transitive: Let R be the binary relation on $\mathbf{Z}$ given by $(x, y) \in R$ iff $x - y = 0, 5$ or -5. Then R is reflexive, since $x - x = 0 \Rightarrow (x, x) \in R$, for all $x \in \mathbf{Z}$. R is symmetric, since $(x, y) \in R$ implies $x - y = 0, 5$ or -5 and hence $y - x = 0, -5$ or 5, so that $(y, x) \in R$. But R is not transitive, since $(12, 7) \in R$ and $(7, 2) \in R$ but $(12, 2) \notin R$.

Example 1.2.7 A relation which is reflexive and transitive but neither symmetric nor antisymmetric: Let $\mathbf{Z}^*$ be the set of all non-zero integers and R be the relation on $\mathbf{Z}^*$ given by $(a, b) \in R$ iff a is a factor b, i.e., iff $a|b$. Since $a|a$ for all $a \in \mathbf{Z}^*$; $a|b$ and $b|c \Rightarrow a|c$, hence R is reflexive and transitive. $2|8$ but $8|2$ is not true; hence R is not symmetric. Again $7|-7$ and $-7|7$ but $7 \neq -7$; hence R is not antisymmetric.

Example 1.2.8 A relation which is symmetric and transitive but not reflexive: Let $\mathbf{R}$ be the set of all real numbers and ρ be the relation on $\mathbf{R}$ given by $(a, b) \in \rho$ iff $ab > 0$. Clearly, ρ is symmetric and transitive. Since $(0, 0) \notin \rho$, ρ is not reflexive.

Definition 1.2.5 Let X be a non-null set. Then a family $\mathcal{P} = \{A_i : i \in I\}$ of subsets of X is said to be a *partition* of X iff

 (i) Each A_i is non-empty, $i \in I$;
(ii) $A_i \cap A_j = \emptyset$ for all $i \neq j, i, j \in I$;
(iii) $\bigcup \{A_i : A_i \in \mathcal{P}\} = X$.

Thus a partition of X is a disjoint class of non-empty subsets of X whose union is the set X itself and a partition of X is the result of splitting the set X into non-empty subsets in such a way that each $x \in X$ belongs to one and only one of the given subsets.

Partition of a set is not unique. For example, if $X = \{1, 2, 3, 4, 5, 6\}$, then $\{1, 3, 5\}, \{2, 4, 6\}$ and $\{1, 2, 3, 4\}, \{5, 6\}$ are two different partitions of X.

We shall show in Theorem 1.2.9 that there is a bijective correspondence between the set of equivalence relations on a set $X \neq \emptyset$ and the set of all partitions of X.

Theorem 1.2.1 *Let ρ be an equivalence relation on a set $X \neq \emptyset$ and let for each $x \in X$, the class (x), denoted by (x) (or $[x]$), be defined by $(x) = \{y \in X : (x, y) \in \rho\}$. Then*

 (i) *For each $x \in X, x \in (x)$;*
 (ii) *If $x, y \in X$, either $(x) = (y)$ or $(x) \cap (y) = \emptyset$;*
(iii) *$\bigcup \{(x) : x \in X\} = X$;*
(iv) *If $\mathcal{P}_\rho = \{(x) : x \in X\}$, then $\mathcal{P}_\rho$ is a partition of X, called the* partition induced *by ρ and denoted by X/ρ.*

Proof (i) Since ρ is reflexive, $(x, x) \in \rho$ for each $x \in X$. Hence by definition of (x), $x \in (x)$.

(ii) Let $(x) \cap (y) \neq \emptyset$. Then there exists some $z \in (x) \cap (y)$. Consequently, $(x, z) \in \rho$ and $(y, z) \in \rho$ by definition of classes. Let w be an arbitrary element such that $w \in (x)$. Then $(x, w) \in \rho$. Now $(x, z) \in \rho \Rightarrow (z, x) \in \rho$ as ρ is symmetric. So, $(z, x) \in \rho$ and $(x, w) \in \rho \Rightarrow (z, w) \in \rho$ as ρ is transitive, and hence

$$(y, z) \in \rho \quad \text{and} \quad (z, w) \in \rho \quad \Rightarrow \quad (y, w) \in \rho \quad \Rightarrow \quad w \in (y) \quad \Rightarrow \quad (x) \subseteq (y) \tag{1.1}$$

as w is an arbitrary element of (x).

Interchanging the role of x and y by symmetric property of ρ, we have

$$(y) \subseteq (x). \tag{1.2}$$

Hence (1.1) and (1.2) show that $(x) = (y)$.

(iii) Since for each $x \in X, x \in (x)$, it follows that

$$X \subseteq \bigcup \{(x) : x \in X\}. \tag{1.3}$$

Again as $(x) \subseteq X$ for each $x \in X$, it follows that

$$\bigcup \{(x) : x \in X\} \subseteq X. \tag{1.4}$$

Hence (1.3) and (1.4) show that $\bigcup \{(x) : x \in X\} = X$.

(iv) By using (ii) and (iii), it follows by definition of a partition that $\mathcal{P}_\rho$ is a partition of X. $\square$

The class (x) associated with ρ is sometimes denoted by $(x)_\rho$ to avoid any confusion. The set (x) is called the *equivalence class* (with respect to ρ) determined by x.

Converse of Theorem 1.2.1 is also true.

Theorem 1.2.2 *Let $\mathcal{P}$ be a given partition of a non-null set X. Define a relation $\rho = \rho_\mathcal{P}$ (depending on $\mathcal{P}$) on X by $(x, y) \in \rho_\mathcal{P}$ iff there exists $A \in \mathcal{P}$ such that $x, y \in A$ (i.e., iff y belongs to the same class as x). Then $\rho_\mathcal{P}$ is an equivalence relation, induced by $\mathcal{P}$. Moreover,*

(a) $\mathcal{P}(\rho_\mathcal{P}) = \mathcal{P}$ *(partition induced by $\rho_\mathcal{P}$)*;
(b) $\rho(\mathcal{P}_\rho) = \rho$ *(equivalence relation induced by partition $\mathcal{P}_\rho$)*.

Proof Let $x \in X$. Since $\mathcal{P}$ is a partition, there exists $A \in \mathcal{P}$, such that $x \in A$. So, $(x, x) \in \rho$. This implies ρ is reflexive. If $(x, y) \in \rho$, then $(y, x) \in \rho$ from the definition of ρ. So, ρ is symmetric. Let $(x, y) \in \rho$ and $(y, z) \in \rho$. Then there exist A, $B \in \mathcal{P}$ such that $x, y \in A$ and $y, z \in B$. Consequently, $y \in A \cap B$. Since $\mathcal{P}$ is a partition, $A = B$. Consequently, $x, z \in A \in \mathcal{P}$; therefore, $(x, z) \in \rho$. So, ρ is transitive. As a result ρ is an equivalence relation.

(a) We now show that $\mathcal{P}(\rho_\mathcal{P}) = \mathcal{P}$. Let $A \in \mathcal{P}$ and $x \in A$. Then for every $y \in A$, $(x, y) \in \rho_\mathcal{P}$. Consequently, $y \in (x)\rho_\mathcal{P} \Rightarrow A \subseteq (x)\rho_\mathcal{P}$. Next let $z \in (x)\rho_\mathcal{P}$. Then there exists some $B \in \mathcal{P}$ such that $x, z \in B$. But $x \in A \Rightarrow x \in A \cap B \Rightarrow A = B$ (by the property of a partition). Consequently, $z \in A \Rightarrow (x)\rho_\mathcal{P} \subseteq A$. As a result,

$$(x)\rho_\mathcal{P} = A, \quad \text{but } (x)\rho_\mathcal{P} \in \mathcal{P}(\rho_\mathcal{P}).$$

Consequently, $\mathcal{P} \subseteq \mathcal{P}(\rho_\mathcal{P})$. Moreover both $\mathcal{P}$ and $\mathcal{P}(\rho_\mathcal{P})$ are partitions of the same set X. Clearly, $\mathcal{P} = \mathcal{P}(\rho_\mathcal{P})$.

(b) To prove $\rho(\mathcal{P}_\rho) = \rho$, let $(x, y) \in \rho$. Then $y \in (x)_\rho \in \mathcal{P}_\rho \Rightarrow (x, y) \in \rho(\mathcal{P}_\rho) \Rightarrow \rho \subseteq \rho(\mathcal{P}_\rho)$. Again $(x, y) \in \rho(\mathcal{P}_\rho) \Rightarrow$ there is an equivalence class $(z)_\rho$ such that $x, y \in (z)_\rho \Rightarrow (z, x) \in \rho$ and $(z, y) \in \rho \Rightarrow (x, z) \in \rho$ and $(z, y) \in \rho \Rightarrow (x, y) \in \rho$ (by transitive property of ρ) $\Rightarrow \rho(\mathcal{P}_\rho) \subseteq \rho$. As a result $\rho = \rho(\mathcal{P}_\rho)$. $\square$

The disjoint classes (x) into which a set X is partitioned by an equivalence relation ρ constitute a set, called the *quotient set* of X by ρ, denoted by X/ρ, where (x) denotes the class containing the element $x \in X$. Each element x of the class (x) is called a representative of (x) and a set formed by taking one element from each class is called a representative system of the partition.

The following example shows that the quotient set of an infinite set may be finite.

Example 1.2.9 (i) *The set of residue classes $\mathbf{Z}_n$*: In Definition 1.2.3, we have defined an equivalence relation R on $\mathbf{Z}$; when an integer a (positive, negative or 0) is divided by a positive integer n, there are n possible remainders, viz., $0, 1, \ldots, n-1$. If r is the remainder for a, then $a - r$ is divisible by n, so that $a \in (r) = \{y \in \mathbf{Z} : y - r$ is divisible by $n\}$. Hence every integer belongs to one and only one of the n classes $(0), (1), \ldots, (n - 1)$. Thus the quotient set $\mathbf{Z}/R$ consists of the n distinct classes $(0), (1), \ldots, (n - 1)$; called the set of *residue classes modulo n* and is denoted by $\mathbf{Z}_n$. The integers $0, 1, \ldots, n - 1$ form a representative system of the partition.

The equivalence relation R on $\mathbf{Z}$ defined in this example is called the *congruence modulo n*. Two integers a and b in the same residue class are said to be congruent, denoted by $a \equiv b \pmod{n}$. The set $\mathbf{Z}_n$ provides very strong different algebraic structures which we shall study in the subsequent chapters.

(ii) *Visual Description of* $\mathbf{Z}_{12}$: We can use a clock to describe $\mathbf{Z}_{12}$ visually. Take the real number line $\mathbf{R}^1$ and wrap it around a circumference bearing the numbers 1 to 12 located as usual on a clock. Since $\mathbf{R}^1$ is infinitely long, it will wrap infinitely many times around the circle, and so each 'hour point' on the clock will coincide with infinitely many integers. The integers located at the hour r for $r = 1, 2, \ldots, 12$ are all integers x such that $x - r$ is divisible by 12 i.e., are all integers congruent to r modulo 12 (see also clock arithmetic in Sect. 2.7.3 of Chap. 2).

Example 1.2.10 Let ρ be the binary relation on $\mathbf{N}^+ \times \mathbf{N}^+$ defined by $(a, b)\rho(c, d) \Leftrightarrow ad = bc$. Thus ρ is an equivalence relation. We assume commutative, associative, and cancellative laws for multiplication on $\mathbf{N}^+$. The relation ρ is clearly reflexive and symmetric. Next suppose $(a, b)\rho(c, d)$ and $(c, d)\rho(e, f)$. Then $ad = bc$ and $cf = de \Rightarrow (ad)(cf) = (bc)(de) \Rightarrow af = be \Rightarrow (a, b)\rho(e, f) \Rightarrow \rho$ is transitive. Consequently, ρ is an equivalence relation on $\mathbf{N}^+ \times \mathbf{N}^+$.

Remark If the ordered pair $(a, b) \in \mathbf{N}^+ \times \mathbf{N}^+$ is written as a fraction $\frac{a}{b}$, then the above relation ρ is the usual definition of equality of two fractions: $\frac{a}{b} = \frac{c}{d} \Leftrightarrow ad = bc$.

1.2.2 Partial Order Relations

We have made so far little use of reflexive, antisymmetric, and transitive laws. We are familiar with the natural ordering $\leq$ between two positive integers. This example suggests the abstract concept of a partial order relation, which is a reflexive, anti-symmetric, and transitive relation. Partial order relations and their special types play an important role in mathematics. For example, partial order relations are essential in Zorn's Lemma, which provides a very powerful tool in mathematics and in lattice theory whose applications are enormous in different sciences.

Definition 1.2.6 A reflexive, antisymmetric, and transitive relation R on a non-empty set P is called a *partial order* relation. Then the pair (P, R) is called a *partially ordered* set or a *poset*.

We adopt the symbol '$\leq$' to represent a partial order relation. So writing $a \leq b$ in place of aRb, from Definition 1.2.6 it follows that

Fig. 1.2 Hasse diagram
representing $a < b$

(i) $a \leq a$ for all $a \in P$;

(ii) $a \leq b$ and $b \leq a$ in $P \Rightarrow a = b$ for $a, b \in P$ and

(iii) $a \leq b$ and $b \leq c$ in $P \Rightarrow a \leq c$ for $a, b, c \in P$.

The following three examples are quite different in nature but possess identical important properties.

Example 1.2.11 (i) $(\mathbf{R}, \leq)$ is a poset, where '$\leq$' denotes the natural ordering in $\mathbf{R}$ (i.e., $a \leq b \Leftrightarrow b - a \geq 0$). Similarly, $(\mathbf{R}, \geq)$ is a poset under natural ordering $\geq$ in $\mathbf{R}$.

(ii) $(\mathcal{P}(A), \leq)$ is a poset under the inclusion relation $\subseteq$ between subsets of A.

(iii) $(\mathbf{N}^+, \leq)$ is a poset under divisibility relation (i.e., $a \leq b \Leftrightarrow a$ divides b in $\mathbf{N}^+$, denoted by $a|b$).

A poset with a finite number of elements can be conveniently represented by means of a diagram, called the *Hasse diagram*, where $a < b$ is represented by means of a diagram of the form as shown in Fig. 1.2, where a, b are represented by small circles, a being written below b and the two circles being joined by a straight line. Thus we have the following examples of posets as described in Fig. 1.3 of Example 1.2.12.

Example 1.2.12 In Fig. 1.3, we have shown six different partial order relations ((i)–(vi)).

In a poset $(P, \leq)$, we say $a < b$ iff $a \leq b$ and $a \neq b$. If $a < b$, we say that a precedes b (or a is a predecessor of b or b succeeds a).

Definition 1.2.7 If a poset $(P, \leq)$ is such that, for every pair of elements a, b of P, exactly one of $a < b, a = b, b < a$ holds [i.e., if any two elements of P are *comparable*], then $(P, \leq)$ is called a *totally (fully* or *linearly) ordered* set or simply an *ordered set* or a *chain*.

Remark The partial order relation in a chain has a fourth property: 'any two elements are comparable', in addition to the properties required in Definition 1.2.6.

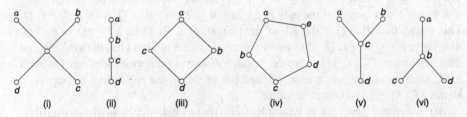

Fig. 1.3 Different partial order relations

Example 1.2.13 (i) The poset $(\mathbf{R}, \leq)$ (in Example 1.2.11(i)) is totally ordered.

(ii) The poset $(\mathbf{N}^+, \leq)$ (in Example 1.2.11(iii)) is not totally ordered, since the integers 5 and 8 are not comparable, as neither divides the other.

(iii) The poset $(\mathcal{P}(X), \leq)$ (in Example 1.2.11(ii)) is not totally ordered, provided that X has at least two elements. For example, if $X = \{1, 2, 3\}$, $A = \{1, 2\}$, $B = \{2, 3\}$, then neither $A \subseteq B$ nor $B \subseteq A$ holds.

Definition 1.2.8 Let $(P, \leq)$ be a poset and $a, b \in P$. An element $x \in P$, where $x \leq a$ and $x \leq b$ is called a *lower bound* of a and b. If x is a lower bound of a, b in P and $y \leq x$ holds for any lower bound y of a and b, then x (uniquely determined) is called the *greatest lower bound* (glb or *infimum* or *meet*) of a and b and is denoted by $a \wedge b$ (or ab).

Similarly, an element $x \in P$, where $a \leq x$ and $b \leq x$ is called an *upper bound* of a and b. If x is an upper bound of a, b in P and $x \leq y$ holds for any upper bound y of a and b, then x (uniquely determined) is called the *least upper bound* (lub or *supremum* or *join*) of a and b and is denoted by $a \vee b$ (or $a + b$).

Definition 1.2.9 A poset in which each pair of elements

 (i) has the lub is called an *upper semilattice*;
 (ii) has the glb is called a *lower semilattice*; and
(iii) has both the lub and the glb are called a *lattice*.

Example 1.2.12(ii)–(iv), as shown in Fig. 1.3, are both upper and lower semilattices and hence are lattices. But Example 1.2.12(i), (v) and (vi) are not so. Indeed, Example 1.2.12(i) is neither an upper semilattice nor a lower semilattice; Example 1.2.12(vi) is an upper semilattice but not a lower semilattice; and Example 1.2.12(v) is a lower semilattice but not an upper semilattice.

The definition of a lower bound and upper bound, glb and lub, can be generalized for any given collection of two or more elements of a poset.

Definition 1.2.10 A lattice $(P, \leq)$ is said to be *complete* iff every non-empty collection of elements of P has both a glb and a lub.

Example 1.2.14 (i) $(\mathcal{P}(A), \leq)$ is a poset (cf. Example 1.2.11(ii)). Let $P, Q \in \mathcal{P}(A)$. Then $P \cap Q \subseteq P$ and $P \cap Q \subseteq Q$, so $P \cap Q$ is a lower bound of the pair P, Q. Also, if R is any lower bound of the pair P, Q, i.e., $R \subseteq P$ and $R \subseteq Q$, then $R \subseteq P \cap Q$. This shows that $P \cap Q$ is the glb of the pair $P, Q \in \mathcal{P}(A)$ i.e., $P \wedge Q = P \cap Q$. Similarly, $P \vee Q = P \cup Q$. Thus every pair of elements of $\mathcal{P}(A)$ has both the glb and the lub. Hence $(\mathcal{P}(A), \leq)$ is a lattice. Again for any non-empty collection of subsets of A, their meet and join are the glb and lub of the given collection, respectively. Hence $(\mathcal{P}(A), \leq)$ is a complete lattice.

(ii) The poset $(\mathbf{N}^+, \leq)$ in Example 1.2.11(iii) is a lattice, called the divisibility lattice, where $x \wedge y = \gcd(x, y)$ and $x \vee y = \mathrm{lcm}(x, y)$, where $\gcd(x, y)$ and

$\text{lcm}(x, y)$ denote, respectively, the greatest common divisor and least common multiple of $x, y \in \mathbf{N}^+$. But this lattice is not complete. For, the subset $X = \{1, 2, 2^2, \ldots\}$ has no lub.

Remark A totally ordered set is a lattice, but its converse is not true in general. From Examples 1.2.14(ii) and 1.2.13(ii) it appears that $(\mathbf{N}^+, \leq)$ is a lattice, but it is not totally ordered.

Definition 1.2.11 A lattice $(L, \leq)$ is said to be *distributive* iff $x \vee (y \wedge z) = (x \vee y) \wedge (x \vee z)$ (or, equivalently, $x \wedge (y \vee z) = (x \wedge y) \vee (x \wedge z)$) for all $x, y, z \in L$.

Definition 1.2.12 A lattice $(L, \leq)$ is said to be *modular* (or *Dedekind*) iff for $x, y, z \in L, x \vee y = z \vee y, x \wedge y = z \wedge y$ and $x \leq z$ imply $x = z$ (or, equivalently, for $x, y, z \in L, x \leq z$, the modular law $(x \vee y) \wedge z = x \vee (y \wedge z)$ holds in L).

Example 1.2.15 (i) The *divisibility* lattice $(\mathbf{N}^+, \leq)$ is modular and distributive.
(ii) Every distributive lattice is modular but the converse is not true.
[*Hint.* (ii) Let $(L, \leq)$ be a distributive lattice and $x, y, z \in L$ be such that $x \leq z$. Then $x \vee (y \wedge z) = (x \vee y) \wedge (x \vee z) = (x \vee y) \wedge z$ (as L is distributive). This shows by Definition 1.2.12 that L is modular.]

Note that its converse is not true. (See Ex. 10 of Exercises-I and Ex. 20 of SE-I.)

Theorem 1.2.3 *If in a poset* $(P, \leq)$, *every subset (including $\emptyset$) has a glb in P, then P is a complete lattice.*

Proof Left as an exercise. □

Remark Any non-void complete lattice contains a least element 0 and a greatest element 1.

Corollary *If a poset P has 1, and every non-empty subset X of P has a glb, then P is a complete lattice.*

Example 1.2.16 The *pentagonal lattice* $(L, \leq)$ is not modular, since $v(u + b) = v$ and $u + vb = u$. As $u < v$ and $u + vb < v(u + b)$, the modular law does not hold in L and L is therefore non-modular as shown in Fig. 1.4.

Theorem 1.2.4 *A lattice is non-modular iff it contains the pentagonal lattice as a sublattice.*

Proof Let $(L, \leq)$ be non-modular lattice. Then in L there exist five elements $u, v, b, u + b = v + b, ub = vb$, which form the pentagonal lattice as shown in Fig. 1.5.
For the validity of the converse, see Example 1.2.16. □

Fig. 1.4 Pentagonal lattice

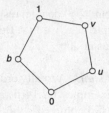

Fig. 1.5 Pentagonal lattice
with $u + b = v + b = 1$ and
$ub = vb = 0$

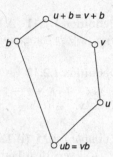

Definition 1.2.13 Let $(P, \leq)$ be a poset. An element $x \in P$ is said to be a *minimal* element of P iff $a \in P$ and $a \leq x$ imply $a = x$. Similarly, an element $x \in P$ is said to be a *maximal* element of P iff $a \in P$ and $x \leq a$ imply $a = x$.

Thus an element $x \in P$ is minimal (or maximal) iff no other element of P precedes (or exceeds) x.

Example 1.2.17 (i) The poset $(\mathbf{R}, \leq)$ (cf. Example 1.2.11(i)) has neither minimal nor maximal elements.

(ii) The minimal elements of the poset $(\mathcal{P}(X) - \emptyset, \leq)$ are the singleton subsets of X.

(iii) In the divisibility lattice, the prime numbers serve as minimal elements.

Note Lattices are used in different areas, such as theoretical computer science, quantum mechanics, social science, biosystems, music etc. For applications of lattices to quantum mechanics, the books (Varadarajan 1968) and (Von Neumann 1955) are referred to.

Our main aim is now to state 'Zorn's Lemma'. A non-constructive criterion for the existence of maximal elements is given by the so called *maximality principle*, which is called *Zorn's Lemma* (the principle goes back to Hausdorff and Kuratowski, but Zorn gave a formulation of it which is particularly suitable to algebra, analysis, and topology).

Lemma 1.2.1 (Zorn's Lemma) *Let $(S, \leq)$ be a non-empty partially ordered set. Suppose every subset $A \subseteq S$ which is totally ordered by $\leq$ has an upper bound (in S). Then S possesses at least one maximal element.*

Remark Zorn's Lemma is indispensable to prove many interesting results of mathematics, one is given in this section and some are given in subsequent chapters.

An abstract Boolean algebra has a close connection with lattices. The following is necessary to understand the concept of Boolean algebra.

By a finite collection of sets we always mean one which is empty or consists of n sets for some positive integer n and by finite unions and finite intersections we mean unions and intersections of finite collection of sets. If we say that a collection $\mathcal{B}$ of sets is closed under the formation of finite unions, we mean that $\mathcal{B}$ contains the union of each of its finite sub-collections; and since the empty sub-collection qualifies as a finite sub-collection of $\mathcal{B}$, we see that its union and the empty set, each is necessarily an element of $\mathcal{B}$. In the same way, a collection of sets, which is closed under the formation of finite intersections is necessarily an element of $\mathcal{B}$.

We assume that universal set $U \neq \emptyset$.

Definition 1.2.14 A *Boolean algebra of sets* is a non-empty collection $\mathcal{B}$ of subsets of U which satisfies the following conditions:

  (i) A and $B \in \mathcal{B}$ imply $A \cup B \in \mathcal{B}$;
 (ii) A and $B \in \mathcal{B}$ imply $A \cap B \in \mathcal{B}$;
(iii) $A \in \mathcal{B}$ implies $A' \in \mathcal{B}$.

Since $\mathcal{B} \neq \emptyset$, it contains at least one set A.

Condition (iii) shows that A' is in $\mathcal{B}$ along with A and since $A \cap A' = \emptyset$ and $A \cup A' = U$, (i) and (ii) show that $\emptyset \in \mathcal{B}$ and $U \in \mathcal{B}$. Clearly, $\{U, \emptyset\}$ is a Boolean algebra of sets and every Boolean algebra of sets contains it. The other extreme is the collection of all subsets $\mathcal{P}(U)$ of U, which is also Boolean algebra of sets.

Let $\mathcal{B}$ be a Boolean algebra of sets. If $\{A_1, A_2, \ldots, A_n\}$ is a non-empty finite sub-collection of $\mathcal{B}$, then $A_1 \cup A_2 \cup \cdots \cup A_n \in \mathcal{B}$ and $A_1 \cap A_2 \cap \cdots \cap A_n \in \mathcal{B}$. Moreover $\emptyset, U \in \mathcal{B}$. Clearly, $\mathcal{B}$ is closed under the formation of finite unions, finite intersections and complements. Conversely, let $\mathcal{B}$ be a collection of sets which is closed under the formation of finite unions, finite intersections, and complements. Then $\emptyset \in \mathcal{B}$ and $U \in \mathcal{B}$. Clearly, $\mathcal{B}$ is a Boolean algebra of sets. Consequently, we may characterize Boolean algebras of sets as collections of sets which are closed under the formation of finite unions, finite intersections, and complements.

Remark An abstract Boolean algebra may be defined by means of lattices (Ex. 7 of Exercises-I).

1.2.3 Operations on Binary Relations

Operations on pairs of binary relations arise in many occasions in the study of modern algebra. Such an operation is defined now.

Let $R : X \to Y$ and $S : Y \to Z$ be two binary relations. Then the *composite* $S \circ R$ of R and S is defined by

$S \circ R = \{(x, z) \in X \times Z:$ if there exists $y \in Y$ such that $(x, y) \in R$ and $(y, z) \in S\} \subseteq X \times Z$.

So, $S \circ R$ is a binary relation from $X \to Z$.

If R is a binary relation from X to Y, the inverse R^{-1} is defined by

$$R^{-1} = \{(y, x) : (x, y) \in R\} \subseteq Y \times X.$$

So, R^{-1} is a binary relation from Y to X.

Proposition 1.2.1 *Let $R : X \to Y$ and $S : Y \to Z$ and $T : Z \to W$ be binary relations. Then*

(i) $(T \circ S) \circ R = T \circ (S \circ R)$ *(associative property)*;
(ii) $(S \circ R)^{-1} = R^{-1} \circ S^{-1}$.

Proof (i) Clearly, both $(T \circ S) \circ R$ and $T \circ (S \circ R)$ are binary relations from X to W. Now $(x, w) \in (T \circ S) \circ R \Leftrightarrow$ there exists $y \in Y$ such that $(x, y) \in R$ and $(y, w) \in T \circ S$, where $x \in X$, $w \in W \Leftrightarrow$ there exists $y \in Y$ such that $(x, y) \in R$ and there exists $z \in Z$ such that $(y, z) \in S$ and $(z, w) \in T \Leftrightarrow$ there exist $y \in Y$ and $z \in Z$ such that $(x, y) \in R$, $(y, z) \in S$ and $(z, w) \in T$. Thus $(x, w) \in (T \circ S) \circ R \Leftrightarrow \exists z \in Z$ such that $(x, z) \in S \circ R$ and $(z, w) \in T \Leftrightarrow (x, w) \in T \circ (S \circ R)$.

Consequently, $(T \circ S) \circ R = T \circ (S \circ R)$.

(ii) Again $S \circ R : X \to Z \Leftrightarrow (S \circ R)^{-1} : Z \to X$.

Clearly, both $(S \circ R)^{-1}$ and $R^{-1} \circ S^{-1}$ are relations from Z to X.

Then for $z \in Z$, $x \in X$, $(z, x) \in (S \circ R)^{-1} \Leftrightarrow (x, z) \in S \circ R \Leftrightarrow \exists y \in Y$ such that $(x, y) \in R$ and $(y, z) \in S \Leftrightarrow \exists y \in Y$ such that $(y, x) \in R^{-1}$ and $(z, y) \in S^{-1} \Leftrightarrow \exists y \in Y$ such that $(z, y) \in S^{-1}$ and $(y, x) \in R^{-1} \Leftrightarrow (z, x) \in R^{-1} \circ S^{-1}$.

Consequently, $(S \circ R)^{-1} = R^{-1} \circ S^{-1}$. $\qquad\qquad\square$

Example 1.2.18 Let R be a relation on a set S. Then R is an equivalence relation on S iff

(i) $\Delta \subseteq R$, where $\Delta = \{(x, x) : x \in S\}$;
(ii) $R = R^{-1}$ and
(iii) $R \circ R \subseteq R$

hold.

1.2.4 Functions or Mappings

The concept of functions (mappings) is perhaps the single most important and universal notion used in all branches of mathematics. Sets and functions are closely related. They have the capacity for vast and intricate development. We are now in a position to define a function in terms of a binary relation.

Definition 1.2.15 Let X and Y be two non-empty sets. A *function* f from X to Y is defined to be a binary relation f such that

$$(x, y) \in f \quad \text{and} \quad (x, z) \in f \quad \text{imply} \quad y = z$$

(i.e., f is single valued).

The *domain of* f denoted by dom f, *range of* f denoted by range f are defined by

$$\text{dom } f = \{x \in X : (x, y) \in f \text{ for some } y \in Y\} \subseteq X;$$
$$\text{range } f = \{y \in Y : (x, y) \in f \text{ for some } x \in X\} \subseteq Y.$$

Remark 1 Definition 1.2.15 means for each $x \in \text{dom } f$, there exists a unique $y \in \text{range } f \subseteq Y$ such that $(x, y) \in f$. Thus a function f from a set X to a set Y is a correspondence which assigns to each $x \in \text{dom } f$ exactly one element $y \in \text{range } f \subseteq Y$. If $(x, y) \in f$, we write $y = f(x)$; y is called the image of x under f and x is called a preimage of y under f.

Remark 2 In elementary calculus, all the functions have the same range, namely, the real numbers, (depicted geometrically as y-axis), in algebra there are many different ranges, so that when we introduce a function it is important to specify both the domain and the range of the function as part of the definition of a function.

The notation $f : X \to Y$ (or $f : X \xrightarrow{f} Y$) is used to mean that dom $f = X$. Sometimes it is convenient to denote the effect of the function f on an element x of X by $x \mapsto f(x)$.

A function $f : X \to Y$ is sometimes called a *map* or *mapping*.

Remark The definition of function identifies with the graph of a function in its usual definition (by means of correspondence).

Definition 1.2.16 Two functions $f, g : X \to Y$ are said to be equal, denoted by $f = g$, iff $f(x) = g(x) \; \forall x \in X$.

Definition 1.2.17 Let $f : X \to Y$ be a binary relation such that if $A \subseteq X$ and $B \subseteq Y$, then the *image* of A under f, denoted by $f(A)$, is defined by

$$f(A) = \{y \in Y : (x, y) \in f \text{ for some } x \in A\} \subseteq Y;$$

the inverse image of B under f denoted by $f^{-1}(B)$ is defined by

$$f^{-1}(B) = \{x \in X : (x, y) \in f \text{ for some } y \in B\} = \{x \in X : f(x) \in B\} \subseteq X.$$

If $B \cap \text{range } f = \emptyset$, $f^{-1}(B) = \emptyset$.

Let $f : X \to Y$ and $A \subseteq X$. Thus $f(A)$ is defined by

$$f(A) = \big\{y \in Y : y = f(x) \text{ for some } x \in A \subseteq X\big\}$$

$$= \text{set of all images of points in } A \text{ under } f.$$

Similarly, $f^{-1}(B) = \bigcup_{y \in B}\{f^{-1}(y)\}$.

Theorem 1.2.5 *Let $f : X \to Y$ be a binary relation and $A, B \subseteq X$ and $C, D \subseteq Y$. Then*

(1) $f(A \cup B) = f(A) \cup f(B)$;
(2) $f(A \cap B) \subseteq f(A) \cap f(B)$;
(3) $f^{-1}(C \cup D) = f^{-1}(C) \cup f^{-1}(D)$;
(4) $f^{-1}(C \cap D) = f^{-1}(C) \cap f^{-1}(D)$.

More generally, if $\{A_i : i \in I\}$ and $\{B_j : j \in J\}$ are two families of subsets X and Y, respectively, then

(1') $f(\bigcup_{i \in I}\{A_i : i \in I\}) = \bigcup_{i \in I} f(A_i)$;
(2') $f(\bigcap\{A_i : i \in I\}) \subseteq \bigcap\{f(A_i) : i \in I\}$;
(3') $f^{-1}(\bigcup\{B_j : j \in J\}) = \bigcup\{f^{-1}(B_j) : j \in J\}$;
(4') $f^{-1}(\bigcap\{B_j : j \in J\}) = \bigcap\{f^{-1}(B_j) : j \in J\}$.

Proof (1') $y \in f(\bigcup\{A_i : i \in I\}) \Leftrightarrow \exists x \in \bigcup A_i$ such that $y = f(x) \Leftrightarrow \exists x \in A_i$ for some $i \in I$ and $y = f(x) \Leftrightarrow$ for some $i \in I, y \in f(A_i) \Leftrightarrow y \in \bigcup\{f(A_i), i \in I\}$. This proves (1').

(2') Clearly, $M = \bigcap\{A_i : i \in I\} \subseteq A_i$ for each $i \in I \Rightarrow f(M) \subseteq f(A_i)$ for each $i \in I \Rightarrow f(M) \subseteq \bigcap\{f(A_i) : i \in I\}$. This proves (2').

(3') $x \in f^{-1}(\bigcup\{B_j : j \in J\}) \Leftrightarrow \exists y \in \bigcup B_j$ such that $f(x) = y \Leftrightarrow \exists y \in B_j$ for some $j \in J$ such that $f(x) = y \Leftrightarrow x \in f^{-1}(B_j)$ for some $j \in J \Leftrightarrow x \in \bigcup\{f^{-1}(B_j) : j \in J\}$. This proves (3').

(4') $x \in f^{-1}(\bigcap\{B_j : j \in J\}) \Leftrightarrow f(x) \in \bigcap\{B_j : j \in J\} \Leftrightarrow f(x) \in B_j$, for each $j \in J \Leftrightarrow x \in f^{-1}(B_j)$, for each $j \in J \Leftrightarrow x \in \bigcap_{j \in J} f^{-1}(B_j)$. This proves (4'). $\square$

Remark Equality in (2) i.e., $f(A \cap B) = f(A) \cap f(B)$ does not occur in general; but occurs iff $x_1 \neq x_2$ in $X \Rightarrow f(x_1) \neq f(x_2)$ for all pairs $x_1, x_2 \in X$ i.e., iff f is 1–1 (see Definition 1.2.18).

Example 1.2.19 Consider the map $f : \mathbf{R}^2 \to \mathbf{R}^2$ by $f(x, y) = (x, 0)$. Let $A = \{(x, y) : x - y = 0\}$ and $B = \{(x, y) : x - y = 1\}$. Then $A \cap B = \emptyset$ and $f(A) \cap f(B) = x$ axis, because geometrically, A, B represent parallel lines and both $f(A)$ and $f(B)$ represent x-axis. Consequently, $f(A \cap B) = \emptyset \neq f(A) \cap f(B)$.

Remark Let there exist a pair $x_1, x_2 \in X$ such that $x_1 \neq x_2$ but $f(x_1) = f(x_2) = y$ (say). Take $A = \{x_1\}$, $B = \{x_2\}$. Then $A \cap B = \emptyset$, $f(A \cap B) = \emptyset$, $f(A) = \{y\} = f(B)$. So, $f(A) \cap f(B) = \{y\} \supsetneq f(A \cap B)$. Equality holds iff $x_1 \neq x_2$ in $X \Rightarrow f(x_1) \neq f(x_2)$ i.e., iff f is 1–1 (see below).

Definition 1.2.18   A map $f : X \to Y$ is said to be

 (i) *injective* (or one–one i.e., 1–1) iff $x, x' \in X$, $x \neq x' \Rightarrow f(x) \neq f(x')$; i.e., iff $\forall x, x' \in X$, $f(x) = f(x') \Rightarrow x = x'$;
 (ii) *surjective* (or a *surjection* or onto) iff $f(X) = Y$; i.e., iff for each $y \in Y$, $\exists$ some $x \in X$ such that $f(x) = y$;
(iii) *bijective* (or a *bijection* or a *one-to-one correspondence*) iff f is both injective and surjective.

Clearly, for any non-null set X, the *identity map* $I_X : X \to X$ defined by $I_X(x) = x$ $\forall x \in X$ is bijective.

Remark Dirichlet observed that for two natural numbers m and n $(m > n)$, there is no injective map $f : \{1, 2, \ldots, m\} \to \{1, 2, \ldots, n\}$ and proved the following principle known as the '*Pigeonhole Principle*'. This principle states that *for two natural numbers m and n $(m > n)$, if m objects are distributed over n boxes, then some box must contain more than one of the objects.* In other words, if n objects are distributed over n boxes in such a way that no box receives more than one object, then each box receives exactly one object.

We now consider some other important concepts related to maps that are frequently used in different branches of mathematics.

Definition 1.2.19   Given a function $f : X \to Y$ and $A \subset X$, the function from A to Y (i.e., the function considered only on A), given by $a \mapsto f(a)$, $\forall a \in A$ is called the *restriction of* f to A and is denoted by $f|A : A \to Y$.

In particular, $I_X|A : A \to X$ is called the *inclusion* map of A into X and is denoted by $i : A \hookrightarrow X$.

To the contrary, given $A \subset X$ and a function $f : A \to Y$, a function $F : X \to Y$ coinciding with f on A (i.e., satisfying $F|A = f$) is called an *extension of* f over X relative to Y.

It is clear from the definitions that restriction of a function is unique but extension of a function is not unique.

If $f : X \to Y$, $A \subset X$, and $g = f|A : A \to Y$, then f is an extension of g over X.

Definition 1.2.20   Let $f : X \to Y$ and $g : Y \to Z$ be two functions. The composite of f and g is the function $X \to Z$ given by $x \mapsto g(f(x))$, $x \in X$.

The composite function is denoted by $g \circ f$ or simply by gf. Clearly $g \circ f$ is meaningful whenever range $f \subseteq$ dom g.

Proposition 1.2.2   *Let $f : X \to Y$ and $g : Y \to Z$ be two maps. Then*

  (i) *f and g are injective imply that $g \circ f$ is injective;*
 (ii) *f and g are surjective imply that $g \circ f$ is surjective;*
(iii) *$g \circ f$ is injective implies that f is injective;*
(iv) *$g \circ f$ is surjective implies that g is surjective;*
 (v) *f and g are bijective imply that $g \circ f$ is bijective;*
(vi) *$g \circ f$ is bijective implies that f is injective and g is surjective.*

Proof (i) Let f and g be injective. Then for each $x, x' \in X$, $(g \circ f)(x) = (g \circ f)(x') \Rightarrow g(f(x)) = g(f(x'))$ (by definition of $g \circ f$) $\Rightarrow f(x) = f(x')$ as g is injective $\Rightarrow x = x'$ as f is also injective $\Rightarrow g \circ f : X \to Z$ is injective.

(ii) Suppose f, g are surjective. Then $(g \circ f)(X) = g(f(X))$ (by definition of $g \circ f$) $= g(Y)$ (as f is surjective) $= Z$ as g is surjective $\Rightarrow g \circ f$ is surjective.

(v) From (i) and (ii) it follows that if f and g are bijective, then $g \circ f$ is also bijective.

(iii), (iv) and (vi) are left as exercises.   □

Remark If $g \circ f$ is bijective, neither f nor g may be bijective.

Example 1.2.20 Let $X = \{1, 2, 3\}$, $Y = \{3, 4, 5, 6\}$. Define $f : X \to Y$ and $g : Y \to X$ by

$$f(1) = 3, \qquad f(2) = 4, \qquad f(3) = 5;$$

$$g(3) = 1, \qquad g(4) = 2, \qquad g(5) = 3, \qquad g(6) = 3.$$

Then $g \circ f = I_X$. As I_X is bijective, $g \circ f$ is bijective but neither f nor g is bijective.

Theorem 1.2.6 *Let $f : X \to Y$, $g : Y \to Z$ and $h : Z \to W$ be maps. Then $(h \circ g) \circ f = h \circ (g \circ f)$ (associative law).*

Proof $((h \circ g) \circ f)(x) = (h \circ g)(f(x)) = h(g(f(x)))$ and $(h \circ (g \circ f))(x) = h((g \circ f)(x)) = h(g(f(x)))$, for all $x \in X$. Consequently, $(h \circ g) \circ f = h \circ (g \circ f)$.   □

Let $f : X \to Y$ be a binary relation. Then f^{-1} is always defined as a binary relation, but it need not be a function.

For example, if $f : \mathbf{R} \to \mathbf{R}^+$ is defined by $f(x) = x^2$, $\forall x \in \mathbf{R}$, then $f^{-1}(x) = \{\pm x^{1/2}\}$ for $x > 0 \Rightarrow f^{-1}$ is not a function as $f^{-1}(x)$ consists of two elements.

In order that $f^{-1} : Y \to X$ is a function it is necessary that $(y, x_1), (y, x_2) \in f^{-1} \Rightarrow x_1 = x_2$ i.e., $(x_1, y), (x_2, y) \in f \Rightarrow x_1 = x_2$ i.e., $f(x_1) = f(x_2) \Rightarrow x_1 = x_2$ i.e., f is injective.

Theorem 1.2.7 *Let $f : X \to Y$ be a map. Then*

(i) *f is injective iff there is a map $g : Y \to X$ such that $g \circ f = I_X$.*
(ii) *f is surjective iff there is a map $h : Y \to X$ such that $f \circ h = I_Y$.*

Proof (i) Let f be injective. Then for each $y \in f(X)$ there is a unique $x \in X$ such that $f(x) = y$. Choose a fixed element $x_0 \in X$.

Define $g : Y \to X$ by

$$g(y) = \begin{cases} x & \text{if } y \in f(X) \text{ and } f(x) = y \\ x_0 & \text{if } y \notin f(X) \end{cases}$$

Then $g \circ f = I_X$.

Conversely, let there be a map $g : Y \to X$ such that $g \circ f = I_X$. Since I_X is bijective, it follows from Proposition 1.2.2(vi) that f is injective.

(ii) Suppose f is surjective. Then $f^{-1}(y) \subseteq X$ is a non-empty set for every $y \in Y$. For each $y \in Y$, choose $x_y \in f^{-1}(y)$. Then the map $h : Y \to X$, $y \mapsto x_y$ is such that $f \circ h = I_Y$.

Conversely, let there be a map $h : Y \to X$ such that $f \circ h = I_Y$. Since I_Y is bijective, $f \circ h$ is bijective and hence by Proposition 1.2.2(vi) it follows that f is surjective.   □

The maps g and h defined in Theorem 1.2.7 are, respectively, called a *left inverse of f* and a *right inverse of f*.

If a map $f : X \to Y$ has both a left inverse g and a right inverse h, then they are equal. This is so because $g = g \circ I_Y = g \circ (f \circ h) = (g \circ f) \circ h = I_X \circ h = h$.

The map $g = h$ is called an *inverse* (or two sided inverse) of f. Thus the inverse of a map f (if it exists) is unique.

Corollary *A map $f : X \to Y$ is bijective $\Leftrightarrow$ f has an inverse $\Leftrightarrow$ there exists $g : Y \to X$ such that $g \circ f = I_X$ and $f \circ g = I_Y$.*

Proof Left as an exercise.   □

The unique inverse of a bijection f is denoted by f^{-1}. Thus if f is bijective, then for each $y \in Y$, there exists a unique element $x \in X$ such that $f(x) = y$ and hence $f^{-1}(y)$ consists of a single element of X. Therefore there exists a correspondence that assigns to each $y \in Y$, a unique element $f^{-1}(y) \in X$. Accordingly, $f^{-1} : Y \to X$ is a function.

A map satisfying any one condition of Corollary of Theorem 1.2.7 is said to be invertible.

Theorem 1.2.8 *Let $f : X \to Y$ and $g : Y \to Z$ have inverse functions $f^{-1} : Y \to X$ and $g^{-1} : Z \to Y$. Then the composite function $g \circ f : X \to Z$ has an inverse function which is $f^{-1} \circ g^{-1} : Z \to X$.*

Proof As $f : X \to Y$ and $g : Y \to Z$ have inverse functions, then $f^{-1} \circ f = I_X$, $f \circ f^{-1} = I_Y$, $g^{-1} \circ g = I_Y$, $g \circ g^{-1} = I_Z$.

Now

$$\left(f^{-1} \circ g^{-1}\right) \circ (g \circ f) = f^{-1} \circ \left(g^{-1} \circ (g \circ f)\right) = f^{-1} \circ \left(\left(g^{-1} \circ g\right) \circ f\right)$$

$$= f^{-1} \circ I_Y \circ f = f^{-1} \circ f = I_X.$$

Similarly, $(g \circ f) \circ (f^{-1} \circ g^{-1}) = I_Z$.

Hence the theorem follows by using the Corollary of Theorem 1.2.7.   □

We now reach the climax of this sequence of theorems.

Theorem 1.2.9 *If X is a non-null set, then the assignment $\rho \mapsto X/\rho$ defines a bijection from the set $E(X)$ of all equivalence relations on X onto the set $\mathcal{P}(X)$ of all partitions of X.*

Proof If ρ is an equivalence relation on X, the set X/ρ of equivalence classes is the partition $\mathcal{P}_\rho$ of X by Theorem 1.2.1 so that $\rho \mapsto X/\rho = \mathcal{P}_\rho$ defines a function

$$f : E(X) \to \mathcal{P}(X).$$

Define a function $g : \mathcal{P}(X) \to E(X)$ as follows:

If $\mathcal{P} = \{X_i : i \in I\}$ is a partition of X, let $g(\mathcal{P})$ be the equivalence relation $\rho_\mathcal{P}$ on X given by:

$$(a, b) \in \rho_\mathcal{P} \quad \Leftrightarrow \quad a \in X_i \text{ and } b \in X_i \text{ for some (unique) } i \in I.$$

Then $\rho_\mathcal{P}$ is an equivalence relation on X by Theorem 1.2.2. Hence g is well defined. It is clear that $g \circ f = I_{E(X)}$ and $f \circ g = I_{\mathcal{P}(X)}$.

Because $g(f(\rho)) = g(\mathcal{P}_\rho) = \rho(\mathcal{P}_\rho) = \rho$ (by Theorem 1.2.2(b)), for all $\rho \in E(X) \Rightarrow g \circ f = I_{E(X)}$.

Again $(f \circ g)(\mathcal{P}) = f(\rho_\mathcal{P}) = \mathcal{P}(\rho_\mathcal{P}) = \mathcal{P}$ (by Theorem 1.2.2(a)), for all $\mathcal{P} \in \mathcal{P}(X) \Rightarrow f \circ g = I_{\mathcal{P}(X)}$.

Then the theorem follows from Theorem 1.2.7.   □

Obviously, the question arises whether or not given two sets have the same number of elements. For sets of finite number of elements the answer is obtained by counting the number of elements in each set. But the difficulty arises if the number of elements of the set is not finite.

The following concept, introduced by German mathematician Georg Cantor (1845–1918), removed this difficulty and led to the theory of sets.

Definition 1.2.21 A set X is *equivalent* (or similar or *equipotent*) to a set Y denoted by $X \sim Y$ iff there is a bijective map $f : X \to Y$.

Theorem 1.2.10 *The relation $\sim$ on the class of all sets is an equivalence relation.*

Proof $X \sim X$ as $I_X : X \to X$ is a bijective map for every X. Again $X \sim Y \Rightarrow Y \sim X$, since a bijective map $f : X \to Y$ has an inverse $f^{-1} : Y \to X$ which is also bijective. Finally, $X \sim Y$ and $Y \sim Z \Rightarrow X \sim Z$, since if $f : X \to Y$ and $g : Y \to Z$ are bijective maps, then $g \circ f : X \to Z$ is also bijective.   □

Example 1.2.21 (i) Consider the concentric circles $C_1 = \{(x, y) \in \mathbf{R} \times \mathbf{R} : x^2 + y^2 = a^2\}$ and $C_2 = \{(x, y) \in \mathbf{R} \times \mathbf{R} : x^2 + y^2 = b^2\}$ with center $O = (0, 0)$, where $0 < a < b$ and the function $f : C_2 \to C_1$, where $f(x)$ is the point of intersection of C_1 and the line segment from the center O to $x \in C_2$. Then f is a bijection.

The bijection of sets yields equivalent sets. Here C_1 and C_2 are *equivalent* as sets as shown in Example-1 of Fig. 1.6.

(ii) Let $\mathbf{C}_\infty = \mathbf{C} \cup \{\infty\}$ be the extended complex plane and $S^2 = \{(x, y, z) \in \mathbf{R}^3 : x^2 + y^2 + z^2 = 1\}$ be the unit sphere in $\mathbf{R}^3$. There exists a bijection $f : S^2 \to \mathbf{C}_\infty$ [Conway 1973] called *stereographic projection*, as shown in Example-2 of Fig. 1.6, and hence S^2 and $\mathbf{C}_\infty$ are equivalent sets. Thus $\mathbf{C}_\infty$ is represented as the sphere S^2 called the *Riemann sphere*.

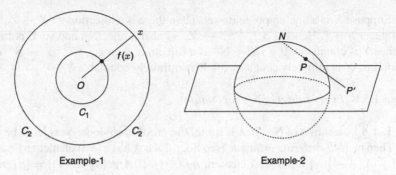

Fig. 1.6 *Example-1* represents the equivalent sets C_1 and C_2 and *Example-2* represents stereographic projection

Definition 1.2.22 A set X is said to be finite iff either $X = \emptyset$ or $X \sim \mathbf{Z}_n$ for some positive integer $n > 1$ i.e., iff X and $\mathbf{Z}_n$ have the same number of elements.

Proposition 1.2.3 *A finite set X cannot be equivalent to a proper subset of X.*

Proof It follows from Definition 1.2.22.   □

Definition 1.2.23 A set which is not finite is said to be infinite. $\mathbf{Z}$, $\mathbf{N}$ are evidently examples of infinite sets.

1.3 Countability and Cardinality of Sets

We are familiar with positive integers $1, 2, 3, \ldots$, and in daily life we use them for counting. They are adequate for counting the elements of a finite set. Beyond the area of mathematics, all sets are generally finite sets. But in mathematics, we come across many infinite sets, such as set of all natural numbers, set of all integers, set of all rational numbers, set of all real numbers, set of all points in a square or in a plane etc. We now describe a method, essentially due to Cantor for counting such infinite sets by introducing the concepts of countability and cardinality. The concept of countability of a set is very important in mathematics, in particular, in algebra, analysis, topology etc. This concept has great aesthetic appeal and develops through natural stages into an excellent structure of thought.

Definition 1.3.1 A set X is said to be *countable* iff either X is finite or there exists a bijection $f : X \to \mathbf{N}^+$. (In the latter case X is said to be *infinitely countable*.)

Proposition 1.3.1 *Let X be a countable set and $f : X \to Y$ be a bijection. Then Y is also a countable set.*

Proof Suppose X is a non-empty finite set. Then there is a bijection $g : X \to \mathbf{Z}_n$ for some integer $n > 1$. Hence $g \circ f^{-1} : Y \to \mathbf{Z}_n$ is also a bijection and so Y is finite and hence Y is countable. If $g : X \to \mathbf{N}^+$ is a bijection, then $g \circ f^{-1} : Y \to \mathbf{N}^+$ is a bijection i.e., X is infinitely countable iff Y is infinitely countable. $\square$

Lemma 1.3.1 *Every subset of* $\mathbf{N}^+$ *is countable.*

Proof Let A be a subset of $\mathbf{N}^+$. If A is finite, the proof is obvious. Next let A be not finite. Then by *well-ordering principle* (see Sect. 1.4), A has a least element n_1 (say). If $\{n_1\} \neq A$, $A - \{n_1\}$ has a least element n_2 (say). If $A \neq \{n_1\} \cup \{n_2\} = \{n_1, n_2\}$, the same process is continued to obtain

$$A = \{n_1, n_2, \ldots\}.$$

Clearly, there exists a bijection $f : A \to \mathbf{N}^+$ by $f(n_r) = r$ for $r = 1, 2, 3, \ldots$. Consequently, A is countable. $\square$

This result can be generalized as follows.

Theorem 1.3.1 *Every subset of a countable set is countable.*

Proof Let X be countable and $Y \subseteq X$. If $Y = \emptyset$ or Y is finite, the proof is obvious. We now consider the other case. As X is infinitely countable, there exists a bijection $f : X \to \mathbf{N}^+$. We define a map $g : Y \to \mathbf{N}^+$ by taking $g = f|Y$. Since f is a bijection and $Y \subseteq X$, g is injective and hence $g : Y \to g(Y) \subseteq \mathbf{N}^+$ is a bijection. As $g(Y)$ is countable by Lemma 1.3.1 and $g^{-1} : g(Y) \to Y$ is a bijection and hence Y is countable by Proposition 1.3.1. $\square$

If a set X is infinitely countable, then there exists a bijection $f : X \to \mathbf{N}^+$. The particular x, for which $f(x) = n$ is called the suffix of the element x denoted by x_n, for $n = 1, 2, 3, \ldots$. Every element of X may thus be suffixed, and the elements of X can be arranged as $x_1, x_2, x_3, \ldots$ in order of the suffixes $1, 2, 3, \ldots$. The set X, thus arranged, is called a *sequence*, and is usually denoted by $\{x_n\}$.

Proposition 1.3.2 *Every infinite set contains a countable subset.*

Proof Let X be an infinite set and $x_1 \in X$. Let $x_2 \in X - \{x_1\}, x_3 \in X - \{x_1, x_2\}, \ldots,$ $x_n \in X - \{x_1, x_2, \ldots, x_{n-1}\}$. Continuing this process, we obtain a sequence of distinct elements $x_1, x_2, \ldots, x_n, \ldots,$ of X. Then X contains a countable subset $\{x_1, x_2, \ldots, x_n, \ldots\}$. $\square$

Theorem 1.3.2 *The union of a countable aggregate of countable sets is countable.*

Proof Let $X_1, X_2, \ldots,$ be a countable aggregate of countable sets X_i. Let $X_i = \{x_{i1}, x_{i2}, \ldots : i \in \mathbf{N}^+\}$. Then the elements of their union $X = X_1 \cup X_2 \cup X_3 \cup \cdots$

can be arranged as a double array:

$$
\begin{array}{cccc}
x_{11}, & x_{12}, & x_{13}, & \cdots \\
x_{21}, & x_{22}, & x_{23}, & \cdots \\
\vdots & \vdots & \vdots & \vdots \\
x_{i1}, & x_{i2}, & x_{i3}, & \cdots \\
\vdots & \vdots & \vdots & \cdots
\end{array}
$$

where the elements of the ith set X_i are arranged in the ith row. Rearranging the elements we obtain a sequence:

$$
x_{11}; \qquad x_{12}, \quad x_{21}; \qquad x_{13}, \quad x_{22}, \quad x_{31}; \qquad \cdots
$$

which consists of 'blocks' separated by semicolons, of elements x_{ij}, where the nth block consists of all x_{ij} for which $i + j = n + 1$, with i increasing. In particular, an element x_{ij} occupies the ith position in the $(i + j - 1)$th block; the position being $[\{1 + 2 + \cdots + (i + j - 2)\} + i]$th from the beginning. Consequently, X is countable.   □

Theorem 1.3.3 *The set $\mathbf{Q}$ of all rationals is countable.*

Proof The set $\mathbf{Q}^+$ of all positive rational numbers is countable by Theorem 1.3.2, since $\mathbf{Q}^+$ is the union of the countable aggregate of countable sets $X_1, X_2, \ldots,$ $X_r, \ldots$, where X_r is the countable set $\{n/r\}$ for $r = 1, 2, \ldots$. Similarly, $\mathbf{Q}^-$ of all negative rational numbers is also countable. Consequently, the set $\mathbf{Q} = \mathbf{Q}^+ \cup \{0\} \cup \mathbf{Q}^-$ is countable.

We now utilize the concept of countability to define a cardinal number. Let $\mathcal{B}$ be the collection of sets. Then the relation $\sim$ (equivalent) is an equivalence relation on $\mathcal{B}$ (by Theorem 1.2.10). Consequently $\sim$ partitions $\mathcal{B}$ into sets of disjoint equivalence classes; each class carries a special name, called a *cardinal number*. Every set belongs to a cardinal number; this is generally expressed by saying that a set has a cardinal number. Finite sets, which are equivalent have the same number of elements; hence the number of elements of a finite set may be taken as its cardinal number. The cardinal numbers of each of the sets $\emptyset, \{1\}, \{1, 2\}, \{a, b, c\}, \ldots$ are $0, 1, 2, 3, \ldots$ respectively.   □

Cantor discovered that the infinite set of all real numbers is not countable. We prove this by using the following theorem.

Theorem 1.3.4 *The set $X = (0, 1]$ of all positive real numbers ≤ 1 is not countable.*

Proof Assume to the contrary, i.e., assume that X is countable. Then $X = \{x_1, x_2, \ldots, x_n, \ldots\}$, say. Now each element $x_n \in X$ can be expressed uniquely as a non-terminating decimal expression: $x_n = 0.x_{n1}x_{n2}x_{n3}\ldots$, where $x_{ni} \in \{0, 1, 2, \ldots, 9\}$, in a non-trivial manner (i.e., when all x_{ni}, for all $i > $ some m are not zero;

a terminating decimal expression is replaced by a non-terminating one, viz., $0.12 = 0.119999\ldots$). Then each element in X can be written as follows:

$$x_1 = 0.x_{11}\, x_{12}\, x_{13} \ldots$$
$$x_2 = 0.x_{21}\, x_{22}\, x_{23} \ldots$$
$$\vdots \qquad \vdots \quad \vdots \quad \vdots$$
$$x_n = 0.x_{n1}\, x_{n2}\, x_{n3} \ldots$$
$$\vdots \qquad \vdots \quad \vdots \quad \vdots \ldots$$

A non-terminating decimal expression $y = 0.y_1 y_2 y_3 \ldots y_n \ldots$, where y_n is taken to be a non-zero digit different from x_{nn} for $n = 1, 2, 3, \ldots$, does not enter in X, since y differs from every x_n at the nth place after decimal. But y determines a real number $\in X$. This contradicts that every real number $r \in X$ occurs in the sequence $\{x_1, x_2, \ldots, x_n, \ldots\}$. Consequently, X must be non-countable. $\qquad\square$

Corollary *The set* **R** *of all real numbers is non-countable.*

We extend the notion of counting by assigning to every set X (finite or infinite) an object $|X|$, called its cardinal number or cardinality, defined in such a way that $|X| = |Y|$ if and only if X is equipotent to Y i.e., $X \sim Y$. This definition carries sense by Theorem 1.2.10. Thus if X is a finite set of n elements then $|X| = n$. The concept of countability accommodates more infinite sets for determination of their cardinality; e.g., $|\mathbf{N}| = |\mathbf{Q}|$. The cardinal number d or c of an infinite set X asserts that the set is countable or not countable, respectively.

The cardinal number of an infinite set is called a *transfinite cardinal number*. In particular, the cardinal number of $\mathbf{N}^+$, and hence also of every infinite countable set, is denoted by the symbol d (or κ_0 i.e., aleph with suffix 0). Let the cardinal number of a set X be denoted by $|X|$. Then $|\mathbf{N}^+| = d$. As $\mathbf{R}$ is non-countable, $|\mathbf{R}| \neq d$ and $|\mathbf{R}|$ is denoted by c (or κ_1 i.e., aleph with suffix 1); and is called the *power* (or *potency*) *of the continuum*. In practice, infinite sets have cardinal number d or c.

We now extend the concept of natural ordering of positive integers to the set of cardinal numbers.

Let A and B be two sets such that $|A| = \alpha$ and $|B| = \beta$. Then we say that $\alpha \leq \beta$ (equivalently $\beta \geq \alpha$) iff the set A is equipotent to a subset of B. Also, we say $\alpha < \beta$ (or $\beta > \alpha$) iff $\alpha \leq \beta$ and $\alpha \neq \beta$ hold.

Proposition 1.3.3 *d is the smallest transfinite cardinal number.*

Proof $d \leq c$ by Proposition 1.3.2. But $d \neq c$ by the Corollary of Theorem 1.3.4. $\square$

The following theorems are used to show that the above relation '$\leq$' is antisymmetric.

Theorem 1.3.5 (Schroeder–Bernstein Equivalence Principle) *Let X, X_1 and X_2 be three sets such that $X \supset X_1 \supset X_2$ and $X \sim X_2$. Then $X \sim X_1$.*

Proof Since $X \sim X_2$, there exists a bijection $f : X \to X_2$. Furthermore, since $X \supset X_1$, $f|X_1$ is injective and hence $X_1 \sim f(X_1) = X_3$ (say) $\subset X_2$. Recursively, let $f(X_r) = X_{r+2}$, for $r = 1, 2, 3, \ldots$. Consequently, we have

$$X \supset X_1 \supset X_2 \supset X_3 \supset \cdots \supset X_r \supset X_{r+1} \supset X_{r+2} \supset \cdots,$$

where $X \sim X_2 \sim X_4 \sim X_6 \sim \cdots$ and $X_1 \sim X_3 \sim X_5 \sim X_7 \sim \cdots$ and hence $X - X_1 \sim X_2 - X_3 \sim X_4 - X_5 \sim X_6 - X_7 \sim \cdots$ holds under f. Thus $(X - X_1) \cup (X_2 - X_3) \cup (X_4 - X_5) \cup (X_6 - X_7) \cup \cdots \sim (X_2 - X_3) \cup (X_4 - X_5) \cup (X_6 - X_7) \cup \cdots$ holds under f.

Let

$$Y = X \cap X_1 \cap X_2 \cap X_3 \cap \cdots.$$

Then

$$X = Y \cup (X - X_1) \cup (X_1 - X_2) \cup (X_2 - X_3) \cup (X_3 - X_4) \cup \cdots,$$

and

$$X_1 = Y \cup (X_1 - X_2) \cup (X_2 - X_3) \cup (X_3 - X_4) \cup (X_4 - X_5) \cup \cdots.$$

Define the map $g : X \to X_1$ as follows:

$$g(x) = \begin{cases} f(x), & \text{if } x \in (X - X_1) \cup (X_2 - X_3) \cup (X_4 - X_5) \cup \cdots, \\ x, & \text{if } x \in (X_1 - X_2) \cup (X_3 - X_4) \cup (X_5 - X_6) \cup \cdots \text{ or } x \in Y. \end{cases}$$

Then g is a bijection. Consequently, $X \sim X_1$. $\qquad\square$

Theorem 1.3.6 (Schroeder–Bernstein Theorem) *If for given two sets X and Y, $X \sim Y_1 \subset Y$ and $Y \sim X_1 \subset X$, then $X \sim Y$.*

Proof Let $f : X \to Y_1$ be a bijection and $f(X_1) = Y_2$. Then $X_1 \sim Y_2$ and $Y_1 \supset Y_2$. Now $Y \supset Y_1 \supset Y_2$. Moreover, $Y \sim X_1$ (by hypothesis) and $X_1 \sim Y_2 \Rightarrow Y \sim Y_2$. Then $Y \sim Y_1$ by Theorem 1.3.5. Hence $X \sim Y_1$ and $Y \sim Y_1 \Rightarrow X \sim Y$. $\qquad\square$

Corollary *If α and β are two cardinal numbers such that $\alpha \leq \beta$ and $\beta \leq \alpha$, then $\alpha = \beta$.*

Applications of Zorn's Lemma and the Schroeder–Bernstein Theorem

A(i) *Let A be a partially ordered set. Then $\exists$ a totally ordered subset of A which is not a proper subset of any other totally ordered subset of A.*

Proof Let $\mathcal{B}$ be the class of all totally ordered subsets of A. Order $\mathcal{B}$ partially by set inclusion. We prove by Zorn's Lemma (Lemma 1.2.1) that $\mathcal{B}$ has a maximal element. Let $\mathcal{A} = \{B_i : i \in I\}$ be a totally ordered subclass of $\mathcal{B}$ and $X = \bigcup\{B_i : i \in I\}$. Then $B_i \subset A \; \forall B_i \in \mathcal{A} \Rightarrow X \subset A$. We claim that X is totally ordered. Let $x, y \in X$. Then $\exists B_j, B_k \in \mathcal{A}$ such that $x \in B_j$, $y \in B_k$ for some $j, k \in I$. As $\mathcal{A}$ is totally ordered by set inclusion, one of them is a subset of other. Without loss of generality, let $B_j \subset B_k$. Then $x, y \in B_k \in \mathcal{A}$, a totally ordered subset of A. Hence either $x \leq y$ or $y \leq x$. Then X is a totally ordered subset of A and hence $X \in \mathcal{B}$. As $B_i \subset X$ $\forall B_i \in \mathcal{A}$, X is an upper bound of $\mathcal{A}$. Thus every totally ordered subset of $\mathcal{B}$ has an upper bound in $\mathcal{B}$. Hence by Zorn's Lemma $\mathcal{B}$ has a maximal element. In other words, $\exists$ a totally ordered subset of A which is not a proper subset of any other totally ordered subset of A. $\square$

A(ii) *Let* $\{A_i\}_{i \in I}$ *be any infinite family of countable sets. If* $A = \bigcup_{i \in I} A_i$, *then card* $A \leq card\ I$, *i.e.,* $|A| \leq |I|$.

Proof If I is countably infinite, then A being the countable union of countable sets is also countable by Theorem 1.3.2. If I is uncountable, then A can be represented by Zorn's Lemma as the union of a disjoint family of countably infinite subsets. Hence card $A \leq$ card I. $\square$

B *Let* $I = [0, 1]$ *and* $I^n = \{(x_1, x_2, \ldots, x_n) : x_i \in I,\ for\ i = 1, \ldots, n\}$ *be the* n-*dimensional unit cube. Then* $I \sim I^n$.

Proof Define $T = \{(x, 0, 0, \ldots, 0) \in I^n\}$. Then $T \subset I^n$. Define a bijection $f : I \to T$ by $f(x) = (x, 0, 0 \ldots, 0)$.

Hence $I \sim T \subset I^n$.

Again for $(x_1, x_2, \ldots, x_n) \in I^n$, x_i can be expressed uniquely as a non-terminating decimal expression:

$$x_i = 0.x_{i1}x_{i2} \ldots, \quad for\ i = 1, 2, \ldots, n.$$

Now define an injective map $g : I^n \to I$ by

$$g(0.x_{11}x_{12} \ldots, 0.x_{21}x_{22} \ldots, \ldots, 0.x_{n1}x_{n2} \ldots) = 0.x_{11}x_{21} \ldots x_{n1}x_{12}x_{22} \ldots x_{n2} \ldots.$$

Then g yields a bijection from I^n onto $g(I^n) = V \subset I$.

Thus $I^n \sim V \subset I$.

By the Schroeder–Bernstein Theorem, (i) and (ii) yield $I^n \sim I$. $\square$

Remark The cardinal numbers of I^n and I are the same.

There is a question arising naturally: does there exist any transfinite cardinal number greater than c? The answer is affirmative. The following theorem prescribes a definite method for obtaining a set, whose cardinal number is greater than the cardinal number of the given set.

Theorem 1.3.7 *If A is a non-empty set and $\mathcal{P}(A)$ is its power set, then $|A| < |\mathcal{P}(A)|$.*

Proof The map $f : A \to \mathcal{P}(A)$ defined by $f(a) = \{a\}$, for all $a \in A$ is injective. Hence, $|A| \leq |\mathcal{P}(A)|$. But there is no bijection $g : A \to \mathcal{P}(A)$ (see Ex. 8 of SE-I). Hence, $|A| \neq |\mathcal{P}(A)|$ implies that $|A| < |\mathcal{P}(A)|$. $\square$

Using this theorem and the result in Ex. 14 of SE-I, the following corollary follows.

Corollary *For any cardinal number* α, $\alpha < 2^{\alpha}$, *where* α *is the cardinality of some set A and* 2^{α} *is the cardinality of the set of all functions from A to* $\{0, 1\}$. *In particular,* $c < 2^c$.

Remark Like natural numbers, the law of trichotomy holds for cardinal numbers.

1.3.1 Continuum Hypothesis

We have shown the existence of three distinct transfinite cardinal numbers d, c and 2^c such that $d < c < 2^c$.

We now state the following natural questions which are still unsolved:

Unsolved Problem 1 Does there exist any cardinal number α such that $d < \alpha < c$?

Unsolved Problem 2 Does there exist any cardinal number β such that $c < \beta < 2^c$?

Cantor's continuum hypothesis stated by Georg Cantor in 1878 says that there is no cardinal number α strictly lying between d and c. Karl Gödel showed in 1939 that if the usual axioms of set theory are consistent, then the introduction of continuum hypothesis does not make any inconsistency [see Cohn (2002)]. Again, the generalized continuum hypothesis asserts that there is no cardinal number β satisfying $\alpha < \beta < 2^{\alpha}$, for any transfinite cardinal α. P.J. Cohen showed in 1963 that the generalized continuum hypothesis is independent of axioms of set theory [see Cohen (1966)].

Supplementary Examples (SE-I)

1 Let A, B, C, and D be non-empty sets. Then

(i) $(A \times B) \cap (C \times D) = (A \cap C) \times (B \cap D)$;
(ii) $(A \times B) \subseteq C \times C \Rightarrow A \subseteq C$;
(iii) $A \times B = B \times A \Leftrightarrow A = B$ (see Proposition 1.1.5);
(iv) $A \neq B \Rightarrow A \times B \neq B \times A$.

[*Hint*. (i) $(a, b) \in (A \times B) \cap (C \times D) \Rightarrow (a, b) \in A \times B$ and $(a, b) \in (C \times D) \Rightarrow a \in A, b \in B, a \in C$ and $b \in D \Rightarrow a \in A \cap C$ and $b \in B \cap D \Rightarrow (a, b) \in (A \cap C) \times (B \cap D) \Rightarrow (A \times B) \cap (C \times D) \subseteq (A \cap C) \times (B \cap D)$.

Similarly, $(a, b) \in (A \cap C) \times (B \cap D) \Rightarrow (a, b) \in (A \times B) \cap (C \times D) \Rightarrow (A \cap C) \times (B \cap D) \subseteq (A \times B) \cap (C \times D)$. Consequently, $(A \times B) \cap (C \times D) = (A \cap C) \times (B \cap D)$.

(ii) Let b be an arbitrary element of B. Then $a \in A \Rightarrow (a, b) \in A \times B \subseteq C \times C \Rightarrow a \in C \Rightarrow A \subseteq C$.

(iii) $(a, b) \in A \times B = B \times A \Rightarrow a \in B$ and $b \in A \Rightarrow A \subseteq B$ and $B \subseteq A \Rightarrow A = B$. The converse is trivial.

(iv) Suppose $A \neq B$. If possible $A \times B = B \times A$. Then (iii) yields $A = B \Rightarrow$ a contradiction.]

2 For each $n \in \mathbf{N}^+$, let $A_n = [0, \frac{1}{n}]$. Then $\bigcap_{n=1}^{\infty} A_n = \{0\}$.

[*Hint.* Let $M = \bigcap_{n=1}^{\infty} A_n$. Clearly $0 \in M$. We claim that $M = \{0\}$. Otherwise, there is a real number $x \in M$ such that $0 < x \leq \frac{1}{n}$ for all $n \in \mathbf{N}^+$. Then $n < \frac{1}{x}$ for all $n \in \mathbf{N}^+ \Rightarrow$ a contradiction.]

3 Let A, B be two given sets such that $A \subset B$. A relation ρ is defined on $\mathcal{P}(B)$ by $(X, Y) \in \rho \Leftrightarrow X \cap A = Y \cap A$. Then

  (i) ρ is an equivalence relation;
 (ii) ρ class $\emptyset\rho$ of $\emptyset$ is $\mathcal{P}(B - A)$;
(iii) $|A| = m \Rightarrow |\mathcal{P}(B)/\rho| = 2^m$.

4 (i) A relation ρ on $\mathbf{R}^2$ is defined by $((a, b), (x, y)) \in \rho \Leftrightarrow a + y = b + x$. Then ρ is an equivalence relation.

(ii) Let ρ and μ be the relations on $\mathbf{R}$ defined by $\rho = \{(x, y) : x^2 + y^2 = 1\}$ and $\mu = \{(y, z) : 2y + 3z = 4\}$. Then $\mu \circ \rho = \{(x, z) : 4x^2 + 9z^2 - 24z + 12 = 0\}$ (by eliminating y from the above two equations).

Remark Considering $\mathbf{R}^2$ as the Euclidean plane, the ρ-class $(a, b)\rho$, in 4(i), represents geometrically the straight line having gradient 1 and passing through the point (a, b).

5 Let $M_n(\mathbf{R})$ denote the set of $n \times n$ real matrices. Define a binary relation ρ on $M_n(\mathbf{R})$ by $(A, B) \in \rho \Leftrightarrow A^r = B^t$ for some $r, t \in \mathbf{N}^+$. Then ρ is an equivalence relation.

6 For any real x, the symbol $[x]$ denotes the greatest integer less than or equal to x (for existence of $[x]$, see Ex. 5 of SE-II).

(a) Let the relation ρ be defined on $\mathbf{R}$ by $(x, y) \in \rho \Leftrightarrow [x] = [y]$. Then ρ is an equivalence relation and for each $n \in \mathbf{N}^+$, $n\rho = [n, n+1)$.
(b) Let $f, g : \mathbf{N}^+ \to \mathbf{N}^+$ be the mapping given by $f(x) = 2x$, $g(x) = [\frac{1}{2}(x + 1)]$. Then f is injective but not surjective, and g is surjective but not injective.
(c) Let $h : \mathbf{N}^+ \to \mathbf{Z}$ be defined by $h(n) = (-1)^n[\frac{n}{2}]$.

   Then h is a bijection.

7 Let the relation ρ be defined on $\mathbf{N}^+ \times \mathbf{N}^+$ by $(a, b)\rho(x, y) \Leftrightarrow a + y = b + x$. Then ρ is an equivalence relation. Define a map $f : (\mathbf{N}^+ \times \mathbf{N}^+)/\rho \to \mathbf{Z}$ by

$$f((a, b)\rho) = a - b, \quad \forall a, b \in \mathbf{N}^+.$$

Then f is well defined and bijective.

8 Let S be a non-empty set. Then there exists no bijection $f : S \to \mathcal{P}(S)$.

[*Hint.* If possible there exists a bijection $f : S \to \mathcal{P}(S)$. Define A by the rule: $A = \{x \in S : x \notin f(x)\}$. Then $A \in \mathcal{P}(S)$. Clearly, f is a bijection $\Rightarrow$ there exists an $s \in S$ such that $f(s) = A$. Now the definition of A shows that if $s \in A$, then $s \notin f(s) = A$ and conversely. This yields a contradiction.]

9 Let $A = \{0, 1\}$ and B, C be non-empty sets and $f : B \to C$ be a non-surjective map. Then there exist distinct maps $g, h : C \to A$ such that $g \circ f = h \circ f$.

[*Hint.* Define $g, h : C \to A$ by $g(x) = 0$ for all $x \in C$ and $h(x) = 0$ if $x \in \mathrm{Im} f \subsetneq C = 1$ if $x \in (C - \mathrm{Im} f)\ (\neq \emptyset)$. Show that $g \neq h$ but $g \circ f = h \circ f$.]

10 If $f : A \to B$, $g : B \to C$ and $h : B \to C$ are maps such that $g \circ f = h \circ f$ and f is surjective. Then $g = h$.

[*Hint.* Let b be an arbitrary element of B. Then there exists some $a \in A$ such that $b = f(a)$, since f is surjective. Now $g(b) = g(f(a)) = (g \circ f)(a) = (h \circ f)(a) = h(f(a)) = h(b) \Rightarrow g = h$ (since $g, h : B \to C$ and b is an arbitrary element of B).]

11 $(A \times B) \times C = A \times (B \times C) \Leftrightarrow$ at least one of the sets A, B, C is empty.

[*Hint.* If one of the sets A, B, C is empty, then by Proposition 1.1.3, $(A \times B) \times C = \emptyset = A \times (B \times C)$. Conversely, suppose $(A \times B) \times C = A \times (B \times C)$. If none of the sets A, B, C is $\emptyset$, then there exists at least one element $(a, c) \in (A \times B) \times C$ where $a \in A \times B$ and $c \in C$. By hypothesis $(a, c) \in A \times (B \times C) \Rightarrow a \in A \Rightarrow$ a contradiction.]

12 (i) Any two open intervals (a, b) and (c, d) are equivalent;

(ii) All the intervals $(0, 1)$, $[0, 1)$ and $(0, 1]$ are equivalent to $[0, 1]$;

(iii) Any open interval is equivalent to $[0, 1]$;

(iv) **R** is equivalent to $[0, 1]$.

[*Hint.* (i) The map $f : (a, b) \to (c, d)$ defined by $f(x) = c + \frac{d-c}{b-a}(x - a)$ is a bijection.

(ii) Let $A = \{x \in (0, 1) : x$ cannot be expressed in the form $\frac{1}{n}$ for integers $n > 1\} = (0, 1) - \{\frac{1}{2}, \frac{1}{3}, \frac{1}{4}, \ldots\}$. Hence $(0, 1) = A \cup \{\frac{1}{2}, \frac{1}{3}, \frac{1}{4}, \ldots\}$.

Then the function $f : [0, 1] \to (0, 1)$ defined by

$$f(x) = \frac{1}{n+2} \quad \text{if } x = \frac{1}{n},\ n \in \mathbf{N}^{+}$$

$$= \frac{1}{2} \quad \text{if } x = 0$$

$$= x \quad \text{if } x \in A$$

is a bijection.

The sets $[0, 1]$ and $[0, 1)$ are equivalent under the bijection $g : [0, 1] \to [0, 1)$ defined by

$$g(x) = x \qquad \text{if } x \neq \frac{1}{n}, \ n \in \mathbf{N}^+$$

$$= \frac{1}{n+1} \quad \text{if } x = \frac{1}{n}, \ n \in \mathbf{N}^+.$$

(iii) Any open interval (a, b) is equivalent to $(0, 1)$ by (i) and $(0, 1)$ is equivalent to $[0, 1]$ by (ii). Hence (a, b) is equivalent to $[0, 1]$ by Theorem 1.2.10.

(iv) Consider the bijection $f : (-\frac{\pi}{2}, \frac{\pi}{2}) \to \mathbf{R}$ defined by $f(x) = \tan x$. Then use (iii) and Theorem 1.2.10.]

13 For any non-empty set A, the collection $\mathcal{B}(A)$ of all characteristic functions on the power set of A is equivalent to the power set $\mathcal{P}(A)$.

[*Hint.* Define the function $f : \mathcal{P}(A) \to \mathcal{B}(A)$ by $f(X) = \chi_X$, the characteristic function of X, given by

$$\chi_X : X \to \{0, 1\};$$

where

$$\chi_X(x) = 1 \quad \text{if } x \in X$$
$$= 0 \quad \text{if } x \in A - X = X'.$$

Then f is a bijection.]

14 Let $f : A \to B$ be one–one. Then f induces a set function $f* : \mathcal{P}(A) \to \mathcal{P}(B)$ such that $f*$ is one–one.

[*Hint.* Define $f* : \mathcal{P}(A) \to \mathcal{P}(B)$ by $f * (X) = f(X)$.

If $A = \emptyset$, $\mathcal{P}(A) = \{\emptyset\}$ and hence $f*$ is one–one.

If $A \neq \emptyset$, $\mathcal{P}(A)$ has at least two elements. Let $X, Y \in \mathcal{P}(A)$ be such that $X \neq Y$. Then $\exists p \in X$ such that $p \notin Y$, (or $p \in Y$, $p \notin X$). Thus $f(p) \in f(X)$ and since f is one–one, $f(p) \notin f(Y)$ (or $f(p) \in f(Y)$ but $f(p) \notin f(X)$). Hence $f(X) \neq f(Y) \Rightarrow f*$ is one–one.]

15 Let ρ be an equivalence relation on a non-empty set A. Then the natural projection map $p : A \to A/\rho$ defined by $p(a) = a\rho = (a)$, the equivalence class of a under ρ is not injective but surjective.

16 Let $f : A \to B$ be a map and ρ be an equivalence relation on A defined by $a\rho b \Leftrightarrow f(a) = f(b)$. Let $\tilde{f} : A/\rho \to f(A)$ be the map defined by $\tilde{f}(a\rho) = f(a)$. Then $\tilde{f}$ is a bijection.

[*Hint.* (i) $\tilde{f}$ is independent of the choice of the representatives of the classes and hence $\tilde{f}$ is well defined. Again for any $x \in f(A)$, $\exists a \in A$ such that $x = f(a)$. Then $\tilde{f}(a\rho) = f(a) = x$ shows that $\tilde{f}$ is surjective. Clearly, $\tilde{f}$ is a bijection.]

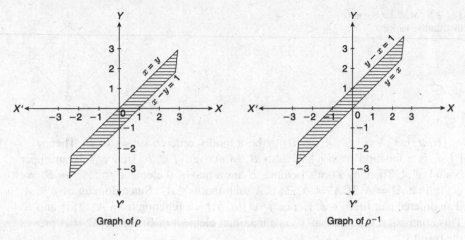

Fig. 1.7 Graphs of ρ and ρ^{-1}

17 Let ρ be the relation on **R** defined by $x\rho y \Leftrightarrow 0 \le x - y \le 1$. Express ρ and ρ^{-1} as subsets of $\mathbf{R} \times \mathbf{R}$ and draw their graphs, where

$$\rho = \{(x, y) : x, y \in \mathbf{R}, \ 0 \le x - y \le 1\} \subset \mathbf{R} \times \mathbf{R}$$

and

$$\rho^{-1} = \{(x, y) : (y, x) \in \rho\} = \{(x, y) : x, y \in \mathbf{R}, \ 0 \le y - x \le 1\} \subset \mathbf{R} \times \mathbf{R}.$$

[*Hint.* See Fig. 1.7.]

18 Let ρ be a relation from a set A to a set B. Then ρ uniquely defines a subset ρ' of $A \times B$ as follows:

$$\rho^* = \{(a, b) \in A \times B : a\rho b\} \subset A \times B.$$

Again any subset ρ^* of $A \times B$ defines a relation ρ from A to B by $a\rho b \Leftrightarrow (a, b) \in \rho^*$.

$$\text{Domain of } \rho = \{a \in A : (a, b) \in \rho \text{ for some } b \in B\},$$

$$\text{Range of } \rho = \{b \in B : (a, b) \in \rho \text{ for some } a \in A\}.$$

Note that this correspondence between the relation ρ from A to B and subsets of $A \times B$ justifies Definition 1.2.1.

Let ρ be a relation from A to B, i.e., $\rho \subset A \times B$ and suppose that the domain of ρ is A. Then $\exists$ a subset f^* of ρ such that f^* is a function from A into B.

Let $\mathcal{B}$ be the class of subsets f of ρ such that $f \subset \rho$ is a function from a subset of A into B. Order $\mathcal{B}$ partially by set inclusion. If $f : A_1 \to B$ is subset of $g : A_2 \to B$, then $A_1 \subset A_2$.

Fig. 1.8 Modular but not
distributive lattice

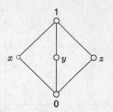

[*Hint.* Let $\mathcal{A} = \{f_i : A_i \to B\}_{i \in I}$ be a totally ordered subset of $\mathcal{B}$. Then $f = \bigcup_i f_i$ is a function from $\bigcup_i A_i$ into B. Moreover, $f \subset \rho$. Hence f is an upper bound of $\mathcal{A}$. Then, by Zorn's Lemma, $\mathcal{B}$ has a maximal element $f^* : A^* \to B$. We claim that $A^* = A$. If $A^* \neq A$, $\exists a \in A$ such that $a \notin A^*$. Since domain of $\rho = A$, $\exists$ an ordered pair $(a, b) \in \rho$. Hence $f^* \cup \{(a, b)\}$ is a function from $A^* \cup \{a\}$ into B. This contradicts the fact that f^* is a maximal element in $\mathcal{B}$. Hence $A^* = A$ proves the result.]

19 Let ρ be the relation on $\mathbf{N}^+$ defined by $(a, b) \in \rho \Leftrightarrow a < b$. Hence $(a, b) \in \rho^{-1} \Leftrightarrow b < a$. Then $\rho \circ \rho^{-1} \neq \rho^{-1} \circ \rho$.
   [*Hint.* $\rho \circ \rho^{-1} = \{(x, y) \in \mathbf{N}^+ \times \mathbf{N}^+ : \exists b \in \mathbf{N}^+ \text{ such that } (x, b) \in \rho^{-1}, (b, y) \in \rho\} = \{(x, y) \in \mathbf{N}^+ \times \mathbf{N}^+ : \exists b \in \mathbf{N}^+ \text{ such that } b < x, b < y\} = (\mathbf{N}^+ - \{1\}) \times (\mathbf{N}^+ - \{1\})$.
   Again

$$\rho^{-1} \circ \rho = \{(x, y) \in \mathbf{N}^+ \times \mathbf{N}^+ : \exists b \in \mathbf{N}^+ \text{ such that } (x, b) \in \rho, (b, y) \in \rho^{-1}\}$$
$$= \{(x, y) \in \mathbf{N}^+ \times \mathbf{N}^+ : \exists b \in \mathbf{N}^+ \text{ such that } x < b, y < b\}$$
$$= \mathbf{N}^+ \times \mathbf{N}^+.]$$

20 The lattice given by the following poset is modular, but not distributive, as shown in Fig. 1.8. Because $x \vee (y \wedge z) = x \vee 0 = x \neq 1 = (x \vee y) \wedge (x \vee z)$.

1.3.2 Exercises

Exercises-I

1. Let U be the universal set and A, B, S, T, X are sets. Prove that

   (i) $A \cap B \subseteq A$ and $A \cap B \subseteq B$;
   (ii) $U \cap A = A$;
   (iii) $A \cap \emptyset = \emptyset$;
   (iv) $(A - B) \subseteq A$;
   (v) $(A - B) \cap B = \emptyset$;
   (vi) $B - A \subseteq A'$;
   (vii) $B - A' = B \cap A$;
   (viii) $A \cap (B \triangle C) = (A \cap B) \triangle (A \cap C)$;

Fig. 1.9 Commutativity of
the diagram

(ix) $(A \cup B) \cap (A \cup B') = A$;

(x) $(A \cap B) \cup (A \cap B') = A$;

(xi) $\underline{(A \cup B)' = A' \cap B' \text{ and } (A \cap B)' = A' \cup B'}$

De Morgan's Law.

(xii) $S \subseteq T \Rightarrow S \cup X \subseteq T \cup X$ and $S \cap X \subseteq T \cap X$ for any set X;

(xiii) $S \subseteq T$ and $T \subseteq X \Rightarrow S \subseteq X$;

(xiv) $S \subseteq T \Leftrightarrow S \cap T = S$;

(xv) $T \subseteq S \Leftrightarrow S = T \cup S$;

(xvi) $T \subseteq S \Rightarrow (S - T) \cup T = S$;

(xvii) $S \times T = T \times S \Leftrightarrow$ either $S = T$ or at least one of S, T is $\emptyset$;

(xviii) $S \times T = S \times A \Leftrightarrow$ either $T = A$ or $S = \emptyset$;

(xix) $(S \times T) \times A = S \times (T \times A) \Leftrightarrow$ at least one of the sets S, T, A is $\emptyset$.

2. Prove the following:

   (i) If S is any set, then $S \times S$ is an equivalence relation on S.

   (ii) If ρ and σ are symmetric relations on a set A, then $\rho \circ \sigma$ is also a symmetric relation on $A \Leftrightarrow \rho \circ \sigma = \sigma \circ \rho$.

   (iii) If a relation ρ on a set A is transitive, then σ^{-1} is also transitive on A.

   (iv) ρ is a reflexive and transitive relation on $A \Rightarrow \rho \cap \rho^{-1}$ is an equivalence relation.

3. If A and B are two sets and $f : A \rightarrow B$ is a surjection, then show that

   (i) f induces an equivalence relation ρ on A;

   (ii) there is a surjection $g : A \rightarrow A/\rho$; and

   (iii) there is a bijection $h : B \rightarrow A/\rho$ such that the diagram as shown in Fig. 1.9 is commutative, i.e., $h \circ f = g$.

4. If A, B, C are sets, then show that

   (a) (i) $A \times B \sim B \times A$;

       (ii) $(A \times B) \times C \sim A \times B \times C \sim A \times (B \times C)$;

   (b) $(0, 1) \sim \mathbf{R}$.

5. Let L be a lattice. If $x, y, z \in L$, then prove that

   (i) $x \wedge x = x$ and $x \vee x = x$;

   (ii) $x \wedge y = y \wedge x$ and $x \vee y = y \vee x$;

   (iii) $x \wedge (y \wedge x) = (x \wedge y) \wedge z$ and $x \vee (y \vee z) = (x \vee y) \vee z$;

   (iv) $(x \wedge y) \vee x = x$ and $(x \vee y) \wedge x = x$.

6. Let L be a non-empty set in which two operations $\wedge$ and $\vee$ are defined and assume that these operations satisfy the conditions of Exercise 5. A binary relation $\leq$ are defined on $L : x \leq y \Leftrightarrow x \wedge y = x$. Then prove that $(L, \leq)$ is a lattice in which $x \wedge y$ and $x \vee y$ are the glb and lub of x and y, respectively.

Fig. 1.10 Diamond lattice

7. A lattice $(L, \leq)$ is said to be complemented iff it contains distinct elements 0 and 1 such that $0 \leq x \leq 1$ for every $x \in L$ and each element x has a complement x' with the property: $x \wedge x' = 0$ and $x \vee x' = 1$. *Boolean algebra* is defined to be a complemented distributive lattice. If B is a Boolean algebra, then prove that

   (i) each element has only one complement;
   (ii) $0' = 1$ and $1' = 0$;
   (iii) $x \leq y \Leftrightarrow y' \leq x'$ for $x, y \in B$;
   (iv) $(x \wedge y)' = x' \vee y'$ and $(x \vee y)' = x' \wedge y'$ for all $x, y \in B$.

8. Prove the following:

   (a) If in a poset P, every subset (including $\emptyset$) has a glb in P, then P is a complete lattice.
   (b) If a poset P has 1, and every non-empty subset X of P has a glb, then P is a complete lattice.

9. If L is a lattice, prove that the following statements are equivalent:

   (i) L is distributive;
   (ii) $ab + c = (a + c)(b + c)$;
   (iii) $ab + bc + ca = (a + b)(b + c)(c + a), \forall a, b, c \in L$, (where $a \wedge b = ab$ and $a \vee b = a + b$).

10. Give an example of a lattice which is modular but not distributive. Consider the diamond as shown in Fig. 1.10. This lattice is modular, as there is no pentagon. But it is not distributive, since $a(b + c) = a1 = a$ and $ab + ac = 0 \Rightarrow a(b + c) \neq ab + ac$.

11. Prove the following:

   (i) If A and B are two finite sets such that $|A| = m$ and $|B| = n$, then $|A \times B| = mn$;
   (ii) If $|X| = n$, then

      (a) the set of all possible mappings $X \rightarrow X$ consists of n^n elements;
      (b) the set of all bijections $X \rightarrow X$ consists of $n!$ elements (each bijection on X is called a permutation of X).

12. Let α and β be cardinal numbers and A, B disjoint sets that $|A| = \alpha$, $|B| = \beta$. Define $\alpha + \beta = |A \cup B|$, $\alpha\beta = |A \times B|$ and $\beta^\alpha = |B^A|$, the cardinal number of the set B^A of all functions from A to B. Prove that for any cardinal numbers α, β and γ,

(i) $(\alpha + \beta) + \gamma = \alpha + (\beta + \gamma)$;
(ii) $(\alpha\beta)\gamma = \alpha(\beta\gamma)$;
(iii) $\alpha + \beta = \beta + \alpha$;
(iv) $\alpha\beta = \beta\alpha$;
(v) $\alpha(\beta + \gamma) = \alpha\beta + \alpha\gamma$;
(vi) $\alpha^\beta \alpha^\gamma = \alpha^{\beta+\gamma}$;
(vii) $(\alpha^\beta)^\gamma = \alpha^{\beta\gamma}$;
(viii) $(\alpha\beta)^\gamma = \alpha^\gamma \beta^\gamma$.

13. Show by examples that the cancellation law for the operations of addition and multiplication of cardinal numbers is not true.

14. Let $\alpha = |A|$ and $\beta = |B|$. Furthermore, let $A \sim B_i \subseteq B$; i.e., suppose there is a mapping $f : A \to B$ which is injective. Then write $A \leq B$ which reads 'A precedes B', and $\alpha \leq \beta$ which reads 'α is less than or equal to β'.

Suppose $A < B$ means $A \leq B$ and $A \neq B$; $\alpha < \beta$ means $\alpha \leq \beta$ and $\alpha \neq \beta$. Then prove the following:

(i) the relation on sets defined by $A \leq B$ is reflexive and transitive, and the relation on the cardinal numbers defined by $\alpha \leq \beta$ is also reflexive and transitive;
(ii) for any cardinal number α, $\alpha < 2^\alpha$ (Cantor's Theorem);
(iii) for cardinal numbers α and β, $\alpha \leq \beta$ and $\beta \leq \alpha \Rightarrow \alpha = \beta$.

1.4 Integers

We are familiar with positive integers as natural numbers from our childhood through the process of counting and start mathematics with them in an informal way by learning to count followed by learning addition and multiplication, the latter being a repeated process of addition. The discovery of positive integers goes back to early human civilization. *Leopold Kronecker* (1823–1891), a great mathematician, once said '*God made the positive integers, all else is due to man*'. The aim of this section is to establish certain elementary properties of integers, which we need in order to develop and illustrate the materials of later chapters. Moreover, some results on theory of numbers are proved in Chap. 10 with an eye to apply them to Basic Cryptography (Chap. 12).

G. *Peano* (1858–1932), an Italian mathematician, first showed in 1889 that the positive integers can be defined formally by a set of axioms, which is now known as *Peano's axioms*, which form the foundation of arithmetic.

Peano's Axiom A non-empty set $\mathbf{N}^+$, having the following properties is called the set of natural numbers and every element of $\mathbf{N}^+$ is called a *natural number*:

(1) to every $n \in \mathbf{N}^+$, there corresponds a unique element $s(n) \in \mathbf{N}^+$, called the successor of n and n is called the predecessor of $s(n)$;
(2) $s(n) = s(m)$ implies $n = m$;

(3) there exists an element of $\mathbf{N}^+$, denoted by 1, which is not the successor of any element of $\mathbf{N}^+$, i.e., 1 has no predecessor;

(4) if $A \subseteq \mathbf{N}^+$ is such that $1 \in A$, and $n \in A$ implies $s(n) \in A$, then $A = \mathbf{N}^+$.

The axiom (4) leads to one of the important principles of mathematics known as the '*Principle of Mathematical Induction*' i.e., if p is a property such that 1 has the property p and whenever a particular natural number n has the property p, implies $s(n)$ has also the property, then every natural number has the property p.

Definition 1.4.1 For $n, m \in \mathbf{N}^+$, we define

(a) *addition* '+' recursively by $n + 1 = s(n)$, $n + s(m) = s(n + m)$;
(b) *multiplication* '·' by $n \cdot 1 = n$, $n \cdot s(m) = n \cdot m + n$;
(c) *order relation* $>$ (or $<$): m is said to be greater than n denoted by $m > n$ (or equivalently, n is said to be less than m denoted by $n < m$) iff $m = n + p$ for some $p \in \mathbf{N}^+$.

Then some basic properties of natural numbers follow:

Proposition 1.4.1 *For all $m, n, p \in \mathbf{N}^+$, we have*

  (i) associative property of addition: $(m + n) + p = m + (n + p)$;
 (ii) associative property of multiplication: $(m \cdot n) \cdot p = m \cdot (n \cdot p)$;
(iii) commutative property of addition: $m + n = n + m$;
 (iv) commutative property of multiplication: $m \cdot n = n \cdot m$;
  (v) cancellation property of addition: $m + n = m + p$ *implies* $n = p$;
 (vi) cancellation property of multiplication: $m \cdot n = m \cdot p$ *implies* $n = p$;
(vii) left distributive property: $m \cdot (n + p) = m \cdot n + m \cdot p$;
      right distributive property: $(n + p) \cdot m = n \cdot m + p \cdot m$;
(viii) property of trichotomy: *Exactly one of the following holds:*

$$m > n, \qquad m < n, \qquad m = n.$$

In respect of subtraction and division in $\mathbf{N}^+$, if $m > n$, then $m - n$ is defined to be the positive integer p such that $m = n + p$. Thus $m - n$ has a meaning in $\mathbf{N}^+$ iff $m > n$. Similarly, if there is a positive integer p such that $m = np$, then m/n is defined to be the positive integer p. Thus m/n has a meaning iff n is a factor of m. Clearly, for any two positive integers m and n we cannot always define $m - n$ or m/n in $\mathbf{N}^+$. These difficulties are overcome by extending the set $\mathbf{N}^+$ to the set of integers $\mathbf{Z}$ and to the set of rational numbers $\mathbf{Q}$, respectively.

Remark If we define $\mathbf{Q}$ as the set $\mathbf{S}$ of all numbers of the form m/n, where m and n are integers and $n \neq 0$, then each element of $\mathbf{Q}$ is represented by an infinite number of elements of $\mathbf{S}$. For example, $3/4 = 6/8 = 12/16 = \cdots$. While doing mathematics, we identify all the elements of $\mathbf{S}$ that represent the same rational number by an equivalence relation $\sim$ on $\mathbf{S}$: $a/b \sim c/d$ if and only if $ad = bc$. Each equivalence class of $\mathbf{S}$ is considered to be a rational number.

We now extend the set $\mathbf{Q}$ of rational numbers to a larger set containing $\mathbf{Q}$ in which all the properties of $\mathbf{Q}$ along with an additional property, called order completeness property (i.e., every bounded set has the greatest lower bound and the least upper bound, which is analogous concepts of the greatest common divisor and least common multiple in the theory of divisibility). This extended set denoted by $\mathbf{R}$ is called the set of real numbers. We have already used the term Euclidean line without any explanation. We use the letter $\mathbf{R}$ (or $\mathbf{R}^1$) to denote an ordinary geometric straight line whose points have been identified with the set $\mathbf{R}$ of real numbers. We use the same letter $\mathbf{R}$ to denote the real line and the set of real numbers and say that a real number corresponds to a unique point on the line and conversely. We assume that the reader is familiar with the properties of the real line, such as inequality of real numbers, usual addition, subtraction, multiplication, division, and lub. We use these properties in subsequent chapters to prove many interesting results. Apart from this approach, there are essentially two methods of constructions of $\mathbf{R}$.

1. Dedekind's Method of completion by cuts and
2. Cantor's Method of sequences.

Clearly, $\mathbf{N}^+ \subsetneq \mathbf{Z} \subsetneq \mathbf{Q} \subsetneq \mathbf{R}$.

In the rest of the section we study the properties of integers only and we discuss some of the important tools that are useful for proving theorems. We assume that the reader is thoroughly familiar with the set $\mathbf{Z}$ of integers, the set $\mathbf{N}^+$ of positive integers and elementary properties of addition, multiplication, and order relation on $\mathbf{Z}$ and $\mathbf{N}^+$. We begin by stating an important axiom, the well-ordering principle.

Principle of Well-Ordering Every non-empty subset S of $\mathbf{N}^+$ contains a least element $b \in S$ (i.e., $\exists$ an element $b \in S$ such that $b \leq a$ for all $a \in S$).

Next we shall show a close connection between the principle of well-ordering and the principle of mathematical induction.

Theorem 1.4.1 *The principle of well-ordering implies the principle of mathematical induction.*

Proof Let S be a non-empty subset of $\mathbf{N}^+$ such that $1 \in S$ and every $n \in S$ implies $n + 1 \in S$. We claim that $S = \mathbf{N}^+$. If $\mathbf{N}^+ - S \neq \emptyset$, then by well-ordering principle, $\mathbf{N}^+ - S$ contains a least positive integer $n > 1$ (as $1 \notin \mathbf{N}^+ - S$). Consequently, $n - 1$ is a positive integer such that $n - 1 \notin \mathbf{N}^+ - S$. As a result $n - 1 \in S$. Then $n = (n-1) + 1 \in S$ by hypothesis. But this is a contradiction. Therefore, $\mathbf{N}^+ - S = \emptyset$ and hence $\mathbf{N}^+ = S$. $\qquad\square$

Remark We now show that the principle of mathematical induction with usual order relation '$\leq$' on $\mathbf{N}^+$ implies the principle of well-ordering.

Proof Let S be a non-empty subset of $\mathbf{N}^+$ and $T = \{n \in \mathbf{N}^+ : n \leq p \text{ for every } p \text{ in } S\}$ (usual order relation '$\leq$' on $\mathbf{N}^+$ is assumed). Then $1 \in T$, so $T \neq \emptyset$. Again $T \neq \mathbf{N}^+$, because $s \in S \Rightarrow s + 1 \notin T$ (since $s + 1 \not\leq s$). Then there exists a natural number

$q \in T$ such that $q + 1 \notin T$, otherwise, it follows by the principle of mathematical induction that $T = \mathbf{N}^+$. Now by definition of $T, q \leq$ every $p \in S$. We claim that $q \in S$. If not, then $q \notin S \Rightarrow q < p, \forall p \in S$. Thus for every $p \in S, p = q + r$, so that $p = q + 1$ (when $r = 1$) or $p = q + s(v)$, where v is the predecessor of r and $s(v)$ denotes the successor of v. Consequently, $q + 1 \leq p$ for every $p \in S \Rightarrow q + 1 \in T$, a contradiction. As a result $q \in S$ and q is a least element in S. Moreover, from the antisymmetric property of the relation '$\leq$' it follows that q is unique. $\square$

We now present "*division algorithm*" for integers. An algorithm means a procedure or a method. The usual method of dividing an integer a by a positive integer b gives a quotient q and a remainder r, which may be 0 or a positive integer less than b. This concept, essentially due to Euclid, is now called 'Division Algorithm'. A geometrical proof is given first, followed by an analytical proof by using the 'well-ordering principle'.

A geometrical proof of division algorithm is given when a and b are positive integers.

Theorem 1.4.2 (Division algorithm) *Let $a, b \in \mathbf{N}^+$ and $b \neq 0$. Then there exist unique integers q and r such that*

$$a = bq + r, \quad and \quad 0 \leq r < b.$$

Geometrical Proof Consider all positive multiples of b and plot them on the real line. Then the integer a lies at some point qb or between two points qb and $(q+1)b$. In the first case, $a - qb = r = 0$. In the second case $qb < a < (q + 1)b$. Hence $0 < r = a - qb < ((q + 1)b - qb) = b$ implies $a = qb + r$, where $0 < r < b$. $\square$

A more general form of Division algorithm is now presented.

Theorem 1.4.3 (Division algorithm) *Let $a, b \in \mathbf{Z}$ and $b \neq 0$. Then there exist unique integers q and r such that*

$$a = bq + r, \quad and \quad 0 \leq r < |b|.$$

Proof We first prove the existence of q and r. If $a = 0$, then $q = 0$ and $r = 0$. So we assume that $a \neq 0$.

Case 1. Suppose $b > 0$. Then $b \geq 1$. If $b = 1$, then $q = a$ and $r = 0$. If $a = bm$ for some $m \in \mathbf{Z}$, then $q = m$ and $r = 0$. For the remaining possibility, consider the set

$$S = \{a - bm : m \in \mathbf{Z} \text{ and } a - bm > 0\}.$$

Since $b > 1, a - (-|a|)b = a + |a|b > 0$ and hence $a - (-|a|)b \in S$. Consequently, $S \neq \emptyset$. Hence by the well-ordering principle S has a least element r. Thus there exists an integer q such that $r = a - bq > 0$. Hence $a = bq + r, 0 < r$. We now prove that $r < b$. Suppose $r > b$. Then $r = b + c$ for some integer c such that

$0 < c < r$. Hence $c = r - b = a - bq - b = a - b(q+1) \in S$. This contradicts the minimality of r in S. Hence $r < b$.

Case 2. Suppose $b < 0$. Since $-b > 0$, by case 1, there exist q', r in $\mathbf{Z}$ such that

$$a = (-b)q' + r, \quad 0 \le r < -b = |b|.$$

Taking $q = -q'$, it follows that $a = bq + r, 0 \le r < |b|$.

From the above two cases we find that there exist q and r in $\mathbf{Z}$ such that

$$a = bq + r, \quad 0 \le r < |b|. \tag{i}$$

Suppose now $a = bq_1 + r_1$, where $0 \le r_1 < |b|$ for some $q_1, r_1 \in \mathbf{Z}$ and $a = bq_2 + r_2$, where $0 \le r_2 < |b|$, for some $q_2, r_2 \in \mathbf{Z}$.

Then $b(q_1 - q_2) = (r_2 - r_1)$. Hence $|b||q_1 - q_2| = |r_2 - r_1|$. If $r_2 - r_1 \ne 0$, then $|b| \le |r_2 - r_1|$. Since $0 \le r_1 < |b|$ and $0 \le r_2 < |b|$, we have $|r_2 - r_1| < |b|$. This contradiction implies that $r_1 = r_2$. Now from $b(q_2 - q_1) = 0$, $(b \ne 0)$, it follows that $q_2 - q_1 = 0$. Hence $r_1 = r_2$ and $q_1 = q_2 \Rightarrow$ uniqueness of q and r in (i). $\qquad\square$

Definition 1.4.2 A non-zero integer b is said to divide (or b is a factor of) an integer a iff there exists an integer c such that $a = bc$. If b divides a, then we write it in symbol $b|a$. In case b does not divide a, we write $b\bar{|}a$.

Some immediate consequences of this definition are listed below:

Theorem 1.4.4 *Let a, b, c be integers.*

(1) *If $0 \ne a$, then $a|0, a|a, 1|b$.*
(2) *If $0 \ne a, 0 \ne c$, then $a|b$ implies $ac|bc$.*
(3) *If $0 \ne a, 0 \ne b$, then $a|b$ and $b|c$ imply $a|c$.*
(4) *If $0 \ne a$, then $a|b$ and $a|c$ imply $a|(bn + cm)$ for every $n, m, \in \mathbf{Z}$.*

Proof Left as an exercise. $\qquad\square$

We now introduce the notion of greatest common divisor.

Definition 1.4.3 Let a, b be two integers, not both zero. A non-zero integer d is said to be a *greatest common divisor* of a and b iff the following hold:

(1) $d|a$ and $d|b$ (d is a common factor of a and b);
(2) if $c \ne 0$ be an integer such that $c|a$ and $c|b$, then $c|d$.

A natural question to ask is whether the elements $a, b \in \mathbf{Z}$ can possess two different greatest common divisors. Suppose d_1 and d_2 are two greatest common divisors of a, b. Then from the definition, $d_1|d_2$ and $d_2|d_1$. Hence $d_1 = \pm d_2$. This shows that one of d_1 and d_2 is a positive integer. Then one will be the negative of the other. The positive one is called the positive greatest common divisor. For two integers a, b, the positive *greatest common divisor*, if it exists, is denoted by $\gcd(a, b)$ or simply by (a, b).

Theorem 1.4.5 *If a and b are integers, not both zero, then* $\gcd(a, b)$ *exists uniquely and* $\gcd(a, b)$ *is expressible in the form*

$$\gcd(a, b) = pa + qb \quad \text{for some } p, q \in \mathbf{Z}.$$

Proof Let $S = \{ma + nb : m, n \in \mathbf{Z}\}$. Since a and b are not both zero, S contains non-zero integers. Suppose $0 \neq ma + nb \in S$. Then $-(ma + nb) = (-m)a + (-n)b \in S$. Since $ma + nb$ and $-(ma + nb)$ both belong to S, we find that S contains positive integers. Let $\mathbf{S}^+$ be the set of all positive integers of S. Since $\mathbf{S}^+ \neq \emptyset$, by the *well-ordering principle*, it follows that $\mathbf{S}^+$ contains a least positive integer d. Hence $d \leq t$, for all $t \in \mathbf{S}^+$ and $d = pa + qb$ for some $p, q \in \mathbf{Z}$. We show that $d = \gcd(a, b)$. From the division algorithm there exist integers c and r such that

$$a = c(pa + qb) + r, \quad 0 \leq r < pa + qb = d.$$

Suppose $r > 0$. Now $r = (1 - cp)a + (-cq)b$ shows that $r \in \mathbf{S}^+$. As $0 < r < d$, it contradicts the fact that d is the least element in $\mathbf{S}^+$. Hence $r = 0$. Consequently, $(pa + qb)|a$. Similarly, we show that $(pa + qb)|b$. Next, assume that u is a non-zero integer such that $u|a$ and $u|b$. Then from Theorem 1.4.4 it follows that $u|(pa + qb)$. Hence $pa + qb$ is a greatest common divisor of a, b. But $pa + qb$ is positive. Consequently $\gcd(a, b)$ exists and $\gcd(a, b) = pa + qb$. The uniqueness of $\gcd(a, b)$ follows from the antisymmetric property of the divisibility relation on $\mathbf{N}^+$ (see Example 1.2.11(iii)). $\square$

Corollary *Let a and b be two integers. Then* $\gcd(a, b) = 1$ *if and only if there exist integers u and v such that* $au + bv = 1$.

Proof First let $\gcd(a, b) = 1$. Then by Theorem 1.4.5, there exist integers u and v such that $au + bv = 1$. Conversely, let there exist integers u and v such that $au + bv = 1$. Let $d = \gcd(a, b)$. Then by definition of greatest common divisor, d divides both a and b and hence $au + bv$. Thus d divides 1, resulting in $d = 1$. $\square$

Remark If $\gcd(a, b) = pa + qb$, for $p, q \in \mathbf{Z}$, then p, q may not be unique. For example, $\gcd(1492, 1066) = -5 \cdot 1492 + 7 \cdot 1066$. So here $p = -5$ and $q = 7$. Now we may take, $p' = p + 1066 = -5 + 1066 = 1061$ and $q' = q - 1492 = 7 - 1492 = -1485$. Then $1492p' + 1066q' = 1492p + 1492 \cdot 1066 + 1066q - 1066 \cdot 1492 = 1492p + 1066q = \gcd(1492, 1066)$.

We now list some of the interesting properties of $\gcd(a, b)$. All these can be proved easily.

Proposition 1.4.2 *If a, b, c are any three non-zero integers, then*

(i) $\gcd(a, \gcd(b, c)) = \gcd(\gcd(a, b), c)$;
(ii) $\gcd(a, 1) = 1$;
(iii) $\gcd(ca, cb) = c \gcd(a, b)$;

(iv) $\gcd(a, b) = 1$ *and* $\gcd(a, c) = 1$ *imply that* $\gcd(a, bc) = 1$;

(v) $a|c, b|c$ *and* $\gcd(a, b) = 1$ *imply that* $ab|c$.

Definition 1.4.4 An integer $p > 1$ is called a *prime integer* iff only factors of p are ± 1 and $\pm p$ in $\mathbf{Z}$, otherwise p is said to be a composite number. Two integers a and b are called relatively prime (or co-prime) iff $\gcd(a, b) = 1$.

The following theorem gives an alternative definition of a prime integer.

Theorem 1.4.6 *An integer* $p > 1$ *is a prime integer iff* $\forall a, b \in \mathbf{Z}$ *with* $p|ab$ *implies either* $p|a$ *or* $p|b$.

Proof Suppose p is a prime integer and $p|ab$. We want to show that either $p|a$ or $p|b$. Suppose $p\nmid a$. Since p is prime, we know that 1 and p are the only positive divisors of p. Hence $\gcd(p, a) = 1$. Then Theorem 1.4.5 asserts that

$$1 = cp + da \quad \text{for some integers } c \text{ and } d.$$

Hence $b = 1 \cdot b = cpb + dab$. Now $p|p$ and $p|ab$. Consequently $p|cpb$ and $p|dab$. Then $p|(cpb + dab)$. This proves that $p|b$.

Conversely, assume that $p > 1$ is an integer such that $p|ab$ $(a, b \in \mathbf{Z})$ implies that $p|a$ or $p|b$. We now prove that p is prime. Suppose $0 \neq m$ and $m|p$. Then $p = mn$ for some $n \in \mathbf{Z}$. Now $p|p$. Hence $p|mn$. Then either $p|m$ or $p|n$.

Suppose $p|m$. Then there exists $d \in \mathbf{Z}$ such that $m = pd$. Then $p = pdn$. This implies $dn = 1$. Since d, n are integers, we find that $d = 1$ and $n = 1$ or $d = -1$ and $n = -1$. Now from $p = mn$ it follows that $m = \pm p$.

Again, if we assume that $p|n$, then proceeding as above we can show that $n = \pm p$ and then $m = \pm 1$. Consequently ± 1 and $\pm p$ are the only factors of p. As a result p is a prime integer. $\qquad\square$

Corollary *Let* n *be a positive integer and* $a_1, a_2, \ldots, a_n$ *be integers such that* $p|a_1 a_2 \cdots a_n$, *where* p *is prime. Then* $p|a_i$ *for some* i *such that* $1 \leq i \leq n$ *i.e.,* p *divides at least one of* $a_1, a_2, \ldots, a_n$.

Proof We prove this corollary by using principle of mathematical induction on n. If $n = 1$, the result is trivially true. Assume that the result is valid for some m such that $1 \leq m < n$. Let $p|a_1 a_2 \cdots a_m a_{m+1}$. Then from Theorem 1.4.6 either $p|a_{m+1}$ or $p|a_1 a_2 \cdots a_m$. If $p|a_1 a_2 \cdots a_m$, then from our hypothesis $p|a_i$ for some i such that $1 \leq i \leq m$. Hence either $p|a_{m+1}$ or p divides at least one of $a_1, a_2, \ldots, a_m$. Now the result follows by the principle of mathematical induction. $\qquad\square$

Remark As a generalization of Theorem 1.4.6 one can prove the following. Let a, b and $m \neq 0$ be integers such that $\gcd(a, m) = 1$ and $m|ab$. Then $m|b$.

The prime numbers form blocks of integers. First we shall show that there exist infinitely many primes. To prove this we use the following lemma.

Lemma 1.4.1 *If $a > 1$ is an integer, then there exists a prime integer p such that $p|a$.*

Proof Let $S = \{b \in \mathbf{Z} : b$ is an integer >1 and $b|a\}$. Now $a \in S$ and then $S \neq \emptyset$. By well-ordering principle, there exists an integer p in S such that $p \leq q$ for all $q \in S$. This clearly shows $p > 1$ and $p|a$. Suppose p is not a prime integer. Then p has a factor m such that $1 < m < p$. Now $m|p$ and $p|a$. Hence $m|a$. As a result $m \in S$. This contradicts the minimality of p in S. Consequently p is a prime integer. □

We are now in a position to prove the following celebrated result.

Theorem 1.4.7 (Euclid) *There are infinitely many prime integers.*

Proof It is true that there are prime integers such as: $2, 3, 5, 7$. Suppose there is only a finite number of distinct primes, say $p_1, p_2, \ldots, p_n$. Consider the positive integer

$$m = p_1 p_2 \cdots p_n + 1.$$

Now $m = p_1(p_2 \cdots p_n) + 1$ shows that $p_1 \overline{|} m$. Similarly we can show that none of $p_2, p_3, \ldots, p_n$ divides m. Since $m > 1$, Lemma 1.4.1 asserts that there exists a prime integer p such that $p|m$. This prime p is clearly different from $p_1, p_2, \ldots, p_n$. Consequently, there is no finite listing of prime integers. In other words, there exist infinite number of prime integers. □

We shall prove a little later another celebrated theorem known as the fundamental theorem of arithmetic which proves that primes are indeed the building blocks for integers. For this purpose we introduce the concept of factorization of integers.

Definition 1.4.5 An integer $n > 1$ is said to be *factorizable* in $\mathbf{N}^+$ iff there exist prime integers $p_1, p_2, \ldots, p_k$ such that $n = p_1 p_2 \cdots p_k$.

An integer $n > 1$ is said to be *uniquely factorizable* in $\mathbf{N}^+$ iff the following two conditions hold:

 (i) n is factorizable;
(ii) if $n = p_1 p_2 \cdots p_r = q_1 q_2 \cdots q_s$ be two factorizations of n (as a product of prime integers), then $r = s$ and the two factorizations differ only in the order of the factors.

We now prove the following basic result.

Theorem 1.4.8 (The Fundamental Theorem of Arithmetic) *Every integer $n > 1$ is uniquely factorizable (up to order).*

We prove this result by using the second principle of mathematical induction which is stated as follows.

Lemma 1.4.2 (Second Principle of Mathematical Induction or Principle of Strong Mathematical Induction)   *Let S be a non-empty subset of* $\mathbf{N}^+$ *such that*

(i) *some fixed positive integer* $n_0 \in S$ *and*
(ii) *for each positive integer* $m > n_0$, *each of* $n_0 + 1, n_0 + 2, \ldots, m - 1 \in S$ *implies* $m \in S$.

 Then $S = \{n \in \mathbf{N}^+ : n \geq n_0\}$.

Proof Let $X = \mathbf{N}^+ - \{1, 2, \ldots, n_0 - 1\}$ and $T = X - S$. If we can show that $T = \emptyset$, then the lemma follows. Suppose $T \neq \emptyset$. Then T is a non-empty subset of $\mathbf{N}^+$. Thus by the well-ordering principle T contains a least element, say m. As $m \in T$, $m \notin S$ and $m \notin \{1, 2, \ldots, n_0 - 1\}$. Now by (i), $m \neq n_0$. For every x satisfying $n_0 \leq x < m$, $x \notin T = X - S = X \cap S'$, where S' is the complement of S in $\mathbf{N}^+$. This implies $x \in X' \cup S$. As $X' = \{1, 2, \ldots, n_0 - 1\}$, this implies $x \in S$. Thus by (ii), $m \in S$. This is a contradiction. Consequently $T = \emptyset$ and then $S = \{n \in \mathbf{N}^+ : n \geq n_0\}$. ☐

Corollary *Let S be a subset of* $\mathbf{N}^+$ *such that*

(i) $1 \in S$, *and*
(ii) *if* $n > 1$ *and* $n \in S$ *for all* $n < m$, *implies* $m \in S$.

 Then $S = \mathbf{N}^+$.

Proof of the Fundamental Theorem of Arithmetic Let $n > 1$. We first show that n is factorizable (as a product of prime integers). We prove this by the second principle of induction. If $n = 2$, then there exists a prime integer $p_1 = 2$ such that $n = p_1$ (=2). Suppose that any integer r, $2 \leq r < n$ can be expressed as a product of primes.

 Consider now the integer n. From Lemma 1.4.1 there exists a prime integer p_1 such that $p_1 | n$. Then we can write $n = p_1 n_1$ for some integer n_1. Since $n > 2$, $p_1 > 1$, we must have $n_1 \geq 1$. If $n_1 = 1$, then $n = p_1$. Suppose $n_1 \neq 1$. Then $2 \leq n_1 < n$ and from the induction hypothesis we can write,

$$n_1 = p_2 p_3 \cdots p_t \quad \text{for some primes } p_2, p_3, \ldots, p_t.$$

Hence $n = p_1 n_1 = p_1 p_2 p_3 \cdots p_t$.

 Thus from the second principle of mathematical induction, it follows that any positive integer $n \geq 2$ can be expressed as a product of prime integers.

 We now show that the factorization of a positive integer $n > 1$ is unique up to the rearrangement of the order of the factors. We prove this also by the method of induction. If $n = 2$, then the factorization of n is unique. Assume that the uniqueness of factorization is true for any positive integer m such that $2 < m < n$. Consider the integer n (>2). Suppose $n = p_1 p_2 \cdots p_s = q_1 q_2 \cdots q_t$ be two factorizations of n (as a product of prime integers). Since $p_1(p_2 \cdots p_s) = q_1 q_2 \cdots q_t$, it follows that $p_1 | q_1 q_2 \cdots q_t$ and hence p_1 divides some q_i $(1 \leq i \leq t)$. But 1 and q_i are the only positive divisors of q_i. Consequently $p_1 = q_i$. We may assume that the q_i's are so arranged such that $q_1 = p_1$. Thus

$$p_1 p_2 \cdots p_s = p_1 q_2 \cdots q_t.$$

Since $p_1 \neq 0$, we may cancel p_1 and get $p_2 p_3 \cdots p_s = q_2 q_3 \cdots q_t = n_1$ (say). But $1 \leq n_1 < n$. Then by induction hypothesis it follows that n_1 is uniquely factorizable. Hence $s - 1 = t - 1$ and that the factorization $q_2 q_3 \cdots q_s$ is just a rearrangement of the p_i's, $i = 2, 3, \ldots, s$. Thus $s = t$ and since $p_1 = q_1$ it follows that n is uniquely factorizable. Hence by induction we find that any integer $n > 1$ is uniquely factorizable.   □

From the fundamental theorem of arithmetic, we can write $n > 1$ as a product of primes and since the prime factors are not necessarily distinct, the result can be written in the form

$$n = p_1^{\alpha_1} p_2^{\alpha_2} \cdots p_r^{\alpha_r},$$

where $p_1, p_2, \ldots, p_r$ are distinct primes and $\alpha_1, \alpha_2, \ldots, \alpha_r$ are positive integers. In turns out that the above representation of n as a product of primes is unique in the sense that, for a fixed n (>1), any other representation is merely a permutation of the factors.

Remark We usually take $p_1 > p_2 > \cdots > p_r$.

Alternative Form of the Second Principle of Mathematical Induction   Let $P(n)$ be a statement which makes sense for any positive integer $n \geq n_0$. If the following two statements are true:

(a) $P(n_0)$ is true and
(b) for all $k \geq n_0$, each of $P(n_0 + 1), P(n_0 + 2), \ldots, P(k - 1)$ is true implies $P(k)$ is true,

then $P(n)$ is true for all $n \geq n_0$.

1.5 Congruences

The language of congruences plays an important role not only in number theory but also an important topic of abstract algebra. It was developed at the beginning of the nineteenth century by Karl Friedrich Gauss (1777–1855). Here we study modular arithmetic, that is, arithmetic of congruence classes where we simplify number theoretic problems by replacing each integer by its remainder when divided by some fixed positive integer n.

Definition 1.5.1 If a and b are integers, we say that a is *congruent* to b modulo a positive integer n, denoted by $a \equiv b \pmod{n}$, iff $n|(a - b)$. If $n\overline{|}(a - b)$ we write $a \not\equiv b \pmod{n}$, and say that a and b are *incongruent* modulo n.

Example 1.5.1 We have $3 \equiv 23 \pmod{10}$, as 10 divides 3–23. Similarly, $2 \equiv 5 \pmod{3}$.

The following proposition provides some important properties of congruences.

Proposition 1.5.1 *Let n be a positive integer. Then the congruence modulo the positive integer n satisfies the following properties*:

(i) reflexive property: *If a is an integer, then $a \equiv a \pmod{n}$*;
(ii) symmetric property: *If a and b are integers such that $a \equiv b \pmod{n}$, then $b \equiv a \pmod{n}$*;
(iii) transitive property: *If a, b, and c are integers with $a \equiv b \pmod{n}$ and $b \equiv c \pmod{n}$, then $a \equiv c \pmod{n}$*.

From Proposition 1.5.1, we see that the set of integers is divided into n different sets called congruence classes modulo n, each containing integers which are mutually congruent modulo n. The reader may also revisit Examples 1.2.3 and 1.2.9.

Arithmetic with congruences is very common in number theory. Congruences ($\equiv$) have many of the properties that "equalities" ($=$) have. The following proposition states that addition, subtraction, or multiplication to both sides of a congruence preserve the congruence.

Proposition 1.5.2 *If a, b, c are integers and n is a positive integer such that $a \equiv b \pmod{n}$, the following properties are satisfied*:

(i) $a + c \equiv b + c \pmod{n}$;
(ii) $a - c \equiv b - c \pmod{n}$;
(iii) $ac \equiv bc \pmod{n}$.

A very natural question that comes to our mind is: what happens when both sides of a congruence are divided by a non-zero integer? Let us consider the following example.

Example 1.5.2 Let us consider 14 modulo 6. We have $7 \cdot 2 \equiv 4 \cdot 2 \pmod 6$. But $7 \not\equiv 4 \pmod 6$.

This example shows that it is not necessarily true that it preserves a congruence when we divide both sides by an integer. However, the following proposition gives a valid congruence when both sides of a congruence are divided by the same integer.

Theorem 1.5.1 *If a, b, c, and n are integers such that $n > 0$, $d = \gcd(c, n)$, and $ac \equiv bc \pmod{n}$, then $a \equiv b \pmod{n/d}$.*

Proof As $ac \equiv bc \pmod{n}$, there exists an integer t such that $c(a - b) = nt$. This implies that $c(a - b)/d = nt/d$. Again, $\gcd(c/d, n/d) = 1$ implies that n/d divides $(a - b)$. Hence $a \equiv b \pmod{n/d}$. $\qquad\square$

Example 1.5.3 Let us take the example of 65 modulo 15. We have $65 \equiv 35 \pmod{15}$. As $\gcd(5, 15) = 5$, by the above theorem we have $65/5 \equiv 35/5 \pmod{15/5}$ which implies $13 \equiv 7 \pmod 3$.

The following corollary is immediate.

Corollary *If a, b, c and n are integers such that $n > 0$, $\gcd(c, n) = 1$ and $ac \equiv bc \pmod{n}$, then $a \equiv b \pmod{n}$.*

The following proposition, which is more general than Proposition 1.5.2, is also useful.

Proposition 1.5.3 *If a, b, c, d are integers and n is a positive integer such that $a \equiv b \pmod{n}$ and $c \equiv d \pmod{n}$, the following properties are satisfied:*

  (i) $a + c \equiv b + d \pmod{n}$;

 (ii) $a - c \equiv b - d \pmod{n}$;

(iii) $ac \equiv bd \pmod{n}$.

The following theorem shows that a congruence is preserved when both sides are raised to the same positive integral power.

Theorem 1.5.2 *If a, b are integers and n and k are positive integers such that $a \equiv b \pmod{n}$, then $a^k \equiv b^k \pmod{n}$.*

Proof As $a \equiv b \pmod{n}$, n divides $a - b$. Now, $a^k - b^k = (a - b)(a^{k-1} + a^{k-2}b + \cdots + ab^{k-2} + b^{k-1})$ implies that $a - b$ divides $a^k - b^k$ and hence n divides $a^k - b^k$. Thus $a^k \equiv b^k \pmod{n}$. $\square$

Supplementary Examples (SE-II)

1 (i) The integer $3^{2n} - 1$ is divisible by 8 $\forall n \geq 1$;

  (ii) $3^{2n} \equiv 1 \pmod{8}$ $\forall n \geq 1$;

 (iii) $3^{2n} - 1 \equiv 0 \pmod{8}$ $\forall n \geq 1$.

[*Hint.* (i) Let $P(n)$ be the statement that the integer $3^{2n} - 1$ is divisible by 8 $\forall n \geq 1$. Then $P(1)$ is true, since $9 - 1 = 8$ is divisible by 8. Next suppose that $P(n)$ is true for some $k \in \mathbf{N}^+$ i.e., $3^{2k} - 1$ is divisible by 8. Now $3^{2(k+1)} - 1 = 3^{2k}3^2 - 3^2 + 3^2 - 1 = 3^2(3^{2k} - 1) + (3^2 - 1)$ is divisible by 8. Hence by the principle of mathematical induction, $P(n)$ is true $\forall n \in \mathbf{N}^+$.

(ii) follows from (i).

(iii) follows from (ii).]

2 (i) $2^n \geq 1 + n$ $\forall n \in \mathbf{N}^+$;

  (ii) $2^n \geq 2n + 1$ for all integers $n \geq 3$.

[*Hint.* (i) Let $P(n)$ be the statement: $2^n \geq 1 + n$ $\forall n \geq \mathbf{N}^+$. Then $P(1)$ is true. Next suppose that $P(k)$ is true for some integer $k \geq 1$ i.e., suppose that $2^k \geq 1 + k$ for some integer $k \geq 1$. Then it follows that $2^{k+1} = 2^k 2 \geq (1 + k)2 = 2 + 2k > 1 + (k + 1) \Rightarrow P(k + 1)$ is also true. Hence by the principle of mathematical induction, $P(n)$ is true $\forall n \in \mathbf{N}^+$.

(ii) Let $P(n)$ be the statement: $2^n \geq 2n + 1$ $\forall$ integers $n \geq 3$. Then $2^3 = 8 > 2.3 + 1 \Rightarrow P(3)$ is true. Next let $P(n)$ be true for some $n \geq 3$ i.e., $2^n \geq 2n + 1$ for some integer $n \geq 3$. Now $2^{n+1} = 2^n \cdot 2 \geq (2n + 1) \cdot 2 \geq 2(n + 1) + 1 \Rightarrow P(n + 1)$ is true. Hence by the principle of mathematical induction, $P(n)$ is true for all integers $n \geq 3$.]

3 Let $m, n \in \mathbf{N}^+$. Then

(i) $n \neq m + n$;

(ii) $n = mn \Leftrightarrow m = 1$.

[*Hint.* (i) Let $P(n)$ be the statement that $n \neq m + n$ for arbitrary $m \in \mathbf{N}^+$ and every $n \in \mathbf{N}^+$. Then by Peano's axiom (3), $1 \neq m + 1 \Rightarrow P(1)$ is true. Next let $P(n)$ be true for some $n \in \mathbf{N}^+$ and arbitrary $m \in \mathbf{N}^+$. Then $n \neq m + n \Rightarrow s(n) \neq s(m + n) \Rightarrow n + 1 \neq m + (n + 1) \Rightarrow P(n + 1)$ is true. Hence by the principle of mathematical induction $P(n)$ is true for arbitrary $m \in \mathbf{N}^+$ and every $n \in \mathbf{N}^+$ i.e., $P(n)$ is true $\forall m, n \in \mathbf{N}^+$.

(ii) $n = 1 \cdot n$ proves the sufficiency of the condition. To prove the necessity of the condition, let $m \neq 1$. Then $\exists$ some $c \in \mathbf{N}^+$ such that $s(c) = m$. Hence $mn = s(c)n = (c + 1)n = cn + n$. Hence for $m \neq 1$, $n = mn$ would imply $n = cn + n$, which contradicts (i).]

4 Show that $\sqrt{n} \in \mathbf{Q} \Rightarrow \sqrt{n} \in \mathbf{N}^+ (n \in \mathbf{N}^+)$.

[*Hint.* Suppose $\sqrt{n} = a/b$ where $a, b \in \mathbf{N}^+$. Then $a^2 = b^2 n$. Now taking prime decomposition, it follows that n is a perfect square $\Rightarrow \sqrt{n} \in \mathbf{N}^+$.]

5 Let $x \in \mathbf{R}$ i.e., x be a real number. We assume that $\exists$ integers which are $\leq x$. So the set S of all such integers is non-empty and is bounded above (x being an upper bound). Hence by Zorn's Lemma there is a greatest integer in S which is less than or equal to x. This integer is called the integral part of x and is noted by $[x]$. Show that $[x]$ satisfies the following properties:

(i) $[x] \leq x < [x] + 1$;

(ii) $x - [x] \in [0, 1)$;

(iii) if $x = n + r$, where $n \in \mathbf{Z}$ and $r \in [0, 1)$, then $n = [x]$;

(iv) let $m \in \mathbf{Z}$ and $n \in \mathbf{N}^+$ and $m = nq + r$ where $q, r \in \mathbf{Z}$ and $0 \leq r \leq n - 1$. Then $q = [\frac{m}{n}]$ and $r = m - n[\frac{m}{n}]$.

[*Hint.* As $\frac{m}{n} = q + \frac{r}{n}$ and $0 \leq \frac{r}{n} \leq \frac{n-1}{n} < 1$.
Hence by (iii), $q = [\frac{m}{n}]$ and so $r = m - n[\frac{m}{n}]$.]

6 If $a \equiv b \pmod{n}$, then $\gcd(a, n) = \gcd(b, n)$ $(a, b, n \in \mathbf{N}^+)$.

[*Hint.* Let $d_1 = \gcd(a, n)$ and $d_2 = \gcd(b, n)$. Now $a \equiv b \pmod{n} \Rightarrow b = a + kn$ for some $k \in \mathbf{Z}$. Hence $d_1 | n$ and $d_1 | a \Rightarrow d_1 | b$ (by Theorem 1.4.4) $\Rightarrow d_1$ is a common factor of $b, n \Rightarrow d_1 \leq d_2$. Similarly, $d_2 \leq d_1$. Hence $d_1 = d_2$.]

7 Let m and n be non-zero integers. The *least common multiple* of m and n, denoted by $\mathrm{lcm}(m, n)$, is defined to be the positive integer l such that in $\mathbf{Z}$,

(i) $m | l$ and $n | l$ and

(ii) $m | c$ and $n | c \Rightarrow l | c$.

Then $\text{lcm}(m, n)$ exists and is unique and $\text{lcm}(m, n)$ has the following relations with $\gcd(m, n)$:

(a) $\gcd(m, n) \cdot \text{lcm}(m, n) = |mn|$;
(b) $\text{lcm}(m, n) = |mn| \Leftrightarrow \gcd(m, n) = 1$.

Supplementary Examples (SE-III)

1 If n is odd, then n^2 is also odd.

[*Hint.* Suppose n is odd and $n = 2m + 1$ for some $m \in \mathbf{Z}$. Then $n^2 = 4(m^2 + m) + 1$ is odd as $4(m^2 + m)$ is even.]

2 For every integer n, neither $n^2 \equiv 2 \pmod 5$ nor $n^2 \equiv 3 \pmod 5$.

[*Hint.* Let n be an arbitrary integer. Then on division n by 5, the only possible remainders r are $0, 1, 2, 3, 4$ i.e., $n \equiv r \pmod 5$ for $r = 0, 1, 2, 3$ and 4 i.e., $n^2 \equiv r^2 \pmod 5$ for $r = 0, 1, 2, 3$ and 4 i.e., $n^2 \not\equiv 2 \pmod 5$ and $n^2 \not\equiv 3 \pmod 5$.]

3 There exists no integral solution of the equation $7x^2 - 15y^2 = 1$.

[*Hint.* Suppose there exist integers n, m such that $7n^2 - 15m^2 = 1$. Then $7n^2 \equiv 1 \pmod 5$, since $15m^2 \equiv 0 \pmod 5$.
Again

$$7 \equiv 2 \pmod 5 \Rightarrow 7n^2 \equiv 2n^2 \pmod 5 \Rightarrow 2n^2 \equiv 1 \pmod 5. \tag{i}$$

By using the above Supplementary Example 2, $\forall n \in \mathbf{Z}$, $n^2 \equiv 0 \pmod 5$ or $n^2 \equiv 4 \pmod 5$ or $n^2 \equiv 1 \pmod 5 \Rightarrow 2n^2 \equiv 0 \pmod 5$ or $2n^2 \equiv 8 \pmod 5 \equiv 3 \pmod 5$ or $2n^2 \equiv 2 \pmod 5 \Rightarrow$ a contradiction if (i).]

4 Every positive integer n is congruent to the sum of its digits modulo 9. Hence deduce a test for divisibility by 9.

[*Hint.* Suppose $n = n_r n_{r-1} \cdots n_2 n_1 n_0$ (written in the customary notation).
Then $n = n_r 10^r + n_{r-1} 10^{r-1} + \cdots + n_2 10^2 + n_1 10 + n_0. \Rightarrow n \equiv (n_r + n_{r-1} + \cdots + n_2 + n_1 + n_0) \pmod 9$ (since $10^m \equiv 1 \pmod 9$ for all integers $m \geq 0$). Clearly, n is divisible by 9 iff the sum of its digits is divisible by 9.]

1.5.1 Exercises

Exercises-II

1. Prove for any positive integer n:

   (i) $3^{3n+1} \equiv 3.5^n \pmod{11}$;
   (ii) $2^{4n+3} \equiv 8.5^n \pmod{11}$;
   (iii) $3^{3n+1} + 2^{4n+3} \equiv 0 \bmod(11)$;

    [*Hint.* (i) $3^3 \equiv 5 \pmod{11} \Rightarrow 3^{3n} \equiv 5^n \bmod(11) \Rightarrow 3^{3n+1} \equiv 3.5^n \pmod{11}$.]

2. Show that there is no solution in integers of the equation

$$x^2 + y^2 = 3z^2$$

except $x = y = z = 0$.

3. Let n be an integer such that n is a perfect square and $n = rt$ where r and t are relatively prime positive integers. Show that r and t are perfect squares.

   [*Hint.* Consider prime decomposition of n. Note that an integer $m \geq 2$ is a perfect square $\Leftrightarrow$ every prime appearing in the decomposition of m is of even power.]

4. Prove that any prime of the form $3n + 1$ is of the form $6r + 1$.

5. Prove that any positive integer of the form $3n + 2$ has a prime factor of the same form.

6. Prove that $\gcd(n, n + 2) = 1$ or 2 for every integer n.

7. Prove that there are infinitely many primes of the form $4n + 3$ or $6n + 5$.

8. Show that $n > 1$ is a prime integer iff for any integer r either $\gcd(r, n) = 1$ or $n | r$.

9. Prove that there are arbitrarily large gaps in the series of primes.

   [*Hint.* See Chap. 10.]

1.6 Additional Reading

We refer the reader to the books (Adhikari and Adhikari 2003; Burton 1989; Halmos 1974; Hungerford 1974; Jones and Jones 1998; Rosen 1993; Simmons 1963) for further details.

References

Adhikari, M.R., Adhikari, A.: Groups, Rings and Modules with Applications, 2nd edn. Universities Press, Hyderabad (2003)

Burton, D.M.: Elementary Number Theory. Brown, Dulreque (1989)

Cohen, P.J.: Set Theory and Continuum Hypothesis. Benjamin, New York (1966)

Cohn, P.M.: Basic Algebra: Groups, Rings and Fields. Springer, London (2002)

Conway, J.B.: Functions of One Complex Variable. Springer, New York (1973)

Halmos, P.R.: Naive Set Theory. Springer, New York (1974)

Hungerford, T.W.: Algebra. Springer, New York (1974)

Jones, G.A., Jones, J.M.: Elementary Number Theory. Springer, London (1998)

Von Neumann, J.: Mathematical Foundations of Quantum Mechanics. Princeton University Press, Princeton (1955)

Rosen, K.H.: Elementary Number Theory and Its Applications. Addison-Wesley, Reading (1993)

Simmons, G.F.: Topology and Modern Analysis. McGraw-Hill, New York (1963)

Varadarajan, V.S.: Geometry of Quantum Theory, I and II. Van Nostrand, Princeton (1968)

Chapter 2
Groups: Introductory Concepts

Groups serve as one of the fundamental building blocks for the subject called today modern algebra. This chapter gives an introduction to the group theory and closely related topics. The idea of group theory was used as early as 1770 by J.L. Lagrange (1736–1813). Around 1830, E. Galois (1811–1832) extended Lagrange's work in the investigation of solutions of equations and introduced the term 'group'. At that time, mathematicians worked with groups of transformations. These were sets of mappings that, under composition, possessed certain properties. Originally, group was a set of permutations (i.e., bijections) with the property that the combination of any two permutations again belongs to the set. Felix Klein (1849–1925) adopted the idea of groups to unify different areas of geometry. In 1870, L. Kronecker (1823–1891) gave a set of postulates for a group. Earlier definitions of groups were generalized to the present concept of an abstract group in the first decade of the twentieth century, which was defined by a set of axioms. The theory of abstract groups plays an important role in the present day mathematics and science. Groups arise in a number of apparently unrelated disciplines. They appear in algebra, geometry, analysis, topology, physics, chemistry, biology, economics, computer science etc. So the study of groups is essential and very interesting. In this chapter, we make an introductory study of groups with geometrical applications along with free abelian groups and structure theorem for finitely generated abelian groups. Moreover, semigroups, homology groups, cohomology groups, topological groups, Lie groups, Hopf groups, and fundamental groups are discussed.

2.1 Binary Operations

Operations on the pairs of elements of a non-empty set arise in several contexts, such as the usual addition of two integers, usual addition of two residue classes in Z_n, usual multiplication of two real or complex square matrices of the same order and similar others. In such cases, we speak of a binary operation. The concept of binary operations is essential in the study of modern algebra. It provides algebraic structures on non-empty sets. The concept of binary operations is now introduced.

M.R. Adhikari, A. Adhikari, *Basic Modern Algebra with Applications*, 55
DOI 10.1007/978-81-322-1599-8_2, © Springer India 2014

Definition 2.1.1 Let S be a non-empty set. A *binary operation* on S is a function $f : S \times S \rightarrow S$.

There are several commonly used notations for the image $f(a, b) \in S$ for $a, b \in S$, such as ab or $a \cdot b$ (multiplicative notation), $a + b$ (additive notation), $a * b$ etc. We may therefore think the usual addition in $\mathbf{Z}$, as being a function: $\mathbf{Z} \times \mathbf{Z} \rightarrow \mathbf{Z}$, where the image of every pair of integers $(m, n) \in \mathbf{Z} \times \mathbf{Z}$ is denoted by $m + n$. Clearly, a binary operation '$\cdot$' on S is equivalent to the statement: if $a, b \in S$, then $a \cdot b \in S$ (known as the *closure axiom*). In this event S is said to be closed under '$\cdot$'. There may be several binary operations on a non-empty set S. For example, usual addition $(a, b) \mapsto a + b$; usual multiplication $(a, b) \mapsto ab$; $(a, b) \mapsto$ maximum $\{a, b\}$, minimum $\{a, b\}$, lcm $\{a, b\}$, gcd $\{a, b\}$ for $(a, b) \in \mathbf{N}^+ \times \mathbf{N}^+$ are binary operations on $\mathbf{N}^+$.

A binary operation on a set S is sometimes called a *composition in S*.

Example 2.1.1 Two binary operations, called addition and multiplication are defined on $\mathbf{Z}_n$ (see Chap. 1) by $((a), (b)) \mapsto (a + b)$ and $((a), (b)) \mapsto (ab)$, respectively. Clearly, each of these operations is independent of the choice of representatives of the classes and hence is well defined.

Example 2.1.2 Let S be any non-empty set. Then

 (i) both union and intersection of two subsets of S define binary operations on $\mathcal{P}(S)$, the power set of S;
(ii) the usual composition '$\circ$' of two mappings of $M(S)$, the set of all mappings of S into S, is a binary operation on $M(S)$ i.e., $(f, g) \mapsto f \circ g$, $f, g \in M(S)$, defines a binary operation on $M(S)$.

   Clearly, $f \circ g \neq g \circ f$, $f, g \in M(S)$ for an arbitrary set S.
   For example, if $S = \{1, 2, 3\}$ and $f, g \in M(S)$ are defined by

$$f(s) = 1 \quad \forall s \in S \quad \text{and} \quad g(s) = 2 \quad \forall s \in S, \quad \text{then } (g \circ f)(s) = g(f(s))$$

$$= g(1) = 2 \quad \text{and} \quad (f \circ g)(s) = f(g(s)) = f(2) = 1, \quad \forall s \in S$$

$$\Rightarrow \quad g \circ f \neq f \circ g.$$

We may construct a table called the *Cayley table* as a convenient way of defining a binary operation only on a finite set S or tabulating the effect of a binary operation on S. Let S be a set with n distinct elements. To construct a table, the elements of S are arranged horizontally in a row, called initial row or 0-row: these are again arranged vertically in a column, called the initial column or 0-column. The element in the ith position in the 0-column determines a horizontal row across it, called the ith row of the table.

Similarly, the element in the jth position in the 0-row determines a vertical column below it, called the jth column of the table. The (i, j)th position in the table is determined by the intersection of the ith row and the jth column.

The (i, j)th positions $(i, j = 1, 2, \ldots, n)$ are filled up by elements of S in any manner. Accordingly, a binary operation on S is determined.

For two elements $u, v \in S$, where u occupies ith position in the 0-column and v occupies jth position in the 0-row, $(u, v) \mapsto u \circ v$, where $u \circ v$ is the element in the (i, j)th position, $i, j = 1, 2, 3, \ldots, n$.

There is also a reverse procedure: given a binary operation f on a finite set S, we can construct a table displaying the effect of f. For example, let $S = \{1, 2, 3\}$ and f a binary operation on S defined by

$$f(1, 1) = 1, \qquad f(1, 2) = 1, \qquad f(1, 3) = 2, \qquad f(2, 1) = 2, \qquad f(2, 2) = 3,$$

$$f(2, 3) = 3, \qquad f(3, 1) = 1, \qquad f(3, 2) = 3, \qquad f(3, 3) = 2.$$

Then the corresponding table is

f :	1	2	3
1	1	1	2
2	2	3	3
3	1	3	2

A table of this kind is termed a *multiplication table or composition table* or *Cayley's table* on S.

Definition 2.1.2   A groupoid is an ordered pair $(S, \circ)$, where S is a non-empty set and '$\circ$' is a binary operation on S and a non-empty subset H of S is said to be a subgroupoid of $(S, \circ)$ iff $(H, \circ_H)$ is a groupoid, where $\circ_H$ is the restriction of '$\circ$' to $H \times H$.

Supplementary Examples (SE-IA)

1   Let $\mathbf{R}$ be the set of all real numbers and $\mathbf{R}^* = \mathbf{R} - \{0\}$.

(i) Define a binary operation $\circ$ on $\mathbf{R}^*$ by $x \circ y = |x|y$. Then $(x \circ y) \circ z = x \circ (y \circ z)$ for all $x, y, z \in \mathbf{R}^*$ but $x \circ y \neq y \circ x$ for some $x, y \in \mathbf{R}^*$.

(ii) Define a binary operation $\circ$ on $\mathbf{R}^*$ by $x \circ y = |x - y|$. Then $(x \circ y) \circ z \neq x \circ (y \circ z)$ for some $x, y, z \in \mathbf{R}^*$ but $x \circ y = y \circ x$ for all $x, y \in \mathbf{R}^*$.

(iii) Define a binary operation $\circ$ on $\mathbf{R}^*$ by $x \circ y = \min\{x, y\}$. Then $(x \circ y) \circ z = x \circ (y \circ z)$ and $x \circ y = y \circ x$ for all $x, y, z \in \mathbf{R}^*$.

2   Let $\mathbf{Q}^*$ be the set of all non-zero rational numbers. Define a binary relation $\circ$ on $\mathbf{Q}^*$ by $x \circ y = x/y$ (usual division).

Then $x \circ y \neq y \circ x$ and $(x \circ y) \circ z \neq x \circ (y \circ z)$ for some $x, y, z \in \mathbf{Q}^*$.

3   A homomorphism f from a groupoid $(G, \circ)$ into a groupoid $(H, *)$ is a mapping $f : G \to H$ such that $f(x \circ y) = f(x) * f(y)$ for all $x, y \in G$. A homomorphism f is said to be a *monomorphism, epimorphism* or *isomorphism* ($\cong$) according as f is injective, surjective or bijective.

(i) Isomorphism relation of groupoids is an equivalence relation.

(ii) If G and H are finite groupoids such that $|G| \neq |H|$. Then G cannot be iso-morphic to H.

(iii) The groupoids $(\mathbf{Z}, \circ)$ and $(\mathbf{Z}, *)$ under the compositions defined by $x \circ y = x + y + xy$ and $x * y = x + y - xy$ are isomorphic.

    [*Hint.* The function $f : (\mathbf{Z}, \circ) \to (\mathbf{Z}, *)$ defined by $f(x) = -x \; \forall x \in \mathbf{Z}$ is an isomorphism.]

(iv) Let $(\mathbf{C}, +)$ be the groupoid of complex numbers under usual addition and $f : (\mathbf{C}, +) \to (\mathbf{C}, +)$ be defined by $f(a + ib) = a - ib$. Then f is an isomorphism.

(v) Let $(\mathbf{N}^+, \cdot)$ and $(\mathbf{R}, +)$ be groupoids under usual multiplication of positive integers and usual addition of real numbers. Then $f : (\mathbf{N}^+, \cdot) \to (\mathbf{R}, +)$ defined by $f(n) = \log_{10} n$ is a monomorphism but not an epimorphism.

    [*Hint.* $0 = \log_{10} 1 < \log_{10} 2 < \log_{10} 3 < \cdots \Rightarrow$ there is no integer $n \in \mathbf{N}^+$ such that $\log_{10} n = -1 \in \mathbf{R} \Rightarrow f$ is not an epimorphism.]

2.2 Semigroups

Our main interest in this chapter is in groups. Semigroups and monoids are con-venient algebraic systems for stating some theorems on groups. Semigroups play an important role in algebra, analysis, topology, theoretical computer science, and in many other branches of science. Semigroup actions unify computer science with mathematics.

Definition 2.2.1 A *semigroup* is an ordered pair $(S, \cdot)$, where S is non-empty set and the dot '$\cdot$' is a binary operation on S, i.e., a mapping $(a, b) \mapsto a \cdot b$ from $S \times S$ to S such that for all $a, b, c \in S$, $(a \cdot b) \cdot c = a \cdot (b \cdot c)$ (*associative law*).

    For convenience, $(S, \cdot)$ is abbreviated to S and $a \cdot b$ to ab. The associative prop-erty in a semigroup ensures that the two products $a(bc)$ and $(ab)c$ are same, which can be denoted as abc.

Definition 2.2.2 A semigroup S is said to be *commutative* iff $ab = ba$ for all $a, b \in S$; otherwise S is said to be non-commutative.

Example 2.2.1 (i) $(\mathbf{Z}, +)$ and $(\mathbf{Z}, \cdot)$ are commutative semigroups, where the binary operations on $\mathbf{Z}$ are usual addition and multiplication of integers.

    (ii) $(\mathcal{P}(S), \cap)$ and $(\mathcal{P}(S), \cup)$ (cf. Example 2.1.2(i)) are commutative semigroups.

    (iii) $(M(S), \circ)$ (cf. Example 2.1.2(ii)) is a non-commutative semigroup.

Definition 2.2.3 Let S be a semigroup. An element $1 \in S$ is called a *left* (*right*) *identity* in S iff $1a = a$ ($a1 = a$) for all $a \in S$ and 1 is called an *identity* iff it is a both left and right identity in S.

Proposition 2.2.1 *If a semigroup S contains a left identity e and a right identity f, then e = f is an identity and there is no other identity.*

Proof ef = f, as *e* is a left identity and also *ef = e*, as *f* is a right identity. Consequently, *e = f* is an identity. Next let 1 ∈ *S* is also an identity. Then 1 is both a left and a right identity. Consequently, *e* = 1 and *f* = 1. □

This proposition shows that a semigroup has at most one identity element and we denote the identity element (if it exists) of a semigroup by 1.

Definition 2.2.4 Let *S* be a semigroup. An element 0 ∈ *S* is called a *left (right) zero* in *S* iff $0a = 0$ ($a0 = 0$) for all *a* ∈ *S* and 0 is called a *zero* iff it is a both left and right zero in *S*.
 If an element *a* ∈ *S* is such that $a \neq 0$, then *a* is called a non-zero element of *S*.

Proposition 2.2.2 *If a semigroup S contains a left zero z and also a right zero w, then z = w is a zero and there is no other zero.*

Proof Similar to proof of Proposition 2.2.1. □

Definition 2.2.5 A semigroup *S* with the identity element is called a *monoid*.

Example 2.2.2 (i) (**Z**, ·) is a monoid under usual multiplication, where 1 is the identity.
 (ii) The set of even integers is not a monoid under usual multiplication.
 If a semigroup *S* has no identity element, it is easy to adjoin an extra element 1 to the set *S*. For this we extend the given binary operation '·' on *S* to $S \cup \{1\}$ by defining $1 \cdot a = a \cdot 1 = a$ for all *a* ∈ *S* and $1 \cdot 1 = 1$. Then $S \cup \{1\}$ becomes a semigroup with identity element 1.
 Analogously, it is easy to adjoin an element 0 to *S* and extend the given binary operation on *S* to $S \cup \{0\}$ by defining $0 \cdot a = a \cdot 0 = 0$ for all *a* ∈ *S* and $0 \cdot 0 = 0$. Then $S \cup \{0\}$ becomes a semigroup with zero element 0.

Definition 2.2.6 Let *S* be a semigroup with zero element 0. If for two non-zero elements *a, b* ∈ *S*, $ab = 0$, then *a* and *b* are called respectively a *left* and a *right divisor of zero*.

Example 2.2.3 $M_2(\mathbf{Z})$, the set of 2×2 matrices over **Z** is semigroup (non-commutative) under usual multiplication of matrices with $\begin{pmatrix} 0 & 0 \\ 0 & 0 \end{pmatrix}$ as zero element. Clearly, $\begin{pmatrix} 2 & 0 \\ 3 & 0 \end{pmatrix}$ and $\begin{pmatrix} 0 & 0 \\ 1 & 5 \end{pmatrix}$ are non-zero elements of $M_2(\mathbf{Z})$ such that $\begin{pmatrix} 2 & 0 \\ 3 & 0 \end{pmatrix}\begin{pmatrix} 0 & 0 \\ 1 & 5 \end{pmatrix} = \begin{pmatrix} 0 & 0 \\ 0 & 0 \end{pmatrix}$. As a result, $\begin{pmatrix} 2 & 0 \\ 3 & 0 \end{pmatrix}$ is a left divisor of zero, having $\begin{pmatrix} 0 & 0 \\ 1 & 5 \end{pmatrix}$ the corresponding right divisor of zero. Again $\begin{pmatrix} -3 & 2 \\ -6 & 4 \end{pmatrix}\begin{pmatrix} 2 & 0 \\ 3 & 0 \end{pmatrix} = \begin{pmatrix} 0 & 0 \\ 0 & 0 \end{pmatrix}$ shows that $\begin{pmatrix} 2 & 0 \\ 3 & 0 \end{pmatrix}$ is also a right divisor of zero, having a different matrix $\begin{pmatrix} -3 & 2 \\ -6 & 4 \end{pmatrix}$ the corresponding left divisor of zero.

If $a_1, a_2, \ldots, a_n$ are elements of a semigroup S, we define the product $a_1 a_2 \cdots a_n$ inductively, as $(a_1 a_2 \cdots a_{n-1}) a_n$.

Theorem 2.2.1 *Let S be a semigroup. Then all the products obtained by insertion of parentheses in the sequence $a_1, a_2, \ldots, a_n, a_{n+1}, a_{n+2}$ of elements of S are equal to the product $a_1 a_2 \cdots a_n a_{n+1} a_{n+2}$. (Generalized associative law.)*

Proof By associative property in S, the theorem holds for $n = 1$. Let the theorem hold for all values of n such that $1 \leq n \leq r$ and let p and p' be the product of an ordered set of $2 + r + 1$ elements $a_1, a_2, \ldots, a_{2+r+1}$, for two different modes of association obtained by insertion of parentheses. Since the theorem holds for all values of $n \leq r$, the parentheses can all be deleted up to the penultimate stage. Then we have $p = (a_1 a_2 \cdots a_t)(a_{t+1} \cdots a_{2+r+1})$, where $1 \leq t \leq 2 + r$, and $p' = (a_1 \cdots a_i)(a_{i+1} \cdots a_{2+r+1})$, where $1 \leq i \leq 2 + r$. If $i = t$, then $p = p'$. Otherwise, we can assume without loss of generality that $i > t$. Then

$$p' = \big((a_1 a_2 \cdots a_t)(a_{t+1} \cdots a_i)\big)(a_{i+1} \cdots a_{2+r+1})$$
$$= (a_1 \cdots a_t)\big((a_{t+1} \cdots a_i)(a_{i+1} \cdots a_{2+r+1})\big), \quad \text{by associativity in } S$$
$$= (a_1 \cdots a_t)(a_{t+1} \cdots a_i a_{i+1} \cdots a_{2+r+1}) = p.$$

Thus $p = p'$ in either case, and the theorem holds for $n = r + 1$ also. Hence by the principle of mathematical induction the theorem holds for all positive integral values of n. $\qquad\square$

Powers of an Element   Let S be a semigroup and $a \in S$. Then the powers of a are defined recursively:

$$a^1 = a, \qquad a^{n+1} = a^n \cdot a \quad \text{for } n = 1, 2, 3, \ldots;$$

n is called the index of a^n. Thus the powers of a are defined for all positive integral values of n.

Proposition 2.2.3 *Let S be a semigroup. Then for any element $a \in S$, $a^r a^t = a^{r+t}$ and $(a^r)^t = a^{rt}$, r, t are positive integers.*

Proof The proof follows from Theorem 2.2.1. $\qquad\square$

Definition 2.2.7 Let S be a semigroup with identity 1 and $a \in S$. An element $a' \in S$ is said to be a left inverse of a with respect to 1 iff $a'a = 1$. Similarly, an element $a^* \in S$ is a right inverse of a iff $aa^* = 1$. An element, which is both a left and right inverse of an element a, with respect to 1, is called a two-sided inverse or an inverse of a.

Theorem 2.2.2 *Let G be a semigroup with an identity element 1. If $a \in G$ has an inverse, then it is unique.*

Proof Let b and b' be inverses of a in G. Then

$$b = b1$$
$$= b(ab'), \quad \text{since } b' \text{ is an inverse of } a.$$
$$= (ba)b' \quad \text{(by associativity in } G)$$
$$= 1b', \quad \text{since } ba = 1$$
$$= b'. \qquad\qquad\qquad\qquad\qquad\qquad\qquad \square$$

We denote by a^{-1} the unique inverse (if it exists) of an element a of a semigroup.

The set $M(X)$ of all mappings of a given non-empty set X into itself, where the binary operation is the usual composition 'o' of mappings, forms an important semigroup.

Proposition 2.2.4 *If X is a non-empty set, $M(X)$ is a monoid.*

Proof Clearly, $(M(X), \circ)$ is a semigroup under usual composition of mappings. Let $1_X : X \to X$ be the identity map defined by $1_X(x) = x$ for all $x \in X$. Then for any $f \in M(X), (1_X \circ f)(x) = 1_X(f(x)) = f(x)$ and $(f \circ 1_X)(x) = f(1_X(x)) = f(x)$ for all $x \in X$ imply $1_X \circ f = f \circ 1_X = f$. Consequently, $M(X)$ is a monoid with 1_X as its identity element. $\qquad\qquad\qquad\qquad\qquad\qquad\qquad \square$

Remark Every element in $M(X)$ may not have an inverse.

For example, if $X = \{1, 2, 3\}$, then $f \in M(X)$ defined by $f(1) = f(2) = f(3) = 1$ has no inverse in $M(X)$. Otherwise, there exists an element $g \in M(X)$ such that $f \circ g = g \circ f = 1_X$. Then $1 = 1_X(1) = (g \circ f)(1) = g(f(1)) = g(1)$ and $2 = 1_X(2) = (g \circ f)(2) = g(f(2)) = g(1)$ imply that $g(1)$ has two distinct values. This contradicts the assumption that g is a map.

We now characterize the elements of $M(X)$ which have inverses.

Theorem 2.2.3 *An element f in the monoid $M(X)$ has an inverse iff f is a bijection.*

Proof The proof follows from Corollary of Theorem 1.2.7 of Chap. 1 (by taking $X = Y$ in particular). $\qquad\qquad\qquad\qquad\qquad\qquad\qquad \square$

Let X be a non-empty set and $B(X)$ denote the set of all binary relations on X. Define a binary operation 'o' on $B(X)$ as follows:

If $\rho, \sigma \in B(X)$, then $\sigma \circ \rho = \{(x, y) \in X \times X : \text{if } \exists z \in X \text{ such that } (x, z) \in \rho \text{ and } (z, y) \in \sigma\}$.

Then $B(X)$ is a semigroup with identity element $\Delta = \{(x, x) : x \in X\}$.

Definition 2.2.8 A relation ρ on a semigroup S is called a *congruence relation* on S iff

(i) ρ is an equivalence relation on S;

(ii) $(a, b) \in \rho \Rightarrow (ca, cb) \in \rho$ and $(ac, bc) \in \rho$ for all $c \in S$.

Let ρ be a congruence relation on a semigroup S and S/ρ the set of all ρ-equivalence classes $a\rho, a \in S$. Define a binary operation 'o' on S/ρ by $a\rho \circ b\rho = (ab)\rho$, for all $a, b \in S$. Clearly, the operation 'o' is well defined.

Then $(S/\rho, \circ)$ becomes a semigroup.

Definition 2.2.9 A mapping f from a semigroup S into a semigroup T is called a homomorphism iff for all $a, b \in S$,

$$f(ab) = f(a)f(b).$$

Let S and T be semigroups. Then a homomorphism $f : S \to T$ is called

(i) an *epimorphism* iff f is surjective;

(ii) a *monomorphism* iff f is injective; and

(iii) an *isomorphism* iff f is bijective.

If $f : S \to T$ is an isomorphism, we say that S is isomorphic to T denoted $S \cong T$.

Proposition 2.2.5 *Let S be a semigroup and ρ a congruence relation on S. Then the map $p : S \to S/\rho$ defined by $p(a) = a\rho$ for all $a \in S$ is a homomorphism from S onto S/ρ (p is called* natural homomorphism).

Proof Trivial.   □

Definition 2.2.10 Let S and T be two semigroups and $f : S \to T$ a homomorphism from S onto T. The kernel of f denoted by $\ker f$ is defined by $\ker f = \{(a, b) \in S \times S : f(a) = f(b)\}$.

Clearly, $\rho = \ker f$ is a congruence relation on S. Then S/ρ is a semigroup.

Proposition 2.2.6 *Let f be a homomorphism from a semigroup S onto a semigroup T. Then the semigroup $S/\ker f$ is isomorphic to T.*

Proof Suppose $\rho = \ker f$. Define $g : S/\rho \to T$ by $g(a\rho) = f(a)$ for all $a \in S$. Clearly, g is an isomorphism.   □

Definition 2.2.11 A non-empty subset I of a semigroup S is called a *left* (*right*) *ideal* of S iff $xa \in I$ ($ax \in I$) for all $x \in S$, for all $a \in I$ and I is called an *ideal* of S iff I is both a left ideal and a right ideal of S.

Example 2.2.4 Let S be a semigroup. Then

(i) S is an ideal of S;

(ii) for each $a \in S$, $Sa = \{sa : s \in S\}$ is a left ideal and $aS = \{as : s \in S\}$ is a right ideal of S.

Definition 2.2.12 A semigroup S is called *left (right) simple* iff S has no left (right) ideals other than S.

Definition 2.2.13 Let S be a semigroup. An element $a \in S$ is called an *idempotent* iff $aa = a^2 = a$.

Clearly if 1 or $0 \in S$, they are idempotents. The set of idempotents of S is denoted by $E(S)$.

If a semigroup S contains 0, then an element $c \in S$ is said to be a *nilpotent* element of rank n for $n \geq 2$ iff $c^n = 0$, but $c^{n-1} \neq 0$ and hence $(c^{n-1})^2 = 0$, since $2(n - 1) \geq n$, so that c^{n-1} is a nilpotent element of rank 2. Thus the following proposition is immediate.

Proposition 2.2.7 *If a semigroup S contains any nilpotent element, then there also exists a nilpotent element of rank 2 in S.*

Definition 2.2.14 A semigroup S is called a *band* iff $S = E(S)$, i.e., iff every element of S is idempotent. S is called a right (left) group, iff S is right (left) simple and left (right) cancellative.

Let S be a band. Define a relation '$\leq$' on S by $e \leq f$, iff $ef = fe = e$. Then clearly, $\leq$ is a partial order on S. A band S is said to be commutative, iff S is a commutative semigroup.

Let X and Y be two sets and define a binary operation on $S = X \times Y$ as follows: $(x, y)(z, t) = (z, t)$, where $x, z \in X$ and $y, t \in Y$. Then S is a non-commutative band. We call S the rectangular band on $X \times Y$.

Definition 2.2.15 Let S be a semigroup. An element $a \in S$ is called *regular* iff $a \in aSa$, i.e., iff $a = axa$ for some $x \in S$. A semigroup S is called regular iff every element of S is regular.

Clearly, if $axa = a$, then $e = ax$ and $f = xa$ are idempotents in S.

Let $a \in S$. An element $b \in S$ is said to be an inverse of a iff $aba = a$ and $bab = b$. Clearly, if a is a regular element of S, then a has at least one inverse.

Example 2.2.5 (i) Every right group is a regular semigroup.

(ii) Every band is a regular semigroup.

(iii) $M(X)$ (cf. Proposition 2.2.4) is a regular semigroup under usual composition of mappings.

(iv) Let $M_2(\mathbf{Q})$ be the set of all 2×2 matrices $\begin{pmatrix} a & b \\ c & d \end{pmatrix}$ over $\mathbf{Q}$, the set of rationals. Then $M_2(\mathbf{Q})$ is a regular semigroup with respect to the usual matrix multiplication.

Definition 2.2.16 A regular semigroup S is called an *inverse semigroup* iff for every element a of S there exists a unique element a^{-1} such $aa^{-1}a = a$ and $a^{-1}aa^{-1} = a^{-1}$.

Clearly, every inverse semigroup is a regular semigroup, but there are regular semigroups which are not inverse semigroups. For example, a rectangular band is a regular semigroup, but it is not an inverse semigroup.

Let $(S, \circ)$ be a semigroup. A non-empty subset T of S is called a sub-semigroup of S iff $\forall a, b \in T$, $a \circ b \in T$ i.e., iff $(T, \circ)$ is also a semigroup where the latter operation '$\circ$' is the restriction of '$\circ$' to $T \times T$.

For example, $(\mathbf{N}, \cdot)$ is a sub-semigroup of $(\mathbf{Z}, \cdot)$, where '$\cdot$' denotes the usual multiplication. Clearly, for any semigroup S, S is a sub-semigroup of itself.

Example 2.2.6 Let $S = M_2(\mathbf{R})$ be the multiplicative semigroup of all 2×2 real matrices and T the subset of all matrices of the form $\left(\begin{smallmatrix} a & 0 \\ 0 & 0 \end{smallmatrix}\right)$, $a \in \mathbf{R}$. Then T is a sub-semigroup of S.

S has an identity $I = \left(\begin{smallmatrix} 1 & 0 \\ 0 & 1 \end{smallmatrix}\right)$ and T has an identity $I' = \left(\begin{smallmatrix} 1 & 0 \\ 0 & 0 \end{smallmatrix}\right)$ but $I \neq I'$.

This example shows that a semigroup S with an identity 1 may have a sub-semigroup with an identity e such that $e \neq 1$.

Proposition 2.2.8 *If $\{A_i\}$ is any collection of sub-semigroups of a semigroup S, then $\bigcap_i A_i$ is also a sub-semigroup of S, provided $\bigcap_i A_i \neq \emptyset$.*

Proof Let $M = \bigcap_i A_i$ $(\neq \emptyset)$ and $a, b \in M$. Then $a, b \in A_i \Rightarrow ab \in A_i$ for each sub-semigroup $A_i \Rightarrow ab \in M \Rightarrow M$ is a sub-semigroup of S, as associative property is hereditary.   □

If X is a non-empty subset of a semigroup S, then $\langle X \rangle$, the sub-semigroup of S generated by X defined by the intersection of all sub-semigroups of S containing X, is the smallest sub-semigroup of S. We say that X generates S iff $\langle X \rangle = S$. Clearly $\langle S \rangle = S$.

Let $a \in S$, then $\langle \{a\} \rangle = \langle a \rangle$ is called the cyclic sub-semigroup of S generated by a. A semigroup S is called *cyclic* iff there exists an element $a \in S$ such that $S = \langle a \rangle$.

Let S be an inverse semigroup and $a, b \in S$. Then the relation '$\leq$' defined by $a \leq b$ iff there exists an idempotent $e \in S$ such that $a = eb$ is a partial order relation.

Let $E(S)$ be the set of all idempotents of an inverse semigroup S. Then $E(S)$ is a sub-semigroup of S. Moreover, $E(S)$ is a semilattice, called the *semilattice of idempotents* of S.

Some Other Semigroups A semigroup S is called fully idempotent iff each ideal I of S is idempotent, i.e., iff $I^2 = I$.

The following statements for a semigroup S are equivalent:

(a) S is fully idempotent;

(b) for each pair of ideals, I, J of S, $I \cap J = IJ$;

(c) for each right ideal R and two-sided ideal I, $R \cap I \subseteq IR$;

(d) for each left ideal L and two-sided ideal I, $L \cap I \subseteq LI$.

Definition 2.2.17 A function $l(r)$ on a semigroup S is said to be *left (right) translation* on S iff $l(r)$ is written as a left (right) operator and satisfies

$$l(xy) = (lx)y \big((xy)r = x(yr)\big) \quad \forall x, y \in S.$$

The pair (l, r) is said to be *linked* iff $x(ly) = (xr)y \ \forall x, y \in S$ and is then called a bitranslation on S.

The set $L(S)$ of all left translations ($R(S)$ of all right translations) on S with the operation of multiplication defined by $(ll_1)x = l(l_1x)(x(rr_1) = (xr)r_1) \ \forall x \in S$ is a semigroup.

The set $B(S)$ of all bitranslations on S with the operation defined by $(l, r)(l_1, r_1) = (ll_1, rr_1)$ is a semigroup and it called the *translational hull* of S.

Two bitranslations (l, r) and (l_1, r_1) are said to be equal iff $l = l_1$ and $r = r_1$.

For any $s \in S$, the functions l_s and r_s given by $l_s x = sx, x r_s = xs \ \forall x \in S$ are called respectively the *inner left and inner right translation* on S induced by s; the pair $T_s = (l_s, r_s)$ is called the *inner bitranslation* on S induced by s.

The set $T(S)$ of all inner bitranslations on S is called the inner part of $B(S)$. If $L_i(S)$ and $R_i(S)$ are two sets of all *inner left* and *inner right translations* on S respectively, then $T(S) \subset B(S), L_i(S) \subset L(S)$, and $R_i(S) \subset R(S)$.

Lemma 2.2.1 *If* $l \in L(S), r \in R(S), l_s \in L_i(S), r_s \in R_i(S), T_s \in T(S), w = (l, r) \in B(S)$, *then for every* $s \in S$,

(i) $ll_s = l_{ls}, l_s l = l_{sr}$;

(ii) $r_s r = r_{sr}, rr_s = r_{ls}$ *and*

(iii) $wT_s = T_{ls}, T_s w = T_{sr}$.

Proof (i) $(ll_s)x = l(l_s x) = l(sx) = (ls)x = l_{ls}x \ \forall x \in S \Rightarrow ll_s = l_{ls} \in L_i(S)$.

Again $(l_s l)x = l_s(lx) = s(lx) = (sr)x = l_{sr}x \ \forall x \in S \Rightarrow l_s l = l_{sr} \in L_i(S)$.

(ii) $x(r_s r) = (xr_s)r = (xs)r = x(sr) = xr_{sr} \ \forall x \in S \Rightarrow r_s r = r_{sr} \in R_i(S)$.

Similarly, $rr_s = r_{ls} \in R_i(S)$.

By using (i) and (ii) it follows that

(iii) $wT_s = (l, r)(l_s, r_s) = (ll_s, rr_s) = (l_{ls}, r_{ls}) = T_{ls} \in T(S)$.

Similarly, $T_s w = T_{sr} \in T(S)$. $\qquad \square$

Corollary 1 *For a semigroup* S,

(i) $L_i(S)$ *is an ideal of* $L(S)$;

(ii) $R_i(S)$ *is an ideal of* $R(S)$; *and*

(iii) $T(S)$ *is an ideal of* $B(S)$.

Corollary 2 *For a semigroup* S,

(i) $L_i(S)$ *is a sub-semigroup of* $L(S)$;

(ii) $R_i(S)$ *is a sub-semigroup of* $R(S)$; *and*
(iii) $T(S)$ *is a sub-semigroup of* $B(S)$.

Definition 2.2.18 A semigroup S is said to be left (right) *reductive*, iff $xa = xb$ $(ax = bx)$ $\forall x \in S \Rightarrow a = b$. If S is both left and right reductive, then S is said to be *reductive*. If $xa = xb$ and $ax = bx$ $\forall x \in S \Rightarrow a = b$, then S is said to be *weakly reductive*.

Clearly, a reductive semigroup is also weakly reductive.

For the application of *category theory* we follow the terminology of Appendix B. It is easy to check that semigroups and their homomorphisms form a category denoted by S_G.

We note that for a semigroup $S \in S_G$, $T(S)$ is also a semigroup $\in S_G$. For $f : S \to W$ in S_G, define $f_* = T(f) : T(S) \to T(W)$ by $f_*((l_r, r_s)) = (l_{f(s)}, r_{f(s)})$. Then for $g : W \to X$ in S_G, $(gf)_* : T(S) \to T(X)$ in S_G is such that

$$(gf)_*((l_s, r_s)) = (l_{(gf)(s)}, r_{(gf)(s)}) = (l_{g(f(s))}, r_{g(f(s))}) = g_*((l_{f(s)}, r_{f(s)}))$$

$$= g_*\big(f_*((l_s, r_s))\big) \quad \forall (l_s, r_s) \in T(S) \quad \Rightarrow \quad (gf)_* = g_* f_*.$$

Also for the identity homomorphism $I_S : S \to S$ in S_G, $I_{S*} : T(S) \to T(S)$ in S_G is such that $I_{S*}((l_s, r_s)) = (l_{I_S(s)}, r_{I_S(s)}) = (l_s, r_s)$ $\forall (l_s, r_s) \in T(S) \Rightarrow I_{S*}$ is the identity homomorphism. The following theorem is immediate from the above discussions.

Theorem 2.2.4 $T : S_G \to S_G$ *is a covariant functor* (*see Appendix B*).

Lemma 2.2.2 *For a semigroup* S, *the function* $f : S \to T(S)$ *defined by* $f(s) = T_s$ $\forall s \in S$ *is a homomorphism from* S *onto* $T(S)$.

Proof $f(st) = T_{st} = T_s T_t = f(s) f(t)$ $\forall s, t \in S \Rightarrow f$ is a homomorphism. Moreover, for any $T_s \in T(S)$, we can find $s \in S$ such that $f(s) = T_s$. □

Theorem 2.2.5 *The function* $f : S \to T(S)$ *defined by* $f(s) = T_s$ *is an isomorphism iff* S *is reductive.*

Proof Suppose S is reductive and $f(s) = f(t)$ for some $s, t \in S$. Then $T_s = T_t \Rightarrow (l_s, r_s) = (l_t, r_t) \Rightarrow l_s = l_t$ and $r_s = r_t \Rightarrow sx = tx$ and $xs = xt$ $\forall x \in S \Rightarrow s = t \Rightarrow f$ is injective. Hence by Lemma 2.2.2, f is an isomorphism.

Conversely, let f be an isomorphism and $ax = bx$, $xa = xb$ hold $\forall a, b \in S$ and $\forall x \in S$. Then $l_a x = l_b x$ and $x r_a = x r_b$ hold $\forall x \in S$. This shows that $l_a = l_b$ and $r_a = r_b$ yielding $f(a) = T_a = (l_a, r_a) = (l_b, r_b) = T_b = f(b)$. Hence $a = b \Rightarrow S$ is reductive. □

Corollary *If* S *is a cancellative semigroup, then the function* $f : S \to T(S)$ *defined by* $f(s) = T_s$ *is an isomorphism.*

For a given semigroup $S \in S_G$, the set $F_S(X)$ of all semigroup homomorphisms from S to X in S_G forms a semigroup under usual multiplication of functions.

We also define for each homomorphism $f : X \to Y$ in S_G, the semigroup homomorphism $f_* = F_S(f) : F_S(X) \to F_S(Y)$ by $f_*(g) = f \circ g \; \forall g : S \to X$ in S_G. Hence the following lemma is immediate

Lemma 2.2.3 $F_S : S_G \to S_G$ *is a covariant functor.*

Let (F_S, T) *denote the set of all natural transformations from the covariant functor F_S to the covariant functor $T : S_G \to S_G$ (see Appendix B).*

Theorem 2.2.6 *For each semigroup $S \in S_G$, the set (F_S, T) admits a semigroup structure such that (F_S, T) is isomorphic to $T(S)$.*

Proof Using the Yoneda Lemma (see Appendix B) for covariant functors F_S and T we find that there is a bijection $f_S : T(S) \to (F_S, T)$ for every $S \in S_G$. As $T(S)$ is a semigroup, f_S induces composition on the set (F_S, T) admitting it a semigroup structure such that (F_S, T) is isomorphic to $T(S)$. $\square$

Theorem 2.2.7 *A semigroup S is isomorphic to the semigroup (F_S, T) iff S is reductive.*

Proof The theorem is immediate by using Theorems 2.2.5 and 2.2.6. $\square$

Remark Theorem 2.2.6 yields a representation of the covariant functor $T : S_G \to S_G$. In this way the problem of classifying semigroups, i.e., of computing $T(S)$ has been reduced to the determination of the set of natural transformations (F_S, T).

Let S be a semigroup. T a sub-semigroup of S and $w \in B(T)$. Then a bitranslation $\tilde{w} \in B(S)$ is said to be an extension of w to S iff $\tilde{w}x = wx$ and $x\tilde{w} = xw \; \forall x \in T$ and we write $\tilde{w}|T = w$.

Theorem 2.2.8 *Let S be a semigroup, T a reductive semigroup, $f : S \to T$ an epimorphism and let $w \in B(S)$. Then there exists a unique element $w' \in B(T)$ such that for each $x \in S$, $w' f(x) = f(wx)$ and $f(x)w' = f(xw)$.*

Proof For $t \in T, \exists s \in S$ such that $f(s) = t$. Now define $w't = f(ws)$ and $tw' = f(xw)$. Then the theorem follows. $\square$

Theorem 2.2.9 *Let S be a reductive semigroup and $l, r : S \to S$ are such that the pair (l, r) is linked, then $(l, r) \in B(S)$.*

Proof It is sufficient to show that $l \in L(S)$ and $r \in R(S)$. Let $x, y \in S$. Then for each $t \in S, t(lx)(y) = (t(lx))y = (tr)(x)y = (tr)(xy) = t(l(xy))$. Since S is reductive, $(lx)y = l(xy) \Rightarrow l \in L(S)$. Similarly, $r \in R(S)$. $\square$

2.2.1 Topological Semigroups

Let S be a semigroup. If A and B are subsets of S, we define $AB = \{ab : a \in A$ and $b \in B\}$. If a subgroup S is a Hausdorff space such that the multiplication function $(x, y) \mapsto xy$ is continuous with the product topology on $S \times S$, then S is called a topological semigroup. The condition that the multiplication on S is continuous is equivalent to the condition that for each $x, y \in S$ and each open set W in S with $xy \in W$, $\exists$ open sets U and V such that $x \in U$, $y \in V$ and $UV \subset W$.

Note that any semigroup can be made into a topological semigroup by endowing it the discrete topology and hence a finite semigroup is a compact semigroup.

Let $\mathbf{C}$ be the space of complex numbers with complex multiplication. Then $\mathbf{C}$ becomes a topological semigroup with a zero and an identity and no other idempotents. The unit disk $D = \{z \in \mathbf{C} : |z| \le 1\}$ is a complex sub-semigroup of $\mathbf{C}$. The circle $S^1 = \{z \in \mathbf{C} : |z| = 1\}$ is also a sub-semigroup of $\mathbf{C}$.

Problems

A *Let A and B be subsets of a topological semigroup S.*

(a) *If A and B are compact, then AB is compact.*
(b) *If A and B are connected, then AB is connected.*
(c) *If B is closed, then $\{x \in S : xA \subset B\}$ is closed.*
(d) *If B is closed, then $\{x \in S : A \subset xB\}$ is closed.*
(e) *If B is compact, then $\{x \in S : xA \subset Bx\}$ is closed.*
(f) *If A is compact and B is open, then $\{x \in S : xA \subset B\}$ is open.*
(g) *If A is compact and B is closed, then $\{x \in S : xA \cap B \ne \emptyset\}$ is closed.*

Proof [See Carruth et al. (1983, 1986)].   □

B *If S is a topological semigroup which is a Hausdorff space, then the set $E(S)$ of all idempotents of S is a closed subset of S.*

Proof $E(S)$ is the set of fixed points of the continuous function $f : S \to S$ defined by $x \mapsto x^2$.

Define $g : S \to S \times S$ by $g(x) = (f(x), x) = (x^2, x)$. Then g is continuous. Since the diagonal of a Hausdorff space is closed in $S \times S$, $g^{-1}(\Delta(S)) = \{x \in S : x^2 = x\}$ is closed.   □

Semigroups of Continuous Self Maps   Let $S(X)$ denote the semigroup of all continuous maps from a topological space X into itself under composition of maps.

At present there are at least four *broad areas* where active researches on $S(X)$ are going on.

These are:

(1) homomorphisms from $S(X)$ into $S(Y)$;
(2) finitely generated sub-semigroups of $S(X)$;

(3) Green's relations and related topics for $S(X)$;
(4) congruences on $S(X)$.

We list some problems:

Problem 1 (a) If X and Y are homeomorphic spaces, then $S(X)$ and $S(Y)$ are iso-morphic semigroups. Is its converse true?
  [*Hint*. Let X be a discrete space with more than one element and Y be the same set, but endowed with the indiscrete topology. Then X and Y are not homeomorphic but $S(X)$ and $S(Y)$ are isomorphic under identity map.]
  (b) If X and Y are two compact 0-dimensional metric spaces, then X and Y are homeomorphic $\Leftrightarrow$ the semigroups $S(X)$ and $S(Y)$ are isomorphic.

Problem 2 Determine Green's relations for elements of $S(X)$ which are not neces-sarily regular.

Remark Some work has been done in this direction but there is much yet to do.

Problem 3 Determine all congruences on $S(X)$.

Remark If X is discrete, this was solved. For the case when X is not discrete [see Magil Jr. (1982)].

Problem 4 If X is Hausdorff and locally compact, then $S(X)$ endowed with com-pact open topology is a topological semigroup. Is the converse true?

Remark The converse is not true [see De Groot (1959)].

2.2.2 Fuzzy Ideals in a Semigroup

Let S be a non-empty set. Consider the closed interval $[0, 1]$. Any mapping from $S \to [0, 1]$ is called a fuzzy subset of S. It is a generalization of the usual concept of a subset of S in the following sense.
  Let A be a subset of S. Define its characteristic function κ_A by

$$\kappa_A(x) = 1 \quad \text{when } x \in A$$

$$= 0 \quad \text{when } x \notin A.$$

So for each subset A of S there corresponds a fuzzy subset κ_A.
  Conversely, suppose $\tilde{A} : S \to [0, 1]$ is a mapping such that $\operatorname{Im} \tilde{A} = \{0, 1\}$.
  Let $A = \{x \in S : \tilde{A}(x) = 1\}$.
  Now A is a subset of S and $\tilde{A}$ is the characteristic function of A.

Definition 2.2.19 Let S be a semigroup. A fuzzy subset $\tilde{A}$ (that is, a mapping from $S \to [0, 1]$) is called a *left* (*right*) *fuzzy ideal* iff

$$\tilde{A}(xy) \geq \tilde{A}(y)\big(\tilde{A}(xy) \geq \tilde{A}(x)\big).$$

Definition 2.2.20 Let $\tilde{A}, \tilde{B}$ be two fuzzy subsets in a semigroup S. Define $\tilde{A} \cup \tilde{B}$, $\tilde{A} \cap \tilde{B}, \tilde{A} \circ \tilde{B}$ by

$$(\tilde{A} \cup \tilde{B})(x) = \max\big(\tilde{A}(x), \tilde{B}(x)\big), \quad \forall x \in S;$$

$$(\tilde{A} \cap \tilde{B})(x) = \min\big(\tilde{A}(x), \tilde{B}(x)\big), \quad \forall x \in S;$$

$$(\tilde{A} \circ \tilde{B})(x) = \max\big\{\min\big(\tilde{A}(u), \tilde{B}(v)\big)\big\}, \quad \text{if } x = uv, \ u, v \in S$$

$$= 0, \quad \text{if } x \text{ cannot be expressed in the form } x = uv.$$

2.2.3 Exercises

Exercises-IA

1. Verify that the following '$*$' are binary operations on the Euclidean plane $\mathbf{R}^2$:
   For $(x, y), (x', y') \in \mathbf{R}^2$,

   (i) $(x, y) * (x', y') = (x, y)$ [if $(x, y) = (x', y')$] = midpoint of the line joining the point (x, y) to the point (x', y') if $(x, y) \neq (x', y')$;

   (ii) $(x, y) * (x', y') = (x + x', y + y')$, where $+$ is the usual addition in $\mathbf{R}$;

   (iii) $(x, y) * (x', y') = (x \cdot x', y \cdot y')$, where '$\cdot$' is the usual multiplication in $\mathbf{R}$.

2. Verify that the following '$\circ$' are binary operations on $\mathbf{Q}$:

   (i) $x \circ y = x - y - xy$;

   (ii) $x \circ y = \frac{x+y+xy}{2}$;

   (iii) $x \circ y = \frac{x+y}{3}$.

   Determine which of the above binary operations are associative and commutative.

3. Determine the number different binary operations which can be defined on a set of three elements.

4. Let S be the set of the following six mappings $f_i : \mathbf{R} \backslash \{0, 1\} \to \mathbf{R}$, defined by

$$f_1 : x \mapsto x; \qquad f_2 : x \mapsto \frac{1}{1-x}; \qquad f_3 : x \mapsto \frac{x-1}{x}; \qquad f_4 : x \mapsto \frac{1}{x};$$

$$f_5 : x \mapsto \frac{x}{x-1}; \qquad f_6 : x \mapsto 1 - x.$$

   Verify that usual composition of mappings is a binary operation on S and construct the corresponding multiplicative table.

Exercises-IB

1. A binary operation $*$ is defined on $\mathbf{Z}$ by $x * y = x + y - xy$, $x, y \in \mathbf{Z}$. Show that $(\mathbf{Z}, *)$ is a semigroup. Let $A = \{x \in \mathbf{Z}; x \le 1\}$. Show that $\{A, *\}$ is a sub-semigroup of $(\mathbf{Z}, *)$.

2. Let A be a non-empty set and $S = \mathcal{P}(A)$ its power set. Show that $(S, \cap)$ and $(S, \cup)$ are semigroups. Do these semigroups have identities? If so, find them.

3. Show that for every element a in a finite semigroup, there is a power of a which is idempotent.

4. Let $(\mathbf{Z}, +)$ be the semigroup of integers under usual addition and $(\mathbf{E}, +)$ be the semigroup of even integers with integer 0 under usual addition. Then the mapping $f : \mathbf{Z} \to \mathbf{E}$ defined by $z \mapsto 2z \ \forall z \in \mathbf{Z}$ is a homomorphism. Find a homomorphism $g : \mathbf{Z} \to \mathbf{E}$ such that $g \ne f$.

5. Show that the following statements are equivalent:

   (i) S is an inverse semigroup;
   (ii) S is regular and its idempotents commute.

6. Prove the following:

   (a) Let I be a left ideal of a semigroup S. Then its characteristic function κ_I is a fuzzy left ideal. Conversely, if for any subset I of S its characteristic function κ_I is a fuzzy left ideal, then I is a left ideal of S;
   (b) If $\tilde{A}, \tilde{B}$ are fuzzy ideals of a semigroup S, then $\tilde{A} \cap \tilde{B}$, $\tilde{A} \cup \tilde{B}$, $\tilde{A} \circ \tilde{B}$ are fuzzy ideals of S.

7. Let S be an inverse semigroup and $E(S)$ the semilattice of idempotents of S. Prove that

   (i) $(a^{-1})^{-1} = a$, $\forall a \in S$
   (ii) $e^{-1} = e$, $\forall e \in E(S)$
   (iii) $(ab)^{-1} = b^{-1}a^{-1}$, $\forall a, b \in S$
   (iv) $aea^{-1} \in E(S) \ \forall a \in S$ and $e \in E(S)$.

8. Let '$\le$' be a partial order on S such that $(S, \circ, \le)$ is a partially ordered (p.o.) semigroup. Define the set $P_l^0(S) = \{p \in S : a \le p \circ a \ \forall a \in S\}$ of all left positive elements of $(S, \circ, \le)$ and correspondingly the set $P_r^0(S)$ of all right positive ones, and $P^0(S) = P_l^0(S) \cap P_r^0(S)$. Let $E^0(S)$ denote the set of all idempotents of (S, o). Prove the following:

   (i) If the p.o. semigroup $(S, \circ, \le)$ has a least element u, i.e., $u \le a \ \forall a \in S$, then u is idempotent, iff $u = x \circ y$ holds for some $x, y \in S$;
   (ii) If $P^0(S) = S$ and $f \in E^0(S)$, then $x \le f \Leftrightarrow x \circ f = f \Leftrightarrow f \circ x = f$ holds $\forall x \in S$.
   (iii) If $P^0(S) = S$ holds and u is the least element of $(S, \le)$, then $S \backslash \{u\}$ is an ideal of (S, o), which is a maximal one. The converse holds if $(S, \le)$ is a fully ordered set, and if, for $a < b \Rightarrow a \circ x = b$ for some $x \in S$.

9. A non-empty subset A of a semigroup S is called a generalized left semi-ideal (gls ideal), iff $x^2 A \subseteq A$ for every $x \in S$.

(a) Let $\mathbf{Z}^*$ denote the set of all non-negative integers. Then $\mathbf{Z}^*$ is a semigroup under usual multiplication. Show that $A = \{x \in \mathbf{Z}^* : x \geq 6\}$ is a gls ideal.

(b) In an idempotent semigroup S, prove that A is a left ideal of $S \Leftrightarrow A$ is a gls ideal.

(c) Let S be a semigroup prove that S is regular iff for each generalized left semi-ideal A of S and for each right ideal B of S, $B \cap A = B^*A$, where $B^* = \{b^2 : b \in B\}$.

2.3 Groups

Historically, the concept of a group arose through the study of bijective mappings $A(S)$ on a non-empty set S. Any mathematical concept comes naturally in a very concrete form from specific sources. We start with two familiar algebraic systems: $(\mathbf{Z}, +)$ (under usual addition of integers) and $(\mathbf{Q}^+, \cdot)$ (under usual multiplication of positive rational numbers). We observe that the system $(\mathbf{Z}, +)$ possesses the following properties:

1. '$+$' is a binary operation on $\mathbf{Z}$;
2. '$+$' is associative in $\mathbf{Z}$;
3. $\mathbf{Z}$ contains an additive identity element, i.e., there is a special element, namely 0, such that $x + 0 = 0 + x = x$ for every x in $\mathbf{Z}$;
4. $\mathbf{Z}$ contains additive inverses, i.e., to each $x \in \mathbf{Z}$, there is an element $(-x)$ in $\mathbf{Z}$, called its negative, such that $x + (-x) = (-x) + x = 0$.

We also observe that the other system $(\mathbf{Q}^+, \cdot)$ possesses similar properties:

1. multiplication '$\cdot$' is a binary operation on $\mathbf{Q}^+$;
2. multiplication '$\cdot$' is associative in $\mathbf{Q}^+$;
3. $\mathbf{Q}^+$ contains a multiplicative identity element, i.e., there is a special element, namely 1, such that $x \cdot 1 = 1 \cdot x = x$ for every x in $\mathbf{Q}^+$;
4. $\mathbf{Q}^+$ contains multiplicative inverses, i.e., to each $x \in \mathbf{Q}^+$, there is an element x^{-1} which is the reciprocal of x in $\mathbf{Q}^+$, such that $x \cdot (x^{-1}) = (x^{-1}) \cdot x = 1$.

If we consciously ignore the notation and terminology, the above four properties are identical in the algebraic systems $(\mathbf{Z}, +)$ and $(\mathbf{Q}^+, \cdot)$. The concept of a group may be considered as a distillation of the common structural forms of $(\mathbf{Z}, +)$ and $(\mathbf{Q}^+, \cdot)$ and of many other similar algebraic systems.

Remark The algebraic system $(A(S), \circ)$ of bijective mappings $A(S)$ on a non-empty set S(under usual composition of mappings) satisfies all the above four properties. This group of transformations is not the only system satisfying all the above properties. For example, the non-zero rationals, reals or complex numbers also satisfy above four properties under usual multiplication. So it is convenient to introduce an abstract concept of a group to include these and other examples.

The above properties lead to introduce the abstract concept of a group. We define a group as a monoid in which every element has an inverse. We repeat the definition in more detail.

Definition 2.3.1 A group G is an ordered pair $(G, \cdot)$ consisting of a non-empty set G together with a binary operation '$\cdot$' defined on G such that

(i) if $a, b, c \in G$, then $(a \cdot b) \cdot c = a \cdot (b \cdot c)$ (associative law);
(ii) there exists an element $1 \in G$ such that $1 \cdot a = a \cdot 1 = a$ for all $a \in G$ (identity law);
(iii) for each $a \in G$, there exists $a' \in G$ such that $a' \cdot a = a \cdot a' = 1$ (inverse law).

Clearly, the identity element 1 is unique by Proposition 2.2.1. The element a' called an inverse of a is unique by Theorem 2.2.2 and is denoted by a^{-1}.

A group G is said to be *commutative* iff its binary operation '$\cdot$' is commutative: $a \cdot b = b \cdot a$, for all $a, b \in G$, otherwise G is said to be non-commutative.

The definition of a group may be re-written using additive notation: We write $a + b$ for $a \cdot b$, $-a$ for a^{-1}, and 0 for 1; $a + b$, $-a$, and 0 are, respectively, called the sum of a and b, negative of a, additive identity or zero and $(G, +)$ is called an additive group.

The binary operation in a group need not be commutative. Sometimes a commutative group is called an *Abelian group* in honor of Niels Henrick Abel (1802–1829); one of the pioneers in the study of groups.

Throughout this section a group G will denote an arbitrary multiplicative group with identity 1 (unless stated otherwise).

Proposition 2.3.1 *If G is a group, then*

(i) *$a \in G$ and $aa = a$ imply $a = 1$;*
(ii) *for all $a, b, c \in G$, $ab = ac$ implies $b = c$ and $ba = ca$ implies $b = c$* (left *and* right cancellation laws);
(iii) *for each $a \in G$, $(a^{-1})^{-1} = a$;*
(iv) *for $a, b \in G$, $(ab)^{-1} = b^{-1}a^{-1}$;*
(v) *for $a, b \in G$, each of the equations $ax = b$ and $ya = b$ has a unique solution in G.*

Proof (i) $aa = a \Rightarrow a^{-1}(aa) = a^{-1}a \Rightarrow (a^{-1}a)a = 1 \Rightarrow a = 1$.

(ii) $ab = ac \Rightarrow a^{-1}(ab) = a^{-1}(ac) \Rightarrow (a^{-1}a)b = (a^{-1}a)c \Rightarrow 1b = 1c \Rightarrow b = c$. Similarly, $ba = ca \Rightarrow b = c$.

(iii) $aa^{-1} = a^{-1}a = 1 \Rightarrow (a^{-1})^{-1} = a$.

(iv) $(b^{-1}a^{-1})(ab) = b^{-1}(a^{-1}(ab)) = b^{-1}((a^{-1}a)b) = b^{-1}(1b) = b^{-1}b = 1$ and $(ab)(b^{-1}a^{-1}) = a(bb^{-1})a^{-1} = a1a^{-1} = aa^{-1} = 1 \Rightarrow (ab)^{-1} = b^{-1}a^{-1}$.

(v) $a^{-1}b = a^{-1}(ax) = (a^{-1}a)x = 1x = x$ and $ba^{-1} = (ya)a^{-1} = y(aa^{-1}) = y1 = y$ are solutions of the equations $ax = b$ and $ya = b$, respectively.

Uniqueness of the two solutions follow from the cancellation laws (ii). $\qquad\square$

Corollary If $a_1, a_2, \ldots, a_n \in G$, then $(a_1 a_2 a_3 \cdots a_n)^{-1} = a_n^{-1} \cdots a_2^{-1} a_1^{-1}$.

Proposition 2.3.2 *Let G be a semigroup. Then G is a group iff the following conditions hold*:

(i) *there exists a left identity e in G*;
(ii) *for each $a \in G$, there exists a left inverse $a' \in G$ of a with respect to e, so that $a'a = e$.*

Proof If G is a group, then conditions (i) and (ii) follow trivially. Next let a semigroup G satisfy conditions (i) and (ii). As $e \in G$, $G \neq \emptyset$. If $a \in G$, then by using (ii) it follows that $aa' = e$. Thus $a' = a^{-1}$ is an inverse of a with respect to e, where $ae = a(a^{-1}a) = (aa^{-1})a = ea = a \ \forall a \in G \Rightarrow e$ is an identity. Therefore G is a group.   □

Proposition 2.3.3 *Let G be a semigroup. Then G is a group iff for all $a, b \in G$ the equations $ax = b$ and $ya = b$ have solutions in G.*

Proof If G is a group, then by Proposition 2.3.1(v), the equations $ax = b$ and $ya = b$ have solutions in G. Conversely, let the equation $ya = a$ have a solution $e \in G$. Then $ea = a$. For any $b \in G$, if t (depending on a and b) be a solution of the equation $ax = b$, then $at = b$.

Now, $eb = e(at) = (ea)t = at = b$.

Consequently, $eb = b, \forall b \in G \Rightarrow e$ is a left identity in G.

Next a left inverse of an element $a \in G$ is given by the solution $ya = e$ and the solution belongs to G. Consequently, for each $a \in G$, there exists a left inverse in G. As a result G is a group by Proposition 2.3.2.   □

Corollary *A semigroup G is a group iff $aG = G$ and $Ga = G$ for all $a \in G$, where $aG = \{ax : x \in G\}$ and $Ga = \{xa : x \in G\}$.*

Definition 2.3.2 The order of a group G is the cardinal number $|G|$ and G is said to be finite (infinite) according as $|G|$ is finite (infinite).

Example 2.3.1 $(\mathbf{Z}, +)$, $(\mathbf{Q}, +)$ and $(\mathbf{R}, +)$ are infinite abelian groups, where $+$ denotes ordinary addition. They are called *additive group of integers*, *additive group of rational numbers* and *additive group of real numbers,* respectively.

Example 2.3.2 $(\mathbf{Q}^*, \cdot)$, $(\mathbf{R}^*, \cdot)$ and $(\mathbf{C}^*, \cdot)$, $(S^1, \cdot)$ $(S^1 = \{z \in \mathbf{C} : |z| = 1\})$ form groups under usual multiplication, where $\mathbf{Q}^*, \mathbf{R}^*, \mathbf{C}^*$ and $\mathbf{C}$ denote respectively the set of all non-zero rational numbers, non-zero real numbers, non-zero complex numbers and all complex numbers. They are called *multiplicative groups of non-zero rationals, multiplicative group of non-zero reals* and *multiplicative group of non-zero complex numbers,* respectively. $(S^1, \cdot)$ is called *circle group* in $\mathbf{C}$.

We now present vast sources of interesting groups.

Example 2.3.3 (Permutation group) Let S be a non-empty set and $A(S)$ be the set of all bijective mappings from S onto itself. Let $f, g \in A(S)$.

Define

$$fg \quad \text{by } (fg)(x) = f\big(g(x)\big), \quad \text{for all } x \in S. \tag{2.1}$$

First we show that $fg \in A(S)$. Suppose $x, y \in S$ and $(fg)(x) = (fg)(y)$. Then $f(g(x)) = f(g(y))$. Since f is injective, it follows that $g(x) = g(y)$. Again this implies $x = y$ as g is injective. Consequently, fg is an injective mapping. Now let $x \in S$. Since f is surjective, there exists $y \in S$ such that $f(y) = x$. Again g is surjective. Hence $g(z) = y$ for some $z \in S$. Then $(fg)(z) = f(g(z)) = f(y) = x$. This implies that fg is surjective and hence $fg \in A(S)$. Since every element of $A(S)$ has an inverse, we find that $A(S)$ is a group under the composition defined by (2.1) This group is called the *permutation group* or the *group of all permutations* on S.

Example 2.3.4 (Symmetric group) In Example 2.3.3, if S contains only n $(n \geq 1)$ elements, say $S = I_n = \{1, 2, \dots n\}$, then the group $A(S)$ is called the *symmetric group S_n* on n elements. If $f \in A(S)$, then f can be described by listing the elements of I_n on a row and the image of each element under f directly below it. According to this assumption we write

$$f = \begin{pmatrix} 1 & 2 & \cdots & n \\ i_1 & i_2 & \cdots & i_n \end{pmatrix}, \quad \text{where } f(t) = i_t \in I_n, \ t \in I_n.$$

Suppose now $i_1, i_2, \dots, i_r$ $(r \leq n)$ are r distinct elements of I_n. If $f \in A(S)$ be such that f maps $i_1 \mapsto i_2, i_2 \mapsto i_3, \dots, i_{r-1} \mapsto i_r, i_r \mapsto i_1$ and maps every other elements of I_n onto itself, then f is also written as $f = (i_1 i_2 \cdots i_r)$. This is called an *r-cycle*. A 2-cycle is called a transposition. The permutations $\alpha_1, \alpha_2, \dots, \alpha_t$ in S_n are said to be disjoint iff for every i, $1 \leq i \leq t$ and every k in I_n, $\alpha_i(k) \neq k$ imply $\alpha_j(k) = k$ for every j, $1 \leq j \leq t$. A permutation $\alpha \in S_n$ is said to be even or odd according as α can be expressed as a product of even or odd number of transpositions. The set of all even permutations in S_n forms a group, called the alternating group of degree n, denoted by A_n. For $n \geq 3$, it is a normal subgroup of S_n (see Ex. 14 of SE-II).

The groups S_n are very useful to the study of finite groups. As n increases the structure of S_n becomes complicated. But we can work out the case $n = 3$ comfortably. The symmetric group S_3 has six elements and is the smallest group whose law of composition is not commutative.

Because of importance of this group, we now describe this group as follows:

For $n = 3$, $I_3 = \{1, 2, 3\}$ and $A(S) = S_3$. This *symmetric groups S_3* consists of exactly six elements:

$$
\begin{array}{cccccc}
e: 1 \mapsto 1, & f_1: 1 \mapsto 2, & f_2: 1 \mapsto 3, & f_3: 1 \mapsto 1, & f_4: 1 \mapsto 3, & f_5: 1 \mapsto 2 \\
2 \mapsto 2 & 2 \mapsto 3 & 2 \mapsto 1 & 2 \mapsto 3 & 2 \mapsto 2 & 2 \mapsto 1 \\
3 \mapsto 3 & 3 \mapsto 1 & 3 \mapsto 2 & 3 \mapsto 2 & 3 \mapsto 1 & 3 \mapsto 3.
\end{array}
$$

We can describe these six mappings in the following way:

$$e = \begin{pmatrix} 1 & 2 & 3 \\ 1 & 2 & 3 \end{pmatrix} = (1), \qquad f_1 = \begin{pmatrix} 1 & 2 & 3 \\ 2 & 3 & 1 \end{pmatrix} = (1 \quad 2 \quad 3),$$

$$f_2 = \begin{pmatrix} 1 & 2 & 3 \\ 3 & 1 & 2 \end{pmatrix} = (1 \quad 3 \quad 2), \qquad f_3 = \begin{pmatrix} 1 & 2 & 3 \\ 1 & 3 & 2 \end{pmatrix} = (2 \quad 3),$$

$$f_4 = \begin{pmatrix} 1 & 2 & 3 \\ 3 & 2 & 1 \end{pmatrix} = (1 \quad 3), \qquad f_5 = \begin{pmatrix} 1 & 2 & 3 \\ 2 & 1 & 3 \end{pmatrix} = (1 \quad 2).$$

In

$$S_3, e = \begin{pmatrix} 1 & 2 & 3 \\ 1 & 2 & 3 \end{pmatrix}$$

is the identity element.

Now

$$f_1 f_3 = \begin{pmatrix} 1 & 2 & 3 \\ 2 & 1 & 3 \end{pmatrix} = f_5 \quad \text{and} \quad f_3 f_1 = \begin{pmatrix} 1 & 2 & 3 \\ 3 & 2 & 1 \end{pmatrix} = f_4.$$

Hence $f_1 f_3 \neq f_3 f_1$. Consequently S_3 is not a commutative group.

Example 2.3.5 (General linear group) Let $GL(2, \mathbf{R})$ denote the set of all 2×2 matrices $\begin{pmatrix} a & b \\ c & d \end{pmatrix}$, where a, b, c, d are real numbers and $ad - bc \neq 0$. Taking usual multiplication of matrices as the group operation, we can show that $GL(2, \mathbf{R})$ is a group where $\begin{pmatrix} 1 & 0 \\ 0 & 1 \end{pmatrix}$ is the identity element and the element

$$\begin{pmatrix} \frac{d}{ad-bc} & \frac{-b}{ad-bc} \\ \frac{-c}{ad-bc} & \frac{a}{ad-bc} \end{pmatrix}$$

is the inverse of $\begin{pmatrix} a & b \\ c & d \end{pmatrix}$ in $GL(2, \mathbf{R})$.

This is a non-commutative group, as $\begin{pmatrix} 1 & 2 \\ 3 & 4 \end{pmatrix}, \begin{pmatrix} 5 & 6 \\ 7 & 8 \end{pmatrix} \in GL(2, \mathbf{R})$ and $\begin{pmatrix} 1 & 2 \\ 3 & 4 \end{pmatrix}\begin{pmatrix} 5 & 6 \\ 7 & 8 \end{pmatrix} = \begin{pmatrix} 19 & 22 \\ 43 & 50 \end{pmatrix}$ and $\begin{pmatrix} 5 & 6 \\ 7 & 8 \end{pmatrix}\begin{pmatrix} 1 & 2 \\ 3 & 4 \end{pmatrix} = \begin{pmatrix} 23 & 34 \\ 31 & 46 \end{pmatrix}$.

This group is called the *general linear group* of order 2 over the set of all real numbers $\mathbf{R}$.

Remark The group $GL(2, \mathbf{C})$ defined in a similar way is called *general linear group of order 2 over* $\mathbf{C}$. In general, $GL(n, \mathbf{R})$ $(GL(n, \mathbf{C}))$, the group of all invertible $n \times n$ real (complex) matrices is called *general linear group of order n over* $\mathbf{R}$ $(\mathbf{C})$.

The concept of congruence of integers (see Chap. 1) is essentially due to Gauss. This is one of the most important concepts in number theory. This suggests the concept of congruence relations on groups, which is important in modern algebra, because every congruence relation on a group produces a new group.

We now introduce the concept of congruence relation on a group. An equivalence relation ρ on a group G is said to be a *congruence* relation on G iff $(a, b) \in \rho$ implies $(ca, cb) \in \rho$ and $(ac, bc) \in \rho$, for all $c \in G$. Let a be an element of a group G and ρ be a congruence relation on G. Then the subset $\{x \in G : (a, x) \in \rho\}$ is called a *congruence class* for the element a. This subset is denoted by (a). The following theorem lists some properties of congruence classes.

Theorem 2.3.1 *Let ρ be a congruence relation on a group G. Then*

(i) $a \in (a)$;
(ii) $(a) = (b)$ iff $(a, b) \in \rho$;
(iii) *if $a, b \in G$, then either $(a) = (b)$ or $(a) \cap (b) = \emptyset$;*
(iv) *If $(a) = (b)$ and $(c) = (d)$, then $(ac) = (bd)$ and $(ca) = (db)$.*

Proof (i) Left as an exercise.

(ii) Left as an exercise.

(iii) Suppose $(a) \cap (b) \neq \emptyset$. Let $c \in (a) \cap (b)$. Then $(a, c) \in \rho$ and $(b, c) \in \rho$. Let $x \in (a)$. Then $(a, x) \in \rho$. From the symmetric and transitive property of ρ and from $(a, c) \in \rho$ and $(a, x) \in \rho$ we find that $(c, x) \in \rho$. Hence $(b, c) \in \rho$ and $(c, x) \in \rho$ imply that $(b, x) \in \rho$. As a result $x \in (b)$ and hence $(a) \subseteq (b)$. Similarly, we can show that $(b) \subseteq (a)$. Consequently $(a) = (b)$.

(iv) From the assumption, $(a, b) \in \rho$ and $(c, d) \in \rho$. Now ρ is a congruence relation. Hence $(ca, cb) \in \rho$ and $(cb, db) \in \rho$. The transitive property of ρ implies that $(ca, db) \in \rho$. Hence $(ca) = (db)$. Similarly $(ac) = (bd)$. $\qquad\square$

The following theorem will show how one can construct a new group from a given group G if a congruence relation ρ on G is given.

Theorem 2.3.2 *Let ρ be a congruence relation on a group G. If G/ρ is the set of all congruence classes for ρ on G, then G/ρ becomes a group under the binary operation given by $(a)(b) = (ab)$.*

Proof Let $(a), (b) \in G/\rho$. Suppose $(a) = (c)$ and $(b) = (d)$ $(c, d \in G)$. Then $(a, c) \in \rho$ and $(b, d) \in \rho$. Hence from (iv) of Theorem 2.3.1, we find that $(ab, cd) \in \rho$. This shows that $(ab) = (cd)$. Therefore the operation defined by $(a)(b) = (ab)$ is well defined. Now, let $(a), (b)$ and (c) be three elements of G. Then $((a)(b))(c) = (ab)(c) = ((ab)c) = (a(bc)) = (a)(bc) = (a)((b)(c))$. Therefore G/ρ is a semi-group. We can show that the congruence class (1) containing the identity element 1 of G is the identity element of G/ρ. If $(a) \in G/\rho$, then $a \in G$ and hence $a^{-1} \in G$ implies $(a^{-1}) \in G/\rho$. Now $(a^{-1})(a) = (a^{-1}a) = (1) = (aa^{-1}) = (a)(a^{-1})$. Hence the inverse of (a) exists in G/ρ. Consequently G/ρ is a group. $\qquad\square$

Corollary 1 *If G is a commutative group and ρ is a congruence on G, then G/ρ is a commutative group.*

Example 2.3.6 (Additive group of integers modulo m)   Let $(\mathbf{Z}, +)$ be the additive group of integers. Let m be a fixed positive integer. Define a relation ρ on $\mathbf{Z}$ by

$$\rho = \{(a, b) \in \mathbf{Z} \times \mathbf{Z} : a - b \text{ is divisible by } m\}.$$

Clearly, ρ is a congruence relation. This relation is usually called *congruence relation modulo m*. Two integers a and b are said to be congruent modulo m, written $a \equiv b \pmod{m}$, iff $a - b$ is divisible by m. For each integer a, the congruence class to which a belongs is $(a) = \{x \in \mathbf{Z} : x \equiv a \pmod{m}\} = \{a + km : k \in \mathbf{Z}\}$. Let $\mathbf{Z}/\rho$ denote the set of all congruence classes modulo m.

From Theorem 2.3.2, we find that $\mathbf{Z}/\rho$ is a group where the group operation + is defined by

$$(a) + (b) = (a + b).$$

This group is a commutative group. In this group the identity element is the congruence class (0) and $(-a)$ is the inverse of (a). Generally we denote this group by $\mathbf{Z}_m$ and this group is said to be the *additive group of integers modulo m*. $(0), (1), \ldots, (m-1)$ exhaust all the elements of $\mathbf{Z}_m$. Given an arbitrary integer a, the division algorithm implies that there exist unique integers q nd r such that

$$a = mq + r, \quad 0 \leq r < m.$$

From the definition of congruence $a \equiv r \pmod{m}$, it follows that $(a) = (r)$. Clearly, there are m possible remainders: $0, 1, \ldots, m - 1$. Consequently, every integer belongs to one and only one of the m different congruence classes: $(0), (1), \ldots, (m-1)$ and $\mathbf{Z}_m$ consists of exactly these elements.

Remark   The above example shows that for every positive integer m, there exists an abelian group G such that $|G| = m$.

Note 1 $\mathbf{Z}_m$ may be considered to be a group consisting of m integers $0, 1, \ldots, m - 1$ together with the binary operation $*$ given by the rule:

$$\text{for } t, s \in \mathbf{Z}, \quad t * s = t + s, \quad \text{if } t + s < m$$
$$= r, \qquad \text{if } t + s \geq m, \text{ where } r \text{ is the remainder}$$
$$\text{when } t + s \text{ is divided by } m.$$

One of the basic problems in group theory is to classify groups up to isomorphisms. For such a classification, we have to either construct an explicit expression for an isomorphism or we have to show that no such isomorphism exists. An isomorphism is a special homomorphism and it identifies groups for their classifications. The concept of a homomorphism is itself very important in modern algebra.

In the context of group theory (like semigroup), the word homomorphism means a mapping from a group to another, which respects binary operations defined on the two groups: If $f : G \to G'$ is a mapping between groups G and G', then f is called a homomorphism iff $f(ab) = f(a)f(b)$ for all $a, b \in G$. The *kernel* of the homomorphism f denoted by ker f is defined by ker $f = \{a \in G : f(a) = 1_{G'} = 1'\}$.

Remark Taking $a = b = 1$, the identity element in G, we find $f(1) = f(1)f(1)$, which shows that $f(1)$ is the identity element $1'$ of G'. Again taking $b = a^{-1}$, we find $f(1) = f(a)f(a^{-1})$. This implies $f(a^{-1}) = [f(a)]^{-1}$. Thus a homomorphism of groups maps identity element into identity element and the inverse element into the inverse element.

Note 2 If $f : G \to G'$ is a homomorphism of groups, then

$$f(a^n) = (f(a))^n, \quad a \in G, \; n \in \mathbf{Z} \; (a^n \text{ is defined in Definition 2.3.4}).$$

Proof It follows by induction on n that for $n = 1, 2, 3, \ldots,$

$$f(a^n) = (f(a))^n.$$

Again $f(a^{-1}) = (f(a))^{-1}$ shows that for $n = -k, k = 1, 2, 3, \ldots,$

$$f(a^n) = f((a^{-1})^k) = (f(a^{-1}))^k = (f(a))^{-k} = (f(a))^n$$

Finally, $f(a^0) = f(1) = 1' = (f(a))^0$. $\qquad\qquad\qquad\qquad\qquad\square$

Definition 2.3.3 A homomorphism $f : G \to G'$ between groups is called

(i) an *epimorphism* iff f is surjective (i.e., onto);
(ii) a *monomorphism* iff f is injective (i.e., 1–1);
(iii) an *isomorphism* iff f is an epimorphism and a monomorphism;

An isomorphism of a group onto itself is called an *automorphism* and a homomorphism of a group G into itself is called an *endomorphism*.

Remark Since two isomorphic groups have the identical properties, it is convenient to identify them to each other. They may be considered as replicas of each other.

If $f : G \to H$ and $g : H \to K$ are homomorphisms of groups, then their usual composition $g \circ f : G \to K$ defined by $(g \circ f)(a) = g(f(a)), a \in G$, is again a homomorphism. Because if $a, b \in G$, then $(g \circ f)(ab) = g(f(ab)) = g(f(a)f(b)) = g(f(a))g(f(b)) = (g \circ f)(a)(g \circ f)(b)$.

Theorem 2.3.3 *Let $f : G \to G'$ be a homomorphism of groups. Then*

(i) *f is a monomorphism iff ker $f = \{1\}$;*
(ii) *f is an epimorphism iff Im $f = G'$;*
(iii) *f is an isomorphism iff there is a homomorphism $g : G' \to G$ such that $g \circ f = I_G$ and $f \circ g = I_{G'}$, where I_G is the identity homomorphism of G.*

Proof (i) Let f be a monomorphism. Then f is injective and hence $\ker f = \{1\}$. Conversely, let $\ker f = \{1\}$. Now if $f(a) = f(b)$, then $f(ab^{-1}) = f(a)f(b)^{-1} = 1$ shows that $ab^{-1} \in \ker f = \{1\}$, and hence $a = b$. Consequently, f is injective.

(ii) and (iii) follow trivially.   □

Theorem 2.3.4 *Let G be a group and* $\text{Aut}\, G$ *be the set of all automorphisms of G. Then* $\text{Aut}\, G$ *is a group, called the automorphism group of G.*

Proof For $f, g \in \text{Aut}\, G$, define $f \circ g$ to be the usual composition. Clearly, $f \circ g \in \text{Aut}\, G$. Then $\text{Aut}\, G$ becomes a group with identity homomorphism I_G of G as identity and f^{-1} as the inverse of $f \in \text{Aut}\, G$.   □

Example 2.3.7 (i) Let $(\mathbf{R}^*, \cdot)$ be the multiplicative group of non-zero reals and $GL(2, \mathbf{R})$ be the general linear group defined in Example 2.3.5. Consider the map $f : GL(2, \mathbf{R}) \to \mathbf{R}^*$ defined by $f(A) = \det A$. Then for $A, B \in GL(2, \mathbf{R}), AB \in GL(2, \mathbf{R})$ and $f(AB) = \det(AB) = \det A \det B = f(A)f(B)$. This shows that f is a homomorphism. As $1 \in \mathbf{R}^*$ is the identity element of $(\mathbf{R}^*, \cdot)$, $\ker f = \{A \in GL(2, \mathbf{R}) : f(A) = \det A = 1\}$.

(ii) Let G be the group under binary operation on $\mathbf{R}^3$ defined by $(a, b, c) * (x, y, z) = (a + x, b + y, c + z + ay)$ and H be the group of matrices

$$\left\{ \begin{pmatrix} 1 & a & c \\ 0 & 1 & b \\ 0 & 0 & 1 \end{pmatrix} : a, b, c \in \mathbf{R} \right\}$$

under usual matrix multiplication. Then the map $f : G \to H$ defined by

$$f(a, b, c) = \begin{pmatrix} 1 & a & c \\ 0 & 1 & b \\ 0 & 0 & 1 \end{pmatrix}$$

is an isomorphism.

(iii) Let G be an abelian group and $f : G \to G$ be the map defined by $f(a) = a^{-1}$. Then f is an automorphism of G. This automorphism is different from the identity automorphism.

(iv) Let $(\mathbf{R}, +)$ and $(\mathbf{R}^+, \cdot)$ denote the additive group of reals and multiplication group of positive reals, respectively. Then the map $f : \mathbf{R} \to \mathbf{R}^+$ defined by $f(x) = e^x$ is an isomorphism.

(v) Let $(\mathbf{Q}, +)$ and $(\mathbf{Q}^+, \cdot)$ denote the additive group of rationals and multiplicative group of positive rationals, respectively. Then these groups cannot be isomorphic. To prove this, let $\exists$ an isomorphism $f : (\mathbf{Q}, +) \to (\mathbf{Q}^+, \cdot)$. Then for $2 \in \mathbf{Q}^+$, $\exists$ a unique element $x \in \mathbf{Q}$ such that $f(x) = 2$. Now $x = \frac{x}{2} + \frac{x}{2}$ and $\frac{x}{2} \in \mathbf{Q}$ show that $2 = f(x) = f(\frac{x}{2} + \frac{x}{2}) = f(\frac{x}{2})f(\frac{x}{2}) = [f(\frac{x}{2})]^2$. This cannot hold as $f(\frac{x}{2})$ is a positive rational. Hence f cannot be an isomorphism.

In any group, the integral powers a^n of a group element a play an important role to study finitely generated groups, in particular, cyclic groups.

Powers of an element have been defined in a semigroup, for positive integral indices. In a group, however, powers can be defined for all integral values of the index, i.e., positive, negative, and zero.

Definition 2.3.4 Let a be an element of a group G. Define

(i) $a^0 = 1$ and $a^1 = a$;
(ii) $a^{n+1} = a^n a$ for any non-negative integer n;

If $n = -m$ ($m > 0$) be a negative integer, then define $a^{-m} = (a^m)^{-1} = (a^{-1})^m$; a^n is said to be the nth power of a.

The exponents so defined have the following properties:

$$a^m a^n = a^{m+n} \quad \text{and} \quad (a^m)^n = a^{mn}.$$

Definition 2.3.5 Let G be a group. An element $a \in G$ is said to be of *finite order* iff there exists a positive integer n such that $a^n = 1$. If a is an element of finite order, then the smallest positive integer of the set $\{m \in \mathbf{N}^+ : a^m = 1\}$ (existence of the smallest positive integer is guaranteed by the well-ordering principle) is called the *order* or *period* of a (denoted by $O(a)$). An element $a \in G$ is said to be of infinite order, iff there is no positive integer n such that $a^n = 1$.

Example 2.3.8 Let $G = \{1, -1, i, -i\}$. Then G is a group under usual multiplication of complex numbers. In the group, -1 is an element of order 2 and i is an element of order 4.

Remark An element a of a group G is of infinite order iff $a^r \neq a^t$ whenever $r \neq t$. This is so because $a^r = a^t \Leftrightarrow a^{r-t} = 1$; that is, a is of finite order, unless $r = t$.

Theorem 2.3.5 *Let G be a group and $a \in G$ such that $O(a) = t$. Then*

(i) *the elements $1 = a^0, a, \ldots, a^{t-1}$ are all distinct;*
(ii) *$a^n = 1$, iff $t \mid n$;*
(iii) *$a^n = a^m$, iff $n \equiv m \pmod{t}$;*
(iv) *$O(a^r) = t / \gcd(t, r)$, $(r, m, n \in \mathbf{Z})$.*

Proof (i) If for $0 < n < m < t$, $a^m = a^n$, then $a^{m-n} = a^m (a^n)^{-1} = 1$. This contradicts the property of the order of a, since $o < m - n < t$ and $O(a) = t$.

(ii) If $n = tm$, then $a^n = a^{tm} = (a^t)^m = 1$. Conversely, if $a^n = 1$, taking $n = tm + r$, where $0 \leq r < t$, we have $1 = a^n = a^{tm+r} = (a^t)^m a^r = a^r$. This implies $r = 0$, since t is the smallest positive integer for which $a^t = 1$ and $0 \leq r < t$. Therefore $n = tm$, and hence t is a factor of n.

(iii) $a^m = a^n \Leftrightarrow a^{n-m} = 1 \Leftrightarrow t$ is a factor of $n - m$ by (ii) $\Leftrightarrow n \equiv m \pmod{t}$.

(iv) Let $\gcd(t, r) = m$ and $t / \gcd(t, r) = n$. Then $t = nm$.

Let $r = ms$, where $\gcd(n, s) = 1$. Now $(a^r)^q = a^{rq} = 1 \Leftrightarrow t \mid rq$ by (ii) $\Leftrightarrow nm \mid msq \Leftrightarrow n \mid sq \Leftrightarrow n \mid q$, since $\gcd(n, s) = 1$. Thus the smallest positive integer q

for which $(a^r)^q = 1$ holds is given by $q = n$. This implies $n = t/\gcd(t,r)$ is the order of a^r. $\square$

Remark Any power a^n is equal to one of the elements given in (i).

Definition 2.3.6 A group G is called a

(i) *torsion group* or a *periodic group* iff every element of G is of finite order;
(ii) *torsion-free* group iff every element of G except the identity element is of infinite order;
(iii) *mixed group* iff some elements of G are of infinite order and some excepting 1 are of finite order.

Example 2.3.9 (i) (Finite torsion group) Consider the multiplicative group $G = \{1, -1, i, -i\}$. Then G is a *torsion group*.

(ii) (Group of rationals modulo 1) Consider the additive group $\mathbf{Q}$ of all rational numbers. Let $\rho = \{(a, b) \in \mathbf{Q} \times \mathbf{Q} : a - b \in \mathbf{Z}\}$. Then ρ is a congruence relation on $\mathbf{Q}$ and hence $\mathbf{Q}/\rho$ is a group under the composition $(a) + (b) = (a + b)$. Now $(\frac{1}{n}), n = 1, 2, 3, \ldots$, are distinct elements of $\mathbf{Q}/\rho$. Then $\mathbf{Q}/\rho$ contains infinite number of elements. Let p/q be any rational number such that $q > 0$. Now $q(p/q) = (p) = (0)$ shows that (p/q) is an element of finite order. Consequently, $\mathbf{Q}/\rho$ is a *torsion group*. This is an infinite abelian group, called the group of rationals modulo 1, denoted by $\mathbf{Q}/\mathbf{Z}$.

Example 2.3.10 (Torsion-free group) The positive rational numbers form a multiplicative group $\mathbf{Q}^+$ under usual multiplication. The integer 1 is the identity element of this group and it is the only element of finite order. Consequently, $\mathbf{Q}^+$ is a *torsion-free group*.

Example 2.3.11 (Mixed group) The multiplicative group of non-zero complex numbers $\mathbf{C}^*$ contains infinitely many elements of finite order, viz. every nth root of unity, for $n = 1, 2, \ldots$. It also contains infinitely many elements $re^{i\theta}$, with $r \neq 1$, of infinite order. Consequently, $\mathbf{C}^*$ is a mixed group.

2.4 Subgroups and Cyclic Groups

Arbitrary subsets of a group do not generally invite any attention. But subsets forming groups contained in larger groups create interest. For example, the group of even integers with 0, under usual addition is contained in the larger group of all integers and the group of positive rational numbers under usual multiplication is contained in the larger group of positive real numbers. Such examples suggest the concept of a subgroup, which is very important in the study of group theory. The cyclic subgroup is an important subgroup and is generated by an element g of a group G. It is the smallest subgroup of G which contains g. In this section we study subgroups and cyclic groups.

Let $(G, \circ)$ be a group and H a non-empty subset of G. If for any two elements $a, b \in H$, it is true that $a \circ b \in H$, then we say that H is closed under this group operation of G. Suppose now H is closed under the group operation '$\circ$' on G. Then we can define a binary operation $\circ_H : H \times H \to H$ by $a \circ_H b = a \circ b$ for all $a, b \in H$. This operation $\circ_H$ is said to be the restriction of '$\circ$' to $H \times H$. We explain this with the help of the following example.

Example 2.4.1 Consider the additive group $(\mathbf{R}, +)$ of real numbers. If $\mathbf{Q}^*$ is the set of all non-zero rational numbers, then $\mathbf{Q}^*$ is a non-empty subset of $\mathbf{R}$. Now for any two $a, b \in \mathbf{Q}^*$, we find that ab (under usual product of rational numbers) is an element of $\mathbf{Q}^*$. But we cannot say that $\mathbf{Q}^*$ is closed under the group operation '$+$' on $\mathbf{R}$. Consider now the set $\mathbf{Z}$ of integers. We can show easily that $\mathbf{Z}$ is closed under the group operation '$+$' on $\mathbf{R}$.

Definition 2.4.1 A subset H of a group $(G, \circ)$ is said to be a subgroup of the group G iff

(i) $H \neq \emptyset$;
(ii) H is closed under the group operation '$\circ$' on G and
(iii) $(H, \circ_H)$ is itself a group, where $\circ_H$ is the restriction of '$\circ$' to $H \times H$.

Obviously, every subgroup becomes automatically a group.

Remark $(\mathbf{Q}^*, \cdot)$ is not a subgroup of the additive group $(\mathbf{R}, +)$. But $(\mathbf{Z}, +)$ is a subgroup of the group $(\mathbf{R}, +)$.

The following theorem makes it easier to verify that a particular subset H of a group G is actually a subgroup of G. In stating the theorem we again use the multiplicative notation.

Theorem 2.4.1 *Let G be a group and H a subset of G. Then H is a subgroup of G iff $H \neq \emptyset$ and $ab^{-1} \in H$ for all $a, b \in H$.*

Proof Let H be a subgroup of G and $a, b \in H$. Then $b^{-1} \in H$ and hence $ab^{-1} \in H$. Conversely, let the given conditions hold in H. Then $aa^{-1} = 1 \in H$ and $1a^{-1} = a^{-1} \in H$ for all $a \in H$. Clearly, associative property holds in H (as it is hereditary). Finally, for all $a, b \in H, ab = a(b^{-1})^{-1} \in H$, since $b^{-1} \in H$. Consequently, H is a subgroup of G.   □

Corollary *A subset H of a group G is a subgroup of G iff $H \neq \emptyset$ and $ab \in H$, $a^{-1} \in H$ for all $a, b \in H$.*

Proof It follows from Theorem 2.4.1.   □

Using the above theorem we can prove the following:

(i) G is a subgroup of the group G.
(ii) $H = \{1\}$ is a subgroup of the group G.

Hence we find that every group G has at least two subgroups: G and $\{1\}$. These are called *trivial subgroups*. Other subgroups of G (if they exist) are called proper subgroups of G.

We now give an interesting example of a subgroup.

Definition 2.4.2 The *center* of a group G, written $Z(G)$ is the set of those elements of G which commute with every element in G, that is, $Z(G) = \{a \in G : ab = ba \text{ for all } b \in G\}$.

This set is extremely important in group theory. In an abelian group G, $Z(G) = G$. The center is in fact a subgroup of G. This follows from the following theorem.

Theorem 2.4.2 *The center $Z(G)$ of a group G is a subgroup of G.*

Proof Since $1b = b = b1$ for all $b \in G$, $1 \in Z(G)$. Hence $Z(G) \neq \emptyset$. Let $a, b \in Z(G)$. Then for all $x \in G$, $ax = xa$ and $bx = xb$. Hence $ax = xa$ and $xb^{-1} = b^{-1}x$. Now $(ab^{-1})x = a(b^{-1}x) = a(xb^{-1}) = (ax)b^{-1} = (xa)b^{-1} = x(ab^{-1})$ for all $x \in G$. Hence $ab^{-1} \in Z(G)$. Consequently $Z(G)$ is a subgroup of G. $\qquad\square$

We now introduce the concepts of conjugacy class and conjugate subgroup.

Let G be a group and $a \in G$. An element $b \in G$ is said to be a *conjugate* of a in G iff $\exists g \in G$, such that $b = gag^{-1}$. Then the relation ρ on G defined by $\rho = \{(a, b) \in G \times G : b \text{ is a conjugate of } a\}$ is an equivalence relation, called *conjugacy* on G; the equivalence class (a) of the relation ρ is called a *conjugacy class* of a in G. Two subgroups H and K of G are said to be conjugate subgroups iff there exists some g in G such that $K = gHg^{-1}$. Clearly, conjugacy is an equivalence relation on the collection of subgroups of G. The equivalence class of the subgroup H is called the conjugacy class or the conjugate class of H.

Theorem 2.4.3 *Let H be a subgroup of a group G and $g \in G$. Then gHg^{-1} is a subgroup of G such that $H \cong gHg^{-1}$.*

Proof Since $1 = g1g^{-1} \in gHg^{-1}$, $gHg^{-1} \neq \emptyset$. For $gh_1g^{-1}, gh_2g^{-1} \in gHg^{-1}$, we have $(gh_1g^{-1})(gh_2g^{-1})^{-1} = gh_1g^{-1}gh_2^{-1}g^{-1} = gh_1h_2^{-1}g^{-1} \in gHg^{-1}$. Hence gHg^{-1} is a subgroup of G. Consider the map $f : H \to gHg^{-1}$ defined by $f(h) = ghg^{-1} \ \forall h \in H$. For $h_1, h_2 \in H, h_1 = h_2 \Rightarrow gh_1g^{-1} = gh_2g^{-1} \Rightarrow f$ is well defined. Also for $a \in gHg^{-1}$, there exists $h = g^{-1}ag \in H$ such that $f(h) = ghg^{-1} = gg^{-1}agg^{-1} = a$. Moreover, $f(h_1) = f(h_2) \Rightarrow gh_1g^{-1} = gh_2g^{-1} \Rightarrow h_1 = h_2$. Finally, $f(h_1h_2) = g(h_1h_2)g^{-1} = (gh_1g^{-1})(gh_2g^{-1}) = f(h_1)f(h_2) \ \forall h_1, h_2 \in H$. Consequently, f is an isomorphism. $\qquad\square$

Let H and K be two subgroups of a group G. Now both subgroups H and K contain the identity 1 of G. Hence $H \cap K \neq \emptyset$. Let $a, b \in H \cap K$. Then $a, b \in H$ and also $a, b \in K$. Since H and K are both subgroups, $ab^{-1} \in H$ and $ab^{-1} \in K$. Hence $ab^{-1} \in H \cap K$ for all $a, b \in H \cap K$. Consequently, $H \cap K$ in a subgroup of G. The more general theorem follows:

Theorem 2.4.4 *Let* $\{H_i : i \in I\}$ *be a family of subgroups of a group* G. *Then* $\bigcap_{i \in I} H_i$ *is a subgroup of* G.

Proof Trivial. □

Theorem 2.4.5 *Let* $f : G \to K$ *be a homomorphism of groups. Then*

(i) $\text{Im} f = \{f(a) : a \in G\}$ *is a subgroup of* K;
(ii) $\ker f = \{a \in G : f(a) = 1_K\}$ *is a subgroup of* G, *where* 1_K *denotes the identity element of* K.

Proof It follows by using Theorem 2.4.1. □

Remark Let $\mathcal{C}$ be a given collection of groups. Define a relation '$\sim$' on $\mathcal{C}$ by $G_1 \sim G_2$ iff $\exists$ an isomorphism $f : G_1 \to G_2$. Then '$\sim$' is an equivalence relation which partitions $\mathcal{C}$ into mutually disjoint classes of isomorphic groups. Two isomorphic groups are abstractly indistinguishable. The main problem of group theory is: given a group G, how is one to determine the above equivalence class containing G, i.e., to determine the class of all groups which are isomorphic to G? In other words, given two groups G_1 and G_2, the problem is to determine whether G_1 is isomorphic to G_2 or not i.e., either to determine an isomorphism $f : G_1 \to G_2$ or to show that there does not exist such an isomorphism. We can solve such problems partially by using the concept of generator, which is important for a study of certain classes of groups, such as cyclic groups, finitely generated groups which have applications to number theory and homological algebra.

Finitely Generated Groups and Cyclic Groups   There are some groups G which are generated by a finite set of elements of G. Such groups are called finitely generated and are very important in mathematics. A cyclic group is, in particular, generated by a single element.

Let G be a group and A be a non-empty subset of G. Consider the family $\mathcal{C}$ of all subgroups of G containing A. This is a non-empty family, because $G \in \mathcal{C}$. Now the intersection of all subgroups of this family is again a subgroup of G. This subgroup contains A and this subgroup is denoted by $\langle A \rangle$.

Definition 2.4.3 If A is a non-empty subset of a group G, then $\langle A \rangle$ is called the *subgroup generated* by A in the group G. If $\langle A \rangle = G$, then G is said to be generated by A, and if, A contains finite number of elements, then G is said to be *finitely generated*.

If A contains a finite number of elements, say $A = \{a_1, a_2, \ldots, a_n\}$, then we write $\langle a_1, a_2, \ldots a_n \rangle$ in place of $\langle \{a_1, a_2, \ldots, a_n\} \rangle$.

The particular case when A consists of a single element of the group is of immense interest. For example, for the circle S of radius 1 in the Euclidean plane, if r is a rotation through an angle $2\pi/n$ radians about the origin, then $r^n = r \circ r \circ \cdots \circ r$ (n times) is the identity. This example leads to the concept of cyclic groups.

Definition 2.4.4 A subgroup H of a group G is called a *cyclic subgroup* of G iff $\exists$ an element $a \in G$ such that $H = \langle a \rangle$ and G is said to be cyclic iff $\exists$ an element $a \in G$ such that $G = \langle a \rangle$, a is called a *generator* of G.

Theorem 2.4.6 *Let G be a group and $a \in G$. Then $\langle a \rangle = \{a^n : n \in \mathbf{Z}\}$.*

Proof Let $T = \{a^n : n \in \mathbf{Z}\}$. Then for $c, d \in T$, there exist $r, t \in \mathbf{Z}$ such that $c = a^r$ and $d = a^t$. Now $cd^{-1} = a^r (a^t)^{-1} = a^r a^{-t} = a^{r-t} \in T \Rightarrow T$ is subgroup of G. Clearly, $\{a\} \subseteq T$. Let H be a subgroup of G such that $\{a\} \subseteq H$. Then $a \in H$ and hence $a^{-1} \in H$. As a result $a^n \in H$ for any $n \in \mathbf{Z}$ and then $T \subseteq H$. Hence T is the intersection of all subgroups of G containing $\{a\}$. This shows that $\langle a \rangle = T = \{a^n : n \in \mathbf{Z}\}$. $\qquad\square$

Remark If a group G is cyclic, then there exists an element $a \in G$ such that any $x \in G$ is a power of a, that is, $x = a^n$ for some $n \in \mathbf{Z}$. Then the mapping $f : \mathbf{Z} \to G$ defined by $f(n) = a^n$ is an epimorphism. Moreover, if ker $f = \{0\}$, f is a monomorphism by Theorem 2.3.3(i), and hence f is an isomorphism, and such G is called an infinite cyclic group or a free cyclic group. For example, $\mathbf{Z}$ is an infinite additive cyclic group with generator 1. Clearly, -1 is also a generator.

Remark Every cyclic group is commutative. Its converse is not true. For example, the Klein four-group (see Ex. 22 of Exercises-III) is a finite commutative group which is not cyclic and $(\mathbf{Q}, +)$ is an infinite commutative group which is not finitely generated (see Ex. 1 of SE-I) and hence not cyclic.

Theorem 2.4.7 *An infinite cyclic group is torsion free.*

Proof Let G be an infinite cyclic group and $1 \neq a \in G$. Then $G = \langle g \rangle$ for some $g \in G$. Consequently, $a = g^t$, where t is a non-zero integer. If a is of finite order $r > 0$, then $g^{tr} = a^r = 1$, a contradiction. Consequently, a is of infinite order. $\qquad\square$

Let G be a group. If G contains a finite number of elements, then the group G is said to be a *finite group*, otherwise, the group G is called an *infinite group*. The number of elements of a finite group is called the order of the group. We write $|G|$ or $O(G)$ to denote the *order* of a group G.

Theorem 2.4.8 *Let G be a group and $a \in G$. Then the order of the cyclic subgroup $\langle a \rangle$ is equal to $O(a)$, when it is finite, and is infinite when $O(a)$ is infinite.*

Proof It follows from Theorem 2.3.5, and Remark of Definition 2.3.5. $\qquad\square$

In particular, if $\gcd(t, r) = 1$ in Theorem 2.3.5(iv), then $O(a) = O(a^r)$ and therefore each of a and a^r generates the same group G. This proves the following theorem.

Theorem 2.4.9 *If G is a cyclic group of order n, then there are $\phi(n)$ distinct elements in G, each of which generates G, where $\phi(n)$ is the number of positive integers less than and relatively prime to n.*

The function $\phi(n)$ is called the Euler ϕ-function *in honor of Leonhard Euler* (1707–1783).

For example, for $\mathbf{Z}_{14}$, $\phi(14) = 6$ and $\mathbf{Z}_{14}$ can be generated by each of the elements (1), (3), (5), (9), (11) and (13), since $r(1)$ is a generator iff gcd $(r, 14) = 1$, where $0 < r < 14$.

Remark If a, b are relatively prime positive integers then $\phi(ab) = \phi(a)\phi(b)$. Moreover, for any positive prime integer p, $\phi(p) = p - 1$, $\phi(p^n) = p^n - p^{n-1}$ for all integers $n \geq 1$ and if $n = p_1^{n_1} p_2^{n_2} \ldots p_r^{n_r}$, then $\phi(n) = p_1^{n_1-1}(p_1 - 1)p_2^{n_2-1}(p_2 - 1) \ldots p_r^{n_r-1}(p_r - 1)$ (see Chap. 10).

Theorem 2.4.10 *Every non-trivial subgroup of the additive group $\mathbf{Z}$ is cyclic.*

Proof Let H be a subgroup of $\mathbf{Z}$, and $H \neq \{0\}$. Then H contains a smallest positive integer a (say). To prove the theorem it is sufficient to show that any integer $x \in H$ is of the form na for some $n \in \mathbf{Z}$. If $x = na + r$ where $0 \leq r < a$, then $r = x - na \in H$ as H is a group. Consequently, $r = 0$, so $x = na$. $\qquad\square$

Theorem 2.4.11 *If G is a cyclic group, then every subgroup of G is cyclic.*

Proof If G is infinite cyclic, then by definition $G \cong \mathbf{Z}$. Then the proof follows from Theorem 2.4.10. Next suppose that G is a finite cyclic group generated by a and H a subgroup of G. Let t be the smallest positive integer such that $a^t \in H$. We claim that a^t is a generator of H. Let $a^m \in H$. If $m = tq + r$, $0 \leq r < t$, then $a^r = a^{m-tq} = a^m(a^t)^{-q} \in H$, since $a^m, a^t \in H$. Since $0 \leq r < t$, it follows by the property of t that $r = 0$. Therefore $m = tq$ and hence $a^m = (a^t)^q$. Thus every element $a^m \in H$ is some power of a^t. Again, since $a^t \in H$, every power of a^t is in H. Hence $H = \langle a^t \rangle$. Thus H is cyclic. $\qquad\square$

The following theorem gives a characterization of the cyclic groups.

Theorem 2.4.12 *Let $(G, \cdot)$ be a cyclic group. Then*

(i) *$(G, \cdot)$ is isomorphic to $(\mathbf{Z}, +)$ iff G is infinite;*
(ii) *$(G, \cdot)$ is isomorphic to $(\mathbf{Z}_n, +)$ iff G is finite and $|G| = n$.*

Proof (i) Let $(G, \cdot)$ be an infinite cyclic group. Then the group $(G, \cdot)$ is isomorphic to $(\mathbf{Z}, +)$. Conversely, suppose $(G, \cdot)$ is isomorphic to $(\mathbf{Z}, +)$. In this case, since $(\mathbf{Z}, +)$ is infinite, $(G, \cdot)$ is also infinite, as an isomorphism is a bijection.

(ii) Suppose $(G, \cdot)$ is a finite cyclic group. Then the case (i) cannot arise and therefore the case (ii) must occur. In this case $(G, \cdot)$ is isomorphic to $(\mathbf{Z}_n, +)$, where

$n = |G|$. Conversely, suppose $(G, \cdot)$ is isomorphic to $(\mathbf{Z}_n, +)$. Then $|\mathbf{Z}_n| = n \Rightarrow$ $|G| = n \Rightarrow G$ is finite.   □

Remark Let G and K be finite cyclic groups such that $|G| = m \neq n = |K|$. Then G and K cannot be isomorphic, i.e., there cannot exist an isomorphism $f : G \to K$.

Theorem 2.4.13 *Let G be a cyclic group of order n and H a cyclic group of order m such that $\gcd(m, n) = 1$. Then $G \times H$ is a cyclic group of order mn. Moreover, if a is a generator of G and b a generator of H, then (a, b) is a generator of $G \times H$.*

Proof Clearly, $G \times H$ is a group under pointwise multiplication defined by $(g, h)(g', h') = (gg', hh')$ for all g, g' in G and h, h' in H. If e_1 and e_2 be identity elements of G and H respectively, then $a^n = e_1$ and $1 \leq r < n \Rightarrow a^r \neq e_1$, $b^m = e_2$ and $1 \leq r < m \Rightarrow b^r \neq e_2$. Let $x = (a, b) \in G \times H$. Then $x^{mn} = (a^{mn}, b^{mn})$ by definition of multiplication (i.e., pointwise multiplication) in

$$G \times H \quad \Rightarrow \quad x^{mn} = (e_1, e_2) \quad \Rightarrow \quad \text{order } x \leq mn \qquad (2.2)$$

Again

$$
\begin{aligned}
x^t = (e_1, e_2) \quad &\Rightarrow \quad (a^t, b^t) = (e_1, e_2) \\
&\Rightarrow \quad a^t = e_1 \quad \text{and} \quad b^t = e_2 \\
&\Rightarrow \quad n | t \quad \text{and} \quad m | t \\
&\Rightarrow \quad mn | t \quad (\text{since } \gcd(m, n) = 1) \\
&\Rightarrow \quad t \geq mn \\
&\Rightarrow \quad \text{order } x \geq mn \qquad (2.3)
\end{aligned}
$$

Consequently, (2.2) and (2.3) $\Rightarrow$ order $x = $ order $(a, b) = mn$. Since $|G \times H| = mn$ and the order of the element $(a, b) = mn$, it follows that (a, b) generates $G \times H$ i.e., $G \times H$ is a cyclic group of order mn.   □

Remark The converse of the last part of Theorem 2.4.13 is also true. Suppose (u, v) is a generator of $G \times H$. Let a be a fixed generator of G and b a fixed generator of H. Then since $\langle (u, v) \rangle = G \times H$, $\exists$ positive integers n and r such that $(u, v)^n = (a, e_2) \in G \times H$ and $(u, v)^r = (e_1, b) \in G \times H$. Then $u^n = a$ and $v^r = b \Rightarrow G = \langle a \rangle \subseteq \langle u \rangle \subseteq G$ and $H = \langle b \rangle \subseteq \langle v \rangle \subseteq H \Rightarrow \langle u \rangle = G$ and $\langle v \rangle = H \Rightarrow u$ is a generator of G and v is a generator of H.

Problems

Problem 1 Let G $(\neq \{1\})$ be a group which has no other subgroup except G and $\{1\}$. Show that G is isomorphic to $\mathbf{Z}_p$ for some prime p.

Solution Let $a \in G - \{1\}$ (since $G \neq \{1\}$). If $T = \langle a \rangle$, then $T = G$ (since $a \in T$ and $a \neq 1$) $\Rightarrow G$ is a cyclic group generated by a. If possible, let G be infinite cyclic.

Then $G = \{a^n : n \in \mathbf{Z}\}$, $a^m \neq a^n$ for $n \neq m$ and $a^0 = 1$. Consider $H = \{a^{2n} : n \in \mathbf{Z}\}$. Then H is proper subgroup of G (as $a \notin H$) and $H \neq \{1\}$. This contradicts the fact that only subgroups of G are G and $\{1\}$. So, G cannot be infinite cyclic. In other words, G must be finite cyclic. Let $|G| = n$. If possible, let $n = rs$, $1 < r < n$, $1 < s < n$. Since $r||G|$ and G is cyclic, G has a cyclic subgroup H of order r. This contradicts the fact that only subgroups of G are G and $\{1\}$. So, n cannot be composite, i.e., n is some prime p, since $G \neq \{1\}$. Thus $(G, \cdot)$ is a cyclic group of order p and hence isomorphic to $(\mathbf{Z}_p, +)$ by Theorem 2.4.12(ii).

Problem 2 Let G be a group which is isomorphic to every proper subgroup H of G with $H \neq \{1\}$ and assume that there exists at least one such proper subgroup $H \neq \{1\}$. Show that G is infinite cyclic. Conversely, if G is an infinite cyclic group, show that G is isomorphic to every proper subgroup H of G such that $H \neq \{1\}$.

Solution There exists an element $a \in G$ such that $a \neq 1$. Let P be the cyclic subgroup generated by a. So, by hypothesis, P is isomorphic to G as $P \neq \{1\}$. Since P is cyclic, G must be also cyclic. Consequently, G is infinite.

Conversely, let G be an infinite cyclic group and H a proper subgroup of G such that $H \neq \{1\}$. Then $H = \langle a^r \rangle (r \geq 1)$, where a is a generator of G. Consider a function $f : G \to H$ by $f(a) = a^r$. Then $f(a^n) = a^{rn}$. Clearly, f is an isomorphism.

2.5 Lagrange's Theorem

One of the most important invariants of a finite group is its order. While studying finite groups Lagrange established a relation between the order of a finite group and the order of its any subgroup. He proved that the order of any subgroup H of a finite group G divides the order of G. This is established by certain decompositions of the underlying set G into a family of subsets $\{aH : a \in G\}$ or $\{Ha : a \in G\}$, called cosets of H.

Let G be a group and H subgroup of G. If $a \in G$, then aH denotes the subset $\{ah \in G : h \in H\}$ and Ha denotes the subset $\{ha \in G : h \in H\}$ of G. Now $a = a1 = 1a$ shows that $a \in aH$ and $a \in Ha$. The set aH is called the *left coset* of H in G containing a and Ha is called the *right coset* of H in G containing a. A subset A of G is called a left (right) coset of H in G iff $A = aH (=Ha)$ for some $a \in G$.

Lemma 2.5.1 *If H is a subgroup of a group G and $a, b \in G$, then each of the following statements is true.*

  (i) $aH = H$ iff $a \in H$;
 (i)' $Ha = H$ iff $a \in H$;
 (ii) $aH = bH$ iff $a^{-1}b \in H$;
(ii)' $Ha = Hb$ iff $ab^{-1} \in H$;
(iii) *Either* $aH = bH$ *or* $aH \cap bH = \emptyset$;
(iii)' *Either* $Ha = Hb$ *or* $Ha \cap Hb = \emptyset$;

(iv) $G = \bigcup_{(a \in G)} aH$;

(iv)' $G = \bigcup_{(a \in G)} Ha$;

(v) *There exists a bijection $aH \to H$ for each $a \in G$;*

(v)' *There exists a bijection $Ha \to H$ for each $a \in G$;*

(vi) *If $\mathcal{L}$ is the set of all left cosets of H in G and $\mathcal{R}$ is the set of all right cosets of H in G, then there exists a bijection from $\mathcal{L}$ onto $\mathcal{R}$.*

Proof (i)–(iv) follow trivially.

(v) If $x \in aH$, then there exists $h \in H$ such that $x = ah$.

Suppose $x = ah_1 = ah_2$ $(h_1, h_2 \in H)$. Then $h_1 = a^{-1}(ah_1) = a^{-1}(ah_2) = h_2$. Hence for each $x \in aH$ there exists a unique $h \in H$ such that $x = ah$. Define $f : aH \to H$ by $f(x) = h$ when $x = ah \in aH$. Now one can show that f is a bijective mapping from aH onto H.

(vi) Define $f : \mathcal{L} \to \mathcal{R}$ by $f(aH) = Ha^{-1}$. One can show that f is a well defined bijective mapping.   □

Example 2.5.1 (i) Consider the group S_3 on the set $I_3 = \{1, 2, 3\}$. In S_3, $H = \{e, (12)\}$ is a subgroup. For this subgroup $eH = \{e, (12)\}$, $(13)H = \{(13), (123)\}$ and $(23)H = \{(23), (132)\}$ are three distinct left cosets and $S_3 = eH \cup (13)H \cup (23)H$. Again $He = \{e, (12)\}$, $H(13) = \{(13), (132)\}$, $H(23) = \{(23), (123)\}$ are three distinct right cosets of H and

$$S_3 = He \cup H(13) \cup H(23).$$

Hence the number of left cosets of H is three and the number of right cosets of H is also three. But $\mathcal{L} \neq \mathcal{R}$.

(ii) Consider the group $\mathbf{Z}_6 = \{0, 1, 2, 3, 4, 5\}$ (see Example 2.3.6). In $\mathbf{Z}_6$, $H = \{0, 3\}$ is a subgroup. Using additive notation, the left cosets $H = \{0, 3\}, 1 + H = \{1, 4\}, 2 + H = \{2, 5\}$ partition $\mathbf{Z}_6$.

We now prove several interesting results about finite groups.

One of the most important invariance of a finite group is its order.

Let G be a finite group. Any subgroup H of G is also a finite group. Lagrange's Theorem (Theorem 2.5.1) establishes a relation between the order of H and the order of G. It plays a very important role in the study of finite groups.

Theorem 2.5.1 (Lagrange's Theorem) *The order of a subgroup of a finite group divides the order of the group.*

Proof Let G be a finite group of order n and H a subgroup of G. Let $|H| = m$. Now $1 \leq m \leq n$. We can write $G = \bigcup_{a \in G} aH$. Since any two left cosets of H are either identical or disjoint, there exist elements $a_1, a_2, \ldots, a_t$ $(t \leq n)$ such that left cosets $a_1 H, a_2 H, \ldots, a_t H$ are all distinct and $G = a_1 H \cup a_2 H \cdots \cup a_t H$. Hence $|G| = |a_1 H| + |a_2 H| + \cdots + |a_t H|$ ($|a_i H|$ denotes the number of elements in $a_i H$).

Let $H = \{h_1, h_2, \ldots, h_m\}$. Then $aH = \{ah_1, ah_2, \ldots, ah_m\}$. The cancellation property of a group implies that $ah_1, ah_2, \ldots, ah_m$ are m distinct elements, so that

$|H| = |aH|$ ($|aH|$ denotes the number of elements in aH). Therefore $n = |G| = |a_1 H| + |a_2 H| + \cdots + |a_t H| = |H| + |H| + \cdots + |H|$ (t times) $= t|H| = tm$. Hence m divides n, where $n = |G|$. $\qquad\square$

Corollary 1 *The order of an element of a finite group divides the order of the group.*

Proof Let G be a finite group and $a \in G$. Then $O(a) = |\langle a \rangle|$ by Theorem 2.4.8. Hence the corollary follows from Lagrange's Theorem 2.5.1. $\qquad\square$

Corollary 2 *A group of prime order is cyclic.*

Proof Let G be a group of prime order p and $1 \neq a \in G$. Then $O(a)$ is a factor of p by Corollary 1. As p is prime, either $O(a) = p$ or $O(a) = 1$. Now $a \neq 1 \Rightarrow O(a) = p \Rightarrow |\langle a \rangle| = p = |G| \Rightarrow G$ is a cyclic group, since $\langle a \rangle \subseteq G$. $\qquad\square$

Remark The converse of the Lagrange Theorem asserts that if a positive integer m divides the order n of a finite group G, then G contains a subgroup of order m. But it is not true in general. For example, the alternating group A_4 of order 12 has no subgroup of order 6 (see Ex. 4 of SE-I). However, the following properties of special finite groups show that the partial converse is true.

1. If G is a finite abelian group, then corresponding to every positive divisor m of n, there exists a subgroup of order m (see Corollary 2 of Cauchy's Theorem in Chap. 3).
2. If $m = p$, a prime integer, then G has a subgroup of order p (see Cauchy's Theorem in Chap. 3).
3. If m is a power of a prime p, then G has a subgroup of order m (see Sylow's First Theorem in Chap. 3).

Let G be a group and H be a subgroup of G. We have proved that there exists a bijective mapping $f : \mathcal{L} \to \mathcal{R}$, where $\mathcal{L}$ is the set of all left cosets of H in G and $\mathcal{R}$ is the set of all right cosets of H in G. Then the cardinal number of $\mathcal{L}$ is the same as the cardinal number of $\mathcal{R}$. This cardinal number is called the index of H in G and is denoted by $[G : H]$. If $\mathcal{L}$ is a finite set, then the cardinal number of $\mathcal{L}$ is the number of left cosets of H in G. Hence in this case the number of left cosets of H (equivalently the number of right cosets of H) in G is the index of H in G.

Theorem 2.5.2 *If H is a subgroup of a finite group G, then*

$$[G : H] = \frac{|G|}{|H|}.$$

Proof Let G be a finite group of order n. Then a subgroup H of G must also be of finite order, r (say). Every left coset of H contains exactly r distinct elements. If $[G : H] = i$, then $n = ri$, since the cosets are disjoint. This proves the theorem. $\qquad\square$

Given two finite subgroups H and K of a group G, $HK = \{hk : h \in H \text{ and } k \in K\}$ may not be a subgroup of G. Hence $|HK|$ may not divide $|G|$. However, Theorem 2.5.3 determines $|HK|$.

Theorem 2.5.3 *If H and K are two finite subgroups of a group G, then*

$$|HK| = \frac{|H||K|}{|H \cap K|}.$$

Proof Let $A = H \cap K$. Then A is a subgroup of G such that $A \subseteq H$ and $A \subseteq K$. Now A is a subgroup of the finite group H. Hence $[H : A]$ is finite. Let $[H : A] = t$. Then there exist elements $x_1, x_2, \ldots, x_t$ in H such that $x_1 A, x_2 A, \ldots, x_t A$ are distinct left cosets of A in H and $H = \bigcup_{i=1}^{t} x_i A$. Now $HK = (\bigcup_{i=1}^{t} x_i A)K = \bigcup_{i=1}^{t} x_i K$, since $A \subseteq K$. We claim that $x_1 K, x_2 K, \ldots, x_t K$ are t distinct left cosets of K.

Suppose $x_i K = x_j K$ for some i and j, where $1 \le i \ne j \le t$. Then $x_j^{-1} x_i \in K$. But $x_j^{-1} x_i \in H$. Hence $x_j^{-1} x_i \in H \cap K = A$. This shows that $x_i A = x_j A$; this is not true as $x_1 A, \ldots, x_t A$ are distinct. As a result $|HK| = |x_1 K| + |x_2 K| + \cdots + |x_t K|$. Also, by Lemma 2.5.1(v) it follows that $|x_i K| = |K|$. Consequently,

$$|HK| = t|K| = [H : A]|K| = \frac{|H||K|}{|A|} = \frac{|H||K|}{|H \cap K|}. \qquad \square$$

Corollary *If H and K are finite subgroups of a group G such that $H \cap K = \{1\}$, then $|HK| = |H||K|$.*

2.6 Normal Subgroups, Quotient Groups and Homomorphism Theorems

In this section we make the study of normal subgroups and develop the theory of quotient groups with the help of homomorphisms. We show how to construct isomorphic replicas of all homomorphic images of a specified abstract group. For this purpose we introduce the concept of normal subgroups. For some subgroups of a group, every left coset is a right coset and conversely, which implies that for some subgroups, the concepts of a left coset and a right coset coincide. Such subgroups are called normal subgroups. Galois first recognized that those subgroups for which left and right cosets coincide are a distinguished one. We shall now study those subgroups H of a group G such that $aH = Ha$ for all $a \in G$. Such subgroups play an important role in determining both the structure of a group and the nature of the homomorphisms with domain G.

Definition 2.6.1 Let H be a subgroup of a group G and $a, b \in G$. Then a is said to be *right (left) congruent* to b modulo H, denoted by $a\rho_r b \pmod{H}$ ($a\rho_l b \pmod{H}$) iff $ab^{-1} \in H$ ($a^{-1}b \in H$).

Theorem 2.6.1 *Let H be a subgroup of a group H. Then the relation ρ_r (ρ_l) on G is an equivalence relation and the corresponding equivalence class of $a \in G$ is the right (left) coset Ha (aH). ρ_r (ρ_l) is called the right (left) congruence relation on G modulo H.*

Proof We write $a\rho b$ for $a\rho_r b$ $(\mathrm{mod}\,H)$ and then prove the theorem only for right congruence ρ and right cosets.

$aa^{-1} = 1 \in H \Rightarrow a\rho a \ \forall a \in G$. Again $a\rho b \Rightarrow ab^{-1} \in H \Rightarrow (ab^{-1})^{-1} \in H \Rightarrow ba^{-1} \in H \Rightarrow b\rho a$. Finally, $a\rho b$ and $b\rho c \Rightarrow ab^{-1} \in H$ and $bc^{-1} \in H \Rightarrow (ab^{-1})(bc^{-1}) \in H \Rightarrow ac^{-1} \in H \Rightarrow a\rho c$. Consequently, ρ is an equivalence relation. Now for each $a \in G$, $a\rho = \{x \in G : x\rho a\} = \{x \in G : xa^{-1} \in H\} = \{x \in G : xa^{-1} = h$ for some $h \in H\} = \{x \in G : x = ha, h \in H\} = \{ha \in G : h \in H\} = Ha$. $\square$

Theorem 2.6.2 *If H is a subgroup of a group G, then the following conditions are equivalent.*

(i) *Left congruence ρ_l and right congruence ρ_r on G modulo H coincide;*
(ii) *every left coset of H in G is also a right coset of H in G;*
(iii) *$aH = Ha$ for all $a \in G$;*
(iv) *for all $a \in G$, $aHa^{-1} \subseteq H$, where $aHa^{-1} = \{aha^{-1} : h \in H\}$.*

Proof (i) $\Leftrightarrow$ (iii) $\rho_l = \rho_r \Leftrightarrow a\rho_l = a\rho_r \ \forall a \in G \Leftrightarrow aH = Ha \ \forall a \in G$ by Theorem 2.6.1.

(iii) $\Leftrightarrow$ (iv) Suppose $aH = Ha \ \forall a \in G$. Then for each $h \in H$, $ah = h'a$ for some $h' \in H$. Consequently, $aha^{-1} = h' \in H \ \forall h \in H$. This shows that $aHa^{-1} \subseteq H$ for every $a \in G$. Again let for every $a \in G$, $aHa^{-1} \subseteq H$. Then $H = a(a^{-1}Ha)a^{-1} \subseteq aHa^{-1}$, since $a^{-1}Ha \subseteq H$ (by hypothesis). Consequently, $aHa^{-1} = H$ for every $a \in G$. This shows that $aH = Ha \ \forall a \in G$.

(ii) $\Leftrightarrow$ (iii) Suppose $aH = Hb$ for $a, b \in G$. Then $a \in Ha \cap Hb$. Since two right cosets of H in G are either disjoint or equal, it follows that $Ha = Hb$ and hence (ii) $\Rightarrow$ (iii). Its converse is trivial. $\square$

Definition 2.6.2 Let G be a group and H a subgroup of G. Then H is said to be a *normal subgroup* of G denoted by $H \triangle G$ iff H satisfies any one of the equivalent conditions of Theorem 2.6.2.

Thus for a normal subgroup, we need not distinguish between the left and right cosets.

In any group G, G and $\{1\}$ are normal subgroups, called trivial normal subgroups. Every subgroup of a commutative group is a normal subgroup.

Example 2.6.1 If S_3 is the symmetric group of order 6, then $H = \{e, (123), (132)\}$ is a subgroup of S_3. Now for any $a \in H$, $aH = H = Ha$. The elements (12), (23), and $(13) \notin H$. Now $(12)H = \{(12), (23), (13)\} = H(12)$. Similarly, $(13)H = H(13)$ and $(23)H = H(23)$. Consequently, H is a normal subgroup of S_3. Clearly, $K = \{e, (12)\}$ is not a normal subgroup of S_3.

Theorem 2.6.3 *Let A and B be two subgroups of a group G.*

(i) *If A is a normal subgroup of a group G, then $AB = BA$ is a subgroup of G.*
(ii) *If A and B are normal subgroups of G, then $A \cap B$ is a normal subgroup of G.*
(iii) *If A and B are normal subgroups of G such that $A \cap B = \{1\}$, then $ab = ba$ $\forall a \in A$ and $\forall b \in B$.*

Proof (i) Let $x = ab \in AB$. Now $bA = Ab \Rightarrow ab = ba_1$ for some $a_1 \in A \Rightarrow x = ba_1 \in BA \Rightarrow AB \subseteq BA$. Similarly, $BA \subseteq AB$. Consequently, $AB = BA$. Again for $b, b' \in B$ and $a, a' \in A$, $(ba)(b'a')^{-1} = ba(a'^{-1}b'^{-1}) = b(aa'^{-1})b'^{-1} = ba_1b'^{-1}(a_1 \in A) = b_1a_2$, where $a_2 \in A$, $b_1 \in B$. This implies $BA = AB$ is a subgroup of G.

(ii) Clearly, $H = A \cap B$ is a normal subgroup of G, since if $h \in H$ and $g \in G$, then ghg^{-1} belongs to both A and B and hence is in H.

(iii) Consider the element $x = aba^{-1}b^{-1}(a \in A, b \in B)$. Now $x = (aba^{-1})b^{-1} \in (aBa^{-1})b^{-1} \subseteq Bb^{-1} = B$ and again $x = a(ba^{-1}b^{-1}) \in a(bAb^{-1}) \subseteq aA = A$. Hence $x \in A \cap B = \{1\}$. Consequently, $aba^{-1}b^{-1} = 1$. This shows that $ab = ba$ $\forall a \in A$ and $\forall b \in B$. $\square$

We now show how to make the set of cosets into a group, called a quotient group.

Theorem 2.6.4 *Let H be a normal subgroup of a group G and G/H the set of all cosets of H in G. Then G/H is a group.*

Proof Define multiplication of cosets by $(aH) \cdot (bH) = (ab)H$, $\forall a, b \in G$. We show that the multiplication is well defined. Let $aH = xH$ and $bH = yH$. Then $a \in aH$ and $b \in bH$ are such that $a = xh_1$ and $b = yh_2$ for some $h_1, h_2 \in H$. Now, $(xy)^{-1}(ab) = y^{-1}x^{-1}ab = y^{-1}x^{-1}xh_1yh_2 = y^{-1}h_1yh_2 \in H$, since $y^{-1}h_1y \in H$ as H is a normal subgroup of G. Consequently, by Lemma 2.5.1, it follows that $(ab)H = (xy)H \Rightarrow (aH) \cdot (bH) = (xH) \cdot (yH)$. Then G/H is a group where the identity element is the coset H, and the inverse of aH is $a^{-1}H$. $\square$

Corollary *If H is a normal subgroup of a finite group G then $|G/H| = |G|/|H|$.*

Proof It follows from Theorem 2.5.2. $\square$

Definition 2.6.3 If H is a normal subgroup of a group G, then group G/H is called the *factor group* (or quotient group) of G by H.

Remark If G is an additive abelian group, the group operation on G/H is defined by

$$(a + H) + (b + H) = (a + b) + H$$

and the corresponding quotient group G/H is sometimes called difference group.

We now show that the construction of quotient groups is closely related to homomorphisms of groups and prove a natural series of theorems.

Theorem 2.6.5 *If $f : G \to T$ is a homomorphism of groups, then the kernel of f is a normal subgroup of G. Conversely, if H is a normal subgroup of G, then the map*

$$\pi : G \to G/H, \quad \text{given by } \pi(g) = gH, \quad g \in G$$

is an epimorphism with kernel H.

Proof Clearly, $1 \in \ker f$. So, $\ker f \neq \emptyset$. Let $x, y \in \ker f$. Then $f(xy^{-1}) = f(x)f(y^{-1}) = f(x)(f(y))^{-1} = 1 \cdot 1 = 1 \; \forall x, y \in \ker f \Rightarrow \ker f$ is a subgroup of G by Theorem 2.4.1. Again if $g \in G$, then $f(gxg^{-1}) = f(g)f(x)(f(g))^{-1} = f(g) \cdot 1 \cdot (f(g))^{-1} = 1 \Rightarrow gxg^{-1} \in \ker f \Rightarrow \ker f$ is a normal subgroup of G.

The map $\pi : G \to G/H$ given by $\pi(g) = gH$ is clearly surjective. Moreover, $\pi(gg') = gg'H = gHg'H = \pi(g)\pi(g') \; \forall g, g' \in G \Rightarrow \pi$ is an epimorphism, called the canonical epimorphism or projection or natural homomorphism. Now $\ker \pi = \{g \in G : \pi(g) = H\} = \{g \in G : gH = H\} = \{g \in G : g \in H\} = H$. $\qquad \square$

Remark Let G be a group and H a subgroup of G. Then H is a normal subgroup of G iff H is the kernel of some homomorphism.

Theorem 2.6.6 (Fundamental Homomorphism Theorem for groups) *Let $f : G \to H$ be a homomorphism of groups and N a normal subgroup of G such that $N \subseteq \ker f$. Then there is a unique homomorphism $\tilde{f} : G/N \to H$ such that $\tilde{f}(gN) = f(g) \; \forall g \in G$, $\operatorname{Im} \tilde{f} = \operatorname{Im} f$ and $\ker \tilde{f} = \ker f/N$. Moreover, $\tilde{f}$ is an isomorphism iff f is an epimorphism and $\ker f = N$.*

Proof If $x \in gN$, then $x = gn$ for some $n \in N$ and $f(x) = f(gn) = f(g)f(n) = f(g)1 = f(g)$, since by hypothesis $N \subseteq \ker f$. This shows that the effect of f is the same on every element of the coset gN. Thus the map $\tilde{f} : G/N \to H$, $gN \mapsto f(g)$ is well defined. Now $\tilde{f}(gN \cdot hN) = \tilde{f}(ghN) = f(gh) = f(g)f(h) = \tilde{f}(gN)\tilde{f}(hN) \; \forall g, h \in G \Rightarrow \tilde{f}$ is a homomorphism. From the definition of $\tilde{f}$, it is clear that $\operatorname{Im} \tilde{f} = \operatorname{Im} f$. Moreover, $gN \in \ker \tilde{f} \Leftrightarrow f(g) = 1 \Leftrightarrow g \in \ker f$. Consequently, $\ker \tilde{f} = \{gN : g \in \ker f\} = \ker f/N$. Since $\tilde{f}$ is completely determined by f, $\tilde{f}$ is unique. Clearly, f is a monomorphism iff $\ker \tilde{f} = \ker f/N$ is a trivial subgroup of G/N. The latter holds iff $\ker f = N$. Moreover, $\tilde{f}$ is an epimorphism iff f is an epimorphism. Consequently, $\tilde{f}$ is an isomorphism iff f is an epimorphism and $\ker f = N$. $\qquad \square$

Corollary 1 (First Isomorphism Theorem) *If $f : G \to H$ is a homomorphism of groups, then f induces an isomorphism $\tilde{f} : G/\ker f \to \operatorname{Im} f$.*

Corollary 2 *If G and H are groups and $f : G \to H$ is an epimorphism, then the group $G/\ker f$ and H are isomorphic.*

Remark The homomorphic images of a given abstract group G are the quotient groups G/N by its different normal subgroups N.

Proposition 2.6.1 *If G and H are groups and $f : G \to H$ is a homomorphism of groups, N is a normal subgroup of G, and T is a normal subgroup of H such that $f(N) \subseteq T$, then there is a homomorphism $\tilde{f} : G/N \to H/T$.*

Proof Define $\tilde{f} : G/N \to H/T$ by $\tilde{f}(gN) = f(g)T$.   □

Corollary *If $f : G \to H$ is a homomorphism of groups with $\ker f = N$, then there is a monomorphism $\tilde{f} : G/N \to H$.*

Proof We have $f(N) = \{1\}$ and $\{1\}$ is a trivial normal subgroup of H. Then by Proposition 2.6.1, f induces a homomorphism $\tilde{f} : G/N \to H$ given by $\tilde{f}(gN) = f(g) \; \forall g \in G$. This shows that $\tilde{f}$ is a monomorphism. Moreover, if $f : G \to H$ is an epimorphism, then $\tilde{f} : G/N \to H$ is also an epimorphism.   □

Remark These results also prove the *First Isomorphism Theorem*.

Theorem 2.6.7 (Subgroup Isomorphism Theorem or the Second Isomorphism Theorem) *Let N be a normal subgroup of a group G and H a subgroup of G. Then $H \cap N$ is a normal subgroup of H, HN is a subgroup of G and $H/H \cap N \cong HN/N$ (or NH/N), where $HN = \{hn : h \in H, n \in N\}$.*

Proof Suppose $x \in H \cap N$ and $h \in H$. Then $h^{-1}xh \in N$ as N is a normal subgroup of G. Moreover, $h^{-1}xh \in H$ as $x \in H$. Consequently, $h^{-1}xh \in H \cap N \Rightarrow H \cap N$ is a normal subgroup of H. Clearly, $HN \neq \emptyset$. Now $x, y \in HN \Rightarrow x = hn$ and $y = h'n'$ for some $n, n' \in N, h, h' \in H \Rightarrow xy^{-1} = hn(h'n')^{-1} = hnn'^{-1}h'^{-1} = hn_1h'^{-1}$ (for some $n_1 \in N$)$= hh'^{-1}h'n_1h'^{-1} = h_1n_2$ (where $hh'^{-1} = h_1 \in H$ and $h'n_1h'^{-1} = n_2 \in N$ as N is a normal subgroup of G) $\in HN \Rightarrow HN$ is a subgroup of G.

Define $f : H \to HN/N$ by $h \mapsto hN$. Then f is an epimorphism. By using First Isomorphism Theorem, $H/\ker f \cong f(H) = HN/N$.

Now $\ker f = \{x \in H : f(x) = N\} = \{x : x \in H \text{ and } xN = N\} = \{x : x \in H \text{ and } x \in N\} = H \cap N$. Consequently, $H/H \cap N \cong HN/N$.   □

Theorem 2.6.8 (Factor of a Factor Theorem or Third Isomorphism Theorem) *If both H and K are normal subgroups of a group G and $K \subseteq H$, then H/K is a normal subgroup of G/K and $G/H \cong (G/K)/(H/K)$.*

Proof Clearly, G/K, G/H and H/K are groups.

Consider the map $f : G/K \to G/H$ given by

$$f(aK) = aH, \quad \forall a \in G.$$

As $aK = bK \Rightarrow a^{-1}b \in K \subseteq H \Rightarrow aH = bH$, the map f is well defined.

Again,

$$f(aK \cdot bK) = f(abK) = abH = aHbH$$
$$= f(aK)f(bK), \quad \text{for all } aK, bK \in G/K$$

shows that f is a homomorphism.

Clearly, f is an epimorphism. We have

$$\ker f = \{aK \in G/K : f(aK) = aH = H\}$$
$$= \{aK : a \in H\} = H/K.$$

Consequently, by the First Isomorphism Theorem, $(G/K)/(H/K) \cong G/H$. □

Theorem 2.6.9 Let $f : G \to K$ be an epimorphism of groups and S a subgroup of K. Then $H = f^{-1}(S)$ is a subgroup of G such that $\ker f \subseteq H$. Moreover, if S is normal in K, then H is normal in G. Furthermore, if $H_1 \supseteq \ker f$ is any subgroup of G such that $f(H_1) = S$, then $H_1 = H$.

Proof It is clear that, as $1 \in H$, $H \neq \emptyset$. For $g, h \in H$, $f(gh^{-1}) = f(g)f(h^{-1}) = f(g)f(h)^{-1} \in S \Rightarrow gh^{-1} \in H \Rightarrow H$ is a subgroup of G. Now $\ker f = \{g \in G : f(g) = 1_K \in S\} \Rightarrow \ker f \subseteq H$. Next let S be normal in K, $g \in G$ and $h \in H$. Then $f(ghg^{-1}) = f(g)f(h)f(g)^{-1} \in S \Rightarrow ghg^{-1} \in H \Rightarrow H$ is normal in G.

Finally, let $H_1 \supseteq \ker f$ be a subgroup of G such that $f(H_1) = S$. We claim that $H_1 = H$. Now $h_1 \in H_1 \Rightarrow f(h_1) \in S \Rightarrow H_1 \subseteq H$. Again $h \in H \Rightarrow f(h) = s \in S$. Choose $h_1 \in H_1$ such that $f(h_1) = s$. Then $f(hh_1^{-1}) = f(h)f(h_1)^{-1} = ss^{-1} = 1_K \Rightarrow hh_1^{-1} \in \ker f \subseteq H_1 \Rightarrow h \in H_1 \Rightarrow H \subseteq H_1$. Consequently, $H = H_1$. □

Corollary 1 (Correspondence Theorem) *If $f : G \to K$ is an epimorphism of groups, then the assignment $H \mapsto f(H)$ defines an one-to-one correspondence between the set $S(G)$ of all subgroups H of G which contain $\ker f$ and the set $S(K)$ of all subgroups of K such that normal subgroups correspond to normal subgroups.*

Proof The assignment $H \mapsto f(H)$ defines a map $\psi : S(G) \to S(K)$. By Theorem 2.6.9 it follows that ψ is a bijection. Moreover, the last part also follows from this theorem. □

Corollary 2 *If N is a normal subgroup of a group G, then every subgroup of G/N is of the form T/N, where T is a subgroup of G that contains N.*

Proof Apply Corollary 1 (Theorem 2.6.9) to the natural epimorphism $\pi : G \to G/N$. □

By using the First Isomorphism Theorem, the structure of certain quotient groups is determined.

Example 2.6.2 (i) Let $G = GL(n, \mathbf{R})$ be the group of all non-singular $n \times n$ matrices over $\mathbf{R}$ under usual matrix multiplication and H the subset of all matrices

$X \in GL(n, \mathbf{R})$ such that $\det X = 1$, i.e., $H = \{X \in GL(n, \mathbf{R}) : \det X = 1\}$. Since the identity matrix $I_n \in H$, $H \neq \emptyset$. Let $X, Y \in H$, so that $\det X = \det Y = 1$. By using the properties of determinants, we have $\det(XY^{-1}) = \det X \cdot \det(Y^{-1}) = \det X \cdot (1/\det Y) = 1$. Thus $X, Y \in H \Rightarrow XY^{-1} \in H \Rightarrow H$ is a subgroup of G by Theorem 2.4.1. Moreover, H is a normal subgroup of G by Theorem 2.6.5, because H is the kernel of the determinant function $\det : GL(n, \mathbf{R}) \to \mathbf{R}^*$ given by $X \mapsto \det X$, where $\mathbf{R}^*$ is the multiplicative group of non-zero reals. As the image set of this homomorphism is $\mathbf{R}^*$, we have $GL(n, \mathbf{R})/H \cong \mathbf{R}^*$.

Remark A similar result holds for $GL(n, \mathbf{C})$ with complex entries.

(ii) Let $f : (\mathbf{R}, +) \to (\mathbf{C}^*, \cdot)$ be the homomorphism defined by $f(x) = e^{2\pi i x} = \cos 2\pi x + i \sin 2\pi x \; \forall x \in \mathbf{R}$. Then $\ker f$ is the additive group $(\mathbf{Z}, +)$ and $\operatorname{Im} f$ is the multiplicative group H of all complex numbers with modulus 1. Hence the additive quotient group $\mathbf{R}/\mathbf{Z}$ is isomorphic to the multiplicative group H.

Problem If H is a non-normal subgroup of a group G, is it possible to make the set $\{aH\}$ of left cosets of H in G a group with the rule of multiplication $aH \cdot bH = abH \; \forall a, b \in G$?

Solution (*The answer is no.*) If it was possible, then the mapping $f : G \to G'$, the group formed by the cosets, given by $x \mapsto xH$, would be a homomorphism with H as kernel. But this is not possible unless H is normal in G.

We now study a very important subgroup defined below which is closely associated with given subgroup.

Definition 2.6.4 If H is a subgroup of a group G, then the set $N(H) = \{a \in G : aHa^{-1} = H\} = \{a \in G : aH = Ha\}$ is called the normalizer of H and H is said to be *invariant* under each $a \in N(H)$.

Proposition 2.6.2 *If $N(H)$ is the normalizer of a subgroup H of a group G, then*

(i) *$N(H)$ is a subgroup of G;*
(ii) *H is a normal subgroup of $N(H)$;*
(iii) *if H and K are subgroups of G, and H is a normal subgroup of K, then $K \subseteq N(H)$;*
(iv) *H is a normal subgroup of $G \Leftrightarrow N(H) = G$.*

Proof (i) Since $1 \in N(H)$, $N(H) \neq \emptyset$. Now $x \in N(H) \Rightarrow xH = Hx \Rightarrow x^{-1}(xH) = x^{-1}(Hx) \Rightarrow H = (x^{-1}H)x \Rightarrow Hx^{-1} = x^{-1}H$.

Then $a, x \in N(H) \Rightarrow aH = Ha$ and $x^{-1}H = Hx^{-1} \Rightarrow (ax^{-1})H = a(x^{-1}H) = a(Hx^{-1}) = (aH)x^{-1} = Hax^{-1} \Rightarrow N(H)$ is a subgroup of G.

(ii) and (iii) follow trivially.

(iv) H is a normal subgroup of $G \Rightarrow$ for each $g \in G$, $gH = Hg \Rightarrow G \subseteq N(H)$. Consequently, $G = N(H)$. The converse part follows from (ii). $\square$

In abelian groups every subgroup is normal but it is not true in an arbitrary non-abelian group. There are some groups in which no normal subgroup exists except its trivial subgroups. Such groups are called simple groups.

Definition 2.6.5 A group G is said to be simple iff G and $\{1\}$ are its only normal subgroups.

Definition 2.6.6 A normal subgroup H of a group G is said to be a maximal normal subgroup iff $H \neq G$ and there is no normal subgroup N of G such that $H \subsetneqq N \subsetneqq G$.

Theorem 2.6.10 *If H is a normal subgroup of a group G, then G/H is simple iff H is a maximal normal subgroup of G.*

Proof Let H be a maximal normal subgroup of G. Consider the canonical homomorphism $f : G \to G/H$. As f is an epimorphism, f^{-1} of any proper normal subgroup of G/H would be a proper normal subgroup of G containing H, which contradicts the maximality of H. Consequently G/H must be simple. Conversely, let G/H be simple. If N is a normal subgroup of G properly containing H, then $f(N)$ is a normal subgroup of G/H and if also $N \neq G$, then $f(N) \neq G/H$ and $f(N) \neq H$, which is not possible. So no such N exists and hence H is maximal. $\square$

Remark There was a long-standing conjecture that a non-abelian simple group of finite order has an even number of elements. This conjecture has been proved to be true by Walker Feit and John Thompson.

2.7 Geometrical Applications

2.7.1 Symmetry Groups of Geometric Figures in Euclidean Plane

The study of symmetry gives the most appealing applications in group theory. While studying symmetry we use geometric reasoning. Symmetry is a common phenomenon in science. In general a symmetry of a geometrical figure is a one-one transformation of its points which preserves distance. Any symmetry of a polygon of n sides in the Euclidean plane is uniquely determined by its effect on its vertices, say $\{1, 2, \ldots, n\}$. The group of symmetries of a regular polygon of n sides is called the *dihedral group* of degree n denoted by D_n which is a subgroup of S_n and contains $2n$ elements. D_n is generated by rotation r of the regular polygon of n sides through an angle $2\pi/n$ radians in its own plane about the origin in anti-clockwise direction and certain reflections s satisfying some relations (see Ex. 19 of Exercises-III).

(a) *Isosceles triangle.* Figure 2.1 is symmetric about the perpendicular bisector 1D. So the symmetry group consists of identity and reflection about the line 1D. In

Fig. 2.1 Group of
symmetries of isosceles
triangle (S_2)

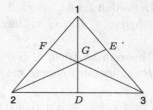

Fig. 2.2 Group of
symmetries of equilateral
triangle (S_3)

terms of permutations of vertices this group is

$$\left\{\begin{pmatrix} 1 & 2 & 3 \\ 1 & 2 & 3 \end{pmatrix}, \begin{pmatrix} 1 & 2 & 3 \\ 1 & 3 & 2 \end{pmatrix}\right\} \cong S_2.$$

(b) *Equilateral triangle*. In Fig. 2.2, the symmetry consists of three rotations of magnitudes $\frac{2\pi}{3}, \frac{4\pi}{3}, \frac{6\pi}{3} = 2\pi$ about the center G denoted by r_1, r_2 and r_3, respectively, together with three reflections about the perpendicular lines $1D, 2E, 3F$ denoted by t_1, t_2, and t_3, respectively. So in terms of permutations of vertices the symmetry group is

$$\left\{r_1 = \begin{pmatrix} 1 & 2 & 3 \\ 2 & 3 & 1 \end{pmatrix}, r_2 = \begin{pmatrix} 1 & 2 & 3 \\ 3 & 1 & 2 \end{pmatrix}, r_3 = \begin{pmatrix} 1 & 2 & 3 \\ 1 & 2 & 3 \end{pmatrix}, \right.$$

$$\left. t_1 = \begin{pmatrix} 1 & 2 & 3 \\ 1 & 3 & 2 \end{pmatrix}, t_2 = \begin{pmatrix} 1 & 2 & 3 \\ 3 & 2 & 1 \end{pmatrix}, t_3 = \begin{pmatrix} 1 & 2 & 3 \\ 2 & 1 & 3 \end{pmatrix}\right\} \cong S_3.$$

(c) *Rectangle* (not a square). In this case, the symmetry group consists of the identity I, reflections r_1, r_2 about DE, FG (lines joining the midpoints of opposite sides), respectively, and reflection r_3 about O (which is actually a rotation about O through an angle π). In terms of permutations of vertices this group is

$$\left\{I = \begin{pmatrix} 1 & 2 & 3 & 4 \\ 1 & 2 & 3 & 4 \end{pmatrix}, r_1 = \begin{pmatrix} 1 & 2 & 3 & 4 \\ 4 & 3 & 2 & 1 \end{pmatrix}, \right.$$

$$\left. r_2 = \begin{pmatrix} 1 & 2 & 3 & 4 \\ 2 & 1 & 4 & 3 \end{pmatrix}, r_3 = \begin{pmatrix} 1 & 2 & 3 & 4 \\ 3 & 4 & 1 & 2 \end{pmatrix}\right\},$$

which is Klein's 4-group (see Ex. 22 of Exercises-III).

(d) *Square*. In Fig. 2.3, the symmetry group consists of four rotations r_1, r_2, r_3, r_4 (= Identity I) of magnitudes $\frac{\pi}{2}, \pi, \frac{3\pi}{2}, 2\pi$ about the circumcenter O respectively and four reflections t_1, t_2, t_3, t_4 about $DE, FG, 13, 24$.

Fig. 2.3 Group of symmetries of the square (D_4)

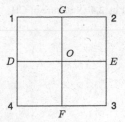

In terms of permutations of vertices involved here, this group is

$$\left\{ r_1 = \begin{pmatrix} 1 & 2 & 3 & 4 \\ 4 & 1 & 2 & 3 \end{pmatrix}, r_2 = \begin{pmatrix} 1 & 2 & 3 & 4 \\ 3 & 4 & 1 & 2 \end{pmatrix}, r_3 = \begin{pmatrix} 1 & 2 & 3 & 4 \\ 2 & 3 & 4 & 1 \end{pmatrix}, \right.$$

$$r_4 = \begin{pmatrix} 1 & 2 & 3 & 4 \\ 1 & 2 & 3 & 4 \end{pmatrix} = I, t_1 = \begin{pmatrix} 1 & 2 & 3 & 4 \\ 4 & 3 & 2 & 1 \end{pmatrix}, t_2 = \begin{pmatrix} 1 & 2 & 3 & 4 \\ 2 & 1 & 4 & 3 \end{pmatrix},$$

$$\left. t_3 = \begin{pmatrix} 1 & 2 & 3 & 4 \\ 1 & 4 & 3 & 2 \end{pmatrix}, t_4 = \begin{pmatrix} 1 & 2 & 3 & 4 \\ 3 & 2 & 1 & 4 \end{pmatrix} \right\}.$$

This group is called the group of symmetries of the square (also called *octic* group) and is the dihedral group D_4 (see Ex. 19 of Exercises-III).

Remark In this group D_4 there are eight (hence octic) mappings of a square into itself that preserve distance between points and distance between their images, map adjacent vertices into adjacent vertices, center into center so that the only distance preserving mappings are rotations of the square about its center, reflections (flips) about various bisectors, and the identity map.

2.7.2 *Group of Rotations of the Sphere*

A sphere with a fixed center O can be brought from a given position into any other position by rotating the sphere about an axis through O. Clearly, the rotations about the same axis have the same result iff they differ by a multiple of 2π. Thus if $r(S)$ denotes the set of all rotations about the same axis, then we call the rotations r and r' in $r(S)$ equal or different iff they differ by a multiple of 2π or not. Clearly, the result of two successive rotations in $r(S)$ can also be obtained by a single rotation in $r(S)$. It follows that $r(S)$ forms a group. (The identity is a rotation in $r(S)$ through an angle 0 and the inverse of a rotation r in $r(S)$ has the same angle but in the opposite direction of r). Thus if any rotation r in $r(S)$ has the angle of rotation θ about the axis, then the map $f : r(S) \to S^1$ defined by $f(r) = e^{i\theta}$ is a group isomorphism from the group $r(S)$ onto the circle group S^1 (see Example 2.3.2).

Remark The rotations of $\mathbf{R}^2$ or $\mathbf{R}^3$ about the origin are the linear operators whose matrices with respect to the natural basis are orthogonal and have determinant 1 (see Chap. 8).

Fig. 2.4 Structure of CH_4

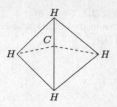

2.7.3 Clock Arithmetic

On a 24-hour clock, let the cyclic group $\langle a \rangle$ of order 24 represent hours. Then each of 24 numerals of the dial serves as representatives of a coset of hours. Here we use the fact that a 10 hour journey starting at 20 hrs (8 o'clock) ends at 30 hrs, i.e., $10 + 20 = 30 \equiv 6$ o'clock (the following day). In shifting from a 24-hour clock to a 12 hour clock, we take the 24 hours modulo the normal subgroup $H = \{1, a^{12}\}$. Then the quotient group $\langle a \rangle / H$ is a group of 12 cosets or a group of representatives: $\{a, a^2, a^3, a^{11}, a^{12}\}$, where $(a) = \{a, a^{13}\}, (a^2) = \{a^2, a^{14}\}, (a^3) = \{a^3, a^{15}\}, (a^{11}) = \{a^{11}, a^{23}\}$ and $(a^{12}) = \{a^{12}, a^{24}\}$. A seven hour journey starting at 9 AM (9 PM) ends at 4 PM (4 AM). On the 12-hour clock we do not distinguish between the congruent elements in the same coset $\{4 \text{ AM}, 4 \text{ PM}\}$.

A Note on Symmetry Group   Group theory is an ideal tool to study symmetries. In this note we call any subset of $\mathbf{R}^2$ or $\mathbf{R}^3$ an object. We study orthogonal maps from $\mathbf{R}^2$ to $\mathbf{R}^2$ or from $\mathbf{R}^3$ to $\mathbf{R}^3$ and these are rotations or reflections or rotation-reflections. As orthogonal maps are only those linear maps which keep lengths and angles invariant, they do not include translations.

Let X be an object in $\mathbf{R}^2$ or $\mathbf{R}^3$. Then the set $S(X)$ of all orthogonal maps g with $g(X) = X$ is a group with respect to the usual composition of maps. This group is called the symmetry group of X. The elements of $S(X)$ are called symmetry operations on X. Let $S_2(X)$ (or $S_3(X)$) correspond to the symmetric group of X in $\mathbf{R}^2$ (or $\mathbf{R}^3$).

Examples (a) If $X =$ regular n-gon in $\mathbf{R}^2$ with center at $(0, 0)$, then $S_2(X) = D_n$, the dihedral group with $2n$ elements and for $S_3(X)$, we also get the reflection about the xy-plane and its compositions with all $g \in S_2(X)$, i.e., $S_3(X) = D_n \times \mathbf{Z}_2$.

(b) If $X =$ regular tetrahedron with center at $(0, 0, 0)$, then $S_3(X) \cong S_4$.

(c) If $X =$ cube, then $S_3(X) = S_4 \times \mathbf{Z}_2$.

(d) If $X =$ the letter M, then $S_2(X) \cong \mathbf{Z}_2$ and $S_3(X) \cong \mathbf{Z}_2 \times \mathbf{Z}_2$.

(e) If $X =$ a circle, then $S_2(X) = O_2(\mathbf{R})$, the whole orthogonal group of all orthogonal maps from $\mathbf{R}^2$ to $\mathbf{R}^2$.

For applications to chemistry and crystallography, X is considered a molecule and $S(X)$ depends on the structural configuration of the molecule X.

For example, methane CH_4 (see Fig. 2.4) has the shape of a regular tetrahedron with H-atoms at the vertices and C-atom at its center $(0, 0, 0)$. Then by (b), $S_3(CH_4) \cong S_4$.

Weyl (1952) first realized the importance of group theory to study symmetry in nature and the application of group theory was boosted by him. Molecules, crystals, and elementary particles are different, but they have very similar theories of symmetry.

Applications of groups to physics are described in Elliot and Dawber (1979) and Sternberg (1994) and those to Chemistry in Cotton (1971) and Farmer (1996).

2.8 Free Abelian Groups and Structure Theorem

If we look at the dihedral group D_n (see Ex. 19 of SE-III) we see that D_n has two generators r and s satisfying some relations other than the associative property. But we now consider groups which have a set of generators satisfying no relations other than associativity which is implied by the group axioms. Such groups are called free groups. In this section we study free abelian groups and prove the 'Fundamental Theorem of Finitely Generated Abelian Groups' which is a structure theorem. This theorem gives notions of 'Betti numbers' and 'invariant factors'. We also introduce the concept of 'homology and cohomology groups' which are very important in homological algebra and algebraic topology.

A free abelian groups is a direct sum of copies of additive abelian group of integers $\mathbf{Z}$. It has some properties similar to vector spaces. Every free abelian group has a basis and its rank is defined as the cardinality of a basis. The rank determines the groups up to isomorphisms and the elements of such a group can be written as finite formal sums of the basis elements.

The concept of free abelian groups is very important in mathematics. It has wide applications in homology theory. Algebraic topology is also used to prove some interesting properties of free abelian groups [see Rotman (1988)]

We consider in this section an additive abelian group G.

Definition 2.8.1 Let G be an additive abelian group and $\{G_i\}_{i \in I}$ be a family of subgroups of G. Then G is said to be generated by $\{G_i\}$ iff every element $x \in G$ can be expressed as

$$x = x_{i_1} + \cdots + x_{in}, \quad \text{where the additive indices } i_t \text{ are distinct.}$$

We sometimes use the following notation:

$$x = \sum_{i \in I} x_i, \quad \text{where we take } x_i = 0, \text{ if } i \text{ is not one of the indices } i_1, i_2, \ldots, i_n.$$

If the subgroups $\{G_i\}$ generate G, we write

$$G = \sum_{i \in I} G_i, \quad \text{in general,} \quad \text{and} \quad G = \sum_{i=1}^{n} G_i, \quad \text{when } I = \{1, 2, \ldots, n\}.$$

Fig. 2.5 Commutativity of
the triangle

Definition 2.8.2 If a group G is generated by the groups $\{G_i\}_{i \in I}$ and every element $x \in G$ has the unique representation as $x = \sum_{i \in I} x_i$, where $x_i = 0$ for all but finitely many i, then G is said to be *direct sum of the groups* G_i, written

$$G = \bigoplus_{i \in I} G_i, \quad \text{in general} \quad \text{and} \quad G = \sum_{i=1}^{n} G_i, \quad \text{if } I = \{1, 2, \ldots, n\}.$$

Remark For $n = 2$, see Ex. 21 of Exercises-III.

A Characterization of Direct Sums Let G be an abelian group and $\{G_i\}$ be a family of subgroups of G. We say that G satisfies the extension condition (EC) for direct sums iff given any abelian group A and any family of homomorphisms $h_i : G_i \to A$, there exists a unique homomorphism $h : G \to A$ extending h_i, for each i, i.e., $h|G_i = h_i$, i.e., making the diagram in Fig. 2.5 commutative, where $k_i : G_i \hookrightarrow G$ is an inclusion map for each i.

Proposition 2.8.1 *Let G be an abelian group and $\{G_i\}$ be a family of subgroups of G.*

(a) *If $G = \bigoplus G_i$, then G satisfies the condition (EC);*
(b) *if G is generated by $\{G_i\}$ and G satisfies the condition (EC), then $G = \bigoplus G_i$.*

Proof (a) Let $G = \oplus G_i$. Then for the given homomorphisms $h_i = G_i \to A$, we define a homomorphism $h : G \to A$ as follows.

If $x = \sum x_i$ (unique finite), then the homomorphism h given by $h(x) = \sum h_i(x_i)$ is well defined and is our required homomorphism.

(b) Let $x = \sum x_i = \sum y_i$. Given an index j, let $A = G_j$.

Define $h_i : G_i \to G_j$ as follows:

$$h_i = id, \quad \text{if } i = j;$$
$$= 0, \quad \text{if } i \neq j.$$

Let $h : G \to G_j$ be the homomorphism extending each h_i by the condition (EC). Then

$$h(x) = \sum_{i \in I} h_i(x_i) = x_j \quad \text{and} \quad h(x) = \sum_{i \in I} h_i(y_i) = y_j.$$

Hence $x_j = y_j \Rightarrow x$ has a unique representation $\Rightarrow G = \bigoplus G_i$. □

Corollary 1 *Let $G = H \oplus K$, where H and K are subgroups of G such that $H = \bigoplus_{i \in I} G_i$ and $K = \bigoplus_{j \in J} G_j$ and $I \cap J = \emptyset$. Then $G = \bigoplus_{t \in I \cup J} G_t$.*

Proof Let A be an abelian group. If $h_i : G_i \to A$ and $h_j : G_j \to A$ are families of homomorphisms, then by Proposition 2.8.1, they can be extended to homomorphisms $f : H \to A$ and $g : K \to A$, respectively. Again f and g can be extended to a homomorphism $h : G \to A$ by Proposition 2.8.1. Hence $G = \bigoplus_{t \in I \cup J} G_t$. $\quad\square$

Corollary 2 $(G_1 \oplus G_2) \oplus G_3 = G_1 \oplus (G_2 \oplus G_3) = G_1 \oplus G_2 \oplus G_3$ *for any subgroups* G_1, G_2 *and* G_3 *of a group* G.

Corollary 3 *For any group* G *and subgroups* G_1 *and* G_2, $G = G_1 \oplus G_2 \Rightarrow G/G_2 \cong G_1$.

Proof Let $A = G_1$ and $h_1 : G_1 \to A = G_1$ be the identity homomorphism and $h_2 : G_2 \to A$ be the zero homomorphism. If $h : G \to A$ is the homomorphism extending h_1 and h_2, then h is an epimorphism with $\ker h = G_2$. Hence $G/G_2 \cong G_1$. $\quad\square$

We are now in a position to study free abelian groups. Let G be an additive group. Then $(m + n)x = mx + nx \; \forall x \in G$ and $\forall m, n \in \mathbf{Z}$.

If the group G is abelian, then $n(x + y) = nx + ny$, $\forall x, y \in G$ and $\forall n \in \mathbf{Z}$.

If $S \; (\neq \emptyset)$ is a subset of G, then the subgroup $\langle S \rangle$ generated by S is given by

$$\langle S \rangle = \{n_1 s_1 + n_2 s_2 + \cdots + n_t s_t : n_i \in \mathbf{Z}, s_i \in S\}.$$

In particular, if $S = \{s\}$, then $\langle s \rangle = \{ns : n \in \mathbf{Z}\}$ is the cyclic group generated by s and if $S = \emptyset$, then $\langle S \rangle = \{0\}$. The subset S is said to be independent iff $\sum n_i s_i = 0 \Rightarrow n_i = 0$, $\forall i$.

Definition 2.8.3 Let G be an additive abelian group. The group G is said to be a free *abelian group* with a basis B iff

 (i) for each $b \in B$, the cyclic subgroup $\langle b \rangle$ is infinite cyclic; and
(ii) $G = \bigoplus_{b \in B} \langle b \rangle$ (direct sum).

Remark A free abelian group G is a direct sum of copies of $\mathbf{Z}$ and every element $x \in G$ can be expressed uniquely as $x = \sum n_b b$, where $n_b \in \mathbf{Z}$ and almost all n_b (i.e., all but a finite number of n_b) are zero.

Using the extension condition (EC) for direct sums, the following characterization of free abelian groups follows.

Proposition 2.8.2 *Let* G *be an abelian group and* $\{b_i\}$ *be a family of elements of* G *that generates* G. *Then* G *is a free abelian group with a basis* $\{b_i\}$ *iff for any abelian group* A *and any family* $\{a_i\}$ *of elements of* A, *there is a unique homomorphism* $h : G \to A$ *such that* $h(b_i) = a_i$ *for each* i.

Proof Let $G_i = \langle b_i \rangle$. Then $\{G_i\}$ is a family of subgroups of G. First suppose that the condition (EC) holds. We claim that each G_i is infinite cyclic. If for some index j, the element b_j generates a finite cyclic subgroup of G, then taking $A = \mathbf{Z}$, there exists no homomorphism $h : G \to A$, mapping each b_i to the number 1. This is so because b_j is of finite order but 1 is not of finite order in $\mathbf{Z}$. Hence by Proposition 2.8.1(b), $G = \bigoplus G_i$.

Conversely, let G be a free group with a basis $\{b_i\}$. Then given elements $\{a_i\}$ of A, $\exists$ homomorphisms $h_i : G_i \to A$ such that $h_i(b_i) = a_i$, because G_i is infinite cyclic. Hence the proof is completed by using Proposition 2.8.1(a). $\square$

Theorem 2.8.1 *If G is a free abelian group with a basis $B = \{b_1, b_2, \ldots, b_n\}$, then n is uniquely determined by G.*

Proof By hypothesis, $G = \mathbf{Z} \oplus \mathbf{Z} + \cdots \oplus \mathbf{Z}$ (n summands). Then $2G$ is a subgroup of G such that

$$2G \cong 2\mathbf{Z} \oplus 2\mathbf{Z} + \cdots + 2\mathbf{Z} \quad (n \text{ summands}).$$

Hence $G/2G \cong \mathbf{Z}/2\mathbf{Z} \oplus \mathbf{Z}/2\mathbf{Z} \oplus \cdots \oplus \mathbf{Z}/2\mathbf{Z}$ (n summands) shows that card $(G/2G) = 2^n$. $\square$

Definition 2.8.4 An abelian group G is said to be a finitely generated free abelian group iff G has a finite basis.

Remark The basis of a finitely generated free abelian group is not unique. For example, $\{(1, 0), (0, 1)\}$ and $\{(-1, 0), (0, -1)\}$ are two different bases of $G = \mathbf{Z} \oplus \mathbf{Z}$.

Corollary 1 *Let G be a finitely generated abelian group. Then any two bases of G have the same cardinal number.*

Proof Let $B = \{b_1, b_2, \ldots, b_n\}$ and $X = \{x_1, x_2, \ldots x_r\}$ be two bases of G. Then by Theorem 2.8.1, card $(G/2G) = 2^n$ and also card $(G/2G) = 2^r$. Hence $n = r$. $\square$

For more general result, like vector spaces, we prove the following theorem.

Theorem 2.8.2 *Any two bases of a free abelian group F have the same cardinality.*

Proof Let B and C be two bases of F. Given a prime integer $p > 1$, the quotient group F/pF is a vector space over the field $\mathbf{Z}_p$ (see Chap. 8 and Example 4.1.5 of Chap. 4). Hence the cosets $\{b + pF : b \in B\}$ form a basis $\Rightarrow \dim_{\mathbf{Z}_p}(F/pF) =$ card B. Similarly, $\dim_{\mathbf{Z}_p}(F/pF) =$ card C. Hence card $B =$ card C. $\square$

Remark If $V = F/pF$ is infinite dimensional, Zorn's Lemma is used to prove the existence of a basis of V and then using the result that the family of all finite subsets of an infinite set B has the same cardinality as B the invariance of the cardinality of a basis is proved.

Definition 2.8.5 Let F be a free abelian group with a basis B. The cardinality of B is called the rank of F, denoted rank F. In particular, if F is finitely generated, then the number of elements in a basis of F is the rank F.

Remark Rank F is well defined by Theorem 2.8.2.

Like vector spaces free abelian groups are characterized by their ranks. More precisely, two free abelian groups are isomorphic iff they have the same rank. Let G be an arbitrary abelian group. We define its rank as follows:

Definition 2.8.6   An abelian group G has rank r (possibly infinite) iff $\exists$ a free abelian subgroup F of G such that

(a) rank $F = r$;
(b) G/F is a torsion group (i.e., every element of G/F is of finite order).

Existence of F: Let B be an independent subset of G. Then the subgroup $\langle B \rangle$ generated by B is abelian and free with a basis B. Let S be a maximal independent subset of G, which exists by Zorn's Lemma. Then $F = \langle S \rangle$ is a free abelian group and G/F is a torsion group.

Remark Rank F precisely depends on G, since rank $G = \dim_Q(Q \otimes G)$, where $Q \otimes G$ is a vector space over Q [see Chaps. 8 and 9]. Hence rank G is well defined.

Sometimes, given an arbitrary family of abelian groups $\{G_i\}$, we can find a group G that contains subgroups H_i isomorphic to the group G_i, such that G is isomorphic to the direct sum of these groups. This leads to the concept of the external direct sum of groups.

Definition 2.8.7   Let $\{G_i\}$ be a family of abelian groups. If G is an abelian group and $f_i : G_i \to G$ is a family of monomorphisms, such that G is the direct sum of the groups $f_i(G_i)$. Then we say that G is the *external direct sum* of the groups G_i, relative to the monomorphisms $f_i : G_i \to G$.

We prescribe a construction of G.

Theorem 2.8.3   *Given a family of abelian groups $\{G_i\}_{i \in I}$, there exists an abelian group G and a family of monomorphisms $f_i : G_i \to G$ such that $G = \bigoplus f_i(G_i)$.*

Proof We consider the cartesian product $\prod_{i \in I} G_i$. It is an abelian group under componentwise addition. Let G be the subgroup of the cartesian product consisting of those tuples $(x_i)_{i \in I}$ such that $x_i = o_i$, for all but a finitely many values of i. Given an index j, define $f_j : G_j \to G$ by $f_j(x)$ be the tuple that has x at its jth coordinate and 0 at its ith coordinate for $i \neq j$. Then f_j is a monomorphism. Since each element $x \in G$ has only finitely many non-zero coordinates, x can be expressed uniquely as a finite sum of elements from the groups $f_j(G_j)$. $\square$

A Characterization of External Direct Sums of Groups   Like extension condition (EC) of direct sums, extension condition (ED) for external direct sums is defined.

Fig. 2.6 Commutativity of
the triangle for external direct
sum

Let $\{G_i\}_{i \in I}$ be a family of abelian groups, G be an abelian group and $f_i : G_i \to$ G be a family of homomorphisms. Then G is said to satisfy the condition (ED) iff given any abelian group A and a family of homomorphisms $h_i : G_i \to A$, $\exists$ a unique homomorphism $h : G \to A$ making the diagram in Fig. 2.6 commutative, i.e., $h \circ f_i = h_i$, $\forall i$.

Proposition 2.8.3 *Let $\{G_i\}$ be a family of abelian groups and A be an abelian group.*

(a) *If each $f_i : G_i \to G$ is a monomorphism and $G = \bigoplus_{i \in I} f_i(G_i)$, then G satisfies the condition (ED).*

(b) *Conversely, if $\{f_i(G_i)\}$ generates G and G satisfies the condition (ED), then each $f_i : G_i \to G$ is a monomorphism and $G = \bigoplus_{i \in I} f_i(G_i)$.*

Proof Left as an exercise. $\qquad\square$

We now prove the following *structure theorem* for finitely generated abelian groups and also prove as its corollary the structure theorem of an arbitrary finite abelian group (also known as fundamental theorem of finite abelian groups).

Theorem 2.8.4 (Fundamental theorem for finitely generated abelian groups) *Every finitely generated abelian group can be expressed uniquely as*

$$G \cong \overbrace{\mathbf{Z} \oplus \mathbf{Z} \oplus \cdots \oplus \mathbf{Z}}^{r \text{ summands}} \oplus \mathbf{Z}_{n_1} \oplus \mathbf{Z}_{n_2} \oplus \cdots \oplus \mathbf{Z}_{n_t}$$

for some integers $r, n_1, n_2, \ldots n_t$ such that

(i) *$r \geq 0$ and $n_j \geq 2$, $\forall j$; and*

(ii) *$n_i | n_{i+1}$, for $1 \leq i \leq t - 1$.*

Proof As finitely generated $\mathbf{Z}$-modules are finitely generated abelian groups, the theorem follows from Theorem 9.6.5 of Chap. 9 by taking $R = \mathbf{Z}$. $\qquad\square$

Definition 2.8.8 The integer r in Theorem 2.8.4 is called the *free rank* or *Betti number* of the group G and the integers $n_1, n_2, \ldots, n_i$ are called the *invariant factors of G*.

Remark Betti numbers are very important in mathematics. For example, two closed surfaces are homeomorphic iff their homology groups have the same Betti numbers in all dimensions [see Massey (1991)].

Theorem 2.8.5 *Two finitely generated abelian groups are isomorphic iff they have the same rank and the same invariant factors (up to units).*

Proof It follows from Theorem 9.6.11 of Chap. 9 by taking $R = \mathbf{Z}$.   □

Remark If we write $F(G) = \mathbf{Z} \oplus \cdots \oplus \mathbf{Z}$ (r summands) and $T(G) = \mathbf{Z}_{n_1} \oplus \cdots \oplus \mathbf{Z}_{n_t}$ in Theorem 2.8.4, then $G \cong F(G) \oplus T(G)$. $F(G)$ is called a *free subgroup* of G and $T(G)$ is called a *torsion subgroup* of G.

Corollary 1 (Structure Theorem for finite abelian groups) *Any finite abelian group G can be expressed uniquely as $G \cong \mathbf{Z}_{n_2} \oplus \cdots \oplus \mathbf{Z}_{n_t}$ such that $n_i | n_{i+1}$, for $1 \leq i \leq t - 1$.*

Proof The finite abelian group G is certainly finitely generated (which is generated by the finite set consisting of all its elements). Hence, in this case, $F(G) = 0$.   □

Corollary 2 *Two finite abelian groups are isomorphic iff they have the same invariant factors.*

Supplementary Examples (SE-IB)

1 For any abelian group G, the following statements are equivalent:

(a) G has a finite basis;
(b) G is the internal direct sum of a family of infinite cyclic groups;
(c) G is isomorphic to a finite direct sum of finite copies of $\mathbf{Z}$.

   [*Hint.* (a) $\Rightarrow$ (b). Let $B = \{b_1, b_2, \ldots, b_t\}$ be a basis of G. Let $nb_i = 0$ for some $n \in \mathbf{Z} \Rightarrow ob_1 + \cdots + nb_i + \cdots ob_t = 0 \Rightarrow n = 0 \Rightarrow b_i$ is of infinite order $\Rightarrow \langle b_i \rangle$ is an infinite cyclic group for $1 \leq i \leq t \Rightarrow G = \bigoplus_{i=1}^{t} \langle b_i \rangle$.
   (b) $\Rightarrow$ (c). Let $G = \bigoplus_{i=1}^{t} G_i$, where each $G_i \cong \mathbf{Z}$. This implies (c).
   (c) $\Rightarrow$ (a). Let $G \cong \mathbf{Z} \oplus \mathbf{Z} + \cdots + \oplus \mathbf{Z} = \mathbf{Z}^t$ (say) be a finite direct sum of t copies of $\mathbf{Z}$ and $f : G \to \mathbf{Z}^t$ be an isomorphism.
   Let $e_i = (0, \ldots, 0, 1, 0 \ldots, 0) \in \mathbf{Z}^t$, where 1 is at the i the place.
   Then $\exists b_i \in G$, such that $f(b_i) = e_i$ for each i.
   Clearly, $B = \langle b_1, b_i, \ldots, b_t \rangle$ forms a basis of G.]

2 Let F be a free abelian group with a basis B and G be an abelian group.

(a) For every group $f : B \to G$, $\exists$ a unique group homomorphism $\tilde{f} : F \to G$ such that $\tilde{f}(b) = f(b)$, $\forall b \in B$.
(b) For every abelian group G, $\exists$ a free abelian group F such that G is isomorphic to a quotient group of F.

   [*Hint.* (a) $x \in F \Rightarrow \exists n_b \in \mathbf{Z}$ such that $x = \sum n_b b$.
   Define $\tilde{f} : F \to G$, by $\tilde{f}(x) = \sum n_b f(b)$. Unique expression of $x \Rightarrow \tilde{f}$ is well defined.

Clearly, $\tilde{f}$ is a homomorphism such that $\tilde{f}$ is unique, because homomorphisms agreeing on the set of basis are equal (like linear transformations of vector spaces).

(b) For each $g \in G$, take an infinite cyclic group generated by b_g (say). Then $F = \bigoplus_{g \in G} \langle b_g \rangle$ is a free abelian group with a basis $B = \{b_g : g \in G\}$. Define a map $f : B \to G, b_g \mapsto g$ and extend it by linearity.

Then by (a), $\exists$ a unique homomorphism $\tilde{f} : F \to G$ such that $\tilde{f}(b_g) = f(b_g) = g$. Hence $\tilde{f}$ is an epimorphism $\Rightarrow G \cong F/\ker \tilde{f}$.]

3 Every finitely generated abelian group G is the homomorphic image of a finitely generated free abelian group F.

[*Hint.* Let $G = \langle X \rangle$, where $X = \{x_1, x_2, \ldots, x_t\}$ and $B = \{b_1, b_2, \ldots, b_t\}$ be a basis of F.

Define a map $f : F \to G, b_i \mapsto x_i$ and extend it by linearity. Then f is well defined and an epimorphism $\Rightarrow G = f(F)$].

4 Let $F = \langle x \rangle$ be a free abelian group. Then $F/\langle nx \rangle \cong \mathbf{Z}_n$, for all integers $n \geq 0$.

[*Hint.* Define map $f : F \to \mathbf{Z}_n, mx \mapsto [m]$, $\forall m \in \mathbf{Z}$. Then f is an epimorphism with $\ker f = \langle nx \rangle \Rightarrow F/\langle nx \rangle \cong \mathbf{Z}_n$.]

2.8.1 Exercises

Exercises-II

1. For a given non-empty set S, define a word to be a finite product $x_1^{n_1} x_2^{n_2} \cdots x_t^{n_t}$, where $x_i \in S$ and $n_i \in \mathbf{Z}$. The word is said to be reduced iff $x_i \neq x_{i+1}$ and n_i's are non-zero. We can make a given word reduced by collecting up powers of the adjacent elements x_i and omitting the zero powers and continue the process according to the necessity.

We consider the word x_1^0 as the empty word. Let $F(S)$ denote the set of all reduced words on S. If $S = \emptyset$, $F(S)$ is taken to be the trivial group. If $S \neq \emptyset$, we define a binary operation on $F(S)$ by concatenation of reduced words, i.e., $(x_1^{n_1} \cdots x_t^{n_t})(y_1^{m_1} \cdots y_r^{m_r}) = x_1^{n_1} \cdots x_t^{n_t} y_1^{m_1} \cdots y_r^{m_r} = w$ (say) (making the word reduced).

Show that

(a) $F(S)$ becomes a group under this operation with the empty word as its identity element and $x_t^{-n_t} \cdots x_1^{-n_1}$ as the inverse of the reduced word $x_1^{n_1} \cdots x_t^{n_t}$;
(b) $|X| = |S| \Rightarrow F(X) \cong F(S)$;
(c) The free group $F(\{x\})$ is the infinite cyclic group having only possible non-empty reduced words are the powers x^r;
(d) If S consists of more than one element, then $F(S)$ is not abelian.

2. Show that the following statements on an abelian group G are equivalent:

(a) G has a non-empty basis;
(b) G is the internal direct sum of a family of infinite cyclic subgroups;

(c) G is isomorphic to a direct sum of copies of $\mathbf{Z}$;

(d) for a subset S and a map $g : S \to G$ such that given an abelian group A and a map $h : S \to A$, $\exists$ a unique homomorphism $f : G \to A$ satisfying $f \circ g = h$.

3. Show that given an abelian group A, $\exists$ a free abelian group G and an epimorphism $f : G \to A$.

   [*Hint*. Show that any abelian group is the homomorphic image of a free abelian group.]

4. Let G and H be free abelian groups on the set S w.r.t. maps $g : S \to G$ and $h : S \to H$ respectively. Show that $\exists$ a unique isomorphism $f : G \to H$ such that $f \circ g = h$.

5. If A is a free abelian group of rank n, then any subgroup B of A is a free abelian group of rank at most n.

   [*Hint*. Let $A = \mathbf{Z} \oplus \mathbf{Z} \oplus \cdots \oplus \mathbf{Z}$ (n summands) and $\pi_i : A \to \mathbf{Z}$ be the projection on the ith coordinate.

   Given $t \leq n$, let $B_t = \{b \in B : \pi_i(b) = 0, \ \forall i > t\}$. Then B_t is a subgroup of $B \Rightarrow \pi_t(B_t)$ is a subgroup of $\mathbf{Z}$. If $\pi_t(B_t) \neq 0$, take $x_t \in B_t$ such that $\pi_t(x_t)$ is a generator of this group. Otherwise, take $x_t = 0$. Then

   (i) $B_t = \langle x_1, x_2, \ldots, x_t \rangle$, for each t;
   (ii) the non-zero elements of $\{x_1, x_2, \ldots, x_t\}$ form a basis of B_t, for each t;
   (iii) $B_t = B$ is a free abelian group with rank B at most n.]

6. *Homology* and *cohomology groups* (see Sect. 9.11 of Chap. 9). A sequence $\{C_n\}$ of abelian groups and a sequence $\{\partial_n\}$ of homomorphisms $\partial_n : C_n \to C_{n-1}$ such that $\partial_{n-1} \circ \partial_n = 0$, $\forall n \in \mathbf{Z}$, are called a *chain complex* $C = (C_n, \partial_n)$ of abelian groups. Given two chain complexes $C = (C_n, \partial_n)$ and $C' = (C_n', \partial_n')$, a sequence $f = \{f_n\}$ of homomorphisms $f_n : C_n \to C_n'$ such that $f_{n-1} \circ \partial_n = \partial_n' \circ f_n$, $\forall n \in \mathbf{Z}$, is called a *chain map* $f : C \to C'$. The elements of $Z_n = \ker \partial_n$ are called *n-cycles*, elements of $B_n = \operatorname{Im} \partial_{n+1}$ are called *n-boundaries*. $f = \{f_n\}$ is said to be an isomorphism iff each f_n is an isomorphism of groups. Then the quotient group Z_n/B_n, denoted $H_n(C)$ exists and is called the nth homology group of C. Moreover each f_n induces a group homomorphism $f_n* : H_n(C) \to H_n(C')$ defined by $f_n * ([z]) = [f(z)]$, $\forall [\mathbf{z}] \in H_n(C)$ and $f_* = \{f_n*\}$ is said to be an isomorphism iff each f_n* is an isomorphism.

   The nth cohomology group $H^n(C)$ is defined dually.

   (a) (*Eilenberg–Steenrod*). Let the groups in the chain complexes C and C' be free and $f : C \to C'$ be a chain map. Let $K = \{K_n = \ker f_n\}$. Show that f_* is an isomorphism iff $H_n(K) = 0$ $\forall n \in \mathbf{Z}$.

   (b) Show that the complex C is exact (see Sect. 9.7 of Chap. 9) iff $H(C) = 0$.

   (c) Establish the dual results of (a) and (b) for $H^n(C)$.

7. Using the techniques of algebraic topology, the following results of free groups can be proved [see Rotman (1988, p. 305)]:

   (a) Every subgroup G of a free group F is itself free;

   (b) a free group F of rank 2 contains a subgroup that is not finitely generated;

(c) let F be a free group of finite rank n, and let G be a subgroup of finite index j. Then G is a free group of finite rank; indeed rank $G = jn - j + 1$.

2.9 Topological Groups, Lie Groups and Hopf Groups

An abstract group endowed with an additional structure having its group operations (multiplication and inversion) compatible with the additional structure, invites further attraction. Some of such groups are studied in this section. For example, we study topological groups, Lie groups and Hopf groups which are very important in topology, geometry, and physics. Topological groups were first considered by S. Lie (1842–1899). A topological group is a topological space whose elements form an abstract group such that the group operations are continuous with respect to the topology of the space. A Lie group is a topological group having the structure of a smooth manifold for which the group operations are smooth functions. On the other hand, a Hopf group (H-group) is a pointed topological space with a continuous multiplication such that it satisfies all the axioms of a group up to homotopy. The concept of an H-group is a generalization of the concept of topological groups. Lie groups are an important branch of group theory. The importance of Lie groups lies in the fact that Lie groups include almost all important groups of geometry and analysis. The theory of Lie groups stands at the crossing point of the theories of differential manifolds, topological groups, and Lie algebras.

2.9.1 Topological Groups

Definition 2.9.1 A *topological group* G is a non-empty set with a group structure and a topology on G such that the function $f : G \times G \to G$, $(x, y) \mapsto xy^{-1}$ is continuous.

The above condition of continuity is equivalent to the statement:
The functions $G \times G \to G$, $(x, y) \mapsto xy$ and $G \to G$, $x \mapsto x^{-1}$ are both continuous.
Some important examples of topological groups.

Example 2.9.1 (i) $(\mathbf{R}, +)$, under usual addition of real numbers and with the topology induced by the Euclidean metric $d(x, y) = |x - y|$.

(ii) The circle group $(S^1, \cdot)$ in $\mathbf{C}$, topologized by considering it as a subset of $\mathbf{R}^2$.

(iii) $(\mathbf{R}^n, +)$, under usual coordinatewise addition and with product topology.

(iv) $(GL(n, \mathbf{R}), \cdot)$, under usual multiplication of real matrices and with the Euclidean subspace topology of $\mathbf{R}^{n^2}$ $(n > 1)$.

(v) The orthogonal group $(O(n), \cdot)$ of real matrices with the Euclidean subspace topology $(n > 1)$. It is a subgroup of $GL(n, \mathbf{R})$.

(vi) The general linear group $(GL(n, \mathbf{C}), \cdot)$ over $\mathbf{C}$ topologized by considering it as a subspace of $\mathbf{R}^{2n^2}$.

(vii) $(U(n), \cdot)$ of all $n \times n$ complex matrices A such that $A\bar{A}^T = I$. It is a subgroup of $GL(n, \mathbf{C})$.

2.9.2 Lie Groups

A Lie group is a topological group which is also a manifold with some compatibility conditions. Lie groups have algebraic structures and they are also subsets of well known spaces and have a geometry, moreover, they are locally Euclidean in the sense that a small portion of them looks like a Euclidean space making it possible to do analysis on them. Thus Lie groups and other topological groups lie at the crossing of different areas of mathematics. Representations of Lie groups and Lie algebras have revealed many beautiful connections with other branches of mathematics, such as number theory, combinatorics, algebraic topology as well as mathematical physics.

Definition 2.9.2   A topological group G is called a *real Lie group* iff

(i) G is a differentiable manifold;
(ii) the group operations $(x, y) \mapsto xy$ and $x \mapsto x^{-1}$ are both differentiable.

Definition 2.9.3   A topological G is called a *complex Lie group* iff

(i) G is a complex manifold;
(ii) the group operations $(x, y) \mapsto xy$ and $x \mapsto x^{-1}$ are both holomorphic.

The name Lie group is in honor of Sophus Lie. The dimension of a Lie group is defined as its dimension as a manifold.

Some Important Examples of Lie Groups

Example 2.9.2   (i) $(\mathbf{R}^n, +)$ is an n-dimensional Lie group over $\mathbf{R}$.

(ii) $(\mathbf{C}^n, +)$ is an n-dimensional Lie group over $\mathbf{C}$, but it is a $2n$-dimensional Lie group over $\mathbf{R}$.

(iii) $GL(n, \mathbf{R})$ is a real Lie group of dimension n^2.

(iv) $GL(n, \mathbf{C})$ is a complex Lie group of dimension n^2.

(v) $SL(n, \mathbf{R})$ is a real Lie group of dimension $n^2 - 1$.

(vi) $SL(n, \mathbf{C})$ is a complex Lie group of dimension $n^2 - 1$.

2.9.3 Hops's Groups or H-Groups

A pointed topological space is a non-empty topological space with a distinguished element.

Let P and Q be pointed topological spaces and $f, g : P \to Q$ be base point preserving continuous maps. Then f and g are said to be homotopic relative to the base point $p_0 \in P$, denoted by $f \simeq g \operatorname{rel} p_0$, iff there exists a continuous one-parameter family of maps

$$f_t : P \to Q,$$

such that $f_0 = f$, $f_1 = g$ and $f_t(p_0) = f(p_0) = g(p_0)$ for all $t \in I = [0, 1]$. f_t is called a *homotopy* between f and g relative to p_0, denoted by $f_t : f \simeq g \operatorname{rel} p_0$.

Since '$\simeq$' is an equivalence relation; one can speak of homotopy class of maps between two spaces. Thus $[P; Q]$ usually denotes the set of all homotopy classes of base point preserving continuous maps $P \to Q$ (all homotopies are relative to the base point).

Definition 2.9.4 A pointed topological space P is called a *Hopf group* or *H-group* iff $\exists$ a continuous multiplication $\mu : P \times P \to P$ such that

(i) $\mu(c, 1_P) \simeq 1_p \simeq \mu(1_P, c)$;
(ii) $\mu(\mu \times 1_P) \simeq \mu(1_P \times \mu)$;
(iii) $\mu(1_P, \phi) \simeq c \simeq \mu(\phi, 1_P)$,

where $c : P \to p_0 \in P$ is the constant map (p_0 is the base point of P), $1_P : P \to P$ is the identity map and $\phi : P \to P$ is a continuous map and ϕ is called a *homotopy inverse* for P and μ. Then the homotopy class $[c]$ of c is a homotopy identity and $[\phi]$ is homotopy inverse for P and μ.

Clearly, any topological group is an H-group with identity element e as a base point.

Theorem 2.9.1 *Let X be an arbitrary pointed topological space and P be an H-group. Then $[X; P]$ is a group.*

Proof Define for every pair of maps $g_1, g_2 : X \to P$, the product $g_1 g_2 : X \to P$ by $(g_1 g_2)(x) = \mu(g_1(x), g_2(x)) = g_1(x)g_2(x)$; where the right hand multiplication is the multiplication μ in the H-group P. This law of composition carries over to give an operation on homotopy classes such that $[g_1][g_2] = [g_1 g_2]$. As the two homotopic maps determine the same homotopy class, the group axioms for $[X; P]$ follow from (i)–(iii) of Definition 2.9.4. $\square$

Theorem 2.9.2 *If $f : X \to Y$ is a base point preserving continuous map, then f induces a group homomorphism*

$$f^* : [Y; P] \to [X; P] \quad \text{for each H-group } P.$$

Proof Define f^* by $f^*[h] = [h \circ f]$. This map is well defined, since $h_0 \simeq h_1 \Rightarrow h_0 o f \simeq h_1 o f$ (cf. Spanier 1966, Th. 6, p. 24). Verify that f^* is a group homomorphism. $\square$

Let P and P' be H-groups with multiplications μ and μ', respectively. Then a continuous map $\alpha : P \to P'$ is called a *homomorphism* of H-groups, iff $\alpha\mu \simeq \mu'(\alpha \times \alpha)$.

Theorem 2.9.3 *If* $\alpha : P \to P'$ *is a homomorphism between H-groups, then* α *induces a group homomorphism* $\alpha_* : [X; P] \to [X; P']$.

Proof Define $\alpha_* : [X; P] \to [X; P']$ by $\alpha_*[f] = [\alpha \circ f]$. Then α_* is well defined and a homomorphism. $\qquad\square$

Remark For further results, see Appendix B.

2.10 Fundamental Groups

Homotopy theory plays an important role in mathematics. In this section, we define fundamental group of a topological space X based at a point x_0 in X. This group is an algebraic invariant in the sense that the fundamental groups of two homeomorphic spaces are isomorphic. By using the fundamental groups many problems of algebra, topology, and geometry are solved by reducing them into algebraic problems. For example, the fundamental theorem of algebra is proved in Chap. 11 by homotopy theory and the Brouwer Fixed Point Theorem for dimension 2 is proved in Appendix B by the fundamental group.

Definition 2.10.1 Let X be a topological space and x_0 be a fixed point of X, called a base point of X. A continuous map $f : I \to X$ (where $I = [0, 1]$) is called a *path* in X; $f(0)$ and $f(1)$ are called the initial point and the terminal point of the path f, respectively. If $f(0) = f(1) = x_0$, the path f is called a *loop* in X based at x_0. A space X is said to be *path connected*, iff any two points of X can be joined by a path. Let $f, g : I \to X$ be two loops based at x_0. Then they are called *homotopic* relative to 0 and 1, denoted by $f \simeq g$ rel$\{0, 1\}$ iff $\exists$ a continuous map $F : I \times I \to X$ such that

$$F(t, 0) = f(t), \qquad F(t, 1) = g(t), \qquad F(0, s) = F(1, s) = x_0 \quad \forall t, s \in I.$$

The above homotopy relation between loops is an equivalence relation, and this gives the set of homotopy classes relative to 0 and 1 of the loops based at x_0 and this is denoted by $\pi_1(X, x_0)$. Given loops $f, g : I \to X$ based at x_0, their product $f * g$ is a loop $f * g : I \to X$ at x_0 defined by

$$(f * g)(t) = \begin{cases} f(2t), & 0 \leq t \leq \frac{1}{2}, \\ g(2t - 1), & \frac{1}{2} \leq t \leq 1. \end{cases}$$

If $[f]$ and $[g]$ are two elements of $\pi_1(X, x_0)$, define their product $[f] \circ [g] = [f * g]$. This product is well defined and $\pi_1(X, x_0)$ is group under this composition. This group is called the *fundamental group* of X based at x_0.

Clearly, if X is path connected, $\pi_1(X, x_0)$ is independent of its base point x_0.

We now find geometrically the fundamental group of the unit circle S^1 which is the boundary of the unit ball or disc E^2 in $\mathbf{R}^2$. Since S^1 is path connected, its fundamental group $\pi_1(S^1, x_0)$ is independent of its base point x_0. A loop f in S^1 based at x_0 is a closed path starting at x_0 and ending at x_0. Then f is either a null-path (constant path) or f is given by one or more complete description (clockwise or anti-clockwise) around the circle. Let f and g be two loops in S^1 based at the same point x_0 such that f describes S^1, m times and g describes S^1, n times. If $m > n$, then $f * g^{-1}$ is a path describing $(m - n)$ times the circle S^1 and such that it is not homotopic to a null path. Thus the homotopy classes of loops in S^1 based at x_0 are in bijective correspondence with $\mathbf{Z}$. Hence $\pi_1(S^1, x_0)$ is isomorphic to $\mathbf{Z}$.

Proceeding as above, for a solid disc E^2 the fundamental $\pi_1(E^2, x_0) = 0$, since the space E^2 is a convex subspace of $\mathbf{R}^2$.

For an analytical proof use the degree function d defined in (v) below or [see Rotman (1988)].

Let $\dot{I} = \{0, 1\}$ be the end point of the closed interval $I = [0, 1]$. Then a loop in S^1 based at 1 is a continuous map $f : (I, \dot{I}) \to (S^1, 1)$. Let $p : (\mathbf{R}, r_0) \to (S^1, 1)$ be the exponential map defined by $p(t) = e^{2\pi i t}$, $\forall t \in \mathbf{R}$, where $r_0 \in \mathbf{Z}$. We now present some interesting properties of fundamental groups (see Rotman (1988)).

(i) There exists a unique continuous map $\tilde{f} : I \to \mathbf{R}$ such that $p \circ \tilde{f} = f$ and $\tilde{f}(0) = 0$.

(ii) If $g : (I, \dot{I}) \to (S^1, 1)$ is continuous and $f \simeq g$ relative to $\dot{I}$, then $\tilde{f} \simeq \tilde{g}$ relative to $\dot{I}$ and $\tilde{f}(1) = \tilde{g}(1)$, where $p \circ \tilde{g} = g$ and $\tilde{g}(0) = 0$.

(iii) If $f : (I, \dot{I}) \to (S^1, 1)$ is continuous, the degree of f denoted by $\deg f$ is defined by $\deg f = \tilde{f}(1)$, where $\tilde{f}$ is the unique lifting of f with $\tilde{f}(0) = 0$. Then $\deg f$ is an integer.

(iv) If $f(x) = x^n$, then $\deg f = n$ and the degree of a constant loop is 0.

(v) The function $d : \pi_1(S^1, 1) \to \mathbf{Z}$, defined by $d([f]) = \deg f$, is an isomorphism of groups.

(vi) Two loops in S^1 at the base point 1 are homotopic relative to $\dot{I}$ if and only if they have the same degree.

2.10.1 A Generalization of Fundamental Groups

Rodes (1966) introduced the fundamental group $\sigma(X, x_0, G)$ of a *transformation group* (X, G, σ) (see Definition 3.1.1), where X is a path connected space with x_0 as base point. Given an element $g \in G$, a path α of order g with base point x_0 is a continuous map $\alpha : I \to X$ such that $\alpha(0) = x_0$, $\alpha(1) = gx_0$. A path α_1 of order g_1 and a path α_2 of order g_2 give rise to a path $\alpha_1 + g\alpha_2$ of order $g_1 g_2$ defined by the equations

$$(\alpha_1 + g\alpha_2)(s) = \begin{cases} \alpha_1(2s), & 0 \leq s \leq \frac{1}{2}, \\ g_1\alpha_2(2s - 1), & \frac{1}{2} \leq s \leq 1. \end{cases}$$

Two paths α and β of the same order g are said to be homotopic iff $\exists$ a continuous map $F : I \times I \to X$ such that

$$F(s, 0) = \alpha(s), \quad 0 \leq s \leq 1;$$
$$F(s, 1) = \beta(s), \quad 0 \leq s \leq 1;$$
$$F(0, t) = x_0, \quad 0 \leq t \leq 1;$$
$$F(1, t) = gx_0, \quad 0 \leq t \leq 1.$$

The homotopy class of a path α of order g is denoted by $[\alpha; g]$. Prove that the set of homotopy classes of paths of prescribed order with rule of composition 'o' form a group where o is defined by

$$[\alpha; g_1] \circ [\beta; g_2] = [\alpha + g_1\beta; g_1g_2].$$

This group denoted by $\sigma(X, x_0, G)$ is called the *fundamental group* of (X, G, σ) with base point x_0.

Problems and Supplementary Examples (SE-I)

1 Show that $(\mathbf{Q}, +)$ is not finitely generated ($+$ denotes usual addition of rational numbers).

Solution If possible, suppose $(\mathbf{Q}, +)$ is finitely generated. Then there exists a finite set

$$S = \left\{ \frac{p_1}{q_1}, \frac{p_2}{q_2}, \ldots, \frac{p_n}{q_n} \right\}$$

of rational numbers such that $\mathbf{Q} = \langle S \rangle$. Now we can find a prime p such that p does not divide $q_1, q_2, \ldots, q_n$. Let $x \in \mathbf{Q}$. Then there exist integers $m_1, m_2, \ldots, m_n$ such that

$$x = m_1 \frac{p_1}{q_1} + m_2 \frac{p_2}{q_2} + \cdots + m_n \frac{p_n}{q_n} = \frac{m}{q_1 q_2 \cdots q_n}$$

for some integer m. Since p does not divide $q_1, q_2, \ldots, q_n$, p does not divide $q_1 q_2 \ldots q_n$. Hence p does not divide the denominator of any rational number (expressed in lowest terms) of $\langle S \rangle$. This shows that $\frac{1}{p} \notin \langle S \rangle = \mathbf{Q}$. This yields a contradiction. Hence $(\mathbf{Q}, +)$ is not finitely generated.

2 Let S_n be the symmetric group on $\{x_1, x_2, \ldots, x_n\}$ with identity e. For $i = 1, 2, \ldots, n-1$, write σ_i for the permutation of S which transposes x_i and x_{i+1} and keep the remaining elements fixed. Then show that

(i) $\sigma_i^2 = e$ for $i = 1, 2, \ldots, n-1$;

(ii) $\sigma_i \sigma_j = \sigma_j \sigma_i$ if $|i - j| > 2$;

(iii) $(\sigma_i \sigma_{i+1})^3 = e$ for $1 \leq i \leq n-2$;

(vi) $\sigma_1, \sigma_2, \ldots, \sigma_{n-1}$ generate the group S_n.

(v) The relations (i)–(iii) between the generators determine the Cayley table of S_n completely (these relations are called the defining relations of S_n).

3 Consider as a conjecture. "Let H be a subgroup of a finite group G. If H is abelian, then H is normal in G." Disprove the conjecture by reference to the Octic group and one of its subgroups.

4 Show by an example that the converse of Lagrange's Theorem is not true for arbitrary finite groups.

Solution We show that the alternating group A_4 of order 12 has no subgroup H of order 6. If possible, let $|H| = 6$. All the following eight 3-cycles (1 2 3), (1 3 2), (1 2 4), (1 4 2), (1 3 4), (1 4 3), (2 3 4), (2 4 3) are in A_4. As $|H| = 6$, H cannot contain all these elements. Let $\beta = (x\ y\ z)$ be a 3-cycle in A_4 such that $\beta \notin H$. Let $K = \{\beta, \beta^2, \beta^3 = 1\}$. Then K is a subgroup of A_4. Hence $H \cap K = \{1\}$ shows that $HK = \frac{|H||K|}{|H\cap K|} = 18$. But $HK \subseteq A_4$ implies a contradiction.

Problems and Supplementary Examples (SE-II)

1 Let G be a group and H be a subgroup of G. Show that $\forall g \in G$, $gH = H \Leftrightarrow g \in H$.

Solution Let $g \in G$ and $gH = H$. Then $g = g \cdot 1 \in gH = H \Rightarrow g \in H$. Thus $\forall g \in G$, $gH = H \Rightarrow g \in H$. Conversely, suppose $g \in H$. Then for $\forall h \in H$, $gh \in H \Rightarrow gH \subseteq H$. Again $g, h \in H \Rightarrow g^{-1}h \in H \Rightarrow h = g(g^{-1}h) \in gH \Rightarrow H \subseteq gH$. Thus for any $g \in H$, $gH = H$.

2 Show that any factor group of a cyclic group is cyclic.

Solution Let $G = \langle a \rangle$ be a cyclic group and N be a normal subgroup of G. Then computation of all powers of aN amounts to computing in G, all powers of the representative a and these powers give all the elements of G. Hence the powers of aN give all the cosets of N and thus $G/N = \langle aN \rangle \Rightarrow G/N$ is cyclic.

3 Find the kernel of the homomorphism: $f : (\mathbf{R}, +) \to (C^*, \cdot)$ defined by $f(x) = e^{ix}$ and hence find the factor group $\mathbf{R}/\ker f$.

Solution

$$\ker f = \left\{x \in \mathbf{R} : e^{ix} = 1\right\} = \{x \in \mathbf{R} : \cos x + i \sin x = 1\}$$
$$= \{x \in \mathbf{R} : x = 2\pi n, n \in \mathbf{Z}\} = \langle 2\pi \rangle.$$

Thus $\mathbf{R}/\ker f = \mathbf{R}/\langle 2\pi \rangle \cong \mathrm{Im}\, f = S^1$ (circle group in C^*, see Example 2.3.2).

4 Show that a group having only a finite number of subgroups is finite.

Solution If $G = \{1\}$, then G is finite. Suppose $G \neq \{1\}$. Then $\exists$ an element $a\ (\neq 1) \in G$. Thus $H = \langle a \rangle$ is a finite subgroup of G.

Otherwise, $(H, \cdot) \cong (\mathbf{Z}, +)$ and $(\mathbf{Z}, +)$ has infinite number of subgroups viz. $\langle n \rangle$ (by Theorem 2.4.10), $n = 0, 1, 2, \ldots$ would imply that H has an infinite number of subgroups contradicting our hypothesis. Thus $O(a)$ is also finite. If $G = \langle a \rangle$, the proof ends. If $G \neq \langle a \rangle$, then $\exists$ only a finite number of elements in $(G - \langle a \rangle)$. Otherwise, for each $g \in (G - \langle a \rangle)$, $\langle g \rangle$ is a subgroup of $G \Rightarrow \exists$ an infinitely many non-trivial subgroups of $G \Rightarrow$ a contradiction of hypothesis. Consequently, G is finite.

5 Let G be a finite group and H be a proper subgroup of G such that for $x, y \in G - H, xy \in H$. Prove that H is a normal subgroup of G such that $|G|$ is an even integer.

Solution $g \in H \Rightarrow ghg^{-1} \in H$, $\forall h \in H$. Again $g \notin H \Rightarrow hg^{-1} \notin H$, $\forall h \in H$. This is so because $hg^{-1} \in H$ and $h, h^{-1} \in H \Rightarrow h^{-1}(hg^{-1}) \in H \Rightarrow g^{-1} \in H \Rightarrow g \in H \Rightarrow$ a contradiction. Thus $g, hg^{-1} \in G - H \Rightarrow g(hg^{-1}) \in H$ by hypothesis $\Rightarrow ghg^{-1} \in H$. Hence it follows that H is a normal subgroup of G. Thus G/H is a group. We now show that $|G|$ is of even order. By hypothesis, $\exists x \in G$ such that $x \notin H$ but $x^2 \in H$. Hence $x^2 H = H \Rightarrow (xH)^2 = eH$. Again $x \notin H \Rightarrow xH \neq eH$. Thus $O(xH) = |xH| = 2$. Now $|xH|$ divides $|G/H| \Rightarrow 2$ divides $|G/H|$. Hence $|G/H|$ is an even integer $\Rightarrow |G| = |G/H| \cdot |H|$ is an even integer.

6 Let G be a finite group and H, K be subgroups of G such that $K \subseteq H$. Then $[G : K] = [G : H][H : K]$.

Solution $H = \bigcup_i x_i K$ and $G = \bigcup_j y_j H$, where both are disjoint union $\Rightarrow G = \bigcup_{i,j} y_j x_i K$ is also a disjoint union $\Rightarrow [G : K] = [G : H][H : K]$.

7 Let G be a finite cyclic group of order n. Show that corresponding to each positive divisor d of n, $\exists$ a unique subgroup of G of order d.

Solution Let $G = \langle g \rangle$ for some $g \in G$ and d a positive divisor of n. Then $n = md$ for some $m \in \mathbf{Z}$.

Now $g^m \in G \Rightarrow O(g^m) = \frac{O(g)}{\gcd(m,n)} = \frac{n}{m} = d$ by Theorem 2.3.5.

Let $H = \langle g^m \rangle$. Then H is subgroup of G of order d.

Uniqueness of H: Let K be a subgroup of G of order d. Let t be the smallest positive integer such that $g^t \in K$. Then $K = \langle g^t \rangle$. Now $|K| = d \Rightarrow O(g^t) = d \Rightarrow d = \frac{O(g)}{\gcd(t,n)} = \frac{n}{\gcd(t,n)}$ by Theorem 2.3.5 $\Rightarrow \gcd(t,n) = \frac{n}{d} = m \Rightarrow m|t$. If $t = ml$ for some $l \in \mathbf{Z}$, then $g^t = g^{ml} = (g^m)^l \in H \Rightarrow K \subseteq H \Rightarrow K = H$, since $|H| = |K|$.

8 Let G be a finite group and $f : G \to \mathbf{Z}_{15}$ be an epimorphism. Show that G has normal subgroups with indices 3 and 5.

Solution $G/\ker f \cong \mathbf{Z}_{15}$ by the First Isomorphism Theorem. Then $\mathbf{Z}_{15}$ being a cyclic group of order 15, $G/\ker f$ is a cyclic group of order $15 \Rightarrow G/\ker f$ has a normal subgroup N_1 of order 5 and a normal subgroup N_2 of order 3 $\Rightarrow \exists$ normal subgroups H_1 and H_2 of G such that $\ker f \subseteq H_1$, $\ker f \subseteq H_2$, $H_1/\ker f = N_1$ and $H_2/\ker f = N_2$ by the Correspondence Theorem. Hence $15 = |G/\ker f| = [G : \ker f] = [G : H_1][H_1 : \ker f]$ (by Ex. 6 of SE-II) $= [G : H_1] \cdot 5 \Rightarrow [G : H_1] = 3$. Similarly, $[G : H_1] = 5$.

9 Prove that the group $\mathbf{Z}_m \times \mathbf{Z}_n$ is isomorphic to the group $\mathbf{Z}_{mm} \Leftrightarrow \gcd(m, n) = 1$ i.e., $\mathbf{Z}_m \times \mathbf{Z}_n \cong \mathbf{Z}_{mn} \Leftrightarrow m, n$ are relatively prime.

Solution $\mathbf{Z}_m$ and $\mathbf{Z}_n$ are cyclic groups of order m and n, respectively. If $\gcd(m, n) = 1$, then $\mathbf{Z}_m \times \mathbf{Z}_n$ is a cyclic group of order mn and hence isomorphic to $\mathbf{Z}_{mn}$ (see Theorems 2.4.12 and 2.4.13). Conversely, let $\mathbf{Z}_{mn} \cong \mathbf{Z}_m \times \mathbf{Z}_n$. If $\gcd(m, n) = d > 1$, then $mn = d \cdot l\,\mathrm{cm}(m, n) \Rightarrow \frac{mn}{d} = l\,\mathrm{cm}(m, n) \Rightarrow \frac{mn}{d} = t$ is divisible by both m and n. Now for $(x, y) \in \mathbf{Z}_m \times \mathbf{Z}_n$, $(x, y) + (x, y) + \cdot + (x, y)$ (t summand) $= (0, 0) \Rightarrow$ no element $(x, y) \in \mathbf{Z}_m \times \mathbf{Z}_n$ can generate the entire group $\mathbf{Z}_m \times \mathbf{Z}_n$. Hence $\mathbf{Z}_m \times \mathbf{Z}_n$ cannot be cyclic and therefore cannot be isomorphic to $\mathbf{Z}_{mn}$. So we reach a contradiction. Hence $\gcd(m, n) = 1$.

10 If $n = p_1^{n_1} p_2^{n_2} \cdots p_n^{n_r}$, where p_i's are distinct primes, $n_i \geq 0$, show that $\mathbf{Z}_n$ is isomorphic to $\mathbf{Z}_{p_1}^{n_1} \times \mathbf{Z}_{p_2}^{n_2} \times \cdots \times \mathbf{Z}_{p_r}^{n_r}$.

Solution The result follows from Ex. 9 of SE-II by an induction argument.

11 Prove that a finitely generated group cannot be expressed as the union of an ascending sequence of its proper subgroups.

Solution Let G be a finitely generated group $G = \langle a_1, a_2, \ldots, a_n \rangle$. If possible, let $G = \bigcup_i A_i$, where $A_1 \subseteq A_2 \subseteq \cdots$ and each A_i is a proper subgroup of $G, i = 1, 2, \ldots$. Then $\exists$ a positive integer t such that $a_1, a_2, \ldots, a_n \in A_t$. Then $G = \langle a_1, a_2, \ldots, a_n \rangle \subseteq A_t$. Again $A_t \subseteq G$. Hence $G = A_t \Rightarrow$ a contradiction, since A_t is a proper subgroup of G.

12 Show that every finitely generated subgroup of $(\mathbf{Q}, +)$ is cyclic.

Solution Let H be a finitely generated subgroup of $(\mathbf{Q}, +)$. Then any element x of H is of the form

$$x = \frac{m}{q_1 q_2 \cdots q_n} \text{ for some integer } m \text{ (proceed as Ex. 1 of SE-I)}$$

$$\Rightarrow \quad H \subseteq \left\langle \frac{1}{q_1 q_2 \cdots q_n} \right\rangle = K \text{ (say)}$$

$$\Rightarrow \quad K \text{ is a cyclic group generated by } \frac{1}{q_1 q_2 \cdots q_n} \in \mathbf{Q}$$

$$\Rightarrow \quad H \text{ is a cyclic group.}$$

13 Prove that every finite subgroup of $(\mathbf{C}^*, \cdot)$ is cyclic.

Solution Let $(H, \cdot)$ be a finite subgroup of $(\mathbf{C}^*, \cdot)$. Let $|H| = n$ and $x \in H$. Then $x^n = 1$.

The nth roots of $x^n = 1$ are given by

$$1, w, w^2, \ldots, w^{n-1}, \quad \text{where } w = \cos\frac{2k\pi}{n} + i\sin\frac{2k\pi}{n}, \; k = 0, 1, 2, \ldots, n-1.$$

Then each of the roots belongs to H and $O(w) = n = |H|$. Hence $H = \langle w \rangle \Rightarrow H$ is cyclic.

14 Prove that the alternating group A_n of order n (≥ 3) is a normal subgroup of S_n (see Ex. 58 of Exercises-III).

Solution Consider the subgroup $G = (\{1, -1\}, \cdot)$ of $(\mathbf{R}^*, \cdot)$. Define a mapping $\psi : S_n \to G$ by

$$\psi(\sigma) = \begin{cases} 1 & \text{if } \sigma \text{ is an even permutation,} \\ -1 & \text{if } \sigma \text{ is an odd permutation.} \end{cases}$$

Then ψ is an epimorphism such that $\ker \psi = \{\sigma \in S_n : \psi(\sigma) = 1\} = A_n$.
Hence A_n is a normal subgroup of S_n by Theorem 2.6.5.

15 Let H be a normal subgroup of a group G such that H and G/H are both cyclic. Show that G is not in general cyclic.

Solution Take $G = S_3$ and $H = \{e, (1\ 2\ 3), (1\ 3\ 2)\}$.
Then $|G/H| = 2 \Rightarrow H$ is a normal subgroup of G. Hence G/H is also a group. Since $|G/H| = 2$ and $|H| = 3$, which are both prime, G/H and H are both cyclic groups. But G is not commutative $\Rightarrow S_3$ is not cyclic.

16 Let G be a non-cyclic group of order p^2 (p is a prime integer) and $g(\neq e) \in G$. Show that $O(g) = p$.

Solution $O(g)||G| = p^2 \Rightarrow O(g) = 1, p$ or p^2. Now $g \neq e \Rightarrow O(g) \neq 1$. If $O(g) = p^2$, then G contains an element g such that $O(g) = |G|$. Hence G is cyclic $\Rightarrow$ a contradiction of hypothesis $\Rightarrow O(g) \neq p^2$. Thus $O(g) \neq 1, O(g) \neq p^2 \Rightarrow O(g) = p$.

2.11 Exercises

Exercises-III

1. Prove that a semigroup S is a group iff for each $a \in S$, $aS = Sa = S$.

2. Let S be a semigroup. A non-empty subset A of S is called a pseudo-ideal (left, right) iff $ab \in A \ \forall a, b \in A$; $x^2 A \subseteq A$ and $Ax^2 \subseteq A (x^2 A \subseteq A, Ax^2 \subseteq A)$ for every $x \in S$. Show that a semigroup S is a group iff $x(A \backslash B)x \subseteq A \backslash B$ for every $x \in S$ when $A \backslash B \neq \emptyset$, where A, B are either both left pseudo-ideals or both right pseudo-ideals of S.

3. Show that (a) in the semigroup $(\mathbf{Z}^*, \cdot)$ (of all non-zero integers under usual multiplication), $A = \{a \in \mathbf{Z}^* : a \geq 6\}$ is a pseudo-ideal but not an ideal of $\mathbf{Z}^*$.

   In the multiplicative group $(\mathbf{Q}^*, \cdot)$ of all non-zero rational numbers, $A = \{r^2 : r \in \mathbf{Q}^*\}$ is a proper pseudo-ideal of $\mathbf{Q}^*$.

   (This example shows that a group may contain a proper pseudo-ideal.)

4. Define a binary operation o on $\mathbf{Z}$ by $a \circ b = a + b - 2$. Show that $(\mathbf{Z}, \circ)$ is a commutative group.

5. Let G be the set of all rational numbers excepting -1. Define a binary operation '$\circ$' on G by $a \circ b = a + b + ab$. Show that $(G, \circ)$ is a group.

6. Define a binary operation '$\circ$' on the set of non-zero rational numbers by $a \circ b = |a|b$. Is $(\mathbf{R}^*, \circ)$ a group? Justify your answer.

7. Let

$$G = \left\{ \begin{pmatrix} a & b \\ c & d \end{pmatrix} : a, b, c, d \in \mathbf{R} \text{ (or } \mathbf{Z}) \text{ and } ad - bc = 1 \right\}.$$

Show that G is a group under usual multiplication. Is this group commutative?

8. (i) A semigroup (X, o) is said to be cancellative iff for each $x, y, z \in X, x \circ y = x \circ z \Rightarrow y = z$ and $y \circ x = z \circ x \Rightarrow y = z$. Show that a finite cancellative semigroup $(X, \circ)$ is a group.

   Does this result hold if the semigroup is not finite?

   (ii) Show that a cancellative semigroup can be extended to a group iff it is commutative.

9. Let $SL(n, \mathbf{R})$ $(SL(n, \mathbf{C}))$ be the set of all $n \times n$ $(n > 1)$ unimodular matrices A (i.e., $\det A = 1$) with real (complex) entries. Show that $(SL(n, \mathbf{R}), \cdot)$ $((SL(n, \mathbf{C}), \cdot))$ is a group under usual multiplication.

10. Let U_n $(n > 1)$ denote the set of all $n \times n$ complex matrices A such that $A\bar{A}^T = I$. Show that $(U_n, \cdot)$ is a non-commutative group and is a subgroup of $GL(n, \mathbf{C})$ (under usual multiplication). ($(U_n, \cdot)$ is called *unitary group*.)

    [*Hint.* $A\bar{A}^T \Rightarrow I \Rightarrow |\det A|^2 = 1$.] Show that in particular, the set SU_n of all unitary matrices A with $\det A = 1$ is a non-commutative group (known as *special unitary group*).

11. Let O_n $(n > 1)$ denote the set of all orthogonal $n \times n$ real matrices A (i.e., $AA^T = I$). Show that $(O_n, \cdot)$ is a non-commutative group (under usual multiplication).

12. Let SO_n be the special *orthogonal group* consisting of all orthogonal matrices A such that $\det A = 1$. Show that $(SO_n, \cdot)$ is a non-commutative group but $(SO_n, +)$ is not a group (where '$\cdot$' and '$+$' denote usual multiplication and addition of matrices respectively).

    [*Hint.* $(SO_n, +)$ does not contain identity.]

13. Let $(G, \cdot)$ be a group and S, T subsets of G. Then define the product $S \circ T$ as

$$S \circ T = \begin{cases} z \in G : z = xy & \text{for some } x \in S, y \in T \\ \emptyset & \text{if either } S \text{ or } T \text{ is empty.} \end{cases}$$

Then the inverse of the set S denoted by S^{-1} is defined by

$$S^{-1} = \begin{cases} z \in G : z = x^{-1} & \text{for some } x \in S \\ \emptyset & \text{iff } S = \emptyset. \end{cases}$$

Show that if S, T, W are subsets of G, then

(a) $(S \circ T) \circ W = S \circ (T \circ W)$;

(b) $(S \circ T)^{-1} = T^{-1} \circ S^{-1}$;

(c) $S \circ T \neq T \circ S$ (in general).

14. Show that a non-empty subset T of a group $(G, \cdot)$ is a subgroup of G iff $T \circ T \subseteq T$ and $T^{-1} \subseteq T$ or equivalently, $T \circ T^{-1} \subseteq T$.

15. Let S, T be subgroups of a group $(G, \cdot)$. Show that $S \circ T$ is a subgroup of G iff $S \circ T = T \circ S$.

16. Let G be a group and T_1, T_2 two proper subgroups of G.

    Show that $T_1 \cup T_2$ is a subgroup of G iff $T_1 \subseteq T_2$ or $T_2 \subseteq T_1$. Hence show that a group cannot be the union of two proper subgroups.

17. Let $(G, \cdot)$ be a group and S a non-empty subset of G.

    Show that the subgroup $\langle S \rangle$ generated by S consists of all elements of the form $x_1 x_2 \cdots x_n$, where $x_i \in S \cup S^{-1}$ for all integers $n \geq 1$ and, moreover, if S is a countable subset of G, then $\langle S \rangle$ is countable.

    [*Hint.* T be the set of all finite products of the given form $x_1 x_2 \cdots x_n$. Taking $n = 1$ and allowing x_1 run over S, it follows that $S \subseteq T$. Suppose $a = x_1 x_2 \cdots x_n$ and $b = y_1 y_2 \cdots y_m \in T$. Now $ab^{-1} = x_1 x_2 \cdots x_n x_{n+1} x_{n+2} \cdots x_{n+m}$, where $x_{n+i} = (y_{m-i+1})^{-1}$.

    But $y_j \in S \cup S^{-1} \Rightarrow y_{j-1} \in S \cup S^{-1}$. Consequently, $x_{n+i} \in S \cup S^{-1} \Rightarrow x_i \in S \cup S^{-1}, i = 1, 2, \ldots, m + n \Rightarrow ab^{-1} \in T \Rightarrow T$ is a subgroup of G. Now $S \subseteq T \Rightarrow \langle S \rangle \subseteq T$. Since $S \subseteq \langle S \rangle$ and $\langle S \rangle$ is a subgroup, $S^{-1} \subseteq \langle S \rangle$.

    $S \cup S^{-1} \subseteq \langle S \rangle$. By closure property and induction, it follows that any finite product of elements of $S \cup S^{-1}$ is also contained in $\langle S \rangle$ i.e., $T \subseteq \langle S \rangle$.]

18. Show that $(\mathbf{R}, +)$ is not finitely generated ($+$ denotes usual addition of reals).

    [*Hint.* If possible, $\exists$ a finite subset $S \subset \mathbf{R}$ such that $\langle S \rangle = \mathbf{R}$. Then S is countable $\Rightarrow \mathbf{R}$ is countable $\Rightarrow$ a contradiction.]

19. The *dihedral group* D_n: The group D_n is completely determined by two generators s, t; and the defining relations $t^n = 1, s^2 = 1, (ts)^2 = 1$, and $t^k \neq 1$ for $0 < k < n$, where 1 denotes the identity element. Find $|D_n|$. Construct the Cayley table for the group D_3. Interpret D_n as a group of symmetries of a regular n-gon.

20. The *quaternion group* H. This group is completely determined by two generators s, t; and the defining relations $s^4 = 1, t^2 = s^2, ts = s^3 t$. Construct its Cayley table.

21. (a) *External direct product of groups.* Let H and K be groups. Show that

   (i) $H \times K$ is a group under binary operation: $(h, k)(h', k') = (hh', kk')$
       $\forall h, h' \in H$ and $k, k' \in K$;
  (ii) If $H \neq \{1\}$ and $K \neq \{1\}$, $H \times K$ is neither isomorphic to H nor to K;
 (iii) $H \times K$ is an abelian group iff H and K are both abelian groups.

   (b) *Internal direct product of subgroups.* Let G be a group and H, K be
normal subgroups of G such that $G = HK$ and $H \cap K = \{1\}$. Then G is called
the internal direct product of H and K. If G is an additive abelian group, then
it is called the internal direct sum of H and K and denoted by $G = H \oplus K$.
   Prove the following:

   (i) $G = H \oplus K$ iff every element $x \in G$ can be expressed uniquely as $x =$
       $h + k$ for some $h \in H$ and $k \in K$.
  (ii) Let G be the internal direct product of normal subgroups H and K. Then
       G is isomorphic to $H \times K$.
 (iii) If $G = H \times K$, then H and K are isomorphic to suitable subgroups $\bar{H}$ and
       $\bar{K}$ of G respectively, such that G is the internal direct product of $\bar{H}$ and
       $\bar{K}$.
 (iv) Let $f : G \rightarrow H$ and $g : H \rightarrow K$ be homomorphisms of abelian groups
      such that $g \circ f$ is an isomorphism. Then $H = \text{Im} \, f \oplus \ker f$.

22. *Klein four-group* $C \times C$: If C is the cyclic group of order 2 generated by g, show
    by using Ex. 21 that $C \times C$ is a group of order 4. Work out the multiplication
    table for $C \times C$. Show that the cyclic group of order 4, $C_4 = \{1, a, a^2, a^3\}$,
    where $a^4 = 1$ and the group $C \times C$ are not isomorphic.

    [*Hint.* All elements other than $(1, 1)$ of $C \times C$ are of order 2 but it is not true
    for C_4.]

    Remark The Klein 4-group is named after Felix Klein (1849–1925). This group
    may also be defined as the set $\{1, a, b, c\}$ together with the multiplication de-
    fined by the following table:

$\bullet$	1	a	b	c
1	1	a	b	c
a	a	1	c	b
b	b	c	1	a
c	c	b	a	1

23. (a) Show that a finite group G of order r is cyclic iff it contains an element g of
    order r.
       (b) Let $\langle a \rangle$ be a finite cyclic group of order n. Show that $\langle a \rangle = \{a, a^2, \ldots,$
    $a^{n-1}\}$.
24. Show that every element a $(\neq 1)$ in a group of prime order p has period p.
25. Show that a group in which every element a $(\neq 1)$ has period 2 is abelian.
26. Show that every group of order <6 is abelian.

27. Let G be a non-abelian group of order 10. Prove that G contains at least one element of period 5.
28. If G is a group of order p^2 (p is prime >1), show that G is abelian (see Ex. 3. of Exercises-I, Chap. 3).
29. Let A and B be different subgroups of a group G such that both A and B are of prime order p. Show that $A \cap B = \{1\}$.
30. Let G be a group of order 27. Prove that G contains a subgroup of order 3.
31. Let G be a finite group and A, B subgroups of G such that $B \subseteq A$. Show that $[G : A]$ is a factor of $[G : B]$.
32. Let f be a homomorphism from G to G', where G and G' are finite groups. Show that $|\operatorname{Im} f|$ divides $|G'|$ and $\gcd(|G|, |G'|)$.
    [*Hint.* Use Theorem 2.4.5(i) and Theorem 2.5.1.]
33. Show that

    (i) two cyclic groups of the same order are isomorphic;
    (ii) every cyclic group of order n is isomorphic to $\mathbf{Z}_n$ (cf. Theorem 2.4.12);
    (iii) $\mathbf{Z}_4$ is cyclic and both 1 and 3 are generators;
    (iv) $\mathbf{Z}_p$ has no proper subgroup if p is a prime integer;
    (v) an infinite cycle group has exactly two generators.

34. Find all the subgroups of $\mathbf{Z}_{18}$ and give their lattice diagram.
35. (a) Let G be a group of order pq, where p and q are primes. Verify that every proper subgroup of G is cyclic.
    (b) Prove that finite groups of rotations of the Euclidean plane are cyclic.
36. Let G be a finite group of order 30. Can G contain a subgroup of order 9? Justify your answer.
37. If the index of a subgroup H in a group G is 2, prove that $aH = Ha \; \forall a \in G$ and G/H is isomorphic to a cyclic group of order 2.
38. Give an example of a subgroup H in a group G satisfying $aH = Ha \; \forall a \in H$ but $[G : H] > 2$.
39. If H is a subgroup of finite index in G, prove that there is only a finite number of distinct subgroups of G of the form aHa^{-1}.
40. Show that the order of an element of finite group G divides the order of G.
    [*Hint.* The order of an element $a \in G$ is the same as the order of $\langle a \rangle$ (by Theorem 2.4.8) and then apply Lagrange's Theorem.]
41. Let G be a group and $x \in G$ be such that x is of finite order m. If $x^r = 1$, show that $m | r$.
42. (i) Let G be a cyclic group of prime order p. Show that G has no proper subgroups.
    (ii) Show that the only non-trivial groups which have no proper subgroups are the cyclic groups of prime order p.
43. Let $(G, \cdot)$, $(H, \cdot)$ and $(K, \cdot)$ be groups and $f : G \to H$ and $g : H \to K$ be group homomorphisms. Show that

    (i) $g \circ f : G \to K$ is also a homomorphism. Further, if f, g are both monomorphisms (epimorphisms), then $g \circ f$ is also a monomorphism (epimorphism);

(ii) in particular, if f and g are both isomorphisms, then $g \circ f$ is also an isomorphism;

(iii) further, if $f : G \to H$ is an isomorphism, then $f^{-1} : H \to G$ is also an isomorphism.

44. Let P be the set of all polynomials with integral coefficients. Show that $(P, +)$ is an abelian group isomorphic to the multiplicative group $(\mathbf{Q}^+, .)$ of positive rationals.

[*Hint.* Let $\{p_n\}_{n=0}^{\infty}$ be an *enumeration of* positive rationals *in increasing order*: $p_0 < p_1 < p_2 < \cdots < p_n < \cdots$.

Define $f : P \to \mathbf{Q}^+$ as follows: for $p(x) = a_0 + a_1 x + a_2 x^2 + \cdots + a_n x^n \in P$, $f(p(x)) = p_0^{a_0} p_1^{a_1} p_2^{a_2} \cdots p_n^{a_n} \in \mathbf{Q}^+$.

Verify that f is a homomorphism with $\ker f = \{a_0 + a_1 x + \cdots + a_n \in P : p_0^{a_0} p_1^{a_1} \cdots p_n^{a_n} = 1\} = \{0\}$. Check that f is an *isomorphism*.]

45. A homomorphism of a group into itself is called an endomorphism. Let $(\mathbf{C}^*, \cdot)$ be the multiplicative group of non-zero complex numbers. Define $f : (\mathbf{C}^*, \cdot) \to (\mathbf{C}^*, \cdot)$ by $f(z) = z^n$ (n is a positive integer). Show that f is an endomorphism. Find $\ker f$.

46. Let G be a group. Show that a mapping $f : G \to G$ defined by $x \mapsto x^{-1}$ is an automorphism of G iff G is abelian.

47. Let G be a group and $a \in G$. Then the mapping $f_a : G \to G$ defined by $f_a(x) = a^{-1}xa$ ($\forall x \in X$) is an automorphism of G (called an inner automorphism induced by a) and $f_1 = 1_G$ is an inner automorphism (called a trivial automorphism). Show that for an abelian group G, the only inner automorphism is 1_G but for a non-abelian group there exist non-trivial automorphisms.

48. Show that

(a) the map $f : (\mathbf{R}^+, \cdot) \to (\mathbf{R}, +)$ defined by $f(x) = \log_e x$ is an isomorphism;
(b) the map $g : (\mathbf{R}, +) \to (\mathbf{R}^+, \cdot)$ defined by $g(x) = e^x$ is an isomorphism.

[*Hint.* Check that f and g are homomorphisms such that $g \circ f = 1$ and $f \circ g = 1$].

49. Show that every homomorphic image of an abelian group is abelian but the converse is not true.

50. Show that

(a) a homomorphism from any group to a simple group is either trivial or surjective;
(b) a homomorphism from a simple group is either trivial or one-to-one.

51. Prove the following:

$$\text{Aut}\, \mathbf{Z}_6 \cong \mathbf{Z}_2 \quad \text{and} \quad \text{Aut}\, \mathbf{Z}_5 \cong \mathbf{Z}_4$$

(see Theorem 2.3.4).

52. (a) If G is a cyclic group and $f : G \to G'$ is a homomorphism of groups, show that $\text{Im}\, f$ is cyclic.

(b) Let G be a group such that every cyclic subgroup of G is normal in G. Show that every subgroup of G is a normal subgroup of G.

53. If G is a cyclic group and a, b are generators of G. Show that there is an automorphism of G which maps a onto b and also any automorphism of G maps a generator onto a generator.

54. Let $A(S)$ be the permutation group on $S(\neq \emptyset)$. If two permutations $f, g \in A(S)$ are such that for any $x \in S$, $f(x) \neq x \Rightarrow g(x) = x$ and for any $y \in S$, $g(y) \neq y \Rightarrow f(y) = y$, (then f and g are called disjoint), examine the validity of the equality $fg = gf$.

55. Show that every permutation can be expressed as a product of disjoint cycles (the number of cycle may be 1), where the component cycles are uniquely determined except for their order of combination.

56. Show that a permutation σ can be expressed as a product of transpositions (i.e., cycles of order 2) in different ways but the number of transpositions in such a decomposition is always odd or always even.

   (σ is said to be an odd or even permutation according as r is odd or even, where r is the number of transpositions in a decomposition.)

57. Show that in a permutation group G, either all permutations are even or exactly half the permutations are even, which form a subgroup of G.

58. Show that all even permutations of a set of n distinct elements ($n \geq 3$) form a normal subgroup A_n of the symmetric group S_n (A_n is called the *alternating group* of degree n).

   Find its order (see Ex. 14 of SE-II).

59. (i) Show that the alternating group A_n is generated by the following cycles of degree 3: $(123), (124), \ldots, (12n)$.

   (ii) If a normal subgroup H of the alternating group A_n contains a cycle of degree 3, show that $H = A_n$.

   (iii) Show that the groups S_3 and $\mathbf{Z}_6$ are not isomorphic but for every proper subgroup G of S_3, there exists a proper subgroup H of $\mathbf{Z}_6$ such that $G \cong H$.

60. Let G be a group and $a, b \in G$. The element $aba^{-1}b^{-1}$ denoted by $[a, b]$ is called a commutator of G. The subgroup of G whose elements are finite products of commutators of G is called the commutator subgroup of G. Denote this subgroup by $C(G)$. Show that

   (i) $[a, b][b, a] = 1$;
   (ii) G is an abelian group iff $C(G) = \{1\}$;
   (iii) $C(G)$ is a normal subgroup of G;
   (iv) the quotient group $G/C(G)$ is abelian;
   (v) if H is any normal subgroup of G such that G/H is abelian, then $C(G) \subset H$;
   (vi) $C(S_3) = A_3$.

61. A group G is called a *quaternion* group iff G is generated by two elements $a, b \in G$ satisfying the relations:

   $$O(a) = 4, \qquad a^2 = b^2 \quad \text{and} \quad ba = a^3 b \quad (\text{see Ex. 20}).$$

   Let G be a subgroup of the General Linear group $GL(2, \mathbf{C})$ of order 2 over $\mathbf{C}$, generated by $A = \begin{pmatrix} 0 & 1 \\ -1 & 0 \end{pmatrix}$ and $B = \begin{pmatrix} 0 & i \\ i & 0 \end{pmatrix}$. Show that G is a quaternion group.

62. Prove the following:

   (a) A semigroup S is a semilattice of groups iff S is a regular semigroup, all of whose one sided ideals are two sided.
   (b) A semigroup S is a semilattice of left groups iff S is a regular semigroup, all of whose left ideals are two sided.
   (c) (i) A semigroup S is a semilattice of left groups iff the condition $I \cap L \cap R = RLI$ holds for any two-sided ideal I, every ideal left L and right ideal R of S;
       (ii) a semigroup S is a semilattice of right groups iff the condition $I \cap L \cap R = IRL$ holds for every two-sided ideal I, every left ideal L and every right ideal R of S.

63. A *Clifford* semigroup is a regular semigroup S in which the idempotents are central i.e., in which $ex = xe$ for every idempotent $e \in S$ and every $x \in S$.

   Prove that a semigroup S is a Clifford semigroup iff S is a semilattice of groups.

64. Prove the following:

   (i) Finite groups of rotations of Euclidean plane are cyclic; any such group of order n consists of all rotations about a fixed point through angles of $2\frac{r\pi}{n}$ for $r = 0, 1, 2, \ldots, n-1$.
   (ii) The cyclic and dihedral groups and the *groups of tetrahedron, octahedron* and *icosahedron* are all the finite subgroups of the group of rotations of Euclidean 3-dimensional space.

   The latter three groups are called the *tetrahedral group* of 12 rotations carrying a regular tetrahedron to itself, the *octahedral group* of order 24 of rotations of a regular octahedron and the *icosahedral group* of 60 rotations of a regular icosahedron, respectively.

65. *Series of Subgroups.* A subnormal (or subinvariant) series of subgroups of a group G is a finite sequence $\{H_i\}$ of subgroups of G such that

$$\{1\} = H_0 \subseteq H_1 \subseteq \cdots \subseteq H_n = G$$

and H_i is a normal subgroup of H_{i+1} for each $i = 0, 1, \ldots, n-1$.

   In addition, if each H_i is a normal subgroup of G, then $\{H_i\}$ is called a *normal* (or *invariant series*) series of G:

   Two subnormal (normal) series $\{H_i\}$ and $\{T_j\}$ of the same group G are isomorphic iff $\exists$ a one-to-one correspondence between the factor groups $\{H_{i+1}/H_i\}$ and $\{T_{j+1}/T_j\}$ such that the corresponding factor groups are isomorphic.

   A subnormal series (normal series) $\{T_j\}$ is a refinement of a subnormal series (normal series) $\{H_i\}$ of a group G iff $\{H_i\} \subseteq \{T_j\}$, i.e., iff each H_i is one of the T_j.

   A subnormal series $\{T_j\}$ of a group G is a composition series iff all the factors groups T_{j+1}/T_j are simple. A normal series $\{N_i\}$ of G is a *principal series* iff all the factor groups N_{i+1}/N_i are simple.

A group G is solvable iff it has a composition series $\{T_j\}$ such that all factor groups T_{j+1}/T_j are abelian.

Prove the following:

(a) (i) *Schreier's Theorem*: Two subnormal (normal)series of a group G have isomorphic refinements.

  (ii) **Z** has neither a composition series nor a principal series.

  (iii) *Jordan–Hölder's Theorem*. Any two composition (principal) series of group G are isomorphic.

  (iv) If G has a composition (principal) series, and if N is a proper normal subgroup of G, then $\exists$ a composition (principal) series containing N.

  (v) A subgroup of a solvable group and the homomorphic image of a solvable group are solvable.

  (vi) If G is a group and N is a normal subgroup of G such that both N and G/N are solvable, then G is solvable.

(b) Let G be a group and A a pseudo-ideal of G. Then G is solvable $\Leftrightarrow$ A is solvable.

66. An abelian group A is said to an *injective group* iff wherever $i : G' \subset G$ (G' is a subgroup of G), for any homomorphism $f : G' \to A$, of abelian groups $\exists$ a homomorphism $\tilde{f} : G \to A$ such that $\tilde{f} \circ i = f$ and A is said to be a *projective group* iff for any epimorphism $\alpha : G \to G''$ of abelian groups, and any homomorphism $f : A \to G''$, $\exists$ a homomorphism $\tilde{f} : A \to G$ such that $\alpha \circ \tilde{f} = f$. An abelian group A is said to be a *divisible group* iff for any element $a \in A$, and any non-zero integer n, $\exists$ an element $b \in A$ such that $nb = a$. Show that

(i) an abelian group A is injective $\Leftrightarrow$ A is divisible;

(ii) any abelian group is a subgroup of an injective group.

2.12 Exercises (Objective Type)

Exercises A   *Identify the correct alternative(s) (there may be more than one) from the following list*:

1. Let G be a group and $a, b \in G$ such that $O(a) = 4$, $O(b) = 2$ and $a^3 b = ba$. Then $O(ab)$ is
   (a) 2   (b) 3   (c) 4   (d) 8.

2. Let G be a cyclic group of infinite order. Then the number of elements of finite order in G is
   (a) 1   (b) 2   (c) infinitely many   (d) 4.

3. Let G be an infinite cyclic group. Then the number of generators of G is
   (a) 1   (b) 3   (c) 2   (d) infinitely many.

4. The number of subgroups of S_3 is
   (a) 4   (b) 5   (c) 6   (d) 3.

5. The order of the permutation $(1\ 2\ 4)\ (3\ 5\ 6)$ in S_6 is
   (a) 2   (b) 4   (c) 3   (d) 9.

6. Let G be a finite group and H, K be two finite subgroups of G such that $|H| = 7$ and $|K| = 11$. Then $|H \cap K|$ is
   (a) 1   (b) 7   (c) 11   (d) 77.

7. Let G be a finite group and H, K be two subgroups of G such that $|H| = 7$, $|K| = 11$. Then $|HK|$ is
   (a) 77   (b) 1   (c) 7   (d) 11.

8. Let G be a finite group of order ≤ 400. If G has subgroups of order 45 and 75, then $|G|$ is
   (a) 400   (b) 225   (c) 300   (d) 75.

9. Let G be a finite group having two subgroups H and K and such that $|H| = 45$, $|K| = 75$. If $|G|$ is n, where $225 < n < 500$, then n is given by
   (a) 230   (b) 475   (c) 450   (d) 460.

10. Let G be a cyclic group of order 25. Then the number of elements of order 5 in G is
    (a) 1   (b) 3   (c) 4   (d) 20.

11. Let G be a non-trivial group of order n. Then G has

    (a) always an element of order 5;
    (b) no element of order 5;
    (c) an element of order 5 if 5 divides n;
    (d) an element of order 30 if 30 divides n.

12. The number of elements of order 5 in the group $\mathbf{Z}_{25} \times \mathbf{Z}_5$ is
    (a) 1   (b) 16   (c) 24   (d) 20.

13. The number of subgroups of order 5 in the group $\mathbf{Z}_5 \times \mathbf{Z}_9$ is
    (a) 1   (b) 2   (c) 24   (d) 5.

14. Automorphism group Aut($\mathbf{Z}_5$) of $\mathbf{Z}_5$ is isomorphic to
    (a) $\mathbf{Z}_5$ (b) $\mathbf{Z}_3$ (c) $\mathbf{Z}_4$ (d) $\mathbf{Z}_2$.

15. Let p be a prime integer. Then $|\mathrm{Aut}(\mathbf{Z}_p)|$ is
    (a) p (b) $p - 1$ (c) 1   (d) $p(p - 1)$.

16. Let $f : \mathbf{Z} \to \mathbf{Z}$ be a group homomorphism such that $f(1) = 1$. Then

    (a) f is an isomorphism;
    (b) f is a monomorphism but not an isomorphism;
    (c) f is not a monomorphism;
    (d) f is not an epimorphism.

17. Let G be an arbitrary group and $f : G \to G$ be an epimorphism. Then

    (a) f is always an isomorphism;
    (b) f is not always an isomorphism;
    (c) f is not always a monomorphism;
    (d) ker f contains more than one element.

18. The number of subgroups of $4\mathbf{Z}/16\mathbf{Z}$ is
    (a) 5   (b) 3   (c) 4   (d) 2.

19. The number of subgroups of $(\mathbf{Z}, +)$ which contains $10\mathbf{Z}$ as a subgroup is
    (a) 4   (b) 10   (c) infinite   (d) 3.

20. Consider the groups $G = \mathbf{Z}_2 \times \mathbf{Z}_3$, and $K = \mathbf{Z}_6$. Then

    (a) $\exists$ no homomorphism from G onto K;
    (b) $\exists$ no monomorphism from G onto K;
    (c) $\exists$ an isomorphism from G onto K;
    (d) $\exists$ an epimorphism from G onto K.

21. Let A_n be the set of all even permutations of the symmetric group S_n. Then

    (a) A_n is a subgroup of S_n but not normal;
    (b) A_n is not a subgroup of S_n;
    (c) A_n is a normal subgroup of S_n;
    (d) A_n is always a non-commutative subgroup of S_n.

22. The group $\mathbf{Z}_3 \times \mathbf{Z}_3$ is

    (a) cyclic;
    (b) not isomorphic to $\mathbf{Z}_9$;
    (c) isomorphic to $\mathbf{Z}_9$;
    (d) simple.

23. The group $\mathbf{Z}_2 \times \mathbf{Z}_2$ is

    (a) cyclic;
    (b) isomorphic to Klein's 4-group;
    (c) not isomorphic to Klein 4-group;
    (d) simple.

24. The group $\mathbf{Z}_8 \times \mathbf{Z}_9$ is

    (a) not cyclic;
    (b) isomorphic to $\mathbf{Z}_8 \times \mathbf{Z}_3 \times \mathbf{Z}_3$;
    (c) isomorphic to $\mathbf{Z}_{72}$;
    (d) simple.

25. The number of homomorphisms from the group $(\mathbf{Q}, +)$ into the group $(\mathbf{Q}^+, \cdot)$ is

    (a) one   (b) two   (c) infinitely many   (d) zero.

26. The number of generators of the group $\mathbf{Z}_p \times \mathbf{Z}_q$, where p, q are relatively prime to each other, is

    (a) pq (b) $(p-1)(q-1)$ ·   (c) $p(q-1)$ (d) $q(p-1)$.

27. Given a group G, let $Z(G)$ be its center. For $n \in \mathbf{N}^+$ (the set of positive integers) define $G_n = \{(g_1, \ldots, g_n) \in Z(G) \times \cdots \times Z(G) : g_1 \cdots g_n = 1\}$. As a subset of the direct product group $G \times \cdots \times G$ (n times direct product of the group G), G_n is

    (a) isomorphic to the direct product $Z(G) \times \cdots \times Z(G)$ $((n-1)$ times);
    (b) a subgroup but not necessarily a normal subgroup;
    (c) a normal subgroup;
    (d) not necessarily a subgroup.

28. Let G be a group of order 77. Then its center $Z(G)$ is isomorphic to
    (a) $\mathbf{Z}_7$ (b) $\mathbf{Z}_1$ (c) $\mathbf{Z}_{11}$ (d) $\mathbf{Z}_{77}$.

29. The number of group homomorphisms from the symmetric group S_3 to the group $\mathbf{Z}_6$ is
    (a) 2   (b) 3   (c) 6   (d) 1.

30. Let G be the group given by $G = \mathbf{Q}/\mathbf{Z}$ and n be a positive integer. Then the statement that there exists a cyclic subgroup of G of order n is

    (a) never true;
    (b) not necessarily true;
    (c) true but not necessarily a unique one;
    (d) true but a unique one.

31. Let $H = \{e, (1\ 2)(3\ 4), (1\ 3)(2\ 4), (1\ 4)(2\ 3)\}$ and $K = \{e, (1\ 2)(3\ 4)\}$ be subgroups of S_4, where e denote the identity element of S_4. Then

    (a) H and K are both normal subgroups of S_4;
    (b) K is normal in A_4 but A_4 is not normal in S_4;
    (c) H is normal in S_4, but K is not;
    (d) K is normal in H and H is normal in A_4.

32. Let n be the orders of permutations σ of 11 symbols such that σ does not fix any symbol. Then n is
    (a) 18   (b) 15   (c) 28   (d) 30.

33. Which of the following groups is (are) cyclic?
    (a) $\mathbf{Z}_8 \oplus \mathbf{Z}_8$ (b) $\mathbf{Z}_8 \oplus \mathbf{Z}_9$ (c) $\mathbf{Z}_8 \oplus \mathbf{Z}_{10}$ (d) a group of prime order.

34. Let G be a finite group and H be a subgroup of G. Let $O(G) = m$ and $O(H) = n$. Which of the following statements is (are) correct?

    (a) If m/n is a prime number, then H is normal in G;
    (b) if $m = 2n$, then H is normal in G;
    (c) if there exist normal subgroups A and B of G such that $H = \{ab \mid a \in A, b \in B\}$, then H is normal in G;
    (d) if $[G : H] = 2$, then H is normal in G.

35. Let G be a finite abelian group of odd order. Which of the following maps define an automorphism of G?

    (a) The map $x \mapsto x^{-1}$ for all $x \in G$;
    (b) the map $x \mapsto x^2$ for all $x \in G$;
    (c) the map $x \mapsto x$ for all $x \in G$;
    (d) the map $x \mapsto x^{-2}$ for all $x \in G$.

36. Let $GL(n, \mathbf{R})$ denote the group of all $n \times n$ matrices with real entries (with respect to matrix multiplication) which are invertible. Which of the following subgroups of $GL(n, \mathbf{R})$ is (are) normal?

    (a) The subgroup of all real orthogonal matrices;
    (b) the subgroup of all matrices whose trace is zero;

(c) the subgroup of all invertible diagonal matrices;

(d) the subgroup of all matrices with determinant equal to unity.

37. Which of the following subgroups of $GL(2, \mathbf{C})$ is (are) abelian?

(a) The subgroup of invertible upper triangular matrices;

(b) the subgroup A defined by $A = \{ \begin{bmatrix} a & b \\ -b & a \end{bmatrix} : a, b \in \mathbf{R} \text{ and } |a|^2 + |b|^2 = 1 \}$;

(c) the subgroup $G = \{ M \in GL_2(C) : \det M = 1 \}$;

(d) the subgroup B defined by $B = \{ \begin{bmatrix} a & b \\ -\bar{b} & \bar{a} \end{bmatrix} : a, b \in \mathbf{C} \text{ and } |a|^2 + |b|^2 = 1 \}$.

38. Let $f : (\mathbf{Q}, +) \to (\mathbf{Q}, +)$ be a non-zero homomorphism. Then

(a) f is always bijective;

(b) f is always surjective;

(c) f is always injective;

(d) f is not necessarily injective or surjective.

39. Consider the element

$$\alpha = \begin{pmatrix} 1 & 2 & 3 & 4 & 5 \\ 2 & 1 & 4 & 5 & 3 \end{pmatrix}$$

of the symmetric group S_5 on five elements. Then

(a) α is conjugate to $\begin{pmatrix} 4 & 5 & 2 & 3 & 1 \\ 5 & 4 & 3 & 1 & 2 \end{pmatrix}$;

(b) the order of α is 5;

(c) α is the product of two cycles;

(d) α commutes with all elements of S_5.

40. Let G be a group of order 60. Then

(a) G has a subgroup of order 30;

(b) G is abelian;

(c) G has subgroups of order 2, 3, and 5;

(d) G has subgroups of order 6, 10, and 15.

41. Let G be an abelian group of order n. Then
  (a) $n = 36$ (b) $n = 65$ (c) $n = 15$ (d) $n = 21$.

42. Which of the following subgroups are necessarily normal subgroups?

(a) The kernel of a group homomorphism;

(b) the center of a group;

(c) a subgroup of a commutative group;

(d) the subgroup consisting of all matrices with positive determinant in the group of all invertible $n \times n$ matrices with real entries (under matrix multiplication).

43. Which of the given subgroup H of a group G is a normal subgroup of G?

(a) G is the group of all 2×2 invertible upper matrices with real entries under matrix multiplication and H is the subgroup of all such matrices (a_{ij}) such that $a_{11} = 1$;

(b) G is the group of all $n \times n$ invertible matrices with real entries under matrix multiplication and H is the subgroup of such matrices with determinant 1;

(c) G is the group of all 2×2 invertible upper matrices with real entries under matrix multiplication and H is the subgroup of all such matrices (a_{ij}) such that $a_{11} = a_{22}$;

(d) G is the group of all $n \times n$ invertible matrices with real entries under matrix multiplication and H is the subgroup of such matrices with positive determinant.

44. Let S_7 denote the symmetric group of all permutations of the symbols $\{1, 2, 3, 4, 5, 6, 7\}$. Then

(a) S_7 has an element of order 10;

(b) S_7 has an element of order 7;

(c) S_7 has an element of order 15;

(d) the order of any element of S_7 is at most 12.

Exercises B (True/False Statements)   Determine which of the following statements are true or false. Justify your answers with a proof or give counter-examples accordingly.

1. The set of all Möbius transformations $S : \mathbf{C} \to \mathbf{C}$, $z \mapsto \frac{az+b}{cz+d}$, $ad - bc \neq 0$; $a, b, c, d \in \mathbf{C}$ form a group under usual composition of maps.

2. The set $SL(2, \mathbf{R}) = \{X \in GL(2, \mathbf{R}) : \det X = 1\}$ is a normal subgroup of the group $GL(2, \mathbf{R})$ of all real non-singular matrices of order 2.

3. The set $G = \{x \in \mathbf{Q} : 0 < x \leq 1\}$ is a group under the usual multiplication.

4. If each element of a group G, except its identity element is of order 2, then the group is abelian.

5. A cyclic group can have more than one generator.

6. Let G be a group of infinite order. Then all the elements of G, which are of the form $a^n, n \in \mathbf{Z}$, for a given $a \neq 1$, are distinct.

7. The octic group D_4 is cyclic.

8. If every proper subgroup of group G is cyclic, then G is also cyclic.

9. Let G be a cyclic group of order n. Then G has $\phi(n)$ distinct generators.

10. If G is an abelian group, then the only inner automorphism of G is the identity automorphism.

11. Let G be the internal direct product of its normal subgroups H and K. Then the groups G/H and K are isomorphic.

12. If g and g^2 are both generators of a cyclic group G of order n, then n is prime.

13. Let G be a finite group of odd order. If $H = \{x^2 : x \in G\}$, then H is a subgroup of G only if G is cyclic.

14. Two permutations in S_n are conjugate to each other $\Leftrightarrow$ they have the same cycle decomposition.

15. $Z(A_4)$, the center of the alternating group A_4 is $\{e\}$.

16. Let H be a normal subgroup of a group G. If H and G/H are both cyclic, then G is cyclic.

17. Every group of order 4 is commutative.

18. Every finitely generated group G can be expressed as the union of an ascending sequence of proper subgroups of G.
19. Let G be a finite group and H be a proper subgroup of G such that $x, y \in G$ and $x, y \notin H \Rightarrow xy \in H$. Then $|G|$ is an even integer.
20. The group $(\mathbf{Q}, +)$ is cyclic.
21. The group $(\mathbf{R}, +)$ is cyclic.
22. Any finitely generated subgroup of $(\mathbf{Q}, +)$ is cyclic.
23. In the group $(\mathbf{Q}, +)$, there exists an ascending sequence of cyclic groups G_n such that $\mathbf{Q} = \bigcup G_n$.
24. Let $(H, \cdot)$ be a subgroup of $(\mathbf{C}^*, \cdot)$ such that $[\mathbf{C}^* : H] = n$ (finite). Then $H \subsetneqq \mathbf{C}^*$.
25. Every finite subgroup $(H, \cdot)$ of $(\mathbf{C}^*, \cdot)$ is cyclic.
26. The groups $(\mathbf{Z}, +)$ and $(\mathbf{R}, +)$ are isomorphic.
27. The groups $(\mathbf{R}^*, \cdot)$ and $(\mathbf{R}, +)$ are isomorphic.
28. The groups $(\mathbf{Z}, +)$ and $(\mathbf{Q}, +)$ are isomorphic.
29. Every proper subgroup of a non-cyclic abelian group is non-cyclic.
30. A group G cannot be isomorphic to a proper subgroup H of G.
31. A group G is commutative $\Leftrightarrow (ab)^n = a^n b^n$, $\forall a, b \in G$ and for any three consecutive integers n.
32. $\mathbf{Z}_9$ is a homomorphic image of $\mathbf{Z}_3 \times \mathbf{Z}_3$.
33. The groups $\mathbf{Z}_6 \times \mathbf{Z}_8$ and $\mathbf{Z}_{48}$ are isomorphic.
34. Let G be a group. For a fixed $a \in G$, a mapping $\psi_a : G \to G$ is defined by $\psi_a(x) = ax, \forall x \in G$. Then ψ_a is a permutation of G.

2.13 Additional Reading

We refer the reader to the books (Adhikari and Adhikari 2003, 2004; Artin 1991; Birkoff and Mac Lane 1965; Chatterjee 1964; Herstein 1964; Howie 1976; Hungerford 1974; Jacobson 1974, 1980; Lang 1965; Mac Lane 1997; Malik et al. 1997; Shafarevich 1997; van der Waerden 1970) for further details.

References

Adhikari, M.R., Adhikari, A.: Groups, Rings and Modules with Applications, 2nd edn. Universities Press, Hyderabad (2003)

Adhikari, M.R., Adhikari, A.: Text Book of Linear Algebra: An Introduction to Modern Algebra. Allied Publishers, New Delhi (2004)

Artin, M.: Algebra. Prentice-Hall, Englewood Cliffs (1991)

Birkoff, G., Mac Lane, S.: A Survey of Modern Algebra. Macmillan, New York (1965)

Carruth, J.H., Hildebrant, J.A., Koch, R.J.: The Theory of Topological Semigroups I. Dekker, New York (1983)

Carruth, J.H., Hildebrant, J.A., Koch, R.J.: The Theory of Topological Semigroups II. Dekker, New York (1986)

Chatterjee, B.C.: Abstract Algebra I. Dasgupta, Kolkata (1964)

Cotton, F.A.: Chemical Application to Group Theory. Wiley, New York (1971)

Elliot, J.P., Dawber, P.G.: Symmetry in Physics. Macmillan, New York (1979)

Farmer, D.W.: Groups and Symmetry. Mathematics World Series, vol. 5, Am. Math. Soc., Providence (1996)

De Groot, J.: Groups represented by Homomorphic Groups I. Math. Ann. **138**, 80–102 (1959)

Herstein, I.: Topics in Algebra. Blaisdell, New York (1964)

Howie, I.M.: An Introduction to Semigroup Theory. Academic Press, New York (1976)

Hungerford, T.W.: Algebra. Springer, New York (1974)

Jacobson, N.: Basic Algebra I. Freeman, San Francisco (1974)

Jacobson, N.: Basic Algebra II. Freeman, San Francisco (1980)

Lang, S.: Algebra, 2nd edn. Addison-Wesley, Reading (1965)

Mac Lane, S.: Category for the Working Mathematician. Springer, Berlin (1997)

Magil, K.D. Jr: Some open problems and directions for further research in semigroups of continuous selfmaps. In: Universal Algebra and Applications, Banach Centre Publications, pp. 439–454. PWN, Warsaw (1982)

Malik, S., Mordeson, J.N., Sen, M.K.: Abstract Algebra, Fundamentals of Abstract Algebra. McGraw Hill, New York (1997)

Massey, W.S.: A Basic Course in Algebraic Topology. Springer, New York (1991)

Rodes, F.: On the fundamental group of a transformation group. Proc. Lond. Math. Soc. **16**, 635–650 (1966)

Rotman, I.J.: An Introduction to Algebraic Topology. Springer, New York (1988)

Shafarevich, I.R.: Basic Notions of Algebra. Springer, Berlin (1997)

Spanier, E.H.: Algebraic Topology. McGraw-Hill, New York (1966)

Sternberg, S.: Group Theory and Physics. Cambridge University Press, Cambridge (1994)

van der Waerden, B.L.: Modern Algebra. Ungar, New York (1970)

Weyl, H.: Symmetry. Princeton University Press, Princeton (1952)

Chapter 3
Actions of Groups, Topological Groups and Semigroups

The actions of groups, topological groups and semigroups are very important concepts. An action of a group (semigroup) on a non-empty set assigns to each element of the group (semigroup) a transformation of the set in such a way that it assigns to the product of the two elements of the group (semigroup) the product of the two corresponding transformations. As a consequence, each element of a group determines a permutation on the set under a group action. For a topological group action on a topological space, this permutation is a homeomorphism and for a Lie group action on a differentiable manifold, it is a diffeomorphism. Group actions are applied to develop the theory of finite groups. More precisely, group actions are used to determine the number of distinct conjugate classes of a subgroup of a finite group and also to prove Cayley's Theorem, Cauchy's Theorem and Sylow's Theorems for groups. The counting principle is applied to determine the structure of a group of prime power order. These groups arise in the Sylow Theorems and in the description of finite abelian groups. The Sylow Theorems give the existence of p-Sylow subgroups of a finite group and describe the subgroups of prime power order of an arbitrary finite group. We discuss the converse of Lagrange's theorem which is not true for arbitrary finite groups. We also study actions of topological groups and Lie groups and obtain some orbit spaces which are very important in topology and geometry. For example, $\mathbf{R}P^n$, $\mathbf{C}P^n$ are obtained as orbit spaces. On the other hand, in this chapter, semigroup actions are applied to theoretical computer science yielding state machines, which unify computer science with mainstream mathematics.

3.1 Actions of Groups

In this section we introduce the concept of an action of a group G on a non-empty set X and study such actions. We show that a group action of G on X assigns to each element of G a permutation on the set X. This leads to a proof of Cayley's theorem that every abstract group is isomorphic to some group of permutations and, in particular, every group of order n is isomorphic to a certain subgroup of S_n. This

M.R. Adhikari, A. Adhikari, *Basic Modern Algebra with Applications*,
DOI 10.1007/978-81-322-1599-8_3, © Springer India 2014

interesting theorem of Cayley establishes the equivalence between the historical concept of a group of transformations and the modern axiomatic concept of a group.

Definition 3.1.1 A group G is said to act on a non-empty set X from the left (or X is said to be a left G-set) iff there is a function $\sigma : G \times X \to X$, denoted by $(g, x) \mapsto g \cdot x$ (or gx) for all $x \in X$, $g \in G$ such that

(i) for any $x \in X$, $1 \cdot x = x$, where 1 is the identity of G;
(ii) for any $x \in X$, $g_1, g_2 \in G$, $(g_1 g_2) \cdot x = g_1 \cdot (g_2 \cdot x)$.

σ is said to be an action of G on X from the left and the ordered triple (X, G, σ) is called a *transformation group*.

Example 3.1.1 Let G be a subgroup of S_n. Then an action of G on the set $X = \{1, 2, \ldots, n\}$ is defined by $\alpha \cdot x = \alpha(x)$ $\forall \alpha \in G$ i.e., $\alpha \cdot x$ is the effect of applying the permutation $\alpha \in G$ to the element $x \in X$.

Example 3.1.2 Let X denote the Euclidean 3-space and G the group of all rotations of X which leave the origin fixed. Then an action of G on X is defined by $g \cdot x = g(x)$, the image of the point $x \in X$ under the rotation g.

There may exist different actions of a group on a given set X.

Example 3.1.3 Let H be a subgroup of a group G. Then for all $h \in H$ and $x \in G$, each of the following is a group action:

(i) $h \cdot x = hx$ (*left translation*);
(ii) $h \cdot x = xh^{-1}$;
(iii) $h \cdot x = hxh^{-1}$ (*conjugation by h*);

Each of the above is an action of H on G. The element hxh^{-1} is said to be a conjugate of x.

(iv) if H is a normal subgroup of G, then $x \cdot h = xhx^{-1}$ is an action of G on H, where the right hand multiplication is the group multiplication.

Remark A right action of a group G on a set X is defined in an analogous manner.

Definition 3.1.2 A group G is said to act on a set X from the right (or X is said to be a right G-set) iff there is a function $\sigma : X \times G \to X$, denoted by $(x, g) \mapsto x \cdot g$ (or xg) for all $x \in X$, $g \in G$ such that

(i)' for any $x \in X$, $x \cdot 1 = x$;
(ii)' for any $x \in X$, $g_1, g_2 \in G$, $x \cdot (g_1 g_2) = (x \cdot g_1) \cdot g_2$.

Remark The essential difference between the right and left G-sets is not whether the elements of G are written on the right or left of those of X. The main point is the difference between respective conditions (ii) and (ii)': If X is a left G-set, then the product $g_1 g_2$ acts on $x \in X$ in such a way that g_2 operates first and g_1 operates on the result, but for the right G-sets, g_1 operates first and g_2 operates on the result.

If X is a left G-set, then for any $x \in X$ and $g \in G$, $x \cdot g = g^{-1} \cdot x$ defines a right G-set structure. Conversely, if X is a right G-set, then $g \cdot x = x \cdot g^{-1}$ defines a left G-set structure. Since there is a bijective correspondence between left and right G-set structures, we need to study only one of them.

Theorem 3.1.1 *Let X be a left G-set. For any $g \in G$, the map $X \to X$ defined by $x \mapsto g \cdot x$ is a permutation on the set X.*

Proof Let $\phi_g : X \to X$ be the map defined by $\phi_g(x) = g \cdot x$. Then

$$\phi_{g^{-1}}(x) = g^{-1} \cdot x \quad \text{for all } g \in G \text{ and } x \in X$$

$$\Rightarrow \quad (\phi_g \phi_{g^{-1}})(x) = \phi_g(g^{-1} \cdot x) = g \cdot (g^{-1} \cdot x) = (gg^{-1}) \cdot x = 1 \cdot x = x$$

$$\forall g \in G \text{ and } \forall x \in X \quad \Rightarrow \quad \phi_g \phi_{g^{-1}} = 1_X.$$

Similarly $\phi_{g^{-1}} \phi_g = 1_X$. Therefore ϕ_g is a bijection on X. Consequently, ϕ_g is a permutation on X. $\qquad \square$

Remark This theorem shows that the notion of a left G-set X is equivalent to the notion of a representation of G by permutations on the set X.

Theorem 3.1.2 *Let X be a left G-set. Then*

(i) *the relation ρ on X defined by $x \rho y$, iff $gx = y$ for some $g \in G$ is an equivalence relation;*
(ii) *for each $x \in X$, $G_x = \{g \in G : gx = x\}$ is subgroup of G.*

Proof (i) $1x = x \ \forall x \in X \Rightarrow x \rho x \ \forall x \in X \Rightarrow \rho$ is reflexive. Again $x \rho y \Rightarrow gx = y$ for some $g \in G \Rightarrow x = g^{-1}y \Rightarrow y \rho x \Rightarrow \rho$ is symmetric. Finally, $x \rho y$ and $y \rho z \Rightarrow gx = y$ and $g'y = z$ for some $g, g' \in G \Rightarrow (g'g)x = g'(gx) = g'y = z \Rightarrow x \rho z$ (since $g'g \in G$) $\Rightarrow \rho$ is transitive. Consequently, ρ is an equivalence relation on X.

(ii) $1 \in G_x \Rightarrow G_x \neq \emptyset$. Let $g, h \in G_x$. Then $gx = x$ and $hx = x \ \forall x \in X$. Now $(gh^{-1})x = g(h^{-1}x) = gx = x \Rightarrow gh^{-1} \in G_x \Rightarrow G_x$ is a subgroup of G. $\qquad \square$

Definition 3.1.3 Let X be a left G-set. Then the equivalence classes x_ρ of Theorem 3.1.2(i) are called the *orbits* of G on X and each x_ρ is called the orbit of $x \in X$, denoted by $\text{orb}(x)$. G is said to act on X transitively iff $\text{orb}(x) = X$ for every $x \in X$. Two orbits of G on X are identical or disjoint and the set of all distinct orbits on X is called the orbit set denoted $X \bmod G$.

Definition 3.1.4 The subgroup G_x is called the *isotropy* group of x or the *stabilizer* of x.

Clearly, the subgroup G_x fixes every $x \in X$. If $G_x = \{1\}$, $\forall x \in X$, then the action of G on X is said to be free.

Lemma 3.1.1 *If a group G acts on a set X, then there is a bijection between the set of cosets of the isotropy group G_x in G onto the* orb(x) *for each $x \in X$.*

Proof Let $\{gG_x\}$ denote the set of all left cosets of G_x in G. Consider the map $f : \{gG_x\} \to$ orb(x) given by $f(gG_x) = gx$. Let $g, g' \in G$. Then $gx = g'x \Leftrightarrow g^{-1}g'x = x \Leftrightarrow g^{-1}g' \in G_x \Leftrightarrow g'G_x = gG_x \Rightarrow f$ is well defined. Clearly, f is a bijection. $\qquad\square$

Theorem 3.1.3 *Let a group G act on a set X. Then $|\mathrm{orb}(x)|$ is the index. $[G : G_x]$ for every $x \in X$. In particular, if G acts transitively on X, then $|X| = [G : G_x]$.*

Proof Since $[G : G_x]$ is the cardinal number of the set $\{gG_x\}$, the theorem follows from Lemma 3.1.1. $\qquad\square$

Theorem 3.1.4 *Let X be a set and $A(X)$ the group of all permutations on X. If a group G acts on X, then this action induces a homomorphism $G \to A(X)$.*

Proof For each $g \in G$, the map $\phi_g : X \to X$ defined by $\phi_g(x) = gx$ is a bijection by Theorem 3.1.1. Consider the map $\psi : G \to A(X)$ defined by $\psi(g) = \phi_g$. Then $\forall g, g' \in G$, $\psi(gg') = \phi_{gg'}$. But $\phi_{(gg')}(x) = (gg')x = g(g'x) = \phi_g(g'x) = (\phi_g\phi_{g'})(x)$ $\forall x \in X \Rightarrow \phi_{gg'} = \phi_g\phi_{g'}$. Thus, $\psi(gg') = \psi(g)\psi(g') \Rightarrow \psi$ is a homomorphism. $\qquad\square$

Lemma 3.1.2 *For every group G, there is a monomorphism $G \to A(G)$.*

Proof By Example 3.1.3, we find that every group G acts on itself by left translation. Then by Theorem 3.1.4, this action induces a homomorphism $\psi : G \to A(G)$ given by $\psi(g) = \phi_g$, where $\phi_g(g') = gg'$ $\forall g, g' \in G$. Now $\psi(g) = 1_{A(G)}$ (identity automorphism of G) $\Leftrightarrow gg' = g'$ $\forall g' \in G \Rightarrow g = 1 \Rightarrow \psi$ is a monomorphism. $\qquad\square$

The following theorem, essentially due to the British mathematician Arthur Cayley (1821–1895), is a direct consequence of Lemma 3.1.2. He observed that every group could be realized as a subgroup of a permutation group.

Theorem 3.1.5 (Cayley's Theorem) *Every group is isomorphic to a group of permutations. In particular every group of order n is isomorphic to a certain subgroup of S_n.*

Proof Let G be any group. Then by Lemma 3.1.2, there is a monomorphism $\psi : G \to A(G)$. Consequently, $G \cong \mathrm{Im}\,\psi$. This implies that G is isomorphic to a group of permutations. If G is, in particular, a finite group of order n, then $A(G)$ becomes the symmetric group S_n. Hence the last statement follows from the first one. $\qquad\square$

Corollary *Let G be a group. Then*

(i) *for each $g \in G$, conjugation by g induces an automorphism of G;*

(ii) *there is a homomorphism $G \to \operatorname{Aut} G$, whose kernel is $C(G) = \{g \in G : gx = xg \; \forall x \in G\}$.*

Proof (i) Let G act on itself by conjugation. Then for each $g \in G$, consider the map $\sigma_g : G \to G$ defined by conjugation: $\sigma_g(x) = gxg^{-1} \; \forall x \in G$. Now $(\sigma_g \sigma_{g^{-1}})(x) = \sigma_g(g^{-1}xg) = g(g^{-1}xg)g^{-1} = x \; \forall x \in G \Rightarrow \sigma_g \sigma_{g^{-1}} = \tilde{1}_G$, where $\tilde{1}_G$ is the identity mapping on G. Similarly, $\sigma_{g^{-1}} \sigma_g = \tilde{1}_G$. Therefore σ_g is a bijection on G. Moreover $\forall x, y \in G, \sigma_g(xy) = g(xy)g^{-1} = gxg^{-1}gyg^{-1} = \sigma_g(x)\sigma_g(y) \Rightarrow \sigma_g$ is a homomorphism. Consequently, for each g, σ_g is an automorphism of G.

(ii) Let G act on itself by conjugation. Then by Theorem 3.1.4, this action induces a homomorphism $\psi : G \to A(G)$ given by $\psi(g) = \sigma_g$ (defined in (i)). Now $g \in \ker \psi \Leftrightarrow \sigma_g = \tilde{1}_G \Leftrightarrow gxg^{-1} = x \; \forall x \in G \Leftrightarrow gx = xg \Leftrightarrow g \in C(G) \Rightarrow \ker \psi = C(G)$. $\qquad \square$

Remark σ_g is called the *inner automorphism* induced by g and the normal subgroup $\ker \psi = C(G)$ is the center of G. The group of symmetries of the square has four distinct inner automorphisms but the cyclic group of order 3 has no inner automorphism except the identity.

Lemma 3.1.3 *Let H be a subgroup of a group G and let G act on the set X of all left cosets of H in G by left translation. Then the kernel of the induced homomorphism $G \to A(X)$ is contained in H.*

Proof The induced *homomorphism* $\psi : G \to A(X)$ is defined by $\psi(g) = \sigma_g$, where $\sigma_g : X \to X$ is given by $\sigma_g(xH) = gxH$. Now $g \in \ker \psi \Rightarrow \sigma_g = \tilde{1}_X \Rightarrow gxH = xH$ $\forall x \in G$. Then for $x = 1$, $g1H = 1H = H \Rightarrow g \in H$. $\qquad \square$

Theorem 3.1.6 *Let H be a subgroup of index n of a group G and no non-trivial normal subgroup of G is contained in H. Then G is isomorphic to a subgroup of S_n.*

Proof Clearly, H is not a normal subgroup of G. Now by using the notation of Lemma 3.1.3, the $\ker \psi$ is a normal subgroup of G contained in H and hence by hypothesis $\ker \psi = \{1\}$. This shows that ψ is a monomorphism. Consequently, G is isomorphic to a subgroup of S_n. $\qquad \square$

Corollary *Let G be a finite group and p be the smallest prime dividing $|G|$. If H is a subgroup of G such $[G : H] = p$, then H is normal in G.*

Proof Let X be the set of all left cosets of H in G. Hence $[G : H] = p \Rightarrow A(X) \cong S_p$. Now consider the map

$$\psi : G \to S_p \quad \text{as defined in Lemma 3.1.3.}$$

As $\ker \psi$ is a normal subgroup in G and contained in H by Lemma 3.1.3 and $G/\ker \psi$ is isomorphic to a subgroup of S_p, $|G/\ker \psi|$ divides $|S_p| = p!$. But every divisor of $|G/\ker\psi| = [G : \ker\psi]$ must divide $|G| = |\ker\psi|[G : \ker\psi]$. Since no positive integer less than p (other than 1) can divide $|G|$, $|G/\ker\psi|$ must be p or 1. Then $|G/\ker\psi| = [G : H][H : \ker\psi] = p[H : \ker\psi] \geq p \Rightarrow |G/\ker\psi| = p$ and $[H : \ker\psi] = 1$. Consequently, $H = \ker\psi \Rightarrow H$ is normal in G.   □

Definition 3.1.5 Let X and Y be two left G-sets. Then a mapping $f : X \rightarrow Y$ is called G-*equivariant* or G-homomorphism or simply a mapping of left G-sets iff $f(g \cdot x) = g \cdot f(x)$ $\forall x \in X$ and $g \in G \cdot$ A G-equivariant map $f : X \rightarrow Y$ is called an isomorphism of left G-sets iff there exists another G-equivariant map $h : Y \rightarrow X$ such that $h \circ f = \tilde{1}_X$ and $f \circ h = \tilde{1}_Y$. As usual, an automorphism of a G-set is a self isomorphism.

Definition 3.1.6 Let X be a left G-set. We say that X is a homogeneous left G-set iff for any elements $x, y \in X$, $\exists g \in G$ such that $g \cdot x = y$, i.e., iff, G acts transitively on X.

Example 3.1.4 Let G be group and H an arbitrary subgroup of G. Define a map $\psi : G \times G/H \rightarrow G/H, (g, g'H) \mapsto gg'H$. Then ψ defines an action of G on G/H such that G/H is a homogeneous left G-set.

Theorem 3.1.7 *Any homogeneous left G-set is isomorphic to some homogeneous left G-set G/H.*

Proof Let X be an arbitrary homogeneous left G-set. Choose an element $x_0 \in X$. Then $H = \{g \in G : gx_0 = x_0\}$ is a subgroup of G. Consider the map $f : G \rightarrow X$ defined by $g \mapsto gx_0$. As X is a homogeneous G-set, f is onto. Now for $g, h \in G$, $gx_0 = hx_0 \Leftrightarrow h^{-1}gx_0 = x_0 \Leftrightarrow h^{-1}g \in H \Leftrightarrow h, g \in$ same coset of H. Consequently, f induces a map $\tilde{f} : G/H \rightarrow X$, which is clearly a bijection. Moreover, $\tilde{f}$ is G-equivariant. Consequently, G/H and X are isomorphic left G-sets.   □

Remark The isomorphism $\tilde{f}$ and the subgroup H defined above depend on the choice of the point $x_0 \in X$. H is called the *isotropy subgroup* corresponding to x_0. A different choice of x_0 gives rise to a conjugate subgroup.

3.2 Group Actions to Counting and Sylow's Theorems

The counting principle and Sylow's Theorems play a basic role in understanding the structure of finite groups. We apply the results of G-sets of Sect. 3.1 in counting and prescribe a method for determining the number of distinct orbits in a G-set. While studying finite groups, it is a natural question: does the converse of Lagrange's Theorem hold for arbitrary finite groups? This means that if a positive integer m divides the order of a finite group G, does G have a subgroup of order m? If m is a prime integer, the answer is shown to be positive by Cauchy's Theorem. But the alternating group A_4 of order 12 has no subgroup of order 6 (see Ex. 4 of SE-I of Chap. 2). This example shows that the converse of Lagrange's Theorem is not true for arbitrary finite groups. Perhaps the best partial converse (not true for all cases) is the First Sylow Theorem which states that the answer is positive whenever m is a power of a prime integer. The Second and Third Sylow Theorems come as a natural consequence of subgroups of maximal prime power order. In this section we begin with the counting principle and prove Cauchy's Theorem and Sylow's Theorems.

We first present some interesting group actions to counting.

Define $X_g = \{x \in X : gx = x\}$ for each $g \in G$ and $G_x = \{g \in G : gx = x\}$ for each $x \in X$, where G is a group and X is a G-set.

Theorem 3.2.1 (Burnside) *Let G be a finite group and X a finite G-set. If r is the number of orbits of G on X, then*

$$r|G| = \sum_{g \in G} |X_g|. \tag{3.1}$$

Proof Consider all ordered pairs (g, x), where $gx = x$, and let N be the number of such pairs. Now, for each $g \in G$, there exist $|X_g|$ ordered pairs having g as the first coordinate. Consequently,

$$N = \sum_{g \in G} |X_g|. \tag{3.2}$$

Again for each $x \in X$, there exist $|G_x|$ [see Theorem 3.1.2(ii)] ordered pairs (g, x) having x as the second coordinate. Consequently, $N = \sum_{x \in X} |G_x|$.

Then by Theorem 3.1.3, $|\text{orb}(x)| = [G : G_x]$. Since $[G : G_x] = |G|/|G_x|$, it follows that $|\text{orb}(x)| = |G|/|G_x|$. So

$$N = \sum_{x \in X} \frac{|G|}{|\text{orb}(x)|} = |G| \sum_{x \in X} \frac{1}{|\text{orb}(x)|}.$$

Let Ω be any orbit. Then

$$\sum_{x \in \Omega} \frac{1}{|\text{orb}(x)|} = 1,$$

since $1/|\text{orb}(x)|$ has the same value for all x in the same orbit.

Consequently,

$$N = |G| \cdot \text{(number of orbits of } G \text{ on } X)$$

$$= |G| \cdot r \Rightarrow r|G| = \sum_{g \in G} |X_g|.$$

$\square$

We now find an equation that counts the number of elements of a finite G-set.

Let X be a finite G-set and r the number of orbits of G on X and $\{x_1, x_2, \ldots, x_r\}$ the representative system of the orbits, i.e., the set containing one element from each orbit of G on X. As every element of X is in exactly one orbit,

$$|X| = \sum_{i=1}^{r} \left|\mathrm{orb}(x_i)\right|. \tag{3.3}$$

Let $X_G = \{x \in X : gx = x \; \forall g \in G\}$ i.e., X_G is precisely the union of the one element orbits of G on X. If $|X_G| = t$, then (3.3) is reduced to

$$|X| = |X_G| + \sum_{i=t+1}^{r} \left|\mathrm{orb}(x_i)\right|. \tag{3.4}$$

We shall come back a little later to (3.4) to deduce an important equation known as class equation. Before that we need another important notion, known as p-groups.

3.2.1 p-Groups and Cauchy's Theorem

The class equation can be effectively utilized when the order of a finite group is a power of a prime. In the rest of the section p denotes a prime integer.

Definition 3.2.1 (p-groups) A finite group G is said to be a prime power group or a *p-group* iff the order of every non-identity element in G is a positive power of the prime p. A subgroup H of a group G is a p-subgroup of G iff H is itself a p-group.

Theorem 3.2.2 *Let G be a group of order p^n for some prime integer p and positive integer n. If X a finite G-set, then $|X| \equiv |X_G| \pmod{p}$.*

Proof Since $|\mathrm{orb}(x_i)| = [G : G_{x_i}]$ by Theorem 3.1.3 and $[G : G_{x_i}]$ divides $|G|$, so, p divides $[G : G_{x_i}]$ and thus p divides $|\mathrm{orb}(x_i)|$ for $t + 1 \le i \le r$ (in the notation of equation (3.4)). This shows that $|X| - |X_G|$ is divisible by p and hence $|X| \equiv |X_G| \pmod{p}$. $\qquad \square$

We now prove Cauchy's Theorem which is essentially due to A.L. Cauchy (1789–1857). This theorem and its Corollary 2 show two different situations (other than condition of the First Sylow Theorem) under which the converse of Lagrange's Theorem becomes also true.

Theorem 3.2.3 (Cauchy) *Let G be a finite group and a prime p divide $|G|$. Then G has an element of order p and consequently a subgroup of order p.*

Proof Consider the set $X = \{(g_1, g_2, \ldots, g_p) : g_i \in G \text{ and } g_1 g_2 \cdots g_p = 1\}$. In forming an element in X, we may take the elements $g_1, g_2, \ldots, g_p$ of G such that $g_p = (g_1 g_2 \cdots g_{p-1})^{-1}$. Thus $|X| = |G|^{p-1}$. Now $p||G| \Rightarrow p||X|$. Let α be the cycle $(123 \cdots p)$ in S_p. Let α act on X by $\alpha \cdot (g_1, g_2, \ldots, g_p) =$

$(g_{\alpha(1)}, g_{\alpha(2)}, \ldots, g_{\alpha(p)}) = (g_2, g_3, \ldots, g_p, g_1) \in X$, since $g_1(g_2 \cdots g_p) = 1 \Rightarrow g_1 = (g_2 g_3 \cdots g_p)^{-1} \Rightarrow (g_2 g_3 \cdots g_p) g_1 = 1$. We consider the subgroup $\langle \alpha \rangle$ of S_p to act on X by iteration. Now $|\langle \alpha \rangle| = p$. Then by using Theorem 3.2.2, $|X| \equiv |X_{\langle \alpha \rangle}|$ (mod p). Hence $p||X| \Rightarrow p||X_{\langle \alpha \rangle}| \Rightarrow$ there must be at least p elements in $X_{\langle \alpha \rangle} \Rightarrow \exists$ some $a \in G$, $a \neq 1$ such that $(a, a, \ldots, a) \in X_{\langle \alpha \rangle} \Rightarrow a^p = 1 \Rightarrow$ order of a is $p \Rightarrow \langle a \rangle$ is a subgroup of G of order p.   □

Corollary 1   *Let G be a finite group. Then G is a p-group iff $|G|$ is a power of p.*

Proof Let G be a finite group such that $|G| = p^r$. Then for each $a \in G$, $O(\langle a \rangle) = O(a)$ and $O(a)|p^r \Rightarrow O(a)$ is a power of p for each $a \in G \Rightarrow G$ is a p-group.

Conversely, let G be a p-group. Then no prime t $(\neq p)$ divides $|G|$, otherwise G would contain an element of order t, a prime number $\neq p$ by Theorem 3.2.3 implying a contradiction, since every element of G has order, a power of p, implying $t = p$. Consequently, $|G|$ is a power of p.   □

We now apply Cauchy's Theorem to prove a partial converse of Lagrange's Theorem.

Corollary 2   *Let G be a finite abelian group of order n and m be a positive integer such that m divides n. Then G has a subgroup of order m.*

Proof If $m = 1$, then $\{1\}$ consisting of the identity element of G is the required subgroup of G. If $n = 1$, then $m = n = 1$ and the result is trivial. So we assume that $m > 1$, $n > 1$ and prove the corollary by induction on order n of G. If $n = 2$, then $m = 2$ and hence G is the required subgroup. We now assume that the corollary is true for all finite abelian groups of order r satisfying $2 \leq r \leq n$. Let p be a prime such that p divides m. Then there exists an integer s such that $m = ps$. Hence G has a subgroup H of order p by Cauchy's theorem. Then G/H is a group, as H is a normal subgroup of the abelian group G. Consequently, $1 \leq |G/H| = |G|/|H| < |G|$ and $|G/H| = n/p$. Again $n = mt$ for some positive integer t. Hence $|G/H| = mt/p = st$ implies that s divides $|G/H|$. This shows by induction hypothesis that the group G/H has a subgroup K/H (say) such that $|K/H| = s$, where K is a subgroup of G. This implies that $|K| = |K/H||H| = sp = m$. Hence K is a subgroup of G of order m.   □

3.2.2 Class Equation and Sylow's Theorems

We are now in a position to define the class equation of a finite group. Let G be a finite group and X a finite G-set. We now consider the special case of equation (3.4) in which $X = G$ and the action of G on itself is by conjugation, i.e., for $g \in G$, $x \mapsto gxg^{-1}$ (cf. Example 3.1.3(iii)). Then $X_G = \{x \in G : gxg^{-1} = x \ \forall g \in G\} = \{x \in G : gx = xg \ \forall g \in G\} = Z(G)$, the center of G.

If $m = |Z(G)|$ and n_i is the number elements in the ith orbit of G under conjugation, i.e., $n_i = |\text{orb}(x_i)|$, then (3.4) becomes

$$|G| = m + n_{m+1} + \cdots + n_r. \tag{3.5}$$

Remark 1 n_i divides $|G|$ for $m + 1 \le i \le r$, since $|\text{orb}(x_i)| = [G : G_{x_i}]$ is a divisor of $|G|$.

Remark 2 Equation (3.5) is called the *class equation of G*. Each orbit of G under conjugation by G is a conjugate or conjugacy class in G.

The class equation is now effectively applied to study a certain class of finite groups. Using the class equation, we are going to prove three celebrated theorems, essentially due to Norwegian mathematician L.M. Sylow (1832–1918), to understand the structure of an arbitrary finite group. There are very few theorems which show the existence of subgroups of prescribed order under specific conditions. Among them the best partial converse to Lagrange's Theorem is the First Sylow Theorem. This is a basic theorem and is largely used. The First Sylow Theorem leads to the Second and Third Sylow Theorems under specific conditions. The First Sylow Theorem describes the subgroups of orders which are some powers of prime integer. The Second Sylow Theorem gives an interesting relation among different Sylow p-subgroups of a finite group. The Third Sylow Theorem prescribes the number of Sylow p-subgroups.

As a first step to prove the Sylow Theorems we now apply group action to determine the number of distinct conjugate classes of a subgroup of a finite group. Let G be a finite group and S the collection of all subgroups of G. We make S into a G-set by the action of G on S by conjugation i.e., if $H \in S$, then $g \cdot H = gHg^{-1}$. Consider the normalizer $N(H) = \{g \in G : gHg^{-1} = H\}$ of H in G. Then $N(H)$ is a subgroup of G and H is a normal subgroup of $N(H)$ such that $N(H)$ is the largest subgroup of G having H as a normal subgroup. Clearly, the element H of S is a fixed point under the above conjugation iff H is normal in G and $\text{orb}(H)$ is precisely the set of all subgroups of G which are conjugate to H, i.e., the conjugate class of H. The isotropy subgroup $G_H = N(H) \Rightarrow |G/N(H)| =$ number of distinct conjugate classes of H in G. Such results are very important in algebraic topology.

Lemma 3.2.1 *Let H be a p-subgroup of a finite group G. Then*

$$\left[N(H) : H\right] \equiv [G : H] \pmod{p}.$$

Proof Let S be the set of all left cosets of H in G and H act on S by left translation: $h \cdot (xH) = (hx)H$. Then S becomes an H-set and $|S| = [G : H]$. Consider $S_H = \{xH \in S : h(xH) = xH \; \forall h \in H\}$. Thus $\forall h \in H$, $xH = h(xH) \Leftrightarrow x^{-1}hx = x^{-1}h(x^{-1})^{-1} \in H \; \forall h \in H \Leftrightarrow x^{-1} \in N(H) \Leftrightarrow x \in N(H) \Rightarrow$ left cosets in S_H are those contained in $N(H)$. The number of such cosets is $[N(H) : H]$ and hence $|S_H| = [N(H) : H]$. Again H is a p-group $\Rightarrow H$ has order a power of p by Corollary 1 of Theorem 3.2.3 shows that $|\mathcal{S}| \equiv |\mathcal{S}_H|$ (mod p) by Theorem 3.2.2. Consequently, $[G : H] \equiv [N(H) : H]$ (mod p) i.e., $[N(H) : H] \equiv [G : H]$ (mod p).   □

Corollary 3 *Let H be a p-subgroup of a finite group G. If $p | [G : H]$, then $N(H) \neq H$.*

Proof Clearly, p divides $[N(H) : H]$ by Lemma 3.2.1. So, $H \neq N(H)$.   □

We are now equipped to prove the First Sylow theorem, which gives the existence of p-subgroups of G for any prime power dividing $|G|$.

Theorem 3.2.4 (First Sylow Theorem) *Let G be a finite group of order $p^n m$, with $m, n \geq 1$, p prime and $\gcd(p, m) = 1$. Then G contains a subgroup of order p^i for each i satisfying $1 \leq i \leq n$ and every subgroup H of G of order p^i is a normal subgroup of a subgroup of order p^{i+1} for $1 \leq i < n$.*

Proof We prove the theorem by induction. $p | |G| \Rightarrow G$ contains a subgroup of order p by Theorem 3.2.3. We assume that H is a subgroup of order p^i $(1 \leq i < n)$. Then $p | [G : H] \Rightarrow p | [N(H) : H]$ by Lemma 3.2.1. Since H is a normal subgroup of $N(H)$, we can form the factor group $N(H)/H$ such that $p | |N(H)/H|$. Again the factor group $N(H)/H$ has a subgroup of order p by Theorem 3.2.3. Then this group is of the form T/H, where T is a subgroup of $N(H)$ containing H. Since H is normal in $N(H)$, H is necessarily normal in T. Finally, $|T| = |H||T/H| = p^i p = p^{i+1}$. Thus we prove that the existence of a subgroup H of order p^i for $i < n$ implies the existence of a subgroup T, of order p^{i+1}, in which H is normal. Thus the theorem follows by an induction argument.   □

Corollary 4 *Let G be a finite group and p a prime. If $p^i | |G|$, then G has a subgroup of order p^i.*

Definition 3.2.2 A p-subgroup P of a finite group G of order $p^n m$, with $m, n \geq 1$ and $\gcd(p, m) = 1$, is said to be a Sylow p-subgroup of G iff P is a maximal p-subgroup of G, i.e., $P \subseteq H \subsetneq G$ with H a p-subgroup of G implies $P = H$.

Remark If G is a finite group such that $|G| = p^n m$ where $m, n \geq 1$ and $\gcd(m, p) = 1$, then Theorem 3.2.4 shows that the Sylow p-subgroups of G are precisely those subgroups of order p^n. Moreover, every conjugate of a Sylow p-subgroup is a Sylow p-subgroup. Its converse is also true by Theorem 3.2.5.

Theorem 3.2.5 (Second Sylow Theorem) *Let P and H be Sylow p-subgroups of a finite group G. Then P and H are conjugate subgroups of G.*

Proof Let $\mathcal{S}$ be the set of all left cosets of P in G and let H act on $\mathcal{S}$ by (left) translation, i.e., $h(xP) = hxP$ $\forall h \in H$. So, $\mathcal{S}$ is a left H-set. Then by Theorem 3.2.2, $|\mathcal{S}_H| \equiv |\mathcal{S}|$ (mod p). Since H is a Sylow p-subgroup of G, $|\mathcal{S}| = [G : H]$ is not

Table 3.1 The number of groups of order ≤ 32, up to isomorphisms

$|G|$	2	3	4	5	6	7	8	9	10	11	12	13	14	15	16	17
Number of groups	1	1	2	1	2	1	5	2	2	1	5	1	2	1	14	1

$|G|$	18	19	20	21	22	23	24	25	26	27	28	29	30	31	32
Number of groups	5	1	5	2	2	1	15	2	2	5	4	1	4	1	51

divisible by p, so $|\mathcal{S}_H| \neq 0$. Hence $\exists x P \in \mathcal{S}_H$. Now $x P \in \mathcal{S}_H \Leftrightarrow h(x P) = x P$ $\forall h \in H \Leftrightarrow x^{-1} h x P = P \ \forall h \in H$. Hence $x^{-1} H x \subseteq P \Rightarrow H \subseteq x P x^{-1}$. Again P and H are Sylow p-subgroups of $G \Rightarrow |H| = |P| = |x P x^{-1}| \Rightarrow H = x P x^{-1} \Rightarrow P$ and H are conjugate subgroups of G. $\qquad\square$

The following theorem prescribes the number of Sylow p-subgroups of a finite group.

Theorem 3.2.6 (Third Sylow Theorem) *If G is a finite group and p divides $|G|$, then the number of Sylow p-subgroups is congruent to 1 modulo p and divides $|G|$.*

Proof Let P be a Sylow p-subgroup of G and $\mathcal{S}$ the set of all Sylow p-subgroups and P act on $\mathcal{S}$ by conjugation. Then by Theorem 3.2.2, $|\mathcal{S}| \equiv |\mathcal{S}_P| \pmod{p}$, where $\mathcal{S}_P = \{Q \in \mathcal{S} : x Q x^{-1} = Q \ \forall x \in P\}$. Since $P \in \mathcal{S}_P$, $\mathcal{S}_P \neq \emptyset$. We claim that $\mathcal{S}_P = \{P\}$. Now $Q \in \mathcal{S}_P \Leftrightarrow x Q x^{-1} = Q \ \forall x \in P \Rightarrow P \subseteq N(Q)$. Both P and Q are Sylow p-subgroups of G and hence of $N(Q)$. Consequently P and Q are conjugates in $N(Q)$ by Theorem 3.2.5. But since Q is normal in $N(Q)$, it is only conjugate in $N(Q)$. This implies $P = Q$. Consequently, $\mathcal{S}_P = \{P\}$. Then

$$|\mathcal{S}| \equiv 1 \pmod{p}.$$

Now let G act on $\mathcal{S}$ by conjugation. Then the last part follows. $\qquad\square$

Remark We now list all the groups G of order ≤ 7 up to isomorphisms.

If $|G| = 2$, then $G \cong \mathbf{Z}_2$.
If $|G| = 3$, then $G \cong \mathbf{Z}_3$.
If $|G| = 4$, then $G \cong \mathbf{Z}_4$ or $\mathbf{Z}_2 \oplus \mathbf{Z}_2$.
If $|G| = 5$, then $G \cong \mathbf{Z}_5$.
If $|G| = 6$, then a non-abelian group appears for the first time and $G \cong S_3$ or $\mathbf{Z}_2 \oplus \mathbf{Z}_3$.
If $|G| = 7$, then $G \cong \mathbf{Z}_7$.

Table 3.1 gives the number of groups of order ≤ 32, up to isomorphisms.

3.2.3 *Exercises*

Exercises-I

1. If G is a group of order p^r, $r > 1$, show that G has a normal subgroup of order p^{r-1}.

2. Suppose G is a group and H is a subgroup of the center $Z(G)$. Show that G is abelian if G/H is cyclic.

3. Show that for a prime number p, every group of order p^2 is abelian.

   [*Hint*. Let G be a group of order p^2 and $Z(G)$ its center. Then $Z(G) \neq \{1\}$. If $Z(G) = G$, then G is abelian. Suppose $Z(G) \neq G$, then $|G/Z(G)| = p \Rightarrow G/Z(G)$ is cyclic $\Rightarrow G$ is abelian by Exercise 2.]

4. Show that a non-abelian group G of order p^3 has a center of order p (p is prime).

   [*Hint*. $Z(G) \neq \{1\}$. Also $Z(G) \neq G$, since G is non-abelian. $|Z(G)| = p^2 \Rightarrow |G/Z(G)| = p \Rightarrow G/Z(G)$ is cyclic $\Rightarrow G$ is abelian (by Exercise 2) $\Rightarrow$ a contradiction $\Rightarrow |Z(G)| = p$.]

5. If P is a p-Sylow subgroup of a finite group G and $x \in G$, show that xPx^{-1} is also a p-Sylow subgroup of G.

   [*Hint*. Suppose $|G| = p^m n$, where p is prime such that $p\nmid n$ and if P is a p-Sylow subgroup of G, then $|P| = p^m$. For $x \in G$, xPx^{-1} is a subgroup of G. Define a mapping $f : P \to xPx^{-1}$, $y \mapsto xyx^{-1}$ $\forall y \in P$. Show that f is bijective. Then $|P| = p^m = |xPx^{-1}| \Rightarrow xPx^{-1}$ is a p-Sylow subgroup of G.]

6. If a finite group G has only one p-Sylow subgroup P, show that P is normal in G. Its converse is also true.

   [*Hint*. Using Exercise 5, xPx^{-1} is a p-Sylow subgroup $\forall x \in G \Rightarrow xPx^{-1} = P$ (by hypothesis) $\Rightarrow P$ is normal in G.]

7. Show that a group G of order 30 is not a simple group.

   [*Hint*. $|G| = 30 = 5 \cdot 3 \cdot 2$. Prove that G has either a normal subgroup of order 5 or a normal subgroup of order 3. Hence G is not a simple group.]

8. Examine whether a group G of order 56 is simple or not.

   [*Hint*. Show that G has at least one non-trivial normal subgroup. So, G is not simple.]

9. Show that a group G of order 108 is not simple.

   [*Hint*. Prove that G has a non-trivial normal subgroup of order 9.]

10. Let G be a group of order pq, where p and q are prime numbers such that $p > q$ and $q\nmid(p - 1)$. Show that G is cyclic.

    [*Hint*. $|G| = pq$. Let n_p be the number of Sylow p-subgroups of G. Then $n_p | pq$ and $n_p = pr + 1$, $(r = 0, 1, 2, \ldots) \Rightarrow pq = sn_p = s(pr + 1)$ for some positive integer $s \Rightarrow s = pq - spr = p(q - sr) = pt$ (say), where $t = q - sr < p$ as $q < p$.

    Hence $pq = s(pr + 1) = pt(pr + 1) \Rightarrow q = t(pr + 1) \Rightarrow r = 0$ (as $q < p$) $\Rightarrow n_p = 1 \Rightarrow G$ contains only one Sylow p-subgroup H (say) such that $|H| = p \Rightarrow H$ is normal in G. Again let n_q be the number of Sylow q-subgroups of G. Then $n_q | pq$ and $n_q = qr' + 1$, $(r' = 0, 1, 2, \ldots) \Rightarrow pq =$

$s' n_q = s'(qr' + 1)$ for some positive integer $s' \Rightarrow s' = q(p - s'r') = qt'$ (say), where $t' = p - s'r'$. Hence $pq = qt'(qr' + 1) \Rightarrow p = t'(qr' + 1)$. Since $q|(p - 1)$, $p = t'(qr' + 1) \Rightarrow r' = 0 \Rightarrow n_q = 1 \Rightarrow G$ contains only one Sylow q-subgroup K (say) such that $|K| = q \Rightarrow K$ is normal in G. Now proceed to show that G is cyclic.]

11. Let G be a group containing an element of finite order $n(> 1)$ and exactly two conjugate classes. Show that $|G| = 2$.

12. Show that any group G of order 35 is cyclic.

13. If the order of a group G is 42, prove that G contains a unique Sylow 7-subgroup which is normal.

14. Show that the center $Z(G)$ of a non-trivial finite p-group G contains more than one element.

15. Prove that there exists no simple group of order 48.

16. Prove that any group of order 15 is cyclic.

17. Prove that a group of order 65 is cyclic.

18. Prove that no group of order 65 is simple.

19. If a non-trivial finite group G has no non-trivial subgroups, then prove that G is a group of prime order.

20. Let G be a finite group. If G has exactly one non-trivial subgroup, then prove that the order of G is p^2 for some prime p.

21. Prove that every group G of order 45 has a unique Sylow 3-group of order 9, which is normal.

22. Let G be a group of order p^n (p is prime and $n > 1$). Then prove that G is not a simple group.

23. Let G be a group of order pq, where p and q are prime numbers. Then prove that G is not a simple group.

24. Prove that any group of order $2p$ (p is a prime) has a normal subgroup of order p.

25. Identify the correct alternative(s) (there may be more than one) from the following list:

   - Let p be a prime number and $GL_{50}(F_p)$ be the group of invertible 50×50 matrices with entries from the finite field F_p. Then the order of a p-Sylow subgroup of the group $GL_{50}(F_p)$ is:

     $$\text{(a) } p^{50} \qquad \text{(b) } p^{1250} \qquad \text{(c) } p^{125} \qquad \text{(d) } p^{1225}.$$

   - If G is the group given by $G = \mathbf{Z}_{10} \times \mathbf{Z}_{15}$, then

     (a) G contains exactly one element of order 2;
     (b) G contains exactly 24 elements of order 5;
     (c) G contains exactly five elements of order 3;
     (d) G contains exactly 24 elements of order 10.

   - If G is the group given by $S_4 \times S_3$, then

     (a) G has a normal subgroup of order 72;
     (b) a 3-Sylow subgroup of G is normal;

(c) a 2-Sylow subgroup of G is normal;

(d) G has a non-trivial normal subgroup.

- Which of the following statement(s) is (are) valid?

(a) Any group of order 15 is abelian;

(b) Any group of order 25 is abelian;

(c) Any group of order 65 is abelian;

(d) Any group of order 55 is abelian.

3.3 Actions of Topological Groups and Lie Groups

The actions of topological groups and Lie groups are very important in topology and geometry. Many well known geometrical objects are obtained as orbit spaces.

Definition 3.3.1 A topological group G is said to act on a topological space X from the left iff there is a continuous function $\sigma : G \times X \to X$, denoted $(g, x) \mapsto g \cdot x$ (or gx), $\forall x \in X$, $\forall g \in G$ such that

(i) for any $x \in X$, $1 \cdot x = x$, where 1 is the identity element of G;

(ii) for any $x \in X$, $g_1, g_2 \in G$, $(g_1 g_2) \cdot x = g_1 \cdot (g_2 \cdot x)$.

σ is said to be a topological action (or just an action) of G on X from the left.

The ordered triple (X, G, σ) is called a *topological transformation group* and X is said to be a *G-space*. A right action of G on X is defined in a similar way. There is a bijective correspondence between the left and right G-space structures.

Example 3.3.1 (i) The space $\mathbf{R}^n$ is a left $GL(n, \mathbf{R})$ space under the usual multiplication of matrices.

(ii) The space $\mathbf{R}^n$ is also a left $O(n)$-space [see Example 2.9.1 of Chap. 2].

Definition 3.3.2 Let X be a left G-space and $X \bmod G$ be the set of all orbits Gx, $\forall x \in X$, with the quotient topology, i.e., the largest topology such that the projection map $p : X \to X \bmod G$, $x \mapsto Gx$ is continuous. The space $X \bmod G$ is called the *orbit space* of the action σ of the transformation group (X, G, σ) and p is also called an *identification map*. In particular, if X is a Hausdorff space and G is a finite topological group acting on X in such a way that $g \cdot x = 1$, for some $x \in X$ implies $g = 1$, then the action is said to be *free*.

Some Important Examples

Example 3.3.2 (Real projective space) Let $S^n = \{x \in \mathbf{R}^{n+1} : \|x\| = 1\}$ be the n-sphere in $\mathbf{R}^{n+1}$, for $n \geq 1$ and $A : S^n \to S^n$, $x \mapsto -x$ be the antipodal map. Then $A^2 = A \circ A = I$ (identity map). The group $G = \{A, I\}$ is a group of homeomorphisms on S^n and acts on S^n. The orbit space $S^n \bmod G$ is called the *real projective n-space*, denoted $\mathbf{R}P^n$. This action is free.

Example 3.3.3 (Complex projective space) Let the circle group S^1 act on $S^{2n+1} = \{(z_0, z_1, \ldots, z_n) \in \mathbf{C}^{n+1} : \sum_{i=0}^{n} |z_i|^2 = 1\}$ continuously under the action

$$z \cdot (z_0, z_1, \ldots, z_n) \mapsto (zz_0, zz_1, \ldots, zz_n).$$

The orbit space S^{2n+1} mod S^1 is called the *complex projective n-space*, denoted $\mathbf{C}P^n$. This action is free.

Example 3.3.4 (Torus) Let $f : \mathbf{R} \to \mathbf{R}$, $x \mapsto x + 1$. Then f is a homeomorphism and for each integer n, $f^n : \mathbf{R} \to \mathbf{R}$ is also a homeomorphism. The infinite cyclic group $\mathbf{Z} = \langle f \rangle$ endowed with the discrete topology acts on a group of homeomorphisms on $\mathbf{R}$. This action is free and the orbit space $\mathbf{R}$ mod $\mathbf{Z}$ is homeomorphic to the circle group S^1. Again consider the action of the discrete group $\mathbf{Z} \times \mathbf{Z}$ on $\mathbf{R} \times \mathbf{R}$ by setting $(f, g)(x, y) = (x + 1, y + 1)$. Then $(f^m, g^n)(x, y) = (x + m, y + n)$ for every pair of integers (m, n). This action is free and the orbit space is $T = S^1 \times S^1$, which is called a 2-*torus* (or simply *torus*). An n-torus is defined similarly as an orbit space of $\mathbf{R}^n$, $\forall n \geq 2$.

Definition 3.3.3 A real Lie group G is said to act on a differentiable manifold M from the left iff there is a C^∞ function $\sigma : G \times M \to M$, $(g, x) \mapsto g \cdot x$ (or gx) $\forall x \in M$ and $\forall g \in G$ such that

(i) for any $x \in M$, $1 \cdot x = x$, where 1 is the identity element of G;
(ii) for any $x \in M$, $g_1, g_2 \in G$, $(g_1g_2) \cdot x = g_1 \cdot (g_2 \cdot x)$.

Definition 3.3.4 A complex Lie group G is said to act on a complex manifold M from the left iff there is a holomorphic function $\sigma : G \times M \to M$, $(g, x) \mapsto gx$ (or gx), $\forall x \in M$ and $\forall g \in G$ such that

(i) for any $x \in M$, $1 \cdot x = x$, where 1 is the identity element of G;
(ii) for any $x \in M$, $g_1g_2 \in G$, $(g_1g_2) \cdot x = g_1 \cdot (g_2 \cdot x)$.

Right actions of G on M are defined dually.

Definition 3.3.5 If a Lie group G acts on manifolds, then G-manifolds (homogeneous), G-homomorphisms, G-isomorphisms, G-automorphisms etc. are defined in usual ways.

3.3.1 Exercises

Exercises-II

1. Let X be a homogeneous left G-set. Prove the following:

   (a) If $\sigma : X \to X$ is an automorphism of X, then the points x and $\sigma(x)$ have the same isotropy subgroup;
   (b) if x and $y \in X$ have the same isotropy subgroup, then there exists an automorphism σ of X such that $\sigma(x) = y$;

(c) if σ and ρ are automorphisms of X such that for some $x \in X$, $\sigma(x) = \rho(x)$, then $\sigma = \rho$;

(d) a group A of automorphisms of X is the entire group of automorphisms iff for any two points $x, y \in X$ which have the same isotropy subgroup, there exists an automorphism $\sigma \in A$ such that $\sigma(x) = y$.

2. Let X be a homogeneous left G-set and H the isotropy subgroup of G corresponding to the point $x_0 \in X$. Prove that $A(X)$, the group of automorphisms of X is isomorphic to $N(H)/H$, where $N(H)$ is the normalizer of H.

3. A *topological group* G is a set G with a group structure and topology on G such that the function $G \times G \to G$, $(s, t) \mapsto st^{-1}$ is continuous. Show that this condition of continuity is equivalent to the statement that the functions $G \times G \to G$, $(s, t) \mapsto st$ and $G \to G$, $s \mapsto s^{-1}$ are both continuous, i.e., the group operations in G are continuous in the topological space G.

For a topological group G, a left G-space is a topological space X together with a map $G \times X \to X$ (The image of $(s, x) \in G \times X$ under the map is sx) such that

(a) for each $x \in X$, $s, t \in G$, $(st)x = s(tx)$;
(b) for each $x \in X$, the relation $1x = x$ holds.

Similarly, a right G-space is defined.

Show that there is a bijective correspondence between the left and right G-space structures.

4. Two elements $x, y \in X$ in a left G-space are called G-equivalent iff $\exists s \in G$ such that $sx = y$. Show that this relation is an equivalence relation. Suppose $Gx = \{$all $sx : s \in G\}$, the equivalence class determined by $x \in X$ and $X \bmod G = \{$all $Gx : x \in X\}$, with the quotient topology, i.e., the largest topology such that the projection $p : X \to X \bmod G$ is continuous.

Prove that the map $X \to X$ (X is a G-space), $x \mapsto xs$ is a homeomorphism and the projection $p : X \to X \bmod G$ is an open map (see Simmons (1963, p. 93)).

5. Let X be a G-space and $X \bmod G$ be the orbit space. Show that the projection map $p : X \to X \bmod G$ is an open map.

[*Hint.* Let V be an open subset of X. Then $p^{-1}p(V) = \bigcup_{g \in G} gV$ is an open set, as it is a union of open sets gV. Hence $p(V)$ is open in $X \bmod G$ for each open set V of X.]

6. Let a compact topological group G act on a space X and $X \bmod G$ be its orbit space. Show that

(a) if X is Hausdorff, then $X \bmod G$ is also so;
(b) if X is regular, then $X \bmod G$ is also so;
(c) if X is normal, then $X \bmod G$ is also so;
(d) if X is locally compact, then $X \bmod G$ is also so.

7. Let X be a G-space, where G is a compact topological group and X is a Hausdorff space and G_x be its isotropy group at x. Show that the continuous map $f : G/G_x \to \text{orb}(x)$, $gG_x \mapsto gx$ is a homeomorphism.

   Hence show that the spaces $U(n)/U(n-1)$ and S^{2n-1} are homeomorphic, where $U(n)$ is the unitary group.

   [*Hint.* Use Lemma 3.1.1 to show that f is a bijection, which is a homeomorphism in this case. The isotropy group at $(0, 0, \ldots, 0, 1)$ is $U(n-1)$.]

8. Show that the map $f : GL(n, \mathbf{R}) \times \mathbf{R}^n \to \mathbf{R}^n$, defined by $(A, X) \mapsto AX$, the product of the $n \times n$ matrix $A \in GL(n, \mathbf{R})$ and $n \times 1$ column matrix $X \in \mathbf{R}^n$ is an action of $GL(n, \mathbf{R})$ on $\mathbf{R}^n$.

   Hence show that the action of the group $O(n)$ of all orthogonal real matrices is transitive on S^{n-1} and the spaces $O(n)/O(n-1)$ and S^{n-1} are homeomorphic.

9. (Irrational flow). Let α be a fixed irrational number, $T = S^1 \times S^1$ be the torus and $\mathbf{R}$ be the additive group of reals. Define an action $\sigma : \mathbf{R} \times T \to T$,

$$r \cdot \left(e^{2\pi i x}, e^{2\pi i y}\right) \mapsto \left(e^{2\pi i (x+r)}, e^{2\pi i (y+\alpha r)}\right).$$

Show that σ is a free action.

10. Let X be a G-space. Show that for each $g \in G$, the map $\phi_g : X \to X$, $x \mapsto gx$ is a homeomorphism and the map $\psi : G \to \text{Homeo}(X)$, $g \mapsto \phi_g$ is a group homomorphism.

    [*Hint.* ϕ_g is a bijection by Theorem 3.1.1. Since the action is continuous, it follows that ϕ_g is a homeomorphism.]

11. If a real Lie group G acts on a differentiable manifold M, show that for each $g \in G$, the function $\phi_g : M \to M$, $x \mapsto gx$ is a diffeomorphism.

12. Let G be a Lie group and M be a homogeneous G-manifold. Show that

    (a) M is G-isomorphic to the G-manifold G/H for some closed Lie subgroup H of G;

    (b) if $\psi : M \to M$ is a G-automorphism, then x and $\psi(x)$ determine the same closed Lie subgroup of G.

13. Let G be a topological group and X a topological space. Suppose the group G acts on the set X. If this action $G \times X \to X$ is continuous, then G is said to act on X.

    Prove the following:

    Let G be a topological group acting on a topological space X. Then

    (i) $H = \{h \in G : hx = x \ \forall x \in X\}$ is a closed normal subgroup of G.

    (ii) If G is compact, H is Hausdorff and G_x is the isotropy group of x, the continuous map $f : G/G_x \to Gx$ defined by $f(gG_x) = gx$ is a homeomorphism.

    [*Hint.* Use Lemma 3.1.1 to show that f is a bijection. Then use the result that if X is compact and Y is Hausdorff and $f : X \to Y$ is continuous and bijective. Hence f is a homeomorphism.]

3.4 Actions of Semigroups and State Machines

Actions of semigroups are important both in mathematics and computer science. The theory of machines developed so far has largely influenced the development of computer science and its associated language. In this section we discuss semigroup actions and their applications to the theory of state machines to unify computer science with the *mainstream mathematics*.

Definition 3.4.1 A semigroup S is said to act on a non-empty set X from the left iff there is a function $\sigma : S \times X \to X$, denoted by $(a, x) \mapsto ax$ (or $a \cdot x$) such that

(i) for any $x \in X$, $a, b \in S$, $(ab)x = a(bx)$;
(ii) if identity $1 \in S$, then for any $x \in X$, $1x = x$.

σ is said to be an *action* of S on X from the left.

Similarly, a right *action* of S on X is defined.

There is a feeling that much of abstract algebra is of little practical use. However, we apply semigroups to the theory of machines and establish some relations between a state machine homomorphism and a homomorphism of the corresponding transformation semigroups. In this section we describe algebraic aspect of finite state machines. We now proceed to unify finite state machines with semigroup actions. So we need the concept of transformation semigroups, which involves an action of a semigroup on a finite set and another condition on the semigroup action as given below.

Definition 3.4.2 A transformation semigroup is a pair (Q, S) consisting of a finite set Q, a finite semigroup S and a semigroup action $\lambda : Q \times S \to Q$, $(q, s) \mapsto qs$ which means

(i) $q(st) = (qs)t$, $\forall q \in Q$, $\forall s, t \in S$; and such that
(ii) $qs = qt$ $\forall q \in Q \Rightarrow s = t$, $s, t \in S$.

Definition 3.4.3 If X and Y are non-empty sets and $R : X \to Y$ is a relation such that $(x, y) \in R$ and $(x, z) \in R \Rightarrow y = z$, where $x \in X$ and $y, z \in Y$, then R is called a partial function. A partial function $R : X \to Y$ is said to be a function iff $\mathrm{dom}(R) = \{x \in X : (x, y) \in R \text{ for some } y \in Y\} = X$.

Definition 3.4.4 A state machine or a *semiautomation* is an ordered triple $\mu = (Q, \Sigma, F)$, where Q and Σ are finite sets and $F : Q \times \Sigma \to Q$ is a partial function.

Definition 3.4.5 A state machine $\mu = (Q, \Sigma, F)$ is said to complete iff $F : Q \times \Sigma \to Q$ is a function.

Such systems can be successfully investigated by using the algebraic techniques, an important achievement of modern algebra.

Example 3.4.1 (Cyclic state machine) Let m, n be positive integers and $Q = \{0, 1, 2, \ldots, m + n - 1\}$, then the diagram of a cyclic state machine is given below:

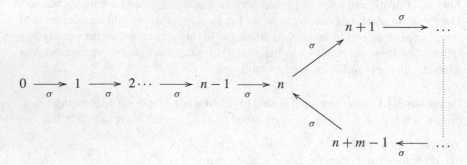

where $F(0, \sigma) = 1$, $F(1, \sigma) = 2, \ldots, F(n-1, \sigma) = n$, $F(n, \sigma) = n+1, \ldots, F(n + m - 2, \sigma) = n + m - 1$ and $F(n + m - 1, \sigma) = n$.

We now associate a semigroup with a state machine.

Let $\mu = \{Q, \Sigma, F\}$ be a state machine with a symbol $\sigma \in \Sigma$ when it is in some state q (say) $\in Q$. The machine then moves to the state $F(q, \sigma) \in Q$. We may equally define $F_\sigma : Q \to Q$ by $F_\sigma(q) = F(q, \sigma)$, $\forall q \in Q$.

Let $\mu = (Q, \Sigma, F)$ be a state machine. Consider the set Σ^+ of all words of length ≥ 1 in the alphabet Σ. Define an equivalence relation ρ on Σ^+ by $\alpha \rho \beta \Leftrightarrow F_\alpha = F_\beta$. An equivalence relation ρ on a semigroup S is said to be a congruence relation on S iff $(a, b) \in \rho \Rightarrow (ca, cb) \in \rho$ and $(ac, bc) \in \rho$ $\forall c \in S$.

Proposition 3.4.1 Σ^+/ρ *is a semigroup.*

Proof Σ^+ admits a semigroup structure by *concatenation* of words. Moreover, ρ is a congruence relation on Σ^+ and hence Σ^+/ρ, the set of all congruence classes (a) becomes a semigroup under the binary operation given by

$$(a)(b) = (ab). \qquad \square$$

Definition 3.4.6 The quotient semigroup Σ^+/ρ is called the semigroup of the state machine μ and is denoted by $S(\mu)$.

Let $\mu = (Q, \Sigma, F)$ be a state machine and $PF(Q)$ be the semigroup of all partial functions from Q to Q, under the usual composition of relations on Q.

Each $\sigma \in \Sigma$ defines a partial function $F_\sigma : Q \to Q \Rightarrow \exists$ a natural function $F : \Sigma \to PF(Q)$, $\sigma \mapsto F_\sigma$. Let $\langle F(\mu) \rangle$ denote the subsemigroup of $PF(Q)$ generated by the set of partial functions $\{F_\sigma : \sigma \in \Sigma\}$.

Corresponding to a state machine $\mu = (Q, \Sigma, F)$, there is a transformation semigroup $(Q, S(\mu))$, denoted by $TS(\mu)$, called the transformation semigroup of μ. Conversely, each transformation semigroup determines a state machine. Suppose $T = (Q, S)$ is a transformation semigroup. We define a state machine $\mu = (Q, S, F)$, where $F : Q \times S \to Q$ is defined by $F(q, s) = qs$, $\forall q \in Q$ and $\forall s \in S$. μ is called the state machine of T, denoted by $SM(T)$.

Definition 3.4.7 Let $\mu = (Q, \Sigma, F)$ and $\mu' = (Q', \Sigma', F')$ be state machines. A pair of maps $(\alpha, \beta) : \mu \to \mu'$ is said to be a state machine homomorphism iff $\alpha : Q \to Q'$ and $\beta : \Sigma \to \Sigma'$ is a pair of maps such that $\alpha \circ F_\sigma = F'_{\beta(\sigma)} \circ \alpha$, $\forall \sigma \in \Sigma$, where $F_\sigma : Q \to Q$ is defined by $F_\sigma(q) = F(q, \sigma)$, $\forall q \in Q$.

Definition 3.4.8 Let $\mu = (Q, \Sigma, F)$ and $\mu' = (Q', \Sigma', F')$ be state machines. A state machine homomorphism $(\alpha, \beta) : \mu \to \mu'$ is said to be

  (i) a monomorphism iff α and β are both injective;
 (ii) an epimorphism iff α and β are both surjective;
(iii) an isomorphism iff (α, β) is both a monomorphism and an epimorphism (written $\mu \cong \mu'$).

3.4.1 Exercises

Exercises-III

1. Let $\mu = (Q, \Sigma, F)$ be a state machine. Show that $\langle F(\mu) \rangle \cong S(\mu)$, where $S(\mu)$ is the semigroup of the state machine μ.
2. Let (Q, S) be a transformation semigroup. Show that there is a natural embedding $\theta : S \to PF(Q)$ and conversely, given any set Q and a subsemigroup $S \subseteq PF(Q)$, (Q, S) is a transformation semigroup.
   [*Hint*. Define $\theta_s : Q \to Q$, $q \mapsto qs$ for each $q \in Q$ and each $s \in S$. Then $\theta : S \to PF(Q)$, $s \mapsto \theta_s$ is a semigroup monomorphism.]
3. Let $\mu = (Q, \Sigma, F)$ be a state machine. Show that there exists a state machine monomorphism $(\alpha, \beta) : \mu \to SM(TS(\mu))$, where $SM(TS(\mu))$ is the state machine of the transformation semigroup $TS(\mu)$ of μ.
4. Let $\mu = (Q, \Sigma, F)$ and $\mu' = (Q', \Sigma', F')$ be complete state machines and $(\alpha, \beta) : \mu \to \mu'$ a state machine homomorphism such that α is onto. Show that $\exists$ a transformation semigroup homomorphism $(f_\alpha, g_\beta) : TS(\mu) \to TS(\mu')$.
5. Let $\mathcal{A} = (Q, S)$ be a transformation semigroup. Show that $TS(SM(\mathcal{A})) \cong \mathcal{A}$.
   [*Hint*. Suppose $SM(\mathcal{A}) = (Q, S, F)$ is the state machine of $\mathcal{A}$ and B is the semigroup generated by $SM(\mathcal{A})$. Then the map $\theta : S \to B$, $s \mapsto \theta_s$ (see Ex. 2) is a semigroup isomorphism. Hence the pair $(I_Q, \theta) : \mathcal{A} \to TS(SM(\mathcal{A}))$ is a transformation semigroup isomorphism.]

3.5 Additional Reading

We refer the reader to the books (Adhikari and Adhikari 2003; Artin 1991; Bredon 1993; Fraleigh 1982; Ginsburg 1968; Herstein 1964; Holcombe 1982; Hungerford

1974; Jacobson 1974, 1980; Lang 1965; Rotman 1988; Spanier 1966; van der Waerden 1970) for further details.

References

Adhikari, M.R., Adhikari, A.: Groups, Rings and Modules with Applications, 2nd edn. Universities Press, Hyderabad (2003)

Artin, M.: Algebra. Prentice-Hall, Englewood Cliffs (1991)

Bredon, G.E.: Topology and Geometry. Springer, New York (1993)

Fraleigh, J.B.: A First Course in Abstract Algebra. Addison-Wesley, Reading (1982)

Ginsburg, A.: Algebraic Theory of Automata. Academic Press, New York (1968)

Herstein, I.: Topics in Algebra. Blaisdell, New York (1964)

Holcombe, W.: Algebraic Automata Theory. Cambridge University Press, New York (1982)

Hungerford, T.W.: Algebra. Springer, New York (1974)

Jacobson, N.: Basic Algebra I. Freeman, San Francisco (1974)

Jacobson, N.: Basic Algebra II. Freeman, San Francisco (1980)

Lang, S.: Algebra, 2nd edn. Addison-Wesley, Reading (1965)

Rotman, J.J.: An Introduction to Algebraic Topology. Springer, New York (1988)

Simmons, G.F.: Topology and Modern Analysis. McGraw-Hill, New York (1963)

Spanier, E.H.: Algebraic Topology. McGraw-Hill, New York (1966)

van der Waerden, B.L.: Modern Algebra. Ungar, New York (1970)

Chapter 4
Rings: Introductory Concepts

In the earlier chapters we have studied groups and their applications. Another fundamental concept in the study of modern algebra is that of a ring. Rings also serve as one of the fundamental building blocks for modern algebra. A group is endowed with only one binary operation but a ring is endowed with two binary operations connected by distributive laws. Fields form a very important class of rings. Under usual addition and multiplication, the set of integers **Z** is the prototype of the ring structure and the sets **Q**, **R**, **C** (of rational numbers, real numbers, complex numbers) are the prototypes of the field structures. The concept of rings arose through the attempts to prove Fermat's Last Theorem and was initiated by Richard Dedekind (1831–1916) around 1880. David Hilbert (1862–1943) coined the term '*ring*'. Emmy Noether (1882–1935) developed the theory of rings under his guidance. Commutative rings play an important role in algebraic number theory and algebraic geometry, and non-commutative rings are used in non-commutative geometry and quantum groups. This chapter starts with introductory concepts of rings and their properties with illustrative examples. Some important rings such as rings of power series, rings of polynomials, Boolean rings, rings of continuous functions, rings of endomorphisms of abelian groups are also studied in this chapter. Further study of theory of rings is given in Chaps. 5–7.

4.1 Introductory Concepts

The abstract concept of a ring has its origin in the set of integers **Z**. The basic difference between a group and a ring is that a group is of one-operational algebraic system and a ring is of two-operational algebraic system. Despite this basic difference, while studying ring theory, many techniques already used for groups are also applied for rings. For example, we obtain in ring theory the appropriate analogues of group homomorphisms, normal subgroups, quotient groups, homomorphism theorems, etc. Integral domains and fields are special classes of rings.

M.R. Adhikari, A. Adhikari, *Basic Modern Algebra with Applications*,   159
DOI 10.1007/978-81-322-1599-8_4, © Springer India 2014

Definition 4.1.1 A *ring* is an ordered triple $(R, +, \cdot)$ consisting of a non-empty set R and two binary operations '+' (called *addition*) and '·' (called *multiplication*) such that

(i) $(R, +)$ is an abelian group;
(ii) $(R, \cdot)$ is semigroup, i.e., $(a \cdot b) \cdot c = a \cdot (b \cdot c)$ for all $a, b, c \in R$; and
(iii) the operation '·' is *distributive* (on both sides) over the operation '+', i.e.,
$a \cdot (b + c) = a \cdot b + a \cdot c$ and $(b + c) \cdot a = b \cdot a + c \cdot a$ for all $a, b, c \in R$
(*distributive laws*).

We adopt the usual convention of designating a ring $(R, +, \cdot)$ simply by the set symbol R and assume that '+' and '·' are known. If in addition, the operation '·' is commutative i.e., $a \cdot b = b \cdot a \; \forall a, b \in R$, R is said to be a *commutative ring*. Otherwise, R is said to be non-commutative. If there exists an element $1 \in R$ such that $1 \cdot a = a \cdot 1 = a$, $\forall a \in R$, 1 is said to be an identity element of R. The identity element is then unique.

Remark A ring R may or may not contain identity element. If there exists an element $l(r) \in R$ such that $l \cdot a = a \; (a \cdot r = a)$ for all $a \in R$, then $l \; (r)$ is called a left (right) identity element of R. It may happen that a R may have a left (right) identity but no right (left) identity. But if both of them exist, then they are equal and we say that the identity element exists in R. On the other hand, if R contains more than one left (right) identity element, then R cannot contain the identity element.

As $(R, +)$ is an abelian group, R has a *zero* element denoted by 0, and every $a \in R$ has a unique (additive) inverse, $-a$. Following usual convention, we can write ab in place of $a \cdot b$ and $a - b$ in place of $a + (-b)$.

Theorem 4.1.1 *If R is a ring, then for any $a, b, c \in R$,*

(i) $0a = a0 = 0$;
(ii) $a(-b) = (-a)b = -(ab)$;
(iii) $(-a)(-b) = ab$;
(iv) $a(b - c) = ab - ac$, $(b - c)a = ba - ca$.

Proof (i) From $0 = 0 + 0$, it follows that $0a = (0 + 0)a = 0a + 0a \Rightarrow 0a = 0$ by the cancellation law for the additive group $(R, +)$. Similarly, it follows that $a0 = 0$.

(ii) $a(-b) + ab = a(-b + b) = a0 = 0 \Rightarrow a(-b) = -(ab)$ and $(-a)b + ab = (-a + a)b = 0b = 0 \Rightarrow (-a)b = -(ab)$.

(iii) $(-a)(-b) = -(a(-b))$ by (ii). Now $a(-b) + ab = a(-b + b) = a0 = 0 \Rightarrow -(a(-b)) = ab$. Consequently, $(-a)(-b) = ab$.

(iv) follows form (ii). □

For each positive integer n, we define the nth natural multiple na recursively as follows: $1a = a$ and $na = (n - 1)a + a$ when $n > 1$. If it is agreed to let $(-na) = -(na)$, then the definition of na can be extended to all integers.

Theorem 4.1.2 *If R is a ring, then for any $a, b \in R$ and $m, n \in \mathbf{Z}$,*

(i) $(m+n)a = ma + na$;

(ii) $(mn)a = m(na)$;

(iii) $m(a+b) = ma + mb$;

(iv) $m(ab) = (ma)b = a(mb)$, *and* $(ma)(nb) = (mn)(ab)$.

Proof Trivial. □

We do not exclude the possibility: a ring R has an identity $1 = 0$. If so, then for any $a \in R$, $a = a1 = a0 = 0 \Rightarrow R$ has only one element 0. This ring is called the *zero ring*, denoted by $\{0\}$. So we assume that any ring with identity contains more than one element and this will exclude the possibility that $1 = 0$.

Example 4.1.1 (i) The triples $(\mathbf{Z}, +, \cdot)$, $(\mathbf{Q}, +, \cdot)$, $(\mathbf{R}, +, \cdot)$ and $(\mathbf{C}, +, \cdot)$ are all commutative rings, where '$+$' is the usual addition and '$\cdot$' is the usual multiplication, with the integer 1 as the multiplicative identity element. On the other hand the ring $2\mathbf{Z}$ of even integers has no multiplicative identity.

(ii) Given a ring R, let $M_2(R)$ denote the set of all 2×2 matrices over R. Then $(M_2(R), +, \cdot)$ is a non-commutative ring under usual compositions of matrices induced by those of R. The matrix ring $M_n(R)$ is defined analogously.

(iii) Given a non-empty set X, let $\mathcal{P}(X)$ denote the power set of X. Then $(\mathcal{P}(X), +, \cdot)$ is a commutative ring, where $A + B = A \triangle B = (A \setminus B) \cup (B \setminus A)$ and $A \cdot B = A \cap B$ for $\forall A, B \in \mathcal{P}(X)$. Clearly, $\emptyset$ is the zero element and X is the identity element. This ring was introduced by *George Boole* (1815–1864) as a formal notation for assertions in logic (see Ex. 4 of Exercises-I).

Remark Neither $(\mathcal{P}(X), \cup, \cap)$ nor $(\mathcal{P}(X), \cap, \cup)$ forms a ring.

(iv) $(\mathbf{Z}_n, +, \cdot)$ $(n > 1)$ is a commutative ring under usual addition and multiplication of classes. Clearly, (0) is its zero element and (1) is its identity element.

(v) Let X be a non-empty set, $(R, +, \cdot)$ a ring and M the set of all mappings from X into R. In M define the pointwise sum and product denoted by $f + g$ and $f \cdot g$, respectively, of two mappings $f, g \in M$ by $(f + g)(x) = f(x) + g(x)$ and $(f \cdot g)(x) = f(x) \cdot g(x)$ $\forall x \in X$. Then $(M, +, \cdot)$ is a ring having the zero element of the ring the constant map c defined by $c(x) = 0$ for all $x \in X$ and the additive inverse $-f$ of f is characterized by the rule $(-f)(x) = -f(x)$, for all $x \in X$. Moreover, if R has a multiplicative identity 1, M has an identity given by the constant map $1(x) = 1$ $\forall x \in X$.

(vi) If R is a ring, then the opposite ring of R, denoted R^{op}, is the ring that has the same set of elements as R, the same addition as R, and multiplication '$\circ$' is given by $a \circ b = ba$, where ba is the usual multiplication in R.

Definition 4.1.2 A non-zero element a in a ring R is said to be a left (right) zero divisor iff $\exists$ a non-zero element $b \in R$ such that $ab = 0$ ($ba = 0$). A zero divisor is an element of R which is both a left and a right zero divisor i.e., a zero divisor is an element a which divides 0.

Example 4.1.2 (i) In the non-commutative ring $M_2(\mathbf{Z})$,

$$\begin{pmatrix} 2 & 0 \\ 0 & 0 \end{pmatrix}\begin{pmatrix} 0 & 0 \\ 0 & 3 \end{pmatrix} = \begin{pmatrix} 0 & 0 \\ 0 & 0 \end{pmatrix} = \begin{pmatrix} 0 & 0 \\ 0 & 3 \end{pmatrix}\begin{pmatrix} 2 & 0 \\ 0 & 0 \end{pmatrix}$$

shows that $\begin{pmatrix} 0 & 0 \\ 0 & 3 \end{pmatrix}$ is a zero divisor of $\begin{pmatrix} 2 & 0 \\ 0 & 0 \end{pmatrix}$.

Remark The ring $M_2(\mathbf{Z})$ is an example of a ring having infinitely many zero divisors.

(ii) $\mathbf{Z}_n$ is a commutative ring with zero divisors for every composite integer $n > 1$; i.e., if $n = rs$ in $\mathbf{Z}$ $(1 < r, s < n)$, then $(r)(s) = (0)$ in $\mathbf{Z}_n$.

The existence (or absence) of a zero divisor in a ring can be characterized with the help of the cancellation laws for multiplication in the ring.

Theorem 4.1.3 *A ring R has no zero divisors iff it satisfies the cancellation laws for multiplication in R; i.e., for $a, b \in R$, $ab = ac$ and $ba = ca$, where $a \neq 0$, imply $b = c$.*

Proof Suppose R has no zero divisors. Let $ab = ac$, $a \neq 0$. Then $a(b - c) = 0 \Rightarrow b - c = 0 \Rightarrow b = c$. Thus $ab = ac \Rightarrow b = c$. Similarly, $ba = ca \Rightarrow b = c$.

Conversely, let R satisfy the cancellation laws for multiplication. Suppose $ab = 0$, $a \neq 0$. Then $ab = a0 \Rightarrow b = 0$ by cancellation law. Similarly, $ab = 0$, $b \neq 0 \Rightarrow a = 0$. Consequently, R has no zero divisors. $\qquad\square$

Remark According to Definition 4.1.2, 0 is not a zero divisor.

If R contains 1, then 1 is not a zero divisor and even any element of R having a multiplicative inverse is not a zero divisor.

Definition 4.1.3 A commutative ring with 1 which has no zero divisors is called an *integral domain*.

Note The ring of integers $\mathbf{Z}$ is an example of an integral domain, hence the name integral domain has been chosen.

Theorem 4.1.4 *A commutative ring R with 1 is an integral domain iff the cancellation law:*
If $a \neq 0$, then $ab = ac \Rightarrow b = c$ holds $\forall a, b, c \in R$.

Proof It follows from Theorem 4.1.3 and Definition 4.1.3. $\qquad\square$

Proposition 4.1.1 $(\mathbf{Z}_n, +, \cdot)$ $(n > 1)$ *is an integral domain iff n is a prime integer.*

Proof Clearly, $\mathbf{Z}_n$ is a commutative ring with (0) as its zero element and (1) as its identity element. Let n be a prime integer, (a) and (b) be two non-zero elements of $\mathbf{Z}_n$. We claim that $(a)(b) \neq (0)$. If $(a) \neq (0)$, then n cannot divide a, i.e., $n\nmid a$. Similarly, $(b) \neq (0) \Rightarrow n\nmid b$. Thus $(a)(b) = (0) \Rightarrow (ab) = (0) \Rightarrow n|ab \Rightarrow$ either $n|a$ or $n|b \Rightarrow$ a contradiction. Consequently, $(a)(b) \neq (0)$. This shows that $\mathbf{Z}_n$ is an integral domain.

Conversely, let $\mathbf{Z}_n$ be an integral domain. Suppose n is not a prime integer. Then there exist integers p, q such that $n = pq$, $1 < p < n$ and $1 < q < n$. So, $(n) = (pq) = (p)(q) = (0) \Rightarrow$ either $(p) = (0)$ or $(q) = (0)$, which is a contradiction. Hence n must be prime. $\qquad\square$

Definition 4.1.4 Let R be a ring with identity 1. An element $a \in R$ is said to be *left* (*right*) *invertible* iff there exists an element $b \in R$ ($c \in R$) such that $ba = 1$ ($ac = 1$) and a is said to be a *unit* or an *invertible* element iff there exists an element $b \in R$ such that $ab = ba = 1$. The element b is then uniquely determined by a, and is written a^{-1}.

Thus a unit in R is an element a which divides 1. Clearly, the units in R form a (multiplicative) group denoted by $U(R)$.

The integral domain $\mathbf{Q}$ and the integral domain $\mathbf{R}$ enjoy an algebraic advantage over the integral domain $\mathbf{Z}$: every equation $ax = b$ (a is not zero) has a unique solution in them. Integral domains with this property are called fields.

Definition 4.1.5 A ring R with 1 is called a *division ring* or a *skew field* iff every non-zero element of R is invertible.

Thus a ring R with 1 is a division ring iff its non-zero elements form a (multiplicative) group.

Proposition 4.1.2 *A division ring does not contain any zero divisor.*

Proof Let R be a division ring and $ab = 0$, $a \neq 0$. Then $a^{-1} \in R$ and $b = 1b = (a^{-1}a)b = a^{-1}(ab) = a^{-1}0 = 0$. This shows that R does not contain any zero divisor. $\qquad\square$

Theorem 4.1.5 *A ring R with 1 is a division ring iff each of the equations: $ax = b$ and $ya = b$, has a unique solution in R, where $a, b \in R$, and $a \neq 0$.*

Proof Let R be a division ring. If $b = 0$, then $x = 0$ and $y = 0$ are solutions of the given equations. Since R has no zero divisor, each of the solutions is unique. Next, let $b \neq 0$. As the non-null elements of R form a multiplicative group, the given two equations have unique solutions by Proposition 2.3.1(v).

Conversely, let R be a ring with 1, satisfying the given conditions. Suppose $a, b \in R$, where $ab = 0$, and $a \neq 0$. Since the equation $ax = 0$ has a solution $x = 0$, and the solution is unique, it follows that $b = 0$. Thus the product of two non-null elements of R is non-null. Hence $(R \setminus \{0\}, \cdot)$ is a multiplicative semigroup, which is a group by using Proposition 2.3.3. Consequently, R is a division ring. $\qquad\square$

Example 4.1.3 (i) *Quaternion rings.* The quaternions of Hamilton essentially due to Sir Wiliam R. Hamilton (1805–1865) constitute a 4-dimensional vector space over the field of real numbers with its basis $\{1, i, j, k\}$ consisting of four vectors (see Chap. 8). Quaternion ring, discovered by Hamilton in 1843 is an important example of a division ring.

To introduce this ring, let H be the set consisting of all ordered 4-tuples of real numbers, i.e.,

$$H = \{(a, b, c, d) : a, b, c, d \in \mathbf{R}\}.$$

In H we define $+$ and $\cdot$ by the rules:

$$(a, b, c, d) + (x, y, z, t) = (a + x, b + y, c + z, d + t),$$

$$(a, b, c, d) \cdot (x, y, z, t) = (ax - by - cz - dt, ay + bx + ct - dz,$$

$$az - bt + cx + dy, at + bz - cy + dx).$$

Then $(H, +, \cdot)$ is a ring with $(0, 0, 0, 0)$ as the zero element, $(-a, -b, -c, -d)$ as the negative element of (a, b, c, d) and $(1, 0, 0, 0)$ as the identity element.

We introduce the notation by taking

$$1 = (1, 0, 0, 0), \qquad i = (0, 1, 0, 0), \qquad j = (0, 0, 1, 0) \quad \text{and} \quad k = (0, 0, 0, 1).$$

Then 1 is the multiplicative identity of H and

$$i^2 = j^2 = k^2 = -1, \qquad i \cdot j = k, \qquad j \cdot k = i, \qquad k \cdot i = j, \qquad j \cdot i = -k,$$

$$k \cdot j = -i, \qquad i \cdot k = -j.$$

These relations show that the commutative law for multiplication fails to hold in H. The definitions of algebraic operations '$+$' and '$\cdot$' in H show that each element $(a, b, c, d) \in H$ is of the form

$$(a, b, c, d) = (a, 0, 0, 0)1 + (b, 0, 0, 0)i + (c, 0, 0, 0)j + (d, 0, 0, 0)k$$

$$= a + bi + cj + dk,$$

on replacing $(r, 0, 0, 0)$ by r, as the map $f : \{(r, 0, 0, 0) : r \in \mathbf{R}\} \to \mathbf{R}, (r, 0, 0, 0) \mapsto r$ is an isomorphism of rings (see Definition 4.4.1). Thus $H = \{a + bi + cj + dk : a, b, c, d \in \mathbf{R}\}$, where $i^2 = j^2 = k^2 = -1, i \cdot j = k, j \cdot k = i, k \cdot i = j, k \cdot j = -i, j \cdot i = -k$ and $i \cdot k = -j$.

An element of H is called a *real quaternion* and H is called a real *quaternion ring.*

Similarly, the *integral quaternion* ring or *rational quaternion* ring can be defined by taking integral or rational coefficients, respectively. Note that any non-zero real quaternion $q = a + bi + cj + dk$ has a conjugate $\bar{q}$ defined by $\bar{q} = a - bi - cj - dk$. Then $q\bar{q} = \bar{q}q = a^2 + b^2 + c^2 + d^2 = N$ (say) $\neq 0$. Hence $q^{-1} = N^{-1}\bar{q}$ is the multiplicative inverse of q. Consequently, $(H, +, \cdot)$ is a skew field.

Remark The set of all members of H of the form $(a, b, 0, 0) = a + bi$, called the *special quaternions* forms a subring isomorphic to $\mathbf{C}$ (see Example 4.1.1(i)). Again the set of all members of H of the forms $(a, 0, b, 0)$ or $(a, 0, 0, b)$ forms a subring isomorphic to $\mathbf{C}$.

In the above sense, the real quaternions may be viewed as a suitable generalization of the complex numbers.

(ii) All square matrices of order 2, given by

$$\begin{pmatrix} a_1 + a_2 i & a_3 + a_4 i \\ -a_3 + a_4 i & a_1 - a_2 i \end{pmatrix},$$

where $i^2 = -1$ and a_r $(r = 1, 2, 3, 4)$ are rational numbers, form a division ring R under usual addition and multiplication of matrices.

A commutative division ring carries a special name called field. It is customary to introduce the concept of fields abstractly.

Definition 4.1.6 A commutative division ring is called a *field*.

Thus a field is an additively abelian group such that its non-null elements form a multiplicatively commutative group and it satisfies distributive laws.

Remark The division rings in Example 4.1.3 are not fields.

Example 4.1.4 (i) $(\mathbf{R}, +, \cdot)$, $(\mathbf{Q}, +, \cdot)$ and $(\mathbf{C}, +, \cdot)$ (cf. Example 4.1.1(i)) are fields (called *fields* of *real numbers, rational numbers* and *complex numbers,* respectively).

(ii) $(\mathbf{R} \times \mathbf{R}, +, \cdot)$ is a field under '+' and '·' defined by $(a, b) + (c, d) = (a + c, b + d)$, $(a, b) \cdot (c, d) = (ac - bd, ad + bc)$ $\forall a, b, c, d \in \mathbf{R}$. Clearly $(0, 0)$ is its additive zero, $(1, 0)$ is its multiplicative identity and the inverse of (a, b) $(\neq (0, 0))$ is $(a/(a^2 + b^2), -b/(a^2 + b^2))$.

Remark The field $(\mathbf{R} \times \mathbf{R}, +, \cdot)$ defined above is isomorphic to the field of complex numbers $(\mathbf{C}, +, \cdot)$.

The concept of abstract field is relatively harder to grasp than that of a subfield of the field of complex numbers, but an abstract field contains new classes of fields including finite fields. Finite fields form an important class of abstract fields [see Sect. 11.2 of Chap. 11]. We are now going to deduce a relation between a finite integral domain and a field through the following theorem.

Theorem 4.1.6 *A finite integral domain is a field.*

Proof Let D be an integral domain consisting of n distinct elements. Suppose $0 \neq a \in D$. If b and $c \in D$ and $b \neq c$, then $ab \neq ac$, otherwise $ab = ac$ would imply $b = c$ by Theorem 4.1.4. Thus if n distinct elements of D are multiplied by a, we obtain n distinct elements in D, i.e., all the elements of D. Hence for any given $b \in D$, there exists a unique element $u \in D$ such that $au = b$. Consequently, the equation

$ax = b$ has a unique solution in D, for any pair of elements $a, b \in D$, provided that $a \neq 0$. Again, since multiplication is commutative in D, the equation $ya = b$ has also the very same solution. Consequently, D is a division ring by Theorem 4.1.5. Finally, as multiplication is commutative in D, D is a field. ☐

Example 4.1.5 The commutative ring $(\mathbf{Z}_n, +, \cdot)$ $(n > 1)$ is a field iff n is a prime integer.

Proof It follows from Proposition 4.1.1 and Theorem 4.1.6. ☐

Remark For a given prime integer p (>1), there exists a field $(\mathbf{Z}_p, +, \cdot)$ denoted as $GF(p)$ having p elements, called a finite field or Galois field.

Remark Let **0** and **1** denote the property of an integer being even or odd, respectively. Define operations on the symbols **0** and **1** by the tables:

+	0	1		×	0	1
0	0	1		**0**	0	0
1	1	0		**1**	0	1

Note that the operations are defined by analogue of the way in which the corresponding properties of integers behave under usual addition and multiplication. For example, since the sum (product) of an even and an odd integer is odd (even), we write

$$0 + 1 = 1 \quad (0 \times 1 = 0).$$

Now $(\{\mathbf{0}, \mathbf{1}\}, +, \times)$ is a field with **0** as its zero element and **1** as its identity element. We write 0 for **0** and 1 for **1** and the field by $GF(2)$. Consider an arbitrary non-empty set S and the commutative ring R consisting of all maps from S to $GF(2)$ (see Example 4.1.1(v)). Since $x^2 = x$ $\forall x \in GF(2)$, this relation also holds in R.

4.2 Subrings

There is a natural interest to study a subset of a ring, which is also closed under addition, subtraction and multiplication as defined in the ring. We now deal with the situation where a subset of a ring again forms a ring.

Definition 4.2.1 Let $(R, +, \cdot)$ be a ring and S a non-empty subset of R. If the ordered triple $(S, +, \cdot)$ constitutes a ring (using the induced operations) i.e., $(S, +, \cdot)$ is itself a ring under the same compositions of addition and multiplication as in R, then $(S, +, \cdot)$ is said to be a subring of $(R, +, \cdot)$.

For example, $(\mathbf{Z}, +, \cdot)$ is a subring of $(\mathbf{R}, +, \cdot)$, $(2\mathbf{Z}, +, \cdot)$ is a subring of $(\mathbf{Z}, +, \cdot)$.

Remark To find a simpler criterion for a subring, we note that $(S, +, \cdot)$ is a subring of $(R, +, \cdot)$ iff $(S, +)$ is a subgroup of $(R, +)$ and $(S, \cdot)$ is a subsemigroup of $(R, \cdot)$, as the distributive, commutative and associative laws are hereditary.

Theorem 4.2.1 *A subset S ($\neq \emptyset$) of a ring R is a subring of R iff $a - b, ab \in S$ for every pair of elements $a, b \in S$.*

Proof Suppose the given conditions hold. Since $a - b \in S$ for every pair of elements $a, b \in S$, it follows by Theorem 2.4.1 that $(S, +)$ is a subgroup of $(R, +)$ and as $+$ defined on R is commutative, $(S, +)$ is abelian. Again $(S, \cdot)$ is a subsemigroup of $(R, \cdot)$ as by hypothesis, the closure property for multiplication holds in S. Associative and distributive laws hold in S, since they hold for the whole set R. Therefore $(S, +, \cdot)$ is a subring of $(R, +, \cdot)$. Conversely, if $(S, +, \cdot)$ is a subring, then the given conditions hold. $\qquad\square$

Example 4.2.1 Consider $S = \left\{ \begin{pmatrix} x & 0 \\ 0 & 0 \end{pmatrix} : x \in \mathbf{R} \right\}$. Then by Theorem 4.2.1, $(S, +, \cdot)$ is a subring of $(M_2(\mathbf{R}), +, \cdot)$ of all 2×2 real matrices under usual addition and multiplication of matrices. Clearly, $\begin{pmatrix} 1 & 0 \\ 0 & 0 \end{pmatrix}$ is the identity of S but this identity is not the same as the identity $\begin{pmatrix} 1 & 0 \\ 0 & 1 \end{pmatrix}$ of $M_2(\mathbf{R})$.

Remark For a ring R with identity, the identity (if it exists) of a subring of $M_2(R)$ may not be the same as the identity of $M_2(R)$.

Theorem 4.2.2 *The intersection of any aggregate of subrings of a ring R is a subring of R.*

Proof Let $M = \bigcap_i S_i$ be the intersection of a given family $\{S_i\}$ of subrings of R. $M \neq \emptyset$, for $0 \in M$. Then $a, b \in M \Rightarrow a, b \in$ each $S_i \Rightarrow a - b, ab \in$ each $S_i \Rightarrow a - b, ab \in M \Rightarrow M$ is a subring of R by Theorem 4.2.1. $\qquad\square$

Remark The join of two subrings may not be a subring.

Example 4.2.2 The multiples of 2 and multiples of 3 are subrings of the ring of integers. The join of these two subrings contains the elements 2 and 3, but not their sum $2 + 3 = 5$. Consequently, the join is not a subring.

Definition 4.2.2 The *center* of a ring R denoted by cent R is defined by cent $R = \{a \in R : ar = ra, \ \forall r \in R\}$.

Theorem 4.2.3 *For any ring R, cent R is a subring of R.*

Proof Clearly, cent $R \neq \emptyset$, since $0 \in$ cent R. Again $a, b \in$ cent $R \Rightarrow ar = ra$ and $br = rb \ \forall r \in R \Rightarrow (a - b)r = ar - br = ra - rb = r(a - b)$ and $(ab)r = a(br) = a(rb) = (ar)b = (ra)b = r(ab) \ \forall r \in R \Rightarrow$ cent R is a subring of R by Theorem 4.2.1. $\qquad\square$

Remark Every ring R has at least two subrings, viz. $\{0\}$ and R. These two subrings are called *trivial* subrings of R; all other subrings (if they exist) are called *non-trivial*. We use the term proper subring to mean a subring which is different from R.

We now study subrings of a field.

Theorem 4.2.4 *A subring with identity, of a field is an integral domain.*

Proof Let D be a subring of a field F, such that $1 \in D$. Then D is a commutative ring having no zero divisors, since multiplication is commutative in F and F contains no zero divisors. Consequently, D is an integral domain.   □

Definition 4.2.3 Let $(F, +, \cdot)$ be a field, A subset H $(\neq \emptyset)$ of F is called a *subfield* of $(F, +, \cdot)$ iff the triple $(H, +, \cdot)$ is itself a field (under the induced operations).

For example, the field of rational numbers $\mathbf{Q}$ is a subfield of the field of real numbers $\mathbf{R}$. In any field F, F is a subfield of F.

A formulation or a test for a subfield of a given field is given below:

Theorem 4.2.5 *A non-empty subset H $(\neq \{0\})$ of a field F is a subfield of F iff $a - b, ab \in H$ for every pair of elements $a, b \in H$ and $x^{-1} \in H$ $\forall x \in H \setminus \{0\}$.*

Proof Let the given conditions hold in H. Then $(H, +)$ forms a group. Moreover $(H \setminus \{0\}, \cdot)$ forms a multiplicative group. Finally, the commutative properties for addition and multiplication and distributive properties, being hereditary, hold in H. Consequently, H is a subfield of F. Conversely, the necessity of the conditions is obvious.   □

Theorem 4.2.6 *The intersection of a given aggregate $\{H_i\}$ of subfields of a field F is a subfield of F.*

Proof Let $M = \bigcap_i H_i$ be the intersection of the aggregate $\{H_i\}$. Then M is a subfield of F by Theorem 4.2.5.   □

Definition 4.2.4 The intersection of all subfields of a field F is a subfield, called the *prime field* of F.

Thus the *prime field* of F is the smallest subfield of F.

4.3 Characteristic of a Ring

Characteristic of a ring is a non-negative integer (may be 0), which is very important in the study of ring theory. The characteristic of an integral domain (hence of a field) is either 0 or a prime integer

Let $(R, +, \cdot)$ be a ring. For $a \in R$, na already has a meaning in $(R, +)$ for every $n \in \mathbf{Z}$. The order of the additive cyclic subgroup, generated by an element $a \in R$ is called the characteristic of a. Thus a has a non-zero characteristic r iff r is the smallest positive integer such that $ra = 0$. Then for any integer n, $na = 0 \Leftrightarrow n$ is divisible by r. If, on the other hand, the additive cyclic subgroup generated by a contains infinitely many distinct elements, then a is said to have the characteristic infinity or 0; in the case, $ra = na \Rightarrow r = n$, and $ra \neq 0$ for all non-zero integers r. Different elements may have different characteristics; 0 has always the characteristic 1.

Definition 4.3.1 Let R be a ring. If r is the smallest positive integer for which $ra = 0 \; \forall a \in R$, then r is said to be the *characteristic* of R; if no such positive integer exists, then R is said to be of characteristic *infinity* or *zero*. We write char R for the characteristic of R.

Example 4.3.1 (i) The ring $(\mathbf{Z}_{18}, +, \cdot)$ has characteristic 18 and the characteristics of $(6), (9), (2), (4)$ are, respectively, $3, 2, 9, 9$.

(ii) Clearly, char $\mathbf{Z}$ is 0.

Remark Definition 4.3.1 makes an assertion about every element of the ring but char R of a ring R with 1 is completely determined by 1. This is given below.

Theorem 4.3.1 *Let R be a ring with identity 1. Then R has characteristic $n > 0$ iff n is also the characteristic of 1.*

Proof char $R = n \; (>0) \Rightarrow na = 0 \; \forall a \in R \Rightarrow n1 = 0$. Now $m1 = 0$ for some m satisfying $0 < m < n \Rightarrow ma = m(1a) = (m1)a = 0a = 0 \; \forall a \in R \Rightarrow$ char $R < n \Rightarrow$ an impossibility $\Rightarrow n$ is the characteristic of 1. The converse is similar. $\square$

Theorem 4.3.2 *Let R be a ring with identity 1. If R has no zero divisor, then char R is either 0 or a prime integer.*

Proof Let char $R = n \neq 0$. Assume that n is not prime. Then n has a non-trivial factorization $n = rs$, with $1 < r, s < n$. Now $0 = n1 = 1 + 1 + \cdots (rs \text{ terms}) = [1 + 1 + \cdots (r \text{ terms})][1 + 1 + \cdots (s \text{ terms})] = r1 \cdot s1$. But $r1 \neq 0$, $s1 \neq 0$, since $1 < r, s < n$. Then $r1$ and $s1$ are zero divisors, which is impossible by hypothesis. We, therefore, conclude that n must be prime. $\square$

Corollary 1   *Let D be an integral domain. Then char D is 0 or a prime number.*

Corollary 2   *Let F be a field. Then char F is 0 or a prime number.*

Example 4.3.2 (i) The field of rational numbers, the field of real numbers and the field complex numbers have each the characteristic 0 as the integer 1 has the same characteristic.

(ii) char $GF(p)$ is p, as the characteristic of its identity (1) is p.

(iii) Let F be a finite field. Then char F is a prime integer.

[*Hint.* (iii) Since every field is an integral domain, char F is either 0 or a prime p. Suppose char $F = 0$. Then $n1 \neq 0$ for every positive integer n and $n1 \neq m1$, where $n \neq m$. Hence $\{n1 : n$ is a positive integer$\}$ is an infinite subset of F. This contradicts that F is a finite field. Consequently, char F is a prime integer.]

(iv) If an integral domain $(D, +, \cdot)$ has a non-zero characteristic p, then p is the period (order) of every non-zero element in the group $(D, +)$.

[*Hint.* (iv) Let $a \in D \backslash \{0\}$. Then $pa = a + a + \cdots (p$ terms$) = [1 + 1 + \cdots 1 \, (p$ terms$)]a = (p1)a = 0a = 0 \Rightarrow$ period of a is a factor of $p \Rightarrow$ period of a is 1 or p, since p is prime $\Rightarrow$ period of a is p as it is not 1, since $a \neq 0$.]

(v) If an integral domain $(D, +, \cdot)$ has characteristic 0, then in the group $(D, +)$, every non-zero element has infinite period (order).

[*Hint.* Let $a \in D \backslash \{0\}$. Then for every positive integer n, $na = (n1)a$. But $a \neq 0$; and, since char $D = 0$, $n1 \neq 0$. Again since D is an integral domain, $na \neq 0$ for every positive integer n i.e., a has infinite period in the group $(D, +)$.]

4.4 Embedding and Extension for Rings

It is sometimes convenient to study a ring as a subring of a suitable ring having additional properties. This motivates to introduce the concept of an embedding, which is a monomorphism. A homomorphism of a ring is similar to a homomorphism of a group. Consequently, the same terminology, such as, homomorphism, monomorphism, epimorphism, isomorphism, automorphism, etc., is also used for rings. The real significance of homomorphisms, homomorphic images, kernels of homomorphisms, and their basic properties are discussed in detail in this section.

The concept of embedding and extension can be introduced for rings in precisely the same sense as in the case of groups.

Definition 4.4.1 Let R and S be rings. A (ring) homomorphism from R to S means a mapping $f : R \to S$ such that $f(a + b) = f(a) + f(b)$, $f(ab) = f(a) \cdot f(b)$ for every pair of elements $a, b \in R$.

Ring homomorphisms of special character are named like special group homomorphisms:

(i)   an injective ring homomorphism is called a *monomorphism*;

(ii) a surjective ring homomorphism is called an *epimorphism*;

(iii) a bijective ring homomorphism is called an *isomorphism*;

(iv) a ring isomorphism of a ring onto itself is called an *automorphism*;

(v)   a ring homomorphism of a ring into itself is called an *endomorphism*.

The existence of a ring isomorphism from R onto S asserts that R and S, in a significant sense, are the same; i.e., R and S, have the same algebraic structures and we say that R and S are isomorphic and write $R \cong S$.

Let $f : R \to S$ be a ring homomorphism. For the time being, if we forget the multiplications defined on R and S, then f is considered a group homomorphism from the additive group $(R, +)$ to the additive group $(S, +)$. Then it follows:

Proposition 4.4.1 *If $f : R \to S$ is a ring homomorphism, then*

(i) $f(0) = 0'$ *(where 0 and $0'$ are the zero elements of R and S, respectively);*
(ii) $\forall a \in R, \ f(-a) = -f(a)$;
(iii) f *is injective, iff* $\ker f = \{0\}$, *where* $\ker f$ *is the kernel of f defined by* $\ker f = \{a \in R : f(a) = 0'\}$.
(iv) *if f is onto and both R and S contain identity elements 1_R and 1_S, respectively, then $f(1_R) = 1_S$.*

To obtain other information about ring homomorphisms and their kernels, we now bring back into our discussion, the multiplications defined on the rings concerned.

Theorem 4.4.1 *Let $f : R \to S$ be a ring homomorphism $(R, S$ being rings). Then*

(i) *for each subring A of R, $f(A)$ is a subring of S;*
(ii) *for each subring B of S, $f^{-1}(B)$ is a subring of R;*
(iii) $\ker f$ *is a subring of R.*

Proof (i) Let A be a subring of R. Since $A \neq \emptyset$, $f(A) \neq \emptyset$. Let $x, y \in f(A)$. Then $x = f(a)$ and $y = f(b)$ for some $a, b \in A$. Hence $x - y = f(a) - f(b) = f(a) + f(-b)$ (by Proposition 4.4.1) $= f(a - b)$ (since f is a homomorphism) $\in f(A)$, since $a - b \in A$. Similarly, $xy = f(a)f(b) = f(ab) \in f(A)$, since $ab \in A$. Consequently, $f(A)$ is a subring of S by Theorem 4.2.1.

(ii) Proof is similar to that of (i).

(iii) Since $0 \in \ker f$, $\ker f \neq \emptyset$. Let $x, y \in \ker f$. Then $f(x) = f(y) = 0'$ (zero element of S) $\Rightarrow f(x - y) = f(x) + f(-y) = f(x) - f(y) = 0' - 0' = 0'$ and $f(xy) = f(x)f(y) = 0'0' = 0' \Rightarrow$ both $x - y$ and $xy \in \ker f \Rightarrow \ker f$ is a subring of R by Theorem 4.2.1. $\qquad \square$

Remark The kernel of a ring homomorphism $f : R \to S$ describes how far f from being $1-1$. If $\ker f = \{0\}$, then f is $1-1$; two elements of R are sent by f to the same element of $S \Leftrightarrow$ their difference is in $\ker f$.

Simple Examples of Ring Homomorphisms

Example 4.4.1 The most trivial examples are the identity homomorphisms. Let R be a ring, and let $I : R \to R$ be the identity function. Then I is a ring homomorphism which is $1-1$ and onto and hence is an isomorphism.

Example 4.4.2 Let S be a ring and let R be a subring of S. Then the inclusion map $i : R \hookrightarrow S$ is a ring homomorphism. In particular, the inclusion map $i : \mathbf{Q} \hookrightarrow \mathbf{R}$ is a ring homomorphism.

Remark The only difference between the identity function I and i in Examples 4.4.1 and 4.4.2 is their ranges. If $R \subsetneq S$, the function $i : R \to S$ defined by $i(x) = x$ in Example 4.4.2 is a homomorphism but not an epimorphism.

Example 4.4.3 Let R be a commutative ring with 1 of prime characteristic $p \ (>1)$. Then the map $\phi : R \to R$ defined by

$$\phi(a) = a^p, \quad \forall a \in R$$

is a homomorphism (*Frobenius homomorphism*).

Example 4.4.4 If F is a finite field of prime characteristic p, then the homomorphism $\phi : F \to F$, $a \mapsto a^p$ is an automorphism.
   [*Hint.* $\ker \phi = \{0\} \Rightarrow \phi$ is a monomorphism. Again F is finite $\Rightarrow \phi$ is surjective by counting.]

Example 4.4.5 Let F be a field of a prime characteristic p. Give an example to show that the homomorphism $\phi : F \to F$, $a \mapsto a^p$, may not be an automorphism if F is not finite.
   [*Hint.* Consider the field $F = \mathbf{Z}_p\langle x\rangle$, the field of rational functions over the field $\mathbf{Z}_p$, which is the quotient field of $\mathbf{Z}_p[x]$, where x is an indeterminate (see Definition 4.5.5). Then $\phi(F) = \mathbf{Z}_p\langle x^p\rangle$ is a proper subfield of $\mathbf{Z}_p\langle x\rangle \Rightarrow \phi$ is not an automorphism.]

Example 4.4.6 Let F be a field. Determine all the automorphisms of $F\langle x\rangle$.
   [*Hint.* Every homomorphism $f : F\langle x\rangle \to F\langle x\rangle$ mapping x onto some $y = (ax + b)/(cx + d) \in F\langle x\rangle$, $a, b, c, d \in F$ and $ad - bc \neq 0$ is an automorphism and conversely, each such $y \in F\langle x\rangle$ gives rise to an automorphism of $F\langle x\rangle$ mapping x into y.]

Definition 4.4.2 A ring R is said to be *embedded* in a ring S (or S is said to be an *extension* of R) iff there exists a monomorphism $f : R \to S$ from the ring R to the ring S.

Remark If f is a monomorphism, then $f : R \to f(R)(\subseteq S)$ is an isomorphism and f is called an embedding or an identification map. So we can identify R and $f(R)$ and consider R a subring of S.

For example, $\mathbf{R}$ is considered as a subfield of $\mathbf{C}$ under identification map $f : \mathbf{R} \to \mathbf{C}$, $r \mapsto (r, 0)$.

Example 4.4.7 Consider the mapping $f : (\mathbf{Z}, +, \cdot) \to (M_2(\mathbf{Z}), +, \cdot)$ defined by

$$f(a) = \begin{pmatrix} a & 0 \\ 0 & a \end{pmatrix} \quad \forall a \in \mathbf{Z}.$$

Then $f(a + b) = f(a) + f(b)$ and $f(ab) = f(a)f(b)$ $\forall a, b \in \mathbf{Z} \Rightarrow f$ is a ring homomorphism. Clearly, f is a monomorphism. Consequently, $(\mathbf{Z}, +, \cdot)$ is embedded in $(M_2(\mathbf{Z}), +, \cdot)$.

Theorem 4.4.2 (Dorroh Extension Theorem) *Let R be a ring without identity element. Then R can be embedded in a ring S with identity element.*

Proof Let R be a given ring without the identity element and $S = \mathbf{Z} \times R$. Define '+' and '·' in S by $(n, a) + (m, b) = (n + m, a + b)$ and $(n, a) \cdot (m, b) = (nm, nb + ma + ab)$, where nb denotes the nth natural multiple of b and similarly for ma. Under these operations S is a ring with the identity element $(1, 0)$. Define a mapping $f : R \to \mathbf{Z} \times R$ by $f(a) = (0, a)$ $\forall a \in R$. Then f is a homomorphism. Now ker $f = \{a \in R : f(a) = (0, 0_R)\} = \{0_R\} \Rightarrow f$ is a monomorphism by Proposition 4.4.1 (where 0_R is the zero element of R) $\Rightarrow R$ is embedded in $\mathbf{Z} \times R$. $\qquad \square$

Theorem 4.4.3 *Let R be a ring such that* char $R = n$ (>0). *Then R can be embedded in a ring S with identity element such that* char $S = n$.

Proof Let $S = \mathbf{Z}_n \times R$. Define '+' and '·' in S by $((m), a) + ((t), b) = ((m + t), a + b)$ and $((m), a)((t), b) = ((mt), mb + ta + ab)$. Then S forms a ring with $((1), 0)$ as the identity element. Since $n((1), 0) = (n(1), 0) = ((n), 0) = ((0), 0)$, and char $\mathbf{Z}_n = n$, it follows that char $S = n$.

Consider the mapping $f : R \to \mathbf{Z}_n \times R$ by $f(a) = ((0), a)$. Clearly, f is a monomorphism. $\qquad \square$

Corollary 1 *Let R be a ring without identity element. Then R can be embedded in a ring S with identity element (with preservation of characteristic of R).*

Proof If char $R = 0$, then by Theorem 4.4.2, R is embedded in $\mathbf{Z} \times R = S$, where char $S = 0$, as $(1, 0)$ has infinite order in $(S, +)$. Again if char $R = n$ $(n > 0)$, then by Theorem 4.4.3, R is embedded in $\mathbf{Z}_n \times R$ with the requisite properties. $\qquad \square$

Corollary 2 *Let R be a ring without identity. Then R can be extended to a ring with identity (with the preservation of characteristic of R).*

Remark If R is a ring with identity element, then an extension S of R may have a different identity element and in such cases, char $S \neq$ char R.

Example 4.4.8 $(\mathbf{Z}_{10}, +, \cdot)$ is a ring with (1) as its identity. If $S = \{(0), (2), (4), (6), (8)\} \subset \mathbf{Z}_{10}$, then $(S, +, \cdot)$ is also a ring, having (6) as its identity. Thus $\mathbf{Z}_{10}$ is an extension of S, but they have different identities.

Remark A ring with zero divisors cannot be extended to an integral domain or a field, otherwise, we would reach a contradiction (by Theorem 4.2.4). But in case of an integral domain an extension to a field is possible. For example, the standard integral domain $\mathbf{Z}$ is extended to the field of rationals $\mathbf{Q}$. The same technique is used to prove the following theorem for an arbitrary integral domain.

Theorem 4.4.4 *Let D be an integral domain. Then D can be extended to a field.*

Proof Let $T = D \times D \setminus \{0\} = \{(x, y) : x, y \in D, y \neq 0\}$. Define a binary relation ρ on T by $(x, y)\rho(a, b)$ iff $xb = ay$. For any $(x, y) \in T$, $xy = yx \Rightarrow (x, y)\rho(x, y) \Rightarrow \rho$ is reflexive. Again $(x, y)\rho(a, b) \Rightarrow xb = ay \Rightarrow ay = xb \Rightarrow (a, b)\rho(x, y) \Rightarrow \rho$ is symmetric. Finally, $(x, y)\rho(a, b)$ and $(a, b)\rho(c, d) \Rightarrow xb = ay$ and $ad = cb \Rightarrow xbd = ayd$ and $ady = cby \Rightarrow xbd = cby$ (as multiplication is commutative in D) $\Rightarrow (xd)b = (cy)b \Rightarrow xd = cy$ (by cancellation law, since $b \neq 0$) $\Rightarrow (x, y)\rho(c, d) \Rightarrow \rho$ is transitive. Consequently, ρ is an equivalence relation on T. Let $Q(D) = T/\rho$, the set of all ρ-equivalence classes on T. Define addition '+' and multiplication '·' in $Q(D)$ by

 (i) $((a, b)) + ((c, d)) = ((ad + bc, bd))$ and
(ii) $((a, b))((c, d)) = ((ac, bd))$.

As $b \neq 0$, $d \neq 0$ and D is an integral domain, $bd \neq 0$, so $Q(D)$ is closed with respect to both '+' and '·'; clearly, the definitions of '+' and '·' in (i) and (ii) are independent of the choice of the representatives of the classes. Then it is easy to verify that in $Q(D)$,

(1) addition '+' and multiplication '·' are commutative and associative;
(2) $((0, a)) \in Q(D)$ is independent of the choice of $a \in D \setminus \{0\}$, as $((0, a)) = ((0, b))$ and $((0, a))$ is the additive identity;
(3) $((-a, b))$ is the additive inverse of $((a, b))$;
(4) distributive laws hold;
(5) for any $a \in D \setminus \{0\}$, $((a, a))$ is independent of a and is the multiplicative identity;
(6) for $((a, b)) \in Q(D)$ and $((a, b))$ not being the zero element in $Q(D)$ (this means $a \neq 0$), the multiplicative inverse of $((a, b))$ is $((b, a))$.

Consequently, $(Q(D), +, \cdot)$ is a field. We now consider the map $f : D \to Q(D)$ by $f(x) = ((ax, a))$. Note that $((ax, a))$ is independent of the choice of $a \in D \setminus \{0\}$. Then f is a homomorphism. Now ker $f = \{x \in D : f(x) = 0'\}$ (where $0'$ is the zero element of $Q(D)) = \{x \in D : ((ax, a)) = ((0, a))\} = \{0\}$ (as $ax = 0 \Rightarrow x = 0$, since $a \neq 0) \Rightarrow f$ is a monomorphism. Consequently, $Q(D)$ is an extension of D. □

Remark 1 The elements of $Q(D)$ are denoted by $((a, b)) = ab^{-1}$ (or a/b) and $Q(D)$ is called the *quotient field* of D (or field of quotients).

Remark 2 For any integral domain D, the quotient field is unique up to isomorphism. (For proof see Corollary of Theorem 4.4.5 or Theorem 4.4.6.)

Theorem 4.4.5 *The quotient fields of two isomorphic integral domains are isomorphic.*

Proof Let D and C be isomorphic integral domains and $\beta : D \to C$ an isomorphism.

Consider the map $f : Q(D) \to Q(C)$ defined by

$$f(dd'^{-1}) = \beta(d)(\beta(d'))^{-1}, \quad \text{where } (d, d') \in D \times D\backslash\{0\}.$$

Let $cc'^{-1} \in Q(C)$, where $(c, c') \in C \times C\backslash\{0\}$.

Then $c = \beta(d)$ and $c' = \beta(d')$ for some $(d, d') \in D \times D\backslash\{0\}$. So, f is surjective.

Let $(d, d'), (a, a') \in D \times D\backslash\{0\}$.

Then

$$dd'^{-1} = aa'^{-1} \quad \Leftrightarrow \quad da' = ad' \quad \Leftrightarrow \quad \beta(da') = \beta(ad')$$
$$\Leftrightarrow \quad \beta(d)\beta(a') = \beta(a)\beta(d')$$
$$\Leftrightarrow \quad \beta(d)(\beta(d'))^{-1} = \beta(a)(\beta(a'))^{-1}$$
$$\Leftrightarrow \quad f(dd'^{-1}) = f(aa'^{-1}) \quad \Rightarrow \quad f \text{ is injective.}$$

Consequently, f is bijective.

Now

$$f(dd'^{-1} + aa'^{-1}) = f((da' + ad')(d'a')^{-1}) = \beta(da' + ad')(\beta(d'a'))^{-1}$$
$$= (\beta(d)\beta(a') + \beta(a)\beta(d'))(\beta(d')(\beta(a'))^{-1}$$
$$= \beta(d)(\beta(d'))^{-1} + \beta(a)(\beta(a'))^{-1}$$
$$= f(dd'^{-1}) + f(aa'^{-1}).$$

Similarly, $f(dd'^{-1} \cdot aa'^{-1}) = f(dd'^{-1})f(aa'^{-1})$.

Consequently, f is an isomorphism. □

Corollary *For an integral domain D, the quotient field $Q(D)$ is unique (up to isomorphism).*

Proof In Theorem 4.4.5 take $D = C$ and β the identity isomorphism on D. □

Let us now study some properties of the quotient field $Q(D)$.

Theorem 4.4.6 *$Q(D)$ is the intersection of all fields, which contain D as a subring.*

Proof Let F be any field containing D as a subring. Then F contains the solutions of the equations $xb = a$ for all $a, b \in F$, where $b \neq 0$. Hence the quotients ab^{-1} of all pairs $(a, b) \in D \times D\backslash\{0\}$ belong to $F \Rightarrow Q(D) \subseteq F$. Also, $Q(D)$ is itself a field containing D as a subring. Consequently, $Q(D)$ is the intersection of all fields, which contains D as a subring. □

Corollary *Let F be any field containing an integral domain D as a subring. Then F contains $Q(D)$ as a subfield.*

Remark (i) $Q(D)$ is the smallest field containing D as a subring.

(ii) Theorem 4.4.6 also proves the uniqueness of the quotient field of a given integral domain.

(iii) If D is a field, then $Q(D) = D$.

Problem 1 Find the quotient field of the ring of integers **Z**.

Solution Let $T = \mathbf{Z} \times \mathbf{Z} \backslash \{0\}$. Then the equivalence relation ρ on T (defined in Theorem 4.4.4) identifies different pairs, whose ratios are the same. For example, $(5, 8)$, $(-10, -16)$, $(40, 64)$ etc. are ρ-equivalent and hence these pairs constitute a class. As **Z** is an integral domain, the elements of the quotient field $Q(\mathbf{Z})$ of **Z** are abstractly the same as these classes, and are called *rational numbers*; every rational number is of the form mn^{-1}, where m, n are integers and $n \neq 0$. Consequently, $Q(\mathbf{Z}) = \mathbf{Q}$, the field of rational numbers.

Remark Let F be a field. Then by Corollary 1 of Theorem 4.3.2, char $F = 0$ or p (a prime number). We show that in each case F has a subfield isomorphic to a well-known field.

Theorem 4.4.7 *Let F be a field.*

(a) *If char $F = 0$, then F contains a subfield K such that K is isomorphic to **Q**.*

(b) *If char $F = p$, a prime integer, then F contains a subfield K such that K is isomorphic to $\mathbf{Z}_p$.*

Proof Define a map $f : \mathbf{Z} \to F, n \mapsto n1, \forall n \in \mathbf{Z}$. Then f is a ring homomorphism.

(a) Suppose char $F = 0$. Then ker $f = \{n \in \mathbf{Z} : n1 = 0\} = \{0\} \Rightarrow f$ is a monomorphism.

Define a map $\tilde{f} : \mathbf{Q} \to F$ by

$$\tilde{f}(m/n) = f(m)\big(f(n)\big)^{-1}, \quad \forall m/n \in \mathbf{Q}.$$

Clearly, $\tilde{f}$ is well defined and is a monomorphism.

Hence the field **Q** is isomorphic to the subfield $\tilde{f}(\mathbf{Q})$ of $F \Rightarrow$ (a), (taking $K = \tilde{f}(\mathbf{Q})$).

(b) Suppose char $F = p > 0$. Since f is a ring homomorphism, it follows that the rings $\mathbf{Z}/\ker f$ and Im $f = f(\mathbf{Z})$ are isomorphic (see Theorem 5.2.3 of Chap. 5). Now char $F = p > 0 \Rightarrow f(\mathbf{Z}) \neq \{0\} \Rightarrow f(\mathbf{Z})$ is a non-trivial subring with 1, of the field $F \Rightarrow f(\mathbf{Z})$ is an integral domain $\Rightarrow \mathbf{Z}/\ker f$ is an integral domain $\Rightarrow \ker f$ is a prime ideal of **Z** (see Theorem 5.3.1 of Chap. 5) $\Rightarrow \ker f = \langle q \rangle$ for some prime integer $q > 0$ (see Problem 1 of Chap. 5). Now $q \in \ker f \Rightarrow f(q) = 0 \Rightarrow q1 = 0 \Rightarrow p|q \Rightarrow p = q \Rightarrow \mathbf{Z}/\ker f = \mathbf{Z}/\langle p \rangle \cong \mathbf{Z}_p \Rightarrow f(\mathbf{Z}) \cong \mathbf{Z}_p \Rightarrow$ (b). $\quad\square$

Consequently, we have the following theorem.

Theorem 4.4.8 *In every field F, there is a minimal subfield P (i.e., a subfield P contained in every subfield of F), P being isomorphic to $\mathbf{Q}$ or $\mathbf{Z}_p$ according as char $F = 0$ or p.*

Remark 1 A field F cannot have more than one such minimal subfield: for, if H and G are both minimal subfields of the field F, then $G \subseteq H$ and $H \subseteq G$. Consequently $G = H$. The unique minimal subfield in a field F is the *prime subfield* of F.

Remark 2 It follows from Theorem 4.4.8 that the prime subfield of F is the intersection of all subfields of F and it is the smallest subfield of F.

Remark 3 Theorem 4.4.7 shows that the prime subfield of F is the subfield of F generated by the identity element of F.

4.5 Power Series Rings and Polynomial Rings

We are familiar with polynomials and power series in one variable with integral coefficients. We extend these notions having coefficients from an arbitrary ring R. This motivates to define polynomial ring $R[x]$ and power series ring $R[\![x]\!]$ in one variable. An element of $R[x]$ is a formal symbol $a_0 + a_1 x + \cdots + a_n x^n$, where a_i's are in R and that of $R[\![x]\!]$ is a formal expression of the form $a_0 + a_1 x + \cdots + a_n x^n + \cdots$, where a_i's are in R. In this section we also study the influence of the structure of R on the structure of $R[x]$ and $R[\![x]\!]$. The polynomial ring in n variables is defined recursively (see Ex. 29 of Exercises-I).

Let $(R, +, \cdot)$ be a ring. Consider the collection P of all infinite sequences

$$f = (a_0, a_1, a_2, \ldots, a_n, \ldots) = \{a_n\}$$

of elements $a_n \in R$. The elements of P are called *formal power series*, or simply power series over R.

Two power series $f = (a_0, a_1, a_2, \ldots, a_n, \ldots)$ and $g = (b_0, b_1, b_2, \ldots, b_n, \ldots)$ are said to be equal, denoted by $f = g$ iff $a_n = b_n$ $\forall n \geq 0$. Define '+' and '$\cdot$' in P as follows:

$$f + g = (a_0 + b_0, a_1 + b_1, \ldots, a_n + b_n, \ldots),$$

$$fg = (c_0, c_1, c_2, \ldots),$$

where, for each $n \geq 0$, c_n is given by

$$c_n = \sum_{\langle i+j=n \rangle} a_i b_j = a_0 b_n + a_1 b_{n-1} + \cdots + a_{n-1} b_1 + a_n b_0.$$

Then P is closed with respect to '+' and '$\cdot$'. Moreover, $(P, +)$ is an abelian group. Let $f, g, h \in P$, where $f = \{a_n\}$, $g = \{b_n\}$ and $h = \{c_n\}$.

Then $(f \cdot g) = \{d_n\}$, where $d_n = \sum_{\langle i+j=n \rangle} a_i b_j$.
So,

$$(f \cdot g) \cdot h = \{e_n\},$$

where

$$e_n = \sum_{\langle r+s=n \rangle} d_r c_s = \sum_{\langle r+s=n \rangle} \left(\sum_{\langle i+j=r \rangle} a_i b_j \right) c_s = \sum_{\langle i+j+s=n \rangle} (a_i b_j) c_s$$

$$= \sum_{\langle i+j+s=n \rangle} a_i (b_j c_s) = \sum_{\langle i+p=n \rangle} a_i t_p, \quad \text{where } g \cdot h = \{t_n\}$$

$$= f \cdot (g \cdot h).$$

Similarly, we can show that

$$f(g+h) = (a_0, a_1, \ldots, a_n, \ldots)(b_0 + c_0, b_1 + c_1, \ldots) = (e_0, e_1, \ldots),$$

where

$$e_n = \sum_{\langle i+j=n \rangle} a_i (b_j + c_j) = \sum_{\langle i+j=n \rangle} (a_i b_j + a_i c_j) = \sum_{\langle i+j=n \rangle} a_i b_j + \sum_{\langle i+j=n \rangle} a_i c_j$$

$$= f \cdot g + f \cdot h.$$

This shows that the left distributive property is also satisfied in $(P, +, \cdot)$. Similarly, the right distributive property is satisfied in $(P, +, \cdot)$. Thus $(P, +, \cdot)$ is a ring with $0 = (0, 0, 0, \ldots)$ as the zero element of this ring and the additive inverse of an arbitrary element $(a_0, a_1, a_2, \ldots)$ of P is $(-a_0, -a_1, -a_2, \ldots)$.

Then we have the following.

Theorem 4.5.1 *The triple* $(P, +, \cdot)$ *forms a ring, called the* ring of (formal) power series *over R. Moreover, the ring* $(P, +, \cdot)$ *is commutative, iff R is commutative.*

Let $S = \{(a, 0, 0, \ldots) : a \in R\}$. Then $(S, +, \cdot)$ is a subring of $(P, +, \cdot)$ such that S is isomorphic to the ring R (under the isomorphism $S \to R$, given by $(a, 0, 0, \ldots) \mapsto a$). In this sense, P contains the original ring R as a subring, and thus, an element $a \in R$ is identified with the special sequence $(a, 0, 0, \ldots) \in P$. Consequently, the elements of R regarded as power series, are hereafter called *constant series*, or just constants.

We use the symbol ax^n, $n \geq 1$, to designate the sequence $(0, \ldots, 0, a, 0, \ldots) \in P$, where the element a occupies the $(n+1)$th place in this sequence. For example,

$$ax = (0, a, 0, 0, \ldots), \qquad ax^2 = (0, 0, a, 0, \ldots),$$

$$ax^3 = (0, 0, 0, a, 0, \ldots) \quad \text{and so on.}$$

Then each power series $f = \{a_n\} \in P$ can be expressed uniquely in the form:

$$f = (a_0, 0, \ldots) + (0, a_1, 0, \ldots) + \cdots + (0, \ldots, 0, a_n, \ldots) + \cdots$$

$$= a_0 + a_1 x + a_2 x^2 + \cdots + a^n x^n + \cdots$$

(with the identification of $a_0 \in R$ with the sequence $(a_0, 0, 0, \ldots) \in P$). Thus P consists of all formal expressions:

$$f = a_0 + a_1 x + a_2 x^2 + \cdots + a_n x^n + \cdots,$$

where the elements a_i (called the *coefficients* of f) $\in R$ and we write $f = \sum_{r=0}^{\infty} a_r x^r$ or simply $f = \sum a_r x^r$. Then the definitions of addition '+' and multiplication '·' in P assume the form:

$$\sum a_n x^n + \sum b_n x^n = \sum (a_n + b_n) x^n, \qquad \left(\sum a_n x^n \right) \left(\sum b_n x^n \right) = \sum c_n x^n,$$

where $c_n = \sum_{\langle i+j=n \rangle} a_i b_j = \sum_{\langle i=0 \rangle}^{n} a_i b_{n-i}$.

Here x is simply a symbol (called an *indeterminate*), not related to the ring R and in no sense representing an element of R. The monomials x^r are considered independent in the sense that $\sum a_n x^n = \sum b_n x^n$ iff $a_i = b_i$ for all $i = 0, 1, 2, \ldots$.

Following the usual convention, we write $R[\![x]\!]$ for P and $f(x)$ or simply f for any element of $R[\![x]\!]$.

Remark If $1 \in R$, then we can identify the power series $(0, 1, 0, 0, \ldots)$ with x, $(0, 0, 1, 0, \ldots)$ with x^2, $(0, 0, 0, 1, 0, \ldots)$ with x^3 and so on. Thus ax is an actual product of elements of $R[\![x]\!]$ defined by $ax = (a, 0, 0, \ldots)(0, 1, 0, 0, \ldots)$ and so on.

Following the convention we omit terms with zero coefficients and replace $(-a_n) x^n$ by $-a_n x^n$. A formal power series $f(x) = \sum a_i x^i \in R[\![x]\!]$ is said to be a zero power series iff $a_i = 0 \ \forall i$, otherwise it is said to be a non-zero power series.

Definition 4.5.1 If $f(x) = \sum a_r x^r$ is a non-zero power series in $R[\![x]\!]$, then the smallest integer n, such that $a_n \neq 0$ is called the order of $f(x)$ and is denoted by ord f (or simply $O(f)$).

Theorem 4.5.2 *If R is an integral domain, then $R[\![x]\!]$ is also an integral domain.*

Proof Clearly, $R[\![x]\!]$ is a commutative ring with identity. We claim that $R[\![x]\!]$ contains no zero divisors. Let $f(x) \neq 0$, $g(x) \neq 0$ in $R[\![x]\!]$, with $O(f) = n$ and $O(g) = m$, so that

$$f(x) = a_n x^n + a_{n+1} x^{n+1} + \cdots \quad (a_n \neq 0),$$

$$g(x) = b_m x^m + b_{m+1} x^{m+1} + \cdots \quad (b_m \neq 0).$$

Then

$$f(x)g(x) = a_n b_m x^{n+m} + (a_{n+1} b_m + a_n b_{m+1}) x^{n+m+1} + \cdots.$$

As R does not contain zero divisors, $a_n b_m \neq 0$. Hence the product $f(x)g(x)$ cannot be a zero power series. Hence $R[\![x]\!]$ is an integral domain.   □

Lemma 4.5.1 *Let R be a ring with 1. The element $f(x) = \sum a_n x^n$ in $R[\![x]\!]$ is invertible in $R[\![x]\!]$ iff the constant term a_0 is invertible in R.*

Proof Suppose $f(x)$ is invertible in $R[\![x]\!]$. Then there exists $g(x) = \sum b_n x^n \in R[\![x]\!]$ such that $f(x)g(x) = g(x)f(x) = 1$. Now $a_0 b_0 = 1 = b_0 a_0 \Rightarrow a_0$ is invertible in $R \Rightarrow b_0 = a_0^{-1}$.

Conversely, suppose a_0 is invertible in R. Then there exists $b_0 \in R$ such that $a_0 b_0 = b_0 a_0 = 1$. We define inductively the coefficients of power series $\sum b_n x^n$ in $R[\![x]\!]$ which is the inverse of $f(x)$. To start with, we determine the coefficient b_n so that $f(x)g(x) = 1$ i.e., $g(x)$ is a right inverse of $f(x)$. From $f(x)g(x) = 1$, we obtain from the definition of multiplication of power series that if the following equations are satisfied by the coefficients of $g(x)$, then $g(x)$ is a right inverse of $f(x)$:

(i) $a_0 b_0 = 1$
(ii) $a_0 b_1 + a_1 b_0 = 0$
(iii) $a_0 b_2 + a_1 b_1 + a_2 b_0 = 0$
    ⋮
(iv) $a_0 b_n + a_1 b_{n-1} + \cdots + a_n b_0 = 0$
(v) $a_n b_{n+1} + a_1 b_n + \cdots + a_n b_1 + a_{n+1} b_0 = 0$
    ⋮

Since a_0 is invertible, b_0 is uniquely determined as a_0^{-1}. Then b_1 is uniquely determined from equation (ii) above. Suppose $b_0, b_1, \ldots, b_n$ have already been defined. Then b_{n+1} is uniquely determined by equation (v) and so on. So every b_r is determined uniquely by induction, such that $f(x)g(x) = 1$. This shows that $f(x)$ has a right inverse $g(x)$.

Similarly, it can be proved that $f(x)$ has a left inverse. Hence $f(x)$ is invertible in $R[\![x]\!]$ (since in any ring an element with a right inverse and a left inverse has an inverse).   □

Theorem 4.5.3 (i) *A power series $f(x) = \sum a_n x^n$ in $F[\![x]\!]$, where F is a division ring, is invertible in $F[\![x]\!]$ iff its constant term a_0 is non-zero.*

(ii) *If F is a division ring, then the invertible elements of $F[\![x]\!]$ are precisely those power series with non-zero constant term.*

Proof (i) and (ii) follow from Lemma 4.5.1.   □

Let R be a field. Then $R[\![x]\!]$ is an integral domain by Theorem 4.5.2. Let $R(\!(x)\!)$ be the quotient field of $R[\![x]\!]$.

Theorem 4.5.4 *Every non-zero element of $R\langle\langle x\rangle\rangle$ can be uniquely expressed in the form:*

$$x^r\left(a_0 + a_1 x + a_2 x^2 + \cdots\right), \quad \text{where } a_i \in R \text{ and } a_0 \neq 0.$$

Proof Any element f of $R\langle\langle x\rangle\rangle$ can be written as

$$f = \frac{b_0 + b_1 x + b_2 x^2 + \cdots}{x^s(c_s + c_{s+1}x + \cdots)},$$

where s is the least integer such that $c_s \neq 0$. Since $c_s \neq 0$, $c_s + c_{s+1}x + \cdots$ has an inverse $d_0 + d_1 x + d_2 x^2 + \cdots$ in $R[\![x]\!]$.

Therefore,

$$f = \frac{(b_0 + b_1 x + b_2 x^2 + \cdots)(d_0 + d_1 x + d_2 x^2 + \cdots)}{x^s(c_s + c_{s+1}x + \cdots)(d_0 + d_1 x + d_2 x^2 + \cdots)}$$

$$= \frac{a_0 + a_1 x + a_2 x^2 + \cdots}{x^s} = x^{-s}\left(a_0 + a_1 x + a_2 x^2 + \cdots\right)$$

$$= a_0 x^{-s} + a_1 x^{-s+1} + a_2 x^{-s+2} + \cdots.$$

Thus $R\langle\langle x\rangle\rangle$ consists of formal power series with a finite number of terms with negative exponents:

$$f = \sum_{n=N}^{\infty} a_n x^n, \quad \text{where } N \text{ is an integer positive, negative or zero.}$$

If $a_N \neq 0$, N is called the *order of* f and is as usual denoted by $\mathrm{ord}(f)$ (or by $O(f)$). $\square$

Corollary *If R is a field, then $R\langle\langle x\rangle\rangle$ is also a field.*

Proof If follows from Theorem 4.5.4 and Lemma 4.5.1. $\square$

Remark $R\langle\langle x\rangle\rangle$ (sometimes, simply denoted by $R\langle x\rangle$) called the *field of formal power series* in x over the field R.

We now consider a particular class of power series.

Definition 4.5.2 Let R be a ring and $R[x]$ the set of all power series in $R[\![x]\!]$ whose coefficients are 0 from some indEx. onwards (the indEx. varies from series to series). Then $R[x]$ is called a *polynomial ring* (in the indeterminate x) over the ring R. The polynomial 0 such that $a_i = 0\ \forall i$ is called zero polynomial; otherwise, a polynomial is called a non-zero polynomial.

Remark Let R be a ring and $R[x]$ polynomial ring over R. If polynomials $f(x) = \sum a_r x^r$, $g(x) = \sum b_r x^r$ are in $R[x]$, with $a_r = 0 \; \forall r \geq n$ and $b_r = 0 \; \forall r \geq m$, then

$$a_r + b_r = 0 \quad \forall r \geq \max(m, n) \quad \text{and}$$

$$\sum_{\langle i+j=r \rangle} a_i b_j = 0 \quad \forall r \geq m + n.$$

These imply that both the sum $f(x) + g(x)$ and the product $f(x)g(x)$ are in $R[x]$. Clearly, $R[x]$ forms a subring of $R[[x]]$ and is called the *ring of polynomials* over R (in x).

Definition 4.5.3 Given a non-zero polynomial

$$f(x) = a_0 + a_1 x + \cdots + a_n x^n \quad (a_n \neq 0) \text{ in } R[x],$$

we call a_n the leading coefficient of $f(x)$; the integer n, the degree of $f(x)$; and write $\deg f = n$.

Remark The degree of any non-zero polynomial is a non-negative integer; no degree is given to the zero polynomial 0. The polynomials of degree 0 are precisely the non-zero constant polynomials.

Definition 4.5.4 If R is a ring with identity 1, a polynomial $f(x)$ ($\in R[x]$) whose leading coefficient is 1 is said to be a *monic polynomial*.

Proposition 4.5.1 *If R is an integral domain, then $R[x]$ is also an integral domain.*

Proof It follows from Theorem 4.5.2.   □

Definition 4.5.5 If R is a field, the quotient field $R\langle x \rangle$ of $R[x]$ is called the field of rational functions over the field R.

Proposition 4.5.2 *Let R be a ring and $f(x), g(x) \in R[x]$. If $\deg f = n$, $\deg g = m$, $(m, n > 0)$, then*

$$\deg(f + g) \leq \max(n, m) \quad \text{and} \quad \deg fg \leq m + n.$$

If R is an integral domain, $\deg(fg) = m + n$ and the converse is also true.

Proof Let

$$\left. \begin{aligned} f(x) &= a_0 + a_1 x + \cdots + a_n x^n \quad (a_n \neq 0) \quad \text{and} \\ g(x) &= b_0 + b_1 x + \cdots + b_m x^m \quad (b_m \neq 0) \end{aligned} \right\} \tag{A}$$

be polynomials in $R[x]$.

If c_r is the coefficient of x^r in $f(x)g(x)$, then $c_r = \sum_{i+j=r} a_i b_j$. The first part follows from Remark of Definition 4.5.2. Next suppose R is an integral domain. In (A), $a_n \neq 0$ and $b_m \neq 0 \Rightarrow c_{m+n} = \sum_{i+j=m+n} a_i b_j = a_n b_m \neq 0$ as R is an integral domain $\Rightarrow \deg(fg) = m + n$. $\square$

Let R be a ring with identity, S an extension of R, and s be an arbitrary element of S. Then for each polynomial $f(x) = a_0 + a_1 x + \cdots + a_n x^n$ in $R[x]$, define $f(s) = a_0 + a_1 s + \cdots + a_n s^n \in S$.

Then the element $f(s)$ is called the result of substituting s for x in $f(x)$.

Suppose $f(x), g(x) \in R[x]$ and $r \in \text{cent } S$. If $h(x) = f(x) + g(x)$ and $p(x) = f(x)g(x)$, then clearly $h(r) = f(r) + g(r)$ and $p(r) = f(r)g(r)$. Consequently, the map $\sigma_r : R[x] \to S, \sigma_r(f(x)) = f(r)$ is a homomorphism. σ_r is called a *substitution homomorphism induced by r* and its range, denoted by $R[r]$, is thus

$$R[r] = \{f(r) : f(x) \in R[x]\}$$
$$= \{a_0 + a_1 r + \cdots + a_n r^n : a_i \in R : n \geq 0\}.$$

Then $R[r]$ forms a subring of S and $R[r]$ is generated by the set $R \cup \{r\}$, (since $1 \in R \Rightarrow 1x = x \in R[x] \Rightarrow r \in R[r]$). Evidently, $R[r] = R \Rightarrow r \in R$.

We claim that σ_r is unique. If ρ is a substitution homomorphism induced by r, then $\rho(a_i) = a_i$ for each coefficient a_i and $\rho(x^i) = r^i$. Since ρ is a homomorphism, $\rho(f(x)) = \rho(a_0) + \rho(a_1)\rho(x) + \cdots + \rho(a_n)\rho(x^n) = a_0 + a_1 r + \cdots + a_n r^n = f(r) = \sigma_r(f(x)) \; \forall f(x) \in R[x] \Rightarrow \rho = \sigma_r$. Thus the following theorem follows:

Theorem 4.5.5 *Let R be a ring with identity, S an extension of R, and the element $r \in \text{cent } S$. Then there is a unique homomorphism $\sigma_r : R[x] \to S$ such that $\sigma_r(x) = r, \sigma_r(a) = a \; \forall a \in R$.*

Let R be a field and $R\langle x \rangle$ the field of formal power series in x and $t \in R\langle x \rangle$. Then by substitution $x \mapsto t$, we get a homomorphism $\sigma : R\langle x \rangle \to R\langle t \rangle$ defined by

$$\sigma \left(\sum_{i=N}^{\infty} a_i x^i \right) = \sum_{i=N}^{\infty} a_i t^i \tag{4.1}$$

However, if $\text{ord}(t) = r$, then σ is called a substitution of order r.

Theorem 4.5.6 *Every substitution $x \mapsto t$ of order 1 is an automorphism of $R\langle x \rangle$ over R.*

Proof Let $t = ax^r + \cdots \in R\langle x \rangle$, $a \neq 0$, $r \geq 1$, so that $\text{ord}(t) = r \geq 1$.

Let $f(x) = a_N x^N + a_{N+1} x^{N+1} + \cdots \in R\langle x \rangle$, $a_N \neq 0$.

Then $f(x) \neq 0$, and $f(t) = a_N(ax^r + \cdots)^N + a_{N+1}(ax^r + \cdots)^{N+1} + \cdots = a_N a^N x^{rN} + \cdots$ and as $a_N a^N \neq 0$; then by (4.1) we have $\sigma(f(t)) \neq 0$. Thus σ is an isomorphism of $R\langle x \rangle$ onto $R\langle t \rangle$. We shall now show that if $r = 1$, then σ is an automorphism. For this we have to prove that $R\langle t \rangle = R\langle x \rangle$. This will follow, if we can find a sequence of elements $a_1, a_2, \ldots$ in R such that σ maps $a_1 x + a_2 x^2 + \cdots$ onto x. In this case, $a_1(ax + \cdots) + a_2(ax + \cdots)^2 + \cdots = x$.

Comparing the coefficients of $x, x^2, x^3, \ldots, x^i$ from both sides, we have

$$a_1 a = 1,$$

$$a_2 a^2 + a_1(\cdots) = 0,$$

$$a_3 a^3 + a_2(\cdots) + a_1(\cdots) = 0,$$

$$\vdots$$

$$a_i a^i + a_{i-1}(\cdots) + \cdots + a_1(\cdots) = 0.$$

and so on, where the omitted expressions are obtained from the coefficients of $a_1 x + \cdots$, and hence they are known. From these relations we can solve $a_1, a_2, \ldots$, successively. This completes the proof. $\square$

4.6 Rings of Continuous Functions

We now consider another class of important rings, such as rings of functions, in particular, rings of continuous functions.

Definition 4.6.1 If X is a non-empty set and R is a ring, then the set S of all (set) functions $f : X \to R$ is a ring under pointwise addition '+' and pointwise multiplication '$\cdot$' of functions defined by

$$\left. \begin{array}{l} (f + g)(x) = f(x) + g(x), \\ (f \cdot g)(x) = f(x) \cdot g(x) \end{array} \right\} \quad \forall f, g \in S \text{ and } \forall x \in X.$$

This ring S is called a *ring of functions*.

Remark This ring S is commutative if and only if R is commutative. S has an identity (which is a constant function if and only if R has an identity).

Proposition 4.6.1 *Let $S(X, R)$ be the ring of all mappings from a non-empty set X to the ring R. Then given a ring T, every ring epimorphism $k : R \to T$ induces a ring homomorphism $\psi = k_* : S(X, R) \to S(X, T)$ such that if R and T have identity elements, then ψ sends identity element of $S(X, R)$ into the identity element of $S(X, T)$.*

Proof Define $\psi = k_* : S(X, R) \to S(X, T)$ by $\psi(f) = k \circ f$, for all $f \in S(X, R)$. Clearly, $\psi(f + g) = k \circ (f + g) = k \circ f + k \circ g = \psi(f) + \psi(g)$ and $\psi(f \cdot g) = k \circ (f \cdot g) = (k \circ f) \cdot (k \circ g) = \psi(f) \cdot \psi(g)$, for all $f, g \in S(X, R)$.

Suppose R and T have identity elements. Let $c : X \to R$ be the constant map defined by $c(x) = 1_R$ (identity element of R), for all $x \in X$ and $d : X \to T$ be the constant map defined by $d(x) = 1_T$ (identity element of T), for all $x \in X$. Then c and d are, respectively, the identity elements of $S(X, R)$ and $S(X, T)$. Now $(\psi(c))(x) = k(1_R) = 1_T$ (since k is an epimorphism) $= d(x)$, for all $x \in X$. Thus $\psi(c) = d$, proves the second part. $\square$

We now study the particular ring when X is a topological space and R is the field $\mathbf{R}$, the field of real numbers. We now define the ring of real-valued continuous functions on X, denoted by $C(X)$.

Definition 4.6.2 Let X be a topological space and $C(X)$ be the set of all continuous real-valued functions on X. Then $C(X)$ is a commutative ring under point wise operation '+' and '·' defined by

$$\left.\begin{array}{l} (f + g)(x) = f(x) + g(x), \\ (f \cdot g)(x) = f(x) \cdot g(x) \end{array}\right\} \quad \forall f, g \in C(X) \text{ and } \forall x \in X.$$

This ring is called the *ring of real-valued continuous functions on X*.

Proposition 4.6.2 *The ring $C(X)$ contains the identity element $c : C(X) \to \mathbf{R}$ defined by $c(f) = 1$ for all $f \in C(X)$. Further if $\psi : C(X) \to \mathbf{R}$ is a non-zero homomorphism, then it is an epimorphism such that $\psi(tc) = t$ for all $t \in \mathbf{R}$.*

Proof As ψ is a non-zero homomorphism there exists an element $f \in C(X)$ such that $\psi(f) \neq 0$. Then $\psi(f) = \psi(f \cdot c) = \psi(f)\psi(c)$ implies that $\psi(c) = 1$. Proceeding as in Problem 2, we prove that $\psi(nc) = n$ for each integer n, $\psi(rc) = r$ for each rational number r and $\psi(xc) = x$ for each irrational number x. Consequently, ψ is an epimorphism such that $\psi(ct) = t$ for all $t \in \mathbf{R}$. $\square$

Remark 1 Let X and Y be two homeomorphic spaces. Then $C(X)$ and $C(Y)$ are isomorphic as rings (see Ex. 11 of the Exercises in Appendix B).

Remark 2 Let X and Y be two compact Hausdorff spaces. If $C(X)$ and $C(Y)$ are isomorphic as rings, then X and Y are homeomorphic. It follows from a theorem of Gelfand–Kolmogoroff (Dugundji 1980, p. 289).

We now draw attention to the special ring $C([0, 1])$.

Proposition 4.6.3 *Let $C([0, 1])$ be the ring of all real-valued continuous functions on $[0, 1]$. Then $C([0, 1])$ is a commutative ring with identity having zero divisor.*

Proof Clearly $C([0, 1])$ is a commutative ring with identity. Consider a non-zero function $f \in C([0, 1])$ defined by

$$f(x) = \begin{cases} 0, & \text{if } 0 \le x \le \frac{1}{2}, \\ x - \frac{1}{2}, & \text{if } \frac{1}{2} \le x \le 1. \end{cases}$$

Let $g \in C([0, 1])$ be defined by

$$g(x) = \begin{cases} \frac{1}{2} - x, & \text{if } 0 \le x \le \frac{1}{2}, \\ 0, & \text{if } \frac{1}{2} \le x \le 1. \end{cases}$$

Then g is a non-zero function in $C([0, 1])$ such that g is a zero divisor of f.   □

The ring $C([0, 1])$, sometimes written as $C([0, 1])$, has the following properties.

Proposition 4.6.4 *The ring $C([0, 1])$ has neither non-trivial nilpotent elements nor non-trivial idempotent elements.*

Proof If possible, there exists a non-trivial nilpotent element $f \in C([0, 1])$. Then $(f(x))^n = 0$ for some positive integer n. Thus $f(x) = 0$, for all $x \in [0, 1]$, which implies a contradiction.

Next suppose that g is a non-trivial idempotent element in $C([0, 1])$. Then $(g(x))^2 = g(x)$ for all $x \in [0, 1]$ implies $g(x)(g(x) - 1) = 0$ for all $x \in [0, 1]$. This implies $\text{Im } g = \{0, 1\}$. Since the continuous image of $[0,1]$ under g must be a connected subset of $\mathbf{R}$ which is necessarily an interval of $\mathbf{R}$. Hence we reach a contradiction as $\{0, 1\} \subset \mathbf{R}$ is not an interval.   □

In Chap. 5, we shall study ideals of the ring $C([0, 1])$.

4.7 Endomorphism Ring

We now define another important ring with an eye to the additive group of integers. Given an additive abelian group A, there exists a ring, called the endomorphism ring of A, having some interesting properties.

Definition 4.7.1 Let A be an additive abelian group and $\text{End}(A)$ the set of all endomorphisms of A. Then $(\text{End}(A), +, \cdot)$ is a non-commutative ring (in general) with identity $1_A : A \to A$ under usual '+' and '·' defined by $(f + g)(x) = f(x) + g(x)$ and $(f \cdot g)(x) = (f \circ g)(x) = f(g(x))$, $\forall f, g \in \text{End}(A)$ and $\forall x \in A$. $\text{End}(A)$ is called *endomorphism ring of A*.

We now study the special ring $\text{End}(\mathbf{Z})$.

Proposition 4.7.1 *The two rings $\mathbf{Z}$ and $\text{End}(\mathbf{Z})$ are isomorphic.*

Proof Define the mapping $\psi : \mathbf{Z} \to \text{End}(\mathbf{Z})$ by $\psi(n) = f_n$, where $f_n : \mathbf{Z} \to \mathbf{Z}$ is defined by $f_n(x) = nx$, $\forall x \in \mathbf{Z}$. We claim that ψ is a ring isomorphism.

ψ is a *ring homomorphism*: $\psi(m+n) = f_{m+n} : \mathbf{Z} \to \mathbf{Z}$ is defined by $f_{m+n}(x) = (m+n)x = mx + nx = f_m(x) + f_n(x) = (f_m + f_n)(x)$, $\forall x \in \mathbf{Z} \Rightarrow f_{m+n} = f_m + f_n \Rightarrow \psi(m+n) = \psi(m) + \psi(n)$, $\forall m, n \in \mathbf{Z}$. Similarly, $\psi(mn) = \psi(m)\psi(n)$, $\forall m, n \in \mathbf{Z}$. Hence ψ is a ring homomorphism.

ψ is *injective*: Let $\psi(m) = \psi(n)$. Then $f_m = f_n \Rightarrow f_m(x) = f_n(x)$ $\forall x \in \mathbf{Z} \Rightarrow mx = nx$ $\forall x \in \mathbf{Z} \Rightarrow m = n$ (by taking in particular $x = 1$) $\Rightarrow \psi$ is injective.

ψ is *surjective*: Let $f \in \text{End}(\mathbf{Z})$. Then $f : (\mathbf{Z}, +) \to (\mathbf{Z}, +)$ is a group homomorphism. $1 \in \mathbf{Z} \Rightarrow f(1) \in \mathbf{Z} \Rightarrow f(1) = n$ for some $n \in \mathbf{Z} \Rightarrow f_n(x) = nx = f(1)x = xf(1) = f(x \cdot 1) = f(x)$, $\forall x \in \mathbf{Z} \Rightarrow f_n = f \Rightarrow \psi(n) = f \Rightarrow \psi$ is surjective. Consequently, ψ is a ring isomorphism. $\qquad\square$

Remark In general $\text{End}(A)$ is non-commutative. For example, consider the free abelian group $(A, +) = (\mathbf{Z} \oplus \mathbf{Z}, +)$. Let $f, g \in \text{End}(A)$ be defined on the generators $(1, 0)$ and $(0, 1)$ of the group A as follows:

$$f(1,0) = (1,0) \quad \text{and} \quad f(0,1) = (1,0);$$

and

$$g(1,0) = (0,0) \quad \text{and} \quad g(0,1) = (0,1).$$

Then

$$(f \circ g)(n,m) = f\big(g(n,m)\big) = f(0,m) = (m,0)$$

and

$$(g \circ f)(n,m) = g\big(f(n,m)\big) = g(n+m,0) = (0,0)$$

for all $(n,m) \in A \Rightarrow f \circ g \neq g \circ f$.

Any ring with identity can be embedded in an endomorphism ring.

Theorem 4.7.1 *Let R be a ring with 1. Then R is isomorphic to a subring of $\text{End}(R)$.*

Proof Define the mapping $\psi : R \to \text{End}(R)$ by $\psi(a) = f_a$ for all $a \in R$, where $f_a : R \to R$ is defined by $f_a(x) = ax$, for all $x \in R$. Clearly, f_a is an endomorphism of the abelian group R by the distributive property in R. Moreover, as before, ψ is a ring homomorphism. Now $\ker \psi = \{a \in R : f_a = 0\} = \{a \in R : ax = 0 \ \forall x \in R\} = \{0\}$. Hence ψ is a monomorphism. Consequently, R is isomorphic to a subring of $\text{End}(R)$. $\qquad\square$

Corollary *Any ring R can be embedded in the ring $\text{End}(R)$.*

Proof It follows from Theorems 4.4.2 and 4.7.1. $\qquad\square$

We now find the endomorphism ring of any finite cyclic group of prime order.

Proposition 4.7.2 *The two rings* $\text{End}(\mathbf{Z}_p)$ *and* $\mathbf{Z}_p$ *are isomorphic.*

Proof Define $\psi : \text{End}(\mathbf{Z}_p) \to \mathbf{Z}_p$ by $\psi(f) = f((1))$, where $f : \mathbf{Z}_p \to \mathbf{Z}_p$ is a group homomorphism. Clearly, ψ is well defined and an isomorphism. $\qquad\qquad\square$

4.8 Problems and Supplementary Examples

Problem 1 Let $(R, +, \cdot)$ be a ring and d an element of R which is not a left zero divisor. Show that the characteristic of R is finite iff the order of d in the additive group $(R, +)$ is finite. Moreover, $\text{char } R = m$ iff order of d in $(R, +) = m$.

[*Hint.* Let $(R, +, \cdot)$ be a ring and d be not a left zero divisor. Suppose $(R, +, \cdot)$ is of finite characteristic. Then $\exists$ a positive integer $m > 1$ such that $mr = 0 \ \forall r \in R$. Then $md = 0$ and hence d has a finite order in $(R, +)$.

Conversely, let d have finite order in $(R, +)$. Then $\exists$ a positive integer $m > 1$ such that $md = 0$. Let x be an arbitrary element of R. Then $0 = (md)x = d(mx)$ (by distributive property) $\Rightarrow mx = 0$ (as d is not a left zero divisor) $\forall x \in R \Rightarrow \text{char } R$ is finite.

Next let $\text{char } R = m$. Then $md = 0 \Rightarrow$ order of d is a divisor of m. If possible, order of $d = r$, $1 \le r < m$. Now $rd = 0 \Rightarrow$ for any $x \in R$, $0 = (rd)x = d(rx) \Rightarrow rx = 0$ (as d is not a left zero divisor). $\Rightarrow \text{char } R = r < m \Rightarrow$ a contradiction.

Conversely, if order of d in $(R, +)$ is m, then $md = 0$ and $rd \ne 0$ if $1 \le r < m$. Now, for any $x \in R$, $mx = 0 \Rightarrow \text{char } R \le m$. If $1 < r < m$, then $rd \ne 0 \Rightarrow \text{char } R \ge m$. Consequently, $\text{char } R = m$.]

Problem 2 Let $\mathbf{R}$ be the field of reals and $f : \mathbf{R} \to \mathbf{R}$ a ring homomorphism such that $f(1) = 1$. Show that $f(x) = x$ for all $x \in \mathbf{R}$ and f is an automorphism.

[*Hint.* $f(1) = 1 \Rightarrow f(n) = f(1 + 1 + \cdots + 1) = f(1) + f(1) + \cdots + f(1)$ (n times) $= nf(1) = n \ \forall$ integers $n > 0$.

If the integer $n < 0$, $f(n) = -nf(-1) = nf(1) = n$.

Thus $f(n) = n \ \forall$ integers $n \Rightarrow f$ is the identity map on the set of integers.

Let $\frac{p}{q}$ be a rational number such that $\gcd(p, q) = 1$. Now

$$p = \frac{p}{q} + \frac{p}{q} + \cdots + \frac{p}{q} \ (q \text{ times}) \quad \Rightarrow \quad f(p) = qf\left(\frac{p}{q}\right)$$

$$\Rightarrow \quad f\left(\frac{p}{q}\right) = \frac{f(p)}{q} = \frac{p}{q}$$

(since $f(p) = p$, as p is an integer, positive or negative) $\Rightarrow f$ is the identity map on the set of rationals.

Let a and b be real numbers such that $a > b$. Then $a - b > 0$ and hence $\exists$ a real number c such that $c^2 = a - b$. Then $f(a - b) = f(c^2) = (f(c))^2$ (as f is a homomorphism) $> 0 \Rightarrow f(a) - f(b) > 0$. Thus $a > b \Rightarrow f(a) > f(b) \Rightarrow f$ is strictly monotonic.

Let x be any irrational number. Then $\exists$ rational numbers a_n and b_n such that $a_n < x < b_n$ and $a_n - b_n \to 0$ as $n \to \infty$.

So, $f(a_n) < f(x) < f(b_n)$ (as f preserves ordering) $\Rightarrow a_n < f(x) < b_n$ (as f is identity on the set of rationals) $\Rightarrow |x - f(x)| < |a_n - b_n| \Rightarrow x - f(x) \to 0 \Rightarrow f(x) = x$.]

Problem 3 Show that there is at most one isomorphism under which an arbitrary ring R is isomorphic to the ring $\mathbf{Z}$.

[*Hint.* Suppose $f, g : R \to \mathbf{Z}$ are isomorphisms of rings. Then $g^{-1} : \mathbf{Z} \to R$ is also an isomorphism.

Hence $f \circ g^{-1} : \mathbf{Z} \to \mathbf{Z}$ is necessarily a homomorphism.

Then $f \circ g^{-1} =$ identity homomorphism on $\mathbf{Z}$ (by Problem 2) $\Rightarrow f = g$.]

Problem 4 Let F be an arbitrary field. Show that $F[x]$ is not a field.

[*Hint.* Let $f(x) \in F[x]$ be such that $\deg f(x) > 0$. Suppose $\exists$ a non-zero polynomial $g(x) \in F[x]$ such that $f(x)g(x) = 1$.

Then $\deg[f(x)g(x)] = \deg f(x) + \deg g(x) > 0 \Rightarrow$ a contradiction, since $\deg 1 = 0 \Rightarrow F[x]$ is not a field.]

Remark The only invertible elements of $F[x]$ are constant polynomials excepting zero polynomial.

Problem 5 Let D be an integral domain and K be its quotient field. Show that the field of rational functions $D\langle x \rangle$ over D is the same as the field of rational functions over K.

Problem 6 A finite division ring is commutative. This result is known as Wedderburn theorem on finite division rings.

[*Hint.* See Sect. 7.2, Herstein (1964).]

Problems and Supplementary Examples (SE-I)

1 Let F be a finite field of characteristic p (p is prime). Let $f : F \to F$ be the map defined by $f(x) = x^p$. Then f is an automorphism of F.

Solution $\forall a, b \in F, (a+b)^p = a^p + p_{c_1}a^{p-1}b + \cdots + p_{c_r}a^{p-r}b^r + \cdots + b^p$ (see Ex. 12 of Exercises-I) and $(ab)^p = a^p b^p$. Since p divides the binomial coefficient p_{c_r} for $1 \le r \le p-1$, $(a+b)^p = a^p + b^p$ (see Ex. 14 of Exercises-I). Consequently, $f(ab) = f(a)f(b)$ and $f(a+b) = f(a) + f(b)$ $\forall a, b \in F \Rightarrow f$ is a homomorphism. Now $\ker f = \{0\} \Rightarrow f$ is a monomorphism. As F is finite, f is an epimorphism. Consequently, f is an automorphism of F.

2 Let F be a field. Then the groups $(F - \{0\}, \cdot)$ and $(F, +)$ cannot be isomorphic.

Solution Let $f : (F - \{0\}, \cdot) \to (F, +)$ be an isomorphism. Then we claim that characteristic of F is 2. Otherwise, $0 = f(1) = f((-1) \cdot (-1)) = f(-1) +$

$f(-1) \neq 0$ by Theorem 4.3.1 $\Rightarrow$ a contradiction. Let $a \in F - \{0\}$. Then $f(a^2) = f(a \cdot a) = f(a) + f(a) = 0 = f(1) \Rightarrow a^2 = 1$ as f is a monomorphism $\Rightarrow a = 1$ (or -1) in this case $\Rightarrow |F| = 2 \Rightarrow |F - \{0\}| = 1$. Thus f is a bijection from a set of order 1 to a set of order 2 $\Rightarrow$ a contradiction.

3 Let R be a ring such that $|R| > 1$ and for each non-zero $x \in R$ there is a unique $y \in R$ satisfying $xyx = x$. Then

(i) R has no zero divisors;
(ii) $yxy = y$;
(iii) R has an identity;
(iv) R is a division ring.

4 Let $f : R \to D$ be a ring homomorphism, where R is a ring with identity 1 and D is an integral domain with identity $1'$. If $\ker f \neq R$, then $f(1) = 1'$.

Solution Suppose $f(1) = 0$. Then $\forall x \in R$, $f(x) = f(x1) = f(x)f(1) = 0 \Rightarrow$ $\ker f = R \Rightarrow$ a contradiction $\Rightarrow f(1) \neq 0$. Now $1'f(1) = f(1) = f(1 \cdot 1) = f(1)f(1) \Rightarrow f(1) = 1'$.

5 Let $f : R \to S$ be a ring homomorphism, where R is a field and $1_S = 1' \neq 0$ in S. Then f is a monomorphism.

Solution Let $a \neq 0$ be in R. We claim that $f(a) \neq 0$. Since R is a field, a has an inverse a^{-1} in R. Hence $1' = f(1) = f(aa^{-1}) = f(a) \cdot f(a^{-1})$. If $f(a) = 0$, then $1' = 0f(a^{-1}) = 0$, it is a contradiction, since $1' \neq 0$ in S. Thus $\ker f$ contains no element of R except 0 and hence f is a monomorphism.

6 If R and S are two rings, their direct sum denoted by $R \oplus S$ is the ring consisting of the pairs (x, y) with $x \in R$ and $y \in S$, with addition and multiplication given by

$$(x_1, y_1) + (x_2, y_2) = (x_1 + x_2, y_1 + y_2),$$

$$(x_1, y_1) \cdot (x_2, y_2) = (x_1x_2, y_1y_2).$$

The direct sum of any number of rings is defined in a similar way.

(a) $R \oplus S$ is not an integral domain, since $\forall x \in R$, $\forall y \in S$, $(x, 0) \cdot (0, y) = (0, 0)$, which is the zero element of $R \oplus S$.
(b) The direct sum $S = R \oplus R \oplus \cdots \oplus R$ of n-copies of R can be viewed as the ring of functions on a set of n elements viz. $\{1, 2, \ldots, n\}$ with values in R; the element $(x_1, x_2, \ldots, x_n) \in R \oplus R \oplus \cdots \oplus R$ can be identified with the function f given by $f(i) = x_i$. Addition and multiplication of functions are given as usual by operating on their values in R. Clearly, S is a ring.
(c) If $i \in \mathbf{N}^+$, then $P = \sum \oplus R_i$ can be viewed as the set of all infinite sequences $(x_1, x_2, \ldots, x_n, \ldots)$ such that $x_i \in R_i$ for each $i \in \mathbf{N}^+$. Clearly, P is a ring.

7 Let R be a ring isomorphic to a subring A of a ring T. Then there exists an extension S of the ring R such that S is isomorphic to T.

Proof Construct S by $S = R \cup (T - A)$. As R is isomorphic to A, $\exists$ an isomorphism $f : R \to A$. Hence we can define a bijective map $g : S \to T$ given by

$$g(u) = f(u), \quad \text{if } u \in R$$

$$= u, \quad \text{if } u \in T - A = S - R.$$

Define addition $+$ and multiplication $\cdot$ on S by

$$u + v = g^{-1}\big(g(u) + g(v)\big) \quad \text{and} \quad uv = g^{-1}\big(g(u)g(v)\big), \quad \forall u, v \in S \qquad (4.2)$$

Hence $g(u + v) = g(u) + g(v)$ and $g(uv) = g(u)g(v)$ $\forall u, v \in S \Rightarrow g$ is a homomorphism. Consequently, g is an isomorphism. Hence S is a ring. To show that R is a subring of S, take $u, v \in R$. Then $f(u) = g(u)$, $f(v) = g(v)$, $f(u) + f(v) = g(u) + g(v)$ and $f(u)f(v) = g(u)g(v) \Rightarrow u + v = f^{-1}(f(u) + f(v)) = $ the sum of u and v as originally defined in R and $uv = f^{-1}(f(u)f(v)) = $ the product of u and v as originally defined in R.

Hence the original composition $+$ and $\cdot$ defined in R are the restrictions of the corresponding compositions defined in S by (4.2). Consequently, R is a subring of S and so S is an extension of R. $\qquad\qquad\qquad\qquad\qquad\qquad\qquad\qquad\qquad\square$

8 A field F is said to be perfect iff either char F is 0 or its Frobenius homomorphism (see Example (iii) or Ex. 15 of Exercises-I) is an epimorphism. Then any finite field is perfect.

9 Let R be a ring and the elements $a, b, c, d \in R$ satisfy the condition

$$\left.\begin{array}{l} a + b = c + d \\[4pt] a \cdot b = c \cdot d \end{array}\right\} \quad \Rightarrow \quad \text{either } a = c,\ b = d \text{ or } a = d,\ b = c.$$

Then R does not contain any divisor of zero.

Solution If possible, let R contains divisor of zero. Then there exist non-zero elements $a, b \in R$ such that $a \cdot b = 0$. Now by hypothesis

$$\left.\begin{array}{l} a + b = (a + b) + 0 \\[4pt] a \cdot b = (a + b) \cdot 0 \end{array}\right\} \quad \Rightarrow \quad \text{either } a = a + b,\ b = 0 \text{ or } a = 0,\ b = a + b.$$

In both cases, either $a = 0$ or $b = 0$, a contradiction. Hence, R does not contain any divisor of zero.

10 A ring R with 1 is a division ring iff for each $x \ (\neq 1) \in R\ \exists y \in R$ such that $x + y = xy$.

Solution Let for each x ($\neq 1$) $\in R$, $\exists y \in R$ such that $x + y = xy$. Let a ($\neq 0$) $\in R$. Let $a = 1$. Then a is the inverse of itself. Next let $a \neq 1$. Now $1 - a \neq 1$ (since $a \neq 0$). Then by hypothesis, $\exists y \in R$ such that $(1 - a) + y = (1 - a)y \Rightarrow 1 - a + y = y - ay \Rightarrow 1 - a = -ay \Rightarrow 1 = a(1 - y) \Rightarrow a$ has a right inverse in R. Then it follows that a has also a left inverse in R. Thus every non-zero element of R has an inverse in $R \Rightarrow R$ is a division ring. Conversely, let R be a division ring. Let x ($\neq 1$) $\in R$. Then $x - 1 \neq 0 \Rightarrow$ its inverse $(x - 1)^{-1}$ exists in $R \Rightarrow x(x - 1)^{-1} = y$ (say) $\in R$. Now, $(x + y)(x - 1) = x(x - 1) + x(x - 1)^{-1}(x - 1) = x^2$ and $(xy)(x - 1) = x \cdot x(x - 1)^{-1}(x - 1) = x^2 \Rightarrow (x + y)(x - 1) = (xy)(x - 1) \Rightarrow (x + y)(x - 1)(x - 1)^{-1} = (xy)(x - 1)(x - 1)^{-1} \Rightarrow x + y = xy$.

11 Find the non-trivial ring homomorphisms from $(\mathbf{Z}, +, \cdot)$ to $(\mathbf{Q}, +, \cdot)$.

Solution Let $f : \mathbf{Z} \to \mathbf{Q}$ be a non-trivial ring homomorphism. Now $f(1) \in \mathbf{Q} \Rightarrow$ either $f(1) = 0$ or $f(1) \neq 0$. If $f(1) = 0$, then $f(x) = f(x \cdot 1) = f(x) \cdot f(1) = 0$, $\forall x \in \mathbf{Z} \Rightarrow f = 0 \Rightarrow$ a contradiction, since $f \neq 0$ by hypothesis $\Rightarrow f(1) \neq 0$. Let $f(1) = p/q$, $p \neq 0$, $q > 0$. Then $f(1) = f(1 \cdot 1) = f(1)f(1) \Rightarrow p/q = p/q \cdot p/q \Rightarrow p/q = 1 \Rightarrow p = q \Rightarrow f(1) = 1 \Rightarrow f(n) = n$, $\forall n \in \mathbf{Z}$ by Problem 2.

12 Let F be a field of characteristic p ($\neq 0$). Then $\exists$ only one group homomorphism: $f : (F, +) \to (F - \{0\}, \cdot)$.

Solution Let $f, g : (F, +) \to (F - \{0\}, \cdot)$ be two group homomorphisms. Then $f(0) = 1 = g(0)$.
   Again

$$\forall x \in F, \quad px = 0 \text{ by hypothesis}$$
$$\Rightarrow \quad f(px) = f(0) = g(0) = g(px)$$
$$\Rightarrow \quad \big(f(x)\big)^p = \big(g(x)\big)^p$$
$$\Rightarrow \quad \big(f(x)\big)^p - \big(g(x)\big)^p = 0.$$

If p is an odd prime, then $(f(x))^p - (g(x))^p = 0 \Rightarrow (f(x))^p + (-1)^p(g(x))^p = 0 \Rightarrow (f(x))^p + ((-1)g(x))^p = 0 \Rightarrow (f(x) + (-1)g(x))^p = 0$ by Ex. 1 of SE-I $\Rightarrow (f(x) - g(x))^p = 0 \Rightarrow f(x) - g(x) = 0 \Rightarrow f(x) = g(x) \; \forall x \in F \Rightarrow f = g$.
   If $p = 2$, then $(f(x))^2 - (g(x))^2 = 0 \Rightarrow (f(x))^2 + (g(x))^2 = 0$, (since ch $F = 2$, $a = -a \; \forall a \in F$) $\Rightarrow (f(x) + g(x))^2 = 0$, since ch $F = 2 \Rightarrow f(x) + g(x) = 0 \; \forall x \in F \Rightarrow f(x) = -g(x) = g(x) \; \forall x \in F$, since ch $F = 2$.
   Thus $f(x) = g(x) \; \forall x \in F \Rightarrow f = g \Rightarrow \exists$ only one group homomorphism $f : (F, +) \to (F - \{0\}, \cdot)$.

13 Every finite subgroup of the multiplicative group $G = (F - \{0\}, \cdot)$ of a field F is cyclic.

Solution Use the structure theorem of a finitely generated abelian group to show that

$$G \cong \mathbf{Z}_{n_1} \oplus \mathbf{Z}_{n_2} \oplus \cdots \oplus \mathbf{Z}_{n_t},$$

where $n_1 | n_2 | \cdots | n_t$.

We claim that $t = 1$. If $t > 1$, let a prime p be a divisor of n_1. The cyclic group $\mathbf{Z}_{n_1}$ has p elements of order p and similarly, $\mathbf{Z}_{n_2}$ has p elements of order p. Hence G has at least p^2 elements of order p. All the elements of order p in G are the roots of the polynomial $x^p - 1$ over the field F. But it cannot have more than p roots. Hence the contradiction implies that $t = 1$. This shows that G is cyclic.

4.8.1 Exercises

Exercises-I

1. Let $\mathbf{Z}[i] = \{a + bi : a, b \text{ are integers and } i^2 = -1\}$. Show that $\mathbf{Z}[i]$ forms a ring under the compositions $(a + bi) + (c + di) = (a + c) + (b + d)i$, $(a + bi)(c + di) = (ac - bd) + (ad + bc)i$.

   $\mathbf{Z}[i]$ is called Gaussian ring in honor of $C \cdot F$ Gauss (1777–1855). For more result on $\mathbf{Z}[i]$ see Chap. 11.

2. Let G be an (additive) abelian group. Define an operation in G by $ab = 0$ $\forall a, b \in G$. Show that $(G, +, \cdot)$ is a ring.

3. Let F denote the set of all real-valued functions, defined on the closed interval $I = [0, 1]$. Show that F forms a ring under the compositions $(f + g)(x) = f(x) + g(x)$, $(f \cdot g)(x) = f(x)g(x)$, $\forall f, g \in F$ and $x \in I$ (F is called *ring of functions*).

4. A ring B with identity in which $a^2 = a$ $\forall a \in B$ (i.e., in which every element is idempotent) is called a *Boolean ring*. Show that

   (i) in a Boolean ring B, $a + a = 0$ $\forall a \in B$;
   (ii) a Boolean ring is commutative;
   (iii) the power set $\mathcal{P}(A)$ of a given set A forms a Boolean ring under the compositions:

   $$X + Y = (X - Y) \cup (Y - X) \quad \text{and} \quad XY = X \cap Y, \quad \forall X, Y \in \mathcal{P}(A).$$

5. An element a in ring R satisfying $a^n = 0$ for some positive integer n is called a nilpotent element. Show that

   (i) a nilpotent element is a zero divisor (unless $R = \{0\}$) but its converse is not true in general;
   (ii) the residue class ring $\mathbf{Z}_n$ contains nilpotent elements iff n has a square factor.

6. Show that any ring $(R, +, \cdot)$ in which $a + b = ab$ $\forall a, b \in R$ is the zero ring.

7. If the set A contains more than one element, prove that every non-empty proper subset of A is a zero divisor in the ring $\mathcal{P}(A)$ [see Exercise 4(iii)].

8. Let R be a ring with identity element 1 and without zero divisors. Show that for $a, b \in R$

   (i) $ab = 1 \Leftrightarrow ba = 1$;
   (ii) $a^2 = 1 \Rightarrow$ either $a = 1$ or $a = -1$.

9. Find the number of ring homomorphisms from the ring $\mathbf{Z}_{12}$ into $\mathbf{Z}_{28}$.

10. Let $G = \{X_i : i \in I\}$ be any multiplicative group and R any commutative ring with identity. Consider the set $R(G)$ of all formal sums $\sum_{i \in I} a_i x_i$ for $a_i \in R$ and $x_i \in G$, where all but a finite number of a_i's are 0. Two such expressions are said to be equal iff they have the same coefficients. Define $+$ and $\cdot$ in $R(G)$ by

$$\left(\sum_{i \in I} a_i x_i \right) + \left(\sum_{i \in I} b_i x_i \right) = \sum_{i \in I} (a_i + b_i) x_i \quad \text{and}$$

$$\left(\sum_{i \in I} a_i x_i \right) \cdot \left(\sum_{i \in I} b_i x_i \right) = \sum_{i \in I} c_i x_i,$$

where $c_i = \sum a_j b_k$ and summation is extended over all subscripts j and k for which $x_j x_k = x_i$. Show that $(R(G), +, \cdot)$ forms a ring ($R(G)$ is called the *group ring* of G over R).

11. (a) Show that a ring R is commutative iff $(a + b)^2 = a^2 + 2ab + b^2 \; \forall a, b \in R$.
    (b) Let R be a ring with identity and cent R the center of R. Then prove that

        (i) if $x^{n+1} - x^n \in$ cent $R \; \forall x \in R$ and for a fixed positive integer n, then R is commutative;
        (ii) if $x^m - x^n \in$ cent $R \; \forall x \in$ cent R and for fixed relatively prime positive integers m and n, one of which is even, then R is commutative.

12. Let R be a commutative ring with 1. Prove by induction the binomial expansion: $(a + b)^n = a^n + {}^nC_1 a^{n-1} b + \cdots + {}^nC_r a^{n-r} b^r + \cdots + b^n$ for $a, b \in R$ and for every positive integer n, where ${}^nC_r$ has the usual meaning.

13. Let R be a commutative ring. If the elements a and b of R are nilpotent, prove that $(a + b)$ is also nilpotent. Show that the result may not be true if R is not commutative.

14. For any two elements a, b in a field of positive prime characteristic p, show that $(a \pm b)^p = a^p \pm b^p$.

15. Let R be a field of positive characteristic p. Show that

    (i) $\forall a, b \in R, (a + b)^{p^m} = a^{p^m} + b^{p^m}$, where m is a positive integer;
    (ii) the map $R \to R$ given by $r \mapsto r^p$ is a homomorphism of rings (called the Frobenius homomorphism), which is a monomorphism but not an isomorphism in general.

16. Let R be a commutative ring with 1. If x is a nilpotent element of R, show that $1 + x$ is a unit in R. Deduce that the sum of a nilpotent element and a unit is a unit.

    [*Hint.* $(1 + x)^{-1} = 1 - x + x^2 + \cdots + (-1)^{n-1} x^{n-1}$, where $x^n = 0$.]

17. Show that the commutative property of addition is a consequence of other properties of a ring if

 (i) R contains an identity element 1 or
 (ii) R contains at least one element, which is not a divisor of zero.

18. Let D be an integral domain and $a, b \in D$. If m and n are relatively prime positive integers such that $a^n = b^n$ and $a^m = b^m$. Show that $a = b$.

19. Let R be a ring. Prove that for any subset X ($\neq \emptyset$) of R, the set $\langle X \rangle = \bigcap \{S : X \subseteq S; S$ is subring of $R\}$ is the smallest (in the sense of inclusion) subring of R to contain X. ($\langle X \rangle$ is called the *subring generated by X*.)

20. Let R be a ring with identity and S a subring of R. Suppose $a \notin S$ and $\langle S, a \rangle$ is the subring generated by the set $S \cup \{a\}$. If $a \in \operatorname{cent} R$, show that

$$\langle S, a \rangle = \left\{ \sum_{i=0}^{n} r_i a^i : n \text{ is a positive integer}; r_i \in S \right\}.$$

21. Show that every ring with 1 can be embedded in a ring of endomorphisms of some additive abelian group.

22. Let M be the ring of all real matrices of the form $\left(\begin{smallmatrix} a & b \\ -b & a \end{smallmatrix} \right)$ and $\mathbf{C}$ the field of complex numbers. Define $f : \mathbf{C} \to M$ by $f(a + ib) = \left(\begin{smallmatrix} a & b \\ -b & a \end{smallmatrix} \right)$. Show that f is an isomorphism of rings and so M is a field.

23. Let H be the division ring of real quaternions and R the division ring defined in Example 4.1.3(ii). Show that R is isomorphic to H.
    [*Hint*. Define an isomorphism $f : H \to R$ given by $1 \mapsto \left(\begin{smallmatrix} 1 & 0 \\ 0 & 1 \end{smallmatrix} \right), i \mapsto \left(\begin{smallmatrix} i & 0 \\ 0 & -i \end{smallmatrix} \right),$
    $j \mapsto \left(\begin{smallmatrix} 0 & 1 \\ -1 & 0 \end{smallmatrix} \right), k \mapsto \left(\begin{smallmatrix} 0 & i \\ i & 0 \end{smallmatrix} \right).$]

24. Let R be a ring. Prove that the following conditions are equivalent.

 (i) R has no non-zero nilpotent elements.
 (ii) If $a \in R$ and $a^2 = 0$, then $a = 0$.

25. Prove that in a finite field every element can be expressed as a sum of two squares.

26. In this exercise a ring means a commutative ring with identity. Let A be a ring $\neq \{0\}$. Show that A is a field $\Leftrightarrow$ every homomorphism of A into a non-zero ring B is a monomorphism.

27. Let L be a lattice in which sup and inf of two elements a and b are denoted by $a \vee b$ and $a \wedge b$, respectively. L is a Boolean lattice (or Boolean algebra) iff

 (i) L has a least element and a greatest element (denoted by 0 and 1, respectively);
 (ii) each of $\vee$ and $\wedge$ is distributive over the other;
 (iii) each $a \in L$ has a unique complement $a' \in L$ such that $a \vee a' = 1$ and $a \wedge a' = 0$.

(For example, $\mathcal{P}(X)$, the set of all subsets of X, ordered by inclusion is a Boolean lattice.)

(a) Let L be a Boolean lattice. Define addition and multiplication in L by the rules

$$a + b = (a \wedge b') \vee (a' \wedge b) \quad \text{and} \quad ab = a \wedge b.$$

Show that L becomes a Boolean ring (Exercise 4).

Conversely, let A be a Boolean ring. Define an ordering on A as follows: $a \leq b \Rightarrow a = ab$.

Show that $(A, \leq)$ is a Boolean lattice.

Remark The sup and inf are given by $a \vee b = a + b + ab$ and $a \wedge b = ab$ and the complement $a' = 1 - a$.

(b) Show that there is a one-to-one correspondence between (isomorphism classes of) Boolean rings and (isomorphism classes of) Boolean lattices.

28. We consider only rings with identity. Let B be a ring and A be a subring of B such that $1 \in A$. An element $x \in B$ is said to be integral over A, iff x satisfies an equation of the form:

$$x^n + a_1 x^{n-1} + \cdots + a_n = 0, \quad \text{where the } a_i\text{'s are elements of } A.$$

If every element of $x \in B$ is integral over A, then B is said to be integral over A.

Prove the following:

Let $A \subseteq B$ be an integral domain and B be integral over A. Then B is a field $\Leftrightarrow A$ is a field.

29. *Polynomials in several variables (indeterminates).* Let R be a ring and $R_1 = R[x_1]$ be the polynomial ring over R in variable x_1. Consider the polynomial ring $R_1[x_2]$ over R_1, where the variable is x_2. Then $R[x_1, x_2] = R_1[x_2]$ consists of all polynomials:

$$\left(\sum c_{0i} x_1^i \right) + \left(\sum c_{1i} x_1^i \right) x_2 + \cdots + \left(\sum c_{ni} x_1^i \right) x_2^n,$$

$n = 0, 1, 2, \ldots$, where summation is finite in each case. $R[x_1, x_2]$ is called the polynomial ring over R in the two variables x_1 and x_2. The concept can now be generalized. The polynomial ring $R[x_1, x_2, \ldots, x_n]$ over R in the n variables $x_1, x_2, \ldots, x_n$ is defined recursively by $R[x_1, x_2, \ldots, x_n] = R_{n-1}[x_n]$, where $R_{n-1} = R[x_1, x_2, \ldots, x_{n-1}]$ for $n = 2, 3, \ldots$ and $R_0 = R$. Clearly, the polynomial ring remains the same by any permutation of the variables.

Thus the ring $R[x_1, x_2, \ldots, x_n]$ is the set of all functions $f : \mathbf{N}^n \to R$ such that $f(u) \neq 0$ for almost a finite number of elements u of $\mathbf{N}^n$ for each positive integer n together with addition and multiplication defined by

$$(f + g)(u) = f(u) + g(u) \quad \text{and} \quad (fg)(u) = \sum_{\substack{v+w=u \\ v, w \in \mathbf{N}^n}} f(v)g(w),$$

where $f, g \in R[x_1, x_2, \ldots, x_n]$ and $u \in \mathbf{N}^n$.

Prove that

(i) if R is a commutative ring with identity or a ring without zero divisor or an integral domain, then so is $R[x_1, x_2, \ldots, x_n]$;

(ii) the map $R \to R[x_1, x_2, \ldots, x_n]$ given by $r \mapsto f_r$, where $f_r(0, 0, \ldots, 0) = r$ and $f_r(u) = 0$ for all other $u \in \mathbf{N}^n$, is a monomorphism of rings.

Remark R is identified with its isomorphic image under the map (ii) and hence R is considered as a subring of $R[x_1, x_2, \ldots, x_n]$.

30. *Topological rings.* A set R is said to be a topological ring iff

   (i) R is a ring;
   (ii) R is a topological space;
   (iii) the algebraic operations defined on R are continuous.

   In the event that R is a division ring it is said to be a topological division ring iff in addition the following condition is satisfied:

   (iv) for any $a \ (\neq 0) \in R$ and for any arbitrary neighborhood W of a^{-1}, $\exists$ a neighborhood U of a satisfying the condition $U^{-1} \subset W$.

   A commutative topological division ring is called a topological field.

   (a) Show that (i) the field of real numbers, the field of complex numbers, the field of power series over the field of residue classes $\mathbf{Z}_p$ are topological fields.

   (ii) The division ring of quaternions is a topological division ring.

   (b) (Pontryagin 1939) Prove that a continuous algebraic division ring (i.e., a locally compact non-discrete topological division ring), if connected and locally compact, is isomorphic either to the real field, or to the complex field or to the quaternion division ring.

   (c) Using (b) prove that every connected locally compact division ring is isomorphic either to the field of real numbers or to the field of complex numbers or to the division ring of quaternions.

Exercises A (Objective Type)   Identify the correct alternative(s) (may be more than one) from the following list:

1. Let $R = \left\{ \begin{pmatrix} a & -b \\ c & d \end{pmatrix} : a, b \in \mathbf{R} \right\}$. Then under usual matrix addition and usual multiplication

   (a) R is a non-commutative ring.
   (b) R is a commutative ring but not a field.
   (c) R is a field.
   (d) R is an integral domain but not a field.

2. Let $R = (\mathbf{Z}_n, +, \cdot)$ be the ring of integers modulo n. If n is a composite number, then

   (a) R is an integral domain.
   (b) R is a commutative ring but not an integral domain.

(c) R is a field.

(d) R is an integral domain but not a field.

3. Let $R = (M_2(\mathbf{R}), +, \cdot)$ be the ring of real matrices of order 2. If $S = \left\{ \begin{pmatrix} x & 0 \\ 0 & 0 \end{pmatrix} : x \in \mathbf{R} \right\}$, then

   (a) S is a subring of R having no multiplicative identity.

   (b) S is a subring of R having multiplicative identity 1_S such that $1_S = 1_R$, where 1_R is the multiplicative identity element of R.

   (c) S is a subring of R such that $1_S \neq 1_R$.

   (d) S is not a subring of R.

4. Let $(\mathbf{Z}, +, \cdot)$ be the ring of integers and $f : \mathbf{Z} \to \mathbf{Z}$ be a ring homomorphism such that $f(1) = 1$. Then

   (a) f is a monomorphism but not an epimorphism.

   (b) f is an epimorphism but not a monomorphism.

   (c) f is an automorphism.

   (d) f is neither a monomorphism nor an epimorphism.

5. Let $(\mathbf{Q}, +, \cdot)$ be the field of rational numbers and $f : \mathbf{Q} \to \mathbf{Q}$ be a ring homomorphism such that $f(1) \neq 0$. Then

   (a) f is a monomorphism but not an epimorphism.

   (b) f is an epimorphism but not a monomorphism.

   (c) f is an isomorphism.

   (d) f is neither a monomorphism nor an epimorphism.

6. The number of automorphisms of the field of real numbers is

   (a) one   (b) infinitely many   (c) two   (d) zero.

7. Let $(\mathbf{Z}, +, \cdot)$ be the ring of integers and $f : \mathbf{Z} \to \mathbf{Z}$ be defined by $f(x) = -x$, $\forall x \in \mathbf{Z}$. Then

   (a) f is a group homomorphism but not a ring homomorphism.

   (b) f is a ring homomorphism.

   (c) f is neither a ring homomorphism nor a group homomorphism.

   (d) f is a ring isomorphism.

8. Let $\mathbf{R}[x]$ be the polynomial ring with real coefficients. Then

   (a) $\mathbf{R}[x]$ is a field.

   (b) $\mathbf{R}[x]$ is an integral domain but not a field.

   (c) $\mathbf{R}[x]$ is a commutative ring with zero divisors.

   (d) $\mathbf{R}[x]$ is a commutative ring with no multiplicative identity.

9. Let $R = C([0, 1])$ be the set of all real-valued continuous functions on $[0, 1]$. Define $+$ and $\cdot$ on $C([1, 0])$ by

$$\left. \begin{array}{l} (f + g)(x) = f(x) + g(x) \\ (f \cdot g)(x) = f(x) \cdot g(x), \end{array} \right\} \quad \forall f, g \in C([0, 1]) \text{ and } \forall x \in [0, 1].$$

Then

(a) $(R, +, \cdot)$ is not a ring.
(b) $(R, +, \cdot)$ is a ring but not an integral domain.
(c) $(R, +, \cdot)$ is an integral domain.
(d) $(R, +, \cdot)$ is not a non-commutative ring with zero divisors.

10. Let $(M, +)$ be an abelian group and $R = \text{End}(M)$ be the set of all endomorphisms on M, i.e., $\text{End}(M)$ be the set of all group endomorphisms of M into itself. Define $+$ and $\cdot$ on R by

$$\left.\begin{array}{l}(f + g)(x) = f(x) + g(x) \\ (f \cdot g)(x) = f\big(g(x)\big),\end{array}\right\} \quad \forall f, g \in R \text{ and } \forall x \in M.$$

Then

(a) $(R, +, \cdot)$ is not a ring.
(b) $(R, +, \cdot)$ is a commutative ring.
(c) $(R, +, \cdot)$ is a non-commutative ring with identity element.
(d) $(R, +, \cdot)$ is an integral domain.

11. Let $(\mathbf{Z}, +, \cdot)$ be the ring of integers and $R = \text{End}(\mathbf{Z})$ be the set of all endomorphisms of $(\mathbf{Z}, +)$, i.e., $\text{End}(\mathbf{Z})$ be the set of all group homomorphisms of $\mathbf{Z}$ into itself. Define $+$ and $\cdot$ on R by

$$\left.\begin{array}{l}(f + g)(x) = f(x) + g(x) \\ (f \cdot g)(x) = f\big(g(x)\big),\end{array}\right\} \quad \forall f, g \in R \text{ and } \forall x \in \mathbf{Z}.$$

Define $f : \mathbf{Z} \to R$ by $f(n) = f_n$, where $f_n : \mathbf{Z} \to \mathbf{Z}$ is given by $f_n(x) = nx$, $\forall x \in \mathbf{Z}$. Then

(a) f is not a ring homomorphism.
(b) f is a ring homomorphism but not a ring monomorphism.
(c) f is a ring isomorphism.
(d) f is a ring homomorphism but not a ring epimorphism.

12. Let $R = C([0, 1])$ be the ring of all real-valued continuous functions on $[0, 1]$ and $\psi : R \to \mathbf{R}$ (where $\mathbf{R}$ is the set of all real numbers) be the map defined by $\psi(f) = f(1/5)$. Then

(a) ψ is a ring epimorphism.
(b) ψ is a ring homomorphism but not a ring monomorphism.
(c) ψ is a ring monomorphism.
(d) ψ is ring homomorphism but not a ring epimorphism.

13. The ring $\mathbf{Z}$ and $\mathbf{Z} \times \mathbf{Z}$ are not isomorphic. Because

(a) one is commutative but the other is not.
(b) one has zero divisor but the other has not.

(c) their cardinalities are different.

(d) one is countable but the other is not.

14. Let $C = \{f : \mathbf{R} \to \mathbf{R} : f \text{ is continuous on } \mathbf{R}\}$ and $D = \{f \in C : f \text{ is differentiable on } \mathbf{R}\}$ Define $+$ and $\cdot$ on C by

$$\left.\begin{array}{l} (f + g)(x) = f(x) + g(x) \\ (f \cdot g)(x) = f(x) \cdot g(x), \end{array}\right\} \quad \forall f, g \in C \text{ and } \forall x \in \mathbf{R}.$$

Then

(a) C is not a ring.

(b) C is ring but D is not a subring of C.

(c) C is ring and D is a subring of C.

(d) both C and D are integral domains.

15. Let $\mathbf{Q}$ be the field of all rational numbers. Then the number of non-trivial field homomorphisms of $\mathbf{Q}$ into itself is

(a) one   (b) two   (c) zero   (d) infinitely many.

16. Let $(\mathbf{Z}, +, \cdot)$ be the ring of integers and $(\mathbf{E}, +, \cdot)$ be ring of even integers. Then

(a) $\mathbf{E}$ is an integral domain.

(b) $\mathbf{Z}$ and $\mathbf{E}$ are isomorphic rings.

(c) $\mathbf{Z}$ and $\mathbf{E}$ are non-isomorphic rings.

(d) both $\mathbf{Z}$ and $\mathbf{E}$ are integral domains.

17. Let n be the number of non-trivial ring homomorphisms from the ring $\mathbf{Z}_{12}$ to the ring $\mathbf{Z}_{28}$. Then n is

(a) 7   (b) 1   (c) 3   (d) 4.

18. Which of the following rings are integral domains?

(a) $\mathbf{R}[x]$, the ring of all polynomials in one variable with real coefficients.

(b) $D[0, 1]$, the ring of continuously differentiable real-valued functions on the interval $[0, 1]$ (with respect to pointwise addition and pointwise multiplication).

(c) $\mathbf{C}[x]$, the ring of all polynomials in one variable with complex coefficients.

(d) $M_n(\mathbf{R})$, the ring of all $n \times n$ matrices with real entries.

Exercises B (True/False Statements)   *Determine the correct statement with justification*:

1. The ring $\mathbf{Z} \times \mathbf{Z}$ is an integral domain.
2. The rings $(\mathbf{Z}(\sqrt{2}), +, \cdot)$ and $(\mathbf{Z}(\sqrt{3}), +, \cdot)$ are isomorphic, where $\mathbf{Z}(\sqrt{2}) = \{a + b\sqrt{2} : a, b \in \mathbf{Z}\}$ and $\mathbf{Z}(\sqrt{3}) = \{c + d\sqrt{3} : c, d \in \mathbf{Z}\}$.
3. There exists a commutative ring R with 1 such that the rings $(R[x], +, \cdot)$ and $(\mathbf{Z}, +, \cdot)$ are isomorphic.
4. The field of quotient of $(\mathbf{Z}, +, \cdot)$ is isomorphic to $(\mathbf{Q}, +, \cdot)$.

5. Every non-zero element of an integral domain D is a unit in the quotient field $Q(D)$ of D.
6. The quotient fields of two isomorphic integral domains are isomorphic.
7. Let R be a commutative ring such that $a^2 = a$, $\forall a \in R$. Then $a = -a$, $\forall a \in R$.
8. The rings $(\mathbf{Z}_n, +, \cdot)$ and $\text{End}(\mathbf{Z}_n, +)$ are isomorphic.
9. The ring $\text{End}(\mathbf{Z}_p, +)$ is a field for every prime $p > 1$.
10. The rings $(\mathbf{Z}_2 \times \mathbf{Z}_2, +, \cdot)$ and $\text{End}(\mathbf{Z}_2 \times \mathbf{Z}_2, +)$ are isomorphic.
11. Any homomorphism of a field F onto itself is an automorphism of F.
12. A non-zero element $(a) \in \mathbf{Z}_n$ is invertible in the ring $\mathbf{Z}_n \Leftrightarrow \gcd(a, n) = 1$.
13. A commutative ring D with 1 is an integral domain $\Leftrightarrow D$ is a subring of a field.
14. Every field contains a subfield F such that F is isomorphic either to the field $\mathbf{Q}$ or to one of the fields $\mathbf{Z}_p$ for some prime p.
15. The invertible elements of $\mathbf{R}[\![x]\!]$ are precisely the power series in $\mathbf{R}[\![x]\!]$ with non-zero constant terms.

4.9 Additional Reading

We refer the reader to the books (Adhikari and Adhikari 2003, 2004; Artin 1991; Atiya and Macdonald 1969; Birkoff and Mac Lane 1965; Burtan 1968; Chatterjee et al. 2003; Fraleigh 1982; Herstein 1964; Hungerford 1974; Jacobson 1974, 1980; Lang 1965; McCoy 1964; Simmons 1963; van der Waerden 1970) for further details.

References

Adhikari, M.R., Adhikari, A.: Groups, Rings and Modules with Applications, 2nd edn. Universities Press, Hyderabad (2003)

Adhikari, M.R., Adhikari, A.: Text Book of Linear Algebra: An Introduction to Modern Algebra. Allied Publishers, New Delhi (2004)

Artin, M.: Algebra. Prentice-Hall, Englewood Cliffs (1991)

Atiya, M.F., Macdonald, I.G.: Introduction to Commutative Algebra. Addison-Wesley, Reading (1969)

Birkoff, G., Mac Lane, S.: A Survey of Modern Algebra. Macmillan, New York (1965)

Burtan, D.M.: A First Course in Rings and Ideals. Addison-Wesley, Reading (1968)

Chatterjee, B.C., Ganguly, S., Adhikari, M.R.: A Text Book of Topology. Asian Books, New Delhi (2003)

Dugundji, J.: Topology. Universal Book Stall, New Delhi (1980) (Indian Reprint)

Fraleigh, J.B.: A First Course in Abstract Algebra. Addison-Wesley, Reading (1982)

Herstein, I.: Topics in Algebra. Blaisdell, New York (1964)

Hungerford, T.W.: Algebra. Springer, New York (1974)

Jacobson, N.: Basic Algebra I. Freeman, San Francisco (1974)

Jacobson, N.: Basic Algebra II. Freeman, San Francisco (1980)

Lang, S.: Algebra, 2nd edn. Addison-Wesley, Reading (1965)

McCoy, N.: Theory of Rings. Macmillan, New York (1964)

Pontryagin, L.: Topological Groups. Princeton University Press, Princeton (1939)

Simmons, G.F.: Topology and Modern Analysis. McGraw-Hill Book, New York (1963)

van der Waerden, B.L.: Modern Algebra. Ungar, New York (1970)

Chapter 5
Ideals of Rings: Introductory Concepts

In this chapter we continue the study of theory of rings with the help of ideals and we show how to construct isomorphic replicas of all homomorphic images of a specified abstract ring. In ring theory, an ideal is a special subring of a ring. The concept of ideals generalizes many important properties of integers. Ideals and homomorphisms of rings are closely related. Like normal subgroups in the theory of groups, ideals play an analogous role in the study of rings. The real significance of ideals in a ring is that they enable us to construct other rings which are associated with the first in a natural way. Commutative rings and their ideals are closely related. Their relations develop ring theory and are applied in many areas of mathematics, such as number theory, algebraic geometry, topology and functional analysis. In this chapter basic concepts of ideals are introduced with many illustrative examples. Ideals of rings of continuous functions and the Chinese Remainder Theorem for rings with its applications are studied. We study ideals of various important rings and prove different isomorphism theorems. Moreover, applications of ideals to algebraic geometry and the Zariski topology are discussed in this chapter.

5.1 Ideals: Introductory concepts

The concept of ideals came through the work of Cartan, J.H. Wedderburn, E. Noether, E. Artin. Many others also made its significant applications in the theory of rings and algebras. An ideal of a ring R is a subring of R which is closed with respect to multiplication on both sides by every element of R. It is now important to carry over many results of group theory related to normal subgroups to ring theory. Ideals of commutative rings are developed in the study of commutative algebra and applied to many areas of mathematics. The theory of ideals of rings of algebraic integers prescribes methods for solving many problems of number theory. Analogous to a normal subgroup, a subring that is the kernel of a ring homomorphism is what we call an *ideal*.

Unless otherwise stated, R will denote an arbitrary ring having more than one element.

M.R. Adhikari, A. Adhikari, *Basic Modern Algebra with Applications*,
DOI 10.1007/978-81-322-1599-8_5, © Springer India 2014

Definition 5.1.1 Let A be a non-empty subset of a ring R such that A is an additive subgroup of R. Then

(i) A is a *left ideal* of R iff $ra \in A$ for each $a \in A$ and $r \in R$;
(ii) A is a *right ideal* of R iff $ar \in A$ for each $a \in A$ and $r \in R$;
(iii) A is an *ideal* of R iff it is both a left ideal of R and a right ideal of R.

Remark Conditions (i) and (ii) assert that A swallows up multiplication from the left and from the right by arbitrary elements of the ring respectively.

R has two obvious ideals, namely, $\{0\}$ and R itself.

Clearly, if the ring R is commutative, the concepts of left-, right- and two-sided ideals coincide.

Remark Let A be a left ideal (right ideal) of R. Then $a, b \in A \Rightarrow a - b, ab \in A \Rightarrow A$ is a subring of R (by Theorem 4.2.1).

The converse of the result is not true.

Example 5.1.1 (a) Let R be a ring. Consider the matrix ring $M_2(R)$. Then in $M_2(R)$
  (i) the subring $A = \left\{ \left(\begin{smallmatrix} a & b \\ 0 & 0 \end{smallmatrix} \right) : a, b \in R \right\}$ is a right ideal but not a left ideal;
  (ii) the subring $B = \left\{ \left(\begin{smallmatrix} a & 0 \\ b & 0 \end{smallmatrix} \right) : a, b \in R \right\}$ is a left ideal but not a right ideal;
  (iii) the subring $C = \left\{ \left(\begin{smallmatrix} a & 0 \\ 0 & 0 \end{smallmatrix} \right) : a \in R \right\}$ is neither a left ideal nor a right ideal.
  (b) The ring R is not an ideal of $R[x]$, because R does not contain $af(x) \, \forall a \in R$ and $\forall f(x) \in R[x]$.

Theorem 5.1.1 *The intersection of any arbitrary collection of (left, right) ideals of a ring R is a (left, right) ideal of R.*

Proof Let $\{A_i\}$ be an arbitrary collection of ideals of R and $M = \bigcap_i A_i$. Then $M \neq \emptyset$, since $0 \in$ each $A_i \Rightarrow 0 \in M$. Again $a, b \in M$ and $r \in R \Rightarrow a - b \in A_i$, $ar, ra \in A_i$ for each $i \Rightarrow a - b, ar, ra \in M \Rightarrow M$ is an ideal of R. $\qquad\square$

Ideals generated by a finite number of elements or a single element play an important role in the study of rings. We now consider an arbitrary ring R and a non-empty subset A of R. By the symbol $\langle A \rangle$ we mean the set

$$\langle A \rangle = \bigcap \{ I : A \subseteq I; I \text{ is an ideal of } R \}.$$

Clearly, $\langle A \rangle \neq \emptyset$. Thus $\langle A \rangle$ exists and satisfies $A \subseteq \langle A \rangle$. Then by Theorem 5.1.1, $\langle A \rangle$ forms an ideal of R, known as the ideal generated by the set A. Moreover, for any ideal I of R with $A \subseteq I \Rightarrow \langle A \rangle \subseteq I \Rightarrow \langle A \rangle$ is the smallest ideal of R to contain the set A.

If in particular, $A = \{a_1, a_2, \ldots, a_n\}$, then the ideal which A generates is denoted by $\langle a_1, a_2, \ldots, a_n \rangle$. Such an ideal is said to be *finitely generated* with $a_1, a_2, \ldots, a_n$ as its generators. An ideal $\langle a \rangle$ generated by just one element a of R is called a *principal ideal* generated by a.

Remark The forms of the ideal $\langle a \rangle$ in cases of commutative and non-commutative rings are given in Ex. 1 in Exercises-I.

Proposition 5.1.1 *Every ideal of the ring of integers* $(\mathbf{Z}, +, \cdot)$ *is a principal ideal.*

Proof Clearly, $\{0\}$ is an ideal of $(\mathbf{Z}, +, \cdot)$. Let A be a non-zero ideal of $(\mathbf{Z}, +, \cdot)$. Then the positive integers in A have a least element u (say) by well ordering principle (see Chap. 1). Let $v \in A$. By division algorithm, we can express $v = qu + r$, where $0 \leq r < u$. Now $r = v - qu \in A$ and u is the least positive integer in $A \Rightarrow r = 0 \Rightarrow v$ is a multiple of $u \Rightarrow A \subseteq \langle u \rangle = \{nu : n \in \mathbf{Z}\}$ (an ideal) $\subseteq A \Rightarrow A = \langle u \rangle$. Thus every ideal of $(\mathbf{Z}, +, \cdot)$ is a principal ideal. $\square$

Remark Let I be an ideal of $\mathbf{Z}$. Then there exists a unique non-negative integer d such that $I = \langle d \rangle$, i.e., $I = d\mathbf{Z}$. This follows from the following fact:

The case with zero ideal is trivial. For the case of a non-zero ideal, let $d\mathbf{Z} = e\mathbf{Z}$, for some non-negative integers d and e. Then d divides e and e divides d, resulting in $d = \pm e$. As both e and d are non-negative, $e = d$. In particular, $\langle 2 \rangle = \mathbf{Z}_e$, ring of even integers forms an ideal of $\mathbf{Z}$. Clearly, $\langle 0 \rangle = \{0\}$ and $\langle 1 \rangle = \mathbf{Z}$.

Definition 5.1.2 A ring R is said to be a principal ideal ring iff every ideal A of R is of the form $A = \langle a \rangle$ for some $a \in R$.

Proposition 5.5.1 shows that $(\mathbf{Z}, +, \cdot)$ is a principal ideal ring.

Every ring R has at least two ideals viz. $\{0\}$ and R. The ideals $\{0\}$ and R are called *trivial ideals*. Any other ideals (if they exist) are called non-trivial ideals. By a *proper ideal* of R we mean an ideal different from R.

Definition 5.1.3 A ring R is said to be *simple* iff it has no non-trivial ideals.

Theorem 5.1.2 *A division ring has no non-trivial ideals.*

Proof Let R be a division ring and A an ideal of R. If $A \neq \{0\}$, then A contains an element $g \neq 0$. Now $g^{-1}g = 1 \in A$ as $g^{-1} \in R$. Hence $r1 = r \in A \; \forall r \in R \Rightarrow R \subseteq A$. Consequently, $R = A$. $\square$

Corollary *Every division ring is a simple ring.*

Remark The converse is not true (see Ex. 2 of the Worked-Out Exercises).

Remark The join of two ideals of a ring R is not necessarily an ideal, since the join of two subrings is not necessarily a subring.

Definition 5.1.4 The *sum* (or hcf) of two ideals P and Q of a ring R denoted by $P + Q$ is defined by $P + Q = \{p + q : p \in P \text{ and } q \in Q\}$.

Clearly, $P + Q$ is an ideal of R such that it is the smallest ideal of R which contains both P and Q.

Note that the set $\{pq : p \in P$ and $q \in Q\}$ is not an ideal of R.

To avoid this difficulty, we define the product of the ideals P and Q as follows:

Definition 5.1.5 The *product* of two ideals P and Q of a ring R denoted by PQ is defined by

$$PQ = \left\{ \sum_{\text{finite}} p_i q_i : p_i \in P \text{ and } q_i \in Q \right\}.$$

Clearly PQ is an ideal of R such that $PQ \subseteq P \cap Q$.

5.2 Quotient Rings

In this section we continue the discussion of ideals. An ideal of a ring produces a new ring, called a quotient ring, which is closely related to the mother ring in a natural way. The concept of a quotient ring is similar to the concept of a quotient group and hence the method of construction of a quotient ring is borrowed from the corresponding construction of a quotient group. In this section we bring over to rings various results, such as homomorphism and isomorphism theorems, correspondence theorem etc., which are counterparts of the corresponding results of groups and establish the relation of ideals to homomorphisms.

Let $(R, +, \cdot)$ be a ring and I an ideal of R. Then $(I, +)$ is a subgroup of the abelian group $(R, +)$ and hence the quotient group $(R/I, +)$ is defined under usual addition of cosets, i.e., $(a + I) + (b + I) = a + b + I$, $\forall a, b \in R$. In this group, the identity element is the trivial coset I and the inverse of $a + I$ is $-a + I$, $\forall a \in R$.

Theorem 5.2.1 *Given an ideal I of a ring $(R, +, \cdot)$, the natural (usual) addition and multiplication of cosets, namely, $(a + I) + (b + I) = a + b + I$ and $(a + I)(b + I) = ab + I$ $\forall a, b \in R$, make $(R/I, +, \cdot)$ a ring.*

Proof Let $x + I = a + I$ and $y + I = b + I$. Then $x - a \in I$ and $y - b \in I$. Now $xy - ab = xy - xb + xb - ab = x(y - b) + (x - a)b \in I \Rightarrow$ multiplication of cosets is well defined. As I is a normal subgroup of R, $(R/I, +)$ is an abelian group. It can be verified that multiplication in R/I is associative and distributive property holds in R/I. So, $(R/I, +, \cdot)$ is a ring.   □

Definition 5.2.1 Given an ideal I of a ring R, the ring R/I is called the *quotient ring* or *factor ring* of R by I (or modulo I).

Remark If R is a commutative ring with identity, then R/I is also so.

This successful construction of the quotient ring by an ideal enables us to bring over to rings the homomorphism theorems of groups.

Theorem 5.2.2 shows that ideals and homomorphisms of rings are closely related.

Theorem 5.2.2 *If $f : R \to S$ is a homomorphism of rings, then $K = \ker f$ is an ideal of R. Conversely, if I is an ideal of R, then the mapping $\pi : R \to R/I$ given by $\pi(r) = r + I$ is an epimorphism with kernel I.*

Proof $0 \in \ker f \Rightarrow \ker f \neq \emptyset$. Let $x, y \in K$ and $r \in R$. Then $f(x - y) = f(x) - f(y) = 0'$ (zero of S) $\Rightarrow x - y \in K$. Also, $f(rx) = f(r)f(x) = f(r) \cdot 0' = 0' \Rightarrow rx \in K$. Similarly, $xr \in K$. Consequently, K is an ideal of R.

Conversely, let I be an ideal of R. Consider the function $\pi : R \to R/I$ given by $\pi(r) = r + I$. Now $\pi(r + t) = (r + t) + I = (r + I) + (t + I) = \pi(r) + \pi(t)$ and $\pi(rt) = rt + I = (r + I)(t + I) = \pi(r)\pi(t)$, $\forall r, t \in R \Rightarrow \pi$ is a ring homomorphism. Clearly, π is an epimorphism.

Now, $\ker \pi = \{r \in R : \pi(r) = I\} = \{r \in R : r + I = I\} = \{r \in R : r \in I\} = I \Rightarrow I$ is the kernel of the homomorphism $\pi : R \to R/I$. $\quad\square$

Corollary *I is an ideal of a ring R iff it is the kernel of some homomorphism from R onto R/I.*

Note The homomorphism $\pi : R \to R/I$ is called the *canonical* or *natural* epimorphism.

Theorem 5.2.3 (Fundamental Homomorphism Theorem or First Isomorphism Theorem) *Let $(R, +, \cdot)$ and $(S, +, \cdot)$ be rings and $f : R \to S$ be a ring homomorphism. Then*

(i) *$K = \ker f$ is an ideal of R;*
(ii) *the quotient ring $(R/K, +, \cdot)$ is isomorphic to the image $f(R)$ under the mapping $\tilde{f} : R/K \to S$ given by $\tilde{f}(r + K) = f(r)$.*

Proof (i) follows from Theorem 5.2.2.

(ii) First we show that $\tilde{f}$ is well defined and one-one. Now $x + K = y + K \Leftrightarrow x - y \in K \Leftrightarrow f(x - y) = 0' \Leftrightarrow f(x) - f(y) = 0' \Leftrightarrow f(x) = f(y) \Leftrightarrow \tilde{f}(x + K) = \tilde{f}(y + K)$ $\forall x, y \in R \Rightarrow$ (forward implication) shows that $\tilde{f}$ is well defined and $\Leftarrow$ (backward implication) shows that $\tilde{f}$ is one-one.

Again,

$$\tilde{f}\big((x + K) + (y + K)\big) = \tilde{f}(x + y + K) = f(x + y) = f(x) + f(y)$$
$$= \tilde{f}(x + K) + \tilde{f}(y + K)$$

and

$$\tilde{f}\big((x + K)(y + K)\big) = \tilde{f}(xy + K) = f(xy) = f(x)f(y)$$
$$= \tilde{f}(x + K)\tilde{f}(y + K) \quad \forall x, y \in R$$
$$\Rightarrow \quad \tilde{f} \text{ is a ring homomorphism.}$$

Let $x \in f(R)$. Then $\exists r \in R$ such that $f(r) = x$. So

$$\tilde{f}(r + K) = f(r) = x \quad \Rightarrow \quad \tilde{f} : R/K \to f(R) \text{ is surjective.}$$

Consequently, $\tilde{f} : R/K \to f(R)$ is an isomorphism. $\qquad\qquad\square$

Note If in particular, f is an epimorphism, then R/K is isomorphic to S.

Corollary *Let* $f : R \to S$ *be a homomorphism of rings with* $I = \ker f$. *Then there exists a unique monomorphism* $\tilde{f} : R/I \to S$ *such that* $f = \tilde{f} \circ \pi$, *where* π *is the natural epimorphism* $\pi : R \to R/I$, *also* $\tilde{f}$ *is an isomorphism, iff* f *is an epimorphism.*

Proof Define a monomorphism $\tilde{f} : R/I \to S$ by $\tilde{f}(r + I) = f(r) \; \forall r \in R$. Then $(\tilde{f} \circ \pi)(r) = \tilde{f}(r + I) = f(r) \; \forall r \in R \Rightarrow f = \tilde{f} \circ \pi$. Moreover, if f is an epimorphism, then $\tilde{f}$ is an isomorphism by Note after Theorem 5.2.3. Again $\tilde{f}$ is an isomorphism $\Rightarrow \tilde{f}$ is an epimorphism $\Rightarrow f$ is an epimorphism. To prove the uniqueness of $\tilde{f}$, let $\tilde{g} : R/I \to S$ be a monomorphism such that $\tilde{f} \circ \pi = \tilde{g} \circ \pi$. Now $\tilde{g}(r + I) = \tilde{g}(\pi(r)) = (\tilde{g} \circ \pi)(r) = (\tilde{f} \circ \pi)(r) = \tilde{f}(\pi(r)) = \tilde{f}(r + I)$, for all $r + I \in R/I$. This implies $\tilde{f} = \tilde{g}$. $\qquad\square$

Theorem 5.2.4 (Correspondence Theorem) *Let* $(R, +, \cdot)$ *and* $(S, +, \cdot)$ *be rings and* $f : R \to S$ *be an epimorphism. Then there exists a one-to-one correspondence between the collection of ideals of* R *containing* K ($= \ker f$) *and the collection of ideals of* S.

Proof Let $I(R)$ and $I(S)$ denote the collection of all ideals of R containing K and the collection of all ideals of S, respectively. Define $\tilde{f} : I(R) \to I(S)$ by $\tilde{f}(T) = f(T) \in I(S)$ as f is a homomorphism and so, $f(T)$ is an ideal of S. Let $A, B \in I(R)$ and $\tilde{f}(A) = \tilde{f}(B)$. Then $f(A) = f(B)$. We claim that $A = B$. Let $a \in A$. Then $\exists b \in B$ such that $f(a) = f(b) \Rightarrow f(a - b) = O_S \Rightarrow a - b \in \ker f = K \subseteq B \in I(R) \Rightarrow a = (a - b) + b \in B \Rightarrow a \in B \; \forall a \in A \Rightarrow A \subseteq B$. Similarly $B \subseteq A$. Hence $A = B$. Let M be an element of $I(S)$. Then $\tilde{f}^{-1}(M) = f^{-1}(M)$ is an ideal T (say) of R containing K. So, $f(T) = M$ as f is an epimorphism (use $f(f^{-1}(M)) = M \cap f(R) = M$). Consequently, $\tilde{f}$ is surjective. $\qquad\square$

Corollary *Let* $(R, +, \cdot)$ *be a ring and* I *an ideal of* R. *Then the ideals of* R/I *are in one-to-one correspondence with the ideals of* R *containing* I.

Proof Let $\pi : R \to R/I$ be the canonical epimorphism. Then $\ker \pi = I$. Now apply Theorem 5.2.4. $\qquad\square$

Remark Any ideal in R/I is of the form $\pi(A)$ for some ideal A of R containing I.

5.3 Prime Ideals and Maximal Ideals

In this section, we continue discussion of ideals and quotient rings. More precisely, we study quotient rings R/I with the help of prime and maximal ideals I of R. So prime and maximal ideals are ideals of special interest. In this section we introduce the concepts of prime ideals and maximal ideals. These two concepts are different in general but coincide for some particular classes of rings. Their characterizations are also established in this section. We show that prime ideals of $\mathbf{Z}$ and prime integers are closely related.

Throughout this section we assume that R is a commutative ring with identity 1 ($\neq 0$).

Definition 5.3.1 A proper ideal P of R is said to be a *prime* ideal iff for any $a, b \in R$, $ab \in P \Rightarrow a \in P$ or $b \in P$.

Example 5.3.1 (i) Zero ideal $\{0\}$ of an integral domain D is a prime ideal, since for $a, b \in D$, $ab = 0 \Leftrightarrow a = 0$ or $b = 0$.

(ii) Consider the ring of integers $(\mathbf{Z}, +, \cdot)$. If p is a prime integer, then the principal ideal $\langle p \rangle$ in $\mathbf{Z}$ is a prime ideal, since $mn \in \langle p \rangle \Rightarrow p|mn \Rightarrow p|m$ or $p|n \Rightarrow m \in \langle p \rangle$ or $n \in \langle p \rangle$. The ideal $I = \langle 6 \rangle = \{6n : n \in \mathbf{Z}\}$ is not prime, since $3 \cdot 2 = 6 \in I$ but neither $3 \in I$ nor $2 \in I$.

Prime ideals of rings may be characterized in terms of their quotient rings.

Theorem 5.3.1 *A proper ideal I of a commutative ring R with identity 1 is prime iff the quotient ring R/I is an integral domain.*

Proof Since R is a commutative ring with 1, R/I is also a commutative ring with identity element $1 + I$ and zero element $0 + I = I$.

Let I be prime. Clearly, $1 + I \neq I$, otherwise $1 + I = I$ would imply $1 \in I$ and hence $I = R$, a contradiction, since I is a proper ideal.

We claim that R/I has no zero divisors. Let $a + I$ and $b + I \in R/I$. Then $(a + I)(b + I) = I \Rightarrow ab + I = I \Rightarrow ab \in I \Rightarrow a \in I$ or $b \in I \Rightarrow a + I = I$ or $b + I = I \Rightarrow R/I$ has no zero divisors. Consequently, R/I is an integral domain.

Conversely, let R/I be an integral domain and $a, b \in R$ such that $ab \in I$. Then $ab + I = I \Rightarrow (a + I)(b + I) = I \Rightarrow a + I = I$ or $b + I = I \Rightarrow a \in I$ or $b \in I \Rightarrow I$ is a prime ideal. $\square$

Definition 5.3.2 A proper ideal I of R is said to be a *maximal* ideal of R iff there is no ideal J of R such that $I \subsetneq J \subsetneq R$.

Thus I is a maximal ideal of R, iff for every ideal N such that $I \subseteq N \subsetneq R \Rightarrow I = N$ (or $I \subsetneq N \subseteq R \Rightarrow N = R$).

Example 5.3.2 (i) $\langle p \rangle$ is a maximal ideal of $\mathbf{Z}$ for each prime integer p. To show this, suppose there exists an ideal J of $(\mathbf{Z}, +, \cdot)$ such that $\langle p \rangle \subsetneqq J$. Then $\exists$ an integer $n \in J$ such that $n \notin \langle p \rangle \Rightarrow n$ is not divisible by $p \Rightarrow 1 = \gcd(n, p)$. So we can write $1 = nr + pt$ for some $r, t \in \mathbf{Z}$. Now $n \in J$ and $p \in \langle p \rangle \subsetneqq J$ show that $1 \in J \Rightarrow J = \mathbf{Z} \Rightarrow \langle p \rangle$ is a maximal ideal of $\mathbf{Z}$.

(ii) The ideal $\{0\}$ is a prime ideal of $\mathbf{Z}$ but not a maximal ideal, since $\{0\} \subsetneqq \langle 2 \rangle \subsetneqq \mathbf{Z}$.

Problem 1 A proper ideal $I = \langle n \rangle$ in $(\mathbf{Z}, +, \cdot)$ is prime iff $n = 0$ or n is a prime integer.

Solution Let I be a prime ideal of $(\mathbf{Z}, +, \cdot)$. Then by Proposition 5.1.1, I can be expressed as $I = \langle n \rangle$ for some $n \geq 0$. If $n = 0$, then I is prime by Example 5.3.1. Since $\langle 1 \rangle = \mathbf{Z}$, I is not proper for $n = 1$. Consider $n > 1$. If possible, let n be composite, i.e., $n = rs$, where $1 < r < n$, $1 < s < n$. Then $rs = n \in \langle n \rangle$. But neither r nor $s \in \langle n \rangle$, otherwise $n | r$ or $n | s$, which is not possible. So, n cannot be composite. In other words n must be prime. Its converse follows from Example 5.3.1.

Problem 2 A commutative ring with identity 1 is a field iff $\{0\}$ is a maximal ideal.

Solution Suppose R is a field. Then by Theorem 5.1.2, R has only two ideals $\{0\}$ and R. Since $\{0\}$ is the only proper ideal of R, $\{0\}$ must be a maximal ideal of R.

Conversely, suppose $\{0\}$ is a maximal ideal of R. Let $a \ (\neq 0) \in R$. Now $\{0\} \subsetneqq Ra \Rightarrow Ra = $ by maximality of $\{0\}$. Then $1 \in Ra \Rightarrow ta = 1$ for some $t \in R \Rightarrow t = a^{-1} \in R$. Consequently, R is a field.

In the next theorem, we shall show that maximal ideals of rings may be characterized in terms of their quotient rings.

Theorem 5.3.2 *Let R be a commutative ring with identity 1. A proper ideal I of R is a maximal ideal of R iff the quotient ring R/I is a field.*

Proof Suppose I is a proper ideal of R such that the quotient ring R/I is a field. We claim that I is a maximal ideal of R. Suppose there exists an ideal J of R such that $I \subsetneqq J \subseteq R$. Then $\exists$ an element $a \in J$ such that $a \notin I$. Hence $a + I \neq I \Rightarrow a + I$ is a non-zero element of $R/I \Rightarrow \exists$ an element $b + I \in R/I$ such that $(b + I)(a + I) = 1 + I$ (since R/I is a field) $\Rightarrow ba + I = 1 + I \Rightarrow 1 - ba \in I \Rightarrow 1 - ba \in J$ (since $I \subsetneqq J$) $\Rightarrow 1 \in J$ (since $a \in J \Rightarrow ba \in J$) $\Rightarrow J = R \Rightarrow I$ is maximal.

Conversely, assume that I is a maximal ideal of R. Since R is a commutative ring with 1 and I is a proper ideal, R/I is a commutative ring with identity $1 + I \neq I$. Let $a + I$ be a non-zero element of R/I. Then $a + I \neq I \Rightarrow a \notin I$. Consider the set $A = \{u + ra : u \in I \text{ and } r \in R\}$. Then A is an ideal of R. Now $u = u + 0a \Rightarrow u \in A$. Also, $a = 0 + 1a \Rightarrow a \in A$. Hence $I \subsetneqq A \Rightarrow A = R$ (by maximality of I). Then $\exists$ elements $u \in I$ and $r \in R$ such that $1 = u + ra$. Now $1 + I = (u + ra) + I = (u + I) + (ra + I) = ra + I$ (since $u \in I$, $u + I = I$, the zero element of R/I)$=$

$(r + I)(a + I)$. This shows that the inverse of $a + I$ exists in R/I. Consequently, R/I is a field.   □

Problem 3 Let R be a commutative ring with 1 ($\neq 0$). Show that the following statements are equivalent:

 (i) R is a field;
 (ii) R has no non-trivial ideals;
 (iii) $\{0\}$ is a maximal ideal of R.

  [*Hint.* (i) $\Rightarrow$ (ii) (by Theorem 5.1.2) $\Rightarrow$ (iii) $\Rightarrow$ (i) (by Problem 2).]

Remark A commutative ring R with 1 ($\neq 0$) is a field $\Leftrightarrow$ every non-zero homomorphism $f : R \to T$ of rings is a monomorphism.

Theorem 5.3.3 *Let R be a commutative ring with* 1. *Then every maximal ideal of R is a prime ideal.*

Proof

$$\text{Let } I \text{ be a maximal ideal of } R \text{ and } a, b \in R \text{ such that } ab \in I. \qquad (5.1)$$

We claim that either $a \in I$ or $b \in I$. Suppose $a \notin I$. Consider the set $A = \{u + ra : u \in I \text{ and } r \in R\}$. Then proceeding as in the second part of Theorem 5.3.2, we find that $A = R$. Then $\exists$ elements $u \in I$ and $r \in R$ such that $1 = u + ra$. Hence $b = ub + rab \Rightarrow b \in I$ by (5.1).   □

Alternative Proof In this case, I is a maximal ideal of $R \Leftrightarrow R/I$ is a field $\Rightarrow R/I$ is an integral domain $\Rightarrow I$ is a prime ideal of R.   □

Remark The converse of Theorem 5.3.3 is not true. It follows from Example 5.3.3.

Example 5.3.3 (i) Let $\mathbf{Z}$ be the ring of integers. Consider the ring $R = \mathbf{Z} \times \mathbf{Z}$, where operations are defined componentwise. Then $\mathbf{Z} \times \{0\}$ is a prime ideal of R. Since $\mathbf{Z} \times \langle 2 \rangle$ is an ideal of R such that $\mathbf{Z} \times \{0\} \subsetneq \mathbf{Z} \times \langle 2 \rangle \subset R$, $\mathbf{Z} \times \{0\}$ cannot be maximal in R.

  (ii) See Example 5.3.2(ii).
  (iii) See Ex. 7 of the Worked-Out Exercises.
  In the following theorem, we shall now show that in Boolean rings and principal ideal domains the concepts of prime ideals and maximal ideals coincide.

Theorem 5.3.4 *Let R be a Boolean ring. An ideal I of R is prime iff I is a maximal ideal.*

Proof Let I be a prime ideal of R. We claim that I is a maximal ideal. Let J be an ideal of R, such that $I \subsetneq J \subseteq R$. Since $I \subsetneq J$, $\exists$ an element $a \in J$ such $a \notin I$. Now

$a(1-a) = a - a^2 = a - a = 0 \in I \Rightarrow$ either $a \in I$ or $1 - a \in I \Rightarrow 1 - a \in I \subsetneqq J$
(as $a \notin I$) $\Rightarrow 1 \in J$ (as $a \in J$) $\Rightarrow J = R \Rightarrow I$ is a maximal ideal of R.

Conversely, if I is a maximal ideal, then I is a prime ideal by Theorem 5.3.3. $\square$

Definition 5.3.3 A *principal ideal domain* (PID) R is a principal ideal ring which is an integral domain.

Theorem 5.3.5 *Let R be a principal ideal domain. Then a non-trivial ideal I of R is a prime ideal iff I is a maximal ideal.*

Proof Let I be a non-trivial prime ideal of R. Suppose J is an ideal of R such that $I \subsetneqq J \subseteq R$. Since R is a PID, we write $I = \langle a \rangle$ and $J = \langle b \rangle$ for some $a, b \in R \setminus \{0\}$. Now $a \in \langle a \rangle \subsetneqq \langle b \rangle \Rightarrow a = rb$ for some $r \in R \Rightarrow$ either $r \in \langle a \rangle$ or $b \in \langle a \rangle$ as $\langle a \rangle$ is prime. But $b \notin \langle a \rangle$, otherwise $\langle b \rangle \subseteq \langle a \rangle$ leads to a contradiction. So, $r \in \langle a \rangle \Rightarrow r = ta$ for some $t \in R \Rightarrow a = rb = (ta)b = (tb)a \Rightarrow tb = 1$ (as R is an integral domain and $a \neq 0$) $\Rightarrow 1 \in \langle b \rangle = J \Rightarrow J = R \Rightarrow I$ is a maximal ideal of R.

The converse part follows from Theorem 5.3.3. $\square$

Corollary *A non-trivial ideal of $(\mathbf{Z}, +, \cdot)$ is maximal iff it is prime.*

The following theorem is an interesting application of Zorn's Lemma.

Theorem 5.3.6 *If a ring R is finitely generated, then each proper ideal of R is contained in a maximal ideal.*

Proof Let A be a proper ideal of a finitely generated ring R. Suppose $R = \langle x_1, x_2, \ldots, x_n \rangle$. Let $\mathbf{D}$ be the set of all ideals B of R satisfying $A \subseteq B \subsetneqq R$, partially ordered by set inclusion. Then $A \in \mathbf{D} \Rightarrow \mathbf{D} \neq \emptyset$. Consider a chain $\{B_i : i \in I\}$ in $\mathbf{D}$. Then $T = \bigcup_{i \in I} B_i$ is an ideal of R.

We claim that T is a proper ideal of R. If possible, let $T = R = \langle x_1, x_2, \ldots, x_n \rangle$. Then each generator x_k ($k = 1, 2, \ldots, n$) would belong to some ideal B_{i_k} of the chain $\{B_i\}$. As there are only finitely many B_{i_k}, one contains all others, call it B_i. Thus $x_1, x_2, \ldots, x_n$ all lie in this one B_i. Consequently, $B_i = R$, which is not possible. Again $A \subseteq \bigcup_i B_i = T \subsetneqq R \Rightarrow T \in \mathbf{D}$. Hence by Zorn's Lemma the family $\mathbf{D}$ contains a maximal element M. Then M is an ideal of R with $A \subseteq M \subsetneqq R$. We claim that M is a maximal ideal of R. To show this, let J be an ideal of R such that $M \subsetneqq J \subseteq R$. Since M is a maximal element of the family $\mathbf{D}$, $J \notin \mathbf{D}$. Accordingly, J cannot be a proper ideal of R. Hence $J = R$. This shows that M is a maximal ideal of R. $\square$

Corollary (Krull–Zorn) *Let R be a ring with identity 1. Then each proper ideal of R is contained in a maximal ideal of R.*

Proof The proof is immediate, since $R = \langle 1 \rangle$, by Theorem 5.3.6. $\square$

Remark In Appendix A, we present generalizations of some concepts of rings to the corresponding concepts for semirings.

5.4 Local Rings

The concept of local rings was introduced by W. Krull (1899–1971) in 1938. Rings with a unique maximal ideal are of immense interest.

Definition 5.4.1 A *local ring* is a commutative ring with identity which has a unique maximal ideal.

Since every proper ideal of a ring R with 1 is contained in a maximal ideal of R (see the Corollary of Theorem 5.3.6), it follows that the unique maximal ideal of a local ring R contains every proper ideal of the ring.

All fields are local rings.

Example 5.4.1 Let R be a commutative ring with 1 in which the set of all non-units of R forms an ideal M. Then R is a local ring. This is so because M is a maximal ideal of R and that M contains any proper ideal of R. Such a ring is a local ring. To show this, let I be an ideal of R such that $M \subsetneq I \subseteq R$. Then $\exists$ a unit element a (say) in I. Hence $I = R \Rightarrow M$ is maximal. Next let A be any proper ideal of R. Then every element of A must be non-unit. Hence $A \subseteq M$.

We now search for another example of a local ring.

Theorem 5.4.1 *For any field F, the power series ring $F[\![x]\!]$ is a principal ideal domain.*

Proof Let A be any proper ideal of $F[\![x]\!]$. If $A = \{0\}$, then $A = \langle 0 \rangle$. If $A \neq \{0\}$, then $\exists$ a non-zero power series $f(x) \in A$ of minimum order r (say), so that $f(x) = a_r x^r + a_{r+1} x^{r+1} + \cdots = x^r (a_r + a_{r+1} x + \cdots)$, where $a_r \neq 0$. Then $a_r + a_{r+1} x + \cdots$ is invertible in $F[\![x]\!]$ by Lemma 4.5.1 of Sect. 4.5 of Chap. 4. Hence $f(x) = x^r g(x)$, where $g(x)$ has an inverse in $F[\![x]\!] \Rightarrow x^r = f(x)g(x)^{-1} \in A \Rightarrow \langle x^r \rangle \subseteq A$. For the reverse inclusion, let $t(x)$ be any non-zero power series in A of order n. Then $r \leq n \Rightarrow t(x)$ can be written in the form:

$$t(x) = x^r \left(b_n x^{n-r} + b_{n+1} x^{n-r+1} + \cdots \right) \in \langle x^r \rangle.$$

Then $A \subseteq \langle x^r \rangle$. Consequently, $A = \langle x^r \rangle$. $\qquad \square$

Corollary *For any field F, $F[\![x]\!]$ is a local ring with $\langle x \rangle$ as its maximal ideal.*

Proof As the non-trivial ideals of $F[\![x]\!]$ are given by $\langle x^r \rangle$ for positive integers $r \geq 1$, it follows that $F[\![x]\!] \supsetneq \langle x \rangle \supsetneq \langle x^2 \rangle \supsetneq \cdots \supsetneq \{0\}$. This proves the corollary. $\qquad \square$

5.5 Application to Algebraic Geometry

Classical algebraic geometry arose through the study of the sets of simultaneous solutions of systems of polynomial equations in n variables over an algebraically closed field, like the field of complex numbers $\mathbf{C}$. The theory of Noetherian rings and the Hilbert Basis Theorem (to be discussed in Chap. 7), prescribe methods for solutions of many problems of algebraic geometry. So the ideal theory imposes a considerable influence on algebraic geometry. For example, affine varieties play an important role in algebraic geometry and Zariski topology. Moreover, affine varieties are characterized in this section with the help of prime ideals. Oscar Zariski and André Weil redeveloped the foundation of algebraic geometry in the middle of the 20th century.

5.5.1 Affine Algebraic Sets

Let K be an algebraically closed field (i.e., every polynomial of one variable of degree ≥ 1 over K has a root in K). An affine n-space over K, denoted by K^n is the set of all n-tuples $(x) = (x_1, x_2, \ldots, x_n)$ of elements of K. We call K^1 the affine line and K^2 the affine plane. The n-tuple $(x) = (x_1, x_2, \ldots, x_n)$ is called a point of K^n. The n-tuple (x) is called a zero of a polynomial $f(x) = f(x_1, x_2, \ldots, x_n) \in K[x_1, x_2, \ldots, x_n]$ iff $f(x) = f(x_1, x_2, \ldots, x_n) = 0$.

Given a subset S of polynomials in $K[x_1, x_2, \ldots, x_n]$, the *algebraic set of zeros* of S, denoted by $V(S)$, is the subset of K^n consisting of all common zeros of polynomials in S i.e., $V(S) = \{(x) \in K^n : f(x) = 0 \ \forall f \in S\}$. A subset A of K^n is called an *affine algebraic* set, iff $\exists$ a subset S of $K[x_1, x_2, \ldots, x_n]$ such that $A = V(S)$. An affine algebraic set is sometimes called (simply) an algebraic set. The set of all zeros of a single polynomial $f(x_1, x_2)$ i.e., the set of points of K^2 satisfying the polynomial equation $f(x_1, x_2) = 0$ over K is called an *affine algebraic curve* in the affine plane K^2. Similarly, the set of all zeros of a single polynomial equation $f(x_1, x_2, x_3) = 0$ is called an *affine algebraic surface* in K^3. The affine algebraic set in K^3 determined by two polynomial equations $f(x_1, x_2, x_3) = 0$ and $g(x_1, x_2, x_3) = 0$ is called a *space curve*. If $f(x_1, x_2, \ldots, x_n)$ is not a constant, then the set of zeros of f is called the *hypersurface* defined by f and is denoted by $V(f)$.

We now present some elementary properties of algebraic sets *in affine n-space K^n*.

Proposition 5.5.1 *Let K^n be an affine n-space. Then*

(i) $P \subseteq Q$ in $K[x_1, x_2, \ldots, x_n]$ implies $V(P) \supseteq V(Q)$ in K^n.
(ii) $V(0) = K^n$, $V(a) = \emptyset$, where $a \neq 0$ and $a \in K$.
(iii) If $\{P_\alpha\}$ is a family of ideals in $K[x_1, x_2, \ldots, x_n]$, then $V(\bigcup_\alpha P_\alpha) = \bigcap_\alpha V(P_\alpha)$.
(iv) If P and Q are ideals in $K[x_1, x_2, \ldots, x_n]$ and $A = V(P)$ and $B = V(Q)$, then $A \cup B = V(P \cap Q) = V(PQ)$.
(v) Any finite subset of K^n is an algebraic set.

Proof (i) $(x) \in V(Q) \Leftrightarrow (x)$ is a common zero of polynomials of $Q \Rightarrow (x)$ is also a common zero of polynomials of $P \Rightarrow (x) \in V(P) \Rightarrow V(Q) \subseteq V(P)$.

(ii) As every point of K^n satisfies the zero polynomial, $V(0) = K^n$. As the constant non-null polynomial 'a' has no zero, $V(a) = \emptyset$.

(iii) $(x) \in V(\bigcup_\alpha P_\alpha) \Leftrightarrow f(x) = 0 \ \forall f \in \bigcup_\alpha P_\alpha \Leftrightarrow f(x) = 0 \ \forall f \in P_\alpha$, where P_α is any member of the collection $\{P_\alpha\} \Leftrightarrow (x) \in V(P_\alpha) \ \forall \alpha \Leftrightarrow (x) \in \bigcap V(P_\alpha)$.

(iv) $(x) \in A \cup B = V(P) \cup V(Q) \Rightarrow (x)$ is a zero of all polynomials in P or (x) is a zero of all polynomials in $Q \Rightarrow (x)$ is a zero of all polynomials in $P \cap Q \Rightarrow (x)$ is a zero of all polynomials in PQ (since $PQ \subseteq P \cap Q) \Rightarrow A \cup B \subseteq V(P \cap Q) \subseteq V(PQ)$.

Again $(x) \notin A \cup B \Rightarrow \exists f \in P$ and $g \in Q$ such that $f(x) \neq 0$ and $g(x) \neq 0 \Rightarrow fg \in P \cap Q$ (also to PQ) does not vanish at $(x) \Rightarrow (x) \notin V(P \cap Q)$ and also $(x) \notin V(PQ) \Rightarrow$ the zeros of $P \cap Q$ (as well as PQ) are the points $A \cup B$ and only these.

(v) Any point $(x_1, x_2, \ldots, x_n) \in K^n$ is an algebraic set, because $V(X_1 - x_1, X_2 - x_2, \ldots, X_n - x_n) = \{(x_1, x_2, \ldots, x_n)\}$. Hence any finite subset A of K^n is an algebraic set by (iv). □

Using the fact that any algebraic set is of the form $V(P_\alpha)$ it follows from Proposition 5.5.1 that

Corollary *Let K^n be an affine n-space. Then*

(i) *the intersection of any collection of algebraic sets in K^n is an algebraic set;*
(ii) *the union of any two algebraic sets in K^n is an algebraic set.*

We now determine all the algebraic sets of the affine line K^1.

Proposition 5.5.2 *A subset A of K is algebraic iff either $A = K^1 = K$ or a finite subset of K^1.*

Proof We have seen that the condition is sufficient. To prove the necessity of the condition, note that as $K[x]$ is a PID and a non-zero polynomial $f \in K[x]$ has a finite number of zeros $\leq \deg f$, it cannot vanish on an infinite set. □

Remark Proposition 5.5.2 shows that an *infinite union* of algebraic sets may not be an algebraic set.

5.5.2 Ideal of a Set of Points in K^n

Given a subset A in K^n, define the subset $I(A)$ in $K[x_1, x_2, \ldots, x_n]$ to be the set of all polynomials which vanish on A i.e.,

$$I(A) = \{f \in K[x_1, x_2, \ldots, x_n] : f(x) = 0 \ \forall (x) \in A\}.$$

Then $I(A)$ is an ideal of $K[x_1, x_2, \ldots, x_n]$. To show this, let $f, g \in I(A)$. Then $f(x) = 0$ and $g(x) = 0$ $\forall(x) \in A$. Consequently, $(f + g)(x) = f(x) + g(x) = 0$ $\forall(x) \in A$ and $(hf)(x) = h(x)f(x) = 0$ $\forall(x) \in A$ and $\forall h \in K[x_1, x_2, \ldots, x_n]$.

Remark $I(A)$ is called the *ideal of A*.

Proposition 5.5.3 *Let K^n be an affine n-space. Then*

  (i) $A \subseteq B$ *in* $K^n \Rightarrow I(A) \supseteq I(B)$ *in* $K[x_1, \ldots, x_n]$;
 (ii) $S \subseteq I(V(S))$ *for any set of polynomials S in* $K[x_1, \ldots, x_n]$;
(iii) $A \subseteq V(I(A))$ *for any set of points A in* K^n;
 (iv) $V(S) = V(I(V(S)))$ *for any set of polynomials S in* $K[x_1, x_2, \ldots, x_n]$;
  (v) $I(A) = I(V(I(A)))$ *for any set of points A in* K^n;
 (vi) $I(\emptyset) = K[x_1, x_2, \ldots, x_n]$;
(vii) $I(K^n) = \{0\}$, *if K is an infinite field.*

Proof (i) $f \in I(B) \Leftrightarrow f$ vanishes on $B \Rightarrow f$ vanishes on A (since $A \subseteq B$) $\Leftrightarrow f \in I(A) \Rightarrow I(B) \subseteq I(A)$.

(ii) This follows from the fact that S is a set of some polynomials which vanish on $V(S)$, while $I(V(S))$ is the set of all polynomials which vanish on $V(S)$.

(iii) This follows from the fact that A is a set of some common zeros of polynomials in $I(A)$, while $V(I(A))$ is the set of all common zeros of polynomials in $I(A)$.

(iv) $V(S) \subseteq V(I(V(S)))$ by (ii) (take $V(S)$ for A in (iii)).

Conversely, $V(I(V(S))) \subseteq V(S)$. Because $(x) \in V(I(V(S))) \Leftrightarrow f(x) = 0 \forall f \in I(V(S)) \Rightarrow f(x) = 0 \forall f \in S$ by (ii) $\Leftrightarrow x \in V(S)$.

(v) $I(A) \subseteq I(V(I(A)))$ by (iii) (take $I(A)$ for S in (iii)).

Conversely, $I(V(I(A))) \subseteq I(A)$ by (iii). This is so because $f \in I(V(I(A))) \Leftrightarrow f(x) = 0 \forall(x) \in V(I(A)) \Rightarrow f(x) = 0 \forall(x) \in A$ by (iii) $\Rightarrow f \in I(A)$.

(vi) $V(1) = \emptyset \Rightarrow I(V(1)) = I(\emptyset)$. Now by (ii), $1 \in I(V(1)) \Rightarrow I(\emptyset) = K[x_1, x_2, \ldots, x_n]$.

(vii) This part says that if K an infinite field and $f \in K[x_1, x_2, \ldots, x_n]$, then f vanishes on K^n implies $f = 0$. We shall prove this by induction on n. If $n = 1$, a non-zero $f \in K[x_1]$ has a finite number of zeros $\leq \deg f$, by the fundamental theorem of algebra (see Chap. 11), and hence f cannot vanish on the infinite set K, unless $f = 0$. Next, suppose that the result is true for $(n - 1)$ variables. Let $f \in K[x_1, x_2, \ldots, x_n]$ vanish on K^n. We write $f(x_1, x_2, \ldots, x_n) = \sum_i f_i(x_1, x_2, \ldots, x_{n-1})x_n^i$ as a polynomial in x_n with coefficients in $K[x_1, \ldots, x_{n-1}]$. If $\exists$ a point $(a_1, a_2, \ldots, a_{n-1}) \in K^{n-1}$ such that for some j, $f_j(a_1, \ldots, a_{n-1}) \neq 0$, then the non-zero polynomial $f(a_1, a_2, \ldots, a_{n-1}, x_n) \in K[x_n]$ vanishes on K, since $f(x_1, \ldots, x_n)$ vanishes on K^n. This implies, by the fundamental theorem of algebra, that $f(a_1, a_2, \ldots, a_{n-1}, x_n) = 0$, which is impossible, since $f_j(a_1, \ldots, a_{n-1}) \neq 0$. Therefore f_j must vanish on K^{n-1} $\forall j$, and hence $f_j = 0 \forall j$, by induction hypothesis. Consequently, $f = 0$. $\square$

Remark If A is an algebraic set and $A = V(S)$, then $A = V(I(A))$; and if P is an ideal of an algebraic set A and $P = I(A)$, then $P = I(V(P))$.

5.5.3 Affine Variety

An algebraic set A in K^n is called *irreducible* or an *affine variety* iff $A \neq B \cup C$, where B and C are algebraic sets in K^n and $A \neq B$, $A \neq C$ i.e., iff A is not the union of two algebraic sets B and C distinct from A. Otherwise A is called reducible.

Theorem 5.5.1 *An algebraic set A is an affine variety iff $I(A)$ is a prime ideal.*

Proof Suppose A is an algebraic set and P is its ideal, i.e., $P = I(A)$. Then $A = V(P)$ by Proposition 5.5.3(iv). If P is not prime, we can find $f, g \in K[x_1, x_2, \ldots, x_n]$ such that $f \notin P$, $g \notin P$ but $fg \in P$. Now by Proposition 5.5.1(i) and (iii), we get

$$f \notin P \quad \Rightarrow \quad P \cup \{f\} \supsetneq P \quad \Rightarrow \quad V(P) \cap V(f) = V\big(P \cup \{f\}\big) \subsetneq V(P);$$

$$g \notin P \quad \Rightarrow \quad P \cup \{g\} \supsetneq P \quad \Rightarrow \quad V(P) \cap V(g) = V\big(P \cup \{g\}\big) \subsetneq V(P).$$

Also

$$\big[V(P) \cap V(f)\big] \cup \big[V(P) \cap V(g)\big] = V(P) \cap \big[V(f) \cup V(g)\big]$$

(by distributive laws of sets) $= V(P) \cap V(fg)$ (by Proposition 5.5.1(iv)) $= V(P \cup (fg)) = V(P)$, since $fg \in P$. Thus the algebraic set $A = V(P)$ is the union of two algebraic sets $V(P) \cap V(f)$ and $V(P) \cap V(g)$, which are distinct from A, and hence A is reducible.

Conversely, if A is reducible, then $A = B \cup C$, where B and C are algebraic sets different from A. Then $A \supsetneq B \Rightarrow I(A) \subsetneq I(B)$ and $A \supsetneq C \Rightarrow I(A) \subsetneq I(C)$. Therefore, if $f \in I(B)$, $g \in I(C)$ and $f, g \notin I(A)$, then $fg \in I(A)$. This is so because $A = B \cup C = V(I(B)) \cup V(I(C))$ (by Proposition 5.5.3) $= V(I(B)I(C))$ (by Proposition 5.5.1(iv) and hence $I(A) = I(V(I(B)I(C))) \supseteq I(B)I(C)$ (by Proposition 5.5.3 (ii)) $\Rightarrow I(A)$ is not prime (see Ex. 11(c) of Exercises-I). We have proved that an algebraic set A is reducible $\Leftrightarrow I(A)$ is not a prime ideal. Equivalently, A is an affine variety $\Leftrightarrow I(A)$ is a *prime* ideal. $\qquad \square$

Corollary *An affine algebraic set A in K^n is an affine variety iff the coordinate ring $K[x_1, x_2, \ldots, x_n]/I(A)$ is an integral domain.*

5.5.4 The Zariski Topology in K^n

Zariski topology is a particular topology introduced by Oscar Zariski (1899–1986) around 1950 to study algebraic varieties. This topology in $\mathbf{C}^n$ has an advantage over the metric topology, because Zariski topology has many fewer open sets than the metric topology. A refinement of Zariski topology was made by Alexander Grothendieck in 1960.

Let κ be a subfield of K. If A is an affine algebraic set in K^n such that $I(A)$ admits a set of generators in $\kappa[x_1, x_2, \ldots, x_n] \subseteq K[x_1, x_2, \ldots, x_n]$, then A is called an affine (K, κ) *algebraic set* and κ is called the field of definition of A. Thus an affine (K, κ) algebraic set A is a subset in K^n consisting of all common zeros of a subset of polynomials (or an ideal) in $\kappa[x_1, x_2, \ldots, x_n]$. If $\kappa = K$, we call A an *absolute affine algebraic set* (or simply an affine algebraic set) in K^n. Define a topology in K^n, called the *Zariski topology* or κ-topology in the following way:

A subset U of K^n is an open set iff $K^n - U$ is an affine κ-algebraic set. Thus the affine κ-algebraic sets in K^n are the closed sets in K^n. As the results of Propositions 5.5.1(ii–iv) for affine algebraic sets are also true for affine (K, κ) algebraic sets, it follows that the empty set and the whole set are closed sets; the intersection of any arbitrary family of closed sets is a closed set; and the union of two closed sets is a closed set. From all this the Zariski topology in K^n follows.

Any subset A of K^n is given the induced topology from Zariski topology in K^n. Thus if A is an affine κ-variety in K^n, then a subset of A is closed in A iff it is an affine κ-algebraic set.

Proposition 5.5.4 (i) *The Zariski topology in K^n is not Hausdorff.*

(ii) *The Zariski topology in K^n may not be T_1 unless $K = \kappa$.*

Proof (i) Let A be a κ-variety in K^n. Then for any two non-empty open sets U_1 and U_2 in A, we have $U_1 \cap U_2 \neq \emptyset$. This is so because if $U_1 \cap U_2 = \emptyset$, then $A = (A - U_1) \cup (A - U_2) \cup (U_1 \cap U_2) = (A - U_1) \cup (A - U_2)$, and hence A is the union of two affine κ-algebraic sets: $A - U_1$ and $A - U_2$, which are different from A. In other words, A is not a κ-variety in K^n, which implies a contradiction. Therefore, if x and y are distinct points of A, it is not possible to find disjoint neighborhoods U_1 and U_2 of x and y respectively. This means that A is not a Hausdorff space. Therefore, K^n cannot be a Hausdorff space, because every subspace of a Hausdorff space is Hausdorff space.

(ii) If $\kappa = K$, then by Proposition 5.5.1(v), any point of K is an algebraic set and hence closed in the Zariski topology. If $\kappa \neq K$, and if $(\alpha_1, \alpha_2, \ldots, \alpha_n) \in K^n$ is not a zero of any polynomial in $k[x_1, x_2, \ldots, x_n]$, then the point $(\alpha_1, \alpha_2, \ldots, \alpha_n)$ is not a κ-algebraic set and hence not closed. $\square$

Remark The above Proposition 5.5.4 shows that any open set U of a κ-variety A is dense in A, i.e., $\overline{U} = A$. Also, any open set of a κ-variety is connected, because it is not the union of two disjoint non-empty open sets.

Problem 4 (a) Let $s(I)$ be the set of all ideals of $K[x_1, x_2, \ldots, x_n]$ and $s(A)$ be the set of all algebraic sets in K^n. Then the mapping $V : s(I) \to s(A)$, $I \mapsto V(I)$ satisfies the following properties:

(i) $V(0) = K^n$.

(ii) $V(K[x_1, x_2, \ldots, x_n]) = \emptyset$.

(iii) $V(\sum_{\alpha} I_{\alpha}) = \bigcap V(I_{\alpha})$.
(iv) $V(I \cap J) = V(IJ) = V(I) \cup V(J)$.

(b) By using (a), define the Zariski topology on K^n.

(c) Hence deduce that if K is an infinite field, then K^1 endowed with the Zariski topology is not a Hausdorff space.

Solution (a) follows from Proposition 5.5.1.

(b) As the properties (i)–(iv) satisfy the axioms for closed sets of a topology, the Zariski topology is defined on K^n, with algebraic sets as the only closed sets.

(c) If M is a maximal ideal of $K[x_1, x_2, \ldots, x_n]$ of the form $(x_1 - t_1, x_2 - t_2, \ldots, x_n - t_n)$, $t = (t_1, t_2, \ldots, t_n) \in K^n$, then $V(M) = \{t\}$. Thus finite sets of points are closed sets in the Zariski topology in K^n. They are the only closed sets in K^1 except K^1 and empty set $\emptyset$. If the field K is infinite, K^1 endowed with the Zariski topology is not Hausdorff.

Problem 5 If K is not algebraically closed, for a proper ideal I of $K[x_1 - t_1, x_2 - t_2, \ldots, x_n - t_n]$, can $V(I)$ be empty?

Solution $V(I)$ may be empty. For example, if $K = \mathbf{R}$ (which is not algebraically closed), $\langle x^2 + 1 \rangle$ is an ideal of $K[x]$ such that $\langle x^2 + 1 \rangle$ is proper but $V(x^2 + 1) = \emptyset$.

Problem 6 Let $s(A)$ denote the set of all algebraic sets in K^n and $s(I)$ denote the set of all ideals of $K[x_1, x_2, \ldots, x_n]$ and $s_R(I)$ denote the set of all radical ideals of $K[x_1, \ldots, x_n]$.

(a) Then the mapping $I : s(A) \to s(I)$, $A \mapsto I(A)$ is injective.
(b) Then the mapping $I \circ V : s_R(I) \to s_R(I)$, $J \mapsto I(V(J))$ is the identity function, where K is algebraically closed.
(c) If K is an algebraically closed field and J is an ideal of the ring $K[x_1, x_2, \ldots, x_n]$, then $I(V(J)) = \operatorname{rad}(J)$, i.e., if a polynomial vanishes at all common zeros of polynomials in J, then some power of f belongs to J.
(d) If K is an algebraically closed field, then there exists a bijective correspondence between the algebraic sets contained in K^n and the radical ideals of the ring $K[x_1, x_2, \ldots, x_n]$ (i.e., ideals J such that $\operatorname{rad}(J) = J$).

Solution (a) It follows from Proposition 5.5.3(iv) that

$$A = V\big(I(A)\big) = (V \circ I)(A), \quad \forall A \in s(A)$$

$\Rightarrow V \circ I$ is the identity function on $s(A) \Rightarrow I$ is injective.

(b) Take in particular, $I(A) = J$. Then J is a radical ideal in $K[x_1, x_2, \ldots, x_n]$ and hence by Ex. 7 of Exercises-I, $s_R(I) = \{J : J = I(A) \text{ for some } A \in s(A)\}$. Then Proposition 5.5.3 $(v) \Rightarrow J = I(V(J)) = (I \circ V)(J), \forall J \in s_R(I) \Rightarrow I \circ V$ is the identity function on $s_R(I) \Rightarrow I$ is a surjection.

(c) Since $I(A) = J$ is a radical ideal (see Ex. 7 of Exercises-I), (c) follows from Proposition 5.5.3(v).

(d) follows from (a) and (b).

We summarize the above discussion in the basic and important result.

Theorem 5.5.2 (Hilbert's Nullstellensatz) *Let K be an algebrically closed field. Then $I(V(J)) = \mathrm{rad}\,(J)$ for every ideal J in $K[x_1, x_2, \ldots, x_n]$. Moreover, there exists a bijective correspondence between the set of affine algebraic sets in K^n and the set of radical ideals in $K[x_1, x_2, \ldots, x_n]$.*

Remark 1 Hilbert's Nullstellensatz theorem is the first result in representing statements about geometry in the language of algebra.

Remark 2 Hilbert's Nullstellensatz theorem identifies the set of maximal ideals of the polynomial rings $\mathbf{C}[x_1, x_2, \ldots, x_n]$ with the points in the affine space $\mathbf{C}^n$.

Remark 3 Hilbert's Nullstellensatz theorem may be considered as a multidimensional version of the fundamental theorem algebra.

Remark 4 Hilbert's Nullstellensatz theorem fails to be true over the field $\mathbf{R}$ (which is not algebraically closed) [see Smith et al. 2000.]

5.6 Chinese Remainder Theorem

It is so named for the reason that it generalizes the result in part (c) of Theorem 5.6.1, which is a classical result of number theory, proved by Chinese mathematicians during the first century AD. It provides a representation of $\mathbf{Z}_n$ and many interesting applications (see Chaps. 10 and 12). This theorem has several forms according to the context. Throughout this section we assume R is a commutative ring with 1. This theorem is used to prove structure theorems for a class of modules.

Definition 5.6.1 Two ideals A and B of R are said to be coprime (comaximal) iff $A + B = R$.

Proposition 5.6.1 *If ideals A and B are coprime in R, then $A \cap B = AB$.*

Proof Clearly, $AB \subseteq A \cap B$. Since $A + B = R$, $\exists a \in A$, $b \in B$ such that $a + b = 1$. Let $c \in A \cap B$. Then $c = c1 = c(a + b) = ca + cb \Rightarrow A \cap B \subseteq AB$. Hence $A \cap B = AB$. $\square$

Example 5.6.1 Let A and B be two distinct maximal ideals of R. Then $A + B = R$.
 [*Hint.* $A + B \supsetneq A$, since A and B are distinct. Hence $A + B = R$.]

Theorem 5.6.1 (Chinese Remainder Theorem for rings) *Let R be a commutative ring with* 1.

(a) *Let A and B be coprime ideals in R. Then the map $f : R \to R/A \times R/B$, defined by $r \mapsto (r + A, r + B)$ is a ring epimorphism with $\ker f = A \cap B = AB$ and the rings R/AB and $R/A \times R/B$ are isomorphic.*

(b) *(General form) Let $A_1, A_2, \ldots, A_n$ be pairwise coprime ideals in R. Then the map $f : R \to R/A_1 \times R/A_2 \times \cdots \times R/A_n$, defined by $r \mapsto (r + A_1, r + A_2, \ldots, r + A_n)$ is a ring epimorphism with $\ker f = A_1 \cap A_2 \cap \cdots \cap A_n = A_1 A_2 \cdots A_n$ and the rings $R/(A_1 A_2 \cdots A_n)$ and $R/A_1 \times R/A_2 \times \cdots \times R/A_n$ are isomorphic.*

(c) *Let $n_1, n_2, \ldots, n_t$ be positive integers such that every pair of integers n_i, n_j are relatively prime for $i \neq j$. If $a_1, a_2, \ldots, a_t$ are t integers, then the system of congruences $x \equiv a_1 \pmod{n_1}$, $x \equiv a_2 \pmod{n_2}, \ldots, x \equiv a_t \pmod{n_t}$ has a simultaneous solution, which is unique up to modulo $n = n_1 n_2 \cdots n_t$.*

Proof (a) A and B are coprime ideals in $R \Rightarrow \exists a \in A$ and $b \in B$ such that $1 = a + b$. Clearly, f is a ring homomorphism.

To show that f is surjective, let $(x + A, y + B) \in R/A \times R/B$. Then

$$f(xb + ya) = (xb + ya + A, xb + ya + B)$$

$$= (x + A, y + B), \quad \text{since } 1 = a + b.$$

Clearly, $\ker f = A \cap B$.

Hence f is an epimorphism with $\ker f = A \cap B = AB \Rightarrow$ the rings R/AB and $R/A \times R/B$ are isomorphic.

(b) Consider the map $f : R \to R/A_1 \times R/A_2 \times \cdots \times R/A_n$, $r \mapsto (r + A_1, r + A_2, \ldots, r + A_n)$. Then f is a ring epimorphism.

Hence the first part of (b) follows from (a) by induction.

The last part of (b) follows from the first part.

(c) It follows from the last part of (b) by taking $R = \mathbf{Z}$, $A_i = \langle n_i \rangle$. $\qquad\square$

Corollary 1 *Let $n_1, n_2, \ldots, n_t$ be integers > 1 such that every pair of integers n_i, n_j are relatively prime for $i \neq j$, $i, j \in \{1, 2, \ldots, t\}$. If $n = n_1 n_2 \cdots n_t$, then ring $\mathbf{Z}_n$ is isomorphic to the ring $\mathbf{Z}_{n_1} \times \mathbf{Z}_{n_2} \times \cdots \times \mathbf{Z}_{n_t}$.*

Corollary 2 *If n is a positive integer and $p_1^{n_1} p_2^{n_2} \cdots p_t^{n_t}$ is its factorization into powers of distinct primes ($n_i \geq 1$), then*

$$\mathbf{Z}_n \cong \mathbf{Z}_{p_1^{n_1}} \times \mathbf{Z}_{p_2^{n_2}} \times \cdots \times \mathbf{Z}_{p_t^{n_t}}.$$

Corollary 3 *If n is a positive integer and $p_1^{n_1} p_2^{n_2} \cdots p_t^{n_t}$ is its factorization into powers of distinct primes ($n_i \geq 1$), then $\mathbf{Z}_n^* \cong \mathbf{Z}_{p_1^{n_1}}^* \times \mathbf{Z}_{p_2^{n_2}}^* \times \cdots \times \mathbf{Z}_{p_t^{n_t}}^*$, where $\mathbf{Z}_n^*$ denotes the multiplicative group of all units of the ring $\mathbf{Z}_n$ (see Chap. 10).*

5.7 Ideals of $C(X)$

Let $C(X)$ be the ring of real valued continuous functions on a *compact topological space* X having the identity element c defined by the constant function $c(x) = 1$ for all $x \in X$. Then the maximal ideals in $C(X)$ correspond to the points of X in a natural way. It follows from the Theorem 5.7.4.

Theorem 5.7.1 *Let $C(X)$ be the ring of real valued continuous functions on a compact topological space X.*[1]

(a) *If M is a proper ideal of $C(X)$, then there exists a point $x_0 \in X$ such that $f(x_0) = 0$ for all $f \in A$.*
(b) *If M is a maximal ideal of $C(X)$, then there exists a point $x_0 \in X$ such that $M_{x_0} = \{ f \in C(X) : f(x_0) = 0 \} = M$.*

Proof (a) Let $f \in A$. Since f is continuous, $f^{-1}(0)$ is a closed set in X. Then $f^{-1}(0) \neq \emptyset$ for each $f \in A$. Otherwise, if possible, there is some $f \in A$ such that $f^{-1}(0) = \emptyset$. Then $f(x) \neq 0$ for all $x \in X$. So, $1/f \in C(X)$ and hence the unit element $c = f \cdot 1/f \in A \subseteq C(X)$ implies that $A = C(X)$. This contradicts the fact that A is a proper ideal of $C(X)$. We now consider the family $\mathcal{F} = \{ f^{-1}(0) : f \in A \}$ of closed sets in X. We shall show that it has the finite intersection property, i.e., every finite sub-family of $\mathcal{F}$ has a non-empty intersection. If $f_1, f_2, \ldots, f_n \in A$, then each $f_i^2 = f_i \cdot f_i \in A$ and hence $f = \sum_{i=1}^{n} f_i^2 \in A$. We claim that $\bigcap_{i=1}^{n} f_i^{-1}(0) = f^{-1}(0)$. Let $x \in \bigcap_{i=1}^{n} f_i^{-1}(0)$. Then each $f_i(x) = 0$ implies each $f_i^2(x) = 0$, resulting in $f(x) = 0$. Thus $x \in f^{-1}(0)$. Hence $\bigcap_{i=1}^{n} f_i^{-1}(0) \subseteq f^{-1}(0)$. Conversely, let $y \in f^{-1}(0)$. Then $f(y) = 0$ implies that each of $f_i^2(y) = 0$, i.e., each $f_i(y) \cdot f_i(y) = 0$, resulting in each $f_i(y) = 0$, as $\mathbf{R}$ is an integral domain. This implies $y \in f_i^{-1}(0)$ for each i, resulting in $y \in \bigcap_{i=1}^{n} f_i^{-1}(0)$, which implies $f^{-1}(0) \subseteq \bigcap_{i=1}^{n} f_i^{-1}(0)$. Thus $\bigcap_{i=1}^{n} f_i^{-1}(0) = f^{-1}(0) \neq \emptyset$. Since X is compact, $\bigcap \{ f^{-1}(0) : f \in A \} \neq \emptyset$. Hence there exists a point $x_0 \in X$ such that $f(x_0) = 0$ for all $f \in A$.

(b) Let M be a maximal ideal of $C(X)$. Then M is a proper ideal of $C(X)$ and hence by (a) there exists a point $x_0 \in X$ such that $f(x_0) = 0$ for all $f \in M$. Let $M_{x_0} = \{ f \in C(X) : f(x_0) = 0 \}$. Since the identity element which is the constant function $c \notin M_{x_0}$, M_{x_0} is a proper ideal of $C(X)$ such that $M \subseteq M_{x_0}$. As M is a maximal ideal, it follows that $M = M_{x_0}$. $\square$

We now study the particular case when $X = [0, 1]$.

Theorem 5.7.2 *Let R be the ring of all real valued continuous functions on $[0, 1]$ and $M_x = \{ f \in R : f(x) = 0 \}$, where $x \in [0, 1]$. Then M_x is a maximal ideal of R.*

---

[1] A topological space X is compact if and only if every collection of closed sets in X possessing the finite intersection property, the intersection of the entire collection is non-empty (see [Chatterjee et al. (2003)]).

Proof Let $f, g \in M_x$ and $h \in R$. Then $f(x) = 0 = g(x)$. Now $(f + g)(x) = f(x) + g(x) = 0$ and $(f \cdot h)(x) = f(x) \cdot h(x) = 0 = (h \cdot f)(x)$. Hence M_x is an ideal of R. Let N be an ideal of R such that $M_x \subsetneq N \subseteq R$. Then $\exists$ a function $g \in N$ such that $g \notin M_x$. Hence $g(x) = r \neq 0$. Consider a function $h \in R$ defined by $h(y) = g(y) - r$. Then h is a continuous function such that $h(x) = g(x) - r = r - r = 0 \Rightarrow h \in M_x \subsetneq N$. Thus $g, h \in N \Rightarrow g - h \in N$, since N is an ideal of $R \Rightarrow r \in N$. Again $r \neq 0 \Rightarrow r^{-1}$ exists $\Rightarrow rr^{-1} \in N \Rightarrow$ the constant function $c(x) = 1 \in N \Rightarrow N = R$ (this is so because $1 \in N \Rightarrow f \cdot 1 \in N \, \forall f \in R \Rightarrow R \subseteq N$. Again $N \subseteq R$. Hence $N = R$). Consequently, M_x is a maximal ideal of R. $\qquad\square$

Corollary *For every $x \in [0, 1]$, the ideal M_x is also prime.*

We now show that the maximal ideals in the ring $R = C([0, 1])$ correspond to the points in $[0, 1]$.

Theorem 5.7.3 *Let $\mathcal{M}$ be the set of all maximal ideals of the ring $R = C([0, 1])$. Then there exists a bijective correspondence between the elements of $\mathcal{M}$ and the points of $[0, 1]$.*

Proof To prove the theorem, it is sufficient to show that the mapping $\psi : [0, 1] \to \mathcal{M}$ defined by $\psi(x) = M_x = \{f \in R : f(x) = 0\}$ is a bijection, where $\mathcal{M}$ is the set of all maximal ideals of the ring R. By Theorem 5.7.1, ψ is well defined. We claim that ψ is injective. As $[0, 1]$ is compact and Hausdorff, by the Urysohn Lemma[2] for each $y \, (\neq x)$ in $[0, 1]$, there is a function f in R such that $f(x) = 0$ but $f(y) \neq 0$. This shows that the mapping is injective. To show that the mapping is onto, let M be a maximal ideal in R. We claim that for this M, there is a point y in $[0, 1]$ at which every function in M vanishes. If possible, for each point x in $[0, 1]$ there exists a function f in M such that $f(x) \neq 0$. Since f is continuous, x has a neighborhood such that at no point f vanishes. Varying x, we obtain an open cover of $[0, 1]$. As $[0, 1]$ is compact, this open cover has a finite subcover and hence we obtain corresponding functions $f_1, f_2, \ldots, f_n$ in M. Again as f_i is in M, each $f_i f_i = f_i^2 \in M$. Then the function $g = \sum_{i=1}^{n} f_i^2 \in M$ is such that $g(x) > 0$ for all $x \in [0, 1]$. Hence g can not lie in a proper ideal of R. It contradicts the fact that g lies in the proper ideal M. This contradiction implies that there exists a point $x \in [0, 1]$ such that every function in M vanishes at x. Hence $M = M_x$ shows that ψ is onto. Consequently, ψ is a bijection. $\qquad\square$

Corollary 1 *The maximal ideals in the ring R of all real valued continuous functions on $[0, 1]$ correspond to the points in $[0, 1]$.*

---

[2]**Urysohn Lemma** *Let X be a normal space, and A and B be disjoint closed subspaces of X. Then there exists a continuous real function f defined on X, all of whose values lie in $[0, 1]$, such that $f(A) = 0$ and $f(B) = 1$. [Simmons 1963]*

Corollary 2 *Given a maximal ideal M in R, there exists a real number $x \in [0, 1]$ such that $M = M_x$.*

Proof It follows from Theorems 5.7.3 (also from Theorem 5.7.1 when $X = [0, 1]$).   □

Proceeding as above, we can generalize Theorem 5.7.3 for an arbitrary compact topological space X.

Theorem 5.7.4 *Let X be a compact topological space and $R = C(X)$ be the ring of all real valued continuous function on X. Then there exists a bijective correspondence between the maximal ideals of R and the points of X.*

Proof Let X be a compact topological space. Then for each $x \in X$, there is a maximal ideal $M_x \in C(X)$ defined by $M_x = \{f \in C(X) : f(x) = 0\}$. Conversely, given a maximal ideal $M \in C(X)$ there exists a point $x \in X$ at which every function in M vanishes, i.e., $M = M_x$. Hence there is a bijective correspondence between the set of maximal ideals of $C(X)$ and the points of X.   □

Corollary *The maximal ideals in $C(X)$ correspond to the points of X. Moreover, given a maximal ideal M in $C(X)$ there exists a point $x \in X$ such that $M = M_x$.*

Problems and Supplementary Examples (SE-I)

1 (i) Let R be a ring and

$$I_1 \subseteq I_2 \subseteq I_3 \subseteq \cdots \subseteq I_n \cdots \tag{5.2}$$

be an ascending chain of ideals of R. If $I = \bigcup I_n$, then I is an ideal of R.

(ii) Deduce that in a principal ideal domain there cannot exist an infinite sequence of ideals:

$$I_1 \subsetneqq I_2 \subsetneqq \cdots \text{ with } I_n \subsetneqq I_{n+1} \text{ for all } n \in \mathbf{N}^+. \tag{5.3}$$

For part (i), suppose $I = \bigcup_n I_n$. Then $I \neq \emptyset$. Let $a, b \in I$. Then $\exists$ integers s, t such that $a \in I_s$, $b \in I_t$. Hence either $s \leq t$ or $t < s$. For definiteness, let $s \leq t$. Then $a - b \in I_t$ and $ar, ra \in I_t$ $\forall r \in R \Rightarrow a - b \in I, ar, ra \in I \Rightarrow I$ is an ideal of R.

For part (ii), suppose there exists an infinite sequence of ideals of type (5.3) in a principal ideal domain R. Then $I = \bigcup I_n$ is an ideal of R and hence $I = \langle x \rangle$ for some $x \in I$. Consequently, $x \in I_m$ for some $m \in \mathbf{N}^+$. This shows that $I \subseteq I_m$. Thus $I_{m+1} \subsetneqq I \subseteq I_m \Rightarrow$ a contradiction.

2 If A and B are non-null ideals of an integral domain R, then $A \cap B \neq \{0\}$.

Solution Let $a \in A$ and $b \in B$ be two non-zero elements. Then R is an integral domain $\Rightarrow ab \neq 0$ and A, B are ideals of $R \Rightarrow ab \in A$ and $ab \in B \Rightarrow ab \in A \cap B$. Consequently, $A \cap B \neq \{0\}$.

3 Let A be an ideal of a ring R. Then the ring R/A is commutative $\Leftrightarrow xy - yx \in A$ $\forall x, y \in R$. Deduce that if B and C are ideals of R and both R/B and R/C are commutative rings, then $R/(B \cap C)$ is also commutative.

Solution The quotient ring R/A is commutative $\Leftrightarrow (x + A)(y + A) = (y + A)(x + A) \forall x, y \in R \Leftrightarrow (xy + A) = yx + A \Leftrightarrow xy - yx \in A \forall x, y \in R$.

For the second part, the hypothesis implies by the first part that $\forall x, y \in R$, $xy - yx \in B$ and also $xy - yx \in C$ and hence $xy - yx \in B \cap C \Rightarrow R/(B \cap C)$ is commutative by the first part.

4 Let A and B be prime ideals in a commutative ring R. If $A \cap B$ is prime in R, then $A \subseteq B$ or $B \subseteq A$.

Solution If $A \not\subseteq B$, we claim that $B \subseteq A$. $A \not\subseteq B \Rightarrow \exists a \in A$ such that $a \notin B$. Let $b \in B$. Then $ab \in A \cap B \Rightarrow a \in A \cap B$ or $b \in A \cap B$, since $A \cap B$ is prime. Since $a \notin B, b \in A \cap B \subset A \Rightarrow B \subseteq A$.

5 (a) Let D be an integral domain and $a (\neq 0) \in D$ be not invertible. Then $\langle a^{n+1} \rangle \subsetneq \langle a^n \rangle$, for all $n \in \mathbf{N}^+$.

Solution Clearly, $\langle a^{n+1} \rangle \subseteq \langle a^n \rangle$. If possible $\langle a^{n+1} \rangle = \langle a^n \rangle$, then $a^n \in \langle a^{n+1} \rangle \Rightarrow a^n = a^{n+1} r$ for some $r \in D \Rightarrow 1 = ar$ by Theorem 4.1.4 $\Rightarrow a$ is invertible in $D \Rightarrow$ a contradiction.

(b) Deduce from (a) that

(i) there are infinitely many different ideals in an integral domain D which is not a field;
(ii) every finite integral domain is a field (cf. Theorem 4.1.6).

Solution (i) As D is not a field, there is a non-zero non-invertible element a in D. Using (a), $\langle a \rangle, \langle a^2 \rangle, \langle a^3 \rangle, \ldots$ are all different ideals of D.

(ii) follows from (i).

6 Let A, B and C be ideals in a ring R. Then

(i) $A + B + C$ and ABC are ideals;
(ii) $(A + B) + C = A + (B + C)$;
(iii) $(AB)C = ABC = A(BC)$;
(iv) $A(B + C) = AB + AC$ and $(A + B)C = AC + BC$.

[*Hint.* Use Definitions 5.1.4 and 5.1.5.]

7 Let $A = \{f(x) \in \mathbf{Z}[x] : f(0) \equiv 0 \pmod 2\}$. Then A is an ideal of $\mathbf{Z}[x]$ and $A = \langle 2, x \rangle$.

Worked-Out Exercises

1. Let R be a ring with 1 such that R has no non-trivial left (right) ideals. Then R is a division ring.

Solution Let $x\ (\neq 0) \in R$. Then $I = Rx = \{rx : r \in R\}$ is a left ideal of R. Now $x = 1 \cdot x \Rightarrow I \neq \{0\} \Rightarrow I = R$ by hypothesis $\Rightarrow 1 = yx$ for some $y \in R \Rightarrow x$ has a left inverse in $R \Rightarrow$ every non-zero element of R has a left inverse in $R \Rightarrow$ non-zero elements of R form a multiplicative group (see Proposition 2.3.2 of Chap. 2) $\Rightarrow R$ is a division ring.

A similar result holds if R has no non-trivial right ideals.

2. Let $R = M_2(\mathbf{Q})$ be the ring of 2×2 matrices over $\mathbf{Q}$ under usual addition and multiplication. Then R is a simple ring but not a division ring.

Solution Since R contains divisors of zero, R cannot be a division ring. On the other hand, let $I \neq \{0\}$ be an ideal of R. Then there exists a non-zero matrix $M = \begin{pmatrix} a & b \\ c & d \end{pmatrix} \in I$. As $M \neq \begin{pmatrix} 0 & 0 \\ 0 & 0 \end{pmatrix}$, without loss of generality, suppose $a \neq 0$. Consider the matrices in R:

$$E_{11} = \begin{pmatrix} 1 & 0 \\ 0 & 0 \end{pmatrix}, \qquad E_{12} = \begin{pmatrix} 0 & 1 \\ 0 & 0 \end{pmatrix}, \qquad E_{21} = \begin{pmatrix} 0 & 0 \\ 1 & 0 \end{pmatrix}, \qquad E_{22} = \begin{pmatrix} 0 & 0 \\ 0 & 1 \end{pmatrix}.$$

Now, $M \in I \Rightarrow E_{11} M E_{11} = \begin{pmatrix} a & 0 \\ 0 & 0 \end{pmatrix} \in I \Rightarrow E_{11} = \begin{pmatrix} 1 & 0 \\ 0 & 0 \end{pmatrix} = (a^{-1} I_2) \begin{pmatrix} a & 0 \\ 0 & 0 \end{pmatrix} \in I$, where I_2 is the identity matrix in R. Similarly, $E_{12}, E_{21}, E_{22} \in I$. Thus any element $A = \begin{pmatrix} x & y \\ z & t \end{pmatrix} \in R$, can be expressed as $A = x E_{11} + y E_{12} + z E_{21} + t E_{22}$.

Hence $A \in I \Rightarrow R \subseteq I$. But $I \subseteq R$.

Thus $I = R \Rightarrow R$ has no non-trivial ideals $\Rightarrow R$ is a simple ring.

3. The ring $\mathbf{Z} \times \mathbf{Z}$ is not an integral domain but it has a quotient ring which is an integral domain.

Solution $R = \mathbf{Z} \times \mathbf{Z}$ is a ring under component-wise addition and multiplication: $(a, b) + (c, d) = (a + c, b + d), (a, b) \cdot (c, d) = (ac, bd), \forall a, b, c, d \in \mathbf{Z}$. Then $(0, 1)$ and $(1, 0)$ are non-null elements of R such that $(0, 1)(1, 0) = (0, 0) \Rightarrow R$ contains divisors of zero $\Rightarrow R$ is not an integral domain. Consider $I = \{(0, n) : n \in \mathbf{Z}\}$. Then I is an ideal of $R \Rightarrow R/I$ is a quotient ring. Then the map $f : \mathbf{Z} \to R/I$ defined by $f(n) = (n, 0) + I, \forall n \in \mathbf{Z}$ is a ring isomorphism. Hence $\mathbf{Z}$ is an integral domain $\Rightarrow f(\mathbf{Z}) = R/I$ is an integral domain.

4. The rings $(\mathbf{Z}_n, +, \cdot)$ and $(\mathbf{Z}/\langle n \rangle, +, \cdot)$ are isomorphic.

Solution Define a map $\psi : \mathbf{Z}_n \to \mathbf{Z}/\langle n \rangle$ by $\psi((m)) = m + \langle n \rangle = m + n\mathbf{Z}$. Then ψ is well defined and a bijective mapping such that $\psi((m + r)) = \psi((m)) + \psi((r))$ and $\psi((mr)) = \psi((m))\psi((r)), \forall (m), (r) \in \mathbf{Z}_n$. Hence ψ is a ring isomorphism.

5. Let R be the ring $R = \left\{ \begin{pmatrix} m & n \\ 0 & p \end{pmatrix} : m, n, p \in \mathbf{Z} \right\}$ and I be an ideal given by $I = \left\{ \begin{pmatrix} 0 & n \\ 0 & p \end{pmatrix} : n, p \in \mathbf{Z} \right\}$. Then the quotient ring R/I is isomorphic to $\mathbf{Z}$.

Solution I is an ideal of $R \Rightarrow R/I$ is a ring. Consider the mapping $f : R \to \mathbf{Z}$ defined by $f\left(\begin{pmatrix} m & n \\ 0 & p \end{pmatrix} \right) = m$. Then f is a ring epimorphism.

$$\ker f = \left\{ \begin{pmatrix} m & n \\ 0 & p \end{pmatrix} \in R : f\left(\begin{pmatrix} m & n \\ 0 & p \end{pmatrix} \right) = m = 0 \right\}$$

$$= \left\{ \begin{pmatrix} 0 & n \\ 0 & p \end{pmatrix} \in R \right\} = I \neq \{0\}.$$

Thus f is not a monomorphism $\Rightarrow f$ is not an isomorphism. But by the First Isomorphism Theorem for rings, $\exists$ an isomorphism $\tilde{f} : R/I \to \mathbf{Z}$ defined by $\tilde{f}(r + I) = f(r)$.

6. Let R be a commutative ring with 1 and I be an ideal of R. If $\langle I, x \rangle = I[x]$ denotes the ideal of $R[x]$ generated by I and x, then show that rings $R[x]/I[x]$ and $(R/I)[x]$ are isomorphic.

Solution Consider the natural map $\psi : R[x] \to (R/I)[x]$ given by reducing each coefficient r of a polynomial in $R[x]$ to the corresponding coefficient $r + I$ of the polynomial in $(R/I)[x]$. Clearly this map ψ is a ring epimorphism. The result follows from the First Ring Isomorphism Theorem.

7. In the ring $\mathbf{Z}[x]$, $\langle x \rangle$ is a prime ideal but not a maximal ideal.

Solution Consider the mapping $f : \mathbf{Z}[x] \to \mathbf{Z}$ defined by $f(a_0 + a_1 x + \cdots + a_n x^n) = a_0$. Then f is a ring epimorphism such that ker $f = \{a_0 + a_1 x + \cdots + a_n x^n \in \mathbf{Z}[x] : a_0 = 0\} = \{a_1 x + \cdots + a_n x^n \in \mathbf{Z}[x]\} = \langle x \rangle$. Hence by the Epimorphism Theorem (First Isomorphism Theorem), the rings $\mathbf{Z}[x]/\langle x \rangle$ and $\mathbf{Z}$ are isomorphic. So, $\mathbf{Z}$ is an integral domain $\Rightarrow \mathbf{Z}[x]/\langle x \rangle$ is an integral domain $\Rightarrow \langle x \rangle$ is a prime ideal in $\mathbf{Z}[x]$ by Theorem 5.3.1.

To the contrary, $\mathbf{Z}$ is not a field $\Rightarrow \mathbf{Z}[x]/\langle x \rangle$ is not a field $\Rightarrow \langle x \rangle$ is not a maximal ideal in $\mathbf{Z}[x]$ by Theorem 5.3.2.

8. Any integral domain D with only a finite number of ideals is a field.

Solution If possible, let D be not a field. Then $\exists$ a non-zero non-invertible element a in D. Then $\langle a \rangle, \langle a^2 \rangle, \langle a^3 \rangle, \ldots, \langle a^n \rangle, \ldots$ are infinitely many different ideals in D (see Ex. 5(a), (b) of SE-I) $\Rightarrow$ a contradiction of hypothesis. Hence D is a field.

9. Let D be an integral domain and $a, b \in D^* = (D - \{0\})$. Then $\langle a \rangle = \langle b \rangle$ iff $ab^{-1} \in U(D) = \{x : x \text{ is a unit in } D\}$.

Solution Suppose $\langle a \rangle = \langle b \rangle$. Then $a = bc$ for some $c \in D$ and $b = ad$ for some $d \in D$. Hence $b \neq 0$ and $b = bcd \Rightarrow 1 = cd \Rightarrow c \in U(D) \Rightarrow ab^{-1} \in U(D)$.

Conversely, let $ab^{-1} \in U(D)$. Then $a = bc$ for some $c \in U(D)$. Let $x \in \langle a \rangle$. Then $x = ad$ for some $d \in D \Rightarrow x = bcd \Rightarrow x \in \langle b \rangle \Rightarrow \langle a \rangle \subseteq \langle b \rangle$. Again $U(D)$ is a group and $ab^{-1} \in U(D) \Rightarrow ba^{-1} = (ab^{-1})^{-1} \in U(D) \Rightarrow \langle b \rangle \subseteq \langle a \rangle$. Hence $\langle a \rangle = \langle b \rangle$.

5.8 Exercises

Exercises-I

1. Show that the ideal $\langle a \rangle$ generated by a single element a of an arbitrary ring R is given by

$$\langle a \rangle = \left\{ na + ra + as + \sum_{\text{finite}} r_i a s_i : r_i, s_i, r, s \in R; \ n \in \mathbf{Z} \right\}.$$

If R has an identity, show that

$$\langle a \rangle = \left\{ \sum_{\text{finite}} r_i a s_i : r_i, s_i \in R \right\}.$$

If R is a commutative ring with 1, show that $\langle 1 \rangle = R$.

2. For any element a belonging to ring $(R, +, \cdot)$, show that

   (i) $(aR, +, \cdot)$ is a right ideal of $(R, +, \cdot)$;
   (ii) $(Ra, +, \cdot)$ is a left ideal of $(R, +, \cdot)$.

3. Let R be a ring, $A \neq \emptyset$ a subset of R. An element $c \in R$ is said to be a left (right) annihilator of A, iff $ca = 0$ $(ac = 0)$ $\forall a \in A$.

   Show that both annihilators of R form an ideal of R.

4. Prove that

   (a) A ring R with identity is a division ring iff R has no non-trivial left ideals;
   (b) If D is a division ring, then $R = M_{nn}(D)$ has no non-trivial left ideals.

   [*Hint.* (a) For any a $(\neq 0) \in R$, Ra is a left ideal of R. Now use the Corollary of Proposition 2.3.2 of Chap. 2.]

5. (a) Let R be an integral domain in which every proper ideal is prime. Show that R is a field.
   (b) If J and K are non-null ideals of an integral domain, prove that $J \cap K \neq \{0\}$.

6. (a) Let I be an ideal of a commutative ring R. Then the radical of I, denoted $\text{rad}(I)$, is defined by $\text{rad}(I) = \{a \in R : a^n \in I$ for some integer $n > 0\}$. Show that $\text{rad}(I)$ is an ideal of R such that $I \subseteq \text{rad}(I)$.
   (b) Prove that the ideal $I(A)$ of any subset A of an affine n-space K^n (A is not necessarily an algebraic set) is a radical ideal (see Chap. 5 or Ex. 7 of Exercises-I).

   [*Hint.* (a) Since the binomial theorem holds in a commutative ring, use this theorem.
       (b) $f \in I(A) \Leftrightarrow f(x) = 0$ $\forall(x) \in A \Leftrightarrow (f(x))^n = 0$ $\forall(x) \in A \Leftrightarrow f^n \in I(A) \Rightarrow f \in \text{rad}(I(A))$.]

7. An ideal I of a commutative ring is called a radical ideal iff $I = \text{rad}(I)$.

   Let K be an algebraically closed field and K^n the cartesian product of K with itself n times and $A \subseteq K^n$. Define $I(A)$ by $I(A) = \{f \in K[x_1, x_2, \ldots, x_n] : f(x) = 0 \forall(x) \in A\}$. Then $I(A)$ is an ideal of $K[x_1, x_2, \ldots, x_n]$ called the ideal of A. Show that $I(A)$ is a radical ideal.

8. If $f : R \to S$ is a ring homomorphism, I is an ideal of R and J is an ideal of S such that $f(I) \subseteq J$, show that f induces a ring homomorphism $\tilde{f} : R/I \to S/J$, given by $a + I \mapsto f(a) + J$. Prove that $\tilde{f}$ is an isomorphism iff $f(R) + J = S$ and $f^{-1}(J) \subseteq I$. In particular, if f is an epimorphism such that $f(I) = J$ and $\ker f \subseteq J$, then show that $\tilde{f}$ is an isomorphism.

   [Hint. See Proposition 2.6.1 of Chap. 2.]

9. Let I and J be ideals in a ring R. Prove that

(i) (Second Isomorphism Theorem). There is an isomorphism of rings $I/(I \cap J) \cong (I + J)/J$;

(ii) (Third Isomorphism Theorem). If $I \subset J$, then J/I is an ideal of R/I and there is an isomorphism of rings $(R/I)/(J/I) \cong R/J$.

[Hint. See Theorems 2.6.7 and 2.6.8 of Chap. 2.]

10. If I is an ideal of a ring R, prove that there is a one-to-one correspondence between the set of all ideals of R containing I and the set of all ideals of R/I. Hence show that every ideal of R/I is of the form T/I, where T is an ideal of R containing I.

    [Hint. Take $S = R/I$ and $f = \pi : R \to R/I$ in Theorem 5.2.4 or see Corollary or Remark of Theorem 5.2.4.]

11. (a) Let R be a commutative ring with identity and I be a prime ideal of R with the property that R/I is finite. Prove that I is a maximal ideal of R.

    [Hint. I is prime $\Rightarrow R/I$ is an integral domain $\Rightarrow R/I$ is a field (as R/I is finite) $\Rightarrow I$ is maximal.]

    (b) Let R be a commutative ring with identity and I be a proper ideal of R. Show that I is a maximal ideal $\Leftrightarrow \langle I \cup \{a\} \rangle = R$ for every element $a \notin I$.

    (c) Show that an ideal I in a commutative ring R is prime iff for every pair of ideals A and B of R such that $AB \subseteq I \Rightarrow$ either $A \subseteq I$ or $B \subseteq I$.

12. Let R and S be commutative rings with identity. If $f : R \to S$ is a ring homomorphism such that $f(R)$ is a field, show that $\ker f$ is a maximal ideal of R.

    [Hint. Clearly, $\ker f$ is an ideal of R. Let I be an ideal of R such that $\ker f \subsetneq I \subseteq R$.

    Then $\exists$ an element $a \in I$ but $a \notin \ker f$. Then $f(a) \neq 0$ and $f(a) \in f(R)$ which is a field and hence $\exists x \in R$ such that $f(a)f(x) = f(1) \Rightarrow ax - 1 \in \ker f \subset I$. Now $ax \in I \Rightarrow 1 \in I \Rightarrow I = R$.]

13. Prove that a proper ideal M of a commutative ring with 1 is maximal iff for every element $r \notin M$, there exists some $a \in R$ such that $1 + ra \in M$.

    [Hint. Let M be a maximal ideal of R and $r \notin M$. Then $M \subsetneq \langle M, r \rangle = R \Rightarrow 1 = m + rx$ for some $m \in M$ and $x \in R$.]

14. Let R be the ring of all real valued continuous functions defined on the closed interval $[0, 1]$. Let $I = \{f \in R : f(\frac{1}{2}) = 0\}$. Show that I is a maximal ideal of R.

15. Let $\mathbf{Z}[x]$ be the ring of polynomials over the ring of integers $\mathbf{Z}$ and I be the principal ideal of $\mathbf{Z}[x]$ generated by x i.e., $I = \langle x \rangle$. Show that $\langle x \rangle$ is a prime ideal but not maximal.

16. Let R be a commutative ring with 1. Show that there is a one-to-one correspondence between the maximal ideals M of R and the maximal ideals M' of $R[[x]]$ in such a way that M' corresponds to M iff M' is generated by M and x, i.e., $M' = \langle M, x \rangle$.

    [Hint. Let M be a maximal ideal of R. By using Ex. 13 of Exercises-I, show that $M' = \langle M, x \rangle$ is a maximal ideal of $R[[x]]$. Next, let M' be a maximal ideal of $R[[x]]$. Define the set $M = \{a_0 \in R : \sum a_i x^i \in M'\}$ i.e., M consists of the constant terms of power series $\in M'$. Show that M is a maximal ideal of R. To verify that the given correspondence is one-to-one, show that $\langle M, x \rangle = \langle \overline{M}, x \rangle$ for maximal ideals M and $\overline{M}$ of $R \Rightarrow \overline{M} = M$.]

17. If p is a prime and n is an integer ≥ 1, show that $\mathbf{Z}_{p^n}$ is a local ring with unique maximal ideal $\langle p \rangle$.

18. Let R be a commutative ring with 1. Show that the following statements are equivalent:

   (i) R is a local ring;
  (ii) all non-units of R are contained in some ideal $M \neq R$;
 (iii) the non-units of R form an ideal of R.

19. Let R and S be two commutative rings with identity. If $f : R \to S$ is a ring epimorphism, show that

   (i) S is a field $\Leftrightarrow$ ker f is a maximal ideal of R;
  (ii) S is an integral domain $\Leftrightarrow$ ker f is a prime ideal of R.
      [*Hint.* By the Note of Theorem 5.2.3, $R/\ker f \cong S$. Now for $I = \ker f$, R/I is a field $\Leftrightarrow$ ker f is maximal and R/I is an integral domain $\Leftrightarrow$ ker f is prime.]

20. (a) Let R be a commutative ring with identity 1. Then the *nil radical* of an ideal I of R, denoted by $\sqrt{I}$, is the set $\sqrt{I} = \{r \in R : r^n \in I$ for some positive integer n (depending on r)$\}$. Show that $\sqrt{I}$ is an ideal of R and $I \subseteq \sqrt{I}$ (see Ex. 6(a) of Exercises-I).

  (b) A proper ideal I of R is said to be a *semiprime ideal* iff $I = \sqrt{I}$. Prove that a proper ideal I of R is a semiprime ideal iff any one of the following conditions holds:

   (i) for any $a \in R$, $a^2 \in I$ implies $a \in I$;
  (ii) the quotient ring R/I has no non-zero nilpotent element.

  (c) Let R be a commutative ring with 1. A proper ideal I of R is said to be a *primary ideal* iff for $a, b \in R$, $ab \in I$ and $b \notin I \Rightarrow a^n \in I$ for some positive integer n. *Clearly every* prime ideal is a primary ideal. Show that $\langle 4 \rangle$ is a primary ideal of $(\mathbf{Z}, +, \cdot)$; determine all its primary ideals. Prove that a proper ideal I of R is a primary ideal, iff in the quotient ring R/I every non-zero divisor of zero is a nilpotent element.

     Further prove that if I is a primary ideal of R, then $\sqrt{I}$ is a prime ideal of R.

     [*Hint.* Let I be a primary ideal of R and $a, b \in R$ be such that $ab \in \sqrt{I}$. Then $(ab)^n = a^n b^n \in I$ for some integer $n \geq 1$. Then either $a^n \in I$ or $(b^n)^m \in I$ for some integer $m \geq 1$. Hence either $a \in \sqrt{I}$ or $b \in \sqrt{I}$.]

21. Let R be a commutative ring with 1. Show that

   (i) every prime ideal of R is a semiprime ideal but its converse is not true;
  (ii) every prime ideal of R is a primary ideal but its converse is not true;
      [*Hint.* Consider the ring $(\mathbf{Z}, +, \cdot)$. $\langle 6 \rangle$ is a semiprime ideal but not a prime ideal and $\langle 4 \rangle$ is a primary ideal but not a prime ideal.]

22. Let R be a commutative ring with 1. Then the *Jacobson radical* of a ring R, denoted by $\operatorname{rad} R$, is defined by $\operatorname{rad} R = \bigcap \{M : M$ is a maximal ideal of $R\}$. Clearly, $\operatorname{rad} R \neq \emptyset$, since R contains at least one maximal ideal by the Corollary of Theorem 5.3.6 (Krull–Zorn).

    If $\operatorname{rad} R = \{0\}$, then R is called a semisimple ring.

    Show that $(\mathbf{Z}, +, \cdot)$ is a semisimple ring.

    Show that $\operatorname{rad} R[\![x]\!] = \langle \operatorname{rad} R, x \rangle$. Hence show that if F is a field, then $\operatorname{rad} F[\![x]\!] = \langle x \rangle$.

    [*Hint.* For $R = \mathbf{Z}$, the maximal ideals of $\mathbf{Z}$ are precisely $\langle p \rangle$, where p is a prime number. Then $\operatorname{rad} R = \operatorname{rad} \mathbf{Z} = \bigcap \{\langle p \rangle : p$ is a prime number$\} = \{0\}$, since no non-zero integer is divisible by every prime. For the next part, $\operatorname{rad} R[\![x]\!] = \bigcap \{M' : M'$ is maximal ideal of $R[\![x]\!]\} = \bigcap \langle M, x \rangle$ (by Ex. 16 of Exercises-I) $= \langle \operatorname{rad} R, x \rangle$. In particular, if R is a field F, then $\operatorname{rad} F[\![x]\!] = \langle x \rangle$ (by the Corollary of Theorem 5.4.1).]

23. Let R be a commutative ring with 1. Prove the following:

    (i) Let A be an ideal of R. Then $A \subseteq \operatorname{rad} R \Leftrightarrow$ each element of the coset $1 + A$ has an inverse in R;

    (ii) An element $a \in \operatorname{rad} R \Leftrightarrow 1 - ra$ is invertible for each $r \in R$;

    (iii) An element a is invertible in $R \Leftrightarrow$ the coset $a + \operatorname{rad} R$ is invertible in the quotient ring $R / \operatorname{rad} R$;

    (iv) 0 is the only idempotent in $\operatorname{rad} R$.

24. Let I be a primary ideal in a commutative ring R with 1. Show that every zero divisor in R/I is nilpotent in R/I.

    [*Hint.* Let $(a) \in R/I$ be a zero divisor. Then $\exists (b) \neq (0) \in R/I$ such that $(a)(b) = (0)$. Hence $ab \in I \Rightarrow a^n \in I$ for some positive integer n, since $b \notin I$. Hence $(a)^n = (0)$.]

25. Let I, J and K be ideals of a ring R satisfying (i) $J \subseteq K$ (ii) $I \cap J = I \cap K$ and (iii) $J/I = K/I$. Show that $J = K$.

    [*Hint.* To show $K \subseteq J$, take an arbitrary element $k \in K$. Then $\exists$ an element $j \in J$ such that $k + I = j + I$ (by (iii)) $\Rightarrow k - j = i$ for some $i \in I$. Again $k - j \in K$ by (i). Consequently, $i = k - j \in I \cap K = I \cap J \Rightarrow k = i + j \in J \Rightarrow K \subseteq J \Rightarrow J = K$ by (i).]

26. Let I and J be two ideals of a ring R. Show that the sets

    (i) $I :_r J = \{a \in R : aJ \subseteq I\}$ (*right quotient* of I by J) and

    (ii) $I :_l J = \{a \in R : Ja \subseteq I\}$ (*left quotient* of I by J) are ideals of R.

    If R is a commutative ring, then we simply write $I : J$ [see Theorem 7.1.7 (Cohen) of Chap. 7].

Exercises A (True/False Statements) *Determine the correct statements with justification*:

1. Let R be a commutative ring with 1. If R has no non-trivial ideals, then R is a field.

2. Let R be a commutative ring with 1. Then every prime ideal of R is a maximal ideal.

3. Let $f : R \to S$ be a ring homomorphism. Then A is maximal in $R \Rightarrow f(A)$ is maximal in S.

4. Let $f : R \to S$ be a ring epimorphism. Then B is maximal in $S \Rightarrow f^{-1}(B)$ is maximal in R.

5. Let R and S be commutative rings with identity elements and $f : R \to S$ be a ring epimorphism. Then S is an integral domain $\Leftrightarrow \ker f$ is a prime ideal in R.

6. Let A be a two-sided maximal ideal of a ring R. Then R/A is a simple ring.

7. R is a simple ring with $1 \Rightarrow R$ is a division ring.

8. The rings $\mathbf{Z}_n$ and $\mathbf{Z}/n\mathbf{Z} \, \forall n \in \mathbf{N}^+ (n > 1)$ are isomorphic.

9. $\mathbf{Z}/\langle n \rangle$ is a quotient ring of $\mathbf{Z}/\langle mn \rangle$, $\forall m, n \in \mathbf{N}^+$.

10. Let $R = C([0, 1])$ be the ring of real valued continuous functions on $[0, 1]$ and $A = \{f \in R : f(\frac{1}{13}) = 0\}$. Then A is a maximal ideal in R.

11. There exist only a finite number of proper ideals in an integral domain which is not a field.

12. Any integral domain having only a finite number of proper ideals is a field.

13. $(\mathbf{Z}, +, \cdot)$ is an ideal of $(\mathbf{Q}, +, \cdot)$.

14. A ring R has zero divisors $\Rightarrow$ every quotient ring of R has zero divisors.

15. Let R be a commutative ring with 1. Then R is a simple ring $\Rightarrow R$ is a field.

16. Let R be a commutative ring with 1. An element $a \, (\neq 0) \in R$ is invertible in $R \Leftrightarrow a \notin M$ for any maximal ideal M of R.

Exercises B   Identify the correct alternative(s) (there may be more than one) from the following list:

1. Let $R = (C([0, 1]), +, \cdot)$ be the ring of all real valued continuous functions on $[0, 1]$ and $M = \{f \in M : f(1/2) = 0\}$. Then

   (a) M is not an ideal of R.
   (b) M is an ideal of R but not a maximal ideal or R.
   (c) M is a maximal ideal or R.
   (d) M is an ideal of R but not a prime ideal or R.

2. Let $(\mathbf{Z}, +, \cdot)$ be the ring of integers. Then $\mathbf{Z}$ has

   (a) Only one proper ideal.
   (b) Finitely many proper ideals.
   (c) Countably infinite many proper ideals.
   (d) Uncountably many proper ideals.

3. Let R be a commutative ring with identity and F be a field. If $f : R \to F$ be a (ring) epimorphism, then $K = \ker f = \{x \in R : f(x) = 0_F\}$ is

   (a) A maximal ideal of R but not a prime ideal of R.
   (b) A prime ideal of R but not a maximal ideal of R.
   (c) A both maximal and prime ideal of R.
   (d) Not a maximal ideal of R.

4. Let $(\mathbf{Z}[x], +, \cdot)$ be the ring of polynomials over $\mathbf{Z}$ and $I = \langle x \rangle$. Then

   (a) I is a prime ideal of $\mathbf{Z}[x]$ but not a maximal ideal.
   (b) I is a maximal ideal of $\mathbf{Z}[x]$ but not a prime ideal.
   (c) I is a both maximal and prime ideal of $\mathbf{Z}[x]$.
   (d) I is not a prime ideal of $\mathbf{Z}[x]$.

5. Let R be an arbitrary commutative ring with identity and I be a prime ideal of R such that R/I is finite. Then

   (a) I is a maximal ideal of R.
   (b) $I = R$.
   (c) R/I is an integral domain but not a field.
   (d) R/I is not an integral domain.

6. Let $f : (\mathbf{Z}, +, \cdot) \to (\mathbf{Z}/\langle n \rangle, +, \cdot)$ be the ring homomorphism defined by $f((m)) = m + \langle n \rangle$. Then

   (a) f is a monomorphism but not an epimorphism.
   (b) f is an epimorphism but not a monomorphism.
   (c) f is an isomorphism.
   (d) f is neither a monomorphism nor an epimorphism.

7. Let

$$R = \left\{ \begin{pmatrix} m & n \\ 0 & p \end{pmatrix} : m, n, p \in \mathbf{Z} \right\} \quad \text{and} \quad A = \left\{ \begin{pmatrix} 0 & n \\ 0 & p \end{pmatrix} : n, p \in \mathbf{Z} \right\}.$$

Then the ring R and its subring A are such that

   (a) A is an ideal of R such that the ring $\mathbf{Z}$ is isomorphic to the quotient ring R/A.
   (b) A is not an ideal of R.
   (c) A is an ideal of R such that the ring $\mathbf{Z}$ is not isomorphic to the quotient ring R/A.
   (d) The quotient ring R/A is a field.

8. Let

$$M = \left\{ \begin{pmatrix} a & b \\ 0 & c \end{pmatrix} : a, b, c \in \mathbf{Z} \right\}$$

be the ring of upper triangular matrices of order two over $\mathbf{Z}$ and

$$A = \left\{ \begin{pmatrix} 0 & b \\ 0 & c \end{pmatrix} : b, c \in \mathbf{Z} \right\}.$$

Then

   (a) A is not an ideal of M.

(b) A is an ideal of M such

$$M/A = \left\{ \begin{pmatrix} a & 0 \\ 0 & 0 \end{pmatrix} + A : a \in \mathbf{Z} \right\}.$$

(c) A is an ideal of M such M/A is isomorphic to $\mathbf{Z}$.
(d) A is an ideal of M such M/A is not an integral domain.

9. Let $\mathbf{Z}[x]$ be the polynomial ring over $\mathbf{Z}$. Then

(a) $\mathbf{Z}$ is a subring of $\mathbf{Z}[x]$ but not an ideal of $\mathbf{Z}[x]$.
(b) $\mathbf{Z}$ is a left ideal of $\mathbf{Z}[x]$ but not a right ideal of $\mathbf{Z}[x]$.
(c) $\mathbf{Z}$ is a right ideal of $\mathbf{Z}[x]$ but not a left ideal of $\mathbf{Z}[x]$.
(d) $\mathbf{Z}$ is an ideal of $\mathbf{Z}[x]$.

10. Let $\langle I(x) \rangle$ be the ideal generated by the polynomial $I(x)$ in $\mathbf{Q}[x]$ and $f(x) = x^3 + x^2 + x + 1$ and $g(x) = x^3 - x^2 + x - 1$ be two polynomials in $\mathbf{Q}[x]$. Then

(a) $\langle f(x) \rangle + \langle g(x) \rangle = \langle x^3 + x \rangle$.
(b) $\langle f(x) \rangle + \langle g(x) \rangle = \langle x^4 - 1 \rangle$.
(c) $\langle f(x) \rangle + \langle g(x) \rangle = \langle x^2 + 1 \rangle$.
(d) $\langle f(x) \rangle + \langle g(x) \rangle = \langle f(x).g(x) \rangle$.

11. Which of the following statement(s) is (are) true?

(a) The set of all 2×2 matrices with rational entries (with the usual operations of matrix addition and matrix multiplication) is a ring which has no non-trivial ideals.
(b) Let $R = C([0, 1])$ be considered as a ring with the usual operations of pointwise addition and pointwise multiplication. Let $I = \{f : [0, 1] \to \mathbf{R} \mid f(1/2) = 0\}$. Then I is a maximal ideal.
(c) $\mathbf{R}[x]$ is a field.
(d) Let R be a commutative ring and let P be prime ideal of R. Then R/P is an integral domain.

12. Consider the ring $\mathbf{Z}_n$ for $n \geq 2$. If

(a) $\mathbf{Z}_n$ is a field, then n is a composite integer.
(b) $\mathbf{Z}_n$ is a field iff n is a prime integer.
(c) $\mathbf{Z}_n$ is an integral domain, then n is prime integer.
(d) There is an injective ring homomorphism of $\mathbf{Z}_5$ to $\mathbf{Z}_n$, then n is prime.

13. Let $C([0, 1])$ be the ring of continuous real-valued functions on $[0, 1]$, with addition and multiplication defined pointwise. For any subset S of $C([0, 1])$ let $K(S) = \{f \in C([0, 1]) \mid f(x) = 0 \text{ for all } x \in S\}$. If

(a) $K(S)$ is an ideal in $C([0, 1])$, then S is closed in $[0, 1]$.
(b) S has only one point, then $K(S)$ is a prime ideal but not a maximal ideal.
(c) $K(S)$ is a maximal ideal, then S has only one point.
(d) S has only one point, then $K(S)$ is a maximal ideal.

14. Let $C([0, 1])$ denote the ring of all continuous real-valued functions on $[0, 1]$ with respect to pointwise addition and pointwise multiplication. Then

(a) $C([0, 1])$ contains divisors of zero.
(b) For an element $a \in [0, 1]$, the set $K = \{f \in C([0, 1]) | f(a) = 0\}$ is an ideal in $C([0, 1])$.
(c) $C([0, 1])$ is an integral domain.
(d) For any proper ideal M in $C([0, 1])$, there exists at least one point $a \in [0, 1]$ such that $f(a) = 0$ for all $f \in M$.

15. Let R be a (commutative) ring (with identity). Let A and B be ideals in R. Then

(a) $A \cup B$ is an ideal in R.
(b) $A \cap B$ is an ideal in R.
(c) $AB = \{xy : x \in A, y \in B\}$ is an ideal in R.
(d) $A + B = \{x + y : x \in A, y \in B\}$ is an ideal in R.

5.9 Additional Reading

We refer the reader to the books (Adhikari and Adhikari 2003, 2004; Artin 1991; Atiya and Macdonald 1969; Birkoff and Mac Lane 1965; Burtan 1968; Chatterjee et al. 2003; Dugundji 1989; Fraleigh 1982; Fulton 1969; Herstein 1964; Hungerford 1974; Jacobson 1974, 1980; Lang 1965; McCoy 1964; Simmons 1963; van der Waerden 1970; Zariski and Samuel 1958, 1960) for further details.

References

Adhikari, M.R., Adhikari, A.: Groups, Rings and Modules with Applications, 2nd edn. Universities Press, Hyderabad (2003)
Adhikari, M.R., Adhikari, A.: Text Book of Linear Algebra: An Introduction to Modern Algebra. Allied Publishers, New Delhi (2004)
Artin, M.: Algebra. Prentice-Hall, Englewood Cliffs (1991)
Atiya, M.F., Macdonald, I.G.: Introduction to Commutative Algebra. Addison-Wesley, Reading (1969)
Birkoff, G., Mac Lane, S.: A survey of Modern Algebra. Macmillan, New York (1965)
Burtan, D.M.: A First Course in Rings and Ideals. Addison-Wesley, Reading (1968)
Chatterjee, B.C., Ganguly, S., Adhikari, M.R.: A Text Book of Topology. Asian Books, New Delhi (2003)
Dugundji, J.: Topology. Brown, Dufauque (1989)
Fraleigh, J.B.: A First Course in Abstract Algebra. Addison-Wesley, Reading (1982)
Fulton, W.: Algebraic Curves. Benjamin, New York (1969)
Herstein, I.: Topics in Algebra. Blaisdell, New York (1964)
Hungerford, T.W.: Algebra. Springer, New York (1974)
Jacobson, N.: Basic Algebra I. Freeman, San Francisco (1974)
Jacobson, N.: Basic Algebra II. Freeman, San Francisco (1980)

Lang, S.: Algebra, 2nd edn. Addison-Wesley, Reading (1965)
McCoy, N.: Theory of Rings. Macmillan, New York (1964)
Simmons, G.F.: Introduction to Topology and Modern Analysis. McGraw-Hill, New York (1963)
Smith, K.E., Kahanpaaää, L., Kekäläinen, P., Traves, W.: An Invitation to Algebraic Geometry. Springer-Verlag, New York (2000)
van der Waerden, B.L.: Modern Algebra. Ungar, New York (1970)
Zariski, O., Samuel, P.: Commutative Algebra I. Van Nostrand, Princeton (1958)
Zariski, O., Samuel, P.: Commutative Algebra II. Van Nostrand, Princeton (1960)

Chapter 6
Factorization in Integral Domains and in Polynomial Rings

In Chap. 1 we have already discussed divisibility and factorization of integers. It is natural to ask whether an abstract ring can have such factorization. Relatively few such rings exist. In this chapter we are able to extend the concepts of divisibility, gcd, lcm, division algorithm, and Fundamental Theorem of Arithmetic for integers to different classes of integral domains. The main aim of this chapter is to study the problem of factoring the elements of an integral domain as products of irreducible elements. We also generalize the concept of division algorithm for integers to arbitrary integral domains by introducing the concept of Euclidean domains. We also study the polynomial rings over a certain class of important rings and prove Eisenstein's irreducibility criterion, the Gauss Lemma and related topics. Our study culminates in proving the Gauss Theorem, which provides an extensive class of uniquely factorizable domains.

6.1 Divisibility

Ideals of integral domains play an important role in the study of factorization theory.

> *Throughout this chapter R denotes an integral domain unless otherwise stated and R' denotes $R \setminus \{0\}$.*

The concept of divisibility has been developed with an eye on generalizing a similar concept already discussed for integers. This concept introduces the notions of divisors, prime and irreducible elements, associated elements, lcm and gcd properties, and unique factorization domains (UFD). They are studied in this section with the help of the theory of ideals. For example, prime and irreducible elements in an integral domain are characterized by prime and maximal ideals, respectively. Moreover, the Fundamental Theorem of Arithmetic is generalized for a PID (Theorems 6.1.5 and 6.1.6).

Definition 6.1.1 Let R be an integral domain and let $R' = R \setminus \{0\}$. A non-zero element $a \in R$ is said to divide an element $b \in R$ (denoted by $a|b$) in R iff there

M.R. Adhikari, A. Adhikari, *Basic Modern Algebra with Applications*, DOI 10.1007/978-81-322-1599-8_6, © Springer India 2014

exists an element $c \in R$ such that $b = ac$. The elements a and $b \in R'$ are said to be associates (or associated elements) iff $a|b$ and $b|a$. If $a|b$, we sometimes say that b is divisible by a in R or a is a divisor of b in R.

Example 6.1.1 In $(\mathbf{Z}, +, \cdot)$, $3|-3$ and $-3|3 \Rightarrow 3$ and -3 are associates.

Theorem 6.1.1 *Let R be an integral domain and $a, b, u \in R'$. Then*

　(i) $a|a$;
　(ii) $a|b$ and $b|c \Rightarrow a|c$ $(c \in R)$;
　(iii) $a|b$ and $a|c \Rightarrow a|(bx + cy)$ *for every* $x, y \in R$;
　(iv) u *is an invertible element in* $R \Rightarrow u|d$ *for every* $d \in R$;
　(v) u *is an invertible element in* $R \Leftrightarrow u|1$;
　(vi) $a|c$ $(c \in R) \Leftrightarrow \langle c \rangle \subseteq \langle a \rangle$ *(i.e., $Rc \subseteq Ra$)*;
　(vii) a *and b are associates* $\Leftrightarrow a = bv$ *for some invertible element $v \in R$.*

Proof (i) $a = a1 \Rightarrow a|a$ (since $1 \in R$).

　(ii) $a|b$ and $b|c \Rightarrow \exists x, y \in R$ such that $b = ax$ and $c = by \Rightarrow c = axy \Rightarrow a|c$ (since $xy \in R$).

　(iii) Trivial.

　(iv) $d = d1 = d(u^{-1}u)$ (as u is invertible in R) $= u(du^{-1}) = uc$ (where $c = du^{-1} \in R$) $\Rightarrow u|d$ for every $d \in R$.

　(v) $1 = uu^{-1} = uc$ (where $c = u^{-1} \in R$) $\Rightarrow u|1$. Again $u|1 \Rightarrow 1 = uc$ for some $c \in R \Rightarrow$ inverse of u is $c \in R$. So, u is invertible in R.

　(vi) $a|c \Rightarrow c = ra$ (for some $r \in R$) $\in Ra \Rightarrow Rc \subseteq Ra \Rightarrow \langle c \rangle \subseteq \langle a \rangle$. Again, $\langle c \rangle \subseteq \langle a \rangle \Rightarrow c \in \langle a \rangle \Rightarrow c \in Ra \Rightarrow c = ra$ for some $r \in R \Rightarrow a|c$.

　(vii) a and b are associates $\Rightarrow a|b$ and $b|a \Rightarrow b = ac$ and $a = bd$ for some $c, d \in R \Rightarrow a = acd \Rightarrow 1 = cd$ (by the cancellation law in R) $\Rightarrow c$ and d are both invertible $\Rightarrow a = bv$, where $v = d$. Again $a = bv$ (where v is invertible in R) $\Rightarrow b|a$. Again v is invertible in $R \Rightarrow v^{-1} \in R \Rightarrow av^{-1} = bvv^{-1} \Rightarrow b = av^{-1} \Rightarrow a|b$. □

Remark 1 a and b are associates in $R \Leftrightarrow \langle a \rangle = \langle b \rangle$.

Remark 2 u is a unit in $R \Leftrightarrow \langle u \rangle = R$.

Remark 3 For arbitrary commutative ring with 1, the above results (i)–(vi) are also valid but the result (vii) is not so.

Problem 1 Find the invertible elements of the Gaussian ring $\mathbf{Z}[i]$.

Solution Clearly, $\mathbf{Z}[i]$ is a commutative ring with identity having no zero divisor. So, $\mathbf{Z}[i]$ is an integral domain. Again $u = a + bi$ is an invertible element of $\mathbf{Z}[i] \Rightarrow (a+bi)|1 \Rightarrow 1 = (a+bi)(c+di)$ for some $c+di \in \mathbf{Z}[i] \Rightarrow 1 = (a-bi)(c-di) \Rightarrow 1 = (a^2 + b^2)(c^2 + d^2) \Rightarrow a^2 + b^2 = 1 \Rightarrow$ either $a \Rightarrow \pm 1$ and $b = 0$ or $a = 0$ and $b = \pm 1 \Rightarrow 1, -1, i$, and $-i$ are the only invertible elements of $\mathbf{Z}[i]$.

Definition 6.1.2 Let R be a commutative ring with 1. An element $q \in R$ is said to be *irreducible* in R iff

(i) q is a non-zero non-unit;
(ii) whenever $q = bc$ for $b, c \in R$, then b or c is a unit.

An element $p \in R$ is said to be *prime* in R iff

(i) p is a non-zero non-unit;
(ii) whenever $p|ab$ for $a, b \in R$, then $p|a$ or $p|b$.

Example 6.1.2 (i) In the integral domain $(\mathbf{Z}, +, \cdot)$, 6 is not an irreducible element, but 7 is an irreducible element. If p is a prime integer, then both p and $-p$ are irreducible elements and prime (according to Definition 6.1.2).

(ii) In $(\mathbf{Z}_6, +, \cdot)$ (which is not an integral domain), (2) is prime but not irreducible, since $(2) = (2) \cdot (4)$ and neither (2) nor (4) are units in $\mathbf{Z}_6$.

In a principal ideal domain R, prime elements and irreducible elements are characterized by the corresponding prime ideals and maximal ideals of R, respectively.

Theorem 6.1.2 *Let p be a non-zero non-unit element in an integral domain R. Then*

(i) *p is prime iff $\langle p \rangle$ is a non-zero prime ideal of R;*
(ii) *p is irreducible iff $\langle p \rangle$ is maximal in the set S of all proper principal ideals of R;*
(iii) *every prime element of R is irreducible;*
(iv) *if R is a PID, then p is prime iff p is irreducible;*
(v) *every associate of an irreducible (prime) element of R is irreducible (prime);*
(vi) *the only divisors of an irreducible element of R are its associates and the units of R.*

Proof (i) As p is a non-zero non-unit element of R, $\langle p \rangle$ is non-zero and not R. Let p be prime in R. Then for $b, c \in R$, if $bc \in \langle p \rangle$, then $\exists$ an element $r \in R$ such that $bc = rp \Rightarrow p|bc \Rightarrow p|b$ or $p|c \Rightarrow b \in \langle p \rangle$ or $c \in \langle p \rangle \Rightarrow \langle p \rangle$ is a prime ideal of R, such that $\langle p \rangle$ is non-zero.

Conversely, let $\langle p \rangle$ be a non-zero prime ideal of R. Then $p|bc$ (where $b, c \in R$) $\Rightarrow bc \in \langle p \rangle \Rightarrow b \in \langle p \rangle$ or $c \in \langle p \rangle \Rightarrow p|b$ or $p|c \Rightarrow p$ is prime in R.

(ii) Let p be irreducible in R and $\langle a \rangle$ be a principal ideal of R such that $\langle p \rangle \subsetneq \langle a \rangle \subseteq R$. Now $p \in \langle a \rangle \Rightarrow p = ra$ for some $r \in R \Rightarrow$ either r or a is invertible, since p is irreducible. Suppose r is invertible, then $p = ra \Rightarrow a = r^{-1}p \in \langle p \rangle \Rightarrow \langle a \rangle \subseteq \langle p \rangle \Rightarrow$ a contradiction $\Rightarrow a$ is an invertible $\Rightarrow \langle a \rangle = R \Rightarrow \langle p \rangle$ is a maximal principal ideal of R. Conversely, let $\langle p \rangle$ be a maximal principal ideal of R. If p is not irreducible, then $p = bc$ for some $b, c \in R$, where neither b nor c is invertible (the possibility that p is a unit $\Rightarrow \langle p \rangle = R$ is not tenable). Now $b \in \langle p \rangle \Rightarrow b = rp$ for some $r \in R \Rightarrow p = bc = (rp)c = (rc)p \Rightarrow 1 = rc \Rightarrow c$ is invertible in $R \Rightarrow$

a contradiction $\Rightarrow b \notin \langle p \rangle \Rightarrow \langle p \rangle \subsetneqq \langle b \rangle$. Again $\langle b \rangle = R \Rightarrow b$ is a unit $\Rightarrow$ a contradiction $\Rightarrow \langle b \rangle \neq R$. Thus $\langle p \rangle \subsetneqq \langle b \rangle \subsetneqq R \Rightarrow$ a contradiction as $\langle p \rangle$ is a maximal principal ideal of R. So, p must be an irreducible element of R.

(iii) Let p be a prime element of R. Then $p = ab$ for some $a, b \in R \Rightarrow p|a$ or $p|b$. Suppose $p|a$. Then $px = a$ (for some $x \in R$) $\Rightarrow p = p(xb) \Rightarrow 1 = xb \Rightarrow b$ is a unit $\Rightarrow p$ is irreducible.

(iv) $\Rightarrow$ (implication part) follows from (iii).

To prove $\Leftarrow$ part, suppose p is an irreducible element such $p|ab$ $(a, b \in R)$. Then $ab = pc$ for some $c \in R$. As R is a PID, $\langle p, a \rangle = \langle d \rangle$ for some $d \in R \Rightarrow p = rd$ for some $r \in R \Rightarrow$ either r or d is an invertible element. Now d is invertible $\Rightarrow$ $\langle p, a \rangle = R \Rightarrow 1 = tp + ua$ for some $t, u \in R \Rightarrow b = b1 = btp + bua = btp + puc = p(tb + uc) \Rightarrow p|b$. For the other possibility, r is invertible in $R \Rightarrow d = r^{-1}p \in \langle p \rangle \Rightarrow \langle d \rangle \subseteq \langle p \rangle \Rightarrow a \in \langle p \rangle \Rightarrow p|a$. Thus $p|ab \Rightarrow p|a$ or $p|b \Rightarrow p$ is prime.

(v) If p is irreducible and d is an associate of p, then $p = du$, where u is an invertible element in R (by Theorem 6.1.1 (vii)). Now $d = ab \Rightarrow p = abu \Rightarrow a$ is a unit or bu is a unit. But bu is a unit $\Rightarrow b$ is a unit. Consequently, d is irreducible.

(vi) p is irreducible and $a|p \Rightarrow \langle p \rangle \subseteq \langle a \rangle \Rightarrow \langle p \rangle = \langle a \rangle$ or $\langle a \rangle = R$ by (ii) $\Rightarrow a$ is either an associate of p or a unit. $\square$

Remark The converse of (iii) is not true in an arbitrary integral domain (see Ex. 15 of Exercises-I).

We now introduce the concept of a *greatest common divisor* (gcd).

Definition 6.1.3 Let $a_1, a_2, \ldots, a_n$ be elements (not all zero) of an integral domain R. An element $d \in R$ is said to be a greatest common divisor of $a_1, a_2, \ldots, a_n$ iff

(i) $d|a_i$ for $i = 1, 2, \ldots, n$ (d is a common divisor);
(ii) $c|a_i$ (for $i = 1, 2, \ldots, n$) $\Rightarrow c|d$.

Suppose d and $x \in R$ satisfy conditions (i) and (ii). Then $d|x$ and $x|d \Rightarrow d$ and x are associates $\Rightarrow d$ is unique (if it exists) up to arbitrary invertible factors. So we write any greatest common divisor of $a_1, a_2, \ldots, a_n$ by $\gcd(a_1, a_2, \ldots, a_n)$.

Remark Given $a_1, a_2, \ldots, a_n \in R'$, their gcd may not exist. For example, in $R = \mathbf{Z}[\sqrt{5}i]$, the elements $2(1 + \sqrt{5}i)$ and 6 have no gcd.

Theorem 6.1.3 *Let $a_1, a_2, \ldots, a_n$ be non-zero elements of an integral domain R. Then $a_1, a_2, \ldots, a_n$ have a greatest common divisor d, expressible in the form:*

$$d = r_1 a_1 + r_2 a_2 + \cdots + r_n a_n (r_i \in R) \quad \text{iff } \langle a_1, a_2, \ldots, a_n \rangle \text{ is a principal ideal.}$$

Proof Suppose $d = \gcd(a_1, a_2, \ldots, a_n)$ exists and is written in the form

$$d = \sum_{i=1}^{n} r_i a_i \quad (r_i \in R).$$

Then $d \in \langle a_1, a_2, \ldots, a_n \rangle = A$ (say) $\Rightarrow \langle d \rangle \subseteq A$. Now $d|a_i$ (for each i) $\Rightarrow a_i = x_i d$ ($x_i \in R$) $\Rightarrow$ each element of $A \in \langle d \rangle \Rightarrow A \subseteq \langle d \rangle$. Consequently, $A = \langle d \rangle$.

Conversely, let $\langle a_1, a_2, \ldots, a_n \rangle = A$ (say) be a principal ideal of R. Then $A = \langle d \rangle$ for some $d \in R$. So each $a_i \in \langle d \rangle \Rightarrow d|a_i$ for each i. We now show that for any common divisor c of a_i's, $c|d$. Now $a_i = t_i c$, for suitable $t_i \in R$ and $d = \sum_{i=1}^{n} r_i a_i$, $r_i \in R \Rightarrow d = \sum_{i=1}^{n} (r_i t_i) c \Rightarrow c|d \Rightarrow d$ is a gcd($a_1, a_2, \ldots, a_n$), expressed in the desired form. $\qquad \square$

Corollary *Any finite set of non-zero elements $a_1, a_2, \ldots, a_n$ of a principal ideal domain R, has a greatest common divisor in the form:* gcd($a_1, a_2, \ldots, a_n$) $= \sum_{i=1}^{n} r_i a_i$ *for suitable choices of $r_i \in R$.*

Definition 6.1.4 If gcd($a_1, a_2, \ldots, a_n$) $= 1$ ($a_i \in R$), we say that $a_1, a_2, \ldots, a_n$ are relatively prime.

Corollary (Bezout's Identity) *If $a_1, a_2, \ldots, a_n$ are non-zero elements of a principal ideal domain R, then $a_1, a_2, \ldots, a_n$ are relatively prime iff $\sum_{i=1}^{n} r_i a_i = 1$, for some $r_i \in R$.*

We now study the concept of a *least common multiple* (lcm) which is the dual to that of gcd.

Definition 6.1.5 Let $a_1, a_2, \ldots, a_n$ be non-zero elements of an integral domain R. An element $d \in R$ is said to be a *least common multiple* (lcm) of $a_1, a_2, \ldots, a_n$ iff

(i) $a_i|d$ for $i = 1, 2, \ldots, n$ (d is a common multiple);
(ii) $a_i|c$ (for $i = 1, 2, \ldots, n$) implies $d|c$.

Remark An lcm (if it exists) is unique up to associates but lcm may not exist. For example, lcm of $2(i + \sqrt{5}i)$ and 6 does not exist in $\mathbf{Z}[\sqrt{5}i]$.

We write lcm($a_1, a_2, \ldots, a_n$) to denote their any lcm (if it exists).

Theorem 6.1.4 *Let $a_1, a_2, \ldots, a_n$ be non-zero elements of an integral domain R. Then $a_1, a_2, \ldots, a_n$ have a least common multiple iff the ideal $\bigcap \langle a_i \rangle$ is a principal ideal in R.*

Proof Suppose $d = $ lcm($a_1, a_2, \ldots, a_n$) exists. Then $d \in \langle a_i \rangle$ for each $i \Rightarrow \langle d \rangle \subseteq \bigcap \langle a_i \rangle$. Again $r \in \bigcap \langle a_i \rangle \Rightarrow a_i|r$ for each $i \Rightarrow d|r \Rightarrow r \in \langle d \rangle \Rightarrow \bigcap \langle a_i \rangle \subseteq \langle d \rangle$. So, $\bigcap \langle a_i \rangle = \langle d \rangle$.

Conversely, $\bigcap \langle a_i \rangle$ is a principal ideal of $R \Rightarrow \bigcap \langle a_i \rangle = \langle d \rangle$ (for some $d \in R$) $\Rightarrow \langle d \rangle \subseteq \langle a_i \rangle$ for each $i \Rightarrow a_i|d$ for each i. Again for any other common multiple c of $a_1, a_2, \ldots, a_n$ in R, $a_i|c$ for each $i \Rightarrow \langle c \rangle \subseteq \langle a_i \rangle$ for each $i \Rightarrow \langle c \rangle \subseteq \bigcap \langle a_i \rangle = \langle d \rangle \Rightarrow d|c \Rightarrow d = $ lcm($a_1, a_2, \ldots, a_n$). $\qquad \square$

Definition 6.1.6 A ring R is said to have the gcd property (lcm property), iff any finite number of non-zero elements of R admits a gcd (lcm).

Remark Every principal ideal domain satisfies both gcd property and lcm property. This follows from Theorems 6.1.3 and 6.1.4.

Definition 6.1.7 Let R be an integral domain and a be a non-zero non-invertible element of R. Then a is said to have a factorization in R iff a can be expressed as a finite product of irreducible elements of R. If every non-zero non-invertible element of R has a factorization in R then R is said to be a *factorization domain*.

Definition 6.1.8 An integral domain R is a *unique factorization domain* iff the following two conditions hold:

(i) Every non-zero non-invertible element of R can be factored into a finite product of irreducible elements;

(ii) if $a = p_1 p_2 \cdots p_n$ and $a = q_1 q_2 \cdots q_m$ (p_i, q_i are irreducible), then $n = m$ and for some permutation σ of $\{1, 2, \ldots, n\}$; p_i and $q_{\sigma(i)}$ are associates for each i, in R.

Example 6.1.3 Consider $(\mathbf{Z}, +, \cdot)$. By Fundamental Theorem of Arithmetic (Theorem 1.4.8 of Chap. 1), every non-zero non-unit element in $\mathbf{Z}$ is a product of a finite number of irreducible elements (prime integers or their negatives) and this factorization is unique (except for the order of the irreducible factors). Consequently, $(\mathbf{Z}, +, \cdot)$ is a unique factorization domain.

Theorem 6.1.5 *Let R be a PID. Then every non-zero non-invertible element has a factorization into a finite product of primes.*

Proof Let a be a non-zero non-invertible element of R. Then $\exists$ a prime $p_1 \in R$ such that $p_1 | a$ (see Ex. 4 of Exercises-I) $\Rightarrow a = p_1 a_1$ for some non-zero $a_1 \in R \Rightarrow \langle a \rangle \subseteq \langle a_1 \rangle$. Suppose $\langle a \rangle = \langle a_1 \rangle$. Now $\langle a \rangle = \langle a_1 \rangle \Rightarrow a_1 = ra$ for some $r \in R \Rightarrow a = p_1 a_1 = p_1 ra \Rightarrow 1 = p_1 r \Rightarrow p_1$ is invertible $\Rightarrow$ a contradiction $\Rightarrow \langle a \rangle \neq \langle a_1 \rangle \Rightarrow \langle a \rangle \subsetneq \langle a_1 \rangle$. Thus we obtain an increasing chain of principal ideals:

$$\langle a \rangle \subsetneq \langle a_1 \rangle \subsetneq \langle a_2 \rangle \subsetneq \cdots \subsetneq \langle a_n \rangle \subsetneq \cdots \tag{6.1}$$

with $a_{n-1} = p_n a_n$ for some prime $p_n \in R$. This process is continued as long as a_n is not an invertible element of R. But this chain terminates (see Ex. 2 of Exercises-I) and for some n, a_n must have an inverse and then $\langle a_n \rangle = R$. Thus (6.1) gives $a = p_1 p_2 \cdots p_{n-1} p'_n$, where $p'_n = p_n a_n$ is prime (as it is an associate of a prime). $\qquad \square$

Theorem 6.1.6 *Every principal ideal domain R is a unique factorization domain.*

Proof Let a be a non-zero non-invertible element of R. Then a has a prime factorization by Theorem 6.1.5. To prove the unique factorization, suppose $a = p_1 p_2 \cdots p_n = q_1 q_2 \cdots q_m$ ($n \neq m$), where p_i and q_i are primes in R. Then $p_i |$ ($q_1 q_2 \cdots q_m$). Hence p_i divides some q_j ($j = 1, 2, \ldots, m$). Without loss of generality, suppose $p_1 | q_1$ and $n < m$. Then $q_1 = p_1 u_1$ for some unit $u_1 \in R$. Canceling p_1, we find that $p_2 \cdots p_n = u_1 q_2 \cdots q_m$. Continuing this process we find (after n steps) that

$$1 = u_1 u_2 \cdots u_n q_{n+1} \cdots q_m.$$

As q_i are not invertible, we get a contradiction, resulting in $m = n$. Thus every p_i has some q_j as an associate and conversely. Hence the two prime factorizations are identical (up to order and associates of factors).   □

Remark The converse of Theorem 6.1.6 is not true. Consider $\mathbf{Z}[x]$. It is a uniquely factorizable domain but not a principal ideal domain (see Ex. 3 of SE-I).

6.2 Euclidean Domains

We now generalize Division Algorithm for integers to arbitrary integral domains by defining Euclidean domains. The ring of integers, Gaussian rings and polynomial rings are the main sources for the study of Euclidean domains with the help of norm functions.

Definition 6.2.1 An integral domain R is said to be a *Euclidean domain* iff there exists a function $\delta : R \setminus \{0\} \to \mathbf{N}$ (non-negative integers) such that

(i) $\delta(ab) \geq \delta(a)$ (and also $\geq \delta(b)$) for any $a, b \in R$ with $a \neq 0$ and $b \neq 0$;
(ii) *(Division algorithm)* for any $a, b \in R$ with $b \neq 0$, there exist $q, r \in R$ (quotient and remainder) such that $a = qb + r$, where either $r = 0$ or $\delta(r) < \delta(b)$.
     δ is called the *Euclidean valuation* or *Euclidean norm function* or *simply norm function*. We also call the pair (R, δ) a *Euclidean domain*.

We do not assign a value to $\delta(0)$.

Example 6.2.1 (i) Every field F is a Euclidean domain with Euclidean norm function defined by

$$\delta(a) = 1 \quad \forall a \, (\neq 0) \in F$$

and division algorithm is defined by $a = (ab^{-1})b + 0 \, \forall a \in F$ and $\forall b \, (\neq 0) \in F$.

(ii) $(\mathbf{Z}, +, \cdot)$ is a Euclidean domain with Euclidean norm function δ defined by $\delta(a) = |a|, a \, (\neq 0) \in \mathbf{Z}$. Now $\delta(ab) = |ab| = |a||b| = \delta(a)\delta(b)$.

Let $a, b \in \mathbf{Z}$ and $b \neq 0$. Then $\exists$ integers q and r (by Theorem 1.4.3 of Chap. 1) such that $a = bq + r$, where either $r = 0$ or $\delta(r) < \delta(b)$. So division algorithm holds.

(iii) The ring $D = \mathbf{Z}[i]$ of Gaussian integers is a Euclidean domain with Euclidean norm function δ defined by $\delta(m + in) = m^2 + n^2 \, \forall m, n$ (not both zero) $\in \mathbf{Z}$. To define division algorithm let $\alpha, \beta \in D$ and $\beta \neq 0$. Then $\alpha/\beta = a + ib$, where a, b are rational numbers. Thus there exist integers m_0, n_0 such that $|m_0 - a| \leq \frac{1}{2}$ and $|n_0 - b| \leq \frac{1}{2}$.

Hence $a + ib = (m_0 + in_0) + \varepsilon_1 + i\varepsilon_2$, where $|\varepsilon_1| \leq \frac{1}{2}$ and $|\varepsilon_2| \leq \frac{1}{2} \Rightarrow \alpha = (m_0 + in_0)\beta + (\varepsilon_1 + i\varepsilon_2)\beta$. Now $\alpha, \beta \in D$ and $m_0 + in_0 \in D \Rightarrow \alpha - (m_0 + in_0)\beta \in D \Rightarrow (\varepsilon_1 + i\varepsilon_2)\beta \in D \Rightarrow \alpha = q\beta + r$, where $r = (\varepsilon_1 + i\varepsilon_2)\beta$ and $q = (m_0 + in_0)$. Now $\delta(r) = \delta((\varepsilon_1 + i\varepsilon_2)\beta) = |\varepsilon_1^2 + \varepsilon_2^2||\beta|^2 \leq (\frac{1}{4} + \frac{1}{4})|\beta|^2 = \frac{1}{2}\delta(\beta) < \delta(\beta) \Rightarrow \delta : D \to \mathbf{N}$ is a Euclidean function.

Problem 2 Let (R, δ) be a Euclidean domain. Prove that

(i) for each non-zero $a \in R$, $\delta(a) \geq \delta(1)$;
(ii) if two non-zero elements $a, b \in R$ are associates, then $\delta(a) = \delta(b)$;
(iii) an element a ($\neq 0$) $\in R$ is invertible iff $\delta(a) = 1$.

Solution (i) $a = a1 \Rightarrow \delta(a) \geq \delta(1)$;

(ii) a and b are associates $\Rightarrow a = bu$, for some invertible element $u \in R \Rightarrow b = au^{-1} \Rightarrow \delta(a) = \delta(bu) \geq \delta(b)$ and $\delta(b) = \delta(au^{-1}) \geq \delta(a) \Rightarrow \delta(a) = \delta(b)$;

(iii) a ($\neq 0$) has an inverse in $R \Rightarrow ab = 1$, for some $b \in R \Rightarrow \delta(a) \leq \delta(ab) = \delta(1) \leq \delta(a) \Rightarrow \delta(a) = 1$. Conversely, for the element a ($\neq 0$) $\in R$, $\delta(a) = 1 \Rightarrow \exists q, r \in R$ such that $1 = qa + r$, where $r = 0$ or $\delta(r) < \delta(a) \Rightarrow r = 0$ or $\delta(r) < 1 \Rightarrow r = 0$ (since the alternative is not possible) $\Rightarrow 1 = qa \Rightarrow a$ is invertible in R.

Problem 3 Prove that the quotient and remainder in condition (ii) of Definition 6.2.1 are unique iff $\delta(a + b) \leq \max\{\delta(a), \delta(b)\}$ for any $a, b \in R$.

Solution Suppose there exist non-zero elements $a, b \in R$ such that $\delta(a + b) > \max\{\delta(a), \delta(b)\}$. Then $b = 0(a + b) + b = 1(a + b) - a$ with both possibilities $\delta(-a) = \delta(a) < \delta(a + b)$ and $\delta(b) < \delta(a + b)$ shows non uniqueness of quotient and remainder in condition (ii) of Definition 6.2.1. Conversely, if the given condition holds and $a \in R$ and the element $a \in R$ has two representations,

$$\left. \begin{array}{ll} a = qb + r & (r = 0 \text{ or } \delta(r) < \delta(b)), \\ a = q'b + r' & (r' = 0 \text{ or } \delta(r') < \delta(b)). \end{array} \right\} \tag{A}$$

If $r \neq r'$ and $q \neq q'$, then $\delta(b) \leq \delta((q' - q)b) = \delta(r - r') \leq \max\{\delta(r), \delta(-r')\} < \delta(b)$, a contradiction. This shows that $r = r'$ and $q = q'$ by (A), as each of these relations implies the other $\Rightarrow$ uniqueness of the representation.

Remark Division algorithm for $\mathbf{Z}$ holds by defining $\delta(x) = |x|$, for all $x \in \mathbf{Z} \setminus \{0\}$.

Theorem 6.2.1 *Every Euclidean domain is a principal ideal domain.*

Proof Let (R, δ) be a Euclidean domain and A an ideal of R such that $A \neq \{0\}$. Consider the set $\mathbf{S}$ defined by $\mathbf{S} = \{\delta(a) : a \in A; a \neq 0\} \subseteq \mathbf{N}^+$. Then $\mathbf{S} \neq \emptyset$ and $\mathbf{S}$ has a least element by the well-ordering principle. Then $\exists$ an element $b \in A$ such that $\delta(b)$ is least in $\mathbf{S}$. We claim that $A = \langle b \rangle$. For any $a \in A$, $\exists q, r \in R$ such that $a = qb + r$, where $r = 0$ or $\delta(r) < \delta(b) \Rightarrow 0 = a - qb \in A$ (since A is an ideal) or

$\delta(r) < \delta(b) \Rightarrow a = qb \in \langle b \rangle$ (since the other alternative contradicts the minimality of $\delta(b)$) $\Rightarrow A \subseteq \langle b \rangle$. Again $b \in A \Rightarrow \langle b \rangle \subseteq A$. Consequently, $A = \langle b \rangle$. $\qquad\square$

Remark 1 Every ideal of a Euclidean ring R is of the form bR, where $b \in R$.

Remark 2 The converse of Theorem 6.2.1 is not true. For example, $R = \{a + b\sqrt{23} : a, b \in \mathbf{Z}\}$ forms a principal ideal domain under usual addition and multiplication. But R is not a Euclidean domain.

Corollary *Every Euclidean domain is a unique factorization domain.*

Proof It follows from Theorems 6.2.1 and 6.1.6. $\qquad\square$

Theorem 6.2.2 *If F is a field, then $F[x]$ is a Euclidean domain.*

Proof If F is a field, then $F[x]$ is an integral domain by Proposition 4.5.1. For any $f(x) \in F[x]$, define $\delta : F[x] \setminus \{0\} \to \mathbf{N}$ (non-negative integers) by

$$\delta(f) = \text{degree of } f.$$

Then $\delta(f)$ is a non-negative integer such that $\deg(fg) = \deg f + \deg g$ by Proposition 4.5.2 and hence

$$\delta(fg) \geq \delta(f) \quad \left(\text{and also } \geq \delta(g)\right), \quad \forall f(x), g(x) \in F[x] \setminus \{0\}.$$

To verify the division algorithm, let $f(x), g(x) \in F[x]$ with $g(x) \neq 0$. If $\deg f < \deg g$, then $f = 0g + f$, with $\delta(f) < \deg g$. Next assume that $\deg f \geq \deg g$. We apply induction on $\deg f = n$. If $n = 0$, then $m = \deg g = 0$ and hence there is nothing to prove. We assume that the division algorithm holds $\forall f, g \in F[x]$ with $\deg f < n$ and $\deg f \geq \deg g$. Let

$$f(x) = a_0 + a_1 x + a_2 x^2 + \cdots + a_n x^n, \quad a_n \neq 0 \quad \text{and}$$

$$g(x) = b_0 + b_1 x + b_2 x^2 + \cdots + b_m x^m, \quad b_m \neq 0$$

be polynomials in $F[x]$ such that $\deg f = n \geq m = \deg g$.

Then $h(x) = f(x) - a_n b_m^{-1} x^{n-m} g(x)$ is a polynomial in $F[x]$ with the coefficient of x^n is $a_n - a_n b_m^{-1} b_m = 0$. Hence $\deg h \leq n - 1 \Rightarrow h(x) = q(x)g(x) + r(x)$, with $r = 0$ or $\delta(r) < \delta(g)$ by induction hypothesis $\Rightarrow f(x) = a_n b_m^{-1} x^{n-m} g(x) + q(x)g(x) + r(x) = q_1(x)g(x) + r(x)$, where $q_1(x) = a_n b_m^{-1} x^{n-m} + q(x) \in F[x]$ with $r = 0$ or $\delta(r) < \delta(g)$.

Hence $F[x]$ is a Euclidean domain. $\qquad\square$

Corollary 1 *If F is a field, then $F[x]$ is a principal ideal domain.*

Proof It follows from Theorems 6.2.1 and 6.2.2. $\qquad\square$

Corollary 2 *If F is a field, then F[x] is a unique factorization domain.*

Proof It follows from Theorem 6.2.2 and Corollary of Theorem 6.2.1.   □

6.3 Factorization of Polynomials over a UFD

As the polynomial ring over an integral domain is also an integral domain, we can extend the factorization theory to such polynomial rings. In this section we study polynomials over unique factorization domains (UFD). We first give sufficient conditions for irreducibility of polynomials over a UFD, called the Eisenstein criterion. One of the most important applications of Eisenstein criterion is in the proof of the irreducibility of the cyclotomic polynomial $\phi_p(x)$ in $\mathbf{Q}[x]$ (see Ex. 4, SE-I). The property that $\mathbf{Z}[\mathbf{x}]$ is a UFD is generalized for $R[x]$, where $R[x]$ is an arbitrary UFD (see Theorem 6.3.2). This theorem is essentially due to Gauss and provides an extensive class of UFD's.

We start with the concept of primitive polynomials.

Definition 6.3.1 Let R be a UFD and $f(x) = a_0 + a_1x + \cdots + a_nx^n \in R[x]$. Then the content of $f(x)$, denoted $C(f)$ is defined by $C(f) = \gcd(a_0, a_1, \ldots, a_n)$.

For example, for $f(x) = 4x^3 + 2x^2 + 3x + 7 \in \mathbf{Z}[x], C(f) = 1$.

Definition 6.3.2 Let R be a UFD. Then $f(x)$ is said to be a primitive polynomial in $R[x]$ iff $C(f) = 1$.

Proposition 6.3.1 *Let R be a UFD. The product of two primitive polynomials in R[x] is also a primitive polynomial.*

Proof Left as an exercise.   □

The concepts of irreducibility and units in any commutative ring R with 1 are closely related. In particular, it is necessary to look into the units of a polynomial ring $R[x]$ for the study of irreducibility of polynomials in $R[x]$. The units in $R[x]$ are precisely the non-zero constant polynomials which are units in R. We now study the irreducibility of polynomials over UFD.

Proposition 6.3.2

(i) *If c is an irreducible element of a UFD R, then the constant polynomial c is also irreducible in R[x].*

(ii) *Every one degree polynomial over a UFD R whose leading coefficient is a unit in R is irreducible in R[x]. In particular, every one degree polynomial over a field is irreducible.*

(iii) *Let f(x) be a polynomial of degree 2 or 3 over a field F. Then f(x) is irreducible over F iff f(x) has no root in F.*

Proof Left as an exercise. □

Lemma 6.3.1 (Gauss Lemma) *Let R be a UFD and $F = Q(R)$, the quotient field of R. If $f(x) \in R[x]$ is a non-constant irreducible polynomial in $R[x]$, then $f(x)$ is also irreducible in $F[x]$.*

Proof Let $f(x)$ be a non-constant irreducible polynomial in $R[x]$. If possible, let $f(x)$ be reducible in $F[x]$. Then $\exists u(x), v(x) \in F[x]$ such that $f(x) = u(x)v(x)$, where $0 < \deg u < \deg f$ and $0 < \deg v < \deg f$. We may take $u(x) = (a/b)u_1(x)$ and $v(x) = (c/d)v_1(x)$, where $a, b \ (\neq 0), c, d \ (\neq 0) \in R$ and $u_1(x), v_1(x) \in R[x]$ and are both primitive polynomials. Hence $f(x) = (ac/bd)u_1(x)v_1(x)$, where $u_1(x)v_1(x)$ is a primitive polynomial by Proposition 6.3.1. Now $f(x) \in R[x]$ is irreducible and $C(f)$ divides $f \Rightarrow C(f) = 1 \Rightarrow f$ is primitive. Consequently, $(bd)f(x) = (ac)u_1(x)v_1(x) \Rightarrow bd = ac$ (comparing the content). Hence $f(x) = u_1(x)v_1(x)$ in $R[x]$, where $\deg u_1 = \deg u < \deg f$ and $\deg v_1 = \deg v < \deg f$. But this contradicts the assumption that $f(x)$ is irreducible in $R[x]$. This leads to the conclusion that $f(x)$ is also irreducible in $F[x]$. □

Theorem 6.3.1 (The Eisenstein criterion) *Let R be a UFD with quotient field $Q(R)$ and $f(x) = a_0 + a_1x + \cdots + a_nx^n \in R[x]$, $a_n \neq 0$ i.e., $f(x)$ be a non-constant polynomial in $R[x]$ of degree n. If there exists an irreducible element $p \in R$ such that*

(i) $p|a_i, i = 0, 1, 2, \ldots, n-1$;
(ii) $p \nmid a_n$ *and*
(iii) $p^2 \nmid a_0$,

then $f(x)$ is irreducible in $Q(R)[x]$.

Moreover, if $f(x)$ is a primitive polynomial in $R[x]$, then $f(x)$ is also irreducible in $R[x]$.

Proof First suppose that $f(x)$ is a primitive polynomial in $R[x]$ satisfying the above divisibility conditions with respect to p. If possible, let $f(x)$ be reducible in $R[x]$. Then $\exists g(x), h(x) \in R[x]$ such that $f(x) = g(x)h(x)$, where $\deg g < n$ and $\deg h < n$.

Let

$$g(x) = g_0 + g_1x + \cdots + g_rx^r, \quad \text{where } g_r \neq 0, \text{ and } 0 < r < n$$

and

$$h(x) = h_0 + h_1x + \cdots + h_tx^t, \quad \text{where } h_t \neq 0, \text{ and } 0 < t < n.$$

Hence

$$f(x) = g(x)h(x) \quad \Rightarrow \quad a_i = g_ih_0 + g_{i-1}h_1 + \cdots + g_0h_i.$$

Now $p|a_0 \Rightarrow p|g_0$ or $p|h_0$, but not both, otherwise, $p^2|a_0$, which is a contradiction. Suppose $p|g_0$ but $p \nmid h_0$. Clearly, p does not divide $C(g)$, otherwise all the coefficients of $f(x)$ would be divisible by p, which is not true, since $f(x)$ is a primitive polynomial in $R[x]$. Let g_s be the first of the g_i's such that g_s is not divisible by p, where $s \leq r < n$. Then p divides all the coefficients $g_0, g_1, \ldots, g_{s-1}$. But $a_s = g_s h_0 + g_{s-1} h_1 + \cdots + g_0 h_s$ and $p | a_s$, $p | g_{s-1}, \ldots, p | g_0 \Rightarrow$ either $p | g_s$ or $p | h_0 \Rightarrow$ a contradiction in either case. Hence $f(x)$ is irreducible in $R[x]$.

Next we suppose $f(x) = C(f) f_1(x)$ with $f_1(x)$ primitive in $R[x]$; (in particular $f_1 = f$ if f is primitive in $R[x]$). If $f_1(x) = b_0 + b_1 x + \cdots + b_n x^n$ and $C(f) = d$, then $a_i = d b_i$ for $i = 0, 1, 2, \ldots, n$. Since p does not divide a_n, p cannot divide d. Thus p divides b_i, $i = 0, 1, 2, \ldots, n - 1$, p does not divide b_n and p^2 does not divide b_0. Hence by the above argument, $f_1(x)$ is irreducible in $R[x]$ and thus it is irreducible in $Q(R)[x]$ by Lemma 6.3.1. As d is non-zero element in the field $Q(R)$, d is a unit in $Q(R)[x]$, showing that $f(x)$ is irreducible in $Q(R)[x]$. $\square$

Corollary 1 *If p is a prime integer and n (>1) is an integer, then $x^n - p \in \mathbf{Z}[x]$ is irreducible in $\mathbf{Z}[x]$.*

Remark The Eisenstein criterion gives only a sufficient condition for irreducibility but not necessary (see Ex. 4 of SE-I).

Corollary 2 (The Eisenstein criterion for $\mathbf{Z}[x]$) *Let $f(x) = a_0 + a_1 x + \cdots + a_n x^n \in \mathbf{Z}[x]$, $a_n \neq 0$ be a primitive polynomial in $\mathbf{Z}[x]$. If there is a prime integer p such that*

   (i) $p | a_i$, $i = 0, 1, 2, \ldots, n - 1$;
   (ii) $p \nmid a_n$ *and*
   (iii) $p^2 \nmid a_0$,

then $f(x)$ is irreducible in $\mathbf{Z}[x]$.

Proof It follows from Theorem 6.3.1 by taking $R = \mathbf{Z}$. $\square$

Theorem 6.3.2 (Gauss Theorem) *If R is a UFD, then $R[x]$ is also a UFD.*

Proof Let $f(x) \in R[x]$ and $F = Q(R)$, the quotient field of R. Then $F[x]$ is a UFD by Corollary 2 of Theorem 6.2.2. Suppose $f(x) = a f_1(x)$, where $a = C(f)$ and $f_1 \in R[x]$ is a primitive polynomial. Now $F[x]$ is a UFD $\Rightarrow f_1(x) = g_1(x) g_2(x) \cdots g_n(x)$, where $g_i(x) \in F[x]$ are irreducible. We may take $g_i(x) = (a_i / b_i) h_i(x)$, where a_i, b_i ($\neq 0$) $\in R$ and $h_i(x) \in R[x]$ are primitive.
Hence

$$f_1(x) = (a_1 a_2 \cdots a_n) / (b_1 b_2 \cdots b_n) h_1(x) h_2(x) \cdots h_n(x)$$

$$\Rightarrow \quad (b_1 b_2 \cdots b_n) f_1(x) = (a_1 a_2 \cdots a_n) h_1(x) h_2(x) \cdots h_n(x)$$

$$\Rightarrow \quad f_1(x) = h_1(x) h_2(x) \cdots h_n(x),$$

since $b_1 b_2 \cdots b_n = a_1 a_2 \cdots a_n$ (comparing the contents), where $h_i(x) \in R[x]$ are primitive and irreducible in $F[x]$ and hence irreducible in $R[x]$ also.

Let $C(f) = a = a_1 a_2 \cdots a_t$, where $a_i \in R$ are irreducible. Then $f(x) = a f_1(x) = (a_1 a_2 \cdots a_t) h_1(x) h_2(x) \cdots h_n(x)$ is a factorization of $f(x)$ as a product of irreducible elements of $R[x]$.

Let

$$f(x) = (a_1 a_2 \cdots a_t) h_1(x) h_2(x) \cdots h_n(x)$$
$$= (b_1 b_2 \cdots b_s) k_1(x) k_2(x) \cdots k_m(x)$$

be two factorizations of $f(x)$ as a product of irreducible elements of $R[x]$. Comparing the contents and using the Gaussian Lemma, the uniqueness of the above factorizations follows. $\qquad\square$

Corollary 1 $\mathbf{Z}[x]$ *is a UFD.*

Corollary 2 *If R is a UFD, then $R[x_1, x_2, \ldots x_n]$ is also a UFD.*

Proof It follows from Gauss Theorem 6.3.2 by inductive argument using the result that $R[x_1, x_2, \ldots, x_n] = R[x_1, x_2, \ldots, x_{n-1}][x_n]$. $\qquad\square$

We now present an important application of irreducibility of polynomials over a UFD in construction of finite fields.

Theorem 6.3.3 *Given a prime integer p and a positive integer n, there exists a finite field of order p^n.*

Proof Consider the polynomial ring $\mathbf{Z}_p[x]$ and an irreducible polynomial $f(x)$ of degree n over $\mathbf{Z}_p$. As $\mathbf{Z}_p$ is a field, $\mathbf{Z}_p$ is PID and $\langle f(x) \rangle$ is a maximal ideal of $\mathbf{Z}_p[x]$. Thus, $\mathbf{Z}_p[x]/\langle f(x) \rangle$ is a field. As $\deg f(x) = n$, an element of $\mathbf{Z}_p[x]/\langle f(x) \rangle$ is of the form

$$a_0 + a_1 x + a_2 x^2 + \cdots + a_{n-1} x^{n-1},$$

where $a_i \in \mathbf{Z}_p$, $i = 0, 1, \ldots, n-1$.

Thus, $|\mathbf{Z}_p[x]/\langle f(x) \rangle| = p^n$. $\qquad\square$

6.4 Supplementary Examples (SE-I)

1 If p is a prime integer, then $\sqrt[n]{p}$ is an irrational number for every positive integer $n > 1$.

[*Hint.* $x^n - p \in \mathbf{Z}[x]$ is irreducible in $\mathbf{Z}[x]$ by Corollary 1 of Theorem 6.3.1. Then by the Gauss Lemma it is also irreducible in $\mathbf{Q}[x]$. Hence it has no rational root.]

2 Let $f(x) = a_0 + a_1 x + \cdots + a_n x^n \in \mathbf{Z}[x]$, $a_n \neq 0$. If there is a prime integer p such that

(i) $p|a_i, i = 0, 1, 2, \ldots, n - 1$;

(ii) $p\!\!\!\!/\,a_n$, and

(iii) $p^2\!\!\!\!/\,a_0$,

then $f(x)$ is irreducible in $\mathbf{Q}[x]$.

   [*Hint*. Use Theorem 6.3.1.]

3 $\mathbf{Z}[x]$ is a UFD but not a PID.

   [*Hint*. $\mathbf{Z}[x]$ is a UFD by the Corollary to Theorem 6.3.2. To show that $\mathbf{Z}[x]$ is not a PID, let $I = \{f = a_0 + a_1 x + \cdots + a_n x^n \in \mathbf{Z}[x] : a_0 = \text{even}\}$. Then I is not a principal ideal, otherwise $I = \langle h(x) \rangle$ for some $h(x) \in \mathbf{Z}[x]$. As $2 \in I$, h must divide 2 and hence $h = \pm 2$. But $x + 2 \in I$ and h does not divide $x + 2$. Hence $\mathbf{Z}[x]$ cannot be a PID.]

4 For any prime integer $p > 1$, the *cyclotomic polynomial* $\phi_p(x) = 1 + x + x^2 + \cdots + x^{p-1}$ is irreducible in $\mathbf{Q}[x]$.

   [*Hint*. $\phi_p(x)$ is irreducible in $\mathbf{Z}[x]$ but there does not exist any prime integer satisfying the Eisenstein criterion for irreducibility of $\phi_p(x)$, though it is irreducible. If ϕ_p is not irreducible in $\mathbf{Z}[x]$, then $\exists$ non-trivial polynomials $f(x), g(x) \in \mathbf{Z}[x]$ such that $\phi_p(x) = f(x)g(x)$. Hence $\phi_p(x + 1) = f(x + 1)g(x + 1)$ is a non-trivial factorization of $\phi_p(x + 1)$ in $\mathbf{Z}[x]$. On the other hand,

$$\phi_p(x) = \frac{x^p - 1}{x - 1} \quad \Rightarrow \quad \phi_p(x + 1) = p + \cdots + \binom{p}{r} x^{r-1} + \cdots + px^{p-2} + x^{p-1}$$

is irreducible in $\mathbf{Z}[x]$ by Theorem 6.3.1. Hence $\phi_p(x)$ is irreducible in $\mathbf{Z}[x] \Rightarrow \phi_p(x)$ is also irreducible in $\mathbf{Q}[x]$ by Gauss Lemma.]

5 If R is a PID, then $R[x]$ is not necessarily so. If $R[x]$ is a PID, then R is a field.

   [*Hint*. $\mathbf{Z}$ is a PID but $\mathbf{Z}[x]$ is not so by Ex. 3. Suppose $R[x]$ is a PID. Consider the map $f : R[x] \to R$ defined by $f(\sum_{i=0}^{n} a_i x_i) = a_0$. Then f is a ring epimorphism with ker $f = \langle x \rangle$ (see W.O. Ex. 7 of Chap. 5).

   Hence $R[x]/\langle x \rangle \cong R \Rightarrow \langle x \rangle$ is a prime ideal in $R[x] \Rightarrow \langle x \rangle$ is a maximal ideal in $R[x]$ by Theorem 5.3.5 of Chap. 5. $\Rightarrow R$ is a field by Theorem 5.3.2.]

6 The ideal $\langle x^2 + 1 \rangle$ is a maximal ideal in $\mathbf{R}[x]$.

   [*Hint*. If possible, let $\exists$ an ideal I in $\mathbf{R}[x]$ such that $\langle x^2 + 1 \rangle \subsetneqq I \subseteq \mathbf{R}[x]$. Then $\exists f(x) \in I$ such that $f(x) \notin \langle x^2 + 1 \rangle$. By division algorithm, $f(x) = (x^2 + 1)q(x) + r(x)$, where $r(x) \neq 0$ and $\deg(r(x)) < 2$. Let $r(x) = ax + b$, $a, b \in \mathbf{R}$ and they are not both 0. Hence $ax + b = f(x) - (x^2 + 1)q(x) \in I \Rightarrow (ax + b)(ax - b) \in I \Rightarrow a^2 x^2 - b^2 \in I$. Moreover, $a^2(x^2 + 1) \in I$.

   Hence $a^2 + b^2 \in I \Rightarrow I$ contains a non-zero real number $\Rightarrow I = \mathbf{R}[x]$.]

6.5 Exercises

Exercises-I R denotes an integral domain and $R' = R \setminus \{0\}$ (unless specified otherwise).

1. Show that the binary relation ρ on R defined by $a\rho b \Leftrightarrow a$ is an associate of b, is an equivalence relation with equivalence classes which are sets of associated elements, and the associates of the identity are precisely the invertible elements of R. Find the associates of an integer n in $(\mathbf{Z}, +, \cdot)$.

2. Let R be a PID. If $\{A_n\}$ is an infinite sequence of ideals satisfying:

$$A_1 \subseteq A_2 \subseteq \cdots \subseteq A_n \subseteq A_{n+1} \subseteq \cdots .$$

   Show that there exists an integer m such that $A_n = A_m \; \forall n > m$.
   [*Hint.* Consider $A = \bigcup A_n$. Then A is a principal ideal. Now $A = \langle a \rangle \Rightarrow a \in A_m$ (for some m) $\Rightarrow \forall n > m$, $A = \langle a \rangle \subseteq A_m \subseteq A_n \subseteq A \Rightarrow A_n = A_m$.]

3. Let R be a PID. Show that the non-trivial ideal $\langle p \rangle$ is a maximal (prime) ideal of $R \Leftrightarrow p$ is an irreducible (prime) element of R.
   [*Hint.* Use Theorem 6.1.2(ii) for the first assertion. For the second assertion, let p be a prime element of R and $ab \in \langle p \rangle$. Then $ab = rp$ for some $r \in R \Rightarrow p|ab \Rightarrow p|a$ or $p|b \Rightarrow a \in \langle p \rangle$ or $b \in \langle p \rangle \Rightarrow \langle p \rangle$ is a prime ideal of R. The converse is similar.]

4. Let R be a PID and $a \, (\neq 0)$ be non-unit (i.e., non-invertible) in R. Show that there exists a prime $p \in R$ such that $p|a$.
   [*Hint.* $a \, (\neq 0)$ is a non-unit in $R \Rightarrow \langle a \rangle$ is a proper ideal of R (see Remark 2 after Theorem 6.1.1) $\Rightarrow \exists$ a maximal ideal M of R such that $\langle a \rangle \subseteq M$ (by the Corollary of Theorem 5.3.6 of Chap. 5) $\Rightarrow \langle a \rangle \subseteq M = \langle p \rangle$ for some prime element $p \in R \Rightarrow p|a$.]

5. Let a, b, c be non-zero elements of a principal ideal domain R. If $c|ab$ and $\gcd(a, c) = 1$, then show that $c|b$.
   [*Hint.* $\gcd(a, c) = 1 \Rightarrow 1 = ra + tc$ (for some $r, t \in R$) $\Rightarrow b = 1b = rab + tcb$. Now $c|ab$ and $c|c \Rightarrow c|(rab + tcb) \Rightarrow c|b$].

6. Let $a_1, a_2, \ldots, a_n$ and r be non-zero elements of an integral domain R. Prove that

   (a) if $\mathrm{lcm}(a_1, a_2, \ldots, a_n)$ exists, then $\mathrm{lcm}(ra_1, ra_2, \ldots, ra_n)$ also exists and $\mathrm{lcm}(ra_1, ra_2, \ldots, ra_n) = r \, \mathrm{lcm}(a_1, a_2, \ldots, a_n)$;
   (b) if $\gcd(ra_1, ra_2, \ldots, ra_n)$ exists, then $\gcd(a_1, a_2, \ldots, a_n)$ also exists and $\gcd(ra_1, ra_2, \ldots, ra_n) = r \, \gcd(a_1, a_2, \ldots, a_n)$.

7. Let $a_1, a_2, \ldots, a_n$ and $b_1, b_2, \ldots, b_n$ be non-zero elements of an integral domain R such that $a_1 b_1 = a_2 b_2 = \cdots = a_n b_n = x$; show that

   (a) if $\mathrm{lcm}(a_1, a_2, \ldots, a_n)$ exists, then $\gcd(b_1, b_2, \ldots, b_n)$ also exists and satisfies: $\mathrm{lcm}(a_1, a_2, \ldots, a_n) \gcd(b_1, b_2, \ldots, b_n) = x$;
   (b) if $\gcd(ra_1, ra_2, \ldots, ra_i, \ldots, ra_n)$ exists $\forall r \, (\neq 0) \in R$, then $\mathrm{lcm}(b_1, b_2, \ldots, b_n)$ also exists and satisfies: $\gcd(a_1, a_2, \ldots, a_n) \mathrm{lcm}(b_1, b_2, \ldots, b_n) = x$.

8. Prove that an integral domain R has the gcd property $\Leftrightarrow$ R has the lcm property.

9. In a principal ideal domain R, prove that every non-trivial ideal is the product of a finite number of prime (maximal) ideals.

   [*Hint.* $a \in R'$ is non-invertible $\Rightarrow a = p_1 p_2 \cdots p_n$, where each p_i is a prime element of $R \Rightarrow \langle a \rangle = \langle p_1 p_2 \cdots p_n \rangle = \langle p_1 \rangle \langle p_2 \rangle \cdots \langle p_n \rangle$, where each $\langle p_i \rangle$ is a prime ideal of R.]

10. Prove the *Euclid Theorem*: There are an infinite number of primes in $\mathbf{Z}$.

    [*Hint.* Suppose there are only a finite number of primes, $p_1, p_2, \ldots, p_n$ (say). Then $x = (p_1 p_2 \cdots p_n) + 1$ is an integer > 1 such that x is not divisible by any of these primes. But Theorem 6.1.5 $\Rightarrow x$ must have a prime factor $\Rightarrow x$ is divisible by a prime different from the above primes $\Rightarrow \exists$ an infinite number of primes in $\mathbf{Z}$.]

11. Let F be a field and $F[x]$ the polynomial extension of F and $f, g \in F[x]$ such that $g \neq 0$. Show that there exist unique polynomials $q, r \in F[x]$ such that $f = gq + r$, where either $r = 0$ or $\deg r < \deg g$.

12. Let F be a field. Show that $F[x]$ is a Euclidean domain with a Euclidean norm function δ such that δ satisfies the additional condition:

$$\delta(f + g) \leq \max\big(\delta(f), \delta(g)\big).$$

    [*Hint.* Define $\delta : F[x] \setminus \{0\} \to \mathbf{N}$ by

$$\delta(f) = 2^{\deg f}.$$

Then

$$\delta(f + g) = 2^{\deg(f+g)} \leq 2^{\max(\deg f, \deg g)} \quad \text{(by Proposition 4.5.2)}$$

$$= \max\big(2^{\deg f}, 2^{\deg g}\big) = \max\big(\delta(f), \delta(g)\big).]$$

13. Let F be a Euclidean domain with norm function δ satisfying the additional condition $\delta(a + b) \leq \max(\delta(a), \delta(b))$. Prove that either F is a field K or $F \subseteq K[x]$ for some field K.

    [*Hint.* Use the fact a Euclidean domain contains an identity e.

    Then $0 \neq \delta(e) = \delta(e^2) = \delta(e)\delta(e) \Rightarrow \delta(e) = 1$. Define $K = \{a \in F : \delta(a) \leq 1\}$. Now $a, b \in K \Rightarrow \delta(a - b) \leq \max(\delta(a), \delta(b)) \leq 1$ and $\delta(ab) = \delta(a)\delta(b) \leq 1 \Rightarrow K$ is a subring of F.]

14. Let (R, δ) be a Euclidean domain and A any ideal of R. Show that there exists an element $a_0 \in A$ such that $A = \langle a_0 \rangle$.

    [*Hint.* Let $A \neq \{0\}$ and $B = \{\delta(x) : x \in A \setminus \{0\}\} \subseteq \mathbf{N}^+ \Rightarrow B$ has least element $m_0 > 0$ and let $a_0 \in A \setminus \{0\}$ be such that $\delta(a_0) = m_0$. For $v \in A, \exists q, r \in R$ such that $v = a_0 q + r$, where $0 \leq \delta(r) < \delta(a_0) \Rightarrow r = 0$ (proceed as in Theorem 6.2.1) $\Rightarrow A \subseteq \langle a_0 \rangle$.]

15. Show that the *quadratic domain* $\mathbf{Z}[\sqrt{-5}] = \{a + b\sqrt{-5} : a, b \in \mathbf{Z}\}$ with usual addition and multiplication of complex numbers does not admit unique factorization. Also show that 3 is irreducible but not prime in $\mathbf{Z}[\sqrt{-5}]$.

[*Hint.* $2, 1 + \sqrt{-5}, 1 - \sqrt{-5}$ are irreducible elements in $\mathbf{Z}[\sqrt{-5}]$. Now $3|6 \Rightarrow 3|(1 + \sqrt{-5})(1 - \sqrt{-5})$ in $\mathbf{Z}[\sqrt{-5}]$. But $3\overline{|}(1 + \sqrt{-5})$ or $3\overline{|}(1 - \sqrt{-5}) \Rightarrow 3$ is not prime.]

16. (*The division algorithm in polynomial ring.*) Let R be a commutative ring with identity and $f, g \in R[x]$ non-zero polynomials such that the leading coefficient of g is an invertible element. Show that there exist unique polynomials $q, r \in R[x]$ such that $f = qg + r$, where either $r = 0$ or $\deg r < \deg g$.

17. (*Remainder Theorem.*) Let R be a commutative ring with identity. If $f(x) \in R[x]$ and $a \in R$, show that there exists a unique polynomial $q(x)$ in $R[x]$ such that $f(x) = (x - a)q(x) + f(a)$.

18. An element $a \in R$ is said to be a root of a polynomial $f(x) \in R[x]$ iff $f(a) = 0$. Show that $f(x)$ is divisible by $x - a \Leftrightarrow a$ is a root of $f(x)$.

19. Let R be an integral domain and $f(x) \in R[x]$ a non-zero polynomial of degree n. Show that $f(x)$ can have at most n distinct roots.

20. Let R be an integral domain and $f(x), g(x) \in R[x]$ be two non-zero polynomials of degree n. If there exist $n + 1$ distinct elements $a_i \in R$ such that $f(a_i) = g(a_i)$ for $i = 1, 2, \ldots, n + 1$, show that $f(x) = g(x)$.

21. Let R be an integral domain and A be any infinite subset of R. If $f(a) = 0$ $\forall a \in A$, show that $f \equiv 0$.

22. (a) Let R be a commutative ring with 1. Show that the following statements are equivalent:

    (i) R is a field;
    (ii) $R[x]$ is a Euclidean domain;
    (iii) $R[x]$ is a principal ideal domain.

    (b) Using (a) show that $\mathbf{Z}[x]$ is not a principal ideal domain.

23. Show that $\mathbf{Z}[\sqrt{n}]$ is a Euclidean domain for $n = -1, -2, 2, 3$, where $\mathbf{Z}[\sqrt{n}] = \{a + b\sqrt{n} : a, b \in \mathbf{Z}\}$.

24. (a) Let D be the integral domain given by $D = \mathbf{Z}[i\sqrt{3}] = \{a + bi\sqrt{3} : a, b \in \mathbf{Z}\}$ and $v : D \to \mathbf{N}$ be the map defined by $v(a + bi\sqrt{3}) = a^2 + 3b^2$. Show that v is not a Euclidean valuation on D.

    (b) If $a^2 + 3b^2$ is a prime integer, show that $a + bi\sqrt{3}$ is an irreducible element in $\mathbf{Z}[i\sqrt{3}]$.

25. Let R be a UFD with quotient field F. Show that

    (a) two primitive polynomials in $R[x]$ are associates in $R[x]$ iff they are associates in $F[x]$;

    (b) a primitive polynomial of positive degree in $R[x]$ is irreducible in $R[x]$ iff it is irreducible in $F[x]$.

26. Identify the correct alternative(s) (there may be more than one) from the following list:

    (a) Let I be the ideal generated by $x^2 + 1$ and J be the ideal generated by $x^3 - x^2 + x - 1$ in $\mathbf{Q}[x]$. If $R = \mathbf{Q}[x]/I$ and $S = \mathbf{Q}[x]/J$, then

        (i) R and S are both fields;
        (ii) R is an integral domain, but S is not so;

(iii) R is a field but S is not so;

(iv) R and S are not integral domains.

(b) Which of the following integral domains are Euclidean domains?

  (i) $\mathbf{R}[x^2, x^3] = \{f = \sum_{i=0}^{n} a_i x^i \in \mathbf{R}[x] : a_1 = 0\}$;

  (ii) $\mathbf{Z}[\sqrt{-3}] = \{a + b\sqrt{-3} : a, b \in \mathbf{Z}\}$;

  (iii) $\mathbf{Z}[x]$;

  (iv) $(\frac{\mathbf{Z}[x]}{\langle 2, x \rangle}[y])$, where x, y are independent variables and $\langle 2, x \rangle$ is the ideal generated by 2 and x.

(c) Let I be the ideal generated by $1 + x^2$ in $\mathbf{Q}[x]$ and R be the ring given by $R = \mathbf{Q}[x]/I$. If y is the coset of x in R, then

  (i) $y^2 + 1$ is irreducible over R;

  (ii) $y^2 - y + 1$ is irreducible over R;

  (iii) $y^3 + y^2 + y + 1$ is irreducible over R;

  (iv) $y^2 + y + 1$ is irreducible over R.

(d) Let $f_n(x) = x^{n-1} + x^{n-2} + \cdots + x + 1$ be a polynomial over $\mathbf{Q}$ of degree $n > 1$. Then

  (i) $f_n(x)$ is an irreducible polynomial in $\mathbf{Q}[x]$ for every positive integer n;

  (ii) $f_{p^e}(x)$ is an irreducible polynomial in $\mathbf{Q}[x]$ for every prime integer p and every positive integer e;

  (iii) $f_p(x)$ is an irreducible polynomial in $\mathbf{Q}[x]$ for every prime integer p;

  (iv) $f_p(x^{p^{e-1}})$ is an irreducible polynomial in $\mathbf{Q}[x]$ for every prime integer p and every prime integer e.

(e) Consider the element $\alpha = 3 + \sqrt{-5}$ in the ring $R = \mathbf{Z}[\sqrt{-5}] = \{a + b\sqrt{-5} : a, b \in \mathbf{Z}\}$. Then

  (i) R is an integral domain;

  (ii) α is irreducible;

  (iii) α is prime;

  (iv) R is not a unique factorization domain.

(f) Let F_p be the field $\mathbf{Z}_p$, where p is a prime. Let $F_p[x]$ be the associated polynomial ring. Then

  (i) $F_5[x]/<x^2 + x + 1>$ is a field;

  (ii) $F_2[x]/<x^3 + x + 1>$ is a field;

  (iii) $F_3[x]/<x^3 + x + 1>$ is a field;

  (iv) $F_7[x]/<x^2 + 1>$ is a field.

(g) Which of the following statement(s) is (are) false?

  (i) A homomorphic image of a UFD (unique factorization domain) is again a UFD;

  (ii) units of the ring $\mathbf{Z}[\sqrt{-5}]$ are the units of $\mathbf{Z}$;

  (iii) the element $2 \in \mathbf{Z}[\sqrt{-5}]$ is irreducible in $\mathbf{Z}[\sqrt{-5}]$;

  (iv) the element 2 is a prime element in $\mathbf{Z}[\sqrt{-5}]$.

(h) Let R be the polynomial ring $\mathbf{Z}_2[x]$ and write the elements of $\mathbf{Z}_2$ as $\{0, 1\}$.
If $f(x) = x^2 + x + 1$, then the quotient ring $R/\langle f(x) \rangle$ is

(i) a ring but not an integral domain;
(ii) a finite field of order 4;
(iii) an integral domain but not a field;
(iv) an infinite field.

(i) Which of the following ideal(s) is(are) maximal?

(i) The ideal $17\mathbf{Z}$ in $\mathbf{Z}$;
(ii) the ideal $I = \{f : f(0) = 0\}$ in the ring $C([0, 1])$ of all continuous real valued functions on the interval $[0, 1]$;
(iii) the ideal $25\mathbf{Z}$ in $\mathbf{Z}$;
(iv) the ideal generated by $x^3 - x + 1$ in the ring of polynomials $\mathbf{F}_3[x]$, where $\mathbf{F}_3$ is the field of three elements.

6.6 Additional Reading

We refer the reader to the books (Adhikari and Adhikari 2003, 2004; Artin 1991; Atiya and Macdonald 1969; Birkoff and Mac Lane 1965; Burtan 1968; Fraleigh 1982; Herstein 1964; Hungerford 1974; Jacobson 1974, 1980; Lang 1965; McCoy 1964; van der Waerden 1970; Zariski and Samuel 1958, 1960) for further details.

References

Adhikari, M.R., Adhikari, A.: Groups, Rings and Modules with Applications, 2nd edn. Universities Press, Hyderabad (2003)
Adhikari, M.R., Adhikari, A.: Text Book of Linear Algebra: An Introduction to Modern Algebra. Allied Publishers, New Delhi (2004)
Artin, M.: Algebra. Prentice-Hall, Englewood Cliffs (1991)
Atiya, M.F., Macdonald, I.G.: Introduction to Commutative Algebra. Addison-Wesley, Reading (1969)
Birkoff, G., Mac Lane, S.: A Survey of Modern Algebra. Macmillan, New York (1965)
Burtan, D.M.: A First Course in Rings and Ideals. Addison-Wesley, Reading (1968)
Fraleigh, J.B.: A First Course in Abstract Algebra. Addison-Wesley, Reading (1982)
Herstein, I.: Topics in Algebra. Blaisdell, New York (1964)
Hungerford, T.W.: Algebra. Springer, New York (1974)
Jacobson, N.: Basic Algebra I. Freeman, San Francisco (1974)
Jacobson, N.: Basic Algebra II. Freeman, San Francisco (1980)
Lang, S.: Algebra, 2nd edn. Addison-Wesley, Reading (1965)
McCoy, N.: Theory of Rings. Macmillan, New York (1964)
van der Waerden, B.L.: Modern Algebra. Ungar, New York (1970)
Zariski, O., Samuel, P.: Commutative Algebra I. Van Nostrand, Princeton (1958)
Zariski, O., Samuel, P.: Commutative Algebra II. Van Nostrand, Princeton (1960)

Chapter 7
Rings with Chain Conditions

In this chapter we continue to study theory of rings by finiteness conditions (chain conditions) on ideals. The motivation of chain conditions for ideals of a ring came from an interesting property of $\mathbf{Z}$. In the ring of integers $\mathbf{Z}$, let I_1 be a non-zero ideal and I be an ideal such that $I_1 \subseteq I$. Then $\exists$ non-zero integers m and n such that $I_1 = \langle m \rangle$, $I = \langle n \rangle$ and $n|m$, as $\mathbf{Z}$ is a PID. Since m contains only a finitely many divisors n in $\mathbf{Z}$ such that $I_1 \subseteq I$, there cannot exist infinitely many ideals I_t ($t = 2, 3, \ldots$) such that $I_1 \subsetneq I_2 \subsetneq I_3 \subsetneq \cdots \subsetneq I_t \subsetneq \cdots$. This interesting property of $\mathbf{Z}$ was first recognized by the German mathematician Emmy Noether (1882–1935). On the other hand, $\mathbf{Z}$ has an infinite descending chain of ideals $\langle 3 \rangle \supsetneq \langle 6 \rangle \supsetneq \langle 12 \rangle \supsetneq \cdots$. This property of $\mathbf{Z}$ leads to the concept of Artinian rings, the name is in honor of Emil Artin (1898–1962). In this chapter we study special classes of rings and obtain deeper results on ideal theory. For this we need to impose some finiteness conditions and introduce *Noetherian rings* which are versatile. The most convenient equivalent formulation of the Noetherian requirement is that the ideals of the ring satisfy the ascending chain condition. We establish a connection between a Noetherian domain and a factorization domain. We prove the Hilbert Basis Theorem which gives an extensive class of Noetherian rings. Its application to algebraic geometry is discussed. We also prove Cohen's theorem. Our study culminates in rings with descending chain condition for ideals, which determines their ideal structure.

In this chapter a ring R means a commutative ring with identity, unless otherwise stated.

7.1 Noetherian and Artinian Rings

The basic concepts of modern theory of rings arose through the work of Enmy Noether and Emil Artin during 1920s.

Noetherian and Artinian rings form a special class of rings. In this section we study such rings with the help of their ideals and determine their ideal structures.

M.R. Adhikari, A. Adhikari, *Basic Modern Algebra with Applications*, DOI 10.1007/978-81-322-1599-8_7, © Springer India 2014

This study develops ring theory. In this section we prove Hilbert Basis Theorem which has made a revolution in modern algebra and algebraic geometry. We further show that any Noetherian domain is a factorization domain and any Artinian domain is a field.

Definition 7.1.1 A ring R is said to satisfy the *ascending* (*descending*) *chain condition* denoted by acc (dcc) for ideals iff given any sequence of ideals $I_1, I_2, \ldots$ of R with $I_1 \subseteq I_2 \subseteq \cdots \subseteq I_n \subseteq \cdots$, $(I_1 \supseteq I_2 \supseteq \cdots \supseteq I_n \supseteq \cdots)$, there exists an integer n (depending on the sequence) such that $I_m = I_n \; \forall m \geq n$.

For non-commutative rings, the definitions of acc (dcc) for left ideals or for right ideals are similar.

Definition 7.1.2 A ring R is said to be a *Noetherian ring* iff it satisfies the ascending chain condition for ideals of R. (This name is in honor of *Emmy Noether*.)

For non-commutative rings the definition of a left (right) Noetherian ring is similar.

Definition 7.1.3 A ring R is said to satisfy the *maximal condition* (for ideals) iff every non-empty set of ideals of R, partially ordered by inclusion, has a maximal element (i.e., an ideal which is not properly contained in any other ideal of the set).

Theorem 7.1.1 *Let R be a ring. Then the following statements are equivalent*:

 (i) *R is Noetherian.*
 (ii) *The maximal condition (for ideals) holds in R.*
(iii) *Every ideal of R is finitely generated.*

Proof (i) $\Rightarrow$ (ii) Let $\mathbf{F}$ be any non-empty collection of ideals of R and $I_1 \in \mathbf{F}$. If I_1 is not a maximal element, $\exists$ an element $I_2 \in \mathbf{F}$ such that $I_1 \subsetneq I_2$. Again, if I_2 is not a maximal element, then $I_2 \subsetneq I_3$ for some $I_3 \in \mathbf{F}$. If $\mathbf{F}$ has no maximal element, then continuing the above process, we obtain the infinite strictly ascending chain of ideals of R:

$$I_1 \subsetneq I_2 \subsetneq I_3 \subsetneq \cdots .$$

But this contradicts the assumption that R is Noetherian.

(ii) $\Rightarrow$ (iii) Let I be an ideal of R and $\mathbf{F} = \{A : A \text{ is an ideal of } R; \; A \text{ is finitely generated and } A \subseteq I\}$.

$\{0\} \subseteq I \Rightarrow \{0\} \in \mathbf{F} \Rightarrow \mathbf{F} \neq \emptyset$. Then (ii) $\Rightarrow \mathbf{F}$ has a maximal element, say M. We show that $M = I$. Suppose $M \neq I$. Then $\exists$ an element $a \in I$ such that $a \notin M$. Now M is finitely generated. Suppose $M = \langle a_1, a_2, \ldots, a_t \rangle$. Then $A = \langle a_1, a_2, \ldots, a_t, a \rangle \in \mathbf{F} \Rightarrow M \subsetneq A \Rightarrow$ a contradiction (since M is a maximal element in $\mathbf{F}$) $\Rightarrow M = I \Rightarrow I$ is finitely generated.

(iii) $\Rightarrow$ (i) Let $I_1 \subseteq I_2 \subseteq I_3 \subseteq \cdots$ be an ascending chain of ideals of R. Then $I = \bigcup_{n=1}^{\infty} I_n$ is an ideal of R, hence is finitely generated; suppose that

$I = \langle a_1, a_2, \ldots, a_r \rangle$. Now each generator $a_t \in$ some ideal I_{i_t} of the given chain. Let n be the maximum of the indices i_t.

Then each $a_t \in I_n$. Consequently, for $m \geq n$,

$$I = \langle a_1, a_2, \ldots, a_r \rangle \subseteq I_n \subseteq I_m \subseteq I \quad \Rightarrow \quad I_m = I_n$$

$\Rightarrow$ the given chain of ideals is stationary at some point $\Rightarrow R$ is Noetherian. $\qquad \square$

Definition 7.1.4 An integral domain (ring) that satisfies any one of the equivalent conditions of Theorem 7.1.1 is called a Noetherian domain (ring).

We now show that a Noetherian domain is a natural generalization of a PID.

Proposition 7.1.1 *Every principal ideal domain is a Noetherian domain.*

Proof Let D be a principal ideal domain and $I_1 \subseteq I_2 \subseteq \cdots \subseteq I_n \subseteq \cdots$ be an ascending chain of ideals of D. Let $I = \bigcup_r I_r$. Then I is an ideal of D. Now $I = \langle a \rangle$ for some $a \in I \Rightarrow a \in I_t$ of the given chain for some $t \Rightarrow I = \langle a \rangle \subseteq I_t \Rightarrow I \subseteq I_t \subseteq I_m \subseteq I \ \forall m \geq t \Rightarrow I_m = I_t \ \forall m \geq t.$ $\qquad \square$

Most of the rings which we have been studying are Noetherian.

Example 7.1.1 (i) $(\mathbf{Z}, +, \cdot)$ is a Noetherian ring by Proposition 7.1.1.
(ii) Every field is a Noetherian ring.
(iii) Every finite ring is a Noetherian ring.
(iv) If F is a field, then the integral domain $F[x_1, x_2, \ldots]$ (in infinite indeterminates $x_1, x_2, \ldots$) is not Noetherian as it contains the infinite ascending chain of ideals $\langle x_1 \rangle \subsetneqq \langle x_1, x_2 \rangle \subsetneqq \langle x_1, x_2, x_3 \rangle \subsetneqq \cdots$. But $F[x_1, x_2, \ldots, x_n]$ (in finite number of indeterminates $x_1, x_2, \ldots, x_n$) is Noetherian by Corollary 3 of Theorem 7.1.2.

We now study *Hilbert Basis Theorem* which provides an extensive class of Noetherian rings. David Hilbert (1863–1941) asserted the Theorem in 1890. This theorem revolutioned algebraic geometry. The word 'basis' is used in the sense of finite generation.

Theorem 7.1.2 (Hilbert Basis Theorem) *If R is a Noetherian ring with identity, then $R[x]$ is also a Noetherian ring.*

Proof Let I be an arbitrary ideal of $R[x]$. To prove this theorem it is sufficient to show that I is finitely generated. For each integer $t \geq 0$, define the set $I_t = \{r \in R : a_0 + a_1 x + \cdots + rx^t \in I\} \cup \{0\}$ (i.e., consisting of zero and those $r \in R$ such that r appear as the leading non-zero coefficient of some polynomial in I of degree t). Then I_t is an ideal of R such that $I_t \subseteq I_{t+1} \ \forall t \geq 0 \Rightarrow I_0 \subseteq I_1 \subseteq I_2 \subseteq \cdots$. Since R is a Noetherian ring, $\exists$ an integer n such that $I_k = I_n \ \forall k \geq n$. Also each ideal I_i of R is finitely generated and suppose $I_i = \langle a_{i1}, a_{i2}, \ldots, a_{im_i} \rangle \ \forall i = 0, 1, 2, \ldots, n$, where a_{ij} is the leading coefficient of a polynomial $f_{ij} \in I$, of degree i.

We claim that I is generated by the $m_0 + \cdots + m_n$ polynomials $f_{01}, \ldots, f_{0m_0} \cdots$, $f_{n1}, \ldots, f_{nm_n}$. Let $J = \langle f_{01}, \ldots, f_{0m_0}, \ldots, f_{n1}, f_{n2}, \ldots, f_{nm_n} \rangle$. Now each $f_{ij} \in I$ (by our choice) implies

$$J \subseteq I. \tag{7.1}$$

Let $f \ (\neq 0) \in R[x]$ be such that $f \in I$ and of degree t (say): $f = b_0 + b_1 x + \cdots + b_{t-1} x^{t-1} + b x^t$. We now apply induction on t. For $t = 0$, $f = b_0 \in I_0 \subseteq J$. Next assume that any polynomial $\in I$ of degree less than or equal to $t - 1$ also belongs to J.

Case 1. $t > n \Rightarrow$ the leading coefficient b (of f) $\in I_t = I_n$ (as $I_t = I_n \ \forall t \geq n$) $\Rightarrow$ $b = a_{n1} c_1 + a_{n2} c_2 + \cdots + a_{nm_n} c_{m_n}$ for suitable $c_i \in R \Rightarrow g = f - (f_{n1} c_1 + f_{n2} c_2 + \cdots + f_{nm_n} c_{m_n}) x^{t-n} \in I$ having degree $< t$ (since the coefficient of x^t in g is $b - \sum_{i=1}^{m_n} a_{ni} c_i = 0$) $\Rightarrow g \in J$ by induction hypothesis $\Rightarrow f \in J$.

Case 2. $t \leq n \Rightarrow b \in I_t \Rightarrow b = a_{t1} d_1 + a_{t2} d_2 + \cdots + a_{tm_t} d_{m_t}$ for some $d_1, d_2, \ldots, d_{m_t} \in R \Rightarrow h = f - (f_{t1} d_1 + f_{t2} d_2 + \cdots + f_{tm_t} d_{m_t}) \in I$ having degree $< t$ (since the coefficient of x^t in h is $b - \sum_{i=1}^{m_t} a_{ti} d_i = 0$) $\Rightarrow h \in J$ by induction hypothesis $\Rightarrow f \in J$.

Consequently, in either case $I \subseteq J$ and hence $I = J$ by (7.1). Thus I is finitely generated and hence $R[x]$ is Noetherian. $\qquad\square$

By induction, Hilbert Basis Theorem can be extended to polynomials in several indeterminates:

Corollary 1 *If R is a Noetherian ring with identity, then the polynomial ring $R[x_1, x_2, \ldots, x_n]$ in a finite number of indeterminates $x_1, x_2, \ldots, x_n$ is also a Noetherian ring.*

Corollary 2 $\mathbf{Z}[x_1, x_2, \ldots, x_n]$ *is a Noetherian ring.*

Corollary 3 $F[x_1, x_2, \ldots, x_n]$ *is a Noetherian ring for every field F.*

A Noetherian ring need not satisfy dcc on ideals. We now consider rings which satisfy dcc on ideals.

Definition 7.1.5 A ring R is said to be an *Artinian ring* iff it satisfies the descending chain condition for ideals of R. (This ring is named after *Emil Artin* (1898–1962).)

Definition 7.1.6 A ring R is said to satisfy the *minimal condition* (for ideals) iff every non-empty set of ideals of R, partially ordered by inclusion, has a minimal element.

Theorem 7.1.3 *Let R be a ring. Then R is Artinian iff R satisfies the minimal condition (for ideals).*

Proof Let R be Artinian and $\mathbf{F}$ a non-empty set of ideals of R. Let $I_1 \in \mathbf{F}$. If I_1 is not a minimal element, we can find another ideal $I_2 \in \mathbf{F}$ such that $I_1 \supsetneqq I_2$. If $\mathbf{F}$ has no minimal element, the repetition of this process indefinitely yields the infinite strictly descending chain: $I_1 \supsetneqq I_2 \supsetneqq I_3 \supsetneqq \cdots$ of ideals of $R \Rightarrow$ a contradiction to the fact that R is Artinian $\Rightarrow \mathbf{F}$ satisfies the minimal condition. Conversely, suppose R satisfies the minimal condition. Let $I_1 \supseteq I_2 \supseteq I_3 \supseteq \cdots$ be a descending chain of ideals of R. Consider $\mathbf{F} = \{I_t : t = 1, 2, 3, \ldots\}$. Then $I_1 \in \mathbf{F} \Rightarrow \mathbf{F} \neq \emptyset$. Again by hypothesis, $\mathbf{F}$ has a minimal element I_n for some positive integer $n \Rightarrow I_m \subseteq I_n$ $\forall m \geq n$. Now $I_m \neq I_n \Rightarrow I_m \notin \mathbf{F}$ (by the minimality of I_n) $\Rightarrow$ an impossibility $\Rightarrow I_m = I_n \ \forall m \geq n \Rightarrow R$ is Artinian. $\qquad \square$

Remark For non-commutative rings, the definitions of left (right) Artinian rings are obvious.

Theorem 7.1.4 *A homomorphic image of a Noetherian (Artinian) ring is also Noetherian (Artinian).*

Proof Let f be a homomorphism of the Noetherian ring R onto the ring S. Consider the ascending chain of ideals of S:

$$J_1 \subseteq J_2 \subseteq \cdots \subseteq J_n \cdots . \tag{7.2}$$

Suppose $I_r = f^{-1}(J_r)$, for $r = 1, 2, \ldots$. Then

$$I_1 \subseteq I_2 \subseteq \cdots \subseteq I_n \subseteq \cdots . \tag{7.3}$$

Relation (7.3) becomes an ascending chain of ideals of R. Then by hypothesis, $\exists$ some index n such that $I_m = I_n \ \forall m \geq n$. This shows that $J_m = J_n \ \forall m \geq n$. Hence the chain (7.2) becomes stationary at some point. Thus S is Noetherian.

For the Artinian ring, the proof is similar. $\qquad \square$

Corollary *If I is an ideal of a Noetherian (Artinian) ring R, then the quotient ring R/I is Noetherian (Artinian).*

Proof The corollary follows from Theorem 7.1.4 by taking in particular, $S = R/I$ and $f : R \to R/I$ the natural homomorphism. $\qquad \square$

Theorem 7.1.5 *Let I be an ideal of a ring R. If I and R/I are both Noetherian (Artinian) rings, then R is also Noetherian (Artinian).*

Proof Let $I_1 \subseteq I_2 \subseteq \cdots \subseteq I_n \subseteq \cdots$ be an ascending chain of ideals of R. Let $p : R \to R/I$ be the natural homomorphism of R onto R/I. Then $p(I_1) \subseteq p(I_2) \subseteq \cdots \subseteq p(I_n) \subseteq \cdots$ is an ascending chain of ideals in R/I. Since R/I is Noetherian, there exists a positive integer n such that $p(I_n) = p(I_{n+i})$ for all $i \geq 1$. Also, $I_1 \cap I \subseteq I_2 \cap I \subseteq \cdots \subseteq I_n \cap I \subseteq \cdots$ is an ascending chain of ideals in I. Since I is Noetherian, there exists a positive integer m such that $I_m \cap I = I_{m+i} \cap I$ for all $i \geq 1$.

Let $r = \max(m, n)$. Then $p(I_r) = p(I_{r+i})$ and $I_r \cap I = I_{r+i} \cap I$ for all $i \geq 1$. Let $a \in I_{r+i}$. Then there exists an element $x \in I_r$ such that $p(a) = p(x)$, i.e., $a + I = x + I$. Hence $a - x \in I$ and also $a - x \in I_{r+i}$. This shows that $a - x \in I_{r+i} \cap I = I_r \cap I$. Hence $a - x \in I_r$. Again as $x \in I_r$, then $a \in I_r$. Consequently, $I_r = I_{r+i}$ for all $r \geq 1$. This proves that R is Noetherian. $\square$

Definition 7.1.7 An Artinian domain R is an integral domain which is also an Artinian ring.

We now prove some interesting properties of Noetherian and Artinian rings.

Theorem 7.1.6 (a) *Any Noetherian domain is a factorization domain.*
(b) *Any Artinian domain is a field.*

Proof (a) Let R be a Noetherian domain. Suppose R is not a factorization domain. Then there exists at least one non-zero, non-invertible element a in R such that a is not a finite product of irreducible elements of R. Let X be the set of all such elements $x \in R$. As $a \in X$, $X \neq \emptyset$. Consider the set $S = \{\langle x \rangle : x \in X\}$. Then S is a non-empty collection of ideals of R. As R is Noetherian, S has a maximal element, say $\langle y \rangle$. Clearly, $y \in X$ and y is not irreducible. This implies that $y = cd$ for some non-zero non-invertible elements c and d in R. Hence $\langle y \rangle \subsetneq \langle c \rangle$ and $\langle y \rangle \subsetneq \langle d \rangle$. Then by maximality of $\langle y \rangle$ in S, it follows that $\langle c \rangle$ and $\langle d \rangle$ are not elements of S. This shows that c and d are finite products of irreducible elements of R. Hence $y = cd$ is also a finite product of irreducible elements of R, a contradiction as $y \in X$.

(b) Let R be an Artinian domain and a ($\neq 0$) $\in R$. Then for descending chain of ideals: $\langle a \rangle \supseteq \langle a^2 \rangle \supseteq \langle a^3 \rangle \cdots \supseteq \cdots$, there exists an index n such that, $\langle a^n \rangle = \langle a^{n+1} \rangle = \langle a^{n+2} \rangle = \cdots$. Hence $\langle a^n \rangle = \langle a^{n+1} \rangle \Rightarrow a^n = ra^{n+1}$ for some $r \in R \Rightarrow 1 = ra$ (by the cancellation law in R as $a^n \neq 0$) $\Rightarrow a$ is invertible in $R \Rightarrow R$ is a field. $\square$

Corollary 1 *Every principal ideal domain is a factorization domain.*

Proof It follows from Proposition 7.1.1 and Theorem 7.1.6(a). $\square$

Corollary 2 *An integral domain with only a finite number of ideals is a field.*

Proof It follows from Theorem 7.1.6(b). $\square$

We recall that a ring R is Noetherian iff every ideal of R is finitely generated. To show that R is Noetherian it is sufficient to consider just the prime ideals of R (this result is due to I.S. Cohen (1917–1955)).

Theorem 7.1.7 (Cohen) *A ring R is Noetherian iff every prime ideal of R is finitely generated.*

Proof $\Rightarrow$ follows from Theorem 7.1.1.

Next assume that every prime ideal of R is finitely generated, but R fails to be Noetherian. This shows that the collection $\mathbf{F}$ of ideals of R which are not finitely generated is non-empty.

Order $\mathbf{F}$ by inclusion. Now consider an arbitrary chain $\{B_i\}$ in $\mathbf{F}$. Then $B = \bigcup_i B_i$ is an ideal of R such that B cannot be finitely generated. Hence $B \in \mathbf{F}$ is an upper bound in $\mathbf{F}$. Then by Zorn's Lemma $\mathbf{F}$ has a maximal element I (say). So, by hypothesis, I cannot be a prime ideal of R. Then $\exists a, b \in R$ such that $a \notin I, b \notin I$ but $ab \in I$. Clearly, both the ideals $\langle I, b \rangle$ and $I : \langle b \rangle$ (see Ex. 26 of Exercise-I, Chap. 5) properly contain I and $a \in I : \langle b \rangle$. So maximality of I in $\mathbf{F} \Rightarrow \langle I, b \rangle \notin \mathbf{F}$ and so, $I : \langle b \rangle \notin \mathbf{F} \Rightarrow \langle I, b \rangle = \langle c_1, c_2, \ldots, c_n \rangle$ and $I : \langle b \rangle = \langle d_1, d_2, \ldots, d_m \rangle$ for suitable $c_i, d_j \in R (i = 1, 2, \ldots, n; j = 1, 2, \ldots, m) \Rightarrow c_i = a_i + b r_i$, for some $a_i \in I$ and $r_i \in R \Rightarrow \langle I, b \rangle = \langle a_1, a_2, \ldots, a_n, b \rangle$.

Now consider the ideal $J = \langle a_1, a_2, \ldots, a_n, b d_1, \ldots, b d_m \rangle$. Now $b d_j \in I \; \forall j \Rightarrow J \subseteq I$. To show the reverse inclusion, let $x \in I$.

Then

$$x \in \langle I, b \rangle \quad \Rightarrow \quad x = \sum_{i=1}^{n} a_i x_i + b y \quad (x_i, y \in R) \quad \Rightarrow \quad b y \in I \text{ (as each } a_i \in I)$$

$$\Rightarrow \quad y \in I : \langle b \rangle \quad \Rightarrow \quad y = \sum_{i=1}^{m} d_j t_j \quad (t_j \in R)$$

$$\Rightarrow \quad x = \sum_{i=1}^{n} a_i x_i + \sum_{j=1}^{m} (b d_j) t_j \in J \quad \Rightarrow \quad I \subseteq J.$$

Consequently, $I = J \Rightarrow I$ is itself finitely generated $\Rightarrow$ an impossibility as $I \in \mathbf{F}$. This contradiction proves the theorem. $\qquad \square$

The "finiteness condition" for Noetherian (Artinian) rings has an advantage over arbitrary rings which makes the study of Noetherian (Artinian) rings more attractive and interesting. We prove the following theorems for commutative cases and non-commutative cases are left as an exercise.

Theorem 7.1.8 *Let R be a commutative Artinian ring with 1. Then every prime ideal of R is a maximal ideal.*

Proof Let A be a prime ideal of R. Then the quotient ring R/A forms an integral domain. Again as R is an Artinian ring, then R/A is also an Artinian ring. Hence by Theorem 7.1.6, R/A is a field. Consequently, A is a maximal ideal of R. $\qquad \square$

We can determine more ideal structure of an Artinian ring R.

Theorem 7.1.9 *Every Artinian commutative ring R with 1 has only a finite number of prime ideals, each of which is maximal.*

Proof If possible, there exists an infinite sequence $\{A_i\}$ of distinct prime ideals of R. Then we can form a descending chain of ideals:

$$A_1 \supseteq A_1 A_2 \supseteq A_1 A_2 A_3 \supseteq \cdots .$$

Since R is Artinian, $\exists$ a positive integer n for which

$$A_1 A_2 \cdots A_n = A_1 A_2 \cdots A_n A_{n+1}.$$

This shows that $A_1 A_2 \cdots A_n \subseteq A_{n+1}$. Then $A_t \subseteq A_{n+1}$ for some $t \leq n$. But A_t is a maximal ideal of R by Theorem 7.1.8. So we have $A_t = A_{n+1}$, which contradicts the fact that each A_i is distinct.

Consequently, R has only a finite number of prime ideals and each of them is maximal by Theorem 7.1.8. $\square$

We recall that the Jacobson radical of a ring R denoted by $J(R)$ (or rad R) is the intersection of all maximal ideals of R. We now study the Jacobson radical of an Artinian ring.

Theorem 7.1.10 *The Jacobson radical of a commutative Artinian ring is the intersection of finitely many maximal ideals of the ring.*

Proof Let R be an Artinian ring and $\mathbf{S}$ be the set of all maximal ideals of R. Let $J = J(R)$ be the Jacobson radical of R. Then J is the intersection of all maximal ideals of R. Let $\mathbf{F}$ be the set of all ideals of R such that each of which is an intersection of finitely many maximal ideals of R. Then $\mathbf{F}$ is non-empty, since $\mathbf{S} \subseteq \mathbf{F}$. As R is Artinian, $\mathbf{F}$ has a minimal element M_0 (say). Suppose $M_0 = M_1 \cap M_2 \cap \cdots \cap M_n$, where M_i are in $\mathbf{S}$. Then $J \subseteq M_0$. Again if M is in $\mathbf{S}$, then $M_0 \cap M$ is in $\mathbf{F}$. Hence by minimality of M_0, it follows that $M_0 \cap M = M_0$. Thus $M_0 \subseteq M$ for all M in $\mathbf{S}$. Hence $J \subseteq M_0 \subseteq \cap_{M \in \mathbf{S}} M = J$ shows that $J = M_0$. $\square$

Theorem 7.1.11 *The Jacobson radical of a commutative Artinian ring with 1 is nilpotent.*

Proof Let R be a commutative Artinian ring with 1 and $J = J(R)$ be its Jacobson radical. We claim that $J^n = \{0\}$ for some positive integer n. Consider the descending chain of ideals of R: $J \supseteq J^2 \supseteq \cdots \supseteq J^n \cdots$. As R is Artinian, there exists some positive integer n such that $J^n = J^m$ for all $m \geq n$. Suppose $I = J^n$. Hence $I = I^2$ and $IJ = I$. If possible, suppose $I \neq \{0\}$. Let $\mathbf{F}$ be the set of all ideals A of R such that $A \subseteq I$ and $IA \neq \{0\}$. Then $\{0\} \neq I = I^2$ shows that $I \in \mathbf{F}$. Hence $\mathbf{F}$ is non-empty. Since R is Artinian and $\{0\}$ is not in $\mathbf{F}$, it follows that $\mathbf{F}$ has a minimal element, say $M \neq \{0\}$, such that $IM \neq \{0\}$ and $M \subseteq I$. Consequently, there exists a non-zero element $y \in M$ such that $Iy \neq \{0\}$. Then Iy is a non-zero ideal of R. Moreover, $I(Iy) = I^2 y = Iy \neq \{0\}$ and $Iy \subseteq M \subseteq I$ show that $Iy \in \mathbf{F}$. Then by minimality of M, it follows that $Iy = M$. Hence there exists an element $x \in I$ such that $y = xy$. Then $y(1 - x) = 0$. Now $x \in I = J^n \subseteq J$ implies that $1 - rx$ is a unit for all $r \in R$ (see Ex. 23 of Exercises-I of Chap. 5). In particular, $1 - x$ is a unit in R. Hence $y(1 - x) = 0$ implies $y = 0$. This is a contradiction. Hence $I = J^n = \{0\}$. □

We recall that if every element of an ideal I of a ring is nilpotent, then the ideal I is called a *nil ideal*.

Corollary *Every nil ideal of a commutative Artinian ring R with 1 is nilpotent.*

Proof Let A be a nil ideal of R. Then every element of A is nilpotent and hence $\forall a \in A, r \in R, ra$ is nilpotent. This implies that $1 + ra$ is a unit in R $\forall r \in R$. This shows that $a \in \text{rad } R$ (by Ex. 23, Exercises-I of Chap. 5). Hence $A \subseteq \text{rad } R = J(R)$.

Since a nil ideal of a ring is contained in its Jacobson radical, the corollary follows from Theorem 7.1.11. □

The following theorem is the same as Theorem 7.1.9. We now give an alternative proof.

Theorem 7.1.12 *Every commutative Artinian ring with 1 contains only finitely many maximal ideals.*

Proof Let R be a commutative Artinian ring 1. Then its Jacobson radical $J = J(R)$ is an intersection of finitely many maximal ideals of R by Theorem 7.1.10. Let $J = M_1 \cap M_2 \cap \cdots \cap M_n \supseteq M_1 M_2 \cdots M_n$. Again by Theorem 7.1.11, J is nilpotent. Then $J^t = \{0\}$ for some positive integer t. Consequently, $\{0\} = J^t \supseteq (M_1 M_2 \cdots M_n)^t = M_1{}^t M_2{}^t \cdots M_n{}^t$. Let M be a maximal ideal of R. Then $M \supseteq \{0\} = M_1{}^t M_2{}^t \cdots M_n{}^t$ implies that $M \supseteq M_i{}^t$, for some i such that $1 \leq i \leq n$. This implies that $M_i \subseteq M$, since M is a maximal ideal and hence it is prime. Again since both M and M_i are maximal ideals of R, it follows that $M = M_i$. This concludes that the only maximal ideals of R are $M_1, M_2, \ldots, M_n$. □

Remark (i) Ex. 4 of Exercises-I shows that there are rings which are right Noetherian but not left Noetherian;

(ii) Ex. 5 of Exercises-I shows that there are rings which are right Artinian but not left Artinian;

(iii) Ex. 7 of Exercises-I shows that a subring of a Noetherian (or Artinian) ring may not be Noetherian (or Artinian);

(iv) $\mathbf{Z}$ is Noetherian but not Artinian.

So the concepts of left Noetherian and right Noetherian rings are different but these two concepts coincide if the ring is commutative. Again $\mathbf{Z}$ is Noetherian but not Artinian. Hence these two classes of rings are different in general.

We now search for suitable situations under which a Noetherian ring is Artinian and conversely.

Theorem 7.1.13 *Let R be a commutative local ring with 1 such that its maximal ideal in R is nilpotent. Then R is Artinian if and only if R is Noetherian.*

Proof Let M be the maximal ideal of R. Then, by hypothesis, $M^t = \{0\}$ for some positive integer t. Hence $F = R/M$ is a field (by Theorem 5.3.2). Clearly, R is Artinian (or Noetherian) iff M is Artinian (or Noetherian). Again M is Artinian (Noetherian) $\Leftrightarrow M/M^2$ and M^2 are Artinian (or Noetherian) $\Leftrightarrow \cdots \Leftrightarrow M^i/M^{i+1}$ and M^{i+1} are Artinian (or Noetherian). But M^i/M^{i+1} is Artinian (or Noetherian) $\Leftrightarrow M^i/M^{i+1}$ is a finite dimensional vector space over F for all $i = 1, 2, \ldots, t-1$ (see Chaps. 8 and 9). If R is Artinian (or Noetherian), then each M^i/M^{i+1} is Artinian (or Noetherian). Hence each is a finite dimensional vector space over the field F. Now each is Noetherian or Artinian. Thus $M^{t-1} = M^{t-1}/M^t$ and M^{t-2}/M^{t-1} are both Noetherian (or Artinian) show that M^{t-2} is Noetherian (or Artinian). Proceeding in this way we prove that M is Noetherian (or Artinian). This proves the theorem. □

Theorem 7.1.14 *Every ideal in a commutative Noetherian ring with 1 contains a product of prime ideals.*

Proof Let R be a commutative Noetherian ring with 1 and $\mathcal{F}$ be the set of all ideals A of R such that A does not contain any product of prime ideals. If $\mathcal{F}$ is not empty, then $\mathcal{F}$ has a maximal element, say M. This M cannot be prime. Hence there exist elements $x, y \in R$ such that $xy \in M$ but $x, y \notin M$. Consider $A = M + Rx$ and $B = M + Ry$. Then $M \subseteq A \cap B$ and $M \neq A$ and $M \neq B$. Hence by maximality of M in $\mathcal{F}$, $A, B \notin \mathcal{F}$. This shows that both A and B contain some product of prime ideals of R. Again $AB = (M + Rx)(M + Ry) \subseteq M + Rxy = M$, as $xy \in M$. This implies that AB contains a product of prime ideals of R and hence M contains a product of prime ideals of R. This is a contradiction. Consequently, $\mathcal{F}$ must be empty. This proves the theorem. □

Corollary *Let R be a commutative Noetherian ring with 1. Then $\{0\} = P_1^{t_1} P_2^{t_2} \cdots P_n^{t_n}$, where P_i's are distinct prime ideals of R and $t_1, t_2, \ldots, t_n$ are some positive integers.*

Theorem 7.1.15 *A commutative Artinian ring R with 1 is Noetherian. Conversely, if R is a Noetherian ring with 1 in which every prime ideal is maximal, then R is also Artinian.*

Proof Suppose R is a commutative ring with 1 which is Artinian. Then by Theorem 7.1.12, R contains only finitely many maximal ideals: $M_1, M_2, \ldots, M_n$ (say) such that its Jacobson radical J satisfies the relation $\{0\} = J^t = M_1^t M_2^t \cdots M_n^t$, where t is the rank of nilpotency of J. Maximal ideals $M_1, M_2, \ldots, M_n$ are pairwise co-prime and hence by Chinese Remainder Theorem 5.6.1 of Chap. 5 it follows that

$$R \cong R/M_1^t \times R/M_2^t \times \cdots \times R/M_n^t. \tag{7.4}$$

Again each R/M_i^t is an Artinian local ring whose maximal ideal M_i/M_i^t is nilpotent. Hence each R/M_i^t is Noetherian by Theorem 7.1.13. Consequently, R is Noetherian by (7.4).

Next suppose R is a Noetherian ring with 1 in which every prime ideal is maximal. Then each maximal ideal of R is also a prime ideal. Hence it follows by Corollary to Theorem 7.1.14 that $\{0\} = M_1{}^{t_1} M_2{}^{t_2} \cdots M_n{}^{t_n}$ for some maximal ideals M_i and some positive integers t_i. Consequently, $R \cong R/M_1{}^{t_1} \times R/M_2{}^{t_2} \times \cdots \times R/M_n{}^{t_n}$. Proceeding as before, the converse part follows. $\square$

Problem Let R be a Noetherian ring and $f : R \to R$ is an epimorphism. Then f is an isomorphism.

[*Hint.* For each $n \in \mathbf{N}^+$, f^n is an epimorphism and $\ker f \subseteq \ker f^2 \subseteq \ker f^3 \subseteq \cdots$ is an ascending chain of ideals in R. Now R is Noetherian $\Rightarrow \exists m \in \mathbf{N}^+$ such that $\ker f^m = \ker f^{m+t} \ \forall t \in \mathbf{N}^+ \Rightarrow f$ is a monomorphism.]

7.2 An Application of Hilbert Basis Theorem to Algebraic Geometry

The Hilbert Basis Theorem plays a very important role in algebraic geometry. We present in this section an application of this theorem to algebraic geometry.

We follow the notation of Sect. 5.5 of Chap. 5.

Theorem 7.2.1 *If K^n is an affine n-space, then every algebraic set in K^n is determined by a finite set of polynomials in $K[x_1, x_2, \ldots, x_n]$.*

Proof Let $A \subseteq K^n$ be an affine algebraic set. Then $\exists$ a subset S of $K[x_1, x_2, \ldots, x_n]$ such that, $A = V(S)$. If P is the ideal generated by S, then $V(P) = V(S)$. Now P being an ideal of the Noetherian ring $K[x_1, x_2, \ldots, x_n]$ (as every field K is a Noetherian ring), P is generated by a finite set of polynomials in $K[x_1, x_2, \ldots, x_n]$. Then $A = V(S) = V(P)$ is determined by a finite set of polynomials in $K[x_1, x_2, \ldots, x_n]$. $\square$

Remark Every ascending sequence of ideals $P_1 \subseteq P_2 \subseteq \cdots$ in $K[x_1, x_2, \ldots, x_n]$ is stationary i.e., $\exists n$ such that $P_n = P_{n+1} = \cdots$. This theorem implies in view of Proposition 5.5.1(i) Chap. 5 the following descending chain condition for algebraic sets.

Every decreasing sequence of algebraic sets in $K^n : V(P_1) \supseteq V(P_2) \supseteq \cdots$ in K^n must terminate, i.e., $\exists n$ such that $V(P_n) = V(P_{n+1}) = \cdots$.

Then it follows that every non-empty collection Σ of algebraic sets in K^n has a member M such that no member of Σ is properly contained in M, otherwise we can construct inductively a non-terminating decreasing sequence of algebraic sets in K^n. Such a member M of Σ is called a minimal member.

Theorem 7.2.2 *Any algebraic set A can be expressed uniquely as a finite union of affine varieties: $A = A_1 \cup A_2 \cup \cdots \cup A_r$, so that there is no inclusion relation among the A_i's i.e., $A_i \not\subset A_j \ \forall i \neq j$.*

Proof (*Existence*) Let Σ be the collection of all those algebraic sets in K^n which cannot be expressed as a finite union of affine varieties. We claim that $\Sigma = \emptyset$: If $\Sigma \neq \emptyset$, Σ has a minimal member M by the descending chain condition for algebraic sets. This M cannot be an affine variety, otherwise M would be a finite union of irreducible algebraic sets: $M = A_1 \cup A_2$ and hence M would not be a member of Σ. Then $\exists$ two algebraic sets $A_1 \neq M$ and $A_2 \neq M$ such that $M = A_1 \cup A_2$. Since M is a minimal member of Σ, $A_1 \notin \Sigma$ and $A_2 \notin \Sigma$. Consequently, each of A_1 and A_2 can be expressed as a finite union of affine varieties, so M is a finite union of affine varieties. This means $M \notin \Sigma$ implies a contradiction. Therefore Σ must be empty. In other words, any algebraic set A can be expressed as a finite union of affine varieties: $A = A_1 \cup A_2 \cup \cdots \cup A_r$. Moreover, if there is an inclusion relation among the A_i's, we can omit the superfluous terms so that $A_i \not\subset A_j \ \forall i \neq j$.

(*Uniqueness*) Let $A = \bigcup_{i=1}^{r} A_i = \bigcup_{j=1}^{s} B_j$, where each A_i, B_j is irreducible and $A_i \subseteq A_k \Leftrightarrow i = k$, $B_j \subseteq B_k \Leftrightarrow j = k$. Each B_j can be expressed as $B_j = B_j \cap A = B_j \cap (A_1 \cup A_2 \cup \cdots \cup A_r) = (B_j \cap A_1) \cup (B_j \cap A_2) \cup \cdots \cup (B_j \cap A_r)$. Since each $B_j \cap A_i$ is an algebraic set and B_j is irreducible, we must have $B_j = B_j \cap A_m \subseteq A_m$ for some m, $1 \leq m \leq r$. By similar arguments, $A_m \subseteq B_k$ for some k, $1 \leq k \leq s$. Therefore $B_j \subseteq B_k$ which implies $j = k$. Consequently, each $B_j = A_m$ for some m, $1 \leq m \leq r$.

Similarly, each $A_m = B_j$ for some j, $1 \leq j \leq s$. This proves that the representation is unique.

The representation $A = A_1 \cup A_1 \cup \cdots \cup A_r$ is called a *decomposition of A*, and the A_i's are called the *irreducible components* of A. $\square$

7.3 An Application of Cohen's Theorem

In this section we show that Cohen's Theorem (proved in the earlier section) offers a sufficient condition for a ring to be Noetherian.

Theorem 7.3.1 *If R is a ring in which every maximal ideal is generated by an idempotent element, then R is Noetherian.*

Proof Let I be a primary ideal of R (see Ex. 20 of Exercise-I, Chap. 5). We claim that I is a maximal ideal. Otherwise, there exists a maximal ideal M such that $I \subsetneq M$. Then, by hypothesis, $M = \langle e \rangle$ where e is an idempotent element in R such that $e \neq 0$ or $e \neq 1$, since $e = 0 \Rightarrow R$ is a field and the proof is trivial. Then $e(1 - e) = 0 \in I$ and I is a primary ideal $\Rightarrow (1 - e)^n \in I \subsetneq M$ for some positive integer $n \Rightarrow 1 - e \in M = \langle e \rangle \Rightarrow 1 \in M \Rightarrow$ a contradiction $\Rightarrow I$ is a maximal ideal. Since every primary ideal of R is maximal, the concepts of maximal, prime and primary ideals coincide. Hence by our hypothesis, every maximal (hence, every prime) ideal is finitely generated $\Rightarrow R$ is necessarily Noetherian by Cohen's Theorem. $\square$

7.4 Supplementary Examples (SE-I)

1 Every Euclidean domain is always a Noetherian domain.

[*Hint.* D is a Euclidean domain $\Rightarrow$ D is a PID by Theorem 6.2.1 $\Rightarrow$ D is a Noetherian domain by Proposition 7.1.1.]

2 Let D be an integral domain which is not a field but D is Noetherian. Then D contains elements which are irreducible.

[*Hint.* D is not a field $\Rightarrow$ $\exists$ a non-zero non unit element $a \in D \Rightarrow a$ is irreducible. Otherwise, if a is reducible in D, then $\exists$ a non-zero non unit element $a_1 \in D$ such that $a_1 | a$ and a_1 is not an associate of a. Hence $\langle a \rangle \subsetneq \langle a_1 \rangle$. If a_1 is not irreducible, repeat the process to obtain an ascending chain of principal ideals:

$$\langle a \rangle \subsetneq \langle a_1 \rangle \subsetneq \langle a_2 \rangle \subsetneq \cdots$$

which contradicts the fact that D is Noetherian.]

3 Every Noetherian domain is a factorization domain.

[*Hint.* Let D be a Noetherian domain which is not a factorization domain. Then $\exists$ at least one non-zero, non unit element a (say) in D such that a is not a product of irreducible elements. Let $\mathbf{A}$ be the set of all such elements a of D. Hence $\mathbf{A} \neq \emptyset$. Consider the set $\mathbf{S} = \{\langle a \rangle : a \in A\}$. Then $\mathbf{S} \neq \emptyset$. Now D is a Noetherian domain $\Rightarrow \mathbf{S}$ has a maximal element $\langle b \rangle$ (say) $\Rightarrow b$ is not irreducible in $D \Rightarrow \exists$ non-zero non unit elements $c, d \in D$ such that $b = cd$ and c, d are not associates of $b \Rightarrow \langle b \rangle = \langle cd \rangle \subsetneq \langle c \rangle$ and $\langle b \rangle \subsetneq \langle d \rangle \Rightarrow \langle c \rangle \notin \mathbf{S}$ and $\langle d \rangle \notin \mathbf{S}$ (by maximality of $\langle b \rangle$) $\Rightarrow c$ and d are both products of irreducible elements of $D \Rightarrow$ a contradiction.]

4 Let R be a commutative Artinian ring with 1 such that $|R| > 1$ and R does not contain zero divisors. Then R is a field.

[*Hint.* Let $a \ (\neq 0) \in R$. Consider the chain $\langle a \rangle \supseteq \langle a^2 \rangle \supseteq \langle a^3 \rangle \supseteq \cdots$.

R is Artinian $\Rightarrow \exists$ a positive integer m such that $\langle a^m \rangle = \langle a^{m+1} \rangle$. Then $a^m \ (\neq 0) \in \langle a^{m+1} \rangle \Rightarrow \exists$ some $r \in R$ such that $a^m = r a^{m+1} \Rightarrow r a = 1$.]

7.5 Exercises

Exercises-I

1. Show that the ring of integers $(\mathbf{Z}, +, \cdot)$ is Noetherian but not Artinian.

    [*Hint.* For any positive integer n, the strictly descending chain $\langle n \rangle \supsetneq \langle 2n \rangle \supsetneq \langle 4n \rangle \supsetneq \cdots$ of ideals of $\mathbf{Z}$ does not terminate.]

2. Let p be a fixed prime integer. Consider the group $\mathbf{Z}(p^\infty) = \{m/p^n : 0 \leq m < p^n; \ m \in \mathbf{Z}, n = 0, 1, 2, \ldots\}$ under the operation of addition modulo 1. Then $\mathbf{Z}(p^\infty)$ is a ring (without identity) by defining the product ab to be zero $\forall a, b \in \mathbf{Z}(p^\infty)$. Show that $(\mathbf{Z}(p^\infty), +, \cdot)$ is an Artinian ring but not Noetherian.

3. Let F be the ring of all real valued functions on $\mathbf{R}$. For any positive real number r define $I_r = \{f \in F : f(x) = 0 \text{ for } -r \leq x \leq r\}$. Then show that I_r is an ideal

of F such that

$$\cdots I_3 \subsetneqq I_2 \subsetneqq I_1 \subsetneqq I_{1/2} \subsetneqq I_{1/3} \subsetneqq \cdots .$$

Also, show that F is neither Noetherian nor Artinian.

4. Consider the ring $R = \left\{ \begin{pmatrix} a & b \\ 0 & c \end{pmatrix}, a \in \mathbf{Z}, b, c \in \mathbf{Q} \right\}$ under usual addition and multiplication. Show that R is right Noetherian but not left Noetherian.

   [*Hint.* For any non-negative integer n, the set $I_n = \left\{ \begin{pmatrix} 0 & m/2^n \\ 0 & 0 \end{pmatrix} ; m \in \mathbf{Z} \right\}$ is a left ideal of R such that $I_1 \subsetneqq I_2 \subsetneqq \cdots$.

   To show that R is right Noetherian, prove that every non-zero right ideal of R is finitely generated.]

5. Consider the ring $R = \left\{ \begin{pmatrix} a & b \\ 0 & c \end{pmatrix} : a \in \mathbf{Q}; \ b, c \in \mathbf{R} \right\}$ under usual addition and multiplication. Show that R is right Artinian but not left Artinian.

6. (a) Let R be a commutative Noetherian ring with 1. Show that every ideal of R contains a finite product of prime ideals.
   (b) Show that a commutative Artinian ring R with 1 is a field $\Leftrightarrow R$ is an integral domain.

7. Show by examples that a subring of a Noetherian (Artinian) ring may not be Noetherian (Artinian).

   [*Hint.* $\mathbf{Q}[x]$ is Noetherian by Theorem 7.1.2 but its subring $R = \{f \in \mathbf{Q}[x] :$ constant term of f belong to $\mathbf{Z}\}$ is not Noetherian, since the strictly ascending chain $\langle x \rangle \subsetneqq \langle x/2 \rangle \subsetneqq \langle x/2^2 \rangle \subsetneqq \cdots$ of ideals of R does not stabilize at any point. Again $\mathbf{Z}$ is a subring of the field $\mathbf{Q}$, but $\mathbf{Z}$ is not Artinian by Exercise 1.]

8. Is the statement of the Hilbert Basis Theorem true on replacement of Noetherian rings by Artinian rings? Justify your answer.

   [*Hint.* For a field F, the strictly descending chain of principal ideals of $F[x]$: $\langle x \rangle \supsetneqq \langle x^2 \rangle \supsetneqq \langle x^3 \rangle \supsetneqq \cdots$ will never terminate.]

9. An ideal I of a ring R is said to be a nil ideal iff each element x in I is nilpotent; I is said to be nilpotent iff $I^n = \{0\}$ for some positive integer n. Let R be an Artinian ring. Examine the validity of the following statements:

   (i) rad R is a nilpotent ideal of R;
   (ii) every nil ideal of R is nilpotent.

10. Let R be a left Artinian ring with identity 1 and G the group of units of R. An element $a \in R$ is said to be left quasi-regular iff $\exists r \in R$ such that $r + a + ra = 0$. In this case, the element r is called a left *quasi-inverse* of a. Let J denote the *Jacobson radical* of R.

    Prove the following:

    (a) Let G^* be the group of units of R/J. Then $g \in G \Leftrightarrow g + J \in G^*$;
    (b) G is a finite group $\Rightarrow R$ is finite;
    (c) G is an abelian group and a, b are quasi-regular elements of $R \Rightarrow ab = ba$ (in particular, J is commutative).

11. Examine the validity of the statement that every Artinian ring is Noetherian but its converse is not true.

12. Examine the validity of the statement that a ring R is Artinian iff R is Noetherian and every prime ideal of R is maximal.
13. For any commutative ring R, the Krull dimension of R is the maximum possible length of the chain $I_1 \subseteq I_2 \subseteq \cdots \subseteq I_n$ ($I_1 \supseteq I_2 \supseteq \cdots \supseteq I_n$) of distinct prime ideals of R. For example, a field has dimension zero and a PID (which is not a field) has dimension 1. The Krull dimension of R is said to be infinite iff R has an arbitrary chain of distinct prime ideals.

    Show that a ring R is Artinian iff R is Noetherian and it has the Krull dimension zero.
14. Show that any ring which is a quotient of a polynomial ring over the integers or over a field is Noetherian.

    [*Hint*. Use Corollary of Theorem 7.1.4 and Hilbert Basis Theorem.]

7.6 Additional Reading

We refer the reader to the books (Adhikari and Adhikari 2003, 2004; Artin 1991; Atiya and Macdonald 1969; Birkoff and Mac Lane 1965; Burtan 1968; Fraleigh 1982; Fulton 1969; Herstein 1964; Hungerford 1974; Jacobson 1974, 1980; Lang 1965; McCoy 1964; Musuli 1992; van der Waerden 1970; Zariski and Samuel 1958, 1960) for further details.

References

Adhikari, M.R., Adhikari, A.: Groups, Rings and Modules with Applications, 2nd edn. Universities Press, Hyderabad (2003)
Adhikari, M.R., Adhikari, A.: Text Book of Linear Algebra: An Introduction to Modern Algebra. Allied Publishers, New Delhi (2004)
Artin, M.: Algebra. Prentice-Hall, Englewood Cliffs (1991)
Atiya, M.F., Macdonald, I.G.: Introduction to Commutative Algebra. Addison-Wesley, Reading (1969)
Birkoff, G., Mac Lane, S.: A survey of Modern Algebra. Macmillan, New York (1965)
Burtan, D.M.: A First Course in Rings and Ideals. Addison-Wesley, Reading (1968)
Fraleigh, J.B.: A First Course in Abstract Algebra. Addison-Wesley, Reading (1982)
Fulton, W.: Algebraic Curves. Benjamin, New York (1969)
Herstein, I.: Topics in Algebra. Blaisdell, New York (1964)
Hungerford, T.W.: Algebra. Springer, New York (1974).
Jacobson, N.: Basic Algebra I. Freeman, San Francisco (1974)
Jacobson, N.: Basic Algebra II. Freeman, San Francisco (1980)
Lang, S.: Algebra, 2nd edn. Addison-Wesley, Reading (1965)
McCoy, N.: Theory of Rings. Macmillan, New York (1964)
Musuli, C.: Introduction to Rings and Modules. Narosa Publishing House, New Delhi (1992)
van der Waerden, B.L.: Modern Algebra. Ungar, New York (1970)
Zariski, O., Samuel, P.: Commutative Algebra I. Van Nostrand, Princeton (1958)
Zariski, O., Samuel, P.: Commutative Algebra II. Van Nostrand, Princeton (1960)

Chapter 8
Vector Spaces

In the earlier chapters we introduced mainly two algebraic systems, such as groups and rings, which involve only internal operations. In this chapter, we introduce another algebraic system which involves both internal and external operations connected by some relations. A vector space is a combination of both an additive abelian group and a field interlinked by an external law of operation. In this chapter, we study vector spaces and their closely related fundamental concepts, such as linear independence, basis, dimension, linear transformation & its matrix representation, eigenvalue, inner product space, etc. Such concepts form an integral part of linear algebra. Vector spaces have multifaceted applications. Such spaces over finite fields play an important role in computer science, coding theory, design of experiments, combinatorics. Vector spaces over the infinite field $\mathbf{Q}$ of the rationals are important in number theory and design of experiments and vector spaces over $\mathbf{C}$ are essential for the study of eigenvalues. In this chapter, we also show how to construct isomorphic replicas of all homomorphic images of a specified abstract vector space. As the concept of a vector provides a geometric motivation, vector spaces facilitate the study of many areas of mathematics and integrate the abstract algebraic concepts with geometric ideas.

8.1 Introductory Concepts

A vector space is just an algebraic system whose elements combine, under vector addition and multiplication by scalars from a suitable field satisfying certain conditions.

The concept of vector spaces arose through the study of geometry and physics. $\mathbf{R}^2$ endowed with the usual distance function is called 'Euclidean plane'. It is also known as 'Cartesian plane' (or coordinate plane) in honor of René Descartes (1596–1650). He first identified the ordered pairs of real numbers with the points in the coordinate plane. We identify the point $X = (x_1, x_2) \in \mathbf{R}^2$ with the arrow $\overrightarrow{OX}$ starting from the origin $O = (0, 0)$ and ending at the point X. The length of the line segment

M.R. Adhikari, A. Adhikari, *Basic Modern Algebra with Applications*,   273
DOI 10.1007/978-81-322-1599-8_8, © Springer India 2014

OX is denoted by $|\overrightarrow{OX}|$. We define addition '+' and scalar multiplication '·' on $\mathbf{R}^2$ as follows.

Addition on $\mathbf{R}^2$ is defined by the usual parallelogram law. If $\overrightarrow{OX}$ is the arrow corresponding to the point $X = (x_1, x_2) \in \mathbf{R}^2$, then for every non-zero real number r, the arrow $r \cdot \overrightarrow{OX}$ has length $|r|$ times of the length $|\overrightarrow{OX}|$ and whose direction is the same or opposite according to $r > 0$ or $r < 0$. If $r = 0$, then $r \cdot \overrightarrow{OX}$ is identified with $(0, 0)$ and no direction is assigned. For simplicity, we use the symbol $r \cdot X$ (or rX) for $r \cdot \overrightarrow{OX}$. Similarly, the ordered triples $(x_1, x_2, x_3) \in \mathbf{R}^3$ are identified with points in the 3-dimensional Euclidean space $\mathbf{R}^3$.

In physics, there are quantities, such as forces acting at a point in a plane, velocities, and accelerations etc., called vectors. They are also added by parallelogram law and multiplied by real numbers (called scalars) in a similar way.

Using this fact, the operations of pointwise addition and scalar multiplication are defined on $\mathbf{R}^2$ as follows:

$$(x_1, x_2) + (y_1, y_2) = (x_1 + y_1, x_2 + y_2);$$

$$r \cdot (x_1, x_2) = (rx_1, rx_2)$$

for all $(x_1, x_2), (y_1, y_2) \in \mathbf{R}^2$ and $r \in \mathbf{R}$. Then $(\mathbf{R}^2, +)$ is an abelian group.

We consider the above scalar multiplication as a mapping $\mu : \mathbf{R} \times \mathbf{R}^2 \to \mathbf{R}^2$, $(r, X) \mapsto r \cdot X$, satisfying the following properties:

1. $1 \cdot X = X$;
2. $r \cdot (s \cdot X) = (rs) \cdot X$;
3. $(r + s) \cdot X = r \cdot X + s \cdot X$;
4. $r \cdot (X + Y) = r \cdot X + r \cdot Y$,

for all $X, Y \in \mathbf{R}^2$ and $r, s \in \mathbf{R}$.

These properties are called vector properties of $\mathbf{R}^2$. Similar properties hold in the Euclidean n-space $\mathbf{R}^n$ and n-dimensional unitary space $\mathbf{C}^n$ for $n \geq 1$ (see Sect. 8.10). Each point $x = (x_1, x_2, \ldots, x_n)$ of $\mathbf{R}^n$ (or $\mathbf{C}^n$) can be thought to represent a vector.

The above vector properties set up an algebra called algebra of vectors. There are many mathematical systems having such vector properties. These systems introduce the concept of vector spaces over fields. Vector spaces generalize the algebra of vectors.

The vector space structure is one of the most important algebraic structures. The concept of a vector is essential to the study of functions of several variables. Vector spaces play an important role in the solution of specified problems. We now define vector spaces over an arbitrary field with an eye to the model of vector spaces over the field $\mathbf{R}$ of real numbers.

Definition 8.1.1 A *vector space* or a *linear space* over a field F is an additive abelian group V together with an external law of composition (called scalar multiplication).

$\mu : F \times V \to V$, the image of (α, v) under μ is denoted by αv, satisfying the following conditions:

$V(1)$ $1v = v$, where 1 is the multiplicative identity in F;

$V(2)$ $(\alpha\beta)v = \alpha(\beta v)$;

$V(3)$ $(\alpha + \beta)v = \alpha v + \beta v$;

$V(4)$ $\alpha(u + v) = \alpha u + \alpha v$, $\forall \alpha, \beta \in F$ and $u, v \in V$.

Unless otherwise stated, by a vector space V we mean a vector space over a field F. In particular, if $F = \mathbf{R}$, V is called a real vector space and if $F = \mathbf{C}$, V is called a complex vector space. A vector space over a division ring is defined in a similar way. However, in this chapter, we consider vector spaces over a field. The algebraic properties of an arbitrary vector space are similar to the vector properties of $\mathbf{R}^2$, $\mathbf{R}^3$ or $\mathbf{R}^n$. Consequently, an element of a vector space V is called a vector and an element of F is called a scalar.

Example 8.1.1 (i) Let $V = M_{m,n}(F)$ be the set of all $m \times n$ matrices over a field F. Then V forms a vector space over F under usual addition of matrices and multiplication of a matrix by a scalar.

(ii) Let F be a field. Then $F^1 = F$ is itself a vector space over F under usual addition and multiplication in the field F. If $n > 1$, then F^n is also a vector space over F under pointwise addition and scalar multiplication defined by

$$(x_1, x_2, \ldots, x_n) + (y_1, y_2, \ldots, y_n) = (x_1 + y_1, x_2 + y_2, \ldots, x_n + y_n) \quad \text{and}$$

$$\alpha(x_1, x_2, \ldots, x_n) = (\alpha x_1, \alpha x_2, \ldots, \alpha x_n), \quad \forall \alpha \in F \text{ and } (x_1, x_2, \ldots, x_n) \in F^n.$$

In particular, $\mathbf{R}^n$ is a vector space over $\mathbf{R}$ and $\mathbf{C}^n$ is a vector space over $\mathbf{C}$.

Remark An element $(x_1, x_2, \ldots, x_n)$ of $\mathbf{R}^n$ (or $\mathbf{C}^n$) may be considered as the real (or complex) $1 \times n$ matrix, called a row vector or an $n \times 1$ matrix

$$\begin{pmatrix} x_1 \\ x_2 \\ \vdots \\ x_n \end{pmatrix},$$

called a column vector.

(iii) Let K be a subfield of a field F. Then F is a vector space over K, under usual addition and multiplication in F.

(iv) Let $V = F[x]$ be the polynomial ring in x over a field F. Then $F[x]$ is a vector space over F. In particular, if $V_n[x]$ denotes the set of all polynomials in $F[x]$, of degree less than a fixed integer n, along with the zero polynomial, then $V_n[x]$ is also a vector space over F.

(v) (a) Let $V = C([a, b])$ denote the set of all real valued continuous functions on $[a, b]$. Then V is a vector space over $\mathbf{R}$ under the usual pointwise addition and scalar multiplication. In particular, $\mathbf{R}[x]$ is a vector space over $\mathbf{R}$.

(b) The set $V = D([a, b])$ of all real valued differentiable functions on $[a, b]$ is a vector space over $\mathbf{R}$ under the same compositions defined in (a).

(vi) Let V be the set of all sequences of real numbers (or complex numbers). Then V is a vector space over $\mathbf{R}$ (or $\mathbf{C}$).

(vii) Let V be the set of all solutions of a system of m linear homogeneous equations in n variables with real (or complex) coefficients. Then V is a real (or complex) vector space, called the solution space of the system.

(viii) The power set $P(S)$ of a non-empty set S is a vector space over $\mathbf{Z}_2$ under the group addition:

$$x + y = (x - y) \cup (y - x) \quad \text{(symmetric difference), where } x - y = x \setminus y;$$

and scalar multiplication:

$$\alpha x = x, \quad \text{if } \alpha = (1) \in \mathbf{Z}_2,$$
$$= \emptyset, \quad \text{if } \alpha = (0) \in \mathbf{Z}_2.$$

Remark The vector spaces $\mathbf{R}^n$, $\mathbf{C}^n$, $M_{m,n}(\mathbf{R})$ and $M_{m,n}(\mathbf{C})$ are representatives of many vector spaces. They have some additional properties, such as $\mathbf{R}^n$ and $\mathbf{C}^n$ admit inner product, $M_{m,n}(\mathbf{R})$ admits matrix multiplication and transpose.

We use the following notations unless otherwise stated.

Notation

$\mathbf{0}$ represents additive zero in V;
$-v$ represents additive inverse of v in V; and
0 represent the zero element in F.

Proposition 8.1.1 *Let V be a vector space over F. Then*

  (i) $\alpha\mathbf{0} = \mathbf{0}$ *for all α in F*;
 (ii) $0v = \mathbf{0}$ *for all v in V*;
(iii) $(-\alpha)v = \alpha(-v) = -(\alpha v)$ *for all α in F and v in V*;
(iv) $\alpha v = \mathbf{0}$ *implies that either $\alpha = 0$ or $v = \mathbf{0}$*.

Proof (i) and (ii) follow by using the cancellation law of addition in V.
    (iii) follows by uniqueness of inverse in V.
    (iv) If $\alpha \neq 0$ then α^{-1} exists in F and hence (iv) follows.   □

Remark Proposition 8.1.1 says that the multiplication by zero element of V or of F always gives the zero element of V. So we can use the same symbol 0 for both of them.

8.2 Subspaces

In the Euclidean 3-space $\mathbf{R}^3$, the vectors which lie in a fixed plane through the origin form by themselves a 2-dimensional vector space which is a proper subset of the whole space.

We now consider similar subsets of a vector space V which inherit vector space structure from V. Such subsets introduce the concept of subspaces and play an important role in the study of vector spaces. We are interested to show how to construct isomorphic replicas of all homomorphic images of a specified abstract vector space. Subspaces play an important role in determining both the structure of a vector space V and nature of homomorphisms with domain V.

Definition 8.2.1 Let V be a vector space over a field F. A non-empty subset U of V is called a *subspace* of V iff

(i) $(U, +)$ is a subgroup of $(V, +)$;
(ii) for $\alpha \in F$ and $u \in U$, $\alpha u \in U$.

Clearly, for any vector space V, $\{0\}$ and V are subspaces of V, called trivial subspaces. Any other subspace of V (if it exists) is called a proper subspace of V.

An equivalent definition of a subspace can be obtained from the following theorem.

Theorem 8.2.1 *Let V be a vector space over F. Then a non-empty subset U of V is a subspace of V iff for every pair of elements $u, v \in U$, $\alpha u + \beta v \in U$ for all $\alpha, \beta \in F$.*

Proof Let U be a subspace of V. Then for all $\alpha, \beta \in F$ and $u, v \in U$, $\alpha u, \beta v \in U$ imply $\alpha u + \beta v \in U$. Converse part follows by taking $\alpha = 1 = \beta$ and $\beta = 0$ successively in $\alpha u + \beta v$ to show that $u + v \in U$ and $\alpha u \in U$, respectively. $\square$

Example 8.2.1 (i) The set $V_1 = \{(x, 0) \in \mathbf{R}^2\}$, i.e., the x-axis is a proper subspace of $V = \mathbf{R}^2$ over $\mathbf{R}$. Similarly, $V_2 = \{(0, y) \in \mathbf{R}^2\}$, i.e., the y-axis is a proper subspace of V over $\mathbf{R}$.

(ii) All straight lines in the Euclidean plane $\mathbf{R}^2$ passing through the origin $(0, 0)$ are proper subspaces of $\mathbf{R}^2$. Other subspaces of $\mathbf{R}^2$ are $\{0\}$ and $\mathbf{R}^2$ itself.

(iii) All straight lines and planes in the Euclidean space $\mathbf{R}^3$ and passing through the origin $(0, 0, 0)$ are proper subspaces of $\mathbf{R}^3$. Other subspaces of $\mathbf{R}^3$ are $\{0\}$ and $\mathbf{R}^3$ itself. On the other hand, $\mathbf{R}^2$ is not a subspace of $\mathbf{R}^3$, because $\mathbf{R}^2$ is not a subset of $\mathbf{R}^3$.

(iv) $V_n[x]$ is a subspace of $F[x]$ (see Example 8.1.1(iv)).

(v) $D([a, b])$ is a subspace of $C([a, b])$ (see Example 8.1.1(v)).

Proposition 8.2.1 *Let V be a vector space over F and $\{V_i\}_{i \in I}$ be a family of subspaces of V. Then their intersection $M = \bigcap_{i \in I} V_i$ is also a subspace of V.*

Proof Clearly, $0 \in M$ implies that $M \neq \emptyset$. Moreover, for $x, y \in M$ and $\alpha, \beta \in F$, $\alpha x + \beta y$ is in each V_i and hence $\alpha x + \beta y \in M$ shows that M is a subspace of V. $\square$

Remark The union of two subspaces of V may not be a subspace of V. For example, the x-axis X and the y-axis Y are subspaces of $\mathbf{R}^2$ over R, but their union $X \cup Y$

is not a subspace of $\mathbf{R}^2$. Because $u = (1, 0) \in X$ and $v = (0, 1) \in Y$ but $u + v = (1, 1) \notin X \cup Y$.

We now want to define the smallest subspace of $\mathbf{R}^2$ containing $X \cup Y$. This leads to the concept of sum of subspaces of a vector space.

Proposition 8.2.2 *Let V be a vector space over F and V_1, V_2 be subspaces of V. Then $U = V_1 + V_2 = \{v_1 + v_2 : v_1 \in V_1, v_2 \in V_2\}$ is the smallest subspace of V containing both V_1 and V_2.*

Proof Left as an exercise.   □

Definition 8.2.2 Let V be a vector space over F and V_1, V_2 be subspaces of V. Then the subspace $U = V_1 + V_2$ is called the sum of subspaces of V_1 and V_2. In general, if $V_1, V_2, \ldots, V_n$ are subspaces of V, then $\sum_{i=1}^{n} V_i = \{(v_1 + v_2 + \cdots + v_n) : v_i \in V_i\}$ is a subspace of V, called the sum of the subspaces $V_1, V_2, \ldots, V_n$, denoted by $V_1 + V_2 + \cdots + V_n$. In particular, if $V = V_1 + V_2$ and $V_1 \cap V_2 = \{0\}$, then V is said to be a direct sum of V_1 and V_2 and denoted by $V = V_1 \oplus V_2$. The subspace V_1 (or V_2) is said to be a complement of V_2 (or V_1) in V. Similarly, $V = V_1 \oplus V_2 \oplus \cdots \oplus V_n$ iff $V = V_1 + V_2 + \cdots + V_n$ and $(V_1 + V_2 + \cdots + V_i) \cap V_{i+1} = \{0\}$, for $i = 1, 2, \ldots, n - 1$.

Remark A complement of a subspace of V (if it exists) may not be unique. For its existence in a finite dimensional vector space see Corollary 2 to Theorem 8.4.6.

Example 8.2.2 (i) If V_1 is a line through the origin in $\mathbf{R}^2$, then any line through the origin, other than V_1, in $\mathbf{R}^2$ is a complement of V_1 in $\mathbf{R}^2$.
(ii) Let $V_1 = \{(x, 0) \in \mathbf{R}^2\}$, $V_2 = \{(0, y) \in \mathbf{R}^2\}$ and $V_3 = \{(x, x) \in \mathbf{R}^2\}$. Then $\mathbf{R}^2 = V_1 \oplus V_2 = V_2 \oplus V_3$ but $V_1 \neq V_3$.

An equivalent definition of direct sum of subspaces of a vector space is obtained from the following theorem.

Theorem 8.2.2 *A vector space V is a direct sum of its subspaces V_1 and V_2 iff every vector v in V can be expressed uniquely as $v = v_1 + v_2$, where $v_1 \in V_1$ and $v_2 \in V_2$. In general, $V = V_1 \oplus V_2 \oplus \cdots \oplus V_n$ iff every element v of V can be expressed uniquely as $v = v_1 + v_2 + \cdots + v_n$, $v_i \in V_i$, $i = 1, 2, \ldots, n$.*

Proof Let $V = V_1 \oplus V_2$. Then $V = V_1 + V_2$ and $V_1 \cap V_2 = \{0\}$. Let $v = v_1 + v_2 = v_3 + v_4$, where $v_1, v_3 \in V_1$ and $v_2, v_4 \in V_2$. Then $v_1 - v_3 = v_4 - v_2 \in V_1 \cap V_2 = \{0\} \Rightarrow v_1 = v_3$ and $v_2 = v_4$. Conversely, let $v \in V$ be expressed uniquely as $v = v_1 + v_2$, where $v_i \in V_i$, $i = 1, 2$. Then $V = V_1 + V_2$. If $w \in V_1 \cap V_2$, then $w = w + 0 = 0 + w$ show that $w = 0$ by unique expression of w. The proof for the general case is left as an exercise.   □

Example 8.2.3 (i) $\mathbf{R}^4 = V_1 \oplus V_2$, where $V_1 = \{(x, y, 0, 0) \in \mathbf{R}^4\}$ and $V_2 = \{(0, 0, z, t) \in \mathbf{R}^4\}$.

(ii) Let $V_n[x]$ be the vector space over $\mathbf{R}$ defined in Example 8.1.1(iv) for $F = \mathbf{R}$ and $U_i = \{rx^i : r \in \mathbf{R}\}$. Then each U_i is a subspace of $V_n[x]$, for $i = 0, 1, 2, \ldots, n-1$. Moreover, each $f(x) \in V_n[x]$ can be expressed uniquely as

$$f(x) = a_0 + a_1 x + \cdots + a_{n-1} x^{n-1}, \quad a_i \in \mathbf{R}.$$

Consequently, $V_n[x] = U_0 \oplus U_1 \oplus \cdots \oplus U_{n-1}$.

8.3 Quotient Spaces

We continue the study of subspaces of vector spaces.

We are interested to show how to construct isomorphic replicas of all homomorphic images of a specified abstract vector space. For this purpose we introduce the concept of quotient spaces which plays an important role in determining both the structure of a vector space V and nature of homomorphisms with domain V. A subspace of a vector space plays an essential role to construct a quotient space which is associated with the mother vector space in a natural way. Like quotient groups and quotient rings, the concept of quotient spaces is introduced in the following way: Let V be a vector space over F and U a subspace of V. Then $(U, +)$ is a subgroup of the abelian group $(V, +)$ and hence $(V/U, +)$ is also an abelian group. The scalar multiplication

$$\mu : F \times V/U \to V/U, \quad (\alpha, v + U) \mapsto \alpha v + U$$

makes V/U a vector space over F.

Definition 8.3.1 The vector space V/U is called the quotient space of V by U and the map $p : V \to V/U, v \mapsto v + U$ is called a canonical homomorphism.

8.3.1 Geometrical Interpretation of Quotient Spaces

We now present geometrical interpretation of some quotient spaces through Example 8.3.1.

Example 8.3.1 (i) Let $V = \mathbf{R}^2$ be the Euclidean plane and U be the x-axis. Then V/U is the set of all lines in $\mathbf{R}^2$ which are parallel to x-axis. This is so because for $v = (r, t) \in \mathbf{R}^2$, $v + U = (r, t) + U = \{(r + x, t) : x \in \mathbf{R}\}$ is the line $y = t$, which is parallel to x-axis. This line is above or below the x-axis according to $t > 0$ or $t < 0$. If $t = 0$, this line coincides with x-axis.

(ii) If $V = \mathbf{R}^3$ is the Euclidean 3-space and $U = \{(x, y, 0) \in \mathbf{R}^3\}$, then U is the xy-plane and for any $v = (r, t, s) \in \mathbf{R}^3$, the coset $v + U$ represents geometrically the plane parallel to xy-plane through the point $v = (r, t, s)$ at a distance s from the xy-plane (above or below the xy-plane according to $s > 0$ or $s < 0$).

(iii) Let $V = \mathbf{R}^3$ and $U = \{(x, y, z) \in \mathbf{R}^3 : 5x - 4y + 3z = 0 \text{ and } 2x - 3y + 4z = 0\}$. Then for any $v = (a, b, c) \in \mathbf{R}^3$, the coset $v + U \in V/U$ represents geometrically the line parallel to the line determined by the intersection of the two planes:

$$5(x - a) - 4(y - b) + 3(z - c) = 0 \quad \text{and} \quad 2(x - a) - 3(y - b) + 4(z - c) = 0.$$

(iv) Let $V = \mathbf{R}^2$ be the Euclidean plane and $W = \{(x, y) \in \mathbf{R}^2 : ax + by = 0\}$ for a fixed non-zero (a, b) in $\mathbf{R}^2$. The cosets of W in $\mathbf{R}^2$ are given by the lines $ax + by = c$ for $c \in \mathbf{R}$. This is so because if $v \in \mathbf{R}^2$, then the coset $v + W$ is the line parallel to the line $ax + by = 0$ and hence the cosets of W in $\mathbf{R}^2$ are given by the system of parallel lines $ax + by = c, c \in \mathbf{R}$.

8.4 Linear Independence and Bases

We recall that $\mathbf{R}^n$ is a vector space over $\mathbf{R}$ and the n vectors $e_1 = (1, 0, \ldots, 0), e_2 = (0, 1, \ldots, 0), \ldots, e_n = (0, 0, \ldots, 1)$ of $\mathbf{R}^n$ determine every vector of $\mathbf{R}^n$ uniquely. Again if we consider the vector spaces $V = \mathbf{R}[x]$ and $V_n[x]$ (taking $F = \mathbf{R}$ in Example 8.2.1(iv)), we find a finite number of elements, such as $1, x, x^2, \ldots, x^{n-1}$, which determine every element $f(x)$ of $V_n[x]$ uniquely as $f(x) = \alpha_0 + \alpha_1 x + \cdots + \alpha_{n-1} x^{n-1}$. On the other hand, there is no finite set of elements in $\mathbf{R}[x]$, which determines every element $f(x)$ of $\mathbf{R}[x]$ uniquely. Such situations lead to the concept of linear independence and bases in vector spaces. For this purpose we first introduce the concepts of linear combinations and generators in a vector space.

Definition 8.4.1 Let V be a vector space over a field F and S be a non-empty subset (finite or infinite) of V. An element $v \in V$ is said to be a linear combination of elements of S over F iff there exists a finite number of elements $v_1, v_2, \ldots, v_n$ in S and $\alpha_1, \alpha_2, \ldots, \alpha_n$ in F such that

$$v = \alpha_1 v_1 + \alpha_2 v_2 + \cdots + \alpha_n v_n = \sum_{i=1}^{n} \alpha_i v_i.$$

Example 8.4.1 $\mathbf{R}^2$ is a vector space over $\mathbf{R}$. If $S = \{(3, 4), (1, 2)\}$, then the element $(9, 14)$ is a linear combination of elements of S. This is so because $(9, 14) = 2(3, 4) + 3(1, 2)$.

Proposition 8.4.1 *Let V be a vector space over F and S be a non-empty subset of V. Then the set $L(S)$ of all linear combinations of S is a subspace of V.*

Proof As $S \subseteq L(S)$, $L(S) \neq \emptyset$. Let $u, v \in L(S)$. Suppose $u = \sum_{i=1}^{n} \alpha_i u_i$ and $v = \sum_{j=1}^{m} \beta_j v_j, \alpha_i, \beta_j \in F$ and $u_i, v_j \in S$. Then for $\alpha, \beta \in F, \alpha u + \beta v \in L(S)$ implies that $L(S)$ is a subspace of V. $\square$

Definition 8.4.2 $L(S)$ is called the linear space spanned by S in V.

$L(S)$ is the smallest subspace of V containing all the vectors of the given set S.

We now proceed to introduce the concept of "generators" in a vector space V. Let S (may be empty) be a subset of V. Clearly, S is contained in at least one subspace of V, namely, V. Then the intersection of all subspaces of V containing S is a subspace of V by Proposition 8.2.1. It is the smallest subspace of V containing S. This subspace is denoted by $\langle S \rangle$.

Definition 8.4.3 $\langle S \rangle$ is called the subspace generated by S. In particular, if $\langle S \rangle = V$, then V is said to be generated by S and S is called a set of generators for V.

The elements of $\langle S \rangle$ can be obtained from the following theorem.

Theorem 8.4.1 *Let V be a vector space over F and S be an arbitrary subset of V. Then*

$$\langle S \rangle = \{0\}, \quad \text{if } S = \emptyset,$$
$$= L(S), \quad \text{if } S \neq \emptyset.$$

Proof If $S = \emptyset$, the smallest subspace of V containing S is $\{0\}$. Hence $\langle S \rangle = \{0\}$, if $S = \emptyset$. If $S \neq \emptyset$, we claim that $L(S)$ is the smallest subspace of V containing S. Let W be any subspace of V such that $S \subseteq W$. Let $v = \sum_{i=1}^{n} \alpha_i v_i \in L(S), \alpha_i \in F$, $v_i \in S, i = 1, 2, \ldots, n$. Then each $\alpha_i v_i \in W \Rightarrow v \in W \Rightarrow L(S) \subseteq W \Rightarrow L(S)$ is the smallest subspace of V containing $S \Rightarrow L(S) = \langle S \rangle$. $\qquad \square$

Definition 8.4.4 A vector space V is said to be finitely generated iff there exists a finite subset S of V such that $V = \langle S \rangle$ and S is called a set of generators of V. Otherwise, V is said to be not finitely generated; sometimes it is called infinitely generated.

Example 8.4.2 (i) $\mathbf{R}^2, \mathbf{R}^3, \ldots, \mathbf{R}^n$ are finitely generated vector spaces over $\mathbf{R}$. This is so because if $S = \{(1, 0), (0, 1)\}$, then $\langle S \rangle = \mathbf{R}^2$; if $S = \{(1, 0, 0), (0, 1, 0), (0, 0, 1)\}$, then $\langle S \rangle = \mathbf{R}^3$, if $S = \{e_1 = (1, 0, \ldots, 0), e_2 = (0, 1, \ldots, 0), \ldots, e_n = (0, 0, \ldots, 1)\}(\subset \mathbf{R}^n)$, then $\langle S \rangle = \mathbf{R}^n$.

(ii) Set of generators of V may not be unique. For example, $S = \{(1, 0), (0, 1)\}$ and $T = \{(1, 0), (0, 1), (1, 1)\}$ are both generators of $\mathbf{R}^2$.

(iii) If $S = \{1, i\}$, then $\langle S \rangle = \mathbf{C}$ implies that $\mathbf{C}$ is a finitely generated vector space over $\mathbf{R}$.

(iv) $\mathbf{R}[x]$ is not finitely generated. This is so because any linear combination of a finite number of polynomials $f_1(x), f_2(x), \ldots, f_n(x)$ is a polynomial whose degree does not exceed the maximum degree m (say) of the above polynomials. But $\mathbf{R}[x]$ contains polynomials of degree greater than m.

Remark Let V be a vector space over F and S be a non-empty subset of V.

(a) If $S = \{v\}$, then $L(S) = \{\alpha v : \alpha \in F\} = \langle v \rangle$.
(b) If $S = \{u, v\}$, then $L(S) = \{\alpha u + \beta v : \alpha, \beta \in F\} = \langle u, v \rangle$.
(c) For any non-empty subsets A, B of a vector space V,

    (i) A is a subspace of V iff $A = \langle A \rangle$;
    (ii) $A \subseteq B$ implies $L(A) \subseteq L(B)$;
    (iii) $\langle L(A) \rangle = L(A)$;
    (iv) $L(A \cup B) = L(A) + L(B)$.

8.4.1 Geometrical Interpretation

Consider the Euclidean space $\mathbf{R}^3$. It $v \neq 0$ is a vector in $\mathbf{R}^3$, then $L(v) = \{\alpha v : \alpha \in \mathbf{R}\}$ represents geometrically the line in $\mathbf{R}^3$ through the origin and the point v. Again if u, v are two non-zero vectors in $\mathbf{R}^3$ such that $v \neq ru$ for any non-zero real number r (or equivalently, $u \neq tv$ for any non-zero real number t), then for $S = \{u, v\}$, $L(S) = \{\alpha u + \beta v : \alpha, \beta \in \mathbf{R}\}$ represents geometrically the plane through the origin and the points u, v in $\mathbf{R}^3$. Such examples motivate to introduce the concepts of linear independence, basis and dimension of a vector space.

Definition 8.4.5 Let V be a vector space over F.

1. The empty set $\emptyset$ is considered as linearly independent over F.
2. A non-empty finite subset $S = \{v_1, v_2, \ldots, v_n\}$ of V is said to be linearly independent over F iff for $\alpha_1, \alpha_2, \ldots, \alpha_n \in F$, $\alpha_1 v_1 + \alpha_2 v_2 + \cdots + \alpha_n v_n = 0$ implies $\alpha_1 = \alpha_2 = \cdots = \alpha_n = 0$.
3. An infinite subset S of V is said to be linearly independent over F iff every finite subset of S is linearly independent.

Definition 8.4.6 A subset S of a vector space over F is said to linearly dependent over F iff it is not linearly independent over F.

Remark If a subset $S = \{v_1, v_2, \ldots, v_n\}$ is a linearly dependent set in a vector space V over F, this implies that there exist $\alpha_1, \alpha_2, \ldots, \alpha_n$ in F, not all 0, such that $\alpha_1 v_1 + \alpha_2 v_2 + \cdots + \alpha_n v_n = 0$. For example, $S = \{(1, 0), (0, 1)\}$ is a linearly independent subset of $\mathbf{R}^2$ over $\mathbf{R}$, while $T = \{(1, 0), (0, 1), (1, 1)\}$ is a linearly dependent subset of $\mathbf{R}^2$ over $\mathbf{R}$. A linearly independent set cannot contain the vector 0.

Remark Linear independence (or dependence) in a vector space is not a property of an individual vector but a property of a set of vectors.

Example 8.4.3 Consider the vector space $\mathbf{R}^2$ over $\mathbf{R}$.

1. If $v \neq 0$, $\{v\}$ is linearly independent. Because $\alpha v = 0 \Rightarrow \alpha = 0$.
2. If $S = \{(1, 0), (2, 0)\}$, then $1(2, 0) - 2(1, 0) = (0, 0)$ implies that S is linearly dependent.
3. A subset S containing the zero vector 0 is linearly dependent.

Proposition 8.4.2 *If the subset $S = \{v_1, v_2, \ldots, v_n\}$ of a vector space V is linearly independent over F, then every element v is $L(S)$ has a unique expression of the form*

$$v = \alpha_1 v_1 + \alpha_2 v_2 + \cdots + \alpha_n v_n, \quad \alpha_i \in F.$$

Proof Every element $v \in L(S)$ is of the above form by definition. To show its unique expression, let

$$v = \alpha_1 v_1 + \alpha_2 v_2 + \cdots + \alpha_n v_n$$

$$= \beta_1 v_1 + \beta_2 v_2 + \cdots + \beta_n v_n.$$

Then

$$(\alpha_1 - \beta_1)v_1 + (\alpha_2 - \beta_2)v_2 + \cdots + (\alpha_n - \beta_n)v_n = 0$$

$$\Rightarrow \quad \alpha_i = \beta_i, \quad i = 1, 2, \ldots, n. \qquad \square$$

Theorem 8.4.2 *Let $S = \{v_1, v_2, \ldots, v_n\}$ be a finite non-empty ordered set of vectors in a vector space V over F. If $n = 1$, then S is linearly independent iff $v_1 \neq 0$. If $n > 1$ and $v_1 \neq 0$, then either S is linearly independent or some vector v_m $(m > 1)$ is a linear combination of its preceding ones, $v_1, v_2, \ldots, v_{m-1}$.*

Proof If S is linearly independent, there is nothing to prove. So we assume that S is linearly dependent. Then there exist $\alpha_1, \alpha_2, \ldots, \alpha_n$ (not all 0) in F such that

$$\alpha_1 v_1 + \alpha_2 v_2 + \cdots + \alpha_n v_n = 0.$$

Let m be the largest integer such that $\alpha_m \neq 0$. Then $\alpha_i = 0$ for all $i > m$. Hence

$$\alpha_1 v_1 + \alpha_2 v_2 + \cdots + \alpha_{m-1} v_{m-1} + \alpha_m v_m = 0$$

implies that

$$v_m = \alpha_m^{-1}(-\alpha_1 v_1 - \alpha_2 v_2 - \cdots - \alpha_{m-1} v_{m-1})$$

$$= (-\alpha_m^{-1}\alpha_1)v_1 + (-\alpha_m^{-1}\alpha_2)v_2 + \cdots + (-\alpha_m^{-1}\alpha_{m-1})v_{m-1}.$$

This proves the theorem. $\qquad \square$

This theorem gives a series of interesting corollaries.

Corollary 1 *Let V be a vector space over F and $S = \{v_1, v_2, \ldots, v_n\}$ a linearly independent subset of V. If $v \in V$, then $S \cup \{v\}$ is linearly independent iff v is not an element of $L(S)$.*

Proof If $S \cup \{v\}$ is linearly independent, then $v \neq 0$. Suppose $v \in L(S)$. Then $\exists \alpha_i \in F$ such that $v = \alpha_1 v_1 + \alpha_2 v_2 + \cdots + \alpha_n v_n$. Since $v \neq 0, \alpha_1, \alpha_2, \ldots, \alpha_n$ are not all 0. Hence $\alpha_1 v_1 + \alpha_2 v_2 + \cdots + \alpha_n v_n + (-1)v = 0 \Rightarrow S \cup \{v\}$ is linearly dependent over F which is a contradiction. Hence $v \notin L(S)$.

Conversely, suppose $v \notin L(S)$. If possible, suppose $S \cup \{v\}$ is linearly dependent. Then by Theorem 8.4.2, $v \in L(S)$, which is a contradiction. Hence $S \cup \{v\}$ is linearly independent. $\qquad\square$

Corollary 2 Let the subset $S = \{v_1, v_2, \ldots, v_n\}$ of the vector space V be such that $U = L(S)$. If $\{v_1, v_2, \ldots, v_m\}$ $(m \leq n)$ is a linearly independent set, then there exists a linearly independent subset T of S of the form $T = \{v_1, v_2, \ldots, v_m, v_{t1}, v_{t2}, \ldots, v_{tr}\}$ such that $U = L(T)$.

Proof If S is linearly independent, then we take $T = S$. If not, we take out from S the first v_m which is not a linear combination of its preceding ones. Then $v_1, v_2, \ldots, v_t$ are linearly independent for $t < m$. Clearly, $B = \{v_1, v_2, \ldots, v_{m-1}, v_{m+1}, \ldots, v_n\}$ has $n - 1$ elements and $L(B) \subseteq U$. To show the equality, let $u \in U$. Then it can be expressed as a linear combination of elements of S. But in this linear combination, we can replace v_m by a linear combination of $v_1, v_2, \ldots, v_{m-1}$. Thus u is a linear combination of $v_1, v_2, \ldots, v_{m-1}, v_{m+1}, \ldots, v_n$.

Continuing this taking out process we obtain a subset $T = \{v_1, \ldots, v_m, v_{t1}, v_{t2}, \ldots, v_{tn}\}$ (say) of S such that $L(T) = U$ and in which no element is a linear combination of the preceding ones. T is clearly linearly independent by Theorem 8.4.2. $\qquad\square$

Corollary 3 Let V be a finitely generated vector space. Then V contains a linearly independent finite subset T of V such that $L(T) = V$.

Proof Since V is finitely generated, there exists a subset $S = \{v_1, v_2, \ldots, v_n\}$ of V such that $V = L(S)$. By using Corollary 2, we can find a linearly independent subset T of S such that $L(T) = V$. $\qquad\square$

Definition 8.4.7 Let V be a non-zero vector space over F. A non-empty subset B of V is said to be a basis of V over F iff

 (i) B is linearly independent over F; and
(ii) $V = L(B)$.

If $V = \{0\}$, the empty set $\emptyset$ is considered to be a basis of V.

Proposition 8.4.3 If a vector space V is finitely generated and $S = \{v_1, v_2, \ldots, v_n\}$ is a subset of V such that $V = L(S)$, then a subset B of S forms a basis of V.

Proof It follows from Corollary 3. $\qquad\square$

The following theorem proves the existence of a basis of an arbitrary vector space.

Theorem 8.4.3 (Existence Theorem) *Every vector space V over F has a basis.*

Proof If $V = \{0\}$, then $\emptyset$ is considered to be a basis of V. Next suppose $V \neq \{0\}$. Let $\mathcal{S}$ be the family of all linearly independent subsets of V i.e., $\mathcal{S} = \{X \subseteq V :$ X is linearly independent over $F\}$. Since every non-zero element of V forms a linearly independent subset, it is in $\mathcal{S}$ and hence V is non-empty. We order $\mathcal{S}$ partially under set inclusion. By Zorn's Lemma, $\mathcal{S}$ has a maximal element, say B. Then B is linearly independent. We claim that $L(B) = V$. Otherwise, $\exists$ an element $v \in V$ such that $v \notin L(B)$. Then $B \cup \{v\}$ is linearly independent by Corollary 1 to Theorem 8.4.2. This contradicts the maximality of B. Hence B is a basis of V. $\square$

We now present an alternative criterion for a basis.

Theorem 8.4.4 *A non-empty subset B of a vector space V is a basis of V iff every element of V can be expressed uniquely as a linear combination of elements of B.*

Proof First suppose that B forms a basis of V. Then B is linearly independent and $V = L(B)$. Hence every element $v \in V$ has a unique expression in the desired form by Proposition 8.4.2.

Conversely, let every element of V be expressed uniquely as a linear combination of elements of B. Then $V \subseteq L(B) \subseteq V \Rightarrow V = L(B)$. From the unique expression of 0, it follows that B is linearly independent. Hence B is a basis of V. $\square$

Theorem 8.4.5 *Let V be a non-zero vector space over F.*

(a) *Let S be an arbitrary linearly independent subset of V. If S is not a basis of V, then S can be extended to a basis of V. In particular, every singleton set of a non-zero vector of V can be extended to a basis of V.*

(b) *If S is a subset of V such that $V = L(S)$, then S contains a basis of V.*

(c) *If V has a finite basis B consisting of n elements, then any other basis of V is also finite consisting of n elements.*

(d) *Cardinality of every basis of V is the same.*

Proof Let V be a non-zero vector space over F.

(a) Let $\mathcal{F}$ be the family of all linearly independent subsets S of V containing any linearly independent subset A of V i.e., $\mathcal{F} = \{S \subseteq V : S$ is linearly independent and $A \subseteq S$, where A is any linearly independent subset of $V\}$. Then $\mathcal{F}$ is non-empty. This is so because all singleton sets of non-zero elements of V are in $\mathcal{F}$. We order $\mathcal{F}$ partially by set inclusion. By Zorn's Lemma, $\mathcal{F}$ has a maximal element B (say). B is clearly linearly independent. Moreover, $L(B) = V$. Otherwise, we contradict the maximality of B. Hence B is a basis of V. The last part follows immediately as every non-zero vector is linearly independent.

(b) Let S be any subset of V such that $V = L(S)$ and $\mathcal{F}_S = \{B \subseteq S :$ B is linearly independent$\}$. Proceeding as in (a), the maximal element of $\mathcal{F}_S$ forms a basis of V.

(c) Let $B = \{v_1, v_2, \ldots, v_n\} = \{v_i\}$ be a finite basis of V with n elements and $C = \{u_j : j \in J\}$ be any other basis of V, where J is an indexing set. We claim that C is finite and if C has m elements, then $m = n$. If C is not finite, we reach a contradiction, because each v_i is a linear combination of certain u_j's and all the u_j's occurring in this way form a finite subset S of C. If we assume that C is not finite, then there exists an element u_{j0} (say) in C which is not in S. But u_{j0} is a linear combination of the v_i's and hence u_{j0} is also a linear combination of the vectors in S. This implies by Corollary 1 to Theorem 8.4.2 that $S \cup \{u_{j0}\}$ is a linearly dependent subset of C. But this is not possible, since C is a basis of V. Thus one concludes that C is finite.

Suppose $C = \{u_1, u_2, \ldots, u_m\}$ for some positive integer m. Then we prove that $m = n$. Since B is a basis of V, u_1 can be expressed as a linear combination of $v_1, v_2, \ldots, v_n$ and hence the set $S_1 = \{u_1, v_1, v_2, \ldots, v_n\}$ is linearly dependent. Then by Theorem 8.4.2, there is some $v_t \neq u_1$ in B such that v_t is a linear combination of the vectors in S_1 preceding it. If we delete v_t from S_1 then the remaining set $S_2 = \{u_1, v_1, \ldots, v_{t-1}, v_{t+1}, \ldots, v_n\}$ is such that $L(S_2) = V$. Again $u_2 \in C$ is a linear combination of vectors in S_2 and the set $S_3 = \{u_1, u_2, v_1, \ldots, v_{t-1}, \hat{v}_t, v_{t+1}, \ldots, v_n\}$ is linearly dependent and such that $L(S_3) = V$, where $\hat{v}_t$ denotes v_t deleted. In this process, we include one vector u_j from C and delete one vector v_i from B. Continuing this process, we cannot delete all the v_i's before the u_j's are exhausted. This is so because in that case, the remaining u_j's would be linear combinations of vectors of C already used. This contradicts the linear independence of u_j's. This implies that n cannot be less than m. Similarly, m cannot be less than n. So we conclude that $m = n$.

(d) Let $B = \{v_i : i \in I\}$ and $C = \{u_j : j \in J\}$ be two bases of V (finite or infinite). If any one of B or C is finite, then the other is also finite and they have the same number of vectors by (c). We now consider the possibility when both B and C are infinite. Since C is a basis, each $v_i \in B$ can be expressed uniquely as a linear combination with non-zero coefficients of some u_j's (finite in number), say, $v_i = \alpha_1 u_{j1} + \cdots + \alpha_n u_{jn}$. Moreover, every u_j appears in at least one such expression, because, if some u_{j0} does not appear in any such expression, then the basis B shows that u_{j0} is a linear combination of some v_i's and hence of some u_j's each of which is different from u_{j0}. This is not possible, as the u_j's are linearly independent. Consequently, corresponding to each $v_i \in B$, there exists a finite non-empty set S_{v_i} of u_j's such that $C = \bigcup_{v_i \in B} S_{v_i}$. By the property of cardinality (see Chap. 1), it follows that $\operatorname{card} C \leq \operatorname{card} B$. Again interchanging the roles of vectors of B and C, it follows that $\operatorname{card} B \leq \operatorname{card} C$. Hence $\operatorname{card} B = \operatorname{card} C$. $\qquad \square$

The above theorem leads to the definition of the dimension of a vector space.

Definition 8.4.8 Let V be a non-zero vector space over F. The cardinality of every basis of V is the same and this common value is called the dimension of V, denoted by $\dim_F V$ or simply $\dim V$. The vector space V is said to be finite dimensional or infinite dimensional according to V having a finite basis or not. If B has a basis of n elements, then $\dim V = n$. On the other hand, if $V = \{0\}$, then V is said to be 0-dimensional or said to have dimension 0.

Theorem 8.4.5 gives the following series of corollaries.

Corollary 1   *Let V be a vector space over F. If $B = \{v_1, v_2, \ldots, v_n\}$ is a maximal set of linearly independent vectors of V, then B forms a basis of V.*

Corollary 2   *Let V be a finite dimensional vector space over F. If $\dim V = n$ and $B = \{v_1, v_2, \ldots, v_n\}$ is linearly independent, then B forms a basis of V.*

Corollary 3   *Let V be a finite dimensional vector space over F and U be a subspace of V such that $\dim V = \dim U$. Then $U = V$.*

Corollary 4   *Let V be a vector space of dimension n. If $S = \{v_1, v_2, \ldots, v_m\}$ is a linearly independent subset of V and $n > m$, then there are vectors $v_{m+1}, \ldots, v_{m+t}$ in V such that $B = \{v_1, v_2, \ldots, v_m, v_{m+1}, \ldots, v_{m+t}\}$ forms a basis of V, where $t = n - m$.*

Remark   Any linearly independent subset of a finite dimensional vector space V can be extended to a basis of V.

Corollary 5   *Let U be a subspace of a finite dimensional vector space V and $\dim V = n$ $(n \geq 1)$. Then U is also finite dimensional and $\dim U \leq n$. Equality holds iff $U = V$.*

Theorem 8.4.6   *Let A and B be subspaces of a finite dimensional vector space V. Then $A + B$ is also finite dimensional and $\dim(A + B) = \dim A + \dim B - \dim(A \cap B)$.*

Proof   Clearly, $A \cap B$ is finite dimensional. If $A \cap B \neq \{0\}$, then it has a basis $S = \{v_1, v_2, \ldots, v_n\}$, say. This basis can be extended to a basis $S_A = \{v_1, v_2, \ldots, v_n, v_{n+1}, \ldots, v_{n+t}\}$ of A and to a basis $S_B = \{v_1, v_2, \ldots, v_n, w_{n+1}, \ldots, w_{n+q}\}$ of B. Then $\dim A = n + t$, $\dim B = n + q$ and $\dim A \cap B = n$. Let $U = L(\{v_{n+1}, \ldots, v_{n+t}\})$. Then $U \cap B = \{0\}$ and $C = \{v_1, v_2, \ldots, v_n, v_{n+1}, \ldots, v_{n+t}, w_{n+1}, \ldots, w_{n+q}\}$ is linearly independent and $L(C) = A + B$. Consequently, C forms a basis of $A + B$. Hence $\dim(A + B) = n + t + q = (n + t) + (n + q) - n = \dim A + \dim B - \dim(A \cap B)$. $\square$

Corollary 1   *For any two subspaces A and B of a finite dimensional vector space V, $\dim(A + B) \leq \dim A + \dim B$. Equality holds iff $A \cap B = \{0\}$.*

Corollary 2   *Let V be a finite dimensional vector space and A be a subspace of V. Then there exists a subspace B of V such that $V = A \oplus B$ and $\dim V = \dim A + \dim B$.*

Proof   Let $B_A = \{v_1, v_2, \ldots, v_n\}$ be a basis of A. Then B_A can be extended to a basis $B_V = \{v_1, v_2, \ldots, v_n, u_1, u_2, \ldots, u_p\}$ of V. Let $B = L(\{u_1, u_2, \ldots, u_p\})$. Then $V = A + B$ and $A \cap B = \{0\}$. Hence $V = A \oplus B$. Then $\dim V = \dim A + \dim B$. $\square$

If V is a finite dimensional vector space and W is a subspace of V, what is the dimension of V/W? The following theorem gives its answer.

Theorem 8.4.7 *Let W be a subspace of a finite dimensional vector space V. Then* $\dim(V/W) = \dim V - \dim W$.

Proof Let $\dim V = n$, $\dim W = m$ and $B_W = \{w_1, w_2, \ldots, w_m\}$ be a basis of W. We now extend B_W to a basis $B = \{w_1, w_2, \ldots, w_m, v_1, v_2, \ldots, v_t\}$ of V such that $\dim V = m + t = n$. Then $A = \{v_1 + W, v_2 + W, \ldots, v_t + W\}$ forms a basis of V/W. Clearly, $\dim(V/W) = t = n - m$. $\square$

Corollary *Let W be a subspace of a finite dimensional vector space V. Then $\{v_1 + W, v_2 + W, \ldots, v_m + W\}$ constitutes a basis of V/W iff $\{v_1, v_2, \ldots, v_m\}$ constitutes a basis of W.*

8.4.2 Coordinate System

Proposition 8.4.2 leads to extend the concept of usual coordinate system in the Euclidean plane $\mathbf{R}^2$ or Euclidean space $\mathbf{R}^3$ to any finite dimensional vector space.

Definition 8.4.9 Let V be an n-dimensional vector space over F and $B = \{v_1, v_2, \ldots, v_n\}$ be a basis of V. Then the unique representation of v in V as

$$v = \alpha_1 v_1 + \alpha_2 v_2 + \cdots + \alpha_n v_n, \quad \text{with } \alpha_i \in F,$$

determines the unique n-tuple $(\alpha_1, \alpha_2, \ldots, \alpha_n) \in F^n$, called the coordinate of v referred to the basis B i.e., referred to axes $OX_1, OX_2, \ldots, OX_n$ (each extended infinitely on both sides) as coordinate axes, where $X_1, X_2, \ldots, X_n$ are the points corresponding to the vectors $v_1, v_2, \ldots, v_n$ respectively, called reference points along these axes.

Remark The ith coordinate α_i of a point v is determined only from the entire coordinate system and not from v_i alone. The coordinates of v in V depend on the choice of the basis of V like the choice of coordinate axes in $\mathbf{R}^2$ or $\mathbf{R}^3$. Corresponding to different bases of V, we get different coordinates of the same vector in V.

8.4.3 Affine Set

In $\mathbf{R}^3$ all non-trivial subspaces are the lines through the origin and the planes through the origin. But in geometry we also study the planes and lines not necessarily passing through the origin. We now introduce a concept called affine set which is closely related to the concept of a subspace.

Definition 8.4.10 Let V be a vector space over F. A subset A of V is said to be an *affine set* iff either $A = \emptyset$ or $A = v + W$ (a coset) for some subspace W of V and some $v \in V$ and an affine combination of vectors $v_1, v_2, \ldots, v_n$ of V is a linear combination of the form $\alpha_1 v_1 + \alpha_2 v_2 + \cdots + \alpha_n v_n$ such that $\alpha_1 + \alpha_2 + \cdots + \alpha_n = 1$.

Example 8.4.4 A convex combination of x and y in $\mathbf{R}^2$ has the form $tx + (1 - t)y$ for all real $t \geq 0$. An affine combination of points $v_1, v_2, \ldots, v_m$ in $\mathbf{R}^n$ is a point $v = \alpha_1 v_1 + \alpha_2 v_2 + \cdots + \alpha_n v_m$, where $\alpha_i \in \mathbf{R}$ and $\sum_{i=1}^{m} \alpha_i = 1$. A convex combination is an affine combination for which $\alpha_i \geq 0$ for all i.

Remark A subset A of Euclidean space is called an affine set iff for every pair of distinct elements x, y in A, the line determined by x and y lies entirely in A. By default, $\emptyset$ and one-point subsets are affine.

Definition 8.4.11 An affine set $A \subset \mathbf{R}^n$ is said to be spanned by $\{v_0, v_1, \ldots, v_n\} \subset \mathbf{R}^n$ iff A consists of all affine combinations of these vectors.

Definition 8.4.12 An ordered set of vectors $\{v_0, v_1, \ldots, v_m\} \subset \mathbf{R}^n$ is said to be affinely independent iff $\{v_1 - v_0, v_2 - v_0, \ldots, v_m - v_0\}$ is a linearly independent subset of $\mathbf{R}^n$.

Definition 8.4.13 Let $S = \{v_0, v_1, \ldots, v_m\}$ be an affinely independent subset of $\mathbf{R}^n$. Then the convex set spanned by this set S, denoted by $\langle S \rangle$, is the set of all affine combinations of the vectors in S i.e., a point v is in $\langle S \rangle$ iff v can be expressed uniquely as

$$v = \alpha_0 v_0 + \alpha_1 v_1 + \cdots + \alpha_m v_m, \tag{8.1}$$

where α_i's are all non-negative real numbers such that $\alpha_0 + \alpha_1 + \alpha_2 + \cdots + \alpha_m = 1$. This gives a unique $(m + 1)$-tuple $(\alpha_0, \alpha_1, \ldots, \alpha_m)$ with $\sum \alpha_i = 1$ and $v = \sum_{i=0}^{m} \alpha_i v_i$. This $(m + 1)$-tuple is called the barycentric coordinate of v relative to the ordered set S.

Definition 8.4.14 Let $S = \{v_0, v_1, \ldots, v_m\}$ be an affinely independent subset of $\mathbf{R}^n$. The convex set $\langle S \rangle$ is called the (affine) m-simplex with vertices $v_0, v_1, \ldots v_m$, denoted by $\langle v_0, v_1, \ldots, v_m \rangle$.

Example 8.4.5 A 0-simplex is a point, 1-simplex is a closed line segment, 2-simplex is a triangle (with its interior points), 3-simplex is a tetrahedron (solid).

8.4.4 Exercises

Exercises-I

1. Let V be a non-zero vector in $\mathbf{R}^2$. Then the subspace $W = \{rv : r \in \mathbf{R}\}$ is the straight line passing through the origin.

[*Hint.* Let (α, β) represent a vector v is $\mathbf{R}^2$. Then the line passing through the origin $(0, 0)$ and (α, β) is $\frac{x-0}{0-\alpha} = \frac{y-0}{0-\beta}$. Hence any point on the line is given by $(r\alpha, r\beta)$ for $r \in \mathbf{R}$.]

2. Let U be a subspace of a vector space V and $v \in V$ be a fixed vector. Then the set S defined by $S = v + U$ is an affine set.

3. Let $W = \{(x, y) : ax + by = 0\}$ for a fixed non-zero (a, b) in $\mathbf{R}^2$. Then W is a one-dimensional subspace of $\mathbf{R}^2$ and the cosets of W in $\mathbf{R}^2$ are given by the lines $ax + by = c$, for $c \in \mathbf{R}$.

4. Let $W = \{(x, y, z) \in \mathbf{R}^3 : y = z\}$. Then W is a subspace of $\mathbf{R}^3$ such that dim $W = 2$.

5. Let $M_2(\mathbf{R})$ be the vector space of all 2×2 matrices over $\mathbf{R}$ and S be the set of all symmetric matrices in $M_2(\mathbf{R})$. Then dim $S = 3$.

6. Let V be an n-dimensional vector space. Then the following statements are equivalent:

   (a) $B = \{v_1, v_2, \ldots, v_n\}$ is a basis of V;
   (b) $B = \{v_1, v_2, \ldots, v_n\}$ is a maximal linearly independent set;
   (c) $B = \{v_1, v_2, \ldots, v_n\}$ is a minimal generating set.

8.5 Linear Transformations and Associated Algebra

The central problem of linear algebra is to study the algebraic structure of linear transformations.

Like homomorphisms of groups and rings, a linear transformation is a map preserving the algebraic structures of vector spaces. This concept is at the heart of linear algebra. Many problems of mathematics are solved with the help of linear transformations. The importance of vector spaces is based on mainly the linear transformations they carry, because many problems of algebra and analysis, when properly posed, may be reduced to the study of linear transformations of vector spaces. For example, linear transformations play an important role in the study of matrices, differential and integral equations and integration theory etc. The aim of this section is to extend the study of vector spaces with the help of linear transformations.

Definition 8.5.1 Let V and W be vector spaces over the same field F. A transformation T from V to W is a mapping $T : V \to W$ such that

L(i) $T(x + y) = T(x) + T(y)$, for all x, y in V (additivity law);
L(ii) $T(\alpha x) = \alpha T(x)$, for all x in V, and for all α in F (homogeneity law).

Conditions L(i) and L(ii) can be combined together to obtain an equivalent condition:

L(iii) $T(\alpha x + \beta y) = \alpha T(x) + \beta T(y)$, for all α, β in F and for all $x, y \in V$.

In particular, if $V = W$, a linear transformation $T : V \to V$ is called a linear operator.

A linear transformation $T : V \to W$ is called a monomorphism, epimorphism or isomorphism according to T being injective, surjective or bijective.

Remark Every linear transformation is a homomorphism of the corresponding additive groups.

Example 8.5.1 (a) Each of the following examples is a linear transformation (LT):

(i) The identity map $I_d : V \to V, x \mapsto x$, for every vector space V is an LT.

(ii) For $\alpha \in \mathbf{R}, T_\alpha : \mathbf{R}^2 \to \mathbf{R}^2, (x, y) \mapsto (\alpha x, \alpha y)$ is an LT.

(iii) $T_1 : \mathbf{R}^2 \to \mathbf{R}^2, (x, y) \mapsto (y, x)$, i.e., T_1 reflects $\mathbf{R}^2$ about the line $y = x$ is an LT.

(iv) $T_2 : \mathbf{R}^2 \to \mathbf{R}^2, (x, y) \mapsto (x, 0)$, i.e., T_2 projects $\mathbf{R}^2$ onto the x-axis is an LT.

(v) Let $C([0, 1])$ be the vector space of all real valued continuous functions defined on $[0, 1]$. Then the map

$$I : C([0, 1]) \to \mathbf{R}, \quad f \mapsto \int_0^1 f(x)\, dx$$

is an LT.

(vi) For any matrix $M \in M_{m,n}(\mathbf{R})$, the map $T_M : \mathbf{R}^n \to \mathbf{R}^m, X \mapsto MX$, X viewed as a column matrix, is an LT.

The usual dot product may be viewed as a special case of T_M, where $T_M : \mathbf{R}^n \to \mathbf{R}, X \mapsto MX$ for every row matrix M i.e., if

$$M = (a_1, a_2, \ldots, a_n) \quad \text{and} \quad X = \begin{pmatrix} x_1 \\ x_2 \\ \vdots \\ x_n \end{pmatrix},$$

then $T_M(X) = a_1 x_1 + a_2 x_2 + \cdots + a_n x_n$ is an LT.

(vii) Let V be the vector space of all real valued functions in x having derivatives of all orders. Then

$$\frac{d}{dx} = D : V \to V, \quad f \mapsto f'(x) = \frac{df}{dx} \quad \text{is an } LT.$$

(viii) Let V be the vector space of all real valued functions f in n-variables $x_1, x_2, \ldots, x_n$ admitting $\frac{\partial f}{\partial x_i}$ for all i. Then the gradient of f defined by

$$\text{grad } f(X) = \left(\frac{\partial f}{\partial x_1}, \ldots, \frac{\partial f}{\partial x_n} \right) \quad \text{is an } LT.$$

(b) The map $T : \mathbf{R}^2 \to \mathbf{R}^1, (x, y) \mapsto \sin(x + y)$ is not an LT.
This is so because $0 = T(\pi, 0) = T((\frac{\pi}{2}, 0) + (\frac{\pi}{2}, 0)) = 2T(\frac{\pi}{2}, 0) = 2$.

Like groups, we define kernel and image of a linear transformation of vector spaces.

Definition 8.5.2 If $T : V \to W$ is a linear transformation, then $\ker T = \{v \in V : T(v) = 0\} \subseteq V$ is called the kernel of T and $\operatorname{Im} T = \{T(v) : v \in V\} \subseteq W$ is called the range or image of T.

Clearly, $\ker T$ is a subspace of V, called the null space of T and $\operatorname{Im} T$ is a subspace of W, called the range space of T.

Interpretation If T is a left multiplication by a matrix M (see Example 8.5.1(vi)), then $\ker T$ is the set of solutions (in $\mathbf{R}^n$) of the homogeneous linear equation $MX = 0$ and $\operatorname{Im} T$ is the set of all vectors C (in $\mathbf{R}^m$) such that $MX = C$ has a solution.

Example 8.5.2   (i) If $T : \mathbf{R}^3 \to \mathbf{R}^2$, $(x, y, z) \mapsto (x, y)$, then $\ker T = \{(x, y, z) \in \mathbf{R}^3 : T(x, y, z) = (0, 0)\} = \{(0, 0, z) \in \mathbf{R}^3\}$ is the z-axis in $\mathbf{R}^3$ and $\operatorname{Im} T$ is the whole Euclidean plane $\mathbf{R}^2$.

(ii) If $T : M_2(\mathbf{R}) \to \mathbf{R}^4$, $\begin{pmatrix} a & b \\ c & d \end{pmatrix} \mapsto (a, b, c, d)$, then T is a linear map such that $\ker T = \left\{ \begin{pmatrix} 0 & 0 \\ 0 & 0 \end{pmatrix} \right\}$ and $\operatorname{Im} T = \mathbf{R}^4$.

We now establish that an n-dimensional vector space over F and F^n are isomorphic as vector spaces.

Theorem 8.5.1 *Let V be an n-dimensional vector space over F ($n \geq 1$). Then V is isomorphic to the vector space F^n.*

Proof Let $W = F^n$ and $B = \{v_1, v_2, \ldots, v_n\}$ be a basis of V. Then every vector v in V can be expressed uniquely as $v = \alpha_1 v_1 + \cdots + \alpha_n v_n$, $\alpha_i \in F$. Using this coordinate of v referred to B, we define the map $T : V \to F^n$, $v \mapsto (\alpha_1, \alpha_2, \ldots, \alpha_n)$. Then T is an isomorphism of vector spaces. $\square$

Remark The isomorphism T depends on the basis B and also on the order of the elements of B.

Corollary *Any non-zero n-dimensional vector space V over $\mathbf{R}$ (or $\mathbf{C}$) is isomorphic to $\mathbf{R}^n$ (or $\mathbf{C}^n$) as vector spaces.*

For example, the vector space $V_n[x]$ (see Example 8.1.1(iv)) over $\mathbf{R}$ and the vector space $\mathbf{R}^n$ have the same algebraic structure as vector spaces.

For a given subset X of a vector space V over F, the set $M(X)$ of all F-valued functions on X which vanish outside any finite set is a vector space over F under usual addition and scalar multiplication. Using this fact we generalize Theorem 8.5.1.

Theorem 8.5.2 *Let V be a non-zero vector space over F. If B is a basis of V, then V is isomorphic to the vector space $M(B)$.*

Proof Let $B = \{v_i : i \in I\}$ be a basis of V. We now construct an isomorphism of V onto $M(B)$ by assigning to each vector v in V an F-valued function f_v defined on

B as follows: If $v \neq 0$, then v can be expressed uniquely as

$$v = \alpha_1 v_{i_1} + \alpha_2 v_{i_2} + \cdots + \alpha_n v_{i_n}, \quad \text{where } \alpha_1, \alpha_2, \ldots, \alpha_n$$

are non-zero elements in F, $S = \{v_{i_1}, v_{i_2}, \ldots, v_{i_n}\}$ is a finite subset of B and f_v is defined by

$$f_v(v_{ij}) = 0, \quad \text{if } v_{ij} \text{ lies outside the set } S;$$
$$= \alpha_{ij}, \quad \text{if } v_{ij} \text{ lies inside the set } S.$$

Then the map $\psi : V \to M(B)$, $v \mapsto f_v$ is an isomorphism of vector spaces.   □

We now give another extension of Theorem 8.5.1.

Theorem 8.5.3 *Let V be a non-zero vector space over F. Then V is isomorphic to the direct sum of copies of F.*

Proof Let $B = \{v_i : i \in I\}$ be a basis of V. It is sufficient to prove that $V = \bigoplus_{i \in I} F v_i$, where

$$F v_i = \{\alpha v_i : \alpha \in F\} \quad \text{and} \quad F v_i \cong F \quad \text{for every } i \in I.$$

Clearly, $F v_i$ is a non-empty subset of V, since $v_i = 1 \cdot v_i \in F v_i$. Again $F v_i$ is a subspace of V for each $v_i \in B$. Let $v \in V$. Then v can be expressed uniquely as

$$v = \sum_{i \in J} \alpha_i v_i, \quad \text{where } J \text{ is a finite subset of } I.$$

Since $\alpha_i v_i \in F v_i$, $\forall i \in I$ (hence $\forall i \in J$), v can be expressed uniquely as $v = \sum_{i \in J} u_i$, where $u_i \in F v_i$. Thus

$$v = \sum_{i \in J} u_i, \quad \text{for all } i \in J$$
$$= 0, \quad \text{for all } i \text{ outside } J.$$

Hence $V = \bigoplus_{i \in I} F v_i$ where each $T_i : F \to F v_i$, $\alpha \mapsto \alpha v_i$ is an isomorphism.   □

Remark Theorems 8.5.1–8.5.3 show that an abstract vector space V seems to have a nice representation for study. For example, an n-dimensional vector space over **R** may be studied through $\mathbf{R}^n$, which is easier to visualize. But such representation loses much of its importance, because it depends on the choice of a basis B of V. Moreover, almost all the vector spaces of greatest importance carry additional structure (algebraic or topological), which need not be related to the above isomorphisms. For example, the product of two polynomials in the vector space $V_n[x]$ is defined in a natural way but there is no similar concept in $\mathbf{R}^n$, although $V_n[x]$ and $\mathbf{R}^n$ are isomorphic as vector spaces. We now give some interesting properties of linear transformations.

Proposition 8.5.1 *Let $T : U \to V$ be a linear transformation. Then*

(i) $T(0) = 0$;

(ii) $T(-u) = -T(u)$ *for all u in U*;

(iii) *T is injective iff* $\ker T = \{0\}$.

Proof Left as an exercise.   □

Theorem 8.5.4 *Let U and V be finite dimensional vector spaces over F and $B = \{u_1, u_2, \ldots, u_n\}$ be a basis of U.*

(i) *If $T : U \to V$ is a linear transformation, then*

$$T(B) = \big\{ T(u_1), T(u_2), \ldots, T(u_n) \big\} \quad \text{generates} \ \operatorname{Im} T.$$

Moreover, if $\ker T = \{0\}$, *then $T(B)$ constitutes a basis of* $\operatorname{Im} T$.

(ii) *For arbitrary non-zero elements $v_1, v_2, \ldots, v_n$ in V, there exists a unique linear transformation $T : U \to V$ such that $T(u_i) = v_i$, $i = 1, 2, \ldots, n$.*

(iii) *T defined in (ii) is injective iff the vectors $v_1, v_2, \ldots, v_n$ are linearly independent.*

Proof (i) If $v \in \operatorname{Im} T$, $\exists$ some $u \in U$ such that $T(u) = v$. As B is a basis of U, $\exists \alpha_1, \alpha_2, \ldots, \alpha_n \in F$ such that $u = \sum_{i=1}^{n} \alpha_i u_i$. This shows by linearity of T that $\operatorname{Im} T$ is generated by $T(B)$. If $\ker T = \{0\}$, then T is a monomorphism and hence $T(B)$ is linearly independent and forms a basis of $\operatorname{Im} T$.

(ii) Left as an exercise.

(iii) Let $v_1, v_2, \ldots, v_n$ be linearly independent and $u \in \ker T$. Then $u = \sum_{i=1}^{n} \alpha_i u_i$ for some $\alpha_i \in F$ shows that $T(u) = \sum_{i=1}^{n} \alpha_i T(u_i) = 0$. Hence $\sum_{i=1}^{n} \alpha_i v_i = 0 \Rightarrow \alpha_i = 0$, $\forall i \Rightarrow \ker T = \{0\} \Rightarrow T$ is injective. Its converse part follows from (i).   □

Remark Let U be a finite dimensional vector space. Then any linear transformation $T : U \to V$ is completely determined by the action of T on a basis of U.

Like groups and rings, the homomorphism theorems are obtained for vector spaces.

Theorem 8.5.5 (First Isomorphism Theorem) *Let $T : U \to V$ be a linear transformation. Then the vector spaces $U / \ker T$ and $\operatorname{Im} T$ are isomorphic. Conversely, if U is a vector space and W is a subspace of U, then there is a linear transformation from U onto U / W.*

Proof Proceed as in groups.   □

Corollary 1 *If a linear transformation $T : U \to V$ is onto, then the vector spaces $U / \ker T$ and V are isomorphic.*

Corollary 2   *Let V be a finite dimensional vector space and $T : V \to V$ a linear operator. Then T is injective iff T is surjective.*

Proof By the First Isomorphism Theorem, $V/\ker T \cong \operatorname{Im} T = U$ (say). Then $\dim U = \dim V - \dim(\ker T)$. Clearly, T is surjective $\Leftrightarrow V = \operatorname{Im} T = U \Leftrightarrow \ker T = \{0\} \Leftrightarrow T$ is injective.   □

Definition 8.5.3   Let $T : U \to V$ be a linear transformation. Then $\dim(\operatorname{Im} T)$ is called the rank of T and $\dim(\ker T)$ is called the nullity of T.

If U is finite dimensional, then $\ker T$ and $\operatorname{Im} T$ are both finite dimensional and their dimensions satisfy the following relation.

Theorem 8.5.6 (Sylvester's Law of Nullity)   *Let U be a finite dimensional vector space and $T : U \to V$ be a linear transformation. Then $\dim U = \dim(\ker T) + \dim(\operatorname{Im} T)$, i.e., $\dim U = $ nullity of $T + $ rank of T.*

Proof Let $\dim U = n$ and $\dim(\ker T) = k \leq n$. We claim that $\dim(\operatorname{Im} T) = n - k$. Clearly, $\dim(\ker T) = k$ implies that there exists a basis $B = \{u_1, u_2, \ldots, u_k\}$ of $\ker T$. We now extend the basis B to a basis $B_U = \{u_1, u_2, \ldots, u_k, w_1, w_2, \ldots, w_{n-k}\}$ of U. Let $B_1 = \{T(w_1), \ldots, T(w_{n-k})\}$. Then B_1 is linearly independent and generates $\operatorname{Im} T$. Consequently, B_1 is a basis of $\operatorname{Im} T$ and hence $\dim(\operatorname{Im} T) = n - k$.   □

This theorem gives a series of corollaries.

Corollary 1   *Let U be an n-dimensional vector space and $T : U \to V$ a linear transformation. Then the following statements are equivalent:*

(i) *T is injective*;
(ii) *rank of $T = n$*;
(iii) *$B = \{u_1, u_2, \ldots, u_n\}$ is a basis of U implies $T(B) = \{T(u_1), T(u_2), \ldots, T(u_n)\}$ is a basis of $\operatorname{Im} T$ i.e., $\dim U = $ rank of T.*

Proof Left as an exercise.   □

Corollary 2   *Let V be a finite dimensional vector space and $T : V \to V$ a linear operator. Then T is injective iff T is surjective. In particular, a linear transformation $T : \mathbf{R}^n \to \mathbf{R}^n$ is injective iff T is surjective.*

Proof Using the First Isomorphism Theorem it follows that $\dim(T(V)) = \dim V - \dim(\ker T)$. Hence T is injective iff T is surjective, because V is finite dimensional. For $V = \mathbf{R}^n$, the last part follows.   □

Corollary 3   *Let U and V be finite dimensional vector spaces and $T : U \to V$ a linear transformation. Then T is an isomorphism iff $\dim U = \dim V$.*

Proof Left as an exercise.   □

Let U and V be two vector spaces over the same field F. We are yet to define any algebraic structure on the set $L(U, V)$ of all linear transformations from the vector space U to the vector space V. The map $0 : U \to V, u \mapsto 0 \in V$ (i.e., it carries every element $u \in U$ to the zero element of V) and the negative of a linear transformation $T : U \to V$, denoted by $(-T)$ defined by $(-T) : U \to V, u \mapsto -T(u)$ are linear transformations.

Proposition 8.5.2 $L(U, V)$ *forms a vector space under the compositions*: $(T + S)(u) = T(u) + S(u)$ *and* $(\alpha T)(u) = \alpha T(u)$, *where* $u \in U$ *and* $T, S \in L(U, V)$.

Proof Left as an exercise.   □

Proposition 8.5.3 *The composite of two linear transformations is a linear transformation.*

Proof Let U, V and W be vector spaces over F; $S : U \to V$ and $T : V \to W$ be linear transformations. It is sufficient to prove that their composite $T \circ S : U \to W$ defined by $(T \circ S)(x) = T(S(x))$ for all x in U is also a linear transformation. Since $(T \circ S)(\alpha x + \beta y) = \alpha T(S(x)) + \beta T(S(y)) \; \forall x, y \in U$, $T \circ S$ is a linear transformation.   □

We now give an alternative definition of a linear isomorphism.

Definition 8.5.4 Let U and V be vector spaces over F. A linear transformation $T : U \to V$ is said to a (linear) isomorphism iff there exists a linear transformation $S : V \to U$ such that $S \circ T = I_U$ (identity map) and $T \circ S = I_V$. In particular, a linear operator $T : V \to V$ is said to be non-singular or invertible iff T is an isomorphism.

We now characterize non-singular transformations.

Proposition 8.5.4 *Let* $T : V \to V$ *be a linear operator. Then the following statements are equivalent*:

(a) T *is non-singular*;
(b) *nullity of* $T = 0$;
(c) *rank of* $T = \dim V$.

Proof Left as an exercise.   □

Theorem 8.5.7 *Let* U *and* V *be finite dimensional vector spaces over the same field* F. *If* $\dim U = m$ *and* $\dim V = n$, *then* $\dim(L(U, V)) = mn$.

Proof Let W be the vector space $L(U, V)$ over F. Suppose $B_U = \{u_1, u_2, \ldots, u_m\}$ is a basis of U and $B_V = \{v_1, v_2, \ldots, v_n\}$ is a basis of V. Define a family of linear transformations $\psi_{ij} : U \to V$ by

$$\psi_{ij}(u_i) = v_j, \quad \text{for } i = 1, 2, \ldots, m; \; j = 1, 2, \ldots, n$$

and

$$\psi_{ij}(u_k) = 0, \quad \text{for } k \neq i.$$

Then ψ_{ij} maps u_i into v_j and other u's into 0.

Clearly, $\{\psi_{ij}\}$ consisting of mn elements constitutes a basis of W. Consequently, $\dim W = mn$. □

Corollary 1 *If V is a finite dimensional vector space over F of dimension n, then $L(V, V)$ is also a finite dimensional vector space over F of dimension n^2.*

Corollary 2 *If V is an n-dimensional vector space over F, then $L(V, F)$ is an also an n-dimensional vector space over F.*

Let V be a vector space over F. Then linear transformations $T : V \to F$ lead to some extremely important concepts and yield results which are not true in general setting.

Definition 8.5.5 For any vector space V over F, the vector space $L(V, F)$ is called is dual space denoted by V^d and an element of V^d is called a linear functional of V into F.

Remark A linear functional transforms a vector to a scalar. If V is an n-dimensional vector space, then $\dim(V^d) = n$.

Example 8.5.3 (i) Let $V = C([0, 1])$ be the vector space of all real valued continuous functions in x on $[0, 1]$. Then $\psi : V \to \mathbf{R}, \; f(x) \mapsto \int_0^1 f(x)\,dx$ is a linear functional and hence $\psi \in V^d$.

(ii) Let $V = \mathbf{R}[x]$ be the vector space (infinite dimensional) of all real polynomials in x. Then for $a \in \mathbf{R}$, $D_a : V \to \mathbf{R}, \; f \mapsto f'(a)$ is a linear functional.

We now consider only finite dimensional vector spaces V, because if V is not finite dimensional, then V^d is too large to invite attention unless an additional structure, such as topology is endowed.

Theorem 8.5.8 *Let V be a finite dimensional vector space over F and $B = \{v_1, v_2, \ldots, v_n\}$ a basis of V. Then the linear functionals $f_1, f_2, \ldots, f_n$ defined by*

$$f_i(v_j) = \delta_{ij} = \begin{cases} 1, & \text{if } i = j \\ 0, & \text{if } i \neq j \end{cases}$$

constitute a basis B^d of the dual space V^d, called a dual basis of B.

Proof Clearly, for $\alpha_1 f_1 + \alpha_2 f_2 + \cdots + \alpha_n f_n = 0$,

$$\sum_{i=1}^{n} \alpha_i f_i(v_i) = 0, \quad \text{since } f_i(\alpha_1 v_1 + \alpha_2 v_2 + \cdots + \alpha_n v_n) = \alpha_i.$$

Hence

$$\sum_{i=1}^{n} \alpha_i \delta_{ij} = 0 \quad \Rightarrow \quad \alpha_i = 0, \quad \forall i = 1, 2, \ldots, n.$$

Consequently, B^d is linearly independent. Again

$$\dim V = \dim(V^d) = n.$$

Hence B^d consisting of n linearly independent elements of V^d constitutes a basis of V^d.   □

Remark δ_{ij} is called Kronecker delta.

Proposition 8.5.5 *If V is a finite dimensional vector space over F and v ($\neq 0$) in V, then there exists a linear functional $f \in V^d$ such that $f(v) \neq 0$.*

Proof Suppose $v \neq 0$. Then it shows that $\{v\}$ is linearly independent in V. Let $\dim V$ be n. We can construct a basis $B = \{v = v_1, v_2, \ldots, v_n\}$ for V. Then for the elements $1, 0, 0, \ldots, 0$ in F, there exists a unique linear transformation $f : V \to F$ such that $f(v) = 1$, $f(v_i) = 0$, $i = 2, 3, \ldots, n$. This implies that $f \in V^d$ such that $f(v) \neq 0$.   □

Example 8.5.4 Let $V = \mathbf{R}^3$, $v_1 = (1, -1, 3)$, $v_2 = (0, 1, -1)$ and $v_3 = (0, 3, -2)$. Then $B = \{v_1, v_2, v_3\}$ is a basis of V. Its dual basis $B^d = \{f_1, f_2, f_3\}$ given by

$$\left. \begin{array}{l} f_1(x, y, z) = \alpha_1 x + \beta_1 y + \gamma_1 z \\ f_2(x, y, z) = \alpha_2 x + \beta_2 y + \gamma_2 z \\ f_3(x, y, z) = \alpha_3 x + \beta_3 y + \gamma_3 z \end{array} \right\} \tag{8.2}$$

is such that $f_1(v_1) = 1$, $f_1(v_2) = 0$ and $f_1(v_3) = 0$;

$$f_2(v_1) = 0, \qquad f_2(v_2) = 1 \quad \text{and} \quad f_2(v_3) = 0;$$

$$f_3(v_1) = 0, \qquad f_3(v_2) = 0 \quad \text{and} \quad f_3(v_3) = 1.$$

Consequently, $f_1(x, y, z) = x$, $f_2(x, y, z) = 7x - 2y - 3z$ and $f_3(x, y, z) = -2x + y + z$. Hence $V^d = L(B^d)$.

We now introduce the concept of annihilators in vector spaces V which is closely related to dual spaces V^d.

Definition 8.5.6 Let X be a non-empty subset of a vector space V. Then the annihilator $A(X)$ of X is defined by

$$A(X) = \{f \in V^d : f(x) = 0 \text{ for all } x \in X\},$$
$$= \{f \in V^d : f(X) = 0\}.$$

$A(X)$ has the following properties.

Proposition 8.5.6 *Let X be any subspace of a finite dimensional vector space V over F. Then $A(X)$ is a subspace of V^d such that $\dim(A(X)) = \dim V - \dim X$.*

Proof Clearly, $A(X)$ is a subspace of V^d. Let $\dim V = n$, $\dim X = m$, $m \le n$ and $\{u_1, u_2, \ldots, u_m\}$ be a basis of X. Then this basis can be extended to a basis $B = \{u_1, u_2, \ldots, u_m, v_1, v_2, \ldots, v_{n-m}\}$ for V. Let $B^d = \{f_1, f_2, \ldots, f_m, g_1, g_2, \ldots, g_{n-m}\}$ be the dual basis for B. If $C = \{g_1, g_2, \ldots, g_{n-m}\}$, then C is a basis of $A(X)$. Hence $\dim(A(X)) = n - m = \dim V - \dim X$. $\square$

Corollary *Let V be a finite dimensional vector space over F and U be a subspace of V. Then the dual space U^d is isomorphic to the quotient space $V^d/A(U)$.*

Proof Clearly, $A(U)$ is a subspace of V^d and $\dim(V^d/A(U)) = \dim(V^d) - \dim(A(U)) = \dim U = \dim U^d$. Hence the corollary follows. $\square$

An Application of Dual Spaces   We now apply the results of dual spaces to obtain the number of linearly independent solutions of a system of linear homogeneous equations over a field F:

$$\left.\begin{aligned}
\alpha_{11}x_1 + \alpha_{12}x_2 + \cdots + \alpha_{1n}x_n &= 0 \\
\alpha_{21}x_1 + \alpha_{22}x_2 + \cdots + \alpha_{2n}x_n &= 0 \\
\cdots \\
\alpha_{m1}x_1 + \alpha_{m2}x_2 + \cdots + \alpha_{mn}x_n &= 0
\end{aligned}\right\} \tag{A}$$

Let $S = \{\alpha_i = (\alpha_{i1}, \alpha_{i2}, \ldots, \alpha_{in}) \in F^n : i = 1, 2, \ldots, m\}$ and $U = L(S)$. Then U is a subspace of F^n.

If $\dim U = r \le m$, we say that the system of (A) is of rank r. Let $V = F^n$, $B = \{e_1, e_2, \ldots, e_n\}$ be the natural (standard basis) of V and $B^d = \{f_1, f_2, \ldots, f_n\}$ be its dual basis. Then any $f \in V^d$ can be expressed uniquely as

$$f = x_1 f_1 + x_2 f_2 + \cdots + x_n f_n, \; x_i \in F.$$

If $f \in A(U)$, then $f(u) = 0$ for all $u \in U$ and hence $0 = f(\alpha_{11}, \alpha_{12}, \ldots, \alpha_{1n})$, since $(\alpha_{11}, \alpha_{12}, \ldots, \alpha_{1n}) \in U$. Consequently,

$$0 = f(\alpha_{11}e_1 + \alpha_{12}e_2 + \cdots + \alpha_{1n}e_n)$$
$$= x_1\alpha_{11} + \cdots + x_n\alpha_{1n}, \quad \text{since } f_i(e_j) = \delta_{ij}.$$

Similar results hold for other equations in (A). Conversely, for every solution $(x_1, x_2, \ldots, x_n)$ of equations in (A), there exists an element $x_1 f_1 + x_2 f_2 + \cdots + x_n f_n$ in $A(U)$. Then the number of independent solutions of the system (A) is $\dim(A(U))$ i.e., $n - r$.

This proves the following theorem.

Theorem 8.5.9 *If the system of equations in* (A), *where* $\alpha_{ij} \in F$, *is of* rank r, *then the number of linearly independent solutions in* F^n *of the system is* $n - r$.

Corollary *If the number of unknowns exceeds the number of equations in* (A), *then there exists a non-trivial solution of the system* (A).

8.5.1 Algebra over a Field

For any two vector spaces U and V over the same field F, the set $L(U, V)$ admits a vector space structure over F (see Proposition 8.5.2). If $V = U$, we can introduce a multiplication on $L(V, V)$ making it a ring. We can combine in a natural way these two twin structures of $L(V, V)$ to provide a very rich structure called an 'algebra' over F.

We now give the formal definition of an algebra with an eye to $L(V, V)$.

Definition 8.5.7 An algebra $\mathcal{A}$ is a vector space over a field F, whose elements can be multiplied in such a way that $\mathcal{A}$ is also a ring in which the scalar multiplication and ring multiplication are interlined by the rule

$$A(1): \text{For all } x, y \in \mathcal{A} \text{ and } \alpha \in F, \alpha(xy) = (\alpha x)y = x(\alpha y).$$

A commutative algebra $\mathcal{A}$ is an algebra such that

$$A(2): xy = yx \text{ holds for all } x, y \text{ in } \mathcal{A}.$$

An identity element in an algebra $\mathcal{A}$ is a non-zero element 1 in $\mathcal{A}$ such that $1x = x1 = x$ for every x in $\mathcal{A}$.

Remark If the identity element exists in an algebra, then it is unique.

We call an algebra $\mathcal{A}$ over F *real* or *complex* according to $F = \mathbf{R}$ or $\mathbf{C}$.

Definition 8.5.8 A non-empty subset $\mathcal{B}$ of an algebra $\mathcal{A}$ is called a subalgebra of $\mathcal{A}$ iff $\mathcal{B}$ is itself an algebra under the operations defined in $\mathcal{A}$ (i.e., $\mathcal{B}$ is a subspace of $\mathcal{A}$ such that for all x, y in $\mathcal{B}, xy \in \mathcal{B}$).

Remark A subalgebra of an algebra is an algebra in its own right. One of the most important examples of an algebra is $L(V, V)$. In general, it is a non-commutative algebra having zero divisors.

We give some other examples.

Example 8.5.5   (i) **R** is a commutative real algebra under usual compositions in **R**.
(ii) The real vector space $C([0, 1])$ is a commutative real algebra with identity.
(iii) The vector space $B(X, \mathbf{C})$ of all bounded complex valued functions on a topological space X is a commutative complex algebra.
(iv) Let $\mathcal{A}$ be an algebra over F and $\mathcal{B} = \{x \in \mathcal{A} : xy = yx \text{ for all } y \in \mathcal{A}\}$. Then $\mathcal{B}$ is a subalgebra of $\mathcal{A}$, called the *center* of $\mathcal{A}$.

Definition 8.5.9   Let $\mathcal{A}$ be an algebra over F. An algebra ideal (or an ideal) I of $\mathcal{A}$ is a non-empty subset of $\mathcal{A}$ such that I is a subspace of $\mathcal{A}$ (when $\mathcal{A}$ is considered as a vector space) and also a ring ideal of $\mathcal{A}$ (when $\mathcal{A}$ is considered as a ring). For any ideal I of $\mathcal{A}$, $\mathcal{A}/I$ is an algebra, called the quotient algebra of $\mathcal{A}$ modulo I.

Like ring homomorphisms, homomorphisms are defined in algebra.

Definition 8.5.10   Let $\mathcal{A}$ and $\mathcal{B}$ be algebras over the same field F. A homomorphism $\psi : \mathcal{A} \to \mathcal{B}$ is a mapping such that

$$\psi(x + y) = \psi(x) + \psi(y);$$
$$\psi(xy) = \psi(x)\psi(y);$$
$$\psi(\alpha x) = \alpha \psi(x),$$

for all x, y in $\mathcal{A}$ and $\alpha \in F$.

Moreover, if the homomorphism ψ is a bijection, then ψ is said to be an isomorphism and $\mathcal{A}$ is said to be isomorphic to $\mathcal{B}$.

Remark An algebra homomorphism f is a ring homomorphism such that f preserves the vectors space structure.

Proposition 8.5.7   *Let $\psi : \mathcal{A} \to \mathcal{B}$ be a homomorphism of algebras. Then $\ker \psi$ is a subalgebra of $\mathcal{A}$ and $\operatorname{Im} \psi$ is a subalgebra of $\mathcal{B}$.*

Proof Left as an exercise. $\square$

We now prove a theorem for an algebra which is the analog of Cayley's Theorem for a group.

Theorem 8.5.10   *Let $\mathcal{A}$ be an algebra with 1, over F. Then $\mathcal{A}$ is isomorphic to a subalgebra of $L(V, V)$ for some vector space V over F.*

Proof $\mathcal{A}$ being an algebra over F, it is a vector space over F. We take $V = \mathcal{A}$ to prove the theorem. For $x \in \mathcal{A}$, we define the map $T_x : \mathcal{A} \to \mathcal{A}$ by $T_x(v) = xv$, $\forall v \in \mathcal{A}$. Then T_x is a linear transformation on $V = \mathcal{A}$ and the map $\psi : \mathcal{A} \to L(V, V)$, $x \mapsto T_x$ is well defined. Moreover, ψ is both a linear transformation of vector spaces and a homomorphism of rings. Hence ψ is a homomorphism of algebras and $\ker \psi = \{x \in \mathcal{A} : \psi(x) = 0\} = \{x \in \mathcal{A} : T_x = 0\} = \{x \in \mathcal{A} : T_x(v) = 0, \forall v \in V\}$. Since $V = \mathcal{A}$ has identity 1, $T_x(1) = 0 \Rightarrow x = 0$. This shows that $\ker \psi = \{0\}$ and hence ψ is a monomorphism of algebras. Consequently, $\mathcal{A}$ is isomorphic to the subalgebra $\operatorname{Im} \psi$ of $L(V, V)$ over F. $\square$

Remark The algebra $L(V, V)$ plays a universal role to obtain isomorphic copies of any algebra.

We now study the algebra $L(V, V)$, where V is restricted to finite dimensional vector spaces over F.

Definition 8.5.11 A non-zero element T in $L(V, V)$ is said to be invertible iff there is an element S in $L(V, V)$ such that $T \circ S = S \circ T = I$. Otherwise, T is said to be singular.

Clearly, if T is invertible, S is unique and it is denoted by T^{-1}.

Theorem 8.5.11 *Let $\mathcal{A}$ be an algebra over F, with 1, such that $\dim \mathcal{A} = n$. Then every element in $\mathcal{A}$ is a root of some non-trivial polynomial $f(x)$ in $F[x]$ such that $\deg f \leq n$.*

Proof For $v \in \mathcal{A}$, the $(n + 1)$ elements $1, v, v^2, \ldots, v^n$ are in $\mathcal{A}$. Since $\dim \mathcal{A} = n$, these $(n + 1)$ elements must be linearly dependent over F. Hence there exist $(n + 1)$ elements $\alpha_0, \alpha_1, \ldots, \alpha_n$ in F (not all 0) such that $\alpha_0 1 + \alpha_1 v + \cdots + \alpha_n v^n = 0$. This shows that v is a root of the non-trivial polynomial $f(x) = \alpha_0 + \alpha_1 x + \cdots + \alpha_n x^n$ in $F[x]$ such that $\deg f \leq n$. $\square$

Corollary *Let V be an n-dimensional vector space over F. Then given a non-zero element T in $L(V, V)$, there exists a non-trivial polynomial $f(x)$ in $F[x]$ such that $\deg f \leq n^2$ and $f(T) = 0$.*

Proof For the algebra $\mathcal{A} = L(V, V)$ over F, $\dim \mathcal{A} = n^2$. Clearly, $\mathcal{A}$ contains the identity, which is the identity operator on V. The corollary follows from Theorem 8.5.11. $\square$

The corollary leads to the following definition.

Definition 8.5.12 Let V be an n-dimensional vector space over F and T be a non-zero element in $L(V, V)$. Then there exists a non-trivial polynomial $m(x)$ of lowest degree with leading coefficient 1 in $F[x]$ such that $m(T) = 0$. We call $m(x)$ minimal polynomial for T over F.

Remark If T satisfies a minimal polynomial $g(x)$ in $F[x]$, then this $m(x)$ divides $g(x)$ in $F[x]$. Since $g(x)$ is monic and $g(T) = 0$, $m(x) = g(x)$. This shows the uniqueness of a monic polynomial for T over F and we use the term the minimal polynomial for T over F.

We now characterize invertible elements in $L(V, V)$.

Theorem 8.5.12 *Let V be an n-dimensional vector space over F and T a non-zero element in $L(V, V)$. Then T is invertible iff the constant term of the minimal polynomial for T over F is not zero.*

Proof Let $f(x) = \alpha_0 + \alpha_1 x + \cdots + \alpha_{m-1} x^{m-1} + x^m \in F[x]$ be the minimal polynomial for T. Then $0 = f(T) = \alpha_0 I + \alpha_1 T + \cdots + \alpha_{m-1} T^{m-1} + T^m$.

If $\alpha_0 \neq 0$, then $\alpha_0 I = -(T^{m-1} + \alpha_{m-1} T^{m-2} + \cdots + \alpha_1 I)T$. Hence $I = [-\alpha_0^{-1}(T^{m-1} + \alpha_{m-1} T^{m-2} + \cdots + \alpha_1 I)]T$. If we take $S = -\alpha_0^{-1}(T^{m-1} + \alpha_{m-1} T^{m-2} + \cdots + \alpha_1 I)$, then $S \circ T = I$. Similarly, $T \circ S = I$. Consequently, T is invertible. Conversely, let T be invertible in $L(V, V)$. If possible, $\alpha_0 = 0$, then $0 = (\alpha_1 I + \alpha_2 T + \cdots + \alpha_{m-1} T^{m-2} + T^{m-1})T$. Multiplying by T^{-1}, we have $\alpha_1 I + \alpha_2 T + \cdots + \alpha_{m-1} T^{m-2} + T^{m-1} = 0$. Hence T satisfies the polynomial $q(x) = \alpha_1 + \alpha_2 x + \cdots + \alpha_{m-1} x^{m-2} + x^{m-1}$ of degree $(m - 1)$ in $F[x]$. But this is not possible as the minimal polynomial $f(x)$ is of degree m. This leads us to conclude that $\alpha_0 \neq 0$. $\qquad\square$

Corollary *Let V be a finite dimensional vector space over F and $T \in L(V, V)$ be an invertible element. Then its inverse T^{-1} is also a polynomial in T over F.*

8.6 Correspondence Between Linear Transformations and Matrices

In this section linear algebra starts to encompass the isomorphic theory of matrices.

Linear transformations and matrices are closely related. In this section we establish their relations and study matrices with the help of linear transformations and vice versa. Let $T : \mathbf{R}^2 \to \mathbf{R}^2$ be the rotation of the plane through an angle θ about the origin defined by $T(x, y) = (x \cos \theta - y \sin \theta, x \sin \theta + y \cos \theta)$ and $B = \{e_1 = (1, 0), \ e_2 = (0, 1)\}$ be the natural basis or standard basis of $\mathbf{R}^2$. Then the actions of T on e_1, e_2 are given by

$$T\big((1, 0)\big) = (\cos \theta, \sin \theta) = \cos \theta (1, 0) + \sin \theta (0, 1)$$

and

$$T\big((0, 1)\big) = (-\sin \theta, \cos \theta) = -\sin \theta (1, 0) + \cos \theta (0, 1).$$

Then the 2×2 matrix

$$A = \begin{pmatrix} \cos \theta & \sin \theta \\ -\sin \theta & \cos \theta \end{pmatrix}$$

is called the *matrix* of *coefficients* associated with T and its transpose

$$A^t = \begin{pmatrix} \cos\theta & -\sin\theta \\ \sin\theta & \cos\theta \end{pmatrix}$$

is called the matrix representation of T with respect to the given basis B. We now extend this concept to an arbitrary finite dimensional vector space.

Remark We also use the notation of the same matrix A as

$$A = \begin{bmatrix} \cos\theta & \sin\theta \\ -\sin\theta & \cos\theta \end{bmatrix}.$$

Definition 8.6.1 Let V and W be finite dimensional vector spaces over F and $T : V \to W$ a linear transformation. Let $B_1 = \{v_1, v_2, \ldots, v_m\}$ and $B_2 = \{w_1, w_2, \ldots, w_n\}$ be their respective ordered bases. For each i, $1 \le i \le m$, $T(v_i)$ in W can be expressed uniquely as $T(v_i) = \sum_{j=1}^{n} a_{ij}w_j$, $a_{ij} \in F$. The matrix $A = (a_{ij})$ is called the *matrix* of *coefficients* associated with T and its transpose A^t is called the matrix representation of T with respect to given bases B_1 and B_2.

The matrix A^t is denoted by $m(T)_{B_1}^{B_2}$ or simply by $m(T)$. In particular, for $V = W$ and $T = I$ (identity), the matrix $m(I)_{B_1}^{B_2}$ is called the transition matrix from B_1 to B_2. Again if $V = W$ and $B_1 = B_2 = B$, then $m(T)_{B_1}^{B_2}$ is simply written by $m(T)_B$.

Remark Some authors call A the matrix representation of T. The matrix $m(T)$ depends not only on T but also depends on the ordered bases B_1 and B_2.

Definition 8.6.2 Let $\sigma : \{1, 2, \ldots, n\} \to \{1, 2, \ldots, n\}$ be a permutation and $\{e_i\}$ be the natural basis B of $\mathbf{R}^n$. Let $f_\sigma : \mathbf{R}^n \to \mathbf{R}^n$ defined by $f_\sigma(e_i) = e_{\sigma(i)}$ be extended linearly over $\mathbf{R}^n$. The matrix $m(f_\sigma)_B$ is called the permutation matrix corresponding to σ.

Example 8.6.1 (i) If $I : \mathbf{R}^n \to \mathbf{R}^n$ is the identity transformation and $B = \{e_i\}$ is the natural basis of $\mathbf{R}^n$, then $m(I)_B = I$ (identity matrix).

(ii) Let $V = V_3[x]$ be the vector space of polynomials in x over $\mathbf{R}$, of degree less than 3 and $D : V \to V$, $f \mapsto f'$ be the differential operator. For the basis $B = \{1, x, x^2\}$,

$$m(D)_B = \begin{pmatrix} 0 & 1 & 0 \\ 0 & 0 & 2 \\ 0 & 0 & 0 \end{pmatrix}.$$

(iii) Let $T : \mathbf{R}^2 \to \mathbf{R}^2$ be the linear transformation defined by $T(x, y) = (2x + y, x - y)$. If $B = \{(1, 0), (0, 1)\}$ and $C = \{(1, 1), (1, -1)\}$, then

$$m(T)_B = \begin{pmatrix} 2 & 1 \\ 1 & -1 \end{pmatrix} \quad \text{and} \quad m(T)_C = \begin{pmatrix} \frac{3}{2} & \frac{3}{2} \\ \frac{3}{2} & -\frac{1}{2} \end{pmatrix}.$$

(iv) The permutation matrix corresponding to

$$\sigma = \begin{pmatrix} 1 & 2 & 3 \\ 3 & 1 & 2 \end{pmatrix} \quad \text{is} \quad \begin{pmatrix} 0 & 1 & 0 \\ 0 & 0 & 1 \\ 1 & 0 & 0 \end{pmatrix}.$$

This is so because by Definition 8.6.2, $f_\sigma(e_1) = e_{\sigma(1)} = e_3 = 0 \cdot e_1 + 0 \cdot e_2 + 1 \cdot e_3$.
Similarly, $f_\sigma(e_2) = e_{\sigma(2)} = e_1 = 1 \cdot e_1 + 0 \cdot e_2 + 0 \cdot e_3$ and $f_\sigma(e_3) = e_{\sigma(3)} = e_2 = 0 \cdot e_1 + 1 \cdot e_2 + 0 \cdot e_3$.

Given an $m \times n$ matrix M over a field F, we can find a linear transformation $T : F^n \to F^m$ such that $m(T) = M$.

Example 8.6.2 Suppose

$$M = \begin{pmatrix} 1 & 2 \\ -1 & 4 \\ 0 & 5 \end{pmatrix}.$$

Then the column vectors $u = (1, -1, 0)$ and $v = (2, 4, 5)$ define a linear transformation $T : \mathbf{R}^2 \to \mathbf{R}^3$ by

$$T(x, y) = xu + yv = x(1, -1, 0) + y(2, 4, 5)$$
$$= (x + 2y, -x + 4y, 5y) \quad \text{for all } (x, y) \text{ in } \mathbf{R}^2.$$

If B is the natural basis of $\mathbf{R}^2$, then $T(e_1) = T((1, 0)) = (1, -1, 0)$ and $T(e_2) = T((0, 1)) = (2, 4, 5)$. Hence $m(T)_B = M$.

The following proposition is a generalization of Example 8.6.2.

Proposition 8.6.1 *Given an $m \times n$ matrix M over $\mathbf{R}$, the map $T_M : \mathbf{R}^n \to \mathbf{R}^m$, $X \mapsto MX$, is a linear transformation where X is viewed as a column vector of $\mathbf{R}^n$.*

Conversely, if $B = \{E_1, E_2, \ldots, E_n\}$ is a set of unit column vectors of $\mathbf{R}^n$, given a linear transformer $T : \mathbf{R}^n \to \mathbf{R}^m$ such that $T(E_j) = M_j$, where M_j is a column vector in $\mathbf{R}^m$, then M is the matrix where column vectors are $M_1, M_2, \ldots, M_n$.

Proof Clearly, T_M is a linear transformation. Conversely, we suppose

$$T(E_j) = \begin{pmatrix} a_{1j} \\ a_{2j} \\ \vdots \\ a_{mj} \end{pmatrix} = M_j.$$

Then we can write every X in $\mathbf{R}^n$ as

$$X = x_1 E_1 + \cdots + x_n E_n = \begin{pmatrix} x_1 \\ x_2 \\ \vdots \\ x_n \end{pmatrix}.$$

Hence $T(X) = x_1 M_1 + x_2 M_2 + \cdots + x_n M_n = MX$, where M is the matrix where column vectors are $M_1, M_2, \ldots, M_n$. Consequently, $T(X) = MX = T_M(X)$ $\forall X \in \mathbf{R}^n \Rightarrow T = T_M$. $\qquad\square$

Proposition 8.6.2 *Let V and W be finite dimensional vector spaces over F. Suppose $\dim V = n$ and $\dim W = m$. If $B_1 = \{e_1, e_2, \ldots, e_n\}$ and $B_2 = \{f_1, f_2, \ldots, f_m\}$ are ordered bases of V and W respectively, then the mapping $\psi : L(V, W) \to M_{mn}(F), T \mapsto m(T)$ is a bijection and also an isomorphism of vector spaces.*

Proof Let $M \in M_{mn}(F)$. Then there exists a matrix $A = (a_{ij})_{n \times m}$ such that $M = A^t$. Suppose $y_i = \sum_{j=1}^m a_{ij} f_j$, $1 \le i \le n$. Then $y_i \in W$ and there exists a unique linear transformation $T : V \to W$ such that $T(e_i) = y_i = \sum_{j=1}^m a_{ij} f_j$. Hence $m(T) = A^t = M$. This shows ψ is onto. Let T, S be in $L(V, W)$ such that $m(T) = m(S) = A^t$, where $A = (a_{ij})$. If $\sum_{j=1}^m a_{ij} f_j = y_i$, $1 \le i \le n$, then $T(e_i) = \sum_{j=1}^m a_{ij} f_j = y_i = S(e_i)$ for all $i = 1, 2, \ldots, n$. Hence it follows that $S = T$. This shows that ψ is injective. Consequently, ψ is a bijection which is clearly an isomorphism of vector spaces. $\qquad\square$

Corollary 1 *Let U, V, W be finite dimensional vector spaces over F. Given ordered bases B_1, B_2 and B_3 of U, V, and W respectively and linear transformations $T_1 : U \to V, T_2 : V \to W, m(T_2 \circ T_1) = m(T_2) \cdot m(T_1)$.*

Corollary 2 *Let V be a finite dimensional vector space and B be a fixed ordered basis of V. A non-zero linear operator $T : V \to V$ is non-singular (i.e., invertible) iff its corresponding matrix $m(T)$ is non-singular.*

Given two bases B_1 and B_2 of a finite dimensional vector space V and a linear operator $T : V \to V$, the corresponding twin matrices $m(T)_{B_1}$ and $m(T)_{B_2}$ are closely related. This introduces the concept of similar matrices.

Definition 8.6.3 Let A, B be two matrices in $M_{nn}(F)$. Then A is said to be similar to B, denoted by $A \approx B$ iff there exists a non-singular matrix P in $M_{nn}(F)$ such that $B = PAP^{-1}$.

Remark Similarity relation on the set $M_{nn}(F)$ is an equivalence relation. We now characterize similar matrices with the help of linear operators.

Theorem 8.6.1 *Two $n \times n$ matrices over a field are similar iff they represent the same linear operator (each relative to a single basis).*

Proof Let V be an n-dimensional vector space over a field F and B_1, B_2 be two bases of V. Suppose A and B are two $n \times n$ matrices represented by the same linear transformation $T : V \to V$ with respect to bases B_1 and B_2, respectively. We claim that $A \approx B$. Let $B_1 = \{v_1, v_2, \ldots, v_n\}$ and $B_2 = \{u_1, u_2, \ldots, u_n\}$ be two bases of V. If $T(v_i) = \sum_{j=1}^{n} a_{ji} v_j$ and $T(u_i) = \sum_{j=1}^{n} b_{ji} u_j$, where $a_{ji}, b_{ji} \in F$ and $1 \le i, j \le n$, then $A = (a_{ij})$ and $B = (b_{ij})$. Again since B_1 and B_2 are both bases of V, we can represent u_i as $u_i = \sum_{j=1}^{n} c_{ji} v_j$ and v_i as $v_i = \sum_{j=1}^{n} d_{ji} u_j$, $1 \le i \le n$. If $C = (c_{ij})$ and $D = (d_{ij})$, then

$$v_i = \sum_{j=1}^{n} d_{ji} \left(\sum_{k=1}^{n} c_{kj} v_k \right) = \sum_{k=1}^{n} \left(\sum_{j=1}^{n} c_{kj} d_{ji} \right) v_k \quad \Rightarrow \quad \sum_{j=1}^{n} c_{kj} d_{ji} = 1, \quad \text{if } i = k$$
$$= 0, \quad \text{if } i \ne k.$$

This shows that $CD = I$. Similarly, $DC = I$. Hence C is a non-singular matrix with its inverse D, i.e., $C^{-1} = D$. Clearly, $BC = CA$ and hence $B = CAC^{-1} \Rightarrow A \approx B$.

Conversely, let A and B be two $n \times n$ similar matrices over F and $T : V \to V$ be a linear operator of a finite-dimensional vector space V such that $A = m(T)_{B_1}$ for some ordered basis $B_1 = \{v_1, v_2, \ldots, v_n\}$ of V. We now show that there exists an ordered basis B_2 of V such that $B = m(T)_{B_2}$. As B is similar to A, there exists a non-singular matrix P such that $B = PAP^{-1}$. If $A = (a_{ij})$, then $T(v_i) = \sum_{j=1}^{n} a_{ji} v_j$, $1 \le i \le n$. Let $B = (b_{ij})$ and $P = (p_{ij})$. We define a linear transformation $S : V \to V$, $v_i \mapsto \sum_{j=1}^{n} p_{ji} v_j$, $1 \le i \le n$. Then $P = m(S)_{B_1}$. As P is non-singular, the set $B_2 = \{S(v_1), S(v_2), \ldots, S(v_n)\}$ is linearly independent. Hence for $d_i \in F$, the relation $\sum_{i=1}^{n} d_i S(v_i) = 0 \Rightarrow S(\sum_{i=1}^{n} d_i v_i) = 0 \Rightarrow \sum_{i=1}^{n} d_i v_i = 0$, since S is injective. This shows that each $d_i = 0$. Hence B_2 also forms a basis of V. Let $m(T)_{B_2} = M$. Then by the first part, $M = PAP^{-1} = B$. $\square$

Remark The matrix P is not unique (see Example 8.6.3).

We now define another relation on $M_{nn}(F)$.

Definition 8.6.4 Let A, B be in $M_{nn}(F)$. Then A is said to be equivalent to B, denoted $A \sim B$ iff there exist non-singular matrices P, Q in $M_{nn}(F)$ such that $B = PAQ$.

Remark Similar matrices are equivalent but its converse is not true (see Example 8.6.3).

Example 8.6.3 (a) Let $T : \mathbf{R}^2 \to \mathbf{R}^2$ be defined by $T(x, y) = (x - y, y - x)$ and $B_1 = \{(1, 0), (0, 1)\}$ and $B_2 = \{(1, 2), (-1, 1)\}$. Then $m(T)_{B_1} = A = \begin{pmatrix} 1 & -1 \\ -1 & 1 \end{pmatrix}$ and $m(T)_{B_2} = B = \begin{pmatrix} 0 & 0 \\ 1 & 2 \end{pmatrix}$ imply that $A \approx B$. If $P = \begin{pmatrix} 1 & 1 \\ -2 & 1 \end{pmatrix}$ and $Q = \begin{pmatrix} 1 & 1 \\ 2 & -3 \end{pmatrix}$, then $B = PAP^{-1} = QAQ^{-1}$ implies that P is not unique.

(b) The matrices $A = \begin{pmatrix} 1 & 0 \\ 0 & 0 \end{pmatrix}$ and $B = \begin{pmatrix} 0 & 1 \\ 0 & 0 \end{pmatrix}$ are equivalent matrices but not similar.

Remark An important application of equivalent matrices is in their Smith normal form (see Ex. 21 of Exercises-II).

We now establish a natural correspondence between linear transformations and matrices which assigns to the sum and composite of two linear transformations the sum and product of two matrices, respectively.

Theorem 8.6.2 *Let V be an n-dimensional vector space over F. Then the two algebras $L_F(V, V)$ of all linear operators on V over F and $M_{nn}(F)$ of all $n \times n$ matrices over F are isomorphic.*

Proof Let $B = \{v_1, v_2, \ldots, v_n\}$ be a basis of V over F and $T \in L_F(V, V)$. If $m(T)$ is the matrix representation of T with respect to the basis B, then the mapping $\psi : L_F(V, V) \to M_{nn}(F)$, $T \mapsto m(T)$ is an isomorphism of algebras. $\square$

Corollary 1 *A non-zero element in $L_F(V, V)$ is invertible iff $m(T)$ has an inverse in $M_{nn}(F)$ i.e., iff $m(T)$ is non-singular.*

Corollary 2 *The group $L_n(F)$ of all non-singular linear operators (i.e., the group of all invertible elements) in $L_F(V, V)$ is isomorphic to the group $GL(n, F)$ of all non-singular matrices in $M_{nn}(F)$.*

Definition 8.6.5 The group $L_n(F)$ is called a full linear group and the group $GL(n, F)$ is called a general linear group.

Similar matrices are closely related to their ranks and determinants.

Definition 8.6.6 Let A be an $m \times n$ matrix over F and $V = R(A)$ be the subspace of F^n generated by the row vectors of A. The dimension of V is called the row rank of A. If $U = C(A)$ is the subspace of F^m generated by the column vectors of A, then dim U is called the column rank of A.

Proposition 8.6.3 *Let A be an $m \times n$ matrix over F. Then*

 (i) *row rank of $A \leq \min\{m, n\}$;*
 (ii) *column rank of $A \leq \min\{m, n\}$;*
 (iii) *row rank of $A = $ column rank of A.*

Proof (i) Let $B = \{R_1, R_2, \ldots, R_m\}$ be the set of all m rows of A. Then B generates the row space $R(A)$ of A. Hence dim$(R(A)) \leq m$. Again $R(A)$ is a subspace of F^n. Hence dim$(R(A)) \leq n$. This proves (i).

 (ii) Similar to (i).

 (iii) Let row rank of $A = r$ and the following r row vectors of A form a basis of $R(A)$:

$$B_1 = (b_{11}, b_{12}, \ldots, b_{1n}),$$

$$B_2 = (b_{21}, b_{22}, \ldots, b_{2n}),$$

$$\cdots$$

$$B_r = (b_{r1}, b_{r2}, \ldots, b_{rn}).$$

Then each row vector of A can be expressed uniquely as

$$R_1 = c_{11}B_1 + c_{12}B_2 + \cdots + c_{1r}B_r,$$

$$R_2 = c_{21}B_1 + c_{22}B_2 + \cdots + c_{2r}B_r,$$

$$\cdots$$

$$R_m = c_{m1}B_1 + c_{m2}B_2 + \cdots + c_{mr}B_r, \quad \text{where } c_{ij} \in F.$$

Equating the jth components from the above relations, we get

$$a_{1j} = c_{11}b_{1j} + c_{12}b_{2j} + \cdots + c_{1r}b_{rj},$$

$$a_{2j} = c_{21}b_{1j} + c_{22}b_{2j} + \cdots + c_{2r}b_{rj},$$

$$\cdots$$

$$a_{mj} = c_{m1}b_{1j} + c_{m2}b_{2j} + \cdots + c_{mr}b_{rj}.$$

The above relations show that

$$\begin{pmatrix} a_{1j} \\ a_{2j} \\ \vdots \\ a_{mj} \end{pmatrix} = b_{1j} \begin{pmatrix} c_{11} \\ c_{21} \\ \vdots \\ c_{m1} \end{pmatrix} + b_{2j} \begin{pmatrix} c_{12} \\ c_{22} \\ \vdots \\ c_{m2} \end{pmatrix} + \cdots + b_{rj} \begin{pmatrix} c_{1r} \\ c_{2r} \\ \vdots \\ c_{mr} \end{pmatrix}.$$

Thus the column space of the matrix A has dimension at most r i.e., column rank of $A \leq r =$ row rank of A. Similarly, row rank of $A \leq$ column rank of A. This proves (iii). $\square$

Definition 8.6.7 Let A be an $m \times n$ matrix over F. Then row rank of $A =$ column rank of $A = r$. This common value is called the rank of A, denoted by rank A.

Proposition 8.6.4 *Let A and B be two similar matrices of order n over F. Then*

(i) rank $A =$ rank B;
(ii) det $A =$ det B.

Proof $A \approx B$ shows that there exists a non-singular matrix P in $M_{nn}(F)$ such that $B = PAP^{-1}$. Then

(i) rank $B =$ rank$(PAP^{-1}) =$ rank A and
(ii) det $B =$ det$(PAP^{-1}) =$ det A. $\square$

8.7   Eigenvalues and Eigenvectors

The algebra of matrices can be applied smoothly to diagonal matrices. The reason
is that for the addition and multiplication of any two diagonal matrices, one simply
adds (or multiplies) the corresponding diagonal entries. So it has become important
to know which matrices are similar to diagonal matrices and which pairs of diagonal matrices are similar to each other. To obtain the answer to these questions, the
concepts of eigenvalue and eigenvector are introduced.

Let V be an n-dimensional vector space over F and $T : V \to V$ a linear operator.
If B_1 is a basis of V and $m(T)_{B_1} = A$, then by Theorem 8.6.1, A is similar to a
diagonal matrix iff there exists a basis $B_2 = \{v_1, v_2, \ldots, v_n\}$ of V such that $m(T)_{B_2}$
is a diagonal matrix, i.e., iff $T(v_i) = \lambda_i v_i$ for some $\lambda_i \in F$, $i = 1, 2, \ldots, n$. Hence
any non-zero vector $v \in V$ for which $T(v) = \lambda v$ for some $\lambda \in F$ plays an important
role in our study and leads to the following concepts.

Definition 8.7.1   Let V be an n-dimensional vector space over F and $T : V \to V$
a linear operator. An element λ in F is called an eigenvalue of T (or characteristic
root of T) iff there exists a non-zero vector $v \in V$ such that $T(v) = \lambda v$. This vector
v (if it exists) is called an eigenvector of T corresponding to the eigenvalue λ. For a
square matrix we have an analogue of this definition.

Definition 8.7.2   Let A be a square matrix of order n over F. An element λ in F
is called an eigenvalue of A iff there exists a non-zero vector $X = (x_1, x_2, \ldots, x_n)$
in F^n such that $AX = \lambda X$ (or $XA = \lambda X$). The vector X (if it exists) is called an
eigenvector of A corresponding to the eigenvalue λ.

Remark 1   If A is an $n \times n$ matrix over $\mathbf{R}$, then $AX = \lambda X$ shows that AX dilates
(lengthens) X if $|\lambda| > 1$ and contracts (shortens) X if $|\lambda| < 1$, in $\mathbf{R}^n$.

Remark 2   Let $T : \mathbf{R}^n \to \mathbf{R}^n$ be a linear operator and $m(T)_B = A$, where B is the
natural basis of $\mathbf{R}^n$. Then for $X = (x_1, x_2, \ldots, x_n) \in \mathbf{R}^n$, $T(X) = XA$. This shows
that λ is an eigenvalue of A iff λ is an eigenvalue of T and X is the corresponding
eigenvector of both T and A.

Remark 3   If v is an eigenvector of T corresponding to an eigenvalue $\lambda \in F$, then
for any non-zero scalar α in F, αv is also an eigenvector of T corresponding to the
same eigenvalue λ. Hence there exist many eigenvectors of T (or A) corresponding
to an eigenvalue.

The eigenvectors have many interesting properties.

Proposition 8.7.1   *Let A be an $n \times n$ matrix over F and $E(\lambda)$ be the set of all
eigenvectors of A corresponding to an eigenvalue λ, together with the zero vector.
Then $E(\lambda)$ forms a subspace of the vector space F^n over F.*

Proof Clearly, $E(\lambda) = \{X \in F^n : AX = \lambda X\} \neq \emptyset$. Then for $X, Y \in E(\lambda)$ and $\alpha, \beta \in F$, $A(\alpha X + \beta Y) = \lambda(\alpha X + \beta Y) \in E(\lambda) \Rightarrow E(\lambda)$ is a subspace of F^n over F. $\square$

Definition 8.7.3 The vector space $E(\lambda)$ is called the eigenspace of A corresponding to the eigenvalue λ. For a linear operator $T : V \rightarrow V$ and an eigenvalue λ, $E(\lambda) = \{v \in V : T(v) = \lambda v\}$ is a subspace of V, called the null space of $T - \lambda I$ and the number of linearly independent eigenvectors in $E(\lambda)$ is called the *dimension* of $E(\lambda)$.

Remark If λ_1 and λ_2 are two distinct eigenvalues of T, then $E(\lambda_1) \cap E(\lambda_2) = \{0\}$ i.e., zero vector is the only vector common in the corresponding eigenspaces.

Theorem 8.7.1 *Let A be an $n \times n$ matrix over F and $\lambda \in F$. Then λ is an eigenvalue of A iff the matrix $(A - \lambda I)$ is singular.*

Proof Let $B = A - \lambda I$. Then λ is an eigenvalue of $A \Leftrightarrow AX = \lambda X$ for some non-zero $X \in F^n \Leftrightarrow (A - \lambda I)X = 0 \Leftrightarrow$ the homogeneous system of equations $BX = 0$ has a non-trivial solution $\Leftrightarrow$ the solution space of the system $BX = 0$ is non-zero and hence has dimension > 0. If rank $B = r$, then $n - r > 0 \Rightarrow n > r$. This situation occurs iff B is singular. $\square$

Corollary λ *is an eigenvalue of A iff $\det(A - \lambda I) = 0$.*

Proof λ is an eigenvalue of $A \Leftrightarrow$ the matrix $(A - \lambda I)$ is singular $\Leftrightarrow \det(A - \lambda I) = 0$. $\square$

We now consider the polynomial $\det(A - xI)$ in x of degree n over F.

Definition 8.7.4 Let A be an $n \times n$ matrix over F. Then the matrix $(A - xI)$ is called the characteristic matrix of A and $\det(A - xI)$ is called the characteristic polynomial of A denoted by $\kappa_A(x)$. The equation $\kappa_A(x) = 0$ is called the characteristic equation of A.

Remark The eigenvalues of A are the roots of its characteristics polynomial.

Example 8.7.1 For the matrix $A = \begin{pmatrix} 1 & 1 \\ 2 & 4 \end{pmatrix}$

(i) the characteristic matrix is $A - xI = \begin{pmatrix} 1-x & 1 \\ 2 & 4-x \end{pmatrix}$;

(ii) the characteristic polynomial is $\kappa_A(x) = x^2 - 5x + 2$;

(iii) the characteristic equation is $x^2 - 5x + 2 = 0$.

Proposition 8.7.2 *Similar matrices have the same eigenvalues.*

Proof Suppose $A \approx B$. Then there exists a non-singular matrix C such that $B = CAC^{-1} \Rightarrow B - xI = CAC^{-1} - xCIC^{-1} = C(A - xI)C^{-1} \Rightarrow A - xI \approx B - xI \Rightarrow \det(A - xI) = \det(B - xI) \Rightarrow A$ and B have the same eigenvalues.   □

Remark A matrix over **R** may not have an eigenvalue but a matrix over **C** has always an eigenvalue. This is so because a polynomial over **R** may not have a root in **R** but a polynomial over **C** has always a root in **C** [see the Fundamental Theorem of Algebra, Chap. 11 of the book or Theorem 11, Herstein (1964, p. 337)].

Proposition 8.7.3 *Let V be an n-dimensional vector space over a field F and $T : V \to V$ be a linear operator. Then the eigenvalues of T are the eigenvalues of any one of its matrix representation (relative to a single basis B).*

Proof Let $m(T)_B = A$, where B is the natural basis of F^n. Then for $X = (x_1, x_2, \ldots, x_n) \in F^n$, $T(X) = XA$. Hence it follows that λ is an eigenvector of $A \Leftrightarrow \lambda$ is an eigenvalue of T. The proof is completed by using Proposition 8.7.2. □

Example 8.7.2 (i) Let $T : \mathbf{R}^2 \to \mathbf{R}^2$, $(x, y) \mapsto (x, -y)$. Then T is a reflection about the x-axis and every multiple of $e_1 = (1, 0)$ is mapped onto itself by T. Hence $T(X) = 1 \cdot X$ for every $X = \alpha e_1$, $\alpha \in \mathbf{R}$ implies that $\lambda = 1$ is an eigenvalue of T and the corresponding eigenspace $E(1)$ is the x-axis. Similarly, for every multiple $Y = \beta e_2$ of $e_2 = (0, 1)$, $\beta \in \mathbf{R}$, $T(Y) = -1 \cdot Y$ implies that -1 is another eigenvalue of T and the corresponding eigenspace $E(-1)$ is the y-axis. Thus geometrically it follows that 1 and -1 are the only eigenvalues of T.

(ii) Let $T : \mathbf{R}^2 \to \mathbf{R}^2$ be the rotation of the plane through an angle θ, $0 < \theta < \pi$. Then T has no eigenvalue. This is so because λX ($\lambda \in \mathbf{R}$) is a scalar multiple of X and no vector X except $(0, 0)$ is mapped into a scalar multiple of itself. Hence $T(X) \neq \lambda X$ unless $X = (0, 0)$ shows that there is no eigenvalue of T and thus there is no eigenvector of T.

(iii) Let $V = D[x]$ be the vector space of real valued functions on **R** with derivatives of all orders. If $\lambda (\neq 0) \in \mathbf{R}$, $f(x) = e^{\lambda x}$ and $D : V \to V$ is the differential operator, then $D(f(x)) = \lambda e^{\lambda x} \Rightarrow D(f) = \lambda f \Rightarrow \lambda$ is an eigenvalue of D. Since the solutions of the differential equation $y' = \lambda y$ are of the form $y = \alpha e^{\lambda x}$, $\alpha \in \mathbf{R}$, the eigenspace $E(\lambda)$ is 1-dimensional with basis $\{f\}$.

(iv) Let

$$A = \begin{pmatrix} 5 & 0 & 0 \\ 0 & 2 & 2 \\ 0 & 7 & 1 \end{pmatrix} \quad \text{and} \quad X = (1 \ \ 0 \ \ 0).$$

Then $XA = 5X \Rightarrow 5$ is an eigenvalue of A and X is the corresponding eigenvector.

We now prove some other interesting properties of eigenvalues and eigenvectors.

Proposition 8.7.4 *If A is a non-singular matrix of order n over F and $\lambda \neq 0$ is an eigenvalue of A, then λ^{-1} is an eigenvalue of A^{-1}.*

Proof A is non-singular (by hypothesis) $\Rightarrow \det A \neq 0$. Then

$$\det\left(A^{-1} - \lambda^{-1}I\right) = \lambda^{-n}\det\left(\lambda A^{-1} - I\right) = \lambda^{-n} \cdot \frac{\det A \cdot \det(\lambda A^{-1} - I)}{\det A}$$

$$= \lambda^{-n}\frac{\det(A(\lambda A^{-1} - I))}{\det A} = \lambda^{-n}\frac{\det(\lambda I - A)}{\det A} = 0,$$

since λ is an eigenvalue of A. This implies that λ^{-1} is an eigenvalue of A^{-1}. $\square$

Proposition 8.7.5 (Uniqueness of eigenvalue) *Given an eigenvector of a square matrix A, the corresponding eigenvalue of A is unique.*

Proof Let X be an eigenvector of A and λ_1, λ_2 be its corresponding eigenvalues. Then $AX = \lambda_1 X = \lambda_2 X \Rightarrow (\lambda_1 - \lambda_2)X = 0 \Rightarrow \lambda_1 = \lambda_2$, since $X \neq 0$. $\square$

Theorem 8.7.2 *Let A be an $n \times n$ matrix over F and $\lambda_1, \lambda_2, \ldots, \lambda_r$ be distinct eigenvalues of A in F, with corresponding eigenvectors $X_1, X_2, \ldots, X_r$, respectively. Then the set $S = \{X_1, X_2, \ldots, X_r\}$ is linearly independent.*

Proof Let $V = F^n$. Then $X_i \in F^n$ and $AX_i = \lambda_i X_i$ for $i = 1, 2, \ldots, r$. We prove the theorem by induction on r. If $r = 1$, then $X_1 \neq \{0\}$ and hence it is linearly independent. We assume that the statement is true for $m < r$.
  Then

$$\alpha_1 X_1 + \alpha_2 X_2 + \cdots + \alpha_m X_m + \alpha_{m+1}X_{m+1} = 0$$

$$\Rightarrow \quad A(\alpha_1 X_1 + \alpha_2 X_2 + \cdots + \alpha_m X_m + \alpha_{m+1}X_{m+1}) = 0$$

$$\Rightarrow \quad \alpha_1 \lambda_1 X_1 + \alpha_2 \lambda_2 X_2 + \cdots + \alpha_m \lambda_m X_m + \alpha_{m+1}\lambda_{m+1}X_{m+1} = 0$$

$$\Rightarrow \quad \alpha_1(\lambda_1 - \lambda_{m+1})X_1 + \alpha_2(\lambda_2 - \lambda_{m+1})X_2 + \cdots + \alpha_m(\lambda_m - \lambda_{m+1})X_m = 0$$

$$\Rightarrow \quad \alpha_{m+1} = 0 \quad \text{by induction hypothesis, since } X_{m+1} \neq 0.$$

Consequently, the set S is linearly independent. $\square$

Corollary *An $n \times n$ matrix over F has at most n distinct eigenvalues.*

Proof Since $\dim F^n$ over F is n, the corollary follows from Theorem 8.7.2. $\square$

Theorem 8.7.3 *Let V be an n-dimensional vector space over F and $T : V \rightarrow V$ be a linear operator with r distinct eigenvalues $\lambda_1, \lambda_2, \ldots, \lambda_r$ and $X_1, X_2, \ldots, X_r$ the corresponding eigen vectors. Then the set $S = \{X_1, X_2, \ldots, X_r\}$ is linearly independent.*

Proof Similar to Theorem 8.7.2. $\square$

Proposition 8.7.6 *Let A be an $n \times n$ matrix over F. If A has n distinct eigenvalues, then A is similar to an $n \times n$ diagonal matrix.*

Proof Let $T : F^n \to F^n$ be the linear operator such that $m(T)_B = A$, where B is the natural basis of F^n and $\lambda_1, \lambda_2, \ldots, \lambda_n$ be n distinct eigenvalues of A with the corresponding eigenvectors $X_1, X_2, \ldots, X_n$. Then the set $S = \{X_1, X_2, \ldots, X_n\}$ is linearly independent and hence S forms a basis of F^n.

Consequently,

$$T(X_1) = \lambda_1 X_1 = \lambda_1 X_1 + 0 \cdot X_2 + \cdots + 0 \cdot X_n$$

$$T(X_2) = \lambda_2 X_2 = 0 \cdot X_1 + \lambda_2 X_2 + \cdots + 0 \cdot X_n$$

$$\cdots$$

$$T(X_n) = \lambda_n X_n = 0 \cdot X_1 + 0 \cdot X_2 + \cdots + \lambda_n \cdot X_n.$$

Hence

$$m(T)_S = D = \begin{pmatrix} \lambda_1 & 0 & \cdots & 0 \\ 0 & \lambda_2 & \cdots & 0 \\ & & \ddots & \\ 0 & 0 & \cdots & \lambda_n \end{pmatrix} \quad \Rightarrow \quad A \approx D$$

by Theorem 8.6.1.   □

Remark Converse of Proposition 8.7.6 is not true. The matrix

$$A = \begin{pmatrix} 2 & 0 & 0 \\ 0 & 2 & 0 \\ 0 & 0 & 3 \end{pmatrix}$$

has only two distinct eigenvalues 2 and 3 but A is similar to itself, which is a diagonal matrix.

We recall that an $n \times n$ matrix A over $\mathbf{C}$ can be expressed as $\mathbf{A} = X + iY$, where X and Y are matrices over $\mathbf{R}$. Then the matrix $\bar{A} = X - iY$ is called the conjugate matrix of A and the elements of $\bar{A}$ are the conjugates of the corresponding elements of A. We use the symbol A^* to denote the conjugate transpose $\bar{A}^t$ of A.

An $n \times n$ matrix A over $\mathbf{C}$ is said to be Hermitian, skew-Hermitian or unitary according to $A^* = A$ or $A^* = -A$ or $AA^* = A^*A = I$.

Proposition 8.7.7 *The eigenvalues of a Hermitian matrix are all real.*

Proof Let A be an $n \times n$ Hermitian matrix and X be an eigenvector corresponding to an eigenvalue λ of A.

Then $AX = \lambda X \Rightarrow \overline{AX} = \bar{\lambda}\bar{X} \Rightarrow (\overline{AX})^t = (\bar{\lambda}\bar{X})^t \Rightarrow \bar{X}^t \bar{A}^t = \bar{\lambda}\bar{X}^t \Rightarrow \bar{X}^t A = \bar{\lambda}\bar{X}^t \Rightarrow \bar{X}^t AX = \bar{\lambda}\bar{X}^t X \Rightarrow \bar{X}^t \lambda X = \bar{\lambda}\bar{X}^t X \Rightarrow \bar{X}^t X (\lambda - \bar{\lambda}) = 0 \Rightarrow \lambda = \bar{\lambda}$, since $X \neq 0 \Rightarrow \lambda$ is real.   □

Corollary *The eigenvalues of a real symmetric matrix are all real.*

Proposition 8.7.8   *The eigenvalues of a skew-Hermitian matrix are 0 or purely imaginary.*

Proof Proceeding as in Proposition 8.7.7, we have $\lambda + \bar{\lambda} = 0$. This shows that λ is 0 or purely imaginary.   □

Corollary *The eigenvalues of a real skew-symmetric matrix A are 0 or purely imaginary (0 is the only possible real eigenvalue of A).*

Proposition 8.7.9   *Each eigenvalue of a unitary matrix has unit modulus.*

Proof Proceeding as in Proposition 8.7.7 we have $\overline{X}^t X = \lambda \bar{\lambda}(\overline{X}^t X)$ and hence $\lambda \bar{\lambda} = 1$, since $\overline{X}^t X \neq 0$. This implies that $|\lambda| = 1$.   □

Corollary *Each eigenvalue of a real orthogonal matrix A has unit modulus (1 and −1 are the only possible real eigenvalues of A).*

We now introduce the concepts of algebraic multiplicity and geometric multiplicity of an eigenvalue. These concepts are important for our further study.

Definition 8.7.5   Let A be an $n \times n$ matrix over F. The algebraic multiplicity of an eigenvalue λ of A is the multiplicity of λ as a root of the characteristic equation $\kappa_A(x) = 0$. The geometric multiplicity of λ is the dimension of the eigenspace $E(\lambda)$.

Definition 8.7.6   An $n \times n$ matrix A over F is said to be diagonalizable over F iff A is similar to a diagonal matrix D over F.

Definition 8.7.7   A linear operator $T : V \to V$ on a finite-dimensional vector space over F is said to be diagonalizable over F iff there exists a basis B of V such that the matrix $m(T)_B$ is diagonalizable.

Remark 1 An $n \times n$ matrix A over $\mathbf{C}$ is diagonalizable over $\mathbf{C}$ iff for each eigenvalue λ of A, the geometric and algebraic multiplicities of λ are the same.

Remark 2 An $n \times n$ matrix A over $\mathbf{R}$ is diagonalizable over $\mathbf{R}$ iff all the eigenvalues of A are real and for each eigenvalue of A, the geometric and algebraic multiplicities are the same.

Remark 3 An $n \times n$ matrix A over F is diagonalizable iff there exists an $n \times n$ non-singular matrix P over F such that $D = PAP^{-1}$ is a diagonal matrix.

Example 8.7.3 (i) The matrix

$$A = \begin{pmatrix} 2 & 0 & 0 \\ 0 & 0 & -1 \\ 0 & 1 & 0 \end{pmatrix}$$

is diagonalizable over $\mathbf{C}$ but not over $\mathbf{R}$. This is so because A has three distinct eigenvalues in $\mathbf{C}$ but its eigenvalue 2 is only in $\mathbf{R}$.

(ii) The linear operator $T : \mathbf{R}^2 \to \mathbf{R}^2$, $(a, b) \mapsto (b, a)$ has its matrix representation $m(T)_B$ with respect to the natural basis $B = \{(1, 0), (0, 1)\}$ of $\mathbf{R}^2$ given by $A = m(T)_B = \begin{pmatrix} 0 & 1 \\ 1 & 0 \end{pmatrix}$. As 1 and -1 are two distinct eigenvalues of A, A is diagonalizable of $\mathbf{R}$ and hence T is diagonizable.

(iii) Let V be a vector space over F and $T : V \to V$ a linear operator with an eigenvalue λ. Then for any polynomial $f(x)$ in $F[x]$, $f(\lambda)$ is an eigenvalue of $f(T)$.

To show this, let v ($\neq 0$) be an eigenvector of T corresponding to the eigenvalue λ. Then $T(v) = \lambda v \Rightarrow T(T(v)) = T(\lambda v) \Rightarrow T^2(v) = \lambda^2 v \Rightarrow \lambda^2$ is an eigenvalue of T^2. In general, $T^n(v) = \lambda^n v$ for every positive integer n.

Let

$$f(x) = a_0 + a_1 x + \cdots + a_n x^n \in F[x].$$

Then

$$f(T) = a_0 I + a_1 T + \cdots + a_n T^n$$

$$\Rightarrow \quad f(T)(v) = \left(a_0 + a_1 \lambda + \cdots + a_n \lambda^n\right) v = f(\lambda) v.$$

Hence

$$f(T)(v) = f(\lambda) v \quad \Rightarrow \quad f(\lambda)$$

is an eigenvalue of $f(T)$. For simplicity, sometimes we write $f(T)v$ in place of $f(T)(v)$, unless there is any confusion.

We now apply the concepts of eigenvalues and eigenvectors to find conditions for diagonalization of a given square matrix.

Theorem 8.7.4 *Let V be an n-dimensional vector space over F and $T : V \to V$ be a linear operator. Then T is represented by a diagonal matrix with respect to some basis iff V has a basis consisting of eigenvectors of T.*

Proof Let T be represented by a diagonal matrix

$$D = \begin{pmatrix} d_1 & 0 & \cdots & 0 \\ 0 & d_2 & \cdots & 0 \\ & & \ddots & \\ 0 & 0 & \cdots & d_n \end{pmatrix}$$

with respect to a basis $B = \{v_1, v_2, \ldots, v_n\}$ of V.

Then $T(v_1) = d_1 v_1$, $T(v_2) = d_2 v_2, \ldots, T(v_n) = d_n v_n$. Consequently, v_i is an eigenvector of T corresponding to the eigenvalue d_i, $i = 1, 2, \ldots, n$. Thus the basis B consists of eigenvectors of T.

Conversely, if the eigenvectors $v_1, v_2, \ldots, v_n$ of T form a basis $B = \{v_1, v_2, \ldots, v_n\}$ of V, then $m(T)_B$ is a diagonal matrix D of the above form, because $T(v_i) = d_i v_i$, $d_i \in F$, $i = 1, 2, \ldots, n$. $\qquad\square$

We now establish the following equivalent conditions for diagonalization of a square matrix over F.

Theorem 8.7.5 *Let V be an n-dimensional vector space over F and $T : V \to V$ be a linear operator with $m(T)_B = A$ with respect to some basis B of V. Then the following statements are equivalent*:

 (i) *A can be diagonalized*;
 (ii) *A is similar to a diagonal matrix*;
(iii) *F^n has a basis consisting of eigenvectors of A*;
(iv) *V has a basis consisting of eigenvectors of T*.

Proof (i) $\Rightarrow$ (ii) It follows from Definition 8.7.6.

   (ii) $\Rightarrow$ (iii) If A is similar to a diagonal matrix, then there exists a non-singular matrix P such that $D = PAP^{-1}$ is a diagonal matrix. If V_j is the jth column of P^{-1}, then the jth column of AP^{-1} is AV_j. Looking at the jth column of each side of $P^{-1}D = AP^{-1}$, we have $AV_j = d_j V_j$, for some $d_j \in F$. Since P^{-1} is non-singular, each V_j is non-zero and hence each V_j is an eigenvector of A. As $\{V_j\}$ is linearly independent, $\{V_j\}$ constitutes a basis of F^n.

   (iii) $\Rightarrow$ (iv) Let $m(T)_B = A$ for some basis of V. We assume that F^n has a basis consisting of eigenvectors of A and $v_1, v_2, \ldots, v_n$ are the elements of V whose coordinate vectors relative to B are the above eigenvectors of A. Then $B_1 = \{v_1, v_2, \ldots, v_n\}$ is a basis of V and each v_j is an eigenvector of T.

   (iv) $\Rightarrow$ (i) Let $B = \{v_1, v_2, \ldots, v_n\}$ be a basis of V consisting of eigenvectors of T. Then $T(v_j) = d_j v_j, d_j \in F$, $j = 1, 2, \ldots, n$. Consequently, T has a diagonal representation and hence A is diagonalizable. $\qquad\square$

8.8 The Cayley–Hamilton Theorem

Cayley–Hamilton Theorem is a famous theorem in linear algebra and gives a relation between a square matrix and its characteristic equation. More precisely, this theorem proves that every square matrix A satisfies its characteristic equation and the minimal polynomial of A divides its characteristic polynomial. This theorem is used to evaluate large powers of the matrix A.

   Let A be a square matrix over F and $f(x) = a_0 + a_1 x + \cdots + a_n x^n$ be a polynomial in $F[x]$. We use the symbol $a_0 I + a_1 A + \cdots + a_n A^n$ to represent the matrix $f(A)$. The following theorem is important in the study of matrices.

Theorem 8.8.1 (Cayley–Hamilton Theorem) *Every square matrix satisfies its own characteristic equation.*

Proof Let A be an $n \times n$ matrix over F and $\kappa_A(x)$ be its characteristic polynomial in $F[x]$. We claim that $\kappa_A(A) = 0$. Let $\kappa_A(x) = \det(A - xI) = c_0 x^n + c_1 x^{n-1} + \cdots + c_n$ and $B(x) = B_0 x^{n-1} + B_1 x^{n-2} + \cdots + B_{n-1}$, be the adjoint of the matrix $(A - xI)$, where B_i is a matrix over F. Then

$$(A - xI)B(x) = \det(A - xI)I$$

$$\Rightarrow \quad (A - xI)\big(B_0 x^{n-1} + B_1 x^{n-2} + \cdots + B_{n-1}\big) = \big(c_0 x^n + c_1 x^{n-1} + \cdots + c_n\big)I.$$

Hence

$$A\big(B_0 x^{n-1} + B_1 x^{n-2} + \cdots + B_{n-1}\big) - \big(B_0 x^n + B_1 x^{n-1} + \cdots + B_{n-1}x\big)$$

$$= c_0 I x^n + c_1 I x^{n-1} + \cdots + c_n I.$$

Equating the coefficients of like powers of x, we have

$$-B_0 = c_0 I$$

$$AB_0 - B_1 = c_1 I$$

$$AB_1 - B_2 = c_2 I$$

$$\cdots$$

$$AB_{n-2} - B_{n-1} = c_{n-1} I$$

$$AB_{n-1} = c_n I.$$

Multiplying the above equations by $A^n, A^{n-1}, \ldots, A, I$, respectively, and then adding we have

$$c_0 A^n + c_1 A^{n-1} + \cdots + c_{n-1} A + c_n I = 0.$$

This shows that $\kappa_A(A) = 0$. $\square$

Remark The definition of a minimal polynomial of a matrix is similar to that of a linear transformation (see Definition 8.5.12).

Definition 8.8.1 Let A be an $n \times n$ non-zero matrix over F. A polynomial $m(x)$ in $F[x]$ of least degree with leading coefficient 1 is said to be a minimal polynomial of A iff $m(A) = 0$.

Proposition 8.8.1 *Let A be a square matrix over F with $m(x)$ its minimal polynomial. Then A is a root of a polynomial $f(x)$ in $F[x]$ iff $f(x)$ is divisible by $m(x)$ in $F[x]$.*

Proof As F is a field, $F[x]$ is a Euclidean domain. Hence $\exists q(x)$ and $r(x)$ in $F[x]$ such that $f(x) = m(x)q(x) + r(x)$, where $r(x) = 0$ or $\deg r < \deg m$ (by division algorithm). If $f(A) = 0$, the $r(A) = 0$, since $m(A) = 0$. If $r(x) \neq 0$, then $r(x)$ is a polynomial of degree less than the degree of $m(x)$, which has A as a root. Then we reach a contradiction as $m(x)$ is a minimal polynomial of A. Thus $r(x) = 0$ shows that $f(x)$ is divisible by $m(x)$. On the other hand if $f(x)$ is divisible by $m(x)$ in $F[x]$, then $r(x) = 0$. Hence $r(A) = 0 \Rightarrow f(A) = 0 \Rightarrow A$ is a root of $f(x)$.   □

Remark A minimal polynomial $m(x)$ of A over F may not be irreducible in $F[x]$. For example, the matrix $A = \begin{pmatrix} 2 & 1 \\ 0 & 1 \end{pmatrix}$ over **R** has its minimal polynomial $m(x) = x^2 - 3x + 2$, but it is not irreducible in $\mathbf{R}[x]$.

Corollary 1 *Let A be a square matrix over F. Then its characteristic polynomial is divisible by its minimal polynomial.*

Proof A is a root of its characteristic polynomial $\kappa_A(x)$ by Cayley–Hamilton Theorem. Then the corollary follows by Proposition 8.8.1.   □

Corollary 2 *The minimal polynomial of a square matrix over F is unique.*

Proof Let A be a square matrix over F and $m_1(x)$, $m_2(x)$ be its minimal polynomials. Then A is a root of $m_1(x)$ and by Proposition 8.8.1, $m_1(x) = g(x)m_2(x)$ for some polynomial $g(x) \in F[x]$. Hence $\deg m_1 = \deg g + \deg m_2$. Again by minimality of degrees, $\deg m_1 \leq \deg m_2$ and $\deg m_2 \leq \deg m_1$. Hence $\deg m_1 = \deg m_2$ shows that $\deg g = 0$. Thus $g(x)$ is a non-zero constant $c \in F$. Let $m_2(x) = x^m + \cdots + a_1 x + a_0$. Then $m_1(x) = cx^m + \cdots + ca_1 x + ca_0$. But the leading coefficient of m_1 is 1. Hence $c = 1$ proves that $m_1(x) = m_2(x)$.   □

Theorem 8.8.2 *Let A be a non-zero $n \times n$ matrix over F. Then A is non-singular iff the constant term of the minimal polynomial of A is not zero.*

Proof Proceed as in Theorem 8.5.12.   □

Corollary *If A is a non-singular square matrix over F, then its inverse is also polynomial in A over F.*

Remark Let V be a finite-dimensional vector space over F and $T : V \to V$ a linear operator. If B is any basis of V, then the minimal polynomial of T is the same as the minimal polynomial of $m(T)_B$. This is so because similar matrices have the same minimal polynomial (see Ex. 20 of Exercises-II).

Theorem 8.8.3 *Let V be an n-dimensional vector space over F and $T : V \to V$ be a linear operator with its minimal polynomial $m(x)$. Then*

(i) *an element λ in F is an eigenvalue of T iff $m(\lambda) = 0$;*
(ii) *the characteristic polynomial $\kappa(x)$ of T and its minimal polynomial $m(x)$ have the same roots.*

Proof (i) If $m(x)$ divides $\kappa(x)$, then $\kappa(x) = q(x)m(x)$ for some $q(x) \in F[x]$. Hence $\kappa(\lambda) = q(\lambda)m(\lambda)$. If λ is a root of $m(x)$, then λ is also a root of $\kappa(x)$ and thus λ is an eigenvalue of T.

Conversely, let λ be an eigenvalue of T and $x \in V$ be an eigenvector corresponding to λ. Then $m(T)x = m(\lambda)x$ by Example 8.7.3(iii) and $m(T) = 0$. Since $x \neq 0$, $m(\lambda) = 0$. Hence λ is a root of $m(x)$.

(ii) It follows from (i).   □

Remark The multiplicities of the roots of $\kappa(x)$ and $m(x)$ may be different. For example, for the matrix

$$A = \begin{pmatrix} 2 & 1 & 0 \\ -4 & -2 & 0 \\ 2 & 1 & 0 \end{pmatrix}, \qquad \kappa(x) = -x^3 \quad \text{and} \quad m(x) = x^2,$$

the multiplicities of roots of $\kappa(x)$ and $m(x)$ are different.

Corollary *Let V be an n-dimensional vector space over F and $T : V \to V$ a linear operator with minimal polynomial $m(x)$ and characteristic polynomial $\kappa(x)$. If $\kappa(x) = (\lambda_1 - x)^{n_1}(\lambda_2 - x)^{n_2} \cdots (\lambda_t - x)^{n_t}$, then there exist integers m_i such that $1 \leq m_i \leq n_i$, $i = 1, 2, \ldots, t$ and $m(x) = (x - \lambda_1)^{m_1}(x - \lambda_2)^{m_2} \cdots (x - \lambda_t)^{m_t}$.*

8.9 Jordan Canonical Form

Every square matrix is not similar to a diagonal matrix but every square matrix is similar to an upper triangular matrix over the complex field **C** (see Ex. 9 and Ex. 24 of Exercises-II). A linear transformation on a vector space represents matrices which differ depending on different bases.

Given a finite-dimensional vector space V over a field F, there exist linear transformations on V such that each associated matrix in a similarity class of square matrices over F, takes a simple form in some basis, called canonical form. There are many canonical forms of matrices, namely, the triangular form, Jordan normal form, rational canonical form etc. But in this section we study only the Jordan canonical form. Since the Jordan canonical form is determined by the set of elementary divisors, we need basic ideas of determinant divisors, invariant factors, elementary divisors etc., of a square matrix. We first introduce these concepts as background material. For other canonical forms see Exs. 21–23 of Exercises-II.

Definition 8.9.1 Let A be a square matrix of order n over a field F and $d_t(A)$ denote the highest common factor of the set of all minors of order t of A, for $t = 1, 2, \ldots, r$, where r is the rank of A. Then $d_t(A)$ is called the tth determinant divisor of A.

Definition 8.9.2 If we take $d_0(A) = 1$ and $e_i = d_i(A)/d_{i-1}(A)$, for $i = 1, 2, \ldots, r$, where rank $A = r$, then $e_1, e_2, \ldots, e_r$ are called the invariant factors of A.

Remark $d_i(A) = e_1 e_2 \cdots e_i$, for $i = 1, 2, \ldots, r$ and hence the invariant factors of A determine uniquely its determinant divisors and conversely.

Definition 8.9.3 Let $x_1, x_2, \ldots, x_m$ be irreducible elements in F, each of which is a divisor of at least one of the invariant factors $e_1, e_2, \ldots, e_r$ of a square matrix A over F. Then the prime powers $x_1^{p_{11}}, \ldots, x_m^{p_{1m}}, \ldots, x_1^{p_{r1}}, \ldots, x_m^{p_{rm}}$ for which the indices are positive, are called the elementary divisors of A over F.

Example 8.9.1   For the matrix

$$A = \begin{pmatrix} x^2 - x & x & 0 \\ x - 2 & x & x + 4 \\ x & 0 & 3x \end{pmatrix},$$

the determinant divisors d_i, invariant factors e_i and elementary divisors are respectively:

$d_1 = hcf$ of $\{x^2 - x, x, x - 2, x, x + 4, x, 3x\} = 1$;
$d_2 = hcf$ of $\{x(x^2 - 2x + 2), x(x - 1)(x + 4), -x^2, 3x^2, 3x(x^2 - x), x(x + 4),$
$-x^2, 2x(x - 5), 3x^2\} = x$;
$d_3 = x^2(3x^2 - 5x + 10)$;
$e_1 = d_1 = 1$, $e_2 = d_2/d_1 = x$, $e_3 = d_3/d_2 = x(3x^2 - 5x + 10)$.

The elementary divisors of A in $\mathbf{R}[x]$ are $x, x; 3x^2 - 5x + 10$ and these in $\mathbf{C}[x]$ are $x, x; x - \alpha; x - \beta$, where $\alpha, \beta = (5 \pm \sqrt{95}i)/6$.

We recall that if V is a finite-dimensional vector space over an algebraically closed field F, such as complex field $\mathbf{C}$, then the characteristic polynomial of every linear operator $T : V \to V$ decomposes into linear factors in $F[x]$. If $\lambda_1, \lambda_2, \ldots, \lambda_n$ (not necessarily distinct) are all eigenvalues of T, then T is diagonalizable iff there is an ordered basis of V consisting of eigenvectors of T. If $B = \{v_1, v_2, \ldots, v_n\}$ is such a basis in which v_j is an eigenvector corresponding to the eigenvalue λ_j, then

$$m(T)_B = \begin{pmatrix} \lambda_1 & 0 & \cdots & 0 \\ 0 & \lambda_2 & \cdots & 0 \\ & & \ddots & \\ 0 & 0 & \cdots & \lambda_n \end{pmatrix} = (\lambda_j).$$

We take $F = \mathbf{C}$, unless otherwise stated.

Definition 8.9.4 Corresponding to an element λ in $\mathbf{C}$, an elementary Jordan matrix or a Jordan block $J(\lambda)$ of order r is a square matrix of order r over $\mathbf{C}$ that has λ in each diagonal position, 1's in each position just above the main diagonal and 0's elsewhere.
Thus

$$J(\lambda) = \begin{pmatrix} \lambda & 1 & 0 & \cdots & 0 \\ 0 & \lambda & 1 & \cdots & 0 \\ 0 & 0 & \lambda & \cdots & 0 \\ & & & \ddots & \\ 0 & 0 & 0 & \cdots & \lambda \end{pmatrix}$$

is a *Jordan block* corresponding to λ. A Jordan block is an upper triangular matrix.

For example,

$$(5), \qquad \begin{pmatrix} 3 & 1 \\ 0 & 3 \end{pmatrix}, \qquad \begin{pmatrix} 7 & 1 & 0 \\ 0 & 7 & 1 \\ 0 & 0 & 7 \end{pmatrix}$$

are Jordan blocks of order $1, 2$ and 3 corresponding to the eigenvalues 5, 3, and 7 respectively.

Definition 8.9.5 The matrix

$$\begin{pmatrix} J_1(\lambda_1) & 0 & 0 & \cdots & 0 \\ 0 & J_2(\lambda_2) & 0 & \cdots & 0 \\ 0 & 0 & J_3(\lambda_3) & \cdots & 0 \\ & & & \ddots & \\ 0 & 0 & 0 & \cdots & J_r(\lambda_r) \end{pmatrix} \qquad (8.3)$$

where each $J_i(\lambda_i)$ is a Jordan block corresponding to the eigenvalue λ_i of a square matrix A is said to be a Jordan canonical form of A, where $\lambda_1, \lambda_2, \ldots, \lambda_r$ may not be distinct and orders of the Jordan blocks may be different.

Jordan canonical form is one of the most generally used canonical forms connecting linear transformation and matrices. But its serious inconvenience is the condition that all eigen values must lie in the ground field. We study some other canonical form which needs nothing about the location of eigenvalues. Such a canonical form is the rational canonical form which is described in Ex. 23 of Exercises-II.

Theorem 8.9.1 (Jordan Form Theorem) *A square matrix M, over a field F such that the eigenvalues of M lie in F, is similar to a Jordan matrix of the form* (8.3), *which is uniquely determined, except for rearrangements of its diagonal blocks.*

Proof Since the eigenvalues of M are in F, its characteristic polynomial $\kappa_M(x)$ decomposes into linear factors in $F[x]$:

$$\kappa_M(x) = (\lambda_1 - x)^{n_1} (\lambda_2 - x)^{n_2} \cdots (\lambda_m - x)^{n_t},$$

where $n_1 + n_2 + \cdots + n_t = n$ and λ_i's are distinct eigenvalues of M.

Again $\kappa_M(x)$ is the product of invariant factors, and also of the elementary divisors of the characteristic matrix $M - xI$. As the prime factors of $\kappa_M(x)$ in $F[x]$ are linear polynomials $\lambda_i - x$, it follows that the elementary divisors of $M - xI$ are of the form $(\lambda_i - x)^{r_i}$. We suppose that $\{(\lambda_1 - x_1)^{r_1}, \ldots, (\lambda_m - x)^{r_m}\}$ is the complete set of elementary divisors of $M - xI$, where $\lambda_1, \lambda_2, \ldots, \lambda_m$ may not be distinct. Let $J_i(\lambda_i)$ be the Jordan block of order r_i corresponding to the eigenvalue λ_i, $i = 1, 2, \ldots, m$. Then $(\lambda_i - x)^{r_i}$ is both the characteristic polynomial and the minimal polynomial of the matrix $J_i(\lambda_i)$. As the minimal polynomial of $J_i(\lambda_i)$ is the last invariant factor of the matrix $J_i(\lambda_i) - xI$ and the characteristic polynomial of $J_i(\lambda_i)$ is the product of all invariant factors of $J_i(\lambda_i) - xI$, it follows that all the invariant factors of $J_i(\lambda_i) - xI$, excepting the last one, is 1. Hence $(\lambda_i - x)^{r_i}$ is the only elementary divisor of $J_i(\lambda_i) - xI$.

Let $J = \mathrm{diag}(J_1(\lambda_1), \ldots, J_m(\lambda_m))$ be the Jordan matrix. Then the set of the elementary divisors of $J - xI$ is given by the union of the elementary divisors of $J_1(\lambda_1) - xI, \ldots, J_m(\lambda_m) - xI$. Thus the elementary divisors of $J - xI$ are given by $(\lambda_1 - x)^{r_1}, \ldots, (\lambda_m - x)^{r_m}$. Again, as the product of the elementary divisors of $M - xI$ is equal to the characteristic polynomial $\kappa_M(x)$, it follows that $r_1 + r_2 + \cdots + r_m = n_1 + n_2 + \cdots + n_t = n$. This shows that $J - xI$ is a matrix of order n. Again $M - xI$ and $J - xI$ are both non-singular. Hence rank M = rank J. Consequently, $M - xI$ and $J - xI$ are both square matrices of order n over F such that they have the same rank and the same set of elementary divisors. This implies that M is similar to the Jordan matrix J.

Again the set of elementary divisors of $M - xI$ consists of the elementary divisors of $J_i(\lambda_i) - xI$, for $i = 1, 2, \ldots, m$, there is one elementary divisor corresponding to each block $J_i(\lambda_i)$. Moreover, the block diagonal matrices which differ only in the arrangement of diagonal blocks are similar. Consequently, the Jordan matrix J is determined uniquely, except for rearrangement of the diagonal blocks, by the given matrix M. $\Box$

Remark   The Jordan matrix J similar to M is called the Jordan normal form (or the classical canonical form) of the matrix M.

Corollary *Any square matrix M over $\mathbf{C}$ is similar to a Jordan matrix, which is determined uniquely, except for rearrangements of its diagonal blocks.*

Proof As all the eigenvalues of M over $\mathbf{C}$ lie in $\mathbf{C}$, the corollary follows from Theorem 8.9.1. $\Box$

Example 8.9.2   (i) The matrix

$$M = \begin{pmatrix} 4 & -1 & 1 \\ 4 & 0 & 2 \\ 2 & -1 & 3 \end{pmatrix}$$

is similar to the Jordan matrix

$$\begin{pmatrix} 2 & 0 & 0 \\ 0 & 2 & 0 \\ 0 & 0 & 3 \end{pmatrix}.$$

For this purpose we prescribe a method by computing as follows:

Step I: $\kappa_M(x) = \det(M - xI) = -(x - 2)^2(x - 3)$.
Step II: $d_1(x) = 1$, $d_2(x) = x - 2$, $d_3(x) = (x - 2)^2(x - 3)$.

Fig. 8.1 Jordan normal form
of M

$$J = \begin{pmatrix} 2 & 1 & 0 \\ 0 & 2 & 0 \\ 0 & 0 & \boxed{2} \end{pmatrix}$$

Step III: $e_1(x) = d_1(x) = 1$, $e_2(x) = d_2(x)/d_1(x) = x - 2 = p_1(x)$. (say) $e_3(x) = (x - 2)(x - 3) = p_1(x)p_2(x)$ (say).

The multiplicities of the eigenvalues 2 and 3 as roots of the invariant factors are tabulated:

λ	$e_3(x)$	$e_2(x)$	$e_1(x)$
2	1	1	0
3	1	0	0

The symbol $[(1, 1), (1)]$ denoting the multiplicities of the eigenvalues as roots of the invariant factors, collected together in the first brackets for each one of the eigenvalues, is called the *Segre characteristic* of M and it represents the Jordan normal form

$$J = \begin{pmatrix} \boxed{2} & 0 & 0 \\ 0 & \boxed{2} & 0 \\ 0 & 0 & \boxed{3} \end{pmatrix}.$$

(ii) The matrix

$$M = \begin{pmatrix} -6 & 6 & 1 \\ -1 & 1 & 1 \\ -8 & 5 & 8 \end{pmatrix}$$

has no Jordan normal form over the field **R**.

$\kappa_M(x) = -(x+5)(x^2 - 3x + 3)$ shows that all the three roots are not in **R**. Hence there does not exist any Jordan normal form of M over **R**.

(iii) The Jordan normal form of the matrix

$$M = \begin{pmatrix} 0 & 4 & 2 \\ -3 & 8 & 3 \\ 4 & -8 & -2 \end{pmatrix}$$

is shown in Fig. 8.1.

Here $\kappa_M(x) = -(x - 2)^3$, $d_1(x) = 1$, $d_2(x) = (x - 2)$, $d_3(x) = -(x - 2)^3$ and $e_1(x) = 1$, $e_2(x) = x - 2$, $e_3(x) = -(x - 2)^2$.

The Segre characteristic of M is $[(2, 1)]$ and it represents J.

(iv) Let M be a 3×3 matrix over **C**. Then all the eigenvalues of M lie in **C**. Find the different forms of Jordan matrices similar to M.

To find the different forms of Jordan matrices similar to M, we consider all possible cases. If e_1, e_2, e_3 are invariant factors of characteristic matrix $M - xI$, then

(a) e_1 is a divisor of e_2 and e_2 is a divisor e_3; and
(b) the product $e_1\, e_2\, e_3 = \kappa_M(x)$.

Fig. 8.2 Jordan normal form for the Case II(ii)

$$J_3 = \begin{pmatrix} \boxed{\begin{matrix} \lambda_1 & 1 \\ 0 & \lambda_1 \end{matrix}} & \begin{matrix} 0 \\ 0 \end{matrix} \\ \begin{matrix} 0 & 0 \end{matrix} & \boxed{\lambda_2} \end{pmatrix}$$

Fig. 8.3 Jordan normal form for the Case III(ii)

$$J_5 = \begin{pmatrix} \boxed{\begin{matrix} \lambda & 1 \\ 0 & \lambda \end{matrix}} & \begin{matrix} 0 \\ 0 \end{matrix} \\ \begin{matrix} 0 & 0 \end{matrix} & \boxed{\lambda} \end{pmatrix}$$

These properties offer different possible cases:

Let $\lambda_1, \lambda_2, \lambda_3$ be the rots of $\kappa_M(x)$.

Case I: Let $\lambda_1, \lambda_2, \lambda_3$ be three distinct roots of $\kappa_M(x)$.

Then $\kappa_M(x) = (\lambda_1 - x)(\lambda_2 - x)(\lambda_3 - x)$ and hence by (a) & (b), $e_1(x) = 1$, $e_2(x) = 1$ and $e_3(x) = (\lambda_1 - x)(\lambda_2 - x)(\lambda_3 - x)$.

Consequently, the Serge characteristic is [(1), (1), (1)] and hence the Jordan normal form J_1 of M is

$$J_1 = \begin{pmatrix} \lambda_1 & 0 & 0 \\ 0 & \lambda_2 & 0 \\ 0 & 0 & \lambda_3 \end{pmatrix}.$$

Case II: Let $\lambda_1 = \lambda_3$ and $\lambda_2 \neq \lambda_1$. Then $\kappa_M(x) = (\lambda_1 - x)^2(\lambda_2 - x)$. There are two possibilities:

(i) $e_1(x) = 1,\ e_2(x) = \lambda_1 - x,\ e_3(x) = (\lambda_1 - x)(\lambda_2 - x)$ or

(ii) $e_1(x) = 1,\ e_2(x) = 1,\ e_3(x) = (\lambda_1 - x)^2(\lambda_2 - x)$.

The Segre characteristic for (i) is [(1, 1), (1)] and for (ii) is [(2), (1)].

Consequently, the Jordan normal form is

$$J_2 = \begin{pmatrix} \lambda_1 & 0 & 0 \\ 0 & \lambda_1 & 0 \\ 0 & 0 & \lambda_2 \end{pmatrix}$$

for (i) and for (ii) it is J_3 as shown in Fig. 8.2.

Case III: $\lambda_1 = \lambda_2 = \lambda_3 = \lambda$ (say). Then there are three possibilities.

(i) $e_1(x) = \lambda - x,\ e_2(x) = \lambda - x,\ e_3(x) = \lambda - x$.

This shows that the Serge characteristic is [(1, 1, 1)] and hence the Jordan normal form

$$J_4 = \begin{pmatrix} \lambda & 0 & 0 \\ 0 & \lambda & 0 \\ 0 & 0 & \lambda \end{pmatrix}.$$

(ii) $e_1(x) = 1,\ e_2(x) = \lambda - x,\ e_3(x) = (\lambda - x)^2$.

This shows that the Segre characteristic is [(2, 1)] and hence the Jordan normal form J_5 is shown in Fig. 8.3.

(iii) $e_1(x) = 1,\ e_2(x) = 1,\ e_3(x) = (\lambda - x)^3$.

This shows that the Segre characteristic is $[(3)]$ and hence the Jordan normal form

$$J_6 = \begin{pmatrix} \lambda & 1 & 0 \\ 0 & \lambda & 1 \\ 0 & 0 & \lambda \end{pmatrix}.$$

8.10 Inner Product Spaces

In our discussion, so far, the concepts of length, angle, and distance have not appeared. In this section we study special vector spaces over $\mathbf{R}$ and $\mathbf{C}$ which admit the above missing concepts. Such vector spaces introduce the concept of Inner Product Spaces. For its motivation, we recall the geometry in $\mathbf{R}^n$ and $\mathbf{C}^n$.

8.10.1 Geometry in $\mathbf{R}^n$ and $\mathbf{C}^n$

Given two vectors $x = (x_1, x_2, \ldots, x_n)$ and $y = (y_1, y_2, \ldots, y_n)$ in $\mathbf{R}^n$, the real number $x_1 y_1 + x_2 y_2 + \cdots + x_n y_n$ is called dot product (or standard inner product) of x and y and is denoted by $x \cdot y = \langle x, y \rangle$. Thus $x \cdot y = \langle x, y \rangle = \sum_{i=1}^{n} x_i y_i \in \mathbf{R}$. The set $\mathbf{R}^n$ endowed with this dot product is called the *Euclidean n-space*. The length of x, denoted by $|x|$, is defined by $|x| = +\sqrt{x \cdot x}$. This definition of inner product is not adequate for defining the length in $\mathbf{C}^n$. For example, for the non-zero $x = (1 - i, 1 + i) \in \mathbf{C}^2$, the above definition of length shows that $|x| = 0$. Similarly, for $x = (1, 4i) \in \mathbf{C}^2$, $|x| = \sqrt{-15}$. So it needs modification of the definition of inner product in $\mathbf{C}^n$ as $\langle x, y \rangle = \sum_{i=1}^{n} x_i \bar{y}_i$, where $\bar{y}_i$ represents the complex conjugate of y_i. The set $\mathbf{C}^n$ endowed with this inner product is called the *n-dimensional unitary space*. The distance between x and y in $\mathbf{R}^n$ (or $\mathbf{C}^n$) is defined by $d(x, y) = \|x - y\|$, where $\| \ \|$ is the norm function (see Example 8.11.3).

Remark The inner product defined in $\mathbf{R}^n$ (or $\mathbf{C}^n$) is a function of x for every fixed y.

8.10.2 Inner Product Spaces: Introductory Concepts

In analytical geometry and vector analysis we study real vector spaces. We now carry over the concept of a dot product in $\mathbf{R}^n$ to a more abstract setting. So the model of the inner product spaces is the Euclidean n-space $\mathbf{R}^n$.

Definition 8.10.1 Let $F = \mathbf{R}$ or $\mathbf{C}$ and V be a vector space over F. An inner product on V is a mapping $\langle , \rangle : V \times V \to F$ such that

(i) $\langle x, x \rangle \geq 0$ and $\langle x, x \rangle = 0$ iff $x = 0$ (positive definiteness);
(ii) for $F = \mathbf{R}$, $\langle x, y \rangle = \langle y, x \rangle$ (symmetry);
     for $F = \mathbf{C}$, $\langle x, y \rangle = \overline{\langle y, x \rangle}$ (conjugate symmetry);

(iii) for $F = \mathbf{R}$, $\langle \alpha x + \beta y, z \rangle = \alpha \langle x, z \rangle + \beta \langle y, z \rangle$ (bilinearity);

for $F = \mathbf{C}$, $\langle z, \alpha x + \beta y \rangle = \bar{\alpha} \langle z, x \rangle + \bar{\beta} \langle z, y \rangle$ (conjugate bilinearity),

for all $\alpha, \beta \in F$ and $x, y, z \in V$.

The pair $(V, \langle , \rangle)$ is called an inner product space over F.

Remark 1 If there is no ambiguity, we write $(V, \langle , \rangle)$ by simply V.

Remark 2 There may exist different inner products on V.

For example, for $x = (x_1, x_2)$ and $y = (y_1, y_2)$ in $\mathbf{R}^2$, $\langle x, y \rangle_1 = x_1 y_1 + x_2 y_2$ and $\langle x, y \rangle = (2x_1 + x_2)y_1 + (x_1 + x_2)y_2$ are both inner products on $\mathbf{R}^2$ over $\mathbf{R}$.

Example 8.10.1 (i) Let $V = C([a, b])$ be the vector space of all real valued continuous functions defined on $[a, b]$. Then $\langle , \rangle$ defined by $\langle f, g \rangle = \int_a^b f(x)g(x)\,dx$ is an inner product and V is an inner product space.

(ii) Let V be the vector space of all continuous complex valued functions defined on $[0, 1]$. Then $\langle , \rangle$ defined by $\langle f, g \rangle = \int_0^1 f(x)\overline{g(x)}\,dx$ is an inner product and V is an inner product space.

(iii) $\mathbf{C}^n$ is an inner product space under the inner product defined by $\langle x, y \rangle = \sum_{i=1}^n x_i \bar{y}_i$.

(iv) Let V be the vector space of all $m \times n$ matrices over $\mathbf{R}$. Then $\langle , \rangle$ defined by $\langle A, B \rangle = \text{trace}(B^t A)$ is an inner product.

We now extend the concept of length or norm defined in $\mathbf{R}^2$ or $\mathbf{R}^3$ to an arbitrary inner product space.

Definition 8.10.2 Let V be an inner product space and x is in V. The norm of x denoted by $\|x\|$ is defined by $\|x\| = +\sqrt{\langle x, x \rangle}$ and the distance between x and y in V is defined by $d(x, y) = \|x - y\|$ for all x, y in V.

Remark For $x = (x_1, x_2, \ldots, x_n)$ in $\mathbf{R}^n$, $\|x\| = +\sqrt{x_1^2 + x_2^2 + \cdots + x_n^2}$ is just the distance of x from the origin.

We now study some properties of norm functions.

Proposition 8.10.1 *Let V be an inner product space over $\mathbf{R}$. Then the norm function $\| \ \| : V \to \mathbf{R}$ satisfies the following properties:*

(i) $\|x\| \geq 0$ *and* $\|x\| = 0$ *iff* $x = 0$;

(ii) $\|\alpha x\| = |\alpha| \|x\|$, $\forall x \in V$ *and* $\alpha \in \mathbf{R}$;

(iii) $\|\langle x, y \rangle\| \leq \|x\| \|y\|$, $\forall x, y \in V$ *(Cauchy–Schwarz Inequality)*;

(iv) *Corresponding to a non-zero vector x in V, there is a vector u in V such that $\|u\| = 1$ and $x = \|x\|u$. (This u is called the unit vector along x.)*

Proof Left as an exercise. $\square$

Remark The Cauchy–Schwarz inequality in $\mathbf{R}^n$ is $(x_1 y_1 + x_2 y_2 + \cdots + x_n y_n)^2 \leq (x_1^2 + x_2^2 + \cdots + x_n^2)(y_1^2 + y_2^2 + \cdots + y_n^2)$, which was stated in 1821 by A.L. Cauchy. The corresponding inequality in $\mathbf{C}^n$ is

$$(x_1 \bar{y}_1 + x_2 \bar{y}_2 + \cdots + x_n \bar{y}_n)^2 \leq (|x_1|^2 + \cdots + |x_n|^2)(|y_1|^2 + \cdots + |y_n|^2).$$

We recall that two non-zero vectors x and y in $\mathbf{R}^n$ are orthogonal iff their dot product $x \cdot y = 0$. This leads to the following definition.

Definition 8.10.3 Two non-zero vectors x, y in an inner product space V are said to be orthogonal iff their inner product $\langle x, y \rangle = 0$ and said to be orthonormal iff $\langle x, y \rangle = 0$ and $\|x\| = 1 = \|y\|$.

Definition 8.10.4 A basis $B = \{v_1, v_2, \ldots, v_n\}$ of an inner product space V is said to be an orthogonal basis (or an orthonormal basis) iff $\langle v_i, v_j \rangle = 0$ for $i \neq j$ (or $\langle v_i, v_j \rangle = 0$ for $i \neq j$ and $\|v_i\| = 1$ for all i).

Example 8.10.2 (i) The unit vectors $e_1 = (1, 0, \ldots, 0), e_2 = (0, 1, 0, \ldots, 0), \ldots,$ $e_n = (0, 0, \ldots, 0, 1)$ in $\mathbf{R}^n$ are pairwise both orthogonal and orthonormal. The set $B = \{e_1, e_2, \ldots, e_n\}$ is an orthonormal basis of $\mathbf{R}^n$.

(ii) Let V be an inner product space. Then any orthonormal generating set $S = \{v_1, v_2, \ldots, v_m\}$ forms a basis of V.

This is so because if $\sum_{i=1}^{m} \alpha_i v_i = 0$ for some $\alpha_i \in F$, then $0 = \sum_{i=1}^{m} \alpha_i \langle v_i, v_k \rangle = \alpha_k$ (by taking inner product with a fixed v_k in S). This is true for all $k = 1, 2, \ldots, m$. Hence S is linearly independent such that $L(S) = V$. Consequently, S forms a basis of V.

We now prescribe a process of orthogonalization to find an orthogonal basis for certain inner product spaces.

Theorem 8.10.1 (Gram–Schmidt Orthogonalization Process) *Every non-zero finite-dimensional inner product space V over F has an orthogonal basis.*

Proof Let $\dim V = n$ and $B = \{v_1, v_2, \ldots, v_n\}$ be a basis of V. We now construct an orthogonal basis $S = \{u_1, u_2, \ldots, u_n\}$. We take $u_1 = v_1$. If $n = 1$, the theorem is proved. If $n > 1$, we take u_2 by defining $u_2 = v_2 - \alpha_1 v_1$, where $\alpha_1 = \frac{\langle v_2, u_1 \rangle}{\|u_1\|^2}, (\alpha_1 \in F$ is called the Fourier coefficient of v_2 with respect to v_1). Then $\langle u_2, u_1 \rangle = 0$. Since $u_2 \neq 0$ and $\langle u_2, u_1 \rangle = 0$, the set $\{u_1, u_2\}$ is linearly independent. Hence for $n = 2$, the theorem is proved. We assume that the set $A = \{u_1, u_2, \ldots, u_{m-1}\}$ is orthogonal and such that $L(A) = L(\{v_1, v_2, \ldots, v_{m-1}\})$, $1 < m < n$. Let u_m be the non-zero vector in V defined by $u_m = v_m + \alpha_1 v_1 + \cdots + \alpha_{m-1} v_{m-1}$ for which $\langle u_m, u_i \rangle = 0$ for each $i = 1, 2, \ldots, m - 1$.

In general,

$$u_m = v_m - \sum_{i=1}^{m-1} \frac{\langle v_m, u_i \rangle}{\|u_i\|^2} u_i, \quad m = 2, 3, \ldots, n.$$

In this way, for given m linearly independent vectors in V, we can construct an orthogonal set of m vectors. As dim $V = n$, in particular, from the given basis B, we can construct the orthogonal set S of n vectors. This S gives the required orthogonal basis of V.   □

Corollary 1 *Every non-zero finite-dimensional inner product space V has an orthonormal basis.*

Proof Let $S = \{u_1, u_2, \ldots, u_n\}$ be an orthogonal basis of V. Then $B = \{\frac{u_1}{\|u_1\|}, \frac{u_2}{\|u_2\|}, \ldots, \frac{u_n}{\|u_n\|}\}$ forms an orthonormal basis of V.   □

Corollary 2 *Any set of non-zero mutually orthogonal vectors of a non-zero finite-dimensional inner product space V is either an orthogonal basis of V or can be extended to an orthogonal basis of V.*

Proof Let dim $V = n$ and $S = \{u_1, u_2, \ldots, u_m\}$ be the given set of non-zero mutually orthogonal vectors of V. If $m = n$, there is nothing to prove. So we assume that $m < n$. We now extend the set S to a basis $B = \{u_1, u_2, \ldots, u_m, u_{m+1}, \ldots, u_n\}$. If we take $v_i = u_i$ for $i = 1, 2, \ldots, m$ and then construct v_{m+i} from u_{m+i} as above, for $i = 1, 2, \ldots, n - m$, then the set $A = \{v_1, v_2, \ldots, v_n\}$ forms an orthogonal basis of V, where the first m vectors of A are the original vectors $u_1, u_2, \ldots, u_m$.   □

Corollary 3 *Any set of orthonormal vectors of V of a non-zero finite-dimensional inner product space V can be extended to an orthonormal basis of V.*

We now introduce the concept of an orthogonal complement in an inner product space.

Proposition 8.10.2 *Let V be a non-zero finite-dimensional inner product space and U be a subspace of V. Then there exists a subspace W of V such that $V = U \oplus W$, where each vector of W is orthogonal to every vector of U.*

Proof Clearly, U is also an inner product space under inner product induced from V. Then U has an orthonormal basis $B = \{e_1, e_2, \ldots, e_n\}$, say. We construct W as defined by $W = \{w \in V : \langle w, e_j \rangle = 0, 1 \leq j \leq n\}$. Then W is a subspace of V and for any element $u \in U$, $\exists$ an element $w \in W$ such that $\langle u, w \rangle = 0$. Again if $v \in V$, then the element

$$v - \sum_{i=1}^{n} \langle v, e_i \rangle e_i \in W, \quad \text{since}$$

$$\left\langle v - \sum_{i=1}^{n} \langle v, e_i \rangle e_i, e_k \right\rangle = \langle v, e_k \rangle - \langle v, e_k \rangle \|e_k\|^2 = 0, \quad \forall k \text{ such that } 1 \leq k \leq n.$$

Hence

$$v = \sum_{i=1}^{n} \langle v, e_i \rangle e_i + \left(v - \sum_{i=1}^{n} \langle v, e_i \rangle e_i \right)$$

$$= u + w \quad \text{for some } u \in U \text{ and some } w \in W.$$

This shows that $V = U + W$. Clearly, for $x \in U \cap W$, $\langle x, x \rangle = 0$ implies $x = 0$. Hence $V = U \oplus W$. □

Definition 8.10.5 The subspace W defined above is called the orthogonal complement of U in V and is denoted by $U^{\perp}$.

Remark If V is an inner product space and U is a subspace of V, then $V = U \oplus U^{\perp}$ and hence any vector v in an inner product space V has the unique representation as

$$v = u + w, \quad \text{for some } u \in U \text{ and some } w \in W = U^{\perp}.$$

Definition 8.10.6 The decomposition $V = U \oplus U^{\perp}$ is called the orthogonal decomposition of V with respect to the subspace U and V is called an orthogonal direct sum of U and $U^{\perp}$.

Example 8.10.3 (i) Let $U = \{(x, y) \in \mathbf{R}^2 : 3x - y = 0\}$. Then $U^{\perp} = \{(x, y) \in \mathbf{R}^2 : x + 3y = 0\}$. Thus the orthogonal complement of the line U in $\mathbf{R}^2$ is the line passing through the origin and perpendicular to the line U in $\mathbf{R}^2$.

(ii) The orthogonal complement of the subspace U generated by the vector $v = (2, 3, 4)$ in $\mathbf{R}^3$ is the plane $2x + 3y + 4z = 0$ in $\mathbf{R}^3$ passing through the origin and perpendicular to the vector v.

8.11 Hilbert Spaces

The concept of Hilbert spaces is not an actual subject of this book but we feel that the readers should have some knowledge of such spaces. So we need the basic ideas of normed linear spaces, Banach spaces etc., as a background.

Definition 8.11.1 Let X be a non-empty set. A metric on X is a real valued function $d : X \times X \to \mathbf{R}$ satisfying the conditions:

$M(1)$ $d(x, y) \geq 0$ and $d(x, y) = 0$ iff $x = y$;
$M(2)$ $d(x, y) = d(y, x)$ (symmetry);
$M(3)$ $d(x, y) \leq d(x, z) + d(z, y)$ (triangle inequality)
            for all x, y, z in X.

$d(x, y)$ is called the distance between x and y. The pair (X, d) is called a metric space.

Example 8.11.1 (i) Let X be an arbitrary non-empty set. Then $d : X \times X \to \mathbf{R}$ defined by

$$d(x, y) = \begin{cases} 0 & \text{if } x = y \\ 1 & \text{if } x \neq y \end{cases}$$

is a metric on X and (X, d) is a metric space.

(ii) Let $\mathbf{R}$ be the real line. Then $|x|$ defined by

$$|x| = \begin{cases} x & \text{if } x > 0 \\ -x & \text{if } x < 0 \\ 0 & \text{if } x = 0 \end{cases}$$

is called the absolute value function. Then $d(x, y) = |x - y|$ is a metric on $\mathbf{R}$.

Definition 8.11.2 Let (X, d) be a metric space. A Cauchy sequence in X is a function $f : \mathbf{N}^+ \to X$ such that for every positive real number ϵ, there exists a positive integer m such that $d(f(i), f(j)) < \epsilon$ for all integers $i > m$ and $j > m$.

Definition 8.11.3 A complete metric space is a metric space in which every Cauchy sequence is convergent.

Example 8.11.2 $[0, 1]$ is a complete metric space but $(0, 1]$ is not so.

In this section we consider vector spaces over $F = \mathbf{R}$ or $\mathbf{C}$.

Definition 8.11.4 A normed linear space is a vector space X on which a real valued function $\| \ \| : X \to \mathbf{R}$ called a norm function is defined satisfying the conditions:

$N(1)$ $\|x\| \geq 0$ and $\|x\| = 0$ iff $x = 0$;
$N(2)$ $\|x + y\| \leq \|x\| + \|y\|$;
$N(3)$ $\|\alpha x\| = |\alpha| \|x\|$,
     for $x, y \in X$ and $\alpha \in \mathbf{R}$ or $\mathbf{C}$.

Remark 1 The non-negative real number $\|x\|$ is considered the length of the vector x and hence the notion of the distance in X from an arbitrary point to the origin is available.

Remark 2 From $N(3)$, it follows that $\| - x\| = \|(-1)x\| = \|x\|$.

Remark 3 A normed linear space is a metric space with respect to the metric induced by the metric defined by $d(x, y) = \|x - y\|$.

Definition 8.11.5 A Banach space X is a normed linear space which is complete as a metric space i.e. every Cauchy sequence in X is convergent.

Example 8.11.3   (i) $\mathbf{R}^n$ is a real Banach space under the norm $\|x\|$ defined by

$$\|x\| = \left(\sum_{i=1}^{n} |x_i|^2 \right)^{\frac{1}{2}}.$$

(ii) $\mathbf{C}^n$ is a complex Banach space under the norm $\|z\|$ defined by

$$\|z\| = \left(\sum_{i=1}^{n} |z_n|^2 \right)^{\frac{1}{2}}.$$

(iii) Let $\mathbf{R}^\infty$ be the set of all sequences $x = (x_1, x_2, \ldots, x_n, \ldots)$ of real numbers such that $\sum_{n=1}^{\infty} |x_n|^2$ converges. Under the norm function defined by

$$\|x\| = \left(\sum_{n=1}^{\infty} |x_n|^2 \right)^{\frac{1}{2}},$$

the vector space $\mathbf{R}^\infty$ becomes a real Banach space, called infinite-dimensional Euclidean space. Similarly, $\mathbf{C}^\infty$ is a complex Banach space, called infinite-dimensional unitary space. The real and complex spaces l_2 are respectively $\mathbf{R}^\infty$ and $\mathbf{C}^\infty$.

(iv) For any real number p such that $1 \leq p < \infty$, the normed linear space l_p^n of all n-tuples $x = (x_1, x_2, \ldots, x_n)$ of scalars in $\mathbf{R}$ or $\mathbf{C}$, with the norm function defined by

$$\|x\|_p = \left(\sum_{i=1}^{n} |x_i|^p \right)^{\frac{1}{p}},$$

is a Banach space.

(v) For any real number p such that $1 \leq p < \infty$, the normed linear space l_p of all sequences $x = \{x_1, x_2, \ldots, x_n, \ldots\}$ of scalars in $\mathbf{R}$ or $\mathbf{C}$ such that $\sum_{n=1}^{\infty} |x_n|^p$ is convergent is a Banach space under the norm function defined by

$$\|x\|_p = \left(\sum_{n=1}^{\infty} |x_n|^p \right)^{\frac{1}{p}}.$$

Clearly, the real l_2 space is the infinite-dimensional Euclidean space $\mathbf{R}^\infty$ and the complex l_2 space is the infinite-dimensional unitary space as defined in (iii).

Banach spaces are vector spaces providing with the idea of the length of a vector. But it fails to provide the concept of the angle between two vectors in an abstract Banach space. A Hilbert space is a special Banach space having an additional structure providing with the concept of orthogonality of two vectors. This type of spaces also fails to provide the main geometric concept of the angle between two vectors in general. But the dot product in $\mathbf{R}^n$ yields both the concepts of the angle and orthogonality of two non-zero vectors. We now proceed to give the definition of a Hilbert space.

Definition 8.11.6 A Hilbert space is a complex *Banach space* X in which a function $\langle , \rangle : X \times X \to \mathbf{C}$ is defined satisfying the following conditions:

$H(1)$: $\langle \alpha x + \beta y, z \rangle = \alpha \langle x, z \rangle + \beta \langle y, z \rangle$;
$H(2)$: $\overline{\langle x, y \rangle} = \langle y, x \rangle$;
$H(3)$: $\langle x, x \rangle = \|x\|^2$.

Then $\langle x, \alpha y + \beta z \rangle = \bar{\alpha} \langle x, y \rangle + \bar{\beta} \langle x, z \rangle$, for all x, y, z in X and α, β in $\mathbf{C}$.

Remark A Hilbert space is a complex Banach space whose norm is defined by an inner product.

Example 8.11.4 The spaces l_2^n and l_2 are the main examples of Hilbert spaces, where $\langle x, y \rangle$ is defined by

$$\langle x, y \rangle = \sum_{i=1}^{n} x_i \bar{y}_i \quad \text{for } x, y \in l_2^n,$$

$$= \sum_{n=1}^{\infty} x_n \bar{y}_n \quad \text{for } x, y \in l_2.$$

We recall the parallelogram law from elementary geometry that the sum of squares of the sides of a parallelogram equals the sum of squares of its diagonals. An analogue of this law holds in a Hilbert space.

Proposition 8.11.1 (Parallelogram law) *Let X be a Hilbert space. For any two vectors x and y in X, $\|x + y\|^2 + \|x - y\|^2 = 2\|x\|^2 + 2\|y\|^2$.*

Proof $\|x + y\|^2 + \|x - y\|^2 = \langle x + y, \ x + y \rangle + \langle x - y, \ x - y \rangle = 2\langle x, x \rangle + 2\langle y, y \rangle = 2\|x\|^2 + 2\|y\|^2$. $\qquad\square$

Corollary 1 *Let X be a Hilbert space. Then for any x, y in X, the following identity*

$$4\langle x, y \rangle = \|x + y\|^2 - \|x - y\|^2 + i\|x + iy\|^2 - i\|x - iy\|^2 \quad holds.$$

Proof By converting the right hand expression into inner products, the identity is verified. $\qquad\square$

Corollary 2 *Let X be a complex Banach space whose norm satisfies the parallelogram law. If an inner product is defined on X by the identity of Corollary 1, then X is a Hilbert space.*

Proof Left as an exercise.   □

Remark Hilbert spaces are precisely the complex Banach spaces in which the parallelogram law is valid.

Definition 8.11.7 Let X be a Hilbert space and x, y be two vectors in X. Then x is said to be orthogonal to y denoted by $x \perp y$ iff $\langle x, y \rangle = 0$.

Remark Let X be a Hilbert space.

(i) $x \perp y \Leftrightarrow y \perp x$, since $\overline{\langle x, y \rangle} = \langle y, x \rangle$.
(ii) $\langle x, x \rangle = \|x\|^2$ and $x \perp 0 = 0$ for every x in X. Clearly, 0 is the only vector orthogonal to itself.
(iii) $x \perp y \Rightarrow \|x + y\|^2 = \|x - y\|^2 = \|x\|^2 + \|y\|^2$ (it represents geometrically Pythagorean Theorem).

Definition 8.11.8 In an inner product space X, for any non-empty set Y of X its *orthogonal complement*, denoted by $Y^\perp$, is defined by $Y^\perp = \{x \in X : x \perp y$ for every y in $Y\}$.

Clearly, $\{0\}^\perp = X$, $X^\perp = \{0\}$ and $X \cap X^\perp = \{0\}$.

Proposition 8.11.2 (Cauchy–Schwarz inequality) *Let X be a Hilbert space and x, y be any two vectors in X. Then $|\langle x, y \rangle| \le \|x\| \|y\|$.*

Proof Let x be a non-zero vector in X. If $y = 0$, then the result is trivial. If $y \ne 0$, it is sufficient to prove that $|\langle x, y \rangle / \|y\|| \le \|x\|$. If $\|y\| = 1$, then $0 \le \|x - \langle x, y \rangle y\|^2 = \langle x, x \rangle - \langle x, y \rangle \langle x, y \rangle = \|x\|^2 - |\langle x, y \rangle|^2 \Rightarrow |\langle x, y \rangle| \le \|x\|$. If $\|y\| \ne 1$, then the result $|\langle x, y \rangle| \le \|x\| \|y\|$ follows immediately.   □

8.12   Quadratic Forms

The theory of quadratic forms arose through the study of central conics in the Euclidean plane $\mathbf{R}^2$ and central conicoids in the Euclidean space $\mathbf{R}^3$. We now carry this study over to quadratic surfaces in the Euclidean n-space $\mathbf{R}^n$. To classify a central conic $ax^2 + 2hxy + by^2 = 1$ in $\mathbf{R}^2$, we reduce it in the form $a_1 X^2 + b_1 Y^2 = 1$ by a non-zero non-singular linear transformation from (x, y) to (X, Y) defined by $x = X \cos\theta - Y \sin\theta$ and $y = X \sin\theta + Y \cos\theta$, which geometrically means a rotation of axes. The rotation of axes implies a change by an orthogonal transformation.

An expression of the form

$$q(x_1, x_2, \ldots, x_n) = \sum_{i=1}^{n} a_i x_i^2 + 2 \sum_{i<j} a_{ij} x_i x_j, \quad a_i, a_{ij} \in \mathbf{R}$$

is called a quadratic form in $\mathbf{R}^n$.

In general, given a field F, by a quadratic form on a vector space F^n over F, we mean a map $q : F^n \to F$, which is a homogeneous polynomial function of degree 2 of the coordinates. For example, the function $f : \mathbf{R}^n \to \mathbf{R}$ defined by $f(x_1, x_2, \ldots, x_n) = x_1^2 + x_2^2 + \cdots + x_n^2$ is a *quadratic form in* $\mathbf{R}^n$.

Let

$$q(x_1, x_2, \ldots, x_n) = \sum_{i,j} b_{ij} x_i x_j, \quad i, j = 1, 2, \ldots, n$$

be a quadratic form over F of characteristic different from 2. By taking $a_{ij} = \frac{b_{ij} + b_{ji}}{2}$, the given quadratic form is transformed into $\sum_{i,j} a_{ij} x_i x_j$, where $a_{ij} = a_{ji}$. It can be expressed as a product XAX^t, where X is the row matrix $(x_1 x_2 \ldots x_n)$ and A is the symmetric matrix $A = (a_{ij})$.

For example, the symmetric matrix A associated with the quadratic form $3x^2 + 4y^2 + 5z^2 + 4xz$ is

$$A = \begin{pmatrix} 3 & 0 & 2 \\ 0 & 4 & 0 \\ 2 & 0 & 5 \end{pmatrix}.$$

8.12.1 Quadratic Forms: Introductory Concepts

Definition 8.12.1 Let V be a real inner product space of finite dimension and $T : V \to V$ be a symmetric linear operator (i.e., $T(x, y) = T(y, x)$ for all x, y in V). Then the mapping $q : V \to \mathbf{R}$ defined by $q(x) = \langle T(x), x \rangle$ is called a *quadratic form* on V.

We now establish a relation between a quadratic form and a symmetric matrix.

Theorem 8.12.1 *Every real symmetric matrix of order n corresponds to a quadratic form on $\mathbf{R}^n$ and conversely, every quadratic form on $\mathbf{R}^n$ corresponds to a real symmetric matrix of order n.*

Proof Let $V = \mathbf{R}^n$ and $A = (a_{ij})$ be a symmetric matrix of order n over $\mathbf{R}$. We take B the usual orthonormal basis of V i.e., $B = \{e_1 = (1, 0, \ldots, 0), \ldots, e_n = (0, 0, \ldots, 1)\}$. Let $T : V \to V$ be the linear operator corresponding to A with respect to the basis B. Then $q : V \to \mathbf{R}$, defined by $q(x) = \langle T(x), x \rangle$ is the quadratic form corresponding to $T : V \to V$. For $x = (x_1, x_2, \ldots, x_n) = \sum_i x_i e_i$, $T(x) = \sum_i x_i T(e_i) = \sum_{i,k} x_i a_{ik} e_k \Rightarrow q(x) = \sum_{i,j} a_{ij} x_i x_j = \sum_i a_{ii} x_i^2 + 2 \sum_{i<j} a_{ij} x_i x_j$. This proves the first part.

Conversely, let q be the quadratic form $q : V \to \mathbf{R}$ corresponding to a symmetric linear operator $T : V \to V$, where $V = \mathbf{R}^n$. Then $q(x) = \langle T(x), x \rangle = \langle \sum_i x_i T(e_i), \sum_j x_j e_j \rangle = \sum_{i,j} (\langle T(e_i), e_j \rangle) x_i x_j = \sum_{i,j} a_{ij} x_i x_j$, where $a_{ij} = \langle T(e_i), e_j \rangle$. Since T is symmetric, $A = (a_{ij})$ is also symmetric. $\quad\square$

This matrix A is called the *matrix* corresponding to the *quadratic form q* on $\mathbf{R}^n$ with respect to the basis B. We now study the effect of change of basis of B of V on A.

Theorem 8.12.2 Let q be a quadratic form on $\mathbf{R}^n$ and $B_1 = \{e_1, e_2, \ldots, e_n\}$, $B_2 = \{f_1, f_2, \ldots, f_n\}$ be two orthonormal bases of $\mathbf{R}^n$. If A, B are the matrices corresponding to q with respect to bases B_1, B_2 respectively and C is the matrix corresponding to the linear transformation $T : V \to V$ mapping the basis B_1 onto B_2, then $B = CAC^t$.

Proof If $A = (a_{ij})$, then by the given condition, $a_{ij} = \langle T(e_i), e_j \rangle$. Hence $q(x) = \langle T(x), x \rangle = \sum_{i,j} a_{ij} x_i x_j$, where $x = \sum_i x_i e_i$. Hence

$$q(x) = X^t A X, \quad \text{where } X = \begin{pmatrix} x_1 \\ x_2 \\ \vdots \\ x_n \end{pmatrix}.$$

As T sends a basis B_1 onto another basis B_2, T is non-singular and hence its corresponding matrix $C = (c_{ij})$ is also non-singular. Then $f_i = \sum_j c_{ij} e_j$. If $(y_1, y_2, \ldots, y_n)$ are the coordinates of x with respect to B_2, then $x = \sum_i y_i f_i$. Consequently,

$$x = \sum_i y_i f_i = \sum_{i,j} y_i c_{ij} e_j \quad \Rightarrow \quad \sum_i x_i e_i = \sum_{i,j} y_i c_{ij} e_j$$

$$\Rightarrow \quad x_j = \sum c_{ij} y_i \quad \Rightarrow \quad X^t = Y^t C;$$

where Y is the column vector

$$\begin{pmatrix} y_1 \\ \vdots \\ y_n \end{pmatrix}.$$

Hence $q(x) = X^t A X = Y^t C A C^T Y \Rightarrow$ the matrix of q with respect to the basis B_2 is $B = CAC^T$. $\qquad\square$

Theorem 8.12.2 introduces the concept of congruent matrices.

Definition 8.12.2 Let A and B be two $n \times n$ symmetric matrices over $\mathbf{R}$. Then B is said to be congruent to A iff there exists a non-singular matrix P such that $B = P^t A P$.

The relation of being congruent on the set of all real $n \times n$ symmetric matrices is an equivalence relation, it partitions the set into congruence classes.

Definition 8.12.3 Two quadratic forms on $\mathbf{R}^n$ are said to be congruent iff their corresponding associated symmetric matrices are congruent.

We now look for a suitable basis for $\mathbf{R}^n$ to obtain a canonical or normal form for a congruence class.

Theorem 8.12.3 (Sylvester's Theorem) *Any real symmetric matrix A of order n is congruent to a matrix of the form*

$$B = \begin{pmatrix} I_p & & \\ & -I_q & \\ & & 0 \end{pmatrix},$$

where I_p, I_q represent unit $p \times p$ and $q \times q$ real matrices and 0 is the matrix of zeros. Moreover, p and q are uniquely determined by A.

Proof As A is a real symmetric matrix, all the characteristic roots (eigenvalues) of A are real. We arrange them in such a way that the first p of them $\lambda_1, \lambda_2, \ldots, \lambda_p$ (say), are positive, the next q of them, $-\lambda_{p+1}, -\lambda_{p+2}, \ldots, -\lambda_{p+q}$ (say), are negative and the last $n - p - q$ of them are all zero. Then there exists an orthogonal matrix C such that

$$C^{-1}AC = \begin{pmatrix} \lambda_1 & & & & & & & & \\ & \lambda_2 & & & & & & & \\ & & \ddots & & & & & & \\ & & & \lambda_p & & & & & \\ & & & & -\lambda_{p+1} & & & & \\ & & & & & \ddots & & & \\ & & & & & & -\lambda_{p+q} & & \\ & & & & & & & 0 & \\ & & & & & & & & \ddots \\ & & & & & & & & & 0 \end{pmatrix}.$$

Let D be the diagonal matrix, where

$$D = \text{diag}\left(\frac{1}{\sqrt{\lambda_1}}, \ldots, \frac{1}{\sqrt{\lambda_p}}, \frac{1}{\sqrt{\lambda_{p+1}}}, \ldots, \frac{1}{\sqrt{\lambda_{p+1}}}, 1, \ldots, 1\right).$$

Then

$$D(C^{-1}AC)D = \begin{pmatrix} I_p & & \\ & -I_q & \\ & & 0 \end{pmatrix} \quad \text{and} \quad P = CD$$

is a non-singular matrix such that $P^t = D^t C^t = DC^{-1}$. Hence

$$P^t A P = \begin{pmatrix} I_p & & \\ & -I_p & \\ & & 0 \end{pmatrix} = B$$

shows that A is congruent to B. Uniqueness p and q is proved after the corollary.

Corollary Any quadratic form $q(x) = \sum_{i,j} a_{ij} x_i x_j$ is congruent to a quadratic form

$$x_1^2 + x_2^2 + \cdots + x_p^2 - x_{p+1}^2 - x_{p+2}^2 - \cdots - x_{p+q}^2,$$

known as the normal (or diagonal) form of $q(x)$, where p and q are determined uniquely by $q(x)$.

Uniqueness of p and q   Let

$$\left.\begin{aligned} q(x) &= \sum_{i=1}^{p} x_i^2 - \sum_{i=p+1}^{p+q} x_i^2 \quad \text{and} \\ q(x) &= \sum_{i=1}^{t} y_i^2 - \sum_{i=t+1}^{t+s} y_i^2 \end{aligned}\right\} \tag{A}$$

be two canonical expressions of the same quadratic form $q(x)$ with respect to bases $B = \{u_1, u_2, \ldots, u_n\}$ and $S = \{v_1, v_2, \ldots, v_n\}$, respectively. We claim that $p = t$ and $q = s$. If possible, $p \neq t$. Without loss of generality, we assume that $p > t$. Let $U = \text{span}\{u_1, u_2, \ldots, u_p\}$ and $W = \text{span}\{v_{t+1}, \ldots, v_n\}$. Then $\dim U = p$ and $\dim W = n - t$. Hence $\dim U + \dim W = p + n - t > n$, as $p > t$ by assumption $\Rightarrow$ the subspaces U and W have non-zero common vectors $\Rightarrow \exists$ a non-zero vector x in $U \cap W$.

This non-zero vector x can be expressed as

$$x = x_1 u_1 + \cdots + x_p u_p = y_{t+1} v_{t+1} + \cdots + y_n v_n.$$

As for this particular x, $x_{p+1} = \cdots = x_n = 0$ and $y_1 = \cdots = y_t = 0$, we have from (A)

$$x_1^2 + x_2^2 + \cdots + x_p^2 \geq 0 \quad \text{and} \quad -y_{t+1}^2 - \cdots - y_n^2 \leq 0.$$

Hence $x_1 = \cdots = x_p = 0$ and $y_{t+1} = \cdots = y_n = 0 \Rightarrow x = 0$.

This contradiction shows that $p \leq t$. Similarly, $t \leq p$. Hence $p = t$. Again $p + q = t + s \Rightarrow q = s$, since $p = t$.

Thus any quadratic form $q(x) = \sum_{i,j} a_{ij} x_i x_j$ is congruent to a quadratic form of the type

$$x_1^2 + x_2^2 + \cdots + x_p^2 - x_{p+1}^2 - \cdots - x_{p+q}^2, \tag{B}$$

where p and q are uniquely determined by $q(x)$. $\square$

Definition 8.12.4 The integer p is called the index of $q(x)$, which is the number of positive terms in the expression (B). If r is the rank of the matrix A associated with the quadratic form $q(x)$, then the *rank* of $q(x)$ is defined to be the rank of A and signature of $q(x)$ is defined to be the integer $p - (r - p) = 2p - r$, which is also called the signature of the matrix A.

Remark If a real symmetric matrix A of rank r is congruent to the matrix

$$B = \begin{pmatrix} I_p & & \\ & -I_q & \\ & & 0 \end{pmatrix},$$

then $r = p + q$ is the rank of A and $p - q$ is the signature of A.

A quadratic form is of several types given in the following definition.

Definition 8.12.5 A quadratic form $q : V \to \mathbf{R}$ is said to be

 (i) *positive definite* provided $q(x) > 0$ for all non-zero vectors x in V;
 (ii) *negative definite* provided $q(x) < 0$ for all non-zero vectors x in V;
 (iii) *positive semidefinite* provided $q(x) \geq 0$ for all x in V and $q(x) = 0$ for some non-zero vectors x in V;
 (iv) *negative semidefinite* provided $q(x) \leq 0$ for all x in V and $q(x) = 0$ for some non-zero vector x in V;
 (v) *indefinite* provided there exist vectors u and v in V such that $q(u) > 0$ and $q(v) < 0$.

The associated real symmetric matrix A is said to be positive definite, negative definite, positive semidefinite etc., according to $q(x)$ being so.

We now characterize quadratic forms with the help of their associated matrices.

Theorem 8.12.4 *Let $q : \mathbf{R}^n \to \mathbf{R}$ be a quadratic form and A be its associated matrix. Then $q(x)$ is positive definite iff the eigenvalues of A are all positive.*

Proof Let the quadratic form $q(x)$ be positive definite and

$$D = \begin{pmatrix} I_p & & \\ & -I_p & \\ & & 0 \end{pmatrix}$$

be a canonical form of A. Then D is the matrix corresponding to $q(x)$ with respect to some basis B of $\mathbf{R}^n$. If $(x_1, x_2, \ldots, x_n)$ is the coordinate of x with respect to B, then

$$q(x) = x_1^2 + x_2^2 + \cdots + x_p^2 - x_{p+1}^2 - \cdots - x_{p+q}^2.$$

If $p < n$, then for $x = (\overbrace{0, \ldots, 0}^{p}, \overbrace{1, 1, \ldots, 1}^{n-p})$, $q(x) < 0$. As q is positive definite, this contradiction shows that $p = n$.

Conversely, let all the eigenvalues of A be positive. Then A is congruent to the identity matrix and hence $q(x) = x_1^2 + x_2^2 + \cdots + x_n^2$ with respect to some basis of $\mathbf{R}^n$. This implies that q is positive definite. $\square$

Corollary *A quadratic form* $q : \mathbf{R}^n \to \mathbf{R}$ *is positive definite if rank of* $q = $ *signature of* $q = n$.

Theorem 8.12.5 *A quadratic form* $q : \mathbf{R}^n \to \mathbf{R}$ *is negative definite if all the eigenvalues of the matrix* A *corresponding to* $q(x)$ *are negative and is indefinite if* A *has both positive and negative eigenvalues.*

Proof Left as an exercise. $\square$

Definition 8.12.6 let V be an n-dimensional inner product space over $\mathbf{R}$ and $T : V \to V$ be a symmetric linear operator. Then T is said to be positive definite iff $\langle T(x), x \rangle > 0$ for all non-zero vectors x in V.

Proposition 8.12.1 *Let* V *be an* n-*dimensional inner product space over* $\mathbf{R}$. *If* $T : V \to V$ *is positive definite, then all eigenvalues of* T *are positive.*

Proof Let $v \neq 0$ be an eigenvector of T corresponding to an eigenvalue λ. Then $\langle T(v), v \rangle = \langle \lambda v, v \rangle = \lambda \langle v, v \rangle$. Since $\langle v, v \rangle > 0$ and $\langle T(v), v \rangle > 0$, it follows that $\lambda > 0$. $\square$

8.13 Exercises

Exercises-II

1. Find a linear operator $T : \mathbf{R}^2 \to \mathbf{R}^4$ corresponding to the matrix

$$A = \begin{pmatrix} 3 & -1 \\ -2 & 4 \\ 0 & 2 \\ 1 & 0 \end{pmatrix}.$$

   [*Hint.* See Example 8.6.2.]

2. Let V be an n-dimensional vector space over $\mathbf{R}$ and B be any basis of V. If $I : V \to V$ is the identity map, show that $m(I)_B = I_n$ (identity matrix).

3. Let V be the vector space generated by three functions $f(x) = 1, g(x) = x, h(x) = x^2$. If $D : V \to V$ is the differential operator, find $m(D)_B$, where $B = \{f(x), g(x), h(x)\}$.

   Ans.:

$$\begin{pmatrix} 0 & 1 & 0 \\ 0 & 0 & 2 \\ 0 & 0 & 0 \end{pmatrix}.$$

4. Let V be a vector space and A, B be subspaces of V. Show that

   (a) the vector spaces $(A + B)/B$ and $A/(A \cap B)$ are isomorphic;
   (b) the vector spaces $(A + B)/A$ and $B/(A \cap B)$ are isomorphic

   (Second Isomorphism Theorem or Quotient of a Sum Theorem).

   [*Hint.* Define the map $T : A \to (A + B)/B$ by $T(x) = x + B$. Then T is a linear transformation and onto. Clearly, $\ker T = A \cap B$. Apply the First Isomorphism Theorem.]

5. Let A and B be subspaces of a vector space V such that $B \subseteq A \subseteq V$. Show that the vector spaces $(V/B)/(A/B)$ and V/A are isomorphic (Third Isomorphism Theorem or Quotient of a Quotient Theorem for Vector Spaces).

   [*Hint.* Define the map $T: V/B \to V/A$ by $T(x + B) = x + A$. Then T is onto and a linear transformation. Clearly, $\ker T = A/B$. Apply the First Isomorphism Theorem.]

6. Let V be the vector space of 2×2 real matrices and $M = \begin{pmatrix} 1 & 2 \\ 0 & 3 \end{pmatrix} \in V$. If $T : V \to V$ is the linear operator defined by $T(A) = AM - MA$, find a basis of $\ker T$.

   [*Hint.*

   $$\ker T = \left\{ \begin{pmatrix} x & y \\ u & v \end{pmatrix} \in V : T\left(\begin{pmatrix} x & y \\ u & v \end{pmatrix}\right) = \begin{pmatrix} 0 & 0 \\ 0 & 0 \end{pmatrix} \right\}$$

   $$= \left\{ \begin{pmatrix} x & y \\ u & v \end{pmatrix} \in V : \begin{pmatrix} -2u & 2x + 2y - 2v \\ -2u & 2u \end{pmatrix} = \begin{pmatrix} 0 & 0 \\ 0 & 0 \end{pmatrix} \right\}.$$

   Hence $u = 0$ and $2x + 2y - 2v = 0 \Rightarrow u = 0$ and $x + y = v \Rightarrow \dim \ker T = 2$. If we take $x = 1$, $y = -1$, then $v = 0$ and if $x = 1$, $y = 0$, then $v = 1$. Hence the matrices $B = \begin{pmatrix} 1 & -1 \\ 0 & 0 \end{pmatrix}$ and $C = \begin{pmatrix} 1 & 0 \\ 0 & 1 \end{pmatrix}$ form a basis of $\ker T$.]

7. Show that there do not exist $n \times n$ real matrices A and B such that $AB - BA = I$ (identity matrix).

   [*Hint.* If $AB - BA = I$, then $\operatorname{tr}(AB) - \operatorname{tr}(BA) = n$. Hence $0 = n$, which is a contradiction.]

8. Let V be the vector space of 2×2 real matrices and $T : V \to \mathbf{R}^4$ be the linear map defined by $T\left(\begin{pmatrix} x & y \\ s & t \end{pmatrix}\right) = (x, y, s, t)$. Verify the Sylvester Law of Nullity.

   [*Hint.* $\ker T = \left\{ \begin{pmatrix} 0 & 0 \\ 0 & 0 \end{pmatrix} \right\}$ and $\operatorname{Im} T = \mathbf{R}^4$.]

9. Show that the matrix $A = \begin{pmatrix} 1 & 1 \\ 0 & 1 \end{pmatrix}$ cannot be similar to a diagonal matrix.

   [*Hint.* Suppose $D = \begin{pmatrix} d_1 & 0 \\ 0 & d_2 \end{pmatrix}$ is such that A is similar to D. Then there exists a non-singular matrix $P = \begin{pmatrix} a & b \\ c & d \end{pmatrix}$ such that $PAP^{-1} = D$. Then it can be shown that $a = 0 = c$. Hence $ad - bc = 0 \Rightarrow$ a contradiction.]

10. Let V be the inner product space $\mathbf{R}[x]$ of all polynomials over $\mathbf{R}$, with inner product defined by $\langle f(x), g(x) \rangle = \int_{-1}^{1} f(x)g(x)\,dx$. Given the sequence $S = \{1, x, x^2, x^3, \ldots\}$ of linearly independent vectors, find an orthogonal sequence of linearly independent vectors.

[*Hint.* Let $B = \{u_1(x), u_2(x), u_3(x), \ldots\}$ be the required orthogonal sequence. Then $u_1(x) = 1$, $u_2(x) = x - \frac{\int_{-1}^{1} x\,dx}{\int_{-1}^{1} dx} \cdot 1 = x$, $u_3(x) = x^2 - \frac{1}{3}$, $u_4(x) = x^3 - \frac{3}{5}x, \ldots$ by the Gram–Schmidt Orthogonalization Process.]

11. Find an orthonormal basis for the space of solutions of the linear equation $3x - 2y + z = 0$.

    [*Hint.* For $z = 1$, the equation $3x - 2y + 1 = 0$ has values $x = 1$, $y = 2$ or $x = 3$, $y = 5$. Clearly, $u = (1, 2, 1)$ and $v = (3, 5, 1)$ are linearly independent. If V is the space of solutions of the equation, then dim $V = 2$; u and v form a basis of V. Using the Gram–Schmidt orthogonalization process $\frac{1}{\sqrt{6}}(1, 2, 1)$ and $\frac{1}{\sqrt{21}}(2, 1, -4)$ form an orthonormal basis of V.]

12. Show that any orthogonal set of non-zero vectors in an inner product space V is linearly independent.

    [*Hint.* Let $B = \{u_i\}$ be an orthogonal set of any non-zero vectors. Suppose $\alpha_1 u_1 + \alpha_2 u_2 + \cdots + \alpha_n u_n = 0$. Then for any u_j, $0 = \langle 0, u_j \rangle = \langle \alpha_1 u_1 + \alpha_2 u_2 + \cdots + \alpha_j u_j + \cdots + \alpha_n u_n, u_j \rangle = \alpha_j \langle u_j, u_j \rangle$ for $j = 1, 2, \ldots, n$.]

Remark A maximal orthonormal set in an inner product space V is called a Hilbert Basis for V.

13. Let V be the vector space of real valued differentiable functions over $\mathbf{R}$ and $n \neq 0$ be an integer. If $D : V \to V$ is the differential operator on V, show that the functions $\sin nx$ and $\cos nx$ are eigenvectors of D^2.

    [*Hint.* $D^2(\sin nx) = -n^2 \sin nx \Rightarrow \sin nx$ is an eigenvector of D^2 corresponding to the eigenvalue $-n^2$.]

14. Find the Jordan canonical form of the matrix

$$A = \begin{pmatrix} 0 & 1 & 0 \\ -4 & 4 & 0 \\ -2 & 1 & 2 \end{pmatrix}.$$

Ans.:

$$\begin{pmatrix} 2 & 0 & 0 \\ 0 & 2 & 1 \\ 0 & 0 & 2 \end{pmatrix}.$$

15. If a quadratic surface is represented by $q(x, y, z) = x^2 + 2y^2 + 2z^2 + 2xy + 2zx = 1$, find its matrix.

    [*Hint.* Consider the matrix

$$A = \begin{pmatrix} 1 & 1 & 1 \\ 1 & 2 & 0 \\ 1 & 0 & 2 \end{pmatrix}$$

associated with q. The eigenvalues of A are $0, 2, 3$. Then q is reduced to $2u^2 + 3v^2 = 1$. It represents an elliptic cylinder in $\mathbf{R}^3$.]

16. Find all quadratic forms $q(x, y)$ of rank 2 which are in canonical forms. Determine the conic representing the equation $q(x, y) = 1$.

   [*Hint.* rank $q = 2$ shows that the possible canonical forms of q are $x^2 + y^2$, $x^2 - y^2$ or $-x^2 - y^2$.]

17. Find the minimal polynomial of $A = \begin{pmatrix} 2 & 2 \\ 2 & -2 \end{pmatrix}$ over **R**.

   [*Hint.* $A^2 = \begin{pmatrix} 8 & 0 \\ 0 & 8 \end{pmatrix} = 8I \Rightarrow A^2 - 8I = 0 \Rightarrow A$ satisfies the polynomial $f(x) = x^2 - 8$. Let $g(x)$ be a polynomial of degree 1 such that $g(A) = 0$. Then $g(x) = a_0 + a_1 x$, $a_1 \neq 0$ and $g(A) = a_0 I + a_1 A = 0 \Rightarrow A = bI$, where $b = -a_1^{-1} a_0 \Rightarrow$ a contradiction, since A is not in the form bI.]

18. Let U, V and W be finite-dimensional vector spaces over F. If $T : U \to V$ and $S : V \to W$ are linear transformations, show that rank$(S \circ T) \leq \min\{$rank T, rank $S\}$.

   [*Hint.* The dimension of a subspace $\leq$ dimension of the whole space $\Rightarrow$ $\dim \text{Im}(S \circ T) \leq \dim \text{Im} \, S$. Hence rank$(S \circ T) \leq$ rank S.]

19. Show that a square matrix A is orthogonally diagonalizable iff A is symmetric.

   [*Hint.* A is orthogonally diagonalizable $\Rightarrow \exists$ an orthogonal matrix P such that $P^{-1}AP = D$ (diagonal matrix). Again $P^t P = I = PP^t \Rightarrow P^{-1} = P^t$. This implies $A = PDP^t$. Hence $A^t = A$. The converse part is similar.]

20. Show that two similar matrices have the same minimal polynomial but its converse is not true.

   [*Hint.* Let A and B be two similar matrices over F. Then there exists a non-singular matrix P of order n over F such that $B = PAP^{-1}$. Let $m(x)$ and $q(x)$ be the minimal polynomials of A and B, respectively. Hence $B^2 = BB = PA^2 P^{-1}$. By induction it follows that $B^n = PA^n P^{-1}$ for all positive integers n. Suppose $m(x) = a_0 + a_1 x + \cdots + x^t$. Then $m(A) = 0$ and hence $m(B) = a_0 I + a_1 B + \cdots + B^t = Pm(A)P^{-1} = 0$. Again by the property of minimal polynomials, $m(x) | q(x)$ and $q(x) | m(x)$; and their leading coefficients are both 1. Hence $m(x) = q(x)$.

   The converse is not true. The matrices $A = \begin{pmatrix} 1 & 1 \\ 0 & 0 \end{pmatrix}$ and $B = \begin{pmatrix} 0 & 1 \\ 1 & 1 \end{pmatrix}$ have the same minimal polynomial $x(x - 1)$. But rank $A = 1$ and rank $B = 2 \Rightarrow A$ is not similar to B by Proposition 8.6.4.]

21. Prove that any square matrix A, of order n, over a Euclidean domain F, is equivalent to a diagonal matrix of the form

$$S = \begin{pmatrix} e_1 & & & & & & \\ & \ddots & & & & & \\ & & e_r & & & & \\ & & & 0 & & & \\ & & & & \ddots & & \\ & & & & & 0 \end{pmatrix},$$

called the Smith normal form of A in honor of H.J.S. Smith (1826–1883), where each e_i is a divisor of e_{i+1}, for $i = 1, 2, \ldots, r - 1$, where $r =$ rank A.

   [*Hint.* See Hazewinkel et al. (2011).]

22. Let $M_n(F)$ be the set of all $n \times n$ matrices over a field F. Then a matrix A in $M_n(F)$ is said to be triangular iff $a_{ij} = 0$ whenever $i > j$ or $a_{ij} = 0$ whenever $i < j$ (i.e., iff all the entries below or above the main diagonal are 0). A matrix of this form is called triangular.

Let $A \in M_n(F)$ be a *triangular matrix*. Show that

(a) $\det A$ is the product of its diagonal entries;
(b) if no entry on the main diagonal is 0, then A is invertible, otherwise A is singular;
(c) if A has all its characteristic roots in F, then there is a non-singular matrix P in $M_n(F)$ such that PAP^{-1} is a *triangular matrix*.

   [*Hint.* See Theorem 6.4.1, Herstein (1964, p. 287).]

23. Let V be a finite-dimensional vector space over F and $T : V \to V$ a linear operator. A subspace U of V is said to be a T-invariant subspace of U iff $T(U) \subseteq U$ and V is said to be a direct sum of (non-zero) T-invariant spaces $V_1, V_2, \ldots, V_r$ iff $V = V_1 \oplus V_2 \oplus \cdots \oplus V_r$ and $T(V_i) \subseteq V_i, i = 1, 2, \ldots, r$. Let T_i be the restriction of T to the invariant subspace V_i of V for each i. Then T is said to be decomposable into the operators T_i or T is said to be the direct sum of T_i's, denoted $T = T_1 \oplus T_2 \oplus \cdots \oplus T_r$.

Let V be a finite-dimensional vector space over F and $T : V \to V$ be linear operator such that $V = V_1 \oplus V_2 \oplus \cdots \oplus V_r$, where each V_i is a T-invariant subspace of V and T is the direct sum of $T_i's$, where T_i is the restriction of T to the invariant subspace V_i. Clearly, the eigenspace of T corresponding to each eigenvalue is a T-invariant subspace of V.

Then the following statements are true:

(a) If $\chi(x)$ is the characteristic polynomial of $T : V \to V$ and $\chi_i(x)$ is the characteristic polynomial of $T_i : V_i \to V_i$, then $\chi(x) = \chi_1(x)\chi_2(x) \cdots \chi_r(x)$.
(b) If A_i is the matrix of $T_i : V_i \to V_i$ relative to some ordered basis B_i of V_i, then $V = A_1 \oplus A_2 \oplus \cdots \oplus A_r$ iff the matrix M of T relative to the ordered basis B which is the union of B_i's arranged in the order $B_1, B_2, \ldots, B_r$, is given by

$$M = \begin{pmatrix} A_1 & 0 & \cdots & 0 \\ 0 & A_2 & \cdots & 0 \\ & & \ddots & \\ 0 & 0 & \cdots & A_r \end{pmatrix} \qquad (8.4)$$

(c) If $m(x) = a_0 + a_1 x + \cdots + a_{t-1}x^{t-1} + x^t$ is in $F[x]$, then $t \times t$ matrix

$$\begin{pmatrix} 0 & 1 & 0 & \cdots & 0 \\ 0 & 0 & 1 & \cdots & 0 \\ & & \ddots & & \\ 0 & 0 & 0 & \cdots & 1 \\ -a_0 & -a_1 & -a_2 & \cdots & -a_{t-1} \end{pmatrix}$$

is called the companion matrix of $m(x)$, denoted by $C(m(x))$. If $T : V \rightarrow V$ has minimal polynomial $m(x) = q_1(x)^{t_1} \cdots q_r(x)^{t_r}$ over F, where $q_1(x), q_2(x), \ldots, q_r(x)$ are irreducible distinct polynomials in $F[x]$, then there exists a basis B of V such that $m(T)_B$ is of the form

$$
\begin{pmatrix}
M_1 & & & \\
& M_2 & & \\
& & \ddots & \\
& & & M_r
\end{pmatrix}, \tag{8.5}
$$

where each matrix M_i is of the form

$$
M_i = \begin{pmatrix}
C(q_1(x))^{t_{i1}} & & \\
& \ddots & \\
& & C(q_i(x))^{t_{ir_i}}
\end{pmatrix},
$$

where $t_i = t_{i1} \geq t_{i2} \geq \cdots \geq t_{ir_i}$. The matrix (8.5) of T is called the rational canonical form of T.

[*Hint*. See Corollary, Herstein (1964, p. 308).]

24. Show that every square matrix is similar to an upper triangular matrix over **C**.
25. Let V be an n-dimensional vector space over F. V is said to be Noetherian (Artinian) iff it satisfies acc (bcc) for its subspaces. Show that V is both Noetherian and Artinian.

    [*Hint*. Let U be a proper subspace of V, then $\dim U < \dim V = n$. Hence any proper ascending (descending) chain of subspaces of V can not contain more than $n + 1$ terms.]

Exercises A (Objective Type)   Identify the correct alternative(s) (there may be more than one) from the following list:

1. Let V be the vector space of all real polynomials of degree at most 3. Define $D : V \rightarrow V$ be the differential operator defined by $(D)(f(x)) = f'(x)$, where f' is the derivative of f. Then matrix of D for the basis $\{1, x, x^2, x^3\}$, considered as column vectors, is given by

   (a) $\begin{pmatrix} 0 & 1 & 0 & 0 \\ 0 & 0 & 2 & 0 \\ 0 & 0 & 0 & 3 \\ 0 & 0 & 0 & 0 \end{pmatrix}$;   (b) $\begin{pmatrix} 0 & 0 & 0 & 0 \\ 1 & 0 & 0 & 0 \\ 0 & 2 & 0 & 0 \\ 0 & 0 & 3 & 0 \end{pmatrix}$;

   (c) $\begin{pmatrix} 0 & 0 & 0 & 0 \\ 0 & 1 & 0 & 0 \\ 0 & 0 & 2 & 0 \\ 0 & 0 & 0 & 3 \end{pmatrix}$;   (d) $\begin{pmatrix} 0 & 1 & 2 & 3 \\ 0 & 0 & 0 & 0 \\ 0 & 0 & 0 & 0 \\ 0 & 0 & 0 & 0 \end{pmatrix}$.

2. Let V be the vector space of all symmetric matrices of order $n \times n$ ($n \geq 2$) with real entries and trace equal to zero. If d is the dimension of V, then d is

(a) $\dfrac{(n^2 + n)}{2} - 1$; (b) $\dfrac{(n^2 - 2n)}{2} - 1$;

(c) $\dfrac{(n^2 + 2n)}{2} - 1$; (d) $\dfrac{(n^2 - n)}{2} - 1$.

3. For the quadratic form $q = x^2 - 6xy + y^2$,

(a) Rank of q is 3; (b) Signature of q is 1;

(c) Rank of q is 2; (d) Signature of q is 0.

4. For a positive integer n, let V_n denote the space of all polynomials $f(x)$ with coefficients in $\mathbf{R}$ such that $\deg f(x) \leq n$, and let B_n denote the standard basis of V_n given by $B_n = \{1, x, x^2, \ldots, x^n\}$. If $T : V_3 \to V_4$ is the linear transformation defined by $T(f(x)) = x^2 f'(x) + \int_0^x f(t) \, dt$ and $A = (a_{ij})$ is the 5×4 matrix of T with respect to standard bases B_3 and B_4, then

(a) $a_{32} = 0$ and $a_{33} = \dfrac{7}{3}$; (b) $a_{32} = \dfrac{3}{2}$ and $a_{33} = 0$;

(c) $a_{32} = \dfrac{3}{2}$ and $a_{33} = \dfrac{7}{3}$; (d) $a_{32} = 0$ and $a_{33} = 0$.

5. Let M be a 3×3 matrix with real entries such that $\det(M) = 6$ and the trace of M is 0. If $\det(M + I) = 0$, where I denotes the 3×3 identity matrix, then the eigenvalues of M are

(a) $1, 2, -3$ (b) $-1, 2, -3$ (c) $-1, 2, 3$ (d) $-1, -2, 3$.

6. If

$$M = \begin{pmatrix} 1 & 1 & 1 \\ 1 & 1 & 1 \\ 1 & 1 & 1 \end{pmatrix},$$

then

(a) 0 and 3 are the only eigenvalues of M;
(b) M is positive semi definite;
(c) M is not diagonalizable;
(d) M is not positive definite.

7. Let the matrix

$$M = \begin{pmatrix} 40 & -29 & -11 \\ -18 & 30 & -12 \\ 26 & 24 & -50 \end{pmatrix}$$

have a non-zero complex eigenvalue λ. Which of the following numbers must be an eigenvalue of M?

     (a) $20 - \lambda$ (b) $\lambda - 20$ (c) $\lambda + 20$ (d) $-20 - \lambda$.

8. A Jordan canonical form of the matrix

$$M = \begin{pmatrix} 0 & 0 & 0 & -4 \\ 1 & 0 & 0 & 0 \\ 0 & 1 & 0 & 5 \\ 0 & 0 & 1 & 0 \end{pmatrix}$$

is

(a) $\begin{pmatrix} -1 & 0 & 0 & 0 \\ 0 & 1 & 0 & 0 \\ 0 & 0 & 2 & 0 \\ 0 & 0 & 0 & -2 \end{pmatrix}$;   (b) $\begin{pmatrix} -1 & 1 & 0 & 0 \\ 0 & -1 & 0 & 0 \\ 0 & 0 & 2 & 0 \\ 0 & 0 & 0 & -2 \end{pmatrix}$;

(c) $\begin{pmatrix} 1 & 1 & 0 & 0 \\ 0 & 1 & 0 & 0 \\ 0 & 0 & 2 & 0 \\ 0 & 0 & 0 & -2 \end{pmatrix}$;   (d) $\begin{pmatrix} -1 & 1 & 0 & 0 \\ 0 & 1 & 0 & 0 \\ 0 & 0 & 2 & 0 \\ 0 & 0 & 0 & -2 \end{pmatrix}$.

9. For a given positive integer n, let M_n be the vector space of all $n \times n$ real matrices $A = (a_{ij})$ such that $a_{ij} = a_{rs}$ whenever $i + j = r + s$ $(i, j, r, s = 1, \ldots, n)$. Then the dimension of M_n, as a vector space over $\mathbf{R}$, is

    (a) $2n + 1$ (b) $n^2 - n + 1$ (c) n^2 (d) $2n - 1$.

10. Let V the vector space of all symmetric matrices $A = (a_{ij})$ of order $n \times n$ $(n \geq 2)$ with real entries, $a_{11} = 0$ and trace zero. Then dimension of V is

    (a) $(n^2 + n - 4)/2$;   (b) $(n^2 - n + 3)/2$;

    (c) $(n^2 + n - 3)/2$;   (d) $(n^2 - n + 4)/2$.

11. For a positive integer n, let $M_n(\mathbf{R})$ be the vector space of all $n \times n$ real matrices and $T : M_n(\mathbf{R}) \to M_n(\mathbf{R})$ be a linear transformation such that $T(A) = 0$, whenever $A \in M_n(\mathbf{R})$ is symmetric or skew-symmetric. Then the rank of T is

    (a) $\dfrac{n(n+1)}{2}$ (b) 0 (c) n (d) $\dfrac{n(n-1)}{2}$.

12. If $S : \mathbf{R}^3 \to \mathbf{R}^4$ and $T : \mathbf{R}^4 \to \mathbf{R}^3$ are two linear transformations such that $T \circ S$ is the identity transformation of $\mathbf{R}^3$, then

(a) $S \circ T$ is surjective but not injective;

(b) $S \circ T$ is injective but not surjective;

(c) $S \circ T$ is the identity map of $\mathbf{R}^4$;

(d) $S \circ T$ is neither injective nor surjective.

13. If V is a 3-dimensional vector space over the field $F_3 = \mathbf{Z}_3$ of three elements, then the number of distinct 1-dimensional subspaces of V is

$$(a)\ 13 \qquad (b)\ 15 \qquad (c)\ 9 \qquad (d)\ 26.$$

14. Which of the following statements is (are) correct?

(a) The eigenvalues of a unitary matrix are all equal to 1;

(b) The eigenvalues of a unitary matrix are all equal to ± 1;

(c) The determinant of real orthogonal matrix is always ± 1;

(d) The determinant of real orthogonal matrix is always -1.

15. Let V be a vector space of dimension $d < \infty$, over $\mathbf{R}$. Let U be a vector subspace of V. Let S be a non-empty subset of V. Which of the following statements is (are) correct?

(a) If S is a basis of V, then $U \cap S$ is a basis of U;

(b) If $U \cap S$ is a basis of U and $\{s + U \in V/U : s \in S\}$ is a basis of V/U, then S is a basis of V;

(c) If $\dim U = n$, then $\dim V/U$ is $d - n$;

(d) If S is a basis of U as well as V, then the dimension of U is d.

16. Let V be the inner product space consisting of linear polynomials, $f : [0, 1] \to \mathbf{R}$ (i.e., V consists of polynomials of the form $f(x) = ax + b, a, b \in \mathbf{R}$), with the inner product defined by $\langle f, g \rangle = \int_0^1 f(x)g(x)\,dx$ for $f, g \in V$. An orthogonal basis of V is then

$$(a)\quad \{1, x\}; \qquad\qquad (b)\quad \left\{1, x - \frac{1}{2}\right\};$$

$$(c)\quad \{1, (2x - 1)\sqrt{3}\}; \qquad (d)\quad \{1, x\sqrt{3}\}.$$

17. If $f(x)$ is the minimal polynomial of the 4×4 matrix

$$M = \begin{pmatrix} 0 & 0 & 0 & 1 \\ 1 & 0 & 0 & 0 \\ 0 & 1 & 0 & 0 \\ 0 & 0 & 1 & 0 \end{pmatrix},$$

then the rank of the 4×4 matrix $f(M)$ is

$$(a)\ 0 \qquad (b)\ 4 \qquad (c)\ 2 \qquad (d)\ 1.$$

18. Let a, b, c be positive real numbers such that $b^2 + c^2 < a < 1$ and M be the matrix given by

$$M = \begin{pmatrix} 1 & b & c \\ b & a & 0 \\ c & 0 & 1 \end{pmatrix}.$$

Then

(a) M can have a positive as well as a negative eigenvalue;
(b) all the eigenvalues of M are positive numbers;
(c) all the eigenvalues of M are negative numbers;
(d) eigenvalues of M can be non-real complex number.

19. If $f : \mathbf{R}^n \to \mathbf{R}$ is a linear map such that $f(0, \ldots, 0) = 0$, then the set $\{f(x_1, x_2, \ldots, x_n) : \sum_{j=1}^n x_j^2 \leq 1\}$ is

(a) $[0, a]$ for some $a \in \mathbf{R}, a \geq 0$;
(b) $[0, 1]$;
(c) $[-a, a]$ for some $a \in \mathbf{R}, a \geq 0$;
(d) $[a, b]$ for some $a, b \in \mathbf{R}, 0 \leq a < b$.

20. Let the system of equations

$$x + y + z = 1$$
$$2x + 3y - z = 5$$
$$x + 2y - kz = 4$$

where $k \in \mathbf{R}$, have an infinite number of solutions. Then the value of k is

(a) 2   (b) 1   (c) 0   (d) 3.

21. Let $A = (a_{ij})$ be an $n \times n$ complex matrix and let A^* denote the conjugate transpose of A. Which of the following statements are necessarily true?

(a) If $\operatorname{tr}(A^*A) \neq 0$, then A is invertible;
(b) If A is invertible, then $\operatorname{tr}(A^*A) \neq 0$ i.e., the trace of A^*A is non-zero;
(c) If $\operatorname{tr}(A^*A) = 0$, then A is the zero matrix;
(d) If $|\operatorname{tr}(A^*A)| < n^2$, then $|a_{ij}| < 1$ for some i, j.

22. For a given positive integer n, let V be an $(n+1)$-dimensional vector space over $\mathbf{R}$ with a basis $B = \{e_1, e_2, \ldots, e_{n+1}\}$. If $T : V \to V$ is the linear transformation such that $T(e_i) = e_{i+1}$ for $i = 1, 2, \ldots, n$ and $T(e_{n+1}) = 0$, then

(a) nullity of T is 1;
(b) rank of T is n;
(c) trace of T is non-zero;
(d) $T^n = T \circ T \circ \cdots \circ T$ is the zero map.

23. Let M be a 3×3 non-zero matrix with the property $M^3 = 0$. Which of the following statements is (are) false?

(a) M has one non-zero eigenvector;
(b) M is similar to a diagonal matrix;
(c) M is not similar to a diagonal matrix;
(d) M has three linearly independent eigenvectors.

24. If T is a linear transformation on the vector space $\mathbf{R}^n$ over $\mathbf{R}$ such that $T^2 = \lambda T$ for some $\lambda \in \mathbf{R}$, then

(a) $T = \lambda I$ where I is the identity transformation on $\mathbf{R}^n$;

(b) if $\|Tx\| = \|x\|$ for some non-zero vector $x \in \mathbf{R}^n$, then $\lambda = \pm 1$;

(c) $\|Tx\| = |\lambda|\|x\|$ for all $x \in \mathbf{R}^n$;

(d) if $\|Tx\| > \|x\|$ for some non-zero vector $x \in \mathbf{R}^n$, then T is necessarily singular.

25. Let M be the 3×3 real matrix all of whose entries are 1. Then:

(a) M is positive semidefinite, i.e., $\langle Mx, x \rangle \geq 0$ for all $x \in \mathbf{R}^3$;

(b) M is diagonalizable;

(c) 0 and 3 are the only eigenvalues of A;

(d) M is positive definite, i.e., $\langle Mx, x \rangle > 0$ for all $x \in \mathbf{R}^3$ with $x \neq 0$.

26. Let $T : \mathbf{R}^7 \to \mathbf{R}^7$ be the linear transformation defined by $T(x_1, x_2, \ldots, x_6, x_7) = (x_7, x_6, \ldots, x_2, x_1)$. Which of the following statements is (are) correct?

(a) There is a basis of $\mathbf{R}^7$ with respect to which T is a diagonal matrix;

(b) $T^7 = I$;

(c) The determinant of T is 1;

(d) The smallest n such that $T^n = I$ is even.

27. Let A be an orthogonal 3×3 matrix with real entries. Which of the following statements is (are) false?

(a) The determinant of A is a rational number;

(b) All the entries of A are positive;

(c) $d(Ax, Ay) = d(x, y)$ for any two vectors x and $y \in \mathbf{R}^3$, where $d(u, v)$ denotes the usual Euclidean distance between vectors u and $v \in \mathbf{R}^3$;

(d) All the eigenvalues of A are real.

28. Which of the following matrices is (are) non-singular?

(a) Every symmetric non-zero real 3×3 matrix;

(b) $I + A$ where $A \neq 0$ is a skew-symmetric real $n \times n$ matrix for $n \geq 2$;

(c) Every skew-symmetric non-zero real 5×5 matrix;

(d) Every skew-symmetric non-zero real 2×2 matrix.

29. Let A be a real symmetric $n \times n$ matrix whose only eigenvalues are 0 and 1. Let the dimension of the null space of $A - I$ be m. Which of the following statements is (are) false?

(a) The characteristic polynomial of A is $(\lambda - 1)^m \lambda^{m-n}$;

(b) The rank of A is $n - m$;

(c) $A^k = A^{k+1}$ for all positive integers k;

(d) The rank of A is m.

30. Let ω be a complex number such that $\omega^3 = 1$, but $\omega \neq 1$. Suppose

$$M = \begin{pmatrix} 1 & \omega & \omega^2 \\ \omega & \omega^2 & 1 \\ \omega^2 & \omega & 1 \end{pmatrix}.$$

Which of the following statements is (are) correct?

(a) 0 is an eigenvalue of M;
(b) $\text{rank}(M) = 2$;
(c) M is invertible;
(d) There exist linearly independent vectors $v, w \in \mathbf{C}^3$ such that $Mv = Mw = 0$.

31. $M = (a_{ij})_{n \times n}$ is a square matrix with integer entries such that $a_{ij} = 0$ for $i > j$ and $a_{ij} = 1$ for $i = 1, \ldots, n$. Which of the following statements is (are) correct?

(a) M^{-1} exists and it has some entries that are not integers;
(b) M^{-1} exists and it has integer entries;
(c) M^{-1} is a polynomial function of M with integer coefficients;
(d) M^{-1} is not a power of M unless M is the identity matrix.

32. Let M be a 4×4 real matrix such that $-1, 1, 2, -2$ are its eigenvalues. Suppose $B = M^4 - 5M^2 + 5I$, where I denotes the 4×4 identity matrix. Which of the following statements is (are) correct?

(a) trace of $(M - B)$ is 0;
(b) $\det(B) = 1$;
(c) $\det(M + B) = 0$;
(d) trace of $(M + B)$ is 4.

33. Let M be a 2×2 non-zero complex matrix such that $M^2 = 0$. Which of the following statements is (are) correct?

(a) PMP^{-1} is a diagonal matrix for some invertible 2×2 matrix P with entries in $\mathbf{R}$;
(b) M has only one eigenvalue in $\mathbf{C}$ with multiplicity 2;
(c) M has two distinct eigenvalues in $\mathbf{C}$;
(d) $Mv = v$ for some $v \in \mathbf{C}^2, v \neq 0$.

34. Let $M_2(\mathbf{R})$ denote the set of 2×2 real matrices. Let $A \in M_2(\mathbf{R})$ be of trace 2 and determinant -3. Identifying $M_2(\mathbf{R})$ with $\mathbf{R}^4$, consider the linear transformation $T : M_2(\mathbf{R}) \to M_2(\mathbf{R})$ defined by $T(M) = AM$. Which of the following statements is (are) correct?

(a) T is invertible;
(b) 2 is an eigenvalue of T;
(c) T is diagonalizable;
(d) $T(B) = B$ for some $0 \neq B$ in $M_2(\mathbf{R})$.

35. Let λ, μ be two distinct eigenvalues of a 2×2 matrix M. Which of the following statements is (are) correct?

(a) $M^3 = \frac{\lambda^3 - \mu^3}{\lambda - \mu} M - \lambda \mu (\lambda + \mu) I$;
(b) M^2 has distinct eigenvalues;
(c) trace of M^n is $\lambda^n + \mu^n$ for every positive integer n;
(d) M^n is not a scalar multiple of identity for any positive integer n.

36. Let T be a non-zero linear transformation on a real vector space V of dimension n. Let the subspace $V_0 \subset V$ be the image of V under T. Let $k = \dim V_0 < n$ and suppose that for some $\lambda \in \mathbf{R}$, $T^2 = \lambda T$. Then

    (a) $\lambda = 1$;
    (b) λ is the only eigenvalue of T;
    (c) $\det M = |\lambda|$, where M is a matrix representation of T;
    (d) there is a non-trivial subspace $V_1 \subset V$ such that $Tx = 0$ for all $x \in V_1$.

37. Let M be a $n \times n$ real matrix and V be the vector space spanned by $\{I, M, M^2, \ldots, M^{2n}\}$. Then the dimension of the vector space V is

    (a) at most n (b) n^2 (c) $2n$ (d) at most $2n$.

38. Let A and B be two real square matrices of order n. Then

    (a) $\mathrm{trace}(A + B) < \mathrm{trace}(A) + \mathrm{trace}(B)$;
    (b) $\mathrm{trace}(A + B) > \mathrm{trace}(A) + \mathrm{trace}(B)$;
    (c) $\mathrm{trace}(A + B) = \mathrm{trace}(A) + \mathrm{trace}(B)$;
    (d) $\mathrm{trace}(A + B) \neq \mathrm{trace}(A^t) + \mathrm{trace}(B^t)$.

39. Let A and B be two real square matrices of order n. Then

    (a) $\mathrm{trace}(AA^t) = $ sum of all entries of A;
    (b) $\mathrm{trace}(AA^t) = $ sum of all diagonal entries of A;
    (c) $\mathrm{trace}(AB) = \mathrm{trace}(BA)$;
    (d) $\mathrm{trace}(AA^t) = 0$ does not imply that A is a null matrix.

Exercises B (True/False Statements)   Determine the correct statement from the following list:

1. Cardinalities of all bases of a vector space are the same.
2. If V is a finite-dimensional vector space and U is a subspace of V with $\dim V = \dim U$, then U may not be same as V.
3. If V is a non-zero vector in $\mathbf{R}^2$, then the subspace $V = \{rv : r \in \mathbf{R}\}$ represents geometrically a straight line passing through the origin in $\mathbf{R}^2$.
4. Let $M_2(\mathbf{R})$ be the vector space of all 2×2 matrices over $\mathbf{R}$ and S be the set of all symmetric matrices in $M_2(\mathbf{R})$. Then $\dim S = 4$.
5. The map $T : \mathbf{R}^2 \to \mathbf{R}^1$ defined by $T(x, y) = \sin(x + y)$ is a linear transformation.
6. Let V be an n-dimensional vector space and $T : V \to U$ be linear transformation. Then T is injective iff rank of $T = n$.
7. The real vector space $C([0, 1])$ is a commutative real algebra with identity.
8. Let A be an algebra with identity 1 over a field F. Then A is isomorphic to a subalgebra of the algebra $L(V, V)$ of all linear transformations from V to V.
9. Let A be an $n \times n$ matrix over the field F. Then the row rank of A is equal to the column rank of A.
10. Let $T : \mathbf{R}^2 \to \mathbf{R}^2$ be the rotation of the plane through an angle θ, $0 < \theta < \pi$. Then T has no eigenvector.

11. The matrix

$$A = \begin{pmatrix} 2 & 0 & 0 \\ 0 & 0 & -1 \\ 0 & 1 & 0 \end{pmatrix}$$

is diagonalizable over **C** but not over **R**.

12. The matrix

$$M = \begin{pmatrix} 4 & -1 & 1 \\ 4 & 0 & 2 \\ 2 & -1 & 3 \end{pmatrix}$$

is similar to the Jordan matrix

$$\begin{pmatrix} 2 & 0 & 0 \\ 0 & 2 & 0 \\ 0 & 0 & 3 \end{pmatrix}.$$

13. The orthogonal complement of the subspace V generated by the vector $v = (2, 3, 4)$ in $\mathbf{R}^3$ is the plane $2x + 3y + 4z = 0$ in $\mathbf{R}^3$ passing through the origin and perpendicular to the vector v.

14. If $U = \{(x, y, 0) \in \mathbf{R}^3\}$, then for any $v = (r, t, s) \in \mathbf{R}^3$, the coset $v + U$ in $\mathbf{R}^3$ represents geometrically the plane parallel to xy-plane passing through the point (r, t, s) and at a distance $|s|$ from the xy-plane.

15. For the quadratic form $q(x) = x_1^2 - 6x_1 x_2 + x_2^2$, rank is 2 and signature is 1.

16. The matrix $A = \begin{pmatrix} 1 & 2 \\ 2 & 1 \end{pmatrix}$ is positive definite but the matrix $B = \begin{pmatrix} 2 & 1 \\ 1 & 2 \end{pmatrix}$ is not positive definite.

17. A symmetric matrix M is positive definite iff there exists a nonsingular matrix P such that $M = P^t P$.

18. Let V be an n-dimensional inner product space over **R**. If $T : V \to V$ is positive definite, then all eigenvalues of T are not necessarily positive.

8.14 Additional Reading

We refer the reader to the books (Adhikari and Adhikari 2003, 2004, 2007; Artin 1991; Birkhoff and Mac Lane 2003; Chatterjee 1966; Hazewinkel et al. 2011; Herstein 1964; Hoffman and Kunze 1971; Hungerford 1974; Janich 1994; Lang 1986; Simmons 1963), for further details.

References

Adhikari, M.R., Adhikari, A.: Groups, Rings and Modules with Applications, 2nd edn. Universities Press, Hyderabad (2003)

Adhikari, M.R., Adhikari, A.: Text Book of Linear Algebra: An Introduction to Modern Algebra. Allied Publishers, New Delhi (2004)

Adhikari, M.R., Adhikari, A.: Introduction to Linear Algebra with Applications to Basic Cryptography. Asian Books, New Delhi (2007)

Artin, M.: Algebra. Prentice-Hall, Englewood Cliffs (1991)

Birkhoff, G., Mac Lane, S.: A Survey of Modern Algebra. Universities Press, Hyderabad (2003)

Chatterjee, B.C.: Linear Algebra. Das Gupta and Company, Calcutta (1966)

Hazewinkel, M., Gubareni, N., Kirichenko, V.V.: Algebras, Rings and Modules vol. 1. Springer, New Delhi (2011)

Herstein, I.: Topics in Algebra. Blaisdell, New York (1964)

Hoffman, K., Kunze, R.: Linear Algebra. Prentice-Hall, Englewood Cliffs (1971)

Hungerford, T.W.: Algebra. Springer, New York (1974)

Janich, K.: Linear Algebra. Springer, New York (1994)

Lang, S.: Introduction to Linear Algebra. Springer, New York (1986)

Simmons, G.F.: Introduction to Topology and Modern Analysis. McGraw-Hill, New York (1963)

Chapter 9
Modules

The concept of a module arose through the study of algebraic number theory. It became an important tool in algebra in the late 1920s essentially due to insight of E. Noether, who was the first mathematician to realize its importance. A module is an additive abelian group whose elements are suitably multiplied by the elements from some ring. Modules exist in any ring. One of the most important topics in modern algebra is module theory. A module over a ring is a generalization of an abelian group (which is a module over $\mathbf{Z}$) and also a natural generalization of a vector space (which is a module over a division ring (field)). Many results of vector spaces are generalized in some special classes of modules, such as free modules and finitely generated modules over PID. Modules are closely related to the representation theory of groups. One of the basic concepts which accelerates the development of the commutative algebra is the module theory, as modules play a central role in commutative algebra. Modules are also widely used in structure theory of finitely generated abelian groups, finite abelian groups and PID, homological algebra, and algebraic topology. In this chapter we study the basic properties of modules. We also consider modules of special classes, such as free modules, modules over PID along with structure theorems, exact sequences of modules and their homomorphisms, Noetherian and Artinian modules, homology and cohomology modules. Our study culminates in a discussion of the topology of the spectrum of modules and rings with special reference to Zariski topology and schemes. In this chapter we also show how to construct isomorphic replicas of all homomorphic images of a specified abstract module.

9.1 Introductory Concepts

The notion of a module is a generalization of the concept of a vector space; where the scalars are restricted to lie in an arbitrary ring (instead of a field for a vector space). So, naturally, modules and vector spaces have some common properties but they differ in many properties.

M.R. Adhikari, A. Adhikari, *Basic Modern Algebra with Applications*,
DOI 10.1007/978-81-322-1599-8_9, © Springer India 2014

Definition 9.1.1 Let R be a ring. A (left) R-module is an additive abelian group M together with an action (called scalar multiplication) $\mu : R \times M \to M$ (the image $\mu(r, x)$ being denoted by rx) such that $\forall r, s \in R$ and $x, y \in M$:

(i) $r(x + y) = rx + ry$;
(ii) $(r + s)x = rx + sx$;
(iii) $r(sx) = (rs)x$.

Moreover, if R has an identity element 1,

(iv) if $1x = x \ \forall x \in M$, then M is said to be a unitary (left) R-module.

If R is a division ring, then a unitary (left) R-module is called a (left) *vector space* over R.

If $R = \mathbf{Z}$, then an R module is merely an additive abelian group.

A ring R itself can be considered as an R-module by taking scalar multiplication to be the usual multiplication of R.

Remark If we write $\mu(r, x)$ as $f_r(x)$, then $f_r : M \to M, x \mapsto rx$ is a group homomorphism by (i) for each $r \in R$.

If $\text{End}(M)$ is the endomorphism ring of the additive abelian group M, then the map $\psi : R \to \text{End}(M), r \mapsto f_r$ is a ring homomorphism by (ii) and (iii).

Analogously, a (right) R-module or a unitary (right) R-module or a (right) vector space is defined by an action $M \times R \to M$ denoted by $(x, r) \mapsto xr$.

Proposition 9.1.1 *Let M be a (left) R-module with additive identity O_M. Then*

(i) $O_R x = r O_M = O_M \ \forall r \in R$ and $x \in M$;
(ii) $(-r)x = -(rx) = r(-x) \ \forall r \in R$ and $x \in M$,

where O_R is the additive identity of the ring R.

Proof Trivial.   □

Remark A given additive abelian group M may have different R-module structures (both left and right). If R is commutative, then every left R-module M can be given the structure of a right R-module by defining

$$xr = rx \quad \forall r \in R, \ x \in M.$$

From now on, unless specified otherwise, the ring R *will mean a commutative ring with identity* 1 and R-module will mean a unitary (left) R-module.

Example 9.1.1 (i) Every additive abelian group G is a unitary $\mathbf{Z}$-module with nx $(n \in \mathbf{Z}, x \in G)$ defined by

$$nx = \begin{cases} x + x + \cdots + x & (n \text{ times if } n > 0) \\ 0 & (\text{if } n = 0) \\ -x - x - \cdots - x & (|n| \text{ times if } n < 0). \end{cases}$$

(ii) An ideal I of a ring R is an R-module with ra ($r \in R$, $a \in I$) being the usual product in R. In particular, R is itself an R-module.

(iii) If R is a division ring F, then an R-module is called an F-vector space.

(iv) If M is a ring and R is a subring of M, then M is an R-module with rx ($r \in R$, $x \in M$) being the usual multiplication in M. In particular $R[\![x]\!]$ and $R[x]$ are R-modules.

(v) If I is an ideal of R, then I is an additive subgroup of R and R/I is an abelian group.

Clearly, R/I is an R-module with $r(t + I) = rt + I$ $\forall r, t \in R$.

(vi) Let $(M, +)$ be an abelian group and $\text{End}(M)$ be the ring of all endomorphisms of $(M, +)$. Taking $R = \text{End}(M)$, M becomes an R-module under the external law of composition $\mu : R \times M \to M$, defined by

$$\mu(f, x) = f \cdot x = f(x), \quad \forall f \in R \text{ and } \forall x \in M.$$

(vii) If R is a unitary ring and n is a positive integer, then the abelian group $(R^n, +)$ (under componentwise addition) is a unitary R-module under the external law of composition $\mu : R \times R^n \to R^n$, defined by

$$\mu(r, x) = r \cdot x = (rx_1, rx_2, \ldots, rx_n), \quad \forall r \in R \text{ and } \forall x = (x_1, x_2, \ldots, x_n) \in R^n.$$

In particular, if $R = F$ is a field, then F^n is an F-vector space.

(viii) Let R be a ring and $(R^{\mathbf{N}^+}, +)$ denote the abelian group of all mappings $f : \mathbf{N}^+ \to R$ (i.e., of all sequences of elements of R) under addition defined by $(f + g)(n) = f(n) + g(n)$, $\forall n \in \mathbf{N}^+$. Then $R^{\mathbf{N}^+}$ is an R-module under the external law of composition $\mu : R \times R^{\mathbf{N}^+} \to R^{\mathbf{N}^+}$, defined by

$$\big(\mu(r, f)\big)(n) = (r \cdot f)(n) = rf(n), \quad \forall n \in \mathbf{N}^+.$$

(ix) Let $P_n(x)$ denote the set of all polynomials of degrees $< n$ with real coefficients, then $P_n(x)$ becomes a real vector space.

(x) Let M be a smooth manifold. Then $C^\infty(M) = \{f : M \to \mathbf{R} \text{ such that } f \text{ is a smooth function}\}$ forms a ring. Then the set S of all smooth vector fields on M forms a module over the ring $C^\infty(M)$.

(xi) Let V be an n-dimensional vector space over F and $T : V \to V$ be a fixed linear operator. Then V is an $F[x]$-module under the external law of composition

$$F[x] \times V \to V, \qquad (f, v) \mapsto f(T)v.$$

We recall the definition of an algebra.

Definition 9.1.2 An algebra consists of a vector space V over a field F together with a binary operation of multiplication on the set V of vectors, such that $\forall r \in F$ and $x, y, z \in V$ the following conditions are satisfied:

(i) $(rx)y = r(xy) = x(ry)$;
(ii) $(x + y)z = xz + yz$;
(iii) $x(y + z) = xy + xz$.

If, moreover, (iv) $(xy)z = x(yz)$, $\forall x, y, z \in V$, then V is said to be an associative algebra over F.

If $\dim_F(V) = n$ as a vector space over F, then the algebra V is said to be n-dimensional.

Example 9.1.2 (i) Let G be a field extension of a field F. Then F is a subfield of G and $V = G$ is an associative algebra over F, where addition and multiplication of elements of V are the usual field addition and multiplication in G, and scalar multiplication by elements of F is again the usual field multiplication in G.

(ii) Let V be a vector space over a field F. Then the set $L(V, V)$ of all linear transformations of V into itself is an algebra over F.

Definition 9.1.3 For a commutative ring R with identity, an R-algebra (or algebra over R) is a ring K such that

 (i) $(K, +)$ is a unitary (left) R-module;
(ii) $r(xy) = (rx)y = x(ry)$, $\forall r \in R$, $x, y \in K$.

Example 9.1.3 (i) Every ring R is a **Z**-algebra.
  (ii) Let R be a commutative ring with 1, then

(a) the group ring $R(G)$ is an R-algebra with R-module structure $r(\sum a_i x_i) = \sum (ra_i)x_i$, $\forall r, a_i \in R$ and $x_i \in G$ (see Ex. 10 of Exercises-I, Chap. 4).
(b) $M_n(R)$, the ring of all square matrices of order n over R, is an R-algebra.

9.2 Submodules

We are interested to show how to construct isomorphic replicas of all homomorphic images of a specified abstract module. For this purpose we introduce the concept submodules which plays an important role in determining both the structure of a module M and nature of homomorphisms with domain M.

Let M be an R-module. Then $(M, +)$ is an abelian group. We now consider its subgroups $(N, +)$ which are stable under the external law of composition defined on the R-module M.

Definition 9.2.1 Let M be an R-module. A non-empty subset N of M is called an R-submodule or (simply) a submodule of M iff

 (i) for $x, y \in N$, $x - y \in N$ (i.e., $(N, +)$ is a subgroup of $(M, +)$);
(ii) for $x \in N$, $r \in R$, $rx \in N$ (i.e., N is stable under the external law of composition on M).

Thus a submodule N of M is closed under both restricted compositions of addition and scalar multiplication (defined on M) on N. Clearly, $\{O_M\}$ and M are submodules of M in their own right, which are called trivial submodules of M.

Remark $\{O_M\}$ is the smallest submodule of M and M is the largest submodule of M. If there exist any other submodules of M, then they are called non-trivial submodules of M. Any submodule of M other than M is called a proper submodule of M.

Theorem 9.2.1 *Let M be a unitary R-module. Then a non-empty subset N of M is a submodule of M iff for all $a, b \in R$ and $x, y \in N$, $ax + by \in N$.*

Proof Let N be a submodule of the unitary R-module M. Then for $a, b \in R$ and $x, y \in N$, $ax \in N$ and $(-by) \in N$ and hence $ax + by = ax - (-(by)) \in N$.

Conversely, let N be a non-empty subset of M such that for all $a, b \in R$ and x, $y \in N$, $ax + by \in N$. Then 1_R and $-1_R \in R \Rightarrow 1_R x + (-1_R)y \in N \Rightarrow x - y \in N \Rightarrow (N, +)$ is a subgroup of $(M, +)$. Again $O_R \in R \Rightarrow ax + O_R y = ax \in N \Rightarrow N$ is stable under external law of composition. Consequently, N is a submodule of M. $\square$

Corollary *Let M be a unitary R-module. Then a non-empty subset N of M is a submodule of M iff*

(i) $x, y \in N \Rightarrow x + y \in N$ *and*
(ii) $a \in R, x \in N \Rightarrow ax \in N$.

Example 9.2.1 (i) If $(G, +)$ is an abelian group, then G is a **Z**-module by Example 9.1.1(i). The submodules of the **Z**-module G are precisely the subgroups of G. In particular, $E = \{0, \pm 2, \pm 4, \ldots\}$ is a submodule of the **Z**-module **Z**.

(ii) Every ring R may be considered as a left R-module by Example 9.1.1(ii). Let I be a submodule of R. Then $I \subseteq R$ is such that $x - y \in I$ and $rx \in I, \forall x, y \in I$ and $\forall r \in R$ by Definition 9.2.1. Consequently, I is a left ideal of R. Conversely, let I be a left ideal of R. Then $I \subseteq R$ is such that $x - y \in I$ and $rx \in I, \forall x, y \in I$ and $\forall r \in R$. Thus I is a submodule of the left R-module R. Hence the submodules of the left R-module R are just the left ideals of R. Consequently, the submodules of the left R-module R are precisely the left ideals of R.

Likewise, considering R as a right R-module, the submodules of R are precisely the right ideals of R.

(iii) Submodules of a commutative ring are precisely its ideals.

Most of the operations considered for groups have their counterparts for modules.

Theorem 9.2.2 *Let M be an R-module and $\{M_i\}_{i \in I}$ be a family of submodules of M. Then their intersection $\bigcap_{i \in I} M_i$ is again a submodule of M.*

Proof Let $A = \bigcap_{i \in I} M_i$. Now $O_M \in M_i$ for each $i \in I$, since every submodule M_i is a subgroup of $M \Rightarrow O_M \in A \Rightarrow A \neq \emptyset$.

Again $x - y \in M_i$ and $rx \in M_i$ for each $i \in I$ and $\forall x, y \in M_i, \forall r \in R \Rightarrow x - y$, $rx \in A \Rightarrow A$ is a submodule of M. $\square$

Remark The union of two submodules of an R-module M is not in general a submodule of M. The reason is that union of two subgroups of a group is not in general a group. We now cite the following examples.

Example 9.2.2 (i) Let $M_1 = \{0, \pm2, \pm4, \pm6, \ldots\}$ and $M_2 = \{0, \pm3, \pm6, \pm9, \ldots\}$. Then M_1 and M_2 are both submodules of the $\mathbf{Z}$-module $\mathbf{Z}$. Now $3 \in M_1 \cup M_2$ and $2 \in M_1 \cup M_2$ but $3 + 2 = 5 \notin M_1 \cup M_2$. Hence $M_1 \cup M_2$ cannot be a submodule of the $\mathbf{Z}$-module $\mathbf{Z}$.

(ii) Let $M_1 = \{(x, 0) \in \mathbf{R}^2\}$ and $M_2 = \{(0, y) \in \mathbf{R}^2\}$. Then $(1, 0) \in M_1$ and $(0, 1) \in M_2$ but $(1, 0) + (0, 1) = (1, 1) \notin M_1 \cup M_2 \Rightarrow M_1 \cup M_2$ is not a subgroup of $\mathbf{R}^2 \Rightarrow M_1 \cup M_2$ cannot be a submodule of the $\mathbf{Z}$-module $\mathbf{R}^2$.

Let M be an R-module. If M_1 and M_2 are submodules of M, then $M_1 \cap M_2$ is the largest submodule of M contained in both M_1 and M_2.

The Theorem 9.2.3 leads to determine the smallest submodule containing a given subset S (including the possibility of $S = \emptyset$) of an R-module M.

Definition 9.2.2 Let M be an R-module and S be a subset of M. Then the submodule generated or spanned by S denoted by $\langle S \rangle$ is defined to be the smallest submodule of M containing S, i.e., $\langle S \rangle$ is the submodule of M obtained by the intersection of all submodules M_i of M containing S.

To determine the elements of $\langle S \rangle$, we introduce the concept of linear combinations of elements of S as defined in vector spaces.

Definition 9.2.3 Let M be an R-module and $S \neq \emptyset$ be a subset of M. Then an element $x \in M$ is said to be a linear combination of elements of S iff $\exists x_1, x_2, \ldots, x_n \in S$ and $r_1, r_2, \ldots, r_n \in R$ such that

$$x = \sum_{i=1}^{n} r_i x_i, \tag{9.1}$$

Let $C(S)$ denote the set of all linear combinations of elements of S in the form (9.1).

Theorem 9.2.3 *Let M be a unitary R-module and S be a subset of M. Then the submodule $\langle S \rangle$ generated by S is given by*

$$\langle S \rangle = \begin{cases} \{O_M\} & \text{if } S = \emptyset; \\ C(S) & \text{if } S \neq \emptyset, \end{cases}$$

where

$$C(S) = \left\{ \sum_{\text{finite sum}} r_i x_i : r_i \in R, \ x_i \in S \right\}.$$

Proof If $S = \emptyset$, then the smallest submodule of M containing S is the zero submodule $\{O_M\}$. Hence $\langle S \rangle = \{O_M\}$ if $S = \emptyset$. Next suppose that $S \neq \emptyset$. Then $C(S) \neq \emptyset$, since $1_R \in R$ and for $x \in S$, $x = 1_R x \in C(S)$. Hence $S \subseteq C(S)$. We claim that $C(S)$ is a submodule of M. Let $x, y \in C(S)$. Then $x = \sum_{i=1}^{n} r_i x_i$ and $y = \sum_{i=1}^{m} t_i y_i$ for some $r_i, t_i \in R$ and $x_i, y_i \in S$. Hence $x + y = \sum_{i=1}^{n} r_i x_i + \sum_{i=1}^{m} t_i y_i \in C(S)$ and $rx = r(r_1 x_1 + r_2 x_2 + \cdots + r_n x_n) = (rr_1)x_1 + (rr_2)x_2 + \cdots + (rr_n)x_n \in C(S) \Rightarrow C(S)$ is a submodule of M containing S. Finally, let N be a submodule of M such that $S \subseteq N$. If $x \in C(S)$, then x can be represented as $x = \sum_{i=1}^{n} r_i x_i, r_i \in R, x_i \in S$. Now $x_i \in S \Rightarrow x_i \in N \Rightarrow r_i x_i \in N, \forall i \Rightarrow \sum_{i=1}^{n} r_i x_i \in N$, since N is a submodule of $M \Rightarrow x \in N \ \forall x \in C(S) \Rightarrow C(S) \subseteq N \Rightarrow C(S)$ is the smallest submodule of M containing S. Again $\langle S \rangle$ is also the smallest submodule of M containing S. Consequently, $\langle S \rangle = C(S)$, if $S \neq \emptyset$. $\qquad \square$

Remark If the ring R does not contain the identity element, then

$$\langle S \rangle = \left\{ \sum_{\text{finite sum}} (n_i + r_i)x_i \text{ if } S \neq \emptyset, \text{ where } n_i \in \mathbf{Z}, r_i \in R \text{ and } x_i \in S \right\}.$$

Definition 9.2.4 An R-module M is said to be generated by a subset S iff $M = \langle S \rangle$ and S is said to be set of generators of M.

In particular, if $S = \{x_1, x_2, \ldots, x_n\}$ is a finite subset of M such that $M = \langle S \rangle$, then M is said to be finitely generated by S and hence if M is a unitary R-module, then

$$M = \left\{ \sum_{i=1}^{n} r_i x_i : r_i \in R \right\}.$$

If $S = \{x\}$, then $\langle \{x\} \rangle = \langle x \rangle$, the submodule generated by $\{x\}$ is given by

$$\langle x \rangle = \begin{cases} \{rx : r \in R\} & \text{if } M \text{ is a unitary } R\text{-module} \\ \{rx + nx : r \in R, \ n \in \mathbf{Z}\}, & \text{otherwise.} \end{cases}$$

$\langle x \rangle$ is called the cyclic submodule of M generated by $x \in M$.

Thus a unitary R-module M is finitely generated iff $M = M_1 + M_2 + \cdots + M_n$, where each M_i is cyclic i.e., $M_i = Rx_i, i = 1, 2, \ldots, n$ and $M_1 + M_2 + \cdots + M_n$ is the sum of submodules of M (see Definition 9.2.10). Then $\{x_1, x_2, \ldots, x_n\}$ is a set of generators for M.

Definition 9.2.5 Let M be an R-module. A subset S of M is said to be linearly dependent over R iff there exist distinct elements $x_1, x_2, \ldots, x_n \in S$ and elements $r_1, r_2, \ldots, r_n$ (not all zero) in R such that $r_1 x_1 + r_2 x_2 + \cdots + r_n x_n = O_M$. Otherwise, S is said to be linearly independent over R.

Remark (i) $\{O_M\}$ and any subset S (of M) containing O_M are linearly dependent over R.

(ii) If S is linearly dependent over R and T is any subset of M such that $S \subseteq T$, then T is also linearly dependent over R, i.e., any subset containing a linearly dependent set is also linearly dependent.

(iii) If S is linearly independent over R and T is any subset of M such that $T \subseteq S$, then T is also linearly independent over R, i.e., any subset contained in a linearly independent set is also linearly independent.

Definition 9.2.6 Let M be an R-module. A submodule N ($\neq M$) of M is said to be maximal iff for a submodule P of M such that $N \subseteq P \subseteq M$, either $P = N$ or $P = M$, i.e., there is no submodule P of M satisfying $N \subsetneq P \subsetneq M$.

Definition 9.2.7 A submodule N ($\neq \{O_M\}$) of M is said to be minimal iff for a submodule P of M such that $P \subseteq N$, either $P = \{O_M\}$ or $P = N$, i.e., the only submodules of M contained in N are $\{O_M\}$ and N.

Definition 9.2.8 A module M ($\neq \{O_M\}$) is said to be simple iff the only submodules of M are $\{O_M\}$ and M.

Theorem 9.2.4 *Let M be a unitary R-module. Then M is simple iff for every nonzero element $x \in M$, $M = Rx = \{rx : r \in R\}$, i.e., iff M is generated by $\{x\}$ for every $x \neq O_M$ in M.*

Proof Let M be a simple unitary R-module. Then for x ($\neq O_M$) in M, $x = 1_R x \in Rx \Rightarrow Rx \neq \emptyset$. Next let $rx, tx \in Rx$, where $r, t \in R$. Then $(rx + tx) = (r + t)x \in Rx$ and $r(tx) = (rt)x \in Rx$. Consequently, Rx is a submodule of M. Since for $x \neq O_M$, $x = 1_R x \in Rx$, $Rx \neq \{O_M\}$. Again, M being a simple R-module, $Rx = M$. Conversely, let $Rx = M$ for every non-zero $x \in M$. Suppose $A \neq \{O_M\}$ is a submodule of M. Then $\exists$ a non-zero element x in A such that $Rx \subseteq A$ i.e., $M \subseteq A$. Since $A \subseteq M$, it follows that $A = M$. Consequently, M is a simple R-module. $\square$

Corollary 1 *If R is a unitary ring, then R is a simple R-module iff R is a division ring.*

Proof Let R be a unitary ring. Then R is a unitary module over itself. Hence by Theorem 9.2.4, R is a simple R-module iff $R = Rx$ for every non-zero $x \in R$. Again $1_R \in R = Rx \Rightarrow 1_R = yx$ for some $y \in R \Rightarrow x$ has a left inverse in R. Thus it follows that every non-zero x in R has an inverse in R (see Proposition 2.3.2 of Chap. 2). Consequently, R is a division ring. Conversely, let R be a division ring. Then $\{O_M\}$ and R are only ideals of R and these are the only submodules of the R-module R. Hence R is simple. $\square$

Corollary 2 *Let V be a vector space over a field F. Then for every non-zero $x \in V$, the subspace $U_x = \{rx : r \in F\}$ is simple. In particular, every F-vector space F is simple.*

Theorem 9.2.5 *Let $M \neq \{O_M\}$ be a finitely generated R-module. Then each proper submodule of M is contained in a maximal submodule of M.*

Proof Similar to the proof of Theorem 5.3.6. □

Definition 9.2.9 Let M, N be (left) R-modules. Then their cartesian product $M \times N$ is a (left) R-module under addition and scalar multiplication defined in the usual way:

$$(x, y) + (s, t) = (x + s, y + t) \quad \text{and}$$

$$r(x, y) = (rx, ry) \quad \forall (x, y), (s, t) \in M \times N \text{ and } \forall r \in R.$$

The R-module $M \times N$ is called the direct product of R-modules M and N.

More generally, if $\{M_i\}_{i \in I}$ is any family of (left) R-modules, then $M = \prod_{i \in I} M_i$ is a (left) R-module in the usual way:

$$R \times M \to M, \quad \left(r, (x_i)_{i \in I}\right) \mapsto (r x_i)_{i \in I}.$$

The R-module $M = \prod_{i \in I} M_i$ is called the direct product of $\{M_i\}_{i \in I}$.
In particular, $R^2 = R \times R, \ldots, R^n = R \times R \times \cdots \times R$ (n factors), $R^\infty = \prod_1^\infty R_i$.

Definition 9.2.10 Let M be an R-module and N_1, N_2 be submodules of M. Then $N_1 + N_2 = \{n_1 + n_2 : n_1 \in N_1, n_2 \in N_2\}$ is called the sum of the submodules N_1 and N_2.

Proposition 9.2.1 *Let M be an R-module and N_1, N_2 submodules of M. Then $N_1 + N_2$ is the submodule generated by $N_1 \cup N_2$.*

Proof $O_M \in N_1 + N_2 \Rightarrow N_1 + N_2 \neq \emptyset$. Let $x, y \in N_1 + N_2$. Then there exist $n_1, n_3 \in N_1$ and $n_2, n_4 \in N_2$ such that $x = n_1 + n_2$ and $y = n_3 + n_4$. Now $x + y = (n_1 + n_2) + (n_3 + n_4) = (n_1 + n_3) + (n_2 + n_4)$ (by commutativity of $+$ in M) and $rx = rn_1 + rn_2 \in N_1 + N_2 \ \forall x, y \in N_1 + N_2$ and $\forall r \in R$. Hence $N_1 + N_2$ is a submodule of M. We claim that $N_1 \cup N_2 \subseteq N_1 + N_2$ and $N_1 + N_2$ is the smallest submodule of M containing $N_1 \cup N_2$. Again $n_1 \in N_1 \Rightarrow n_1 = n_1 + O_{N_2} \in N_1 + N_2 \Rightarrow N_1 \subseteq N_1 + N_2$. Similarly, $N_2 \subseteq N_1 + N_2$. Consequently, $N_1 \cup N_2 \subseteq N_1 + N_2$. Let N be the submodule of M generated by $N_1 \cup N_2$ and let $n_1 + n_2 \in N_1 + N_2$, where $n_1 \in N_1$ and $n_2 \in N_2$. Hence $N_1 \cup N_2 \subseteq N \Rightarrow n_1, n_2 \in N \Rightarrow n_1 + n_2 \in N \Rightarrow N_1 + N_2 \subseteq N \Rightarrow N_1 + N_2$ is the smallest submodule of M containing $N_1 \cup N_2$. Consequently, $\langle N_1 \cup N_2 \rangle = N_1 + N_2$. □

Corollary *If N_1 and N_2 are submodules of M, then $N_1 + N_2$ is the smallest submodule of M containing both N_1 and N_2.*

The concept of sum $N_1 + N_2$ of two submodules of M can be generalized for any family $\{N_i\}_{i \in I}$ of submodules of M:

$$\sum_{i \in I} N_i = \left\{ \sum_{i \in I} x_i : x_i \in N_i, x_i = O_M \text{ except for finitely many i's} \right\}.$$

This is a submodule of M containing each $N_i, i \in I$.

The submodule $\sum_{i \in I} N_i$ is called the sum of the family of submodules $\{N_i\}_{i \in I}$.
In particular, if $I = \{1, 2, \ldots, n\}$, then the submodule $\sum_{i=1}^{n} N_i = N_1 + N_2 + \cdots + N_n$ is called the sum of n submodules $N_1, N_2, \ldots, N_n$ of M.

We now proceed to define direct sum of two submodules of an R module in two different equivalent ways.

Definition 9.2.11 Let M and N be (left) R-modules. Then $P = M \times N$ is again a (left) R-module. The direct sum of the modules M and N denoted by $M \oplus N$ is defined by

$$M \oplus N = \{(x, O_N) + (O_M, y) : x \in M, y \in N\} = \{(x, y) \in P : x \in M, y \in N\}.$$

Definition 9.2.12 Let M be an R-module and A, B be two submodules of M. Then M is said to be the direct sum of A and B denoted by $M = A \oplus B$ iff $M = A + B$ and $A \cap B = \{O_M\}$. If $M = A \oplus B$, then A and B are called the direct summands of M.

Definition 9.2.12 leads to the following proposition.

Proposition 9.2.2 *Let M be an R-module and A, B be its two submodules. Then $M = A \oplus B$ iff every element $x \in M$ can be expressed uniquely as $x = a + b, a \in A$, $b \in B$.*

Proof $M = A \oplus B \Leftrightarrow M = A + B$ and $A \cap B = \{O_M\}$. Then $x \in M$ can be expressed as $x = a + b, a \in A, b \in B$. If $x = a + b = c + d$, where $a, c \in A$ and $b, d \in B$, then $a - c = d - b \Rightarrow a - c \in A$ and $d - b \in B$ are such that $a - c = d - b \in A \cap B = \{O_M\} \Rightarrow a = c$ and $d = b$. The converse part is similar. $\qquad\square$

Corollary *Let M be an R-module and A, B be two submodules of M. Then $A \cap B = \{O_M\}$ iff every element $x \in A + B$ can be expressed uniquely as $x = a + b$, where $a \in A$ and $b \in B$.*

Definition 9.2.13 An R-module M is called the direct sum of a family of submodules $\{M_i\}_{i \in I}$ denoted by $M = \bigoplus_{i \in I} M_i$, iff $M = \sum_{i \in I} M_i$ and every element $x \in M$ can be expressed uniquely as $x = \sum_{i \in I} x_i$, $x_i \in M_i$, $x_i = O_M$ except for finitely many i's. A non-zero module M is said to be semisimple iff it can be decomposed into a direct sum of a family of minimal submodules of M.

Clearly, a local ring is semisimple iff it is a division ring.

Remark 1 In general, $\bigoplus_{i \in I} M_i \subseteq \prod_{i \in I} M_i$. Equality holds iff I is a finite set.

Remark 2 Remark 1 shows that Definition 9.2.11 and Definition 9.2.12 are equivalent.

Definition 9.2.14 A submodule A of an R-module M is said to be a *direct summand* of M iff there exists a submodule B of M such that $M = A \oplus B$. Such a submodule B of M is called a supplement or complement of A in M.

Remark (i) Every subspace of an inner product space is a direct summand.

(ii) Every submodule of a module is not in general a direct summand.

(iii) If a submodule of a module M is a direct summand, its supplement in M may not be unique.

(iv) Consider the **Z**-module **Z**. A non-zero subgroup $\langle n \rangle$ of **Z** is not a direct summand, because a supplement which is infinite cyclic should be isomorphic to the quotient group $\mathbf{Z}/\langle n \rangle$ $(\simeq \mathbf{Z}_n)$, which is not possible.

(v) Consider the vector space $\mathbf{R}^2$ over **R**. Let l, m, n be three distinct lines in $\mathbf{R}^2$ through its origin $(0, 0)$. Then l, m, n are subspaces of $\mathbf{R}^2$ such that both m and n are supplements of l in $\mathbf{R}^2$. Hence $m \neq n \Rightarrow$ supplement of l is not unique.

Theorem 9.2.6 *Let M be an R-module and $s(M)$ be the set of all submodules of M. Then $s(M)$ is a modular lattice under set inclusion.*

Proof Let M be an R-module and P, Q be submodules of M. Then $P + Q$ is the smallest submodule of M containing both P and Q, i.e., $P \subseteq P + Q$ and $Q \subseteq P + Q$. Again $P \cap Q$ is the largest submodule of M contained in both P and Q, i.e., $P \cap Q \subseteq P$ and $P \cap Q \subseteq Q$. Thus in the set $s(M)$ of all submodules of M, partially ordered by set inclusion, every pair of elements P and Q has $P + Q$ as lub (or supremum) and $P \cap Q$ as glb (or infimum), i.e. their join $P \vee Q = P + Q$ and meet $P \wedge Q = P \cap Q$. Hence $s(M)$ is a lattice under set inclusion. Next we claim that this lattice is modular. Let P, Q, N be submodules of M such that $N \subseteq P$. Then we prove that $P \cap (Q + N) = (P \cap Q) + N$. Now $N \subseteq P \Rightarrow P + N = P$. Again, $(P \cap Q) + N \subseteq P + N$ and $(P \cap Q) + N \subseteq Q + N \Rightarrow (P \cap Q) + N \subseteq (P + N) \cap (Q + N) = P \cap (Q + N)$. To prove the reverse inclusion, let $x \in P \cap (Q + N)$. Then $x \in P$ and $x \in Q + N$. Hence $\exists q \in Q$ and $n \in N$ such that $x = q + n$. Since $N \subseteq P$, we have $n \in P$ and hence $q = x - n \in P$. Consequently, $q \in P \cap Q$. Hence $x = q + n \in (P \cap Q) + N$. Thus $P \cap (Q + N) \subseteq (P \cap Q) + N$. Hence it follows that $P \cap (Q + N) = (P \cap Q) + N$. $\square$

Definition 9.2.15 Let M be a (left) R-module and S be a non-empty subset of M. Then the annihilator of S in R denoted by $\mathrm{Ann}(S)$ is defined by

$$\mathrm{Ann}(S) = \{r \in R : rx = O_M, \forall x \in S\}.$$

Proposition 9.2.3 *Let M be a (left) R-module and S be a non-empty subset of M. Then*

(a) $\mathrm{Ann}(S)$ *is a left ideal of R;*
(b) *if S is a submodule of M, then $\mathrm{Ann}(S)$ is a two-sided ideal of R.*

Proof (a) $O_R x = O_M, \forall x \in S \Rightarrow O_R \in \mathrm{Ann}(S) \Rightarrow \mathrm{Ann}(S) \neq \emptyset$. Let $a, b \in \mathrm{Ann}(S)$. Then $a, b \in R$ and $ax = O_M = bx \, \forall x \in S$. Hence $(a - b)x = O_M \, \forall x \in S \Rightarrow a - b \in \mathrm{Ann}(S)$. Again $(ra)x = r(ax) = rO_M = O_M, \forall x \in S, \forall r \in R \Rightarrow ra \in \mathrm{Ann}(S)$, $\forall r \in R$. Consequently, $\mathrm{Ann}(S)$ is a left ideal of R.

(b) Let S be a submodule of M. Then $rx \in S$, $\forall r \in R$ and $\forall x \in S$. Hence for $a \in \mathrm{Ann}(S)$, $a(rx) = O_M \Rightarrow (ar)x = O_M \ \forall x \in S \Rightarrow ar \in \mathrm{Ann}(S) \Rightarrow \mathrm{Ann}(S)$ is also a right ideal of R. Thus $\mathrm{Ann}(S)$ is a two-sided ideal of R by (a). $\qquad \square$

Corollary 1 *If M is an R- module, then $I = \mathrm{Ann}(M)$ (i.e., $IM = 0$) is a two-sided ideal of R.*

Corollary 2 *If R is a commutative ring with 1 and $I = \mathrm{Ann}(M)$ is a maximal ideal of R, then M is a vector space over the field R/I.*

Remark Let $I = \mathrm{Ann}(M)$. Then R/I is a ring, since I is a two-sided ideal of R by (b). Define scalar multiplication $R/I \times M \to M$, given by $(r + I) \cdot x = rx$.

Then M is both an R-module as well as an R/I module, but their scalar multiplications on M coincide. Thus an R-module M is also an R/I-module and vice versa.

Exercise The intersection of all annihilators of all non-zero elements of a simple R-module M is a two-sided ideal of R.

[*Hint.* $\mathrm{Ann}(M)$ is a two-sided ideal of R and $\bigcap_{O_M \neq x \in M} \mathrm{Ann}(x) = \mathrm{Ann}(M)$.]

9.3 Module Homomorphisms

We continue in this section the study of submodules.

Structure preserving mappings, called homomorphisms, play an important role in group theory as well as in ring theory. It is an essential and basic task to extend this concept to module theory.

Definition 9.3.1 Let M and N be (left) R-modules. A mapping $f : M \to N$ is called an R-homomorphism (or R-morphism) iff

(i) $f(x + y) = f(x) + f(y)$;
(ii) $f(rx) = rf(x) \ \forall x, y \in M$ and $\forall r \in R$.

In particular, if R is a field, then an R-homomorphism is as usual called a linear transformation of vector spaces. Module homomorphisms of special character, carry special names, like special group homomorphisms.

Definition 9.3.2 Let $f : M \to N$ be an R-homomorphism for R-modules. Then f is called an

(1) R-monomorphism iff f is injective;
(2) R-epimorphism iff f is surjective;
(3) R-isomorphism iff f is bijective;
(4) R-endomorphism iff $M = N$, i.e., iff f is an R-homomorphism of M into itself;
(5) R-automorphism iff $M = N$ and f is an R-isomorphism of M onto itself.

The existence of an R-isomorphism between R-modules M and N asserts that M and N are, in a significant sense, the same, i.e., M and N have the same module structures. In that case we say that the R-modules M and N are R-isomorphic and write $M \cong N$.

For the time being, if we forget the scalar multiplication on M, then f is considered a group homomorphism $f : (M, +) \to (N, +)$. Hence it preserves additive zero and additive inverse. More precisely:

Proposition 9.3.1 *If $f : M \to N$ is an R-homomorphism, then $f(O_M) = O_N$ and $f(-x) = -f(x)$, $\forall x \in M$.*

Proof $f(O_M) = f(O_M + O_M) = f(O_M) + f(O_M) \Rightarrow f(O_M) = O_N$. Again $f(x) + f(-x) = f(x + (-x)) = f(O_M) = O_N \Rightarrow f(-x) = -f(x), \forall x \in M$. $\square$

Definition 9.3.3 Let $f : M \to N$ be an R-homomorphism. Then kernel of f denoted by ker f is defined to be the kernel of f as a group homomorphism $f : (M, +) \to (N, +)$, i.e., ker $f = \{x \in M : f(x) = O_N\}$ and the image of f denoted by Im f is the set Im $f = f(M) = \{f(x) : x \in M\}$.

It follows from the definitions that

Proposition 9.3.2 *Let $f : M \to N$ be an R-homomorphism. Then*

(a) ker f *is a submodule of M;*
(b) f *is an R-monomorphism iff ker $f = \{O_M\}$;*
(c) *for a submodule A of M, $f(A)$ is a submodule of N;*
(d) Im $f = f(M)$ *is a submodule of N.*

Remark (i) A necessary and sufficient condition for an R-homomorphism $f : M \to N$ to be an R-monomorphism is that ker f is as small as possible as a submodule of M, namely, ker $f = \{O_M\}$.

(ii) On the other hand, a necessary and sufficient condition for an R-homomorphism $f : M \to N$ is an R-epimorphism is that Im f is as large as possible as a submodule of N, namely, Im $f = N$.

Example 9.3.1 (i) Every homomorphism of abelian groups is a $\mathbf{Z}$-module homomorphism.

To show this let M and N be abelian groups. Then M and N are regarded as $\mathbf{Z}$-modules. Let $f : M \to N$ be a group homomorphism. Then $f(nx) = nf(x)$ by induction on $n \in \mathbf{N}^+$. Clearly, $f(nx) = nf(x)$, $\forall n \in \mathbf{Z}$. Hence f is a $\mathbf{Z}$-module homomorphism.

(ii) Let M be an R-module. Then the identity mapping $1_M : M \to M$ is an R-automorphism.

(iii) Let M be an R-module and n be a positive integer. Then the mapping $p_t : M^n \to M$, defined by $p_t(x_1, x_2, \ldots, x_t, \ldots, x_n) = x_t$ is an R-epimorphism, called the tth projection mapping of M^n onto M; on the other hand, the mapping $i_t : M \to$

M^n, defined by $i_t(x) = (0, 0, \ldots, x, \ldots, 0)$ is an R-monomorphism, called the tth injection mapping, for $t = 1, 2, \ldots, n$, where x is at the tth position.

(iv) The mapping $f : M_n(\mathbf{R}) \to \mathbf{R}$, defined by $f(A) = \text{trace } A$, the trace of the square matrix A of order n over $\mathbf{R}$, is an $\mathbf{R}$-homomorphism.

(v) Let $S_n(\mathbf{R}) = \{A \in M_n(\mathbf{R}) : A \text{ is symmetric}\}$. Then the mapping $f : M_n(\mathbf{R}) \to S_n(\mathbf{R})$, defined by $f(A) = A + A^t$ is an $\mathbf{R}$-epimorphism.

Like group homomorphisms, we can define the composite of R-homomorphisms having the following properties:

Proposition 9.3.3 *If $f : M \to N$ and $g : N \to P$ are R-homomorphisms, then their composite mapping $g \circ f : M \to P$, defined by $(g \circ f)(x) = g(f(x))$, $\forall x \in M$ is also an R-homomorphism satisfying the following properties*:

(a) *f and g are R-epimorphisms imply $g \circ f$ is also so*;
(b) *f and g are R-monomorphisms imply $g \circ f$ is also so*;
(c) *$g \circ f$ is an R-epimorphism implies g is also so*;
(d) *$g \circ f$ is an R-monomorphism implies f is also so*.

Proof For $x, y \in M$ and $r \in R$, $(g \circ f)(x + y) = g(f(x + y)) = g(f(x) + f(y)) = g(f(x)) + g(f(y)) = (g \circ f)(x) + (g \circ f)(y)$ and $(g \circ f)(rx) = g(f(rx)) = g(rf(x)) = rg(f(x)) = r(g \circ f)(x)$. Hence $g \circ f : M \to P$ is an R-homomorphism.

(a)–(d): Left as exercises. $\square$

Proposition 9.3.4 *Let M and N be R-modules and $f : M \to N$ be an R-homomorphism. Then for any submodule B of N, the set $f^{-1}(B)$ defined by*

$$f^{-1}(B) = \{x \in M : f(x) \in B\} \quad \text{is a submodule of } M.$$

Proof $f(O_M) = O_N \in B \Rightarrow O_M \in f^{-1}(B) \Rightarrow f^{-1}(B) \neq \emptyset$.

Now $x, y \in f^{-1}(B) \Rightarrow f(x), f(y) \in B \Rightarrow f(x - y) = f(x) - f(y) \in B$, since B is a submodule of $N \Rightarrow x - y \in f^{-1}(B)$. Similarly, for $r \in R$ and $x \in f^{-1}(B)$, $rx \in f^{-1}(B)$. Consequently, $f^{-1}(B)$ is a submodule of M. $\square$

Proposition 9.3.5 *Let $f : M \to N$ be an R-homomorphism.*

(a) *If A is a submodule of M, then $f^{-1}(f(A)) = A + \ker f$*;
(b) *If B is a submodule of N, then $f(f^{-1}(B)) = B \cap \text{Im } f$.*

Proof (a) $x \in A \Rightarrow f(x) \in f(A) \Rightarrow x \in f^{-1}(f(A)) \Rightarrow A \subseteq f^{-1}(f(A))$. Again $y \in \ker f \Rightarrow f(y) = O_N = f(O_M) \in f(A) \Rightarrow y \in f^{-1}(f(A)) \Rightarrow \ker f \subseteq f^{-1}(f(A))$. Since $f^{-1}(f(A))$ is a submodule of M, it follows that $A + \ker f \subseteq f^{-1}(f(A))$. To obtain the reverse inclusion, let $x \in f^{-1}(f(A))$. Then $f(x) \in f(A) \Rightarrow f(x) = f(a)$ for some $a \in A \Rightarrow f(x - a) = f(x) - f(a) = O_N \Rightarrow x - a \in \ker f \Rightarrow x \in a + \ker f \subseteq A + \ker f \Rightarrow f^{-1}(f(A)) \subseteq A + \ker f$. Hence it follows that $f^{-1}(f(A)) = A + \ker f$.

(b) $a \in f(f^{-1}(B)) \Rightarrow a = f(x)$ for some $x \in f^{-1}(B) \Rightarrow f(x) \in B \Rightarrow a \in B \Rightarrow f(f^{-1}(B)) \subseteq B$. Moreover, $f(f^{-1}(B)) \subseteq f(M) \Rightarrow f(f^{-1}(B)) \subseteq \operatorname{Im} f$. Hence it follows that $f(f^{-1}(B)) \subseteq B \cap \operatorname{Im} f$. For the reverse inclusion, let $y \in B \cap \operatorname{Im} f$. Then $y \in B$ and $y \in \operatorname{Im} f$. Now $y \in \operatorname{Im} f \Rightarrow y = f(m)$ for some $m \in M \Rightarrow f(m) = y \in B \Rightarrow m \in f^{-1}(B) \Rightarrow f(m) \in f(f^{-1}(B)) \Rightarrow y \in f(f^{-1}(B)) \Rightarrow B \cap \operatorname{Im} f \subseteq f(f^{-1}(B))$. Hence it follows that $f(f^{-1}(B)) = B \cap \operatorname{Im} f$. $\qquad \square$

Corollary (a) *If $f : M \to N$ is an R-monomorphism and A is a submodule of M, then $f^{-1}(f(A)) = A$;*

(b) *If $f : M \to N$ is an R-epimorphism and B is a submodule of N, then $f(f^{-1}(B)) = B$.*

Theorem 9.3.1 (Schur's Lemma) *Let M, N be R-modules. Then*

(a) *M is a simple R-module implies any non-zero R-homomorphism $f : M \to N$ is an R-monomorphism;*

(b) *N is a simple R-module implies any non-zero R-homomorphism $f : M \to N$ is an R-epimorphism;*

(c) *M is a simple R-module implies $\operatorname{End}(M)$ is a division ring.*

Proof (a) $\ker f$ is a submodule of the R-module M. Hence M is simple $\Rightarrow \ker f = \{O_M\}$ or $\ker f = M$. Since f is non-zero, $\ker f \neq M$. Hence $\ker f = \{O_M\} \Rightarrow f$ is an R-monomorphism.

(b) $\operatorname{Im} f$ is a submodule of N. Hence N is simple $\Rightarrow \operatorname{Im} f = N$ or $\operatorname{Im} f = \{O_N\}$. Since f is non-zero, $\operatorname{Im} f \neq \{O_N\}$. Hence $\operatorname{Im} f = N \Rightarrow f$ is an R-epimorphism.

(c) In particular, let $f : M \to M$ be a non-zero R-homomorphism. Hence M is simple $\Rightarrow f$ is an automorphism by (a) and (b) $\Rightarrow f$ is invertible in the ring $\operatorname{End}(M) \Rightarrow \operatorname{End}(M)$ is a division ring. $\qquad \square$

Definition 9.3.4 Let R be a commutative ring and M be an R-module. Then for each $a \in R$, $\exists$ an R-endomorphism $f_a : M \to M$, defined by $f_a(x) = ax$, called the homothecy defined by a.

Proposition 9.3.6 *The set $H = \{f_a : a \in R$ and f_a is the homothecy defined by $a\}$ is a subring of the ring $\operatorname{End}(M)$.*

Proof Consider the mapping $\psi : R \to \operatorname{End}(M)$, defined by $\psi(a) = f_a$, where $f_a : M \to M$ is defined by $f_a(x) = ax$.

Then $\psi(a+b) = \psi(a) + \psi(b)$ and $\psi(ab) = \psi(a)\psi(b)$, $\forall a, b \in R \Rightarrow \psi$ is a ring homomorphism $\Rightarrow H = \psi(R)$ is a subring of $\operatorname{End}(M)$. $\qquad \square$

Given an R-module M and an R-endomorphism $f : M \to M$, can we express M as the direct sum of $\operatorname{Im} f$ and $\ker f$? The solution of the following problem gives its answer.

Problem Let $f : M \to M$ be an R-endomorphism such that $f \circ f = f$. Is $M = \operatorname{Im} f \oplus \ker f$ true?

Solution $\ker f$ and $\operatorname{Im} f$ are both submodules of $M \Rightarrow \operatorname{Im} f + \ker f \subseteq M$. We now show that $\operatorname{Im} f \oplus \ker f = M$. For $m \in M, m = f(m) + (m - f(m))$. Again $f(m - f(m)) = f(m) - (f \circ f)(m) = f(m) - f(m) = O_M \Rightarrow m - f(m) \in \ker f \Rightarrow m \in \operatorname{Im} f + \ker f \Rightarrow M \subseteq \operatorname{Im} f + \ker f$. Hence $M = \operatorname{Im} f + \ker f$. Again $x \in \operatorname{Im} f \cap \ker f \Rightarrow f(x) = O_M$ and $\exists y \in M$ such that $x = f(y)$. Now $f(f(y)) = f(x) = O_M \Rightarrow (f \circ f)(y) = O_M \Rightarrow f(y) = O_M \Rightarrow x = O_M \Rightarrow \operatorname{Im} f \cap \ker f = \{O_M\}$. Hence $M = \operatorname{Im} f \oplus \ker f$.

9.4 Quotient Modules and Isomorphism Theorems

In this section we continue discussion of submodules and homomorphisms. The construction of quotient modules is analogous to that of quotient groups.

We now prescribe a method of construction for new modules from the old ones. The most significant role played by submodules of a module is to yield other modules known as quotient modules which are associated with the parent module in a natural way. Moreover, we prove isomorphism theorems for modules.

Let M be an R-module and N be a submodule of M. Then $(N, +)$ is a normal subgroup of the abelian group $(M, +)$ and hence the quotient group $(M/N, +)$ is also an abelian group under the composition $+$, defined by

$$(x + N) + (y + N) = (x + y) + N \quad \forall x, y \in M.$$

Definition 9.4.1 Given an R-module M and a submodule N of M, the abelian group $(M/N, +)$ inherits an R-module structure from M:
$R \times M/N \to M/N$, defined by

$$(r, x + N) \mapsto rx + N, \quad \text{i.e.,} \quad r \cdot (x + N) = rx + N.$$

The R-module M/N is called the quotient module of M by N. The natural map $\pi : M \to M/N$, defined by $\pi(x) = x + N \ \forall x \in M$ is an R-epimorphism with $\ker \pi = N$ and is called the canonical R-epimorphism or natural R-homomorphism.

Theorem 9.4.1 (Correspondence Theorem) *Let M be an R-module and N be a submodule of M. Then there exists an inclusion preserving bijection from the set $s(M)$ of all submodules of M containing N to the set $s(M/N)$ of all submodules of M/N.*

Proof Let $A \in s(M)$. Then A is a submodule of M such that $N \subseteq A \subseteq M$. Again $A/N = \{a + N : a \in A\}$ is a submodule of M/N. We now define a mapping $f : s(M) \to s(M/N)$ by

$$f(A) = A/N \quad \forall A \in s(M).$$

Fig. 9.1 Commutativity of
the triangular diagram

Fig. 9.2 Commutativity of
the rectangular diagram

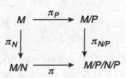

f is injective: Let $A, B \in s(M)$ be such that $f(A) = f(B)$. Then $A/N = B/N \Rightarrow$ given $a \in A$, $\exists b \in B$ such that $a + N = b + N$. Hence $a - b \in N \Rightarrow a - b = n$ for some $n \in N$. Thus $a = b + n \in B$, since B is a submodule of M such that $N \subseteq B$. Hence $A \subseteq B$. Similarly $B \subseteq A$. Thus $A = B \Rightarrow f$ is injective.

f is surjective: Let $P \in s(M/N)$. Then P is a submodule of M/N. Consider $P^* = \{x \in M : x + N \in P\}$. Since $P \neq \emptyset$, $P^* \neq \emptyset$. Again $\forall x \in N$, $x + N = 0 + N \in P \Rightarrow \forall x \in N$, $x \in P^* \Rightarrow N \subseteq P^*$. Clearly, P^* is a submodule of M such that $N \subseteq P^* \Rightarrow P^* \in s(M)$. We claim that $f(P^*) = P$. Now $x + N \in P^*/N \Leftrightarrow x \in P^* \Leftrightarrow x + N \in P \Leftrightarrow P^*/N = P \Rightarrow f(P^*) = P \Rightarrow f$ is surjective.

Consequently, f is a bijection. Finally, let $A, B \in s(M)$ be such that $A \subseteq B$. Now $a + N \in A/N \Rightarrow a \in A \subseteq B \Rightarrow a + N \in B/N \Rightarrow A/N \subseteq B/N \Rightarrow f(A) \subseteq f(B)$. Thus $A \subseteq B \Rightarrow f(A) \subseteq f(B)$. $\qquad\square$

Corollary *Any submodule of M/N is of the form A/N for some submodule A of M such that $N \subseteq A$.*

Theorem 9.4.2 (Fundamental Homomorphism Theorem or First Isomorphism Theorem) *Let M and N be R-modules and $f : M \to N$ be an R-homomorphism with $K = \ker f$. Then there exists a unique R-isomorphism $\tilde{f} : M/K \to \operatorname{Im} f$ such that $f = \tilde{f} \circ \pi$, i.e., such that the diagram in Fig. 9.1 is commutative, where $\pi : M \mapsto M/K$, $x \mapsto x + K$ is the natural R-epimorphism.*

Proof Similar to that of groups. $\qquad\square$

Corollary (Epimorphism Theorem) *Let $f : M \to N$ be an R-epimorphism of R-modules with $\ker f = K$. Then there exists an R-isomorphism $\tilde{f} : M/K \to N$ of R-modules.*

Proof f is an R-epimorphism $\Rightarrow f(M) = N \Rightarrow \operatorname{Im} f = N \Rightarrow \tilde{f}$ is an R-isomorphism by Theorem 9.4.2. $\qquad\square$

Theorem 9.4.3 (Second Isomorphism Theorem or Isomorphism Theorem of Quotient of Quotient) *Let M be an R-module and P, N be submodules of M such that $P \subseteq N$. Then N/P is a submodule of M/P and there exists a unique natural R-isomorphism $\pi : M/N \to M/P/N/P$ such that the diagram in Fig. 9.2 is* *commutative, i.e.,* $\pi_{N/P} \circ \pi_P = \pi \circ \pi_N$, *where*

$$\pi_P : M \to M/P, \quad x \mapsto x + P \quad \forall x \in M,$$

$$\pi_N : M \to M/N, \quad x \mapsto x + N \quad \forall x \in M, \quad and$$

$$\pi_{N/P} : M/P \to M/P/N/P, \quad y \mapsto y + N/P, \quad \forall y \in M/P.$$

Proof N is a submodule of $M \Rightarrow N \neq \emptyset \Rightarrow N/P \neq \emptyset$. Now $x + P, y + P \in N/P \Rightarrow x, y \in N \Rightarrow x + y \in N$ and $rx \in N \ \forall r \in R \Rightarrow (x + y) + P \in N/P$ and $rx + P \in N/P \ \forall r \in R \Rightarrow N/P$ is a submodule of M/P. Define the natural mapping $\pi : M/N \to M/P/N/P$ by the rule

$$\pi(x + N) = (x + P) + N/P.$$

Now

$$x + N = y + N(x, y \in M) \quad \Leftrightarrow \quad x - y \in N$$

$$\Leftrightarrow \quad (x - y) + P \in N/P \quad \Leftrightarrow \quad (x + P) - (y + P) \in N/P$$

$$\Leftrightarrow \quad (x + P) + N/P = (y + P) + N/P \quad \Leftrightarrow \quad \pi(x + N) = \pi(y + N).$$

Thus π is well defined and injective.

Again for $(x + P) + N/P \in M/P/N/P$, $\exists$ an element $x + N \in M/N$ such that $\pi(x + N) = (x + P) + N/P$. Thus π is surjective. Hence π is a bijection. Finally, it is easy to show that π is an R-homomorphism. Consequently, π is a natural R-isomorphism.

Uniqueness of π: Let $\tilde{\pi} : M/N \to M/P/N/P$ be an R-isomorphism (by taking $\tilde{\pi}$ in place of π) with commutating diagram in Fig. 9.2 i.e., $\tilde{\pi} \circ \pi_N = \pi_{N/P} \circ \pi_P$. Then

$$\tilde{\pi}(x + N) = \tilde{\pi}\big(\pi_N(x)\big) = (\tilde{\pi} \circ \pi_N)(x) = (\pi_{N/P} \circ \pi_P)(x)$$

$$= \pi_{N/P}\big(\pi_P(x)\big) = \pi_{N/P}(x + P) = (x + P) + N/P$$

$$= \pi(x + N), \quad \forall x + N \in M/N$$

$$\Rightarrow \quad \tilde{\pi} = \pi \quad \Rightarrow \quad \text{uniqueness of } \pi. \qquad \square$$

Theorem 9.4.4 (Third Isomorphism Theorem or Theorem of Quotient of Quotient of a sum) *Let M be an R-module. If A and B are submodules of M, then*

(i) *the quotient modules $A/A \cap B$ and $(A + B)/B$ are R-isomorphic and*
(ii) *the quotient modules $B/A \cap B$ and $(A + B)/A$ are R-isomorphic.*

Proof (i) Define a mapping $f : A/A \cap B \to (A + B)/B$ by $f(a + A \cap B) = a + B$.

f *is well defined*: For $a, x \in A$, $a + A \cap B = x + A \cap B \Rightarrow a - x \in A \cap B \subseteq B \Rightarrow a - x \in B \Rightarrow a + B = x + B \Rightarrow f$ is independent of the representative a.

f is injective: $f(a + A \cap B) = f(x + A \cap B) \Rightarrow a + B = x + B \Rightarrow a - x \in B$.
Again $a, x \in A \Rightarrow a - x \in A$. Hence $a - x \in A \cap B \Rightarrow a + A \cap B = x + A \cap B$.

f is surjective: From the definition of f, it follows that f is surjective.
Clearly, f is an R-homomorphism. Consequently, f is an R-isomorphism.

(ii) Define a mapping $g : B/A \cap B \rightarrow (A + B)/A$ by $g(b + A \cap B) = b + A$.
Then as in (i), g is well defined and an R-isomorphism. $\qquad\qquad \square$

9.5 Modules of Homomorphisms

In this section R denotes a commutative ring with 1_R and $\mathrm{Hom}_R(M, N)$ denotes
the set of all R-homomorphisms from the R-module M to the R-module N. We
now continue discussion of module homomorphisms and study the structure of
$\mathrm{Hom}_R(M, N)$.

Proposition 9.5.1 *Let R be a commutative ring. Then for R-modules M and N,*
$\mathrm{Hom}_R(M, N)$ *is an R-module.*

Proof For $f, g \in \mathrm{Hom}_R(M, N)$, define $f + g$ by the rule $(f + g)(x) = f(x) + g(x)$,
$\forall x \in M$. Then $f + g : M \rightarrow N$ is an R-homomorphism such that $(\mathrm{Hom}_R(M, N), +)$
is an abelian group. Define an external law of composition $\mu : R \times \mathrm{Hom}_R(M, N) \rightarrow$
$\mathrm{Hom}_R(M, N)$ by $\mu(r, f) = rf$, where $rf : M \rightarrow N$ is defined by $(rf)(x) = rf(x)$
$\forall x \in M$. The composition μ is well defined, because

$$(rf)(x + y) = r\big(f(x + y)\big) = r\big(f(x) + f(y)\big) = rf(x) + rf(y)$$

$$= (rf)(x) + (rf)(y) \quad \forall x, y \in M \quad \text{and}$$

$$(rf)(sx) = r\big(f(sx)\big) = r\big(sf(x)\big) = (rs)f(x) = (sr)f(x),$$

since R is commutative by hypothesis

$$= s(rf(x)) = s(rf)(x) \quad \forall x \in M \text{ and } \forall r, s \in R.$$

Moreover, $r(f + g) = rf + rg$; $(r + s)f = rf + sf$ and $r(sf) = (rs)f$ $\forall r, s \in R$
and $\forall f, g \in \mathrm{Hom}_R(M, N)$. Consequently, $\mathrm{Hom}_R(M, N)$ is an R-module. $\qquad \square$

Remark In absence of commutativity of R, $\mathrm{Hom}_R(M, N)$ fails to be an R-module.
Thus the abelian group $\mathrm{Hom}_R(M, N)$ is not in general an R-module.

We now examine the particular case when $N = R$. We can endow $\mathrm{Hom}_R(M, R)$
the structure of a right R-module as follows: $(f + g)(x) = f(x) + g(x)$ $\forall f, g \in$
$\mathrm{Hom}_R(M, R)$ and $\forall x \in M$; $\mu : \mathrm{Hom}_R(M, R) \times R \rightarrow \mathrm{Hom}_R(M, R)$ is defined by
$\mu(f, r)(x) = (fr)(x) = f(x)r$ $\forall f \in \mathrm{Hom}_R(M, R)$ and $\forall r \in R$. Then $(fr)(x +$
$y) = f(x + y)r = (f(x) + f(y))r = f(x)r + f(y)r = (fr)(x) + f(r)y$ $\forall x, y \in$
M and $(fr)(sx) = f(sx)r = sf(x)r = s(fr)(x)$ $\forall x \in M$ and $\forall r, s \in R$. Hence

the group homomorphism fr is indeed an R-homomorphism, and so belongs to $\mathrm{Hom}_R(M, R)$.

Definition 9.5.1 If M is a left R-module, then the dual module of M is the right R-module $M^d = \mathrm{Hom}_R(M, R)$. The elements of M^d are called linear functionals (or linear forms) on M. Again we can obtain the dual of the right R-module M^d, which is the left R-module $(M^d)^d$. We call it the bidual of M and denote it by M^{dd}.

Proposition 9.5.2 *Given a commutative ring R and a fixed R-module M, the R-homomorphism $f : N \to P$ induces*

(a) *an R-homomorphism $f_* : \mathrm{Hom}_R(M, N) \to \mathrm{Hom}_R(M, P)$ defined by $f_*(\alpha) = f \circ \alpha \; \forall \alpha \in \mathrm{Hom}_R(M, N)$;*
(b) *an R-homomorphism $f^* : \mathrm{Hom}_R(P, M) \to \mathrm{Hom}_R(N, M)$ defined by $f^*(\beta) = \beta \circ f \; \forall \beta \in \mathrm{Hom}_R(P, M)$.*

Proof (a) $f_*(\alpha + \gamma) = f \circ (\alpha + \gamma)$ by definition.
    Now

$$\big(f \circ (\alpha + \gamma)\big)(x) = f\big((\alpha + \gamma)(x)\big) = f\big(\alpha(x) + \gamma(x)\big)$$
$$= f\big(\alpha(x)\big) + f\big(\gamma(x)\big) = (f \circ \alpha)(x) + (f \circ \gamma)(x)$$
$$= (f \circ \alpha + f \circ \gamma)(x) \quad \forall x \in M$$
$$\Rightarrow \quad f \circ (\alpha + \gamma) = f \circ \alpha + f \circ \gamma$$
$$\Rightarrow \quad f_*(\alpha + \gamma) = f_*(\alpha) + f_*(\gamma) \quad \forall \alpha, \gamma \in \mathrm{Hom}_R(M, N).$$

Similarly, $f_*(r\alpha) = rf_*(\alpha) \; \forall r \in R, \; \forall \alpha \in \mathrm{Hom}_R(M, N)$. Hence f_* is an R-homomorphism.
    (b) Proceed as in (a).   □

Corollary *For the identity R-automorphism $I_N : N \to N$,*

(a) $1_{N*} : \mathrm{Hom}_R(M, N) \to \mathrm{Hom}_R(M, N)$ *is the identity R-automorphism.*
(b) $1_N^* : \mathrm{Hom}_R(N, M) \to \mathrm{Hom}_R(N, M)$ *is also the identity R-automorphism.*

Proposition 9.5.3 *Let R be a commutative ring and M, N, P be R-modules. If $f : M \to N$ and $g : N \to P$ are R-homomorphisms, then for any R-module A,*

(i) $(g \circ f)_* : \mathrm{Hom}_R(A, M) \to \mathrm{Hom}_R(A, P)$ *is an R-homomorphism such that $(g \circ f)_* = g_* \circ f_*$;*
(ii) $(g \circ f)^* : \mathrm{Hom}_R(P, A) \to \mathrm{Hom}_R(M, A)$ *is an R-homomorphism such that $(g \circ f)^* = f^* \circ g^*$.*

Proof (i) $g \circ f : M \to P$ is an R-homomorphism $\Rightarrow (g \circ f)_* : \mathrm{Hom}_R(A, M) \to \mathrm{Hom}_R(A, P)$ defined by $(g \circ f)_*(\alpha) = (g \circ f) \circ \alpha$ is an R-homomorphism by

Proposition 9.5.2. Clearly for every $a \in A, ((g \circ f)_*(\alpha))(a) = (g_* \circ f_*)(\alpha)(a) \Rightarrow (g \circ f)_* = g_* \circ f_*$.

(ii) Proceed as in (i). $\qquad\square$

Corollary *If $f : M \to N$ is an R-isomorphism, then for any R-module A, the induced R-homomorphisms*

$$f_* : \operatorname{Hom}_R(A, M) \to \operatorname{Hom}_R(A, N) \quad and \quad f^* : \operatorname{Hom}_R(N, A) \to \operatorname{Hom}_R(M, A)$$

are R-isomorphisms.

Proof f is an R-isomorphism $\Rightarrow \exists$ an R-homomorphism $g : N \to M$ such that $g \circ f = 1_M$ and $f \circ g = 1_N$. Hence $(g \circ f)_* = 1_{M^*} \Rightarrow g_* \circ f_* = 1_{M^*}$ by Corollary to Proposition 9.5.2 and Proposition 9.5.3. Similarly, $f_* \circ g_* = 1_{N^*}, f^* \circ g^* = 1_{M^*}$ and $g^* \circ f^* = 1_N^*$. Hence f_* and f^* are both R-isomorphisms. $\qquad\square$

9.6 Free Modules, Modules over PID and Structure Theorems

In this section we consider a special class of modules which is the most natural generalization of vector space and is also very important in the study of module theory.

Every vector space has a basis, which may be finite or not. But it fails for modules over arbitrary rings. This leads to the concept of free modules, which may be considered as the most natural generalization of the concept of vector spaces. Free modules play an important role in the theory of modules and widely used in algebraic topology. To extend many results of vector spaces, we also study in this section finitely generated modules over principal ideal domains and finally prove the structure theorems.

From now we use the same symbol 0 to represent the zero element of R, zero element of M as well as zero module (unless there is confusion).

9.6.1 Free Modules

We now proceed to study free modules which are very important in the theory of modules and algebraic topology.

Definition 9.6.1 Let R be a ring with 1 and M be an R-module. A subset S of M is said to be a basis of M iff S generates M and S is linearly independent over R. In particular, M is called cyclic iff $M = Rx$ for some $x \in M$.

Remark For a cyclic R-module M, there exists an R-epimorphism $f : R \to M$ such that $R/\ker f \cong M$. Because, $M = Rx$ for some $x \in M$ and R-homomorphism f :

$R \to M, r \mapsto rx$ is an R-epimorphism. In particular, if R is a PID, then ker $f = \langle a \rangle$ for some $a \in R$ and the cyclic R-module M is of the form $R/\langle a \rangle$, where $\langle a \rangle = $ Ann (M).

Proposition 9.6.1 *Let R be a ring with 1. A (left) R-module M is cyclic iff M is isomorphic to R/I for some left ideal I of R.*

Proof M is cyclic $\Rightarrow \exists$ some $x \in M$ such that $M = Rx$. Then the map $f : R \to M$, $r \mapsto rx$, is an R-epimorphism with ker $f = I$ (say), which is a left ideal of R. Hence $M \cong R/I$. Conversely, let $M \cong R/I$ for some left ideal I of R. Then M is cyclic, because $R/I = \langle 1 + I \rangle$. $\qquad\qquad\qquad\qquad\qquad\qquad\qquad\qquad\qquad\qquad\qquad\square$

Proposition 9.6.2 *An R-module M is the direct sum of submodules $M_1, M_2, \ldots, M_n$, denoted $M = M_1 \oplus M_2 \oplus \cdots \oplus M_n$ iff*

(a) $M = M_1 + M_2 + \cdots + M_n$; *and*
(b) $M_{i+1} \cap (M_1 + M_2 + \cdots + M_i) = 0, \forall i, 1 \le i \le n - 1$.

Proof Left as an exercise. $\qquad\qquad\qquad\qquad\qquad\qquad\qquad\qquad\qquad\qquad\qquad\qquad\square$

Definition 9.6.2 A cyclic R-module $M = Rx$ is called free with a basis $\{x\}$ iff every element $y \in M$ can be expressed uniquely as $y = rx$ for some $r \in R$.

Remark A cyclic R-module $M = Rx$ is free with a basis $\{x\}$ iff Ann$(x) = 0$.

Definition 9.6.3 An R-module M is said to be free on a finite basis iff

(a) $M = M_1 \oplus M_2 \oplus \cdots \oplus M_n$; *and*
(b) each M_i is a free cyclic R-module.

If $M_i = Rx_i$, then $B = \{x_1, x_2, \ldots, x_n\}$ is called a basis of the free R-module M.

Remark Every element y of a free R-module M with a finite basis $B = \{x_1, x_2, \ldots, x_n\}$ can be expressed uniquely as

$$y = \sum_{i=1}^{n} r_i x_i, \quad r_i \in R.$$

This definition can be extended.

Definition 9.6.4 An R-module M is said to be free iff M has a basis.

More precisely, M is said to be a free R-module with a basis B iff every element $x \in M$ can be expressed uniquely as $x = \sum_{b \in B} r_b b$, $r_b \in R$ and $r_b = 0$ except for finitely many b's.

Proposition 9.6.3 *If M and N are free R-modules, then $M \oplus N$ is also a free R-module. In general, if $\{M_i\}$ is any family of free R-modules M_i with basis B_i, then $\bigoplus M_i$ is a free R-module with a basis $B = \bigcup_{i \in I} B_i$.*

Proof Let A be a basis of M and B be a basis of N. Then $M \oplus N$ has a basis $A \times \{0\} \cup \{0\} \times B$ and hence $M \oplus N$ is a free R-module.

The proof of the last part is left as an exercise.   □

Example 9.6.1 (i) The zero module is a free module with empty basis.

(ii) If R is a unitary ring, then as a module over itself R admits a basis $B = \{1\}$.

(iii) If R is a unitary ring, then for any positive integer n, R^n is a free R-module with a basis $B = \{e_i\}$, where $e_i = (0, 0, \ldots, 1, 0, \ldots, 0)$, 1 is at the ith place, $i = 1, 2, \ldots, n$.

(iv) Every vector space V is a free module (over a field), since V has a basis.

(v) $M = \mathbf{Z}_n$ is not a free over $\mathbf{Z}$, because $\text{Ann}(x) \neq \{0\}$ for every $x \ (\neq 0) \in \mathbf{Z}_n$.

(vi) Any finite abelian group $G \ (\neq 0)$ is not a free $\mathbf{Z}$-module.

Remark The basis of a free module is not unique. For example, for a unitary ring R, R^3 is a free R-module with different bases $A = \{(1, 0, 0), (0, 1, 0), (0, 0, 1)\}$ and $B = \{(-1, 0, 0), (0, -1, 0), (0, 0, -1)\}$.

Theorem 9.6.1 *Let R be a commutative ring with 1. Any two finite bases of a free R-module M have the same number of elements.*

Proof Let M be a free module with a basis $\{x_1, x_2, \ldots, x_n\}$ and I be a maximal ideal of R. Then $R/I = F$ is field. Since $V = M/IM$ is annihilated by I, V is a vector space over F. If $[x_i] = x_i + IM \ (1 \leq i \leq n)$, then $B = \{[x_1], [x_2], \ldots, [x_n]\}$ is a basis of V over F. Since any two finite bases of a vector space have the same number of elements, the theorem follows.   □

Definition 9.6.5 Let R be a commutative ring with 1. If a free R-module F has a basis with n elements, then n is called the rank of F, denoted rank $F = n$. In general, if F is free module with a basis B, then rank $F = \text{card } B$.

Example 9.6.2 The R-module R^n in Example 9.6.1(iii) is a free module of rank n.

Remark Theorem 9.6.1 fails for modules over arbitrary rings i.e., for an arbitrary ring R, a free R-module may have bases of different cardinalities.

For example, let V be a vector space of countably infinite dimension over a division ring F and $R = \text{End}_F(V) = \{f : V \to V : f \text{ is a linear transformation}\}$. Then R is a free module over itself with a basis $\{1_R\}$. Moreover, $\exists$ a basis $B_n = \{f_1, f_2, \ldots, f_n\}$ for R having $|B_n| = n$, for any given positive integer n, where the f_i's are defined as follows:

Let $B = \{b_1, b_2, \ldots, b_n, \ldots\}$ be a countably infinite basis of V over F. Define $f_1, f_2, \ldots, f_n$ by assigning their values on b_i's is presented in Table 9.1.

Clearly, B_n is a basis of R over itself with card $B_n = n \neq 1$.

Proposition 9.6.4 *Let R be an integral domain and M be a free R-module of finite rank n. Then any $(n + 1)$ elements of M are linearly dependent over R.*

Table 9.1 Table defining the functions $f_1, f_2, \ldots, f_n$

	f_1	f_2	f_3	f_4	$\cdots$	f_n
b_1	b_1	0	0	0	$\cdots$	0
b_2	0	b_1	0	0	$\cdots$	0
$\vdots$	$\vdots$	$\vdots$	$\vdots$	$\vdots$		$\vdots$
b_n	0	0	0	0	$\cdots$	b_1
b_{n+1}	b_2	0	0	0	$\cdots$	0
b_{n+2}	0	b_2	0	0	$\cdots$	0
$\vdots$	$\vdots$	$\vdots$	$\vdots$	$\vdots$		$\vdots$
b_{2n}	0	0	0	0	$\cdots$	b_2
$\vdots$						
b_{mn+1}	b_{m+1}	0	0	0	$\cdots$	0
b_{mn+2}	0	b_{m+1}	0	0	$\cdots$	0
$\vdots$	$\vdots$	$\vdots$	$\vdots$	$\vdots$		$\vdots$
$b_{(m+1)n}$	0	0	0	0	$\cdots$	b_{m+1}
$\vdots$		$\vdots$	$\vdots$	$\vdots$		$\vdots$

Proof By hypothesis, $M \cong R \oplus R \oplus \cdots \oplus R$ (n-summands). Let $Q(R) = F$ be the quotient field of R. Then M is embedded in $F \oplus F \oplus \cdots \oplus F$ (n-summands), which is an n-dimensional vector space over F. Hence any $(n+1)$ elements of M are linearly dependent over F. $\qquad\qquad\qquad\square$

Remark Vector spaces over a field F are unitary F-modules. Free modules and vector spaces are closely related. They have many common properties but they differ in some properties. Some of them are given now.

Like vector spaces it is not true that a linearly independent subset of a free module can always be extended to a basis.

For example, $\mathbf{Z}$ is a $\mathbf{Z}$-module having $\{1\}$ (also $\{-1\}$) as a basis. $\{3\}$ is linearly independent over $\mathbf{Z}$ but $\mathbf{Z} \neq \langle\{3\}\rangle$.

Remark Like vector spaces it is not true that every subset S of a free module M which spans M contains a basis for M.

For example, consider the $\mathbf{Z}$-module $\mathbf{Z}$. Let $S = \{p, q\}$, where p, q are integers such that $\gcd(p, q) = 1$ and they are not units. Then $1 = ap + bq$, for some $a, b \in \mathbf{Z} \Rightarrow x = (xa)p + (xb)q \ \forall x \in \mathbf{Z} \Rightarrow \mathbf{Z} = \langle S \rangle$. But S is linearly dependent over $\mathbf{Z} \Rightarrow S$ is not a basis for $\mathbf{Z}$.

Remark A finitely generated free module M may have a submodule which is neither free nor finitely generated.

For example, let F be a field and $R = F[x_1, x_2, \ldots, x_n, \ldots]$ which is a commutative ring with 1 (in infinite variables). Then $M = R$ is a free module over R with a basis $\{1\}$. The submodules of M are precisely the ideals of R. Consider the ideal N of all polynomials with constant term zero, i.e., $N = \langle x_1, x_2, \ldots, x_n, \ldots \rangle$. Then N is not finitely generated $\Rightarrow N$ is not a principal ideal $\Rightarrow N$ is not free, because the only ideals of R which are free as R-modules are non-zero principal ideals.

Theorem 9.6.2 (a) $\{x_i\}_{i \in I}$ *is a basis of an R-module M iff $M = \bigoplus_{i \in I} Rx_i$, where $Rx_i \cong R$ for every $i \in I$.*

(b) *An R-module M is free iff M is isomorphic to a direct sum of copies of R.*

Proof (a) Let $M = \bigoplus_{i \in I} Rx_i$. Then every element $x \in M$ can be expressed uniquely as $x = \sum_{i \in I} Rx_i$, where $Rx_i = O_M$ except for a finite number of Rx_i's. Hence $\{x_i\}$ forms a basis of $M \Rightarrow M$ is a free R-module.

Conversely, let M be a free R-module with a basis $\{x_i\}$. Then the mapping $f : R \to Rx_i$, defined by $f(r) = rx_i$ is an R-isomorphism.

Moreover, $Rx_i \cap Rx_j = \{O_M\}$ if $i \neq j$. Again $Rx_i \subseteq M$ $\forall i \in I \Rightarrow \bigoplus Rx_i \subseteq M$. Now $x \in M \Rightarrow x = r_1 x_{i_1} + \cdots + r_n x_{i_n}$, where $x_{i_t} \in \{x_i\}_{i \in I} \Rightarrow M \subseteq \bigoplus_{i \in I} Rx_i$. Hence $M = \bigoplus_{i \in I} Rx_i$, where $Rx_i \cong R$ for each $i \in I$.

(b) It follows from (a).   □

Remark In view of Theorem 9.6.2, Definition 9.6.4 can be restated as follows:

Definition 9.6.6 A free R-module is one which is isomorphic to an R-module of the form $\bigoplus_{i \in I} M_i$, where each $M_i \cong R$ (as an R-module).

A finitely generated free R-module is therefore isomorphic to $R \oplus \cdots \oplus R$ (n summands), which is denoted by R^n (R^0 is taken to be the zero module denoted by 0).

Theorem 9.6.3 *Let R be a ring with 1. Then M is a finitely generated R-module iff M is isomorphic to a quotient of R^n for some integer $n > 0$.*

Proof Let $x_1, x_2, \ldots, x_n$ generate M. Define $f : R^n \to M$ by

$$f(a_1, a_2, \ldots, a_n) = \sum_{\langle i=1 \rangle}^{n} a_i x_i.$$

Then f is an R-module homomorphism onto M. Hence $M \cong R^n / \ker f$ by Theorem 9.4.2.

Conversely, let us have an R-module homomorphism f of R^n onto M. If $e_j = (0, 0, \ldots, 1, 0, \ldots, 0)$ ($1 \in R$ being in the jth place), then $\{e_j\}$ $(1 \leq j \leq n)$ generates R^n. Hence $\{f(e_j)\}$ $(1 \leq j \leq n)$ generates M.   □

9.6.2 Modules over PID

Many results of modules are aimed to extend desirable results of vector spaces. For this purpose, we now consider finitely generated modules over principal ideal domains (PID). We now show that every submodule of a free module of finite rank is also a free module and it is possible to choose bases for the two modules which are closely related in a simple way.

Theorem 9.6.4 *Let R be a PID and M be a free module of finite rank r over R and N be a non-zero submodule of M. Then there exists a basis $\{x_1, x_2, \ldots, x_r\}$ of M and an integer t $(1 \le t \le r)$ and non-zero elements $a_1, a_2, \ldots, a_t$ in R such that $\{a_1x_1, a_2x_2, \ldots, a_tx_t\}$ is a basis of N and $a_i|a_{i+1}, 1 \le i \le t - 1$.*

Proof Let $f : M \to R$ be an R-homomorphism. Then $f(N)$ is an ideal of $R \Rightarrow \exists$ some element a_f (say) in R such that $f(N) = \langle a_f \rangle$, since R is a PID. Consider the set $S = \{f(N) = \langle a_f \rangle$ such that $f : M \to R$ is an R-homomorphism$\}$. If $f = 0$, then $\langle a_f \rangle = \langle 0 \rangle \in S \Rightarrow S \ne \emptyset$. Again R is Noetherian. Then the non-empty set S of ideals of R, partially ordered by inclusion, has a maximal element. Hence $\exists$ a maximal element in S, i.e., $\exists$ an R-homomorphism $g : M \to R$ such that the ideal $g(N) = \langle a_g \rangle$ is not properly contained in any other element of S. Let for this g, $a_1 = a_g$ and $g(x) = a_1$ for some $x \in N$. Then $\langle a_1 \rangle$ is a maximal element in $S \Rightarrow a_1 \ne 0$. Let $\{y_1, y_2, \ldots, y_r\}$ be a basis of M and if $x = \sum_{i=1}^{r} d_i y_i$, then $\pi_i : M \to R, x \mapsto d_i$ is the natural R-epimorphism. Hence it follows that $a_1|\pi_i(x)$ $\forall i$. Then $\pi_i(x) = a_1 s_i$ for some $s_i \in R$. Define $x_1 = \sum_{i=1}^{r} s_i y_i$. Then $a_1 x_1 = \sum_{i=1}^{r} a_1 s_i y_i = x \Rightarrow a_1 = g(x) = g(a_1 x_1) = a_1 g(x_1) \Rightarrow g(x_1) = 1$, since $a_1 \ne 0$. Taking this element x_1 as an element in a basis for M and the element $a_1 x_1$ as an element in a basis for N, we show that

(i) $M = Rx_1 \oplus \ker g$ and (ii) $N = Ra_1x_1 \oplus (N \cap \ker g)$.

To prove (i), let $y \in M$. Then $y = g(y)x_1 + (y - g(y)x_1)$.

Now $g(y - g(y)x_1) = g(y) - g(y)g(x_1) = g(y) - g(y)1 = 0 \Rightarrow y - g(y)x_1 \in \ker g \Rightarrow y \in Rx_1 + \ker g \Rightarrow M \subseteq Rx_1 + \ker g$.

Again, $rx_1 \in \ker g \Rightarrow 0 = g(rx_1) = rg(x_1) = r1 = r \Rightarrow Rx_1 \cap \ker g = \{0\}$. Hence $M = Rx_1 \oplus \ker g$.

(ii) is proved similarly.

We now prove the theorem by induction on $t > 0$, the rank of N.

(ii) $\Rightarrow t = \operatorname{rank} N = \operatorname{rank}(Ra_1x_1) + \operatorname{rank}(N \cap \ker g) \Rightarrow N \cap \ker g$ is an R-module of rank $t - 1 \Rightarrow N \cap \ker g$ is free by induction hypothesis $\Rightarrow N$ is a free R-module of rank t having the basis $\{a_1x_1\} \cup B_{t-1}$, where B_{t-1} is any basis of $N \cap \ker g$ (by using (ii)).

To prove the last part we use induction on r, the rank of M. Now (i) $\Rightarrow \operatorname{rank} \ker g = r - 1 \Rightarrow \exists$ a basis $\{x_2, x_3, \ldots, x_r\}$ of $\ker g$ such that $\{a_2x_2, a_3x_3, \ldots, a_tx_t\}$ is a basis of $N \cap \ker g$ by induction hypothesis, for some elements $a_2, a_3, \ldots, a_t$ in R such that $a_2|a_3|\cdots|a_t$. Consequently, (i) and (ii) show that $\{x_1, x_2, \ldots, x_r\}$ is a basis of M and $\{a_1x_1, a_2x_2, \ldots, a_tx_t\}$ is a basis of N.

Finally, maximality of $\langle a_1 \rangle$ in $S \Rightarrow \langle a_2 \rangle \subseteq \langle a_1 \rangle \Rightarrow a_1|a_2$. $\qquad \square$

9.6.3 Structure Theorems

We are now in a position to proceed to determine the structure of a finitely generated module over a PID, in general and in particular, structure of a finitely generated abelian group and finite abelian group. We show in the next theorem that every finitely generated module over a PID is isomorphic to the direct sum of finitely many cyclic modules. This theorem is the main objective of this section. It is often called fundamental theorem or structure theorem for finitely generated modules over principal ideal domains.

Theorem 9.6.5 (Structure Theorem) *Let R be a PID and M be a finitely generated R-module. Then*

$$M \cong \overbrace{R \oplus R \oplus \cdots \oplus R}^{r \text{ copies}} \oplus R/\langle q_1 \rangle \oplus R/\langle q_2 \rangle \oplus \cdots \oplus R/\langle q_t \rangle,$$

for some integer $r \geq 0$ and the non-zero non-unit elements $q_1, q_2, \ldots, q_t$ of R such that $q_1|q_2| \cdots |q_t$, where q_i's are unique up to units.

Proof Let $B = \{y_1, y_2, \ldots, y_m\}$ be a minimal generating set of M. Then R^m is a free R-module of rank m. Define

$$f : R^m \to M, \quad (a_1, a_2, \ldots, a_m) \mapsto \sum_{i=1}^{m} a_i y_i.$$

Then f is an R-epimorphism $\Rightarrow M \cong R^m / \ker f$ (see Theorem 9.4.2).

Again by Theorem 9.6.4 applied to R^m and its submodule $\ker f$, $\exists$ a basis $\{x_1, x_2, \ldots, x_m\}$ of R^m such that $\{q_1 x_1, q_2 x_2, \ldots, q_t x_t\}$, $t \leq m$, is a basis of $\ker f$ for some non-zero non-units q_i of R satisfying the divisibility relation $q_1|q_2| \cdots |q_t$.

Hence $M \cong R^m / \ker f \Rightarrow M \cong (Rx_1 \oplus Rx_2 \oplus \cdots \oplus Rx_m)/(Rq_1 x_1 \oplus Rq_2 x_2 \oplus \cdots \oplus Rq_t x_t)$.

Consider the natural R-epimorphism

$$h : Rx_1 \oplus Rx_2 \oplus \cdots \oplus Rx_m \to R/\langle q_1 \rangle \oplus R/\langle q_2 \rangle \oplus \cdots \oplus R/\langle q_t \rangle \oplus R^{m-t},$$

$$(r_1 x_1 + r_2 x_2 + \cdots + r_m x_m) \mapsto (r_1 + \langle q_1 \rangle) + (r_2 + \langle q_2 \rangle) + \cdots + (r_t + \langle q_t \rangle)$$
$$+ (r_{t+1} x_{t+1} + \cdots + r_m x_m),$$

where q_i divides r_i, $i = 1, 2, \ldots, t$.

Then

$$\ker h = Rq_1 x_1 \oplus Rq_2 x_2 \oplus \cdots \oplus Rq_t x_t,$$

shows that

$$M \cong R^{m-t} \oplus R/\langle q_1 \rangle \oplus \cdots \oplus R/\langle q_t \rangle.$$

If q is a unit, then $R/\langle q \rangle = 0$ and hence any such term will not appear in the above expression.

Taking $m - t = r$, we have $M \cong R^r \oplus R/\langle q_1 \rangle \oplus \cdots \oplus R/\langle q_t \rangle$.   □

Remark We show a little later the invariance of r. If $r = 0$, M is said to be a torsion R-module.

Theorem 9.6.6 *Let R be a PID and M be a torsion R-module, then $M \cong R/\langle q_1 \rangle \oplus R/\langle q_2 \rangle \oplus \cdots \oplus R/\langle q_t \rangle$, where $\langle q_1 \rangle, \langle q_2 \rangle, \ldots, \langle q_t \rangle$ are uniquely determined up to units.*

Proof Taking $r = 0$ in Theorem 9.6.5, let

$$M \cong R/\langle q_1 \rangle \oplus R/\langle q_2 \rangle \oplus \cdots \oplus R/\langle q_t \rangle$$

$$\cong R/\langle a_1 \rangle \oplus R/\langle a_2 \rangle \oplus \cdots \oplus R/\langle a_k \rangle$$

where $q_1 | q_2 | \cdots | q_t$ and $a_1 | a_2 | \cdots | a_k$.

If $x \in M$, then $x = x_1 + x_2 + \cdots + x_t$, for some $x_i \in R/\langle a_i \rangle$. For a prime p in R, let $M(p) = \{x \in M : px = 0\}$. Then $M(p)$ is a vector space over the field $F = R/\langle p \rangle$.

Now $x \in M(p) \Leftrightarrow px = 0 \Leftrightarrow px_i = 0, \forall i \Leftrightarrow M(p)$ is the direct sum of the kernels of the multiplication by p in each component. Hence $\dim M(p)$ over F is the number of terms $R/\langle q_i \rangle$ for which $p | q_i$.

If $p | q_1$, then $p | q_i, \forall i$. Again for the second decomposition of M, p must divide at least t of a_i's. Hence $t \leq k$. Similarly, $k \leq t$. Hence $t = k$. Let $q_i = pb_i$, $b_i \in R$ and $1 \leq i \leq t$. Then

$$pM \cong pR/\langle pb_1 \rangle \oplus pR/\langle pb_2 \rangle \oplus \cdot \oplus pR/\langle pb_t \rangle$$

$$\cong R/\langle b_1 \rangle \oplus R/\langle b_2 \rangle \oplus \cdots \oplus R/\langle b_t \rangle, \quad \text{where } b_1 | b_2 | \cdots | b_t.$$

By induction on the number of irreducible factors of q_1, it follows that $\langle b_1 \rangle, \ldots, \langle b_t \rangle$ are determined uniquely up to units. Hence $\langle q_1 \rangle, \langle q_2 \rangle, \ldots, \langle q_t \rangle$ are uniquely determined up to units.   □

Corollary 1 *If $M \cong F(M) \oplus T(M)$, then $T(M)$ is uniquely determined.*

Remark The torsion part $T(M)$ of M over a PID is unique but $F(M)$ is not unique. Because different bases B and B' for $M/T(M)$ give different free parts F and F' (say). Then

$$M \cong F(M) \oplus T(M) \quad \text{and} \quad M \cong F'(M) \oplus T(M)$$

$$\Rightarrow \quad F(M) \cong M/T(M) \cong F'(M).$$

The Structure Theorem 9.6.5 can also be stated as follows:

Theorem 9.6.7 (Structure Theorem) *Let R be a PID and M be a finitely generated module over R. Then M can be expressed uniquely as*

$$M \cong F \oplus R/\langle q_1 \rangle \oplus R/\langle q_2 \rangle \oplus \cdots \oplus R/\langle q_t \rangle,$$

where F is uniquely determined up to isomorphism called free part of module M and $q_1, q_2, \ldots, q_t$ *are determined uniquely up to multiplication by units such that* $q_1 | q_2 | \cdots | q_t$.

Proof It follows from Theorem 9.6.6 and Corollary 1. $\square$

In particular, for $R = \mathbf{Z}$, the following *structure theorem* for finitely generated abelian groups is obtained.

Corollary (Structure Theorem) *Let G be a finitely generated abelian group. Then* $G \cong F \oplus \mathbf{Z}_{q_1} \oplus \mathbf{Z}_{q_2} \oplus \cdots \oplus \mathbf{Z}_{q_t}, q_1 | q_2 | \cdots | q_t$, *where F is a free abelian group and* $\mathbf{Z}_{q_n}$ *is the cyclic group of order* q_n, $1 \le n \le t$. *In particular, if G is a finite abelian group, then G is isomorphic to the direct sum of the* q_i*-Sylow subgroups of G.*

Definition 9.6.7 The rank of F is called the free rank or Betti number of M and the elements $q_1, q_2, \ldots, q_t$ are called the invariant factors of M.

Remark Torsion modules M are precisely the modules of rank zero and $\text{Ann}(M) = \langle q_t \rangle$. A torsion free module may not be free as a module. For example, the abelian group $\mathbf{Q}$ of rationals is torsion free over $\mathbf{Z}$ but not free over $\mathbf{Z}$ (see Ex. 12 of the Worked-Out Exercises).

Definition 9.6.8 Let R be a PID and M be a finitely generated R-module such that $\text{Ann}(M) = \langle e \rangle$. Then e is called the exponent of M. It is unique up to units.

Existence of e: Let $M = \langle x_1, x_2, \ldots, x_t \rangle$. Choose $r_i \, (\ne 0) \in R$ such that $r_i x_i = 0$. Then $r = \prod r_i \ne 0$. Hence $r M = 0 \Rightarrow \text{Ann}(M) \ne 0 \Rightarrow \text{Ann}(M) = \langle e \rangle, e \, (\ne 0) \in R$.

Proposition 9.6.5 *Let R be a PID and M be a finitely generated torsion module over R with exponent* $e = ab$, *where* $a, b \in R$ *and* $\gcd(a, b) = 1$. *Then* $M = M_1 \oplus M_2$, *where* $M_1 = \{x \in M : ax = 0\}$ *and* $M_2 = \{x \in M : bx = 0\}$.

Proof $\gcd(a, b) = 1 \Rightarrow \exists m, n \in R$ such that $ma + nb = 1 \Rightarrow x = max + nbx$, $\forall x \in M$.
Again, $b(max) = ab(mx) = emx = 0 \Rightarrow max \in M_2$.
Similarly, $nbx \in M_1$. Consequently, $M = M_1 + M_2$.
Again $x \in M_1 \cap M_2 \Rightarrow ax = 0 = bx \Rightarrow max + nbx = 0 \Rightarrow x = 0 \Rightarrow M_1 \cap M_2 = \{0\}$. Hence $M = M_1 \oplus M_2$. $\square$

Using the Chinese remainder theorem we now proceed to decompose further the cyclic modules in Theorem 9.6.5.

Theorem 9.6.8 *Let R be a PID and M be a finitely generated R-module. Let M be a non-zero R-module with exponent e. If* $e = p_1^{n_1} p_2^{n_2} \cdots p_t^{n_t}$ *is the unique factorization* *of e in R, then*

$$M \cong R^r \oplus R/\langle p_1^{n_1} \rangle \oplus R/\langle p_2^{n_2} \rangle \oplus \cdots \oplus R/\langle p_t^{n_t} \rangle,$$

where $r \geq 0$ is an integer and $p_1^{n_1}, p_2^{n_2}, \ldots, p_t^{n_t}$ are positive powers of distinct primes in R.

Proof For $i \neq j$, $\langle p_i^{n_i} \rangle + \langle p_j^{n_j} \rangle = \langle 1 \rangle = R$, since $\gcd(p_i, p_j) = 1 \Rightarrow$ the ideals $\langle p_i^{n_i} \rangle$ are pairwise co-prime, $i = 1, 2, \ldots, t \Rightarrow \bigcap_i \langle p_i^{n_i} \rangle = \langle e \rangle$, since e is the lcm of $p_1^{n_1}, p_2^{n_2}, \ldots, p_t^{n_t} \Rightarrow R/\langle e \rangle \cong R/\langle p_1^{n_1} \rangle \oplus R/\langle p_2^{n_2} \rangle \oplus \cdots \oplus R/\langle p_t^{n_t} \rangle$, by the Chinese Remainder Theorem.

Hence the theorem follows from Theorem 9.6.5.   □

Remark This isomorphism is also an isomorphism both as rings and R-modules.

Definition 9.6.9 Let R be a PID and M be a finitely generated R-module. The prime powers $p_1^{n_1}, p_2^{n_2}, \ldots, p_t^{n_t}$ are called the elementary divisors of M.

Theorem 9.6.9 (Primary Decomposition Theorem) *Let R be a PID and M be a non-zero torsion R-module with exponent e.*

If $e = p_1^{n_1} p_2^{n_2} \cdots p_t^{n_t}$ be the unique factorization in R and $M_i = \{x \in M : p_i^{n_i} x = 0\}$, $1 \leq i \leq t$, then $M = M_1 \oplus M_2 \oplus \cdots \oplus M_t$.

Proof Each M_i is a submodule of M such that the ideals $\langle p_i^{n_i} \rangle$ and $\langle p_j^{n_j} \rangle$ are pairwise co-prime for $i \neq j$. Hence the theorem follows by applying Chinese Remainder Theorem for rings (see Ex. 34 of Exercises-I of Chap. 9).   □

Definition 9.6.10 The submodule $M(p_i) = \{x \in M : p_i^{n_i} x = 0\}$ of M is called the p_i-primary component of M, where p_i is an elementary divisor of M.

Remark The concepts of elementary divisors of a finitely generated module M and the invariant factors of the primary components $M(p_i)$ of $T(M)$ coincide.

Theorem 9.6.10 *Let R be a PID and p be a prime element in R and F be the field $R/\langle p \rangle$.*

(a) *If $M = R^n$, then $M/pM \cong F^n$.*
(b) *If $M = R/\langle q \rangle$ for a non-zero element q in R, then*

$$M/pM \cong \begin{cases} F, & \text{if } p|q \text{ in } R \\ 0, & \text{otherwise.} \end{cases}$$

(c) *If $M \cong R/\langle q_1 \rangle \oplus R/\langle q_2 \rangle \oplus \cdots \oplus R/\langle q_t \rangle$, where $p|q_i$, $\forall i$, then $M/pM \cong F^t$.*

Proof (a) Consider the natural R-epimorphism

$$\pi : R^n \to \left(R/\langle p \rangle \right)^n, \quad (a_1, a_2, \ldots a_n) \mapsto \left(a_1 + \langle p \rangle, a_2 + \langle p \rangle, \ldots, a_n + \langle p \rangle \right).$$

Then $\ker \pi = pR^n$. Hence $R^n/pR^n \cong (R/\langle p \rangle)^n \Rightarrow M/pM \cong F^n$.

(b) Left as an exercise.

(c) Left as an exercise.   □

Theorem 9.6.11 *Let R be a PID. Then two finitely generated R-modules M_1 and M_2 are*

(a) *isomorphic iff they have the same free rank and the same set of invariant factors;*
(b) *isomorphic iff they have the same free rank and the same set of elementary divisors.*

Proof (a) If M_1 and M_2 have the same free rank and the same set of invariant factors, they are isomorphic. Conversely let M_1 and M_2 be isomorphic. Let rank $M_1 = n_1$ and rank $M_2 = n_2$. Then $M_1 \cong M_2 \Rightarrow T(M_1) \cong T(M_2) \Rightarrow M_1/T(M_1) \cong R^{n_1}$ and $M_2/T(M_2) \cong R^{n_2}$. Hence $R^{n_1} \cong R^{n_2}$. Let p ($\neq 0$) be prime in R. Then $M/pR^{n_1} \cong M/pR^{n_2} \Rightarrow F^{n_1} \cong F^{n_2}$ as vector spaces over the field $F = R/pR \Rightarrow n_1 = n_2$.

(b) If M_1 and M_2 have the same free rank and the same set of elementary divisors, then they are isomorphic. We now show that M_1 and M_2 have the same lists of elementary divisors and invariant factors. To do so it is sufficient to consider the isomorphic torsion modules $T(M_1)$ and $T(M_2)$. So we assume that both M_1 and M_2 are torsion R-modules. By using (c) of Theorem 9.6.10, it is proved that the set of elementary divisors of M_1 is the same as the set of elementary divisors of M_2. Let $q_1, q_2, \ldots, q_t$ be a set of invariant factors of M_1 and $a_1, a_2, \ldots, a_k$ be a set of invariant factors of M_2. We can find a set of elementary divisors of M_1 by taking the prime powers of $q_1, q_2, \ldots, q_t$. Then $q_1|q_2|\cdots|q_t \Rightarrow q_t$ is the product of the largest of the prime powers among the elementary divisors, q_{t-1} is the product of the prime powers among these elementary divisors when the factors of q_t have been removed, and so on. Similarly, for $a_1|a_2|\cdots|a_t$. Since the elementary divisors for M_1 and M_2 are the same, it follows that their invariant factors are also the same. If $M_1 \cong M_2$, then it follows that M_1 and M_2 have the same set of invariant factors and the same set of elementary divisors.   □

Corollary 1 *Two finitely generated abelian groups are isomorphic iff they have the same free rank (Betti number) and the same set of invariant factors.*

Corollary 2 *Two finitely generated abelian groups are isomorphic iff they have the same free rank (Betti number) and the same set of elementary divisors.*

9.7 Exact Sequences

Definition 9.7.1 A sequence of R-modules and their homomorphisms

$$\cdots \to M_{n-1} \xrightarrow{f_n} M_n \xrightarrow{f_{n+1}} M_{m+1} \to \cdots \qquad (9.2)$$

is said to be exact at M_n iff Im $f_n = \ker f_{n+1}$.

The sequence (9.2) is exact iff it is exact at each M_n. In particular, an exact sequence of the form given in sequence (9.3),

$$0 \to M' \xrightarrow{f} M \xrightarrow{g} M'' \to 0 \qquad\qquad (9.3)$$

is called a short exact sequence.

Remark 1 Any long exact sequence (9.2) can be split into short exact sequences.

Remark 2 If $f : M \to N$ is a module homomorphism, then $M/\ker f$ is called *co-image* of f and is denoted by Coim f and to the contrary, $N/\operatorname{Im} f$ called *co-kernel* of f is denoted by Coker f.

Note that $0 \to M \xrightarrow{f} N$ is an exact sequence of modules and their homomorphisms $\Leftrightarrow f$ is a module monomorphisms and $N \xrightarrow{g} P \to 0$ is an exact sequence of modules and their homomorphisms $\Leftrightarrow g$ is a module epimorphism.

If $M \xrightarrow{f} N \xrightarrow{g} P$ is exact sequence of modules and their homomorphisms, then $g \circ f = 0$.

Finally, if $M \xrightarrow{f} N \xrightarrow{g} P \to 0$ is an exact sequence of modules and their homomorphisms, then Coker $f = N/\operatorname{Im} f = N/\ker g = \operatorname{Coim} g \cong P$.

Proposition 9.7.1 *For any sequence of R-modules and their homomorphisms of the form* (9.2) *(not necessarily exact),* $f_{n+1} \circ f_n = 0$ *iff* $\operatorname{Im} f_n \subseteq \ker f_{n+1} \; \forall$ *index n.*

Proof $\operatorname{Im} f_n \subseteq \ker f_{n+1} \Rightarrow (f_{n+1} \circ f_n)(x) = f_{n+1}(f_n(x)) = O_{M_{n+1}} \; \forall x \in M_{n-1} \Rightarrow f_{n+1} \circ f_n = 0$.

Conversely, let $f_{n+1} \circ f_n = 0$. Now, for $y \in \operatorname{Im} f_n$, $\exists$ some $x \in M_{n-1}$ such that $y = f_n(x)$. Hence $(f_{n+1} \circ f_n)(x) = f_{n+1}(f_n(x)) = f_{n+1}(y)$. Thus $f_{n+1} \circ f_n = 0 \Rightarrow f_{n+1}(y) = O_{M_{n+1}} \Rightarrow y \in \ker f_{n+1} \Rightarrow \operatorname{Im} f_n \subseteq \ker f_{n+1}$. $\qquad\square$

Corollary *If the sequence* (9.2) *is exact, then* $f_{n+1} \circ f_n = 0$ *for each index n.*

Proof Exactness of (9.2) $\Rightarrow \operatorname{Im} f_n = \ker f_{n+1} \Rightarrow \operatorname{Im} f_n \subseteq \ker f_{n+1} \Rightarrow f_{n+1} \circ f_n = 0$ by Proposition 9.7.1 for each index n. $\qquad\square$

Remark The converse of the corollary is not true is general. Consider the three term sequence:

$$\mathbf{Z} \xrightarrow{f} \mathbf{Z} \xrightarrow{g} \mathbf{Z}_3,$$

where $\mathbf{Z}$ and $\mathbf{Z}_3$ are $\mathbf{Z}$-modules and f, g are defined by $f(n) = 6n$, $g(n) = (n)_{\bmod 3}$, respectively.

Now for $x \in \operatorname{Im} f$, $\exists n \in \mathbf{Z}$ such that $x = f(n) = 6n$. Hence $g(x) = g(6n) = (6n)_{\bmod 3} = (0)_{\bmod 3} \Rightarrow x \in \ker g \Rightarrow \operatorname{Im} f \subseteq \ker g \Rightarrow g \circ f = 0$. Now $3 \in \ker g$ but $3 \notin \operatorname{Im} f \Rightarrow \ker g \neq \operatorname{Im} f \Rightarrow$ the sequence is not exact.

Theorem 9.7.1 *If $f : M \to N$ is an R-homomorphism, $O \to M$ denotes the inclusion map and $N \to O$ denotes zero R-homomorphism, then*

(i) $O \to M \xrightarrow{f} N$ *is exact* $\Leftrightarrow$ *f is an R-monomorphism;*

(ii) $M \xrightarrow{f} N \to O$ *is exact* $\Leftrightarrow$ *f is an R-epimorphism;*

(iii) $O \to M \xrightarrow{f} N \to O$ *is exact* $\Leftrightarrow$ *f is an R-isomorphism.*

Proof (i) $O \to M \xrightarrow{f} N \to O$ is exact $\Leftrightarrow \ker f = \{O_M\} \Leftrightarrow f$ is an R-monomorphism.

(ii) $M \xrightarrow{f} N \to O$ is exact $\Leftrightarrow \operatorname{Im} f = N \Leftrightarrow f$ is an R-epimorphism.

(iii) follows from (i) and (ii). $\qquad\qquad\qquad\qquad\qquad\qquad\qquad\square$

Corollary *If the sequence $O \to M \xrightarrow{f} N \xrightarrow{g} P \to O$ is exact, then f is an R-monomorphism g is an R-epimorphism and g induces an R-isomorphism.*

$$\tilde{g} : N/f(M) \to P.$$

Proof The proof follows from Theorem 9.7.1 and the Epimorphism Theorem for modules. $\qquad\qquad\qquad\qquad\qquad\qquad\qquad\qquad\qquad\qquad\qquad\square$

Example 9.7.1 (a) Let $f : A \to B$ be a homomorphism of abelian groups. Then each of the following sequences is a short exact sequence.

(i) $O \to \ker f \xrightarrow{i} A \xrightarrow{\pi} A/\ker f \to O$, where $i : \ker f \hookrightarrow A$ is the inclusion map and $\pi : A \to A/\ker f$ is the canonical epimorphism defined by $\pi(a) = a + \ker f$.

(ii) $O \to \operatorname{Im} f \xrightarrow{i} B \xrightarrow{\pi} B/\operatorname{Im} f \longrightarrow 0$, where $i : \operatorname{Im} f \hookrightarrow B$ is the inclusion map and $\pi : B \to B/\operatorname{Im} f$ is the canonical epimorphism defined by $\pi(b) = b + \operatorname{Im} f$.

(b) Let A be an R-module and B be a submodule of A. Then the sequence $O \to B \xrightarrow{i} A \xrightarrow{\pi} A/B \to O$ is exact, where $i : B \hookrightarrow A$ is the inclusion map and $\pi : A \to A/B$ is the canonical R-epimorphism defined by $\pi(a) = a + B$.

(c) Let A and B be R-modules and $f : A \to B$ be an R-homomorphism. Then the sequence $O \to \ker f \xrightarrow{i} A \xrightarrow{f} B \xrightarrow{\pi} B/f(A) \to O$ is exact, where $i : \ker f \hookrightarrow A$ is the inclusion map and $\pi : B \to B/f(A)$ is the canonical R-epimorphism defined by $\pi(b) = b + f(A)$.

(d) Given R-modules A and B, there always exists at least one extension of B by A.

Is the extension of B by A unique for arbitrary modules A and B?

[*Hint.* Consider the short exact sequence

$$0 \longrightarrow A \xrightarrow{f} A \oplus B \xrightarrow{g} B \longrightarrow 0,$$

where $f(a) = (a, 0)$ and $g(a, b) = b$. Then $A \oplus B$ is an extension of B by A.

For the second part, consider the short exact sequence by taking $A = \mathbf{Z}$ and $B = \mathbf{Z}_n$ in first part and other by taking

$$0 \longrightarrow \mathbf{Z} \xrightarrow{\ f_n\ } \mathbf{Z} \xrightarrow{\ g\ } \mathbf{Z}_n \longrightarrow 0,$$

where $f_n : \mathbf{Z} \to \mathbf{Z}$, $x \mapsto nx$ and g is the natural projection.]

Definition 9.7.2 An exact sequence of R-modules and their homomorphisms of the form:

(a) $M \xrightarrow{f} N \to O$ is said to split iff $\exists$ an R-homomorphism $g : N \to M$ such that $f \circ g = 1_N$;

(b) $O \to M \xrightarrow{f} N$ is said to split iff $\exists$ an R-homomorphism $h : N \to M$ such that $h \circ f = 1_M$.

Such R-homomorphisms g and h are called splitting R-homomorphisms.

Definition 9.7.3 A short exact sequence of R-modules and their homomorphisms $O \to N \xrightarrow{f} M \xrightarrow{g} P \to O$ is said to split

(a) on the right iff $M \xrightarrow{g} P \to O$ splits;

(b) on the left iff $O \to N \xrightarrow{f} M$ splits.

Theorem 9.7.2 *Let* $O \to M \xrightarrow{f} N \xrightarrow{g} P \to O$ *be an exact sequence of R-modules and their homomorphisms. Then the following statements are equivalent*:

(a) *the sequence splits on the right*;
(b) *the sequence splits on the left*;
(c) $\mathrm{Im}\, f = \ker g$ *is a direct summand of* N.

Proof (a) $\Rightarrow$ (c). Let $\pi : P \to N$ be a right splitting R-homomorphism. Then $g \circ \pi = 1_P$. We now consider $\ker g \cap \mathrm{Im}\, \pi = A$ (say). If $x \in A$, then $g(x) = O_p$ and $x = \pi(p)$ for some $p \in P$. Hence $O_P = g(x) = g(\pi(p)) = (g \circ \pi)(p) = 1_P(p) = p \Rightarrow x = \pi(p) = \pi(O_P) = O_N \Rightarrow A = \{O_N\}$. Again, for every $y \in N$, $g(y - (\pi \circ g)(y)) = g(y) - (g \circ \pi \circ g)(y) = g(y) - (1_P \circ g)(y) = g(y) - g(y) = O_P \Rightarrow y - (\pi \circ g)(y) \in \ker g, \forall y \in N$.

Now, $y = (\pi \circ g)(y) + (y - (\pi \circ g)(y)) \in \mathrm{Im}\, \pi + \ker g \ \forall y \in N \Rightarrow N \subseteq \mathrm{Im}\, \pi + \ker g$. Again $\mathrm{Im}\, \pi + \ker g$ being a submodule of N, $\mathrm{Im}\, \pi + \ker g \subseteq N$. Consequently, $N = \mathrm{Im}\, \pi \oplus \ker g \Rightarrow \mathrm{Im}\, f = \ker g$ is a direct summand of $N \Rightarrow$ (c).

(c) $\Rightarrow$ (a). Let $\mathrm{Im}\, f = \ker g$ be a direct summand of N. Then $N = \ker g \oplus A$ for some submodule A of N. Hence every $n \in N$ has a unique representation as $n = y + a$ for some $y \in \ker g$ and some $a \in A$. Consider the mapping $\psi = g|_A : A \to P$. Then ψ is an R-isomorphism with inverse $\psi^{-1} : P \to A \Rightarrow \psi \circ \psi^{-1} = 1_P \Rightarrow \psi^{-1}$ is a right splitting R-homomorphism $\Rightarrow$ (a).

Thus (a) $\Leftrightarrow$ (c).

Similarly (b) $\Leftrightarrow$ (c). $\square$

Fig. 9.3 The Three Lemma
diagram

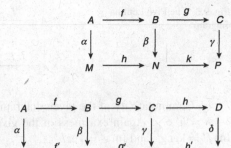

Fig. 9.4 The Four Lemma
diagram

Theorem 9.7.3 (The Three Lemma) *Let the diagram in Fig.* 9.3 *of R-modules and their homomorphisms be commutative with two exact rows. Then*

(i) α, γ, *and h are R-monomorphisms* $\Rightarrow \beta$ *is an R-monomorphism;*
(ii) α, γ, *and g are R-epimorphisms* $\Rightarrow \beta$ *is an R-epimorphism;*
(iii) α, γ *are R-isomorphisms, h is an R-monomorphism and g is an R-epimorphism* $\Rightarrow \beta$ *is an R-isomorphism.*

Proof Commutativity of the given diagram $\Rightarrow h \circ \alpha = \beta \circ f$ and $k \circ \beta = \gamma \circ g$. Again the exactness of the given rows $\Rightarrow \text{Im} f = \ker g$ and $\text{Im} h = \ker k$.

(i) $b \in \ker \beta \Rightarrow \beta(b) = O_N \Rightarrow k(\beta(b)) = k(O_N) \Rightarrow (k \circ \beta)(b) = O_P \Rightarrow (\gamma \circ g)(b) = O_P \Rightarrow \gamma(g(b)) = \gamma(O_C) \Rightarrow g(b) = O_C$, since γ is an R-monomorphism $\Rightarrow b \in \ker g = \text{Im} f \Rightarrow b = f(a)$ for some $a \in A \Rightarrow \beta(b) = \beta(f(a)) \Rightarrow O_N = (\beta \circ f)(a) = (h \circ \alpha)(a) = h(\alpha(a)) \Rightarrow h(O_M) = h(\alpha(a)) \Rightarrow O_M = \alpha(a)$, since h is an R-monomorphism $\Rightarrow \alpha(O_A) = \alpha(a) \Rightarrow O_A = a$, since α is an R-monomorphism $\Rightarrow f(O_A) = f(a) \Rightarrow O_B = b \Rightarrow \ker \beta = \{O_B\} \Rightarrow \beta$ is an R-monomorphism.

(ii) $n \in N \Rightarrow k(n) \in P \Rightarrow k(n) = \gamma(c)$ for some $c \in C$, since γ is an R-epimorphism $\Rightarrow k(n) = \gamma(g(b))$ for some $b \in B$, since g is an R-epimorphism $\Rightarrow k(n) = (\gamma \circ g)(b) = (k \circ \beta)(b) = k(\beta(b)) \Rightarrow k(n - \beta(b)) = O_P \Rightarrow n - \beta(b) \in \ker k \Rightarrow n - \beta(b) \in \text{Im} h \Rightarrow n - \beta(b) = h(m)$ for some $m \in M \Rightarrow n - \beta(b) = h(\alpha(a))$ for some $a \in A$, since α is an R-epimorphism $\Rightarrow n - \beta(b) = (h \circ \alpha)(a) = (\beta \circ f)(a) = \beta(f(a)) \Rightarrow n = \beta(b) + \beta(f(a)) = \beta(b + f(a))$. Thus for each $n \in N$, $\exists$ an element $b + f(a) \in B$ such that $n = \beta(b + f(a)) \Rightarrow \beta$ is an R-epimorphism.

(iii) follows from (i) and (ii). $\square$

Theorem 9.7.4 (The Four Lemma) *Let the diagram in Fig.* 9.4 *of R-modules and R-homomorphisms be commutative with two rows of exact sequences. Then*

(i) α, γ *are R-epimorphisms and δ is an R-monomorphism* $\Rightarrow \beta$ *is an R-epimorphism;*
(ii) α *is an R-epimorphism and β, δ are R-monomorphisms* $\Rightarrow \gamma$ *is an R-monomorphism.*

Fig. 9.5 The Five Lemma diagram

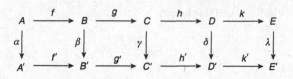

Proof Commutativity of the given diagram $\Rightarrow \beta \circ f = f' \circ \alpha;\ \gamma \circ g = g' \circ \beta$ and $\delta \circ h = h' \circ \gamma$. Again exactness of the given rows $\Rightarrow \operatorname{Im} f = \ker g,\ \operatorname{Im} g = \ker h$, $\operatorname{Im} f' = \ker g'$, and $\operatorname{Im} g' = \ker h'$.

(i) $b' \in B' \Rightarrow g'(b') \in C' \Rightarrow g'(b') = \gamma(c)$ for some $c \in C$, since γ is an R-epimorphism $\Rightarrow h'(g'(b')) = h'(\gamma(c)) \Rightarrow (h' \circ g')(b') = (h' \circ \gamma)(c) \Rightarrow O_{D'} = (\delta \circ h)(c) = \delta(h(c))$, since $h' \circ g' = 0$ by the corollary to Proposition 9.7.1. $\Rightarrow \delta(O_D) = \delta(h(c)) \Rightarrow O_D = h(c)$. since δ is an R-monomorphism $\Rightarrow c \in \ker h = \operatorname{Im} g \Rightarrow c = g(b)$ for some $b \in B \Rightarrow \gamma(c) = \gamma(g(b)) = (\gamma \circ g)(b) \Rightarrow g'(b') = (g' \circ \beta)(b) \Rightarrow g'(b') = g'(\beta(b)) \Rightarrow g'(b' - \beta(b)) = O_{C'} \Rightarrow b' - \beta(b) \in \ker g' = \operatorname{Im} f' \Rightarrow b' - \beta(b) = f'(a')$ for some $a' \in A' \Rightarrow b' - \beta(b) = f'(\alpha(a))$ for some $a \in A$, since α is an R-epimorphism $\Rightarrow (f' \circ \alpha)(a) = (\beta \circ f)(a) = \beta(f(a)) \Rightarrow b' = \beta(b) + \beta(f(a)) = \beta(b + f(a))$. Thus given $b' \in B'$, $\exists$ an element $b + f(a) \in B$ such that $\beta(b + f(a)) = b' \Rightarrow \beta$ is an R-epimorphism.

(ii) $c \in \ker \gamma \Rightarrow \gamma(c) = O_{C'} \Rightarrow h'(\gamma(c)) = (h' \circ \gamma)(c) = O_{D'} \Rightarrow (h' \circ \gamma)(c) = \delta(O_D) \Rightarrow (\delta \circ h)(c) = \delta(O_D) \Rightarrow \delta(h(c)) = \delta(O_D) \Rightarrow h(c) = O_D$, since δ is an R-monomorphism $\Rightarrow c \in \ker h = \operatorname{Im} g \Rightarrow c = g(b)$ for some $b \in B \Rightarrow \gamma(c) = \gamma(g(b)) \Rightarrow \gamma(c) = (\gamma \circ g)(b) \Rightarrow \gamma(c) = (g' \circ \beta)(b) \Rightarrow O_{C'} = g'(\beta(b)) \Rightarrow \beta(b) \in \ker g' = \operatorname{Im} f' \Rightarrow \beta(b) = f'(a')$ for some $a' \in A' \Rightarrow \beta(b) = f'(\alpha(a))$ for some $a \in A$, since α is an R-epimorphism $\Rightarrow \beta(b) = (f' \circ \alpha)(a) \Rightarrow \beta(b) = (\beta \circ f)(a) = \beta(f(a)) \Rightarrow b = f(a)$, since β is an R-monomorphism $\Rightarrow g(b) = g(f(a)) = (g \circ f)(a) = O_C$, since $g \circ f = 0 \Rightarrow c = g(b) = O_C \Rightarrow \ker \gamma = \{O_C\} \Rightarrow \gamma$ is an R-monomorphism. $\qquad \square$

A more frequently used consequence is

Theorem 9.7.5 (The Five Lemma) *Let the diagram in Fig. 9.5 of R-modules and their R-homomorphisms be commutative with two exact rows. If $\alpha, \beta, \delta, \lambda$ are R-isomorphisms, then γ is also an R-isomorphism.*

Proof Apply the Four Lemma to the right hand squares of the given diagram to show that γ is an R-epimorphism. Again apply the same lemma to the left hand squares of the given diagram to show that γ is an R-monomorphism. Hence it follows that under the given conditions, γ is an R-isomorphism. $\qquad \square$

Remark 1 The Short Five Lemma follows from Theorem 9.7.5 as a particular case. However, an independent proof is given in Lemma 9.7.1.

Remark 2 The weaker hypothesis of Theorem 9.7.5 (The Five Lemma) shows in more detail that

Fig. 9.6 Commutativity of diagram of two short exact sequences

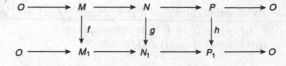

Fig. 9.7 Commutativity of diagram

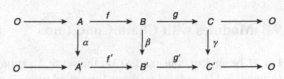

(a) α is an R-epimorphism and β, δ are R-monomorphisms $\Rightarrow \gamma$ is an R-monomorphism.

(b) λ is an R-monomorphism and β, δ are R-epimorphisms $\Rightarrow \gamma$ is an R-epimorphism.

Two short exact sequences $O \to M \to N \to P \to 0$ and $O \to M_1 \to N_1 \to P_1 \to 0$ of R-modules and their homomorphisms are said to be *isomorphic* iff $\exists$ isomorphisms $f : M \to M_1$, $g : N \to N_1$ and $h : P \to P_1$ such that the diagram in Fig. 9.6 is commutative.

Note that

(i) isomorphism of short exact sequences of R-modules and their homomorphisms is an equivalence relation;

(ii) short exact sequences of R-modules and their homomorphisms form a *category* (see Example B.1.1 of Appendix B).

Lemma 9.7.1 (The Short Five Lemma) *Let R be a ring and Fig. 9.7 shows a commutative diagram with rows of short exact sequences of R-modules and their homomorphisms, and if*

(i) *α and γ are monomorphisms, then so is β;*

(ii) *α and γ are epimorphisms, then so is β;*

(iii) *α and γ are isomorphisms, then so is β.*

Proof (i) Suppose $\beta(b) = 0$ for some $b \in B$. We shall show that $b = 0$. Now $\gamma g(b) = g'\beta(b) = 0 \Rightarrow g(b) = 0$, since γ is a monomorphism $\Rightarrow b \in \ker g = \text{Im } f \Rightarrow b = f(a)$ for some $a \in A \Rightarrow f'\alpha(a) = \beta f(a) = \beta(b) = 0 \Rightarrow \alpha(a) = 0$ (since $\ker f' = 0 \Rightarrow f'$ is a monomorphism) $\Rightarrow a = 0$ (since α is a monomorphism) $\Rightarrow \beta$ is a monomorphism (since $b = f(a) = f(0) = 0$).

(ii) Take $b' \in B'$. Then $g'(b') \in C' \Rightarrow g'(b') = \gamma(c)$ for some $c \in C \Rightarrow c = g(b)$ for some $b \in B$ (since $\text{Im } g = C$ by exactness of the top row at C) $\Rightarrow g'\beta(b) = \gamma g(b) = \gamma(c) = g'(b') \Rightarrow g'(\beta(b) - b') = 0 \Rightarrow \beta(b) - b' \in \ker g' = \text{Im } f' \Rightarrow \beta(b) - b' = f'(a')$ for some $a' \in A'$. Clearly, α is an epimorphism $\Rightarrow a' = \alpha(a)$ for some $a \in A$.

Now $b - f(a) \in B$. Then $\beta(b - f(a)) = \beta(b) - \beta f(a)$.

By commutatively of the diagram, $\beta f(a) = f'\alpha(a) = f'(a') = \beta(b) - b' \Rightarrow$ $\beta(b - f(a)) = b' \Rightarrow \beta$ is an epimorphism.

(iii) follows from (i) and (ii).   □

9.8 Modules with Chain Conditions

Let M be an R-module and $S(M)$ the set of all submodules of M. Then analogous to rings, $S(M)$, ordered by the inclusion relation $\subseteq$ is said to satisfy

(i) the ascending chain condition iff every increasing chain of submodules $M_1 \subseteq$ $M_2 \subseteq \cdots$ in $S(M)$ is stationary (i.e., there exists an integer n such that $M_n = M_{n+1} = \cdots$);

(ii) the maximal condition iff every non-empty subset of $S(M)$ has a maximal element.

Remark The conditions (i) and (ii) are equivalent.

If $S(M)$ is ordered by $\supseteq$, then (i) becomes the descending chain condition and (ii) becomes the minimal condition and they become equivalent.

Definition 9.8.1 A module M satisfying either of the equivalent conditions (i) or (ii) for submodules is said to be Noetherian (named after Emmy Noether).

Definition 9.8.2 A module M satisfying either the descending chain condition or the minimal condition for submodules is said to be Artinian (named after Emil Artin).

Theorem 9.8.1 *The following three statements for an R-module M are equivalent*:

 (i) *M satisfies ascending chain condition for submodules.*
 (ii) *Maximal conditions (for submodules) holds in M.*
(iii) *Every submodule of M is finitely generated.*

Proof Proceed as in Theorem 7.1.1.   □

A Noetherian module is now redefined.

Definition 9.8.3 An R-module M satisfying any one of the three equivalent conditions of Theorem 9.8.1 is said to be *Noetherian*.

Remark If R is a PID, then every non-empty set of ideals of R has maximal element and hence R is Noetheiran.

Theorem 9.8.2 *A homomorphic image of a* Noetherian module *is also Noetherian.*

Proof Proceed as in Theorem 7.1.4.   □

Corollary *If A is a submodule of a Noetherian module M, then the quotient module M/A and A are also Noetherian.*

Proof A is Noetherian by Theorem 9.8.1. Again M/A being the homomorphic image of the natural R-epimorphism $\pi : M \to M/A$ is also Noetherian by Theorem 9.8.2.   □

Theorem 9.8.3 *Let M be a module and N a submodule of M. If N and M/N are both Noetherian, then M is also Noetherian.*

Proof Proceed as in Theorem 7.1.5.   □

Corollary 1   *In an exact sequence $O \to M_1 \to M \to M_2 \to O$ of R-modules and their homomorphisms, M is Noetherian iff M_1 and M_2 are both Noetherian.*

Proof It follows from Theorems 9.8.2 and 9.8.3.   □

Corollary 2   *Let M be an R-module and A, B be submodules of M. If $M = A \oplus B$ and if both A, B are Noetherian, then M is Noetherian. A finite direct sum of Noetherian modules is Noetherian.*

Proof $A \times B$ contains A as a submodule whose factor module is isomorphic to B (see Worked-Out Exercises 1). Hence by Theorem 9.8.2, $A \times B$ is a Noetherian module.

Consider the mapping $f : A \times B \to M$, defied by $f(a, b) = a + b$.

Since $M = A \oplus B$, it follows that f is well defined and is an R-epimorphism. Then M being the homomorphic image of the Noetherian module $A \times B, M$ is Noetherian by Theorem 9.8.2.

The last part follows by induction.   □

Proposition 9.8.1 *Let R be a Noetherian ring and let M be a finitely generated R-module. Then M is Noetherian.*

Proof Let $M = \langle x_1, x_2, \ldots, x_n \rangle$. There exists a homomorphism $f : R \times R \times \cdots \times R = R^n \to M$, defined by $f(r_1, r_2, \ldots, r_n) = r_1 x_1 + r_2 x_2 + \cdots + r_n x_n$.

Then f is an epimorphism. The product R^n being Noetherian, M is Noetherian by Theorem 9.8.2.   □

Theorem 9.8.4 *Let M be an R-module. Then the following two statements are equivalent:*

 (i) *M satisfies descending chain condition for submodules.*
(ii) *Minimal condition (for submodules) holds in M.*

Proof Proceed as in Theorem 7.1.3.   □

An Artinian module is now redefined.

Definition 9.8.4 An R-module M satisfying any one of the equivalent conditions of Theorem 9.8.4 is called *Artinian*.

Remark Analogues of Theorems 9.8.2 and 9.8.3 hold for Artinian modules.

Theorem 9.8.5 (Fitting's Lemma) *For any endomorphism f of an Artinian and Noetherian module M, there exists an integer n such that $M = \operatorname{Im} f^n \oplus \ker f^n$.*

This lemma was proved by H. Fitting in 1933.

Proof Left as an exercise.   □

9.9 Representations

A basic tool of ring theory is the representation of a ring R, in the ring $\operatorname{End}(M)$ of endomorphisms of an abelian group M. Representations of R and R-modules are closely related.

If M is an abelian group and $R = \operatorname{End}(M)$, then M can be made into a left R-module by defining an action: $R \times M \to M, (f, x) \mapsto f(x) \,\forall f \in R, x \in M$. On the other hand, if M is a left R-module, then for each $r \in R$, the mapping $I_r : M \to M$, $x \mapsto rx$ is an endomorphism of M. Then the mapping $\rho : R \to \operatorname{End}(M), r \mapsto I_r$ is a ring homomorphism (called a representation of R).

Definition 9.9.1 A homomorphism of a ring R into the ring $\operatorname{End}(M)$ of all endomorphisms of an abelian group M is called a *representation* of R.

In particular, if M is a left R-module, the mapping $\rho : R \to \operatorname{End}(M), \rho \to I_r$ is the representation of R associated with M.

Definition 9.9.2 A left R-module M is said to be faithful iff the representation of R associated with M is injective.

Theorem 9.9.1 *Let R be a ring and I a minimal left ideal of R. If M is a faithful simple left R-module, then M is isomorphic to the left R-module I.*

Proof Let a_0 be a non-zero element of I. Since M is faithful, $a_0 x \neq 0$ for some $x \in M$. Hence the map $\rho_x : I \to M$ defined by $\rho_x(a) = ax (a \in I)$ is a non-zero homomorphism. Hence by Schur's Lemma, ρ_x is a monomorphism and also an epimorphism (since the R-module M is simple). Consequently, ρ_x is an isomorphism.   □

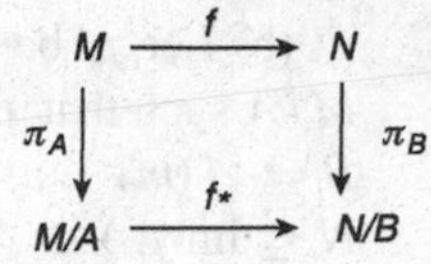

Fig. 9.8 Commutativity of
the rectangular diagram

$$M \xrightarrow{\quad f \quad} N$$
$$\pi_A \downarrow \qquad \qquad \downarrow \pi_B$$
$$M/A \xrightarrow{\quad f_* \quad} N/B$$

9.10 Worked-Out Exercises

1. Let M be an R-module and A, B be submodules of M such that $M = A \oplus B$. Then

   (a) the R-modules M/A and B are R-isomorphic;
   (b) the R-modules M/B and A are R-isomorphic.

   Solution (i) Let $x \in M = A \oplus B$. Then x has the unique representation as $x = a + b$, $a \in A$ and $b \in B$. Consider the mapping

   $$f : M \to B, \quad \text{defined by } f(x) = b.$$

   Then f is an R-epimorphism such that $\ker f = A$. Hence by Epimorphism Theorem for modules, $M/\ker f \cong B \Rightarrow M/A \cong B$.
   (ii) It follows similarly.

2. Let M, N be R-modules and $f : M \to N$ be an R-homomorphism. Let A be a submodule of M and B be a submodule of N such that $f(A) \subseteq B$. Then there exists a unique R-homomorphism $f_* : M/A \to N/B$ making the diagram in Fig. 9.8 commutative, i.e., $\pi_B \circ f = f_* \circ \pi_A$, where π_A and π_B are the respective natural epimorphisms.

   Solution Define $f_* : M/A \to N/B$ by $f_*(m + A) = f(m) + B$.
   Then f_* is well defined and an R-homomorphism, making the given diagram commutative.
   Uniqueness of f_*: Let $g : M/A \to N/B$ be an R-homomorphism, making the given diagram commutative. Then $g \circ \pi_A = \pi_B \circ f$. Now $g(m + A) = g(\pi_A(m)) = (g \circ \pi_A)(m) = (\pi_B \circ f)(m) = \pi_B(f(m)) = f(m) + B = f_*(m + A) \ \forall m + A \in M/A \Rightarrow g = f_* \Rightarrow$ uniqueness of f_*.

3. Consider f_* defined in Worked-Out Exercise 2. Then

   (a) f_* is an R-monomorphism $\Leftrightarrow f^{-1}(B) = A$;
   (b) f_* is a R-epimorphism $\Leftrightarrow \operatorname{Im} f + B = N$.

   Solution (a)

   $$\ker f_* = \left\{ x + A \in M/A : f_*(x + A) = O_N + B \right\}$$
   $$= \left\{ x + A \in M/A : f(x) \in B \right\}$$
   $$= \left\{ x + A \in M/A : x \in f^{-1}(B) \right\}$$
   $$= f^{-1}(B)/A.$$

   Hence f_* is a R-homomorphism $\Leftrightarrow f^{-1}(B) = A$.

(b) Let f_* be an R-epimorphism. Then for every $n + B \in N/B$, $\exists m + A \in M/A$ such that $f_*(m + A) = n + B$. Hence $f(m) + B = n + B \Rightarrow f(m) - n \in B \Rightarrow f(m) - n = b$ for some $b \in B \Rightarrow f(m) - b = n \Rightarrow n \in \operatorname{Im} f + B \Rightarrow N \subseteq \operatorname{Im} f + B \subseteq N$ by hypothesis $\Rightarrow \operatorname{Im} f + B = N$. For the converse, take $n + B \in N/B$. Then $\exists m \in M$ such that $n = f(m) + b \Rightarrow f(m) = n - b$ for some $b \in B \Rightarrow f_*(m + A) = f(m) + B = n - b + B = n + B \Rightarrow f_*$ is surjective $\Rightarrow f_*$ is an R-epimorphism.

4. Let M, N be R-modules and A, B be submodules of M and N, respectively, and $f : M \to N$ be an R-homomorphism. Then the following statements are equivalent:

   (a) $f(A) \subseteq B$;
   (b) There exists a unique R-homomorphism $f_* : M/A \to N/B$ making the diagram of W.O. Ex. 2 commutative.

   Solution Proceed as in Worked-Out Exercise 2.

5. Let M be a unitary R-module. Then M is simple $\Leftrightarrow$ M and R/K are isomorphic for some left ideal K of R.

   Solution M is simple $\Rightarrow M = Rx$ for some x $(\neq O_M) \in M$ by Theorem 9.2.4.
   Consider the mapping $f : R \to M$ defined by $f(r) = rx$, $\forall r \in R$. Then f is an R-epimorphism such that $K = \ker f = \{r \in R : f(r) = O_M\}$ is a left ideal of R. Hence by Epimorphism Theorem M and R/K are isomorphic.

6. Let M be a unitary R-module. If M is simple, then M and R/K are isomorphic for some maximal left ideal K of R.

   Solution By Worked-Out Exercise 5, M and R/K are isomorphic. Since M is simple, R/K is also a simple unitary R-module. Hence by Corollary 1 to Theorem 9.2.4, R/K is division ring $\Rightarrow K$ is a maximal left ideal of R.

7. Let R be a commutative ring and M, N be R-modules. Then

   (a) $\operatorname{Hom}_R(M, N)$ is a left module over the ring $\operatorname{End}(N, +)$;
   (b) $\operatorname{Hom}_R(M, N)$ is a right module over the ring $\operatorname{End}(M, +)$.

   Solution (i) Define a mapping $\mu : \operatorname{End}(N, +) \times \operatorname{Hom}_R(M, N) \to \operatorname{Hom}_R(M, N)$ by $\mu(f, g) = f \circ g$. Then μ is a scalar multiplication.
   (ii) Define a mapping $\mu : \operatorname{Hom}_R(M, N) \times \operatorname{End}(M, +) \to \operatorname{Hom}_R(M, N)$ by $\mu(g, f) = g \circ f$. Then μ is a scalar multiplication.

8. The $\mathbf{Z}$-module $\operatorname{Hom}_{\mathbf{Z}}(\mathbf{Q}, \mathbf{Z}) = \{0\}$.

   Solution Let $f \in \operatorname{Hom}_{\mathbf{Z}}(\mathbf{Q}, \mathbf{Z})$. Then $f : \mathbf{Q} \to \mathbf{Z}$ is a $\mathbf{Z}$-homomorphism. Now $1 \in \mathbf{Q} \Rightarrow f(1) \in \mathbf{Z}$. For any integer q $(\neq 0)$, $1/q \in \mathbf{Q} \Rightarrow f(1) = f(q \cdot \frac{1}{q}) = qf(\frac{1}{q}) \Rightarrow f(\frac{1}{q}) = \frac{1}{q} f(1) \in \mathbf{Z} \Rightarrow f(1)$ is a multiple of $q \Rightarrow q | f(1)$ $\forall q$ $(\neq 0) \in \mathbf{Z} \Rightarrow f(1) = 0$. For any rational number $p/q \in \mathbf{Q}$, $f(p/q) = pf(1/q) = p \cdot \frac{1}{q} f(1) = 0 \Rightarrow f$ is a zero $\mathbf{Z}$-homomorphism $\Rightarrow \operatorname{Hom}_{\mathbf{Z}}(\mathbf{Q}, \mathbf{Z}) = \{0\}$.

9. (a) $\operatorname{Hom}_z(\mathbf{Z}_2, \mathbf{Z}) = \{0\}$.
   (b) $\operatorname{Hom}_z(\mathbf{Z}_2, \mathbf{Q}) = \{0\}$.

Solution (a) $\mathbf{Z}_2 = \{(0), (1)\}$. Then for $f \in \mathrm{Hom}_{\mathbf{Z}}(\mathbf{Z}_2, \mathbf{Z})$, $f : \mathbf{Z}_2 \to \mathbf{Z}$ is a $\mathbf{Z}$-homomorphism. Now $f((1)) \in \mathbf{Z} \Rightarrow f((1)) = n$ for some $n \in \mathbf{Z} \Rightarrow 0 = f((0)) = f((1) + (1)) = f((1)) + f((1)) = n + n = 2n \Rightarrow n = 0 \Rightarrow f((1)) = 0$. Thus $f = 0$. Hence $\mathrm{Hom}_z(\mathbf{Z}_2, \mathbf{Z}) = \{0\}$.

(b) Proceed as in (a).

10. Let R be a commutative ring.

   (a) If $O \to A \xrightarrow{\alpha} B \xrightarrow{\beta} C$ is an exact sequence of R-modules and R-homomorphisms, then for any R-module M, the sequence $O \to \mathrm{Hom}_R(M, A) \xrightarrow{\alpha_*} \mathrm{Hom}_R(M, B) \xrightarrow{\beta_*} \mathrm{Hom}_R(M, C)$ is exact.

   (b) If $A \xrightarrow{\alpha} B \xrightarrow{\beta} C \to O$ is an exact sequence of R-modules and R-homomorphisms, then for any R-module M, the sequence $O \to \mathrm{Hom}_R(C, M) \xrightarrow{\beta^*} \mathrm{Hom}_R(B, M) \xrightarrow{\alpha^*} \mathrm{Hom}_R(A, M)$ is exact (α_* and α^* are defined in Proposition 9.5.2).

   (c) Show by examples that the exactness of the sequence of R-modules and their homomorphisms $O \to B \xrightarrow{f} C \xrightarrow{g} D \to O$ does not in general imply the exactness of any of the following sequences:

   (i) For a given R-module M, the sequence $O \to \mathrm{Hom}_R(M, B) \xrightarrow{f_*} \mathrm{Hom}_R(M, C) \xrightarrow{g_*} \mathrm{Hom}_R(M, D) \to O$ is not in general exact.

   (ii) For a given R-module M, the sequence $O \to \mathrm{Hom}_R(D, M) \xrightarrow{g^*} \mathrm{Hom}_R(C, M) \xrightarrow{f^*} \mathrm{Hom}_R(B, M) \to O$ is not in general exact.

Solution (a) Use the definition of α_* and β_* to show that under the given conditions α_* is an R-monomorphism and $\mathrm{Im}\,\alpha_* = \ker \beta_*$.

(b) It is sufficient to prove that β^* is an R-monomorphism and $\mathrm{Im}\,\beta^* = \ker \alpha^*$.

(c) (i) Consider the exact sequence

$$O \to \mathbf{Z} \xrightarrow{i} \mathbf{Q} \xrightarrow{q} \mathbf{Q}/\mathbf{Z} \to O,$$

where $i : \mathbf{Z} \hookrightarrow \mathbf{Q}$ is the inclusion map and q is the canonical $\mathbf{Z}$-homomorphism defined by $q(x) = x + \mathbf{Z}$ and take $M = \mathbf{Z}_2$. Then the sequence $O \to \mathrm{Hom}_{\mathbf{Z}}(\mathbf{Z}_2, \mathbf{Z}) \xrightarrow{i_*} \mathrm{Hom}_{\mathbf{Z}}(\mathbf{Z}_2, \mathbf{Q}) \xrightarrow{q_*} \mathrm{Hom}_{\mathbf{Z}}(\mathbf{Z}_2, \mathbf{Q}/\mathbf{Z}) \to O$ is not exact. This is because q_* is not surjective, since $\mathrm{Hom}_{\mathbf{Z}}(\mathbf{Z}_2, \mathbf{Q}) = O$ but $\mathrm{Hom}_z(\mathbf{Z}, \mathbf{Q}/\mathbf{Z}) \neq O$.

(ii) Consider the exact sequence (i) and take $M = \mathbf{Z}$. Then the sequence $O \to \mathrm{Hom}_{\mathbf{Z}}(\mathbf{Q}/\mathbf{Z}, \mathbf{Z}) \xrightarrow{q^*} \mathrm{Hom}_{\mathbf{Z}}(\mathbf{Q}, \mathbf{Z}) \xrightarrow{i^*} \mathrm{Hom}_z(\mathbf{Z}, \mathbf{Z}) \to O$ is not exact. This is because $\mathrm{Hom}_{\mathbf{Z}}(\mathbf{Q}, \mathbf{Z}) = O$ and $\mathrm{Hom}_{\mathbf{Z}}(\mathbf{Z}, \mathbf{Z}) = \mathbf{Z}$ and hence i^* is not surjective.

11. A finite abelian group $G \neq \{0\}$ is not a free $\mathbf{Z}$-module.

Solution By Lagrange's Theorem if n (>0) is the number of elements of G, and if $x \in G$, then $nx = 0$, so that $\{x\}$ is not linearly independent over $\mathbf{Z}$ for any

$x \in G$. Hence no non-empty subset of G is linearly independent. Thus G has no basis over $\mathbf{Z} \Rightarrow G$ is not a free $\mathbf{Z}$-module.

12. The module $\mathbf{Q}$ over $\mathbf{Z}$ is not free i.e., $\mathbf{Q}$ is not a free $\mathbf{Z}$-module.

Solution Let p/q $(\neq 0) \in \mathbf{Q}$. Then $np/q = 0 \Rightarrow n = 0 \Rightarrow \{p/q\}$ is linearly independent over $\mathbf{Z}$. Again, if p/q and a/b are two different rational numbers, then $(aq)(p/q) - (pb)(a/b) = 0$, where $aq, pb \in \mathbf{Z}$. Hence p/q and a/b are linearly dependent over $\mathbf{Z}$. We now show that no singleton set can generate $\mathbf{Q}$. To show this, let $\{1/p\}$, where p is prime, generate $\mathbf{Q}$. As $1/2p \in \mathbf{Q}$, $\exists n \in \mathbf{Z}$ such that $n \cdot 1/p = 1/2p$. Then $n = (1/2) \notin \mathbf{Z} \Rightarrow$ a contradiction. In this way we find that $\mathbf{Q}$ admits no basis over $\mathbf{Z}$. In other words $\mathbf{Q}$ is not a free $\mathbf{Z}$-module.

13. Let M be an R-module.

(a) If $a, b \in R$, then $a - b \in \text{Ann}(M)$ in $R \Rightarrow ax = bx \ \forall x \in M$.
(b) If M is endowed with the $R/\text{Ann}(M)$-module structure, then find the annihilator of M in $R/\text{Ann}(M)$.

Solution (i) Let $a, b \in R$ be such that $a - b \in \text{Ann}(M)$ (in R). Then $(a - b)x = O_M \ \forall x \in M \Rightarrow ax = bx \ \forall x \in M$.

(ii) $\text{Ann}(M)$ (in R) is a two-sided ideal in $R \Rightarrow R/\text{Ann}(M)$ is ring. M can be endowed $R/\text{Ann}(M)$-module structure (see Remark of Proposition 9.2.3) under the external law of composition:

$$R/\text{Ann}(M) \times M \to M, \quad (r).m \mapsto rm \quad \forall (r) \in R/\text{ann}(M).$$

This composition is well defined. This is because $t \in (r) \Rightarrow (t) = (r)$ in $R/\text{Ann}(M) = S$ (say) $\Rightarrow t + \text{Ann}(M) = r + \text{Ann}(M) \Rightarrow r - t \in \text{Ann}(M) \Rightarrow rm = tm \ \forall m \in M$ by (i).

Then annihilation of M in

$$S = \left\{ (r) \in S : (r) \cdot m = O_M \ \forall m \in M \right\}$$
$$= \left\{ (r) \in S : rm = O_M \ \forall m \in M \right\}$$
$$= \left\{ (r) \in S : r \text{ belongs to the annihilator of } M \text{ in } R \right\}$$
$$= \{O_S\}.$$

14. (a) Let R be a commutative unitary ring and M be an R-module. Given $r \in R$, rM and M_r are defined by

$$rM = \{rm : m \in M\} \quad \text{and}$$
$$M_r = \{m \in M : rm = O_M\}.$$

Then rM and M_r are both submodules of M.

(b) In particular, if $R = \mathbf{Z}$ and $M = \mathbf{Z}_n$, where $n = rs$ and $\gcd(r, s) = 1$, then $rM = M_s$.

Solution (a) $O_M \in rM \Rightarrow rM \neq \emptyset$. Let $rx, ry \in rM$, where $x, y \in M$. Then $rx + ry = r(x + y) \in rM$, since $x + y \in M$. Again, $\forall a \in R$, $a(rx) = (ar)x =$

$(ra)x$ (since R is commutative) $\Rightarrow r(ax) \in rM$, since $ax \in M$. Consequently, rM is a submodule of M.

Similarly, M_r is a submodule of M.

(b) Suppose $R = \mathbf{Z}$ and $M = \mathbf{Z}_n$, where $n = rs$ and $\gcd(r, s) = 1$.

Now $\gcd(r, s) = 1 \Rightarrow \exists a, b \in \mathbf{Z}$ such that $ra + sb = 1$.

Let $m \in M = \mathbf{Z}_n$. Then $rm \in nM$. Hence $s(rm) = (sr)m = nm = 0$ in $\mathbf{Z}_n \Rightarrow rm \in M_s \Rightarrow rM \subseteq M_s$.

On the other hand, let $m \in M_s$. Then $sm = O_M$.

Again $ra + sb = 1 \Rightarrow m = 1 \cdot m = (ra + sb)m = ram + sbm = r(am) + s(bm) = r(am) + b(sm) = r(am) \in rM$, since $am \in M$ and $sm = O_M$.

Thus $m \in rM \Rightarrow M_s \subseteq rM$. Consequently, $rM = M_s$.

15. Let M be an R-module and A, B, C be submodules of M such that $A \subseteq B$, $A + C = B + C$, $A \cap C = B \cap C$, then $A = B$.

Solution $(S(M), \subseteq)$ forms a modular lattice (see Theorem 9.2.6). Since $A \subseteq B$ by modular law, $(A + C) \cap B = A + (B \cap C)$. Hence $A = A + (A \cap C) = A + (B \cap C) = (B \cap C) + A = B \cap (A + C) = B \cap (B + C) = B$.

9.10.1 Exercises

Exercises-I

1. For an additive abelian group M, let $\text{End}(M)$ be the ring of endomorphisms of M. If R is any ring show that M is an R-module iff there exists a homomorphism $\theta : R \to \text{End}(M)$.

   [*Hint.* Let M be an R-module with an action $R \times M \to M$, denoted by $(r, x) \mapsto rx$. Define $\theta : R \to \text{End}(M)$ by $r \mapsto \theta(r)$, where $\theta(r) : M \to M$ is given by $x \mapsto rx \ \forall x \in M$ and $r \in R$. Then θ is a homomorphism of rings. Conversely, suppose $\theta : R \to \text{End}(M)$ is a ring homomorphism. Define the action $R \times M \to M$, $(r, x) \mapsto rx = (\theta(r))(x) \ \forall r \in R, \ x \in M$. Then M is an R-module.]

2. Show that a subring S of a ring R is a module over the whole ring R only if S is an ideal of R.

3. (a) Let $f : M \to N$ be an R-homomorphism of modules with $P = \ker f$. Show that there exists a unique module monomorphism $\tilde{f} : M/P \to N$ such that $f = \tilde{f} \circ \pi$, where $\pi : M \to M/P$ is the natural module homomorphism.

   (a) Let R be a ring and N a submodule of an R-module M. Show that there is a one-to-one correspondence between the set of all submodules of M containing N and the set of all submodules of M/N, given by $P \mapsto P/N$. Hence prove that every submodule of M/N is of the form P/N, where P is a submodule of M which contains N.

4. Let M be an R-module. Show that the submodules of M form a complete lattice with respect to inclusion.

5. Let L, M, N be R-modules such that $N \subseteq M \subseteq L$. Show that $(L/N)/(M/N) \cong L/M$.

   [*Hint.* Define $f : L/N \to L/M$ by $f(x + N) = x + M \ \forall x \in L$. Check that f is a well-defined R-module homomorphism of L/N onto L/M and ker $f = M/N$. Then apply Theorem 9.4.2.]

6. Let M be an R-module and P, Q submodules of M. Prove that $(P + Q/P \cong Q/(P \cap Q)$.

   [*Hint.* The composite homomorphism $Q \to P + Q \to (P + Q)/P$ is surjective and its kernel is $P \cap Q$. Then apply Theorem 9.4.2.]

7. Let R be a ring, I an ideal of R, M an R-module and A ($\neq \emptyset$) is a subset of M. Show that

   (a) $IA = \{\sum_{\langle i=1 \rangle}^{n} r_i a_i : r_i \in I \text{ and } a_i \in A\}$ is a submodule of M.

   (b) M/IA is an R/I-module under the action: $R/I \times M/IA \to M/IA$, $(r + I)(x + IA) \mapsto rx + IA$.

   (c) $IM = \{\sum_{\text{finite sum}} a_i x_i : a_i \in I \text{ and } x_i \in M\}$ is a submodule of M.

8. Let A, B be submodules of an R-module M. Define $(A : B) = \{r \in R : rB \subseteq A\}$. Show that $(A : B)$ is an ideal of R. In particular, define the annihilator of M denoted by $\text{Ann}(M)$ by $\text{Ann}(M) = (0 : M) = \{r \in R : rM = 0\}$. Show that $\text{Ann}(M)$ is an ideal of R. If I is an ideal of R such that $I \subseteq \text{Ann}(M)$, show that M can be made into an R/I module.

   [*Hint.* Define $R/I \times M \to M$, $((x), m) \mapsto xm \ \forall (x) \in R/I$ and $m \in M$, where (x) is represented by $x \in R$, i.e., $(x) = x + I$.]

9. Let A, B be submodules of an R-module M. Show that

   (i) $\text{Ann}(A + B) = \text{Ann}(A) \cap \text{Ann}(B)$;

   (ii) $(A : B) = \text{Ann}((A + B)/A)$.

10. Let M be a finitely generated R-module and I be an ideal of R and let f be an R-module endomorphism of M such that $f(M) \subseteq IM$. Show that f satisfies an equation of the form

$$f^n + a_1 f^{n-1} + \cdots + a_n = 0 \quad (a_i \in I).$$

   [*Hint.* Let $x_1, x_2, \ldots, x_n$ be a set of generators of M. Then for each $f(x_i) \in IM$,

$$f(x_i) = \sum_{j=1}^{n} a_{ij} x_j \quad (1 \le i \le n; \ a_{ij} \in I) \quad \Rightarrow \quad \sum_{j=1}^{n} (\delta_{ij} f - a_{ij}) x_j = 0,$$

   where δ_{ij} is the Kronecker delta. Multiply on the left by the adjoint of the matrix $(\delta_{ij} f - a_{ij})$. Then it follows that $\det(\delta_{ij} f - a_{ij})$ annihilates each x_i, hence it is the zero endomorphism of M. An equation of the required form is obtained on expansion of the determinant.]

11. Let M be a finitely generated R-module and I an ideal of R such that $IM = M$. Show that there exists $x \equiv 1 \pmod{I}$ such that $xM = 0$.

Fig. 9.9 Commutative
diagram for Five Lemma

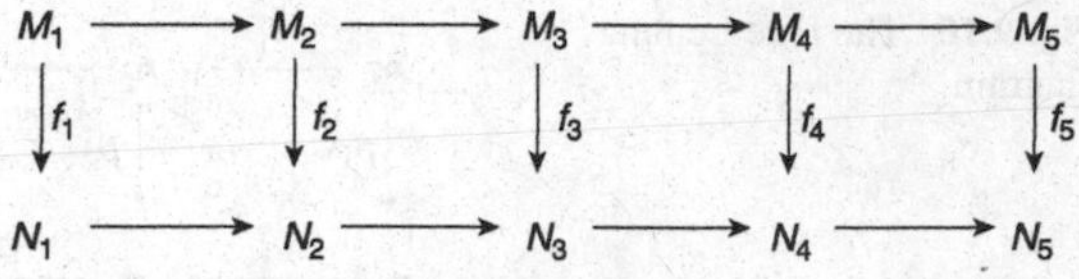

$$
\begin{array}{ccccccccc}
M_1 & \longrightarrow & M_2 & \longrightarrow & M_3 & \longrightarrow & M_4 & \longrightarrow & M_5 \\
\downarrow f_1 & & \downarrow f_2 & & \downarrow f_3 & & \downarrow f_4 & & \downarrow f_5 \\
N_1 & \longrightarrow & N_2 & \longrightarrow & N_3 & \longrightarrow & N_4 & \longrightarrow & N_5
\end{array}
$$

[*Hint*. Take $f =$ identity in Exercise 10. Then $x = 1 + a_1 + \cdots + a_n$ gives the required solution.]

12. (*Nakayama's Lemma*) Let M be a finitely generated R-module and I an ideal of R contained in the Jacobson radical J of R. Then $IM = M$ implies $M = 0$.

    [*Hint*. Using Exercise 11, $xM = 0$ for some $x \equiv 1 \pmod{J} \Rightarrow x$ is invertible in R (cf. Exercise 23(ii), Chap. 5) $\Rightarrow M = x^{-1}xM = 0$.]

13. Let M be a finitely generated R-module, N a submodule of M and I an ideal of R contained in the Jacobson radical J of R. Then show that $M = IM + N \Rightarrow M = N$.

    [*Hint*. Note that $I(M/N) = (IM + N)/N$. Apply Nakayama's Lemma to M/N.]

14. (*The Five Lemma*). Let the diagram of R-modules and their homomorphisms in Fig. 9.9 be a commutative, with exact rows. Prove the following:

    (a) f_1 is an epimorphism and f_2, f_4 are monomorphisms $\Rightarrow f_3$ is a monomorphism.

    (b) f_5 is a monomorphism and f_2, f_4 are epimorphisms $\Rightarrow f_3$ is an epimorphism.

       [The lemma in Exercise 17, which is also called the Five Lemma, follows from this exercise.]

15. (a) Let A, B, C, D be R-modules and $f : C \to A$ and $g : B \to D$ be R-module homomorphisms. Then show that the map $\psi : \mathrm{Hom}_R(A, B) \to \mathrm{Hom}_R(C, D)$ defined by $\psi(\alpha) = g\alpha f$ is a homomorphism of abelian groups.

    (b) Let

$$
A \xrightarrow{f} B \xrightarrow{g} C \to O \tag{9.4}
$$

    be an exact sequence of R-modules and their homomorphisms. Show that the sequence (9.4) is exact $\Leftrightarrow \forall R$-modules N, the sequence of abelian groups

$$
O \to \mathrm{Hom}(C, N) \xrightarrow{g^*} \mathrm{Hom}(B, N) \xrightarrow{f^*} \mathrm{Hom}(A, N) \tag{9.5}
$$

    is exact.

    (c) Let

$$
O \to A \xrightarrow{f} B \xrightarrow{g} C \to \tag{9.6}
$$

Fig. 9.10 The Five Lemma
diagram

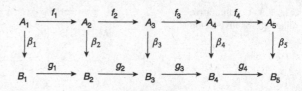

be a sequence of R-modules and their homomorphisms. Show that the sequence (9.6) is exact $\Leftrightarrow \forall R$-modules M, the sequence of abelian groups

$$O \to \mathrm{Hom}(M, A) \xrightarrow{f_*} \mathrm{Hom}(M, B) \xrightarrow{g_*} \mathrm{Hom}(M, C) \to \qquad (9.7)$$

is exact.

16. If N is any submodule of an R-module M, show that the sequence $O \to N \xrightarrow{i} M \xrightarrow{\pi} M/N \to O$ is exact, where i is the inclusion map and π is the natural homomorphism. Conversely, if the sequence of R modules and homomorphisms $O \to A \xrightarrow{f} M \xrightarrow{g} C \to O$ is exact, then $N = \mathrm{Im}\, f = \ker g$ is a submodule of M and $N \cong A$ and $M/N \cong C$.

17. (*The Five Lemma*) If R is a ring and the diagram in Fig. 9.10 is commutative with exact rows of R-modules and their homomorphisms such that each β_i ($i = 1, 2, 4, 5$) is an isomorphism, then show that β_3 is also an isomorphism.

18. Show that M is a Noetherian R-module iff every submodule of M is finitely generated.

19. Let $O \to M' \xrightarrow{f} M \xrightarrow{g} M'' \to O$ be an exact sequence of R-modules and their homomorphisms. Show that

    (a) M is Noetherian $\Leftrightarrow M'$ and M'' are Noetherian;
    (b) M is Artinian $\Leftrightarrow M'$ and M'' are Artinian.

20. If M_i ($1 \le i \le n$) are Noetherian (Artinian) R-modules, show that $\bigoplus_{(i=1)}^{n} M_i$ is also.

21. Let R be a Noetherian (Artinian) ring and M a finitely generated R-module. Show that M is a Noetherian (Artinian) R-module.

22. Prove the following:

    (a) Let M be a Noetherian (Artinian) R-module and I a submodule of M. Then M/I is a Noetherian (Artinian) R-module;
    (b) if an R module M has a submodule I such that both I and M/I are Noetherian (Artinian), then M is Noetherian (Artinian).

23. If M ($\neq 0$) is a finitely generated R-module, show that M has a maximal submodule.

24. (a) Let M be an R-module and N a submodule of M Show that N is maximal in $M \Leftrightarrow R$-module M/N is simple.
    (b) Show that an R-module M is simple $\Leftrightarrow M \cong R/I$ for some maximal ideal I of R.

25. Let M be an R-module and $\{M_i\}$, $\langle i \in J \rangle$ be a family of simple submodules of M such that $M = \sum_{\langle i \in J \rangle} M_i$. Show that for each submodule N of M there is a subset I of J such that $M = \bigoplus_{\langle i \in I \rangle} M_i \oplus N$.

26. Let $M_1 \xrightarrow{f_1} M_2 \xrightarrow{f_2} M_3 \xrightarrow{f_3} M_4 \xrightarrow{f_4} M_5 \xrightarrow{f_5} M_6$ be an exact sequence of R-modules. Show that the following statements are equivalent:

   (a) f_3 is an isomorphism;
   (b) f_2 and f_4 are trivial homomorphisms;
   (c) f_1 is an epimorphism and f_5 is a monomorphism.

27. (a) Let the sequence $O \to M \xrightarrow{f} N \to O$ of R-modules be exact. Show that f is an isomorphism.

   (b) Let the sequence $O \to A_1 \xrightarrow{f} B \xrightarrow{g} A_2 \to O$ of R-modules be exact. Show that the following statements are equivalent:

   (i) There is an R-module homomorphism $h : A_2 \to B$ with $gh = 1_{A_2}$;
   (ii) there is an R-module homomorphism $k : B \to A_1$ such that $kf = 1_{A_1}$;
   (iii) the given sequence is isomorphic (with identity maps on A_1 and A_2) to the short exact sequence $O \to A_1 \xrightarrow{i} A_1 \oplus A_2 \xrightarrow{\pi} A_2 \to O$; in particular, $B \cong A_1 \oplus A_2$.

28. An exact sequence

$$\cdots \to M_n \xrightarrow{f_n} M_{n+1} \xrightarrow{f_{n+1}} M_{n+2} \to \cdots \qquad (9.8)$$

is said to split at the R-module M_{n+1} iff the R-submodule $M = \mathrm{Im}\, f_n = \ker f_{n+1}$ of M_{n+1} is a direct summand of M_{n+1} i.e., iff M_{n+1} is decomposable into the direct sum of M and another R-submodule of M_{n+1}. If the exact sequence (9.8) splits at M_{n+1} $\forall n$, the sequence (9.8) is said to split.

   (a) Prove that the sequence (9.8) splits if either (i) or (ii) holds.

   (i) There exists an R-homomorphism $h : M_{n+1} \to M_n$ such that $hf_n : M_n \to M_n$ is an automorphism.
   (ii) There exists an R-homomorphism $k : M_{n+2} \to M_{n+1}$ such that $f_{n+1}k : M_{n+2} \to M_{n+2}$ is an automorphism.

   (b) Prove that if the sequence (9.8) splits, then

   (i) $M_{n+1} \cong \mathrm{Im}\, f_n \oplus \mathrm{Im}\, f_{n+1} \cong M_n \oplus \mathrm{Im}\, f_{n+1}$
      [*Hint*. Use (a(i)).]
   (ii) $M_{n+1} \cong \mathrm{Im}\, f_n \oplus \mathrm{Im}\, f_{n+1} \cong \mathrm{Im}\, f_n \oplus M_{n+2}$
      [*Hint*. Use a(ii).]

29. Let M, N, and P be given R-modules and $M \times N$ be the product of the sets M and N. A function $f : M \times N \to P$ is said to be bilinear iff

$$f(r_1 m_1 + r_2 m_2, n) = r_1 f(m_1, n) + r_2 f(m_2, n)$$

and $f(m, r_3n_1 + r_4n_2) = r_3 f(m, n_1) + r_4 f(m, n_2)$ hold $\forall r_i \in R$, $m, m_i \in M$ and $n, n_i \in N$.

By a tensor product of M and N, we mean a pair (T, f), where T is an R-module and $f : M \times N \to T$ is a bilinear function such that for every bilinear function $g : M \times N \to P$ there exists a unique homomorphism

$$h : T \to P \quad \text{satisfying} \quad hf = g.$$

Prove the following:

(a) If (T, f) is a tensor product of R-modules M and N, then $f(M \times N)$ generates T.

(b) If (T, f) and (P, g) are tensor products of M and N, then there exists a unique isomorphism $h : T \to P$ of R-modules and that $hf = g$.

(c) Every pair of R-modules M and N determines a unique tensor product (T, f) (up to isomorphism).

   The product is denoted by $M \otimes_R N$ or by simply $M \otimes N$.

(d) For any given R-module M, $M \otimes R \cong M \cong R \otimes M$.

30. (a) For any R-modules M, N and P, prove the following isomorphisms.

   (i) $M \oplus N \cong N \oplus M$;

   (ii) $(M \oplus N) \oplus P \cong M \oplus (N \oplus P)$;

   (iii) $M \otimes N \cong N \otimes M$;

   (iv) $(M \otimes N) \otimes P \cong M \otimes (N \otimes P)$;

   (v) $(M \oplus N) \otimes P \cong M \otimes P \oplus N \otimes P$;

   (vi) $P \otimes (M \oplus N) \cong P \otimes M \oplus P \otimes N$.

(b) Let **S** be the set of all R-modules. Then show that $(\mathbf{S}, \oplus, \otimes)$ forms a commutative semiring.

   [*Hint.* See Appendix A and use (a).]

31. Let G be a free **Z**-module of rank n and H be a subgroup of G of rank $m \le n$. Then the index $[G : H]$ is finite iff $m = n$. For $m = n$, let $\{x_1, x_2, \ldots x_n\}$ be a basis of G and $\{y_1, y_2, \ldots y_n\}$ be a basis of H. Write

$$\begin{pmatrix} y_1 \\ y_2 \\ \vdots \\ y_n \end{pmatrix} = A \begin{pmatrix} x_1 \\ x_2 \\ \vdots \\ x_n \end{pmatrix} \quad \text{with } A \in M_n(\mathbf{Z}).$$

Then show that $[G : H] = |\det A|$.

32. Let R and A be integral domains such that $R \subseteq A$. If R is a Noetherian domain and A is a finitely generated R-module, show that A is a Noetherian domain.

   [*Hint.* Let $I_1 \subseteq I_2 \subseteq \cdots$ be an ascending chain of ideals in A. Then by Proposition 9.8.1, A is a Noetherian R-module. Since each I_1 is an R-submodule of A, then the above chain must terminate.]

33. Let R be a ring with identity, M be an R-module and N be a submodule of M. Show that M is Noetherian iff both N and M/N are Noetherian.

    [*Hint.* Use the Corollary to Theorem 9.8.2 and Theorem 9.8.3.]

34. (*Chinese Remainder Theorem for Modules*) Let R be a commutative ring with 1 and $I_1, I_2, \ldots, I_t$ be ideals in R and M be an R-module.

    (a) The map $M \to M/I_1 M \oplus \cdots \oplus M/I_t M$, $x \mapsto (x + I_1 M, \ldots, x + I_t M)$ is an R-module homomorphism with kernel $I_1 M \cap I_2 M \cap \cdots \cap I_t M$;
    (b) if the ideals $I_1, I_2, \ldots, I_t$ in (a) are pairwise co-prime (i.e., $I_i + I_j = R$ for all $i \neq j$), then

$$M/(I_1 I_2 \cdots I_t)M \cong M/I_1 M \oplus \cdots \oplus M/I_t M.$$

    [*Hint.* The proof is similar to the proof of Chinese Remainder Theorem for rings.]

35. The concept of Lie algebra can be extended for modules over commutative rings with identity element.

    Let M be a module over a commutative ring with identity element. Then a bilinear map $M \times M \to M$, $(a, b) \mapsto [a, b]$ is said to make M a Lie algebra iff it satisfies the condition $[a, a] = 0$, $\forall a \in M$ and the Jacobi identity.

    (a) Let $M_n(R)$ be the ring of square matrices of order n over a commutative ring R with 1. If $x, y \in M_n(R)$, then the bilinear map $M_n(R) \times M_n(R) \to M_n(R)$, $(x, y) \mapsto [x, y] = xy - yx$ makes $M_n(R)$ into a Lie algebra.
    (b) Let $H = \{X \in M_n(R) : \text{trace } X = 0\}$ is also a Lie algebra and a (Lie) subalgebra of $M_n(R)$.

36. A nonzero ring R is said to be right (left) semisimple iff it is semisimple as a right (left) module over itself. Prove that the following statements are equivalent for a ring R with 1.

    (a) R is right semisimple;
    (b) R is isomorphic to a direct sum of a finite number of matrix rings of square matrices over division rings;
    (c) R is left semisimple.

    The result is known as Wedderburn-Artin Theorem.

37. Show that the following statements are equivalent for a semisimple module M:

    (a) M is Artinian;
    (b) M is Noetherian;
    (c) M is a direct sum of a finite number of minimum modules M.

38. Show that a semisimple module is finitely generated iff it is a sum of finitely many minimal submodules.

39. (a) Show that a simple R-module M is free iff R is a division ring and M is of dimension 1 over R;
    (b) Show that

    (i) a matrix ring $R = M_n(D)$ of square matrices of order n over a division
         ring D is semisimple;
   (ii) the matrix ring $R = M_n(\mathbf{Z})$ is not semisimple;
  (iii) a commutative ring is semisimple iff it is a finite direct product of
        fields.

9.11 Homology and Cohomology Modules

In this section we continue discussion of exact sequences.

The concept of homological algebra arose through the study of algebraic topology during the middle of the 20th century. Homological algebra borrows the language of algebraic topology, such as homology groups, homology modules, chains, boundaries, chain complex etc. The concept of cohomology modules is dual to that of homology modules. They are widely used in commutative algebra, algebraic geometry and algebraic topology.

Definition 9.11.1 A sequence $M = \{M_n, \partial_n\}$, $n \in \mathbf{Z}$, of R-modules M_n together with a sequence of R-homomorphisms $\partial_n : M_n \to M_{n-1}$, such that $\partial_n \circ \partial_{n+1} = 0$, $\forall n \in \mathbf{Z}$, is called a chain complex and ∂_n is called a boundary homomorphism.
    More precisely,

$$M : \cdots \to M_{n+1} \xrightarrow{\partial_{n+1}} M_n \xrightarrow{\partial_n} M_{n-1} \to \cdots \tag{9.9}$$

is called a chain complex iff $\partial_n \circ \partial_{n+1} = 0$, $\forall n \in \mathbf{Z}$.

Definition 9.11.2 The elements of $Z_n = \ker \partial_n$ are called n-*cycles*, the elements of $B_n = \operatorname{Im} \partial_{n+1}$ are called n-*boundaries* and the elements of M_n are called n-*chains* of the complex (9.9).

Proposition 9.11.1 B_n *is a submodule of* Z_n, $\forall n \in \mathbf{Z}$.

Proof It follows from the condition of a chain complex M that $\partial_n \circ \partial_{n+1} = 0$, $\forall n \in \mathbf{Z}$.   □

Definition 9.11.3 The quotient module Z_n/B_n for any chain complex M denoted $H_n(M)$ (or simply H_n) is called the n-dimensional homology module of the chain complex M. For $R = \mathbf{Z}$, we get the homology groups $H_n(M)$. The complex M said to be a cyclic iff $H_n(M) = 0$ for $\forall n \in \mathbf{Z}$. The elements of $H_n = Z_n/B_n$ are called *homology classes*, denoted $[z]$ for every $z \in Z_n$.

Remark If the homology module $H_n(C) = 0$, then the sequence (9.9) is exact at M_n. This shows that the homology module of a chain complex measures its deviation from the exactness of the sequence (9.9).

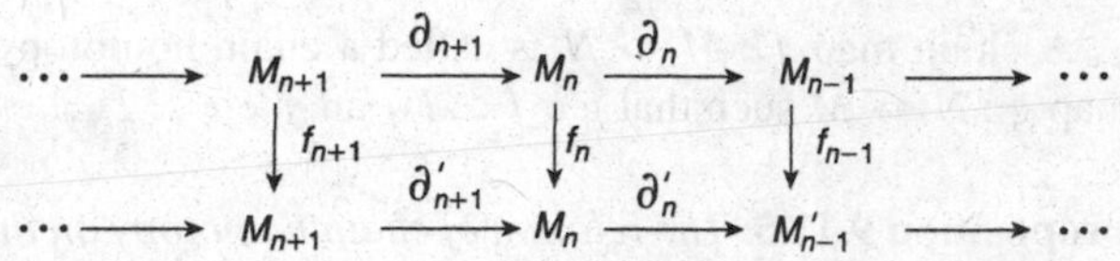

Fig. 9.11 Commutative
diagram of two chain
complexes

$$\cdots \longrightarrow M_{n+1} \xrightarrow{\partial_{n+1}} M_n \xrightarrow{\partial_n} M_{n-1} \longrightarrow \cdots$$
$$\downarrow f_{n+1} \qquad \downarrow f_n \qquad \downarrow f_{n-1}$$
$$\cdots \longrightarrow M_{n+1} \xrightarrow{\partial'_{n+1}} M_n \xrightarrow{\partial'_n} M'_{n-1} \longrightarrow \cdots$$

Definition 9.11.4 Let $M = \{M_n, \partial_n\}$, $n \in \mathbf{Z}$ and $M' = \{M'_n, \partial'_n\}$, $n \in \mathbf{Z}$ be two chain complexes of R-modules. Then a sequence $\{f_n : M_n \to M'_n\}$, $n \in \mathbf{Z}$ of R-homomorphisms is called a chain map from M to M' iff these R-homomorphisms commute with the boundary homomorphisms i.e., iff each square in the diagram of Fig. 9.11 is commutative, i.e., $f_{n-1} \circ \partial_n = \partial'_n \circ f_n$, $\forall n \in \mathbf{Z}$.

We abbreviate the entire collection to $f : M \to M'$ and call f a chain map.

Proposition 9.11.2 *Let $M = \{M_n, \partial_n\}$ and $M' = \{M'_n, \partial'_n\}$ be two chain complexes of R-modules and $\{f_n : M_n \to M'_n\}$ be a chain map. Then f_n maps n-cycles of M into n-cycles of M' and n-boundaries of M into n-boundaries of M', for all $n \in \mathbf{Z}$.*

Proof Left as an exercise. □

Proposition 9.11.3 *Let $M = \{M_n, \partial_n\}$ and $M' = \{M'_n, \partial'_n\}$ be two chain complexes of R-modules and $\{f_n : M_n \to M'_n\}$ be a chain map. Then each $f_n : M_n \to M'_n$ determines an R-homomorphism*

$$H_n(f_n) = f_{n*} : H_n(M) \to H_n(M'), \quad [z] \mapsto [f_n(z)].$$

Proof Left as an exercise. □

Definition 9.11.5 $H_n(f) = f_{n*} : H_n(M) \to H_n(M')$ is called the module homomorphism (or homomorphism) in homology induced by f_n for each $n \in \mathbf{Z}$.

We simply write f and f_* is places of f_n and f_{n*}, respectively, unless there is no confusion.

Proposition 9.11.4 (a) *If $f : M \to M'$ and $g : M' \to M''$ are two chain maps, then their composite $g \circ f : M \to M''$ is a chain map such that $(g \circ f)_* = g_* \circ f_* : H_n(M) \to H_n(M'')$;*

(b) *if $I_M : M \to M$ is the identity chain map, then $(I_M)_* : H_n(M) \to H_n(M)$ is also the identity homomorphism.*

Proof Left as an exercise. □

Definition 9.11.6 Let $M = \{M_n, \partial_n\}$ and $N = \{N_n, \partial'_n\}$ be two chain complexes and $f, g : M \to N$ be two chain maps. Then f is said to be chain homotopic to g, denoted $f \simeq g$, iff $\exists$ a sequence $\{F_n : M_n \to N_{n+1}\}$ of R-homomorphisms such that

$$\partial'_{n+1} F_n + F_{n-1} \partial_n = f_n - g_n : M_n \to N_n, \quad \forall n \in \mathbf{Z}.$$

A chain map $f : M \to N$ is called a chain homotopy equivalence iff $\exists$ a chain map $g : N \to M$ such that $g \circ f \simeq I_M$ an $f \circ g \simeq I_N$.

Proposition 9.11.5 *The relation of chain homotopy on the set $C(M, N)$ of all chain maps from M to N is an equivalence relation.*

Proof Left as an exercise.   □

Proposition 9.11.6 *Two homotopic chain maps $f, g : M \to N$ induce the same homomorphism in the homology, i.e., $f \simeq g : M \to N \Rightarrow f_* = g_* : H_n(M) \to H_n(N)$ $\forall n \in \mathbf{Z}$.*

Proof Let $f \simeq g : M \to N$. Then $\exists$ a chain homotopy $\{F_n : M_n \to N_{n+1}\}$. Let $[z] \in H_n(M)$. Then $\partial_n([z]) = 0 \Rightarrow f_n([z]) - g_n([z]) = \partial_{n+1} F_n([z])$ is a boundary $\Rightarrow [f_n[z]] = [g_n[z]] \Rightarrow f_n * ([z]) = g_n * [z], \forall [z] \in H_n(M) \Rightarrow f_n* = g_n* \Rightarrow f_* = g_*$.   □

A cohomology module is the dual to homology module. A cochain complex, cocycles, coboundaries, cochains, cohomology classes and a cohomology module are defined dually as follows:

Definition 9.11.7 A sequence $M^* = \{M^n, \delta^n\}$, $n \in \mathbf{Z}$, of R-modules M^n together with a sequence of R-homomorphisms $\delta^n : M^n \to M^{n+1}$, such that $\delta^n \circ \delta^{n-1} = 0$, $\forall n \in \mathbf{Z}$ is called a cochain complex and δ^n is called a coboundary homomorphism.
More precisely,

$$M^* : \cdots \to M^{n-1} \xrightarrow{\delta^{n-1}} M^n \xrightarrow{\delta^n} M^{n+1} \to \cdots \tag{9.10}$$

is called a cochain complex iff $\delta^n \circ \delta^{n-1} = 0, \forall n \in \mathbf{Z}$.

Definition 9.11.8 The elements of $Z^n = \ker \delta^n$ are called *n-cocycles*, the elements of $B^n = \text{Im} \, \delta^{n-1}$ are called *n-coboundaries* and the elements of M^n are called *n-cochains* of the cochain complex (9.10).

Proposition 9.11.7 *B^n is a submodule of Z^n, $\forall n \in \mathbf{Z}$.*

Proof It follows from the condition of a cochain complex M^* such that $\delta^n \circ \delta^{n-1} = 0, \forall n \in \mathbf{Z}$.   □

Definition 9.11.9 The quotient module Z^n / B^n for any cochain complex M^*, denoted $H^n(M^*)$ (or simply H^n) is called the *n-dimensional cohomology module* of the cochain complex M^*. For $R = \mathbf{Z}$, we get the cohomology group $H^n(M^*)$. The elements of $H^n = Z^n / B^n$ are called *cohomology classes*, denote $[z]$ for every $[z] \in Z^n$.

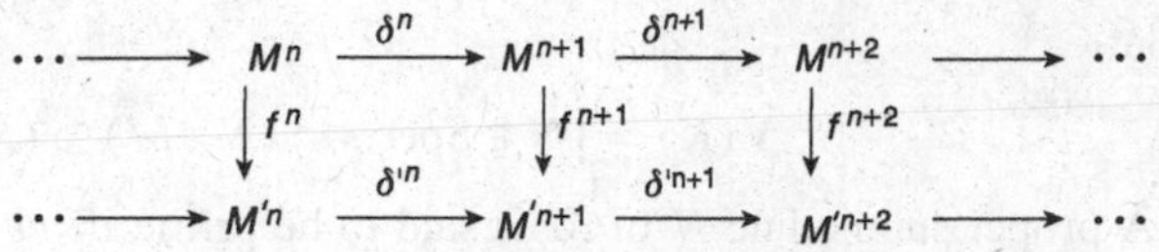

Fig. 9.12 Commutative
diagram of two cochain
complexes

$$
\begin{array}{ccccccc}
\cdots \longrightarrow & M^n & \xrightarrow{\;\delta^n\;} & M^{n+1} & \xrightarrow{\;\delta^{n+1}\;} & M^{n+2} & \longrightarrow \cdots \\
& \downarrow f^n & & \downarrow f^{n+1} & & \downarrow f^{n+2} & \\
\cdots \longrightarrow & M'^n & \xrightarrow{\;\delta'^n\;} & M'^{n+1} & \xrightarrow{\;\delta'^{n+1}\;} & M'^{n+2} & \longrightarrow \cdots
\end{array}
$$

Definition 9.11.10 Let $M^* = \{M^n, \delta^n\}$ and $M'^* = \{M'^n, \delta'^n\}$ be two cochain complexes of R-modules. Then a sequence $\{f^n : M^n \to M'^n\}$, $n \in \mathbf{Z}$ of R-homomorphisms is called a cochain map from M^* to M'^* iff these R-homomorphisms commute with the coboundary homomorphisms i.e., iff each square in the diagram of Fig. 9.12 is commutative, i.e., $f^{n+1} \circ \delta^n = \delta'^n \circ f^n$, $\forall n \in \mathbf{Z}$.

We can define a cochain homotopy between two cochain maps in a similar way as this definition.

9.11.1 Exercises

Exercises-II

1. Show that a chain complex $M = \{M_n, \partial_n\}$ of R-modules and their R-homomorphisms is exact iff $H_n(M) = 0$, $\forall n \in \mathbf{Z}$.
   [*Hint.* $Z_n = B_n \Leftrightarrow \ker \partial_n = \operatorname{Im} \partial_{n+1}$.]
2. Let M and N be two chain complexes of R-modules and R-homomorphisms and $f : M \to N$ is a chain homotopy equivalence. Show that $H_n(f) = f_* : H_n(M) \to H_n(N)$ is an R-isomorphism for all $n \in \mathbf{Z}$.
   [*Hint.* Let $f : M \to N$ be a chain homotopy equivalence. Then $\exists$ a chain homotopy equivalence $g : N \to M$ such that $g \circ f \simeq 1_M$ and $f \circ g \simeq I_N$.
   Then $(g \circ f)_* = g_* \circ f_* = 1d$ and $(f \circ g)_* = f_* \circ g_* = 1d \Rightarrow f_*$ is an isomorphism of R-modules.]
3. Let $f, g : M \to N$ and $h, k : N \to P$ be chain maps. Show that $f \simeq g$ and $h \simeq k \Rightarrow h \circ f \simeq k \circ g : M \to P$, i.e., the composites of chain homotopic maps are chain homotopic.

9.12 Topology on Spectrum of Modules and Rings

In this section we topologize the set of all prime ideals of rings and prime submodules of modules by defining closed sets. Moreover we study the properties of such spaces, with special reference to Zariski topology and schemes.

9.12.1 Spectrum of Modules

Let R be a ring and M an R-module. Define $\operatorname{Spec} M$ as the set of all prime submodules of M and call it as the spectrum of M. For any prime submodule K of M, put

$$V(K) = \{N \in \operatorname{Spec} M : (N : M) \supset (K : M)\}$$

(A proper submodule N of M is said to be prime iff $re \in N$, for $r \in R$, $e \in M \Rightarrow$ either $e \in N$ or $r \in (N : M)$.)

(a) Show that $X = \operatorname{Spec} M$ satisfies the following properties:

  (i) $V(0) = \operatorname{Spec} M$ and $V(M) = \emptyset$;
  (ii) $\bigcap_{t \in I} V(K_t) = V(\sum_{\langle t \in I \rangle}(K_t : M)M)$;
  (iii) $V(K_1) \cup V(K_2) = V(K_1 \cap K_2)$, where K_t's are submodules of M.

  These results show that the sets $V(K)$ satisfy axioms for closed sets in a topological space. Then $\operatorname{Spec} M$ becomes a topological space.

(b) Prove the following:

  (i) Let Y be a closed subset of $\operatorname{Spec} M$. Then Y is irreducible (i.e., every pair of non-empty closed sets in Y intersect) $\Leftrightarrow Y = V(N)$ for some prime submodule N of M (this N is unique in the sense that any other submodule N' of M with $Y = V(N') \Rightarrow (N : M) = (N' : M)$).
  (ii) Let M be a finitely generated R-module. Then $X = \operatorname{Spec} M$ is irreducible $\Leftrightarrow \bigcap_{\langle N \in X \rangle} N$ is a prime submodule of M.
  (iii) M is a Noetherian R-module $\Rightarrow \operatorname{Spec} M$ is a Noetherian space.

  [*Hint.* $X_1 \supseteq X_2 \supseteq \cdots$ be any decreasing sequence of closed subsets of spec M. Then we can write $X_i = V(K_i)$ for every i. Define $K_i' = \bigcap_{\langle N \in X_i \rangle} N$. Then $V(K_i) = V(K_i')$ and $\{K_i'\}$ is an increasing sequence of submodules of M. Since M is a Noetherian R-module, this sequence becomes stationary and hence $\operatorname{Spec} M$ is a Noetherian space.]

9.12.2 Spectrum of Rings and Associated Schemes

The prime spectrum or simply spectrum of a ring R is the set $\operatorname{Spec} R$ of all prime ideals of R, i.e., $\operatorname{Spec} R = \{P : P \text{ is a prime ideal of } R\}$. Let $\operatorname{Max}(R) = \{M : M$ is a maximal ideal of $R\}$. We now consider commutative rings with identity element. Then by Theorem 5.3.3, $\operatorname{Max}(R) \subseteq \operatorname{Spec} R$. $\operatorname{Max}(R)$ is called the maximal spectrum of R.

Let R and T be two commutative rings with identity elements and $f : R \to T$ be a ring homomorphism. Then f induces a function $\operatorname{Spec} f : \operatorname{Spec} T \to \operatorname{Spec} R$, defined by

$$(\operatorname{Spec} f)(Q) = f^{-1}(Q), \quad \forall Q \in \operatorname{Spec} T. \tag{9.11}$$

Q is a prime ideal of $T \Rightarrow T/Q$ is an integral domain by Theorem 5.3.1. Now the ring $R/f^{-1}(Q)$ is isomorphic to a subring of $T/Q \Rightarrow R/f^{-1}(Q)$ in an integral domain $\Rightarrow f^{-1}(Q)$ is a prime ideal of $R \Rightarrow \operatorname{Spec} f$ is well defined.

For any set $X \subseteq R$, define $V(X) = \{P \in \operatorname{Spec} R : X \subseteq P\}$.

If I is the ideal generated by X, then $V(X) = V(I) = V(\operatorname{rad} I)$.

Spec R satisfies the following properties:

(i) $V(O) = \text{Spec } R$ and $V(R) = \emptyset$;
(ii) $\bigcap_{\alpha \in A} V(X_\alpha) = V(\bigcup_{\alpha \in A} X_\alpha)$, where $\{X_\alpha\}_{\alpha \in A}$ is any family of subsets of R.
(iii) $V(I_1) \cup V(I_2) = V(I_1 \cap I_2)$, where I_1 and I_2 are prime ideals of R.

Thus the family of subsets of Spec R which are of the form $V(X)$ is closed with respect to any intersection and finite unions, contains the empty set $\emptyset$ and the entire set Spec R. Hence there exists a unique topology on the set Spec R in which closed sets are of the form $V(X)$ for $X \subseteq R$.

The set Spec R of all prime ideals of R, together with the family $\{V(X)_{X \subseteq R}\}$ as the class of closed sets, forms a topological space. This topology of Spec R is called the *Zariski topology*.

Problem 1 The topological space Spec R endowed with Zariski topology is a Hausdorff space iff Spec $R = \text{Max}(R)$.

Problem 2 The induced function Spec $f : \text{Spec } T \to \text{Spec } R$, defined by (9.11) is a continuous function with respect to Zariski topology.

Solution Given a set $X \subseteq R$, $(\text{Spec } f)^{-1}(V(X)) = \{Q \in \text{Spec } T : \text{Spec } f(Q) \in V(X)\} = \{Q \in \text{Spec } T : f^{-1}(Q) \in V(X)\} = \{Q \in \text{Spec } T : X \subseteq f^{-1}(Q)\} = \{Q \in \text{Spec } T : f(X) \subseteq Q\} = V(f(X)) \Rightarrow \text{Spec } f$ is continuous.

Definition 9.12.1 The spectrum of a ring, endowed with Zariski topology is called a *scheme*.

The name '*scheme*' was given by Alexander Grothendieck. He developed the theory of schemes which made a revolution in algebraic geometry. He was awarded 'Fields Medal' in 1966 in recognition of his contribution to the theorey of schemes. An affine scheme comes with a 'sheaf of rings'.

For further results of this section see Ex. 12 of Appendix B.

9.13 Additional Reading

We refer the reader to the books (Adhikari and Adhikari 2003; Atiyah and MacDonald 1969; Blyth 1977; Herstein 1964; Hilton and Stammbach 1971; Lambek 1966; Lang 1986; Musili 1992) for further details.

References

Adhikari, M.R., Adhikari, A.: Groups, Rings and Modules with Applications, 2nd edn. Universities Press, Hyderabad (2003)
Atiyah, M.F., MacDonald, I.G.: Introduction to Commutative Algebra. Addison-Wesley, Reading (1969)

Blyth, T.S.: Module Theory: An Approach to Linear Algebra. Oxford University Press, London (1977)

Herstein, I.: Topics in Algebra. Blaisdell, New York (1964)

Hilton, P.J., Stammbach, U.: A Course in Homological Algebra. Springer, Berlin (1971)

Lambek, J.: Lectures on Rings and Modules. Blaisdell, Waltham (1966)

Lang, S.: Introduction to Linear Algebra. Springer, New York (1986).

Musili, C.: Introduction to Rings and Modules. Narosa, New Delhi (1992)

Chapter 10
Algebraic Aspects of Number Theory

There is no doubt, number theory has long been one of the favorite subjects not only for the students but also for the teachers of mathematics. It is a classical subject and has a reputation for being the "purest" part of mathematics, yet recent developments in cryptology, coding theory and computer science are based on elementary number theory. Number theory, in a general sense, is the study of numbers and their properties. In Chap. 1, we have already seen some of the basic but important concepts of integers, such as, Peano's Axiom, well ordering principle, division algorithm, greatest common divisors, prime numbers, fundamental theorem of arithmetic etc. In this chapter, we discuss some more interesting properties of integers, in particular properties of prime numbers, and primality testing. In addition, we study the applications of number theory, particularly those directed towards theoretical computer science. Number theory has been used in many ways to devise algorithms for efficient computer and for computer operations with large integers. Both algebra and number theory play an increasingly significant role in computing and communication, as evidenced by the striking applications of these subjects to the fields of coding theory and cryptology. The motivation of this chapter is to provide an introduction to the algebraic aspects of number theory, mainly the study of development of the theory of prime numbers with an emphasis on algorithms and applications, which would be necessary for studying cryptology to be discussed in Chap. 12. In this chapter, we start with the introduction to prime numbers with a brief history. We provide several different proofs of the celebrated theorem by Euclid stating that there exist infinitely many primes. We further discuss Fermat number, Mersenne numbers, Carmichael numbers, quadratic reciprocity, multiplicative functions, such as, Euler ϕ-function, number of divisor functions, sum of divisor functions. This chapter ends with the discussions on primality testing both deterministic and probabilistic, such as, Solovay–Strassen and Miller–Rabin probabilistic primality tests.

10.1 A Brief History of Prime Numbers

Prime numbers belong to an exclusive world of intellectual conceptions. Around 2500 years ago, the ancient Greeks in the school of Pythagoras were interested in

M.R. Adhikari, A. Adhikari, *Basic Modern Algebra with Applications*,
DOI 10.1007/978-81-322-1599-8_10, © Springer India 2014

numbers for their mystical and numerological properties. They made the distinction between composite numbers and prime numbers. They understood the idea of primality and were interested in perfect numbers (a number whose sum of the proper divisors is the number itself, e.g., 6, 28 etc.) and pairs of amicable numbers (i.e., a pair of numbers such that the proper divisors of one number is the sum to the other and vice versa, e.g., 220 and 284).

The book *Elements* written by Euclid, an ancient Greek mathematician, appeared in about 300 BC. In the Book IX of the Elements, Euclid included a proof that there are infinitely many primes. This is considered to be one of the first proofs known which uses the method of contradiction to establish a result. Euclid also came up with a proof of the Fundamental Theorem of Arithmetic which states that "*Every integer can be written as a product of primes in an essentially unique way*". This factorization can be found by trial division of the integer by primes less than its square-root. Euclid also proved that a number $2^{n-1}(2^n - 1)$ is a perfect number if the number $2^n - 1$ is prime. Euler showed that all even perfect numbers are exactly of this form. Till date, it is not known whether there are any odd perfect numbers.

In 17th century, Pierre de Fermat (1601–1665) took a leading role for the advancement of the theory of prime numbers. He proved a speculation of Albert Girard that every prime number of the form $4n + 1$ can be written in a unique way as the sum of two squares. He further proved a theorem which is now known as Fermat's Little theorem. The theorem states that if p is prime and a be an integer such that $\gcd(a, p) = 1$, that is a and p are relatively prime, then, p divides $a^{p-1} - 1$. Fermat's Little theorem is the basis for many results in number theory and is one of the basis for checking primality testing techniques which are still in use on today's electronic computers. He discovered a technique for factorizing large numbers, which he illustrated by factorizing the number $2027651281 = 44021 \times 46061$. Although Fermat claimed that he proved all of his theorems, a very few records of his proofs have survived. Many mathematicians, including Gauss, cast doubts on his several claims, because of the difficulty of some of the problems and the limited mathematical tools available to Fermat. Out of his claims, *Fermat's Last theorem* is the most celebrated one. The theorem was first proposed by Fermat in the form of a note scribbled in the margin of his copy of the ancient Greek text Arithmetica by Diophantus. The theorem states that the Diophantine equation $x^n + y^n = z^n$ has no integer solutions for $n > 2$ and $x, y, z \neq 0$. Fermat left no proof of the conjecture for all n, except for the special case $n = 4$. In the note, Fermat claimed that he discovered a proof of the whole theorem, but due to the small space of the margin he could not write the proof. It was called a "theorem" on the strength of Fermat's statement, despite the fact that mathematicians failed to prove it for hundreds of years. No successful proof was published until 1995, when Andrew Wiles came up with a correct successful proof the theorem. This single theorem simulates many new development of algebraic number theory in the 19th century and the proof of the modularity theorem in the 20th century.

Factorization is considered to be another interesting research area in the field of number theory. Though Euclid came up with a proof of the Fundamental Theorem of Arithmetic, the factorization are found by trial division of the integer by primes less than its square-root. For large enough numbers, clearly this method is inefficient. Fermat, Euler, and many other mathematicians have produced imaginative factorization techniques. However, using the most efficient technique and modern computer facility yet devised, billions of years of computer time may be required to factor a suitably chosen integer even with 300 decimal digits.

From the very early development of prime numbers, mathematicians were very much interested in finding formulas that can generate primes. Fermat conjectured that the numbers $2^n + 1$ are always prime if n is a power of 2. He verified this for $n = 1, 2, 4, 8$, and 16. Numbers of this form are called Fermat numbers. But a century after, his claim was shown to be wrong by the renowned Swiss mathematician Leonard Euler, who discovered that 641 is a factor of $2^{2^5} + 1 = 4294967297$.

The German mathematician Carl Friedrich Gauss, considered to be one of the greatest mathematicians of all time, developed the language of congruences in the early 19th century. When doing certain computations, integers may be replaced by their remainders when divided by a specific integer, using the language of congruences. Many questions can be phrased using the notion of a congruence that can only be awkwardly stated without this terminology. Congruences have diverse applications to computer science, including applications to computer file storage, arithmetic with large integers, and the generation of pseudo-random numbers.

The problem of distinguishing primes from composites, known as primality testing, has been extensively studied. The ancient Greek scholar Eratosthenes discovered a method, now called the sieve of Eratosthenes. This method finds all primes less than a specified limit. Ancient Chinese mathematicians believed that the primes were precisely those positive integers n such that n divides $2^n - 2$. Fermat proved one part of the belief by showing that if n is prime, then n divides $2^n - 2$. The other part of the belief was proven to be wrong, in the early 19th century, by showing the existence of composite integers n such that n divides $2^n - 2$, such as $n = 341$. However, it is possible to develop probabilistic primality tests based on the original Chinese belief. In current days, it is now very well possible to efficiently find primes; in fact, primes with as many as 400 decimal digits can be found in a few minutes of computer time. For example, 1152163959944035801782591958215936605022 9133173385690566201793635642 8602095117223277 57086072102350660443990 98144500755041680003586821792652531770745659087273762673860 8284693 3890552946646764738540931231769626871804922231181205228833608035170 680964347 442368708701479356815648063495583572596496615099354396442 21493 is a prime number having 309 digits. It is generated by using computers through the SAGE software. These large primes numbers are very useful in the context of cryptology. The problem of efficiently determining whether a given integer is prime or not was a longstanding and challenging problem. However, some efficient probabilistic algorithms for primality testing are available, such as Miller–Rabin test, Solovay–Strassen test etc. The breakthrough came in 2002 as Agrawal, Kayal, and Saxena came up with a deterministic efficient algorithm for primality testing.

10.2 Some Properties of Prime Numbers

Theorem 10.2.1 *Every positive integer greater than* 1 *has a prime divisor.*

Proof If possible, let there be a positive integer greater than 1 having no prime divisors. Let $S = \{x : x$ is a positive integer greater than 1 having no prime divisors$\}$. Then $S (\neq \emptyset) \subseteq \mathbf{N}^+$. Thus by the well ordering principle, S must have a least element, say l. Since l has no prime divisors and l divides l, l cannot be a prime integer. Thus $l = ab$, where $1 < a < l$ and $1 < b < l$. As $a < l$, a must have a prime divisor, say x. As, a divides l, x must divide l, contradicting the fact that l does not have a prime divisor. Thus we conclude that every positive integer greater than 1 has a prime divisor.   □

A natural question that comes to our mind: *Is the set of all primes finite?* The answer to this question was given long back by Euclid by proving the following celebrated theorem in his Book IX of *Elements*.

Theorem 10.2.2 *There are infinitely many primes.*

Proof If possible, let there exist only k primes: say $p_1 < p_2 < \cdots < p_k$. Let $N = p_1 p_2 \cdots p_k + 1$. Then $N > 1$ and hence by Theorem 10.2.1, N has a prime factor, say p. Then p must be one of $p_1, p_2, \ldots, p_k$. Thus p divides $p_1 p_2 \cdots p_k = N - 1$. Hence p divides $N - (N - 1) = 1$, a contradiction, proving that there exist infinitely many primes.   □

Corollary 1 *Let p_n denote the nth prime. Then $p_n \leq 2^{2^{n-1}}$, for $n \geq 1$.*

Proof We shall prove the result by strong mathematical induction. Clearly, the result is true for $n = 1$. We assume that the result is true for $n = 2, 3, \ldots, k$. We shall prove the result for $n = k + 1$. From Theorem 10.2.2, we have seen that $N = p_1 p_2 \cdots p_k + 1$ is divisible by a prime p and that prime p is not equal to any of the p_i's, $i \in \{1, 2, \ldots, k\}$. So, the prime p must be greater than equal to p_{k+1}. Thus $p_{k+1} \leq p \leq p_1 p_2 \cdots p_k + 1 \leq 2^{2^0} 2^{2^1} \cdots 2^{2^{k-1}} + 1 = 2^{2^k - 1} + 1 = \frac{1}{2} 2^{2^k} + 1 \leq 2^{2^k}$. So the result is true for $k + 1$. Hence by the strong mathematical induction, the result is true for all $n \geq 1$.   □

Remark The estimation given above is very weak as p_n is much smaller than $2^{2^{n-1}}$. For example, for $n = 5$, $p_5 = 11$ while $2^{2^{5-1}} = 65536$.

Corollary 2 *For $x > 0$, let $\pi(x)$ denote the number of primes less than or equal to x. Then $\pi(x) \geq \lfloor \log_2(\log_2 x) \rfloor + 1$.*

Proof Note that $\lfloor \log_2(\log_2 x) \rfloor + 1$ is the largest integer n such that $2^{2^{n-1}} \leq x$. Then by the above Corollary 1, there are at least n primes, say $p_1, p_2, \ldots, p_n$ such that

$p_i \leq 2^{2^{n-1}}$, for $i = 1, 2, \ldots, n$. All these primes are less than or equal to x and so
$\pi(x) \geq n = \lfloor \log_2(\log_2 x) \rfloor + 1$. $\square$

Remark As before, this bound is also very weak. For example, let $x = 2^{2^4}$. Then
$\lfloor \log_2(\log_2 x) \rfloor + 1 = 5$, while $\pi(x) = 6542$.

In the literature of number theory, there have been many proofs of Theorem 10.2.2. In this chapter, we shall provide several such proof techniques.

In 1878, Kummer came up with a very simple proof.

Alternative Proof of Theorem 10.2.2 Suppose that there exist only finitely many primes, $p_1 < p_2 < \cdots < p_r$. Let $N = p_1 p_2 \cdots p_r > 2$. Let us consider the integer $N - 1 > 1$. Then by Theorem 10.2.1, $N - 1$ must have a prime divisor, say p_i, common with N. So, p_i divides both N and $N - 1$, resulting in: p_i divides $N - (N - 1) = 1$, a contradiction. Hence the number of primes is infinite. $\square$

In 1915, H. Brocard gave another simple proof that was published in the *Intermédiaire des Mathématiciens 22, page 253*, attributed to Hermite.

Alternative Proof of Theorem 10.2.2 Let us consider an integer $s_n = n! + 1$, $n \geq 1$. Then by Theorem 10.2.1, s_n must have a prime divisor, say p_n. We claim that $p_n > n$. If not, i.e., if $p_n \leq n$, then p_n divides $n! = s_n - 1$, leads to the fact that p_n divides 1. Thus we find a prime larger than n, for every positive integer n. This implies that there exist infinitely many primes. $\square$

Alternative Proof of Theorem 10.2.2 If possible, let there exist finitely many primes. Let P be the product of all these primes. Let $P = ab$ be any factorization of P with positive integers a and b. Now for any prime p, either a or b is divisible by p, but not both. This implies that the positive integer $a + b$ cannot have any prime factor, contradicting Theorem 10.2.1. $\square$

In 1917, Métro gave another simple proof.

Alternative Proof of Theorem 10.2.2 If possible let there exist only k primes: say $p_1 < p_2 < \cdots < p_k$. Let P be the product of all these primes, i.e., $P = p_1 p_2 \cdots p_k$. For each $i = 1, 2, \ldots, k$, let $Q_i = P/p_i$. Then p_i does not divide Q_i for each i, while p_i divides Q_j for $i \neq j$. Let $S = Q_1 + Q_2 + \cdots + Q_k$. As $S > 1$, by Theorem 10.2.1, S must have a prime divisor, say q. Then clearly, $q \neq p_i$, for all $i = 1, 2, \ldots, k$, contradicting the fact that there exist only k primes, $p_1, p_2, \ldots, p_k$. Thus there exist infinitely many primes. $\square$

Euler gave an indirect proof of Theorem 10.2.2.

Alternative Proof of Theorem 10.2.2 If possible, let there exist finitely many primes, say $p_1, p_2, \ldots, p_n$. For each $i = 1, 2, \ldots, n$ consider,

$$\sum_{k=0}^{\infty} \frac{1}{p_i^k} = \frac{1}{(1 - \frac{1}{p_i})}.$$

Multiplying these n equalities, we obtain

$$\prod_{i=1}^{n} \sum_{k=0}^{\infty} \frac{1}{p_i^k} = \prod_{i=1}^{n} \frac{1}{(1 - \frac{1}{p_i})},$$

where the left hand side is the sum of the inverses of all the natural numbers, each counted once—this follows from the fundamental theorem of arithmetic that every natural number, greater than 1, can be represented as the product of the primes in a unique way. But the series $\sum_{n=1}^{\infty} \frac{1}{n}$ is divergent. Being a series of positive terms, the order of the summation is irrelevant. So, the left hand side is infinite, while the right hand side is clearly finite. This is absurd.   □

Another famous result of number theory deals with the existence of infinitely many primes in arithmetic progressions. The result is known as Dirichlet's Theorem on primes in arithmetic progressions.

Statement of the Theorem *Let a and b be relatively prime positive integers. Then the arithmetic progression $an + b, n = 1, 2, 3, \ldots$ contains infinitely many primes.*

G. Lejenue Dirichlet proved the theorem in 1837. But the proof technique is out of the scope of this book. However, a special case of Dirichlet's Theorem is illustrated as follows.

Theorem 10.2.3 *There are infinitely many primes of the form $4k + 3$ where $k = 0, 1, 2, \ldots$.*

Proof If possible, let there exist only finitely many primes, say $p_1, p_2, \ldots, p_r$ of the form $4k + 3, k = 0, 1, 2, \ldots$. Let $N = 4p_1 p_2 \cdots p_r - 1$. Then N is of the form $4k + 3$ where k is a non-negative integer. Let p be a prime divisor of N. As N is odd, the form of p is either $4k + 1$ or $4k + 3$ for some integer k. Note that if each of the prime divisors p of N is of the form $4k + 1$, then N must also have the same form, contradicting the fact that N is of the form $4k + 3$. Thus N is divisible by at least one prime p of the form $4k + 3$. So by hypothesis, $p = p_i$ for some $i = 1, 2, \ldots, r$. This implies that p divides $4p_1 p_2 \cdots p_r - N = 1$, a contradiction, proving the fact that there are infinitely many primes of the form $4k + 3$ where $k = 0, 1, 2, \ldots$.   □

10.2.1 Prime Number Theorem

Euclid's Theorem on the infinitude of the primes is considered to be the first result on the distribution of primes. In 1737 Euler went a step further and proved that, in fact, the series $\frac{1}{2} + \frac{1}{3} + \frac{1}{5} + \frac{1}{7} + \frac{1}{11} + \cdots$, i.e., the series of the reciprocals of the primes, diverges. On the other hand, Euler further observed that the rate of divergence of this series is much slower than the rate of divergence of the harmonic series $\frac{1}{1} + \frac{1}{2} + \frac{1}{3} + \frac{1}{4} + \frac{1}{5} + \cdots$. This statement appears to be the earliest attempt to quantify the frequency of the primes among the positive integers.

In 1793, Gauss conjectured that

$$\lim_{n \to \infty} \frac{\pi(n)}{n/\log n} = 1,$$

where $\pi(n)$ denotes the number of primes not exceeding a given positive integer n. This result is called the Prime Number Theorem (PNT) which describes the asymptotic distribution of the prime numbers. Gauss further observed that the logarithmic integral $\mathrm{li}(x) = \int_2^x \frac{dt}{\log t}$ seemed to provide a very good approximation for $\pi(x)$. In 1986, Jacques Hadamard and Charles Jean de la Vallée-Poussin independently came with a proof of the prime number theorem.

Remark Already we have shown that there are infinitely many primes. Now the following theorem will show that there are arbitrary long runs of integers containing no primes.

Theorem 10.2.4 *Given any positive integer n, there exist n consecutive composite integers.*

Proof Consider the integers, $(n+1)! + 2, (n+1)! + 3, \ldots, (n+1)! + n, (n+1)! + n + 1$. Every one of the above integers is composite, since j divides $(n+1)! + j$ if $2 \le j \le n+1$. $\qquad\square$

Example 10.2.1 Note that $8! + 2, 8! + 3, \ldots, 8! + 8$ are 7 consecutive composite integers. However, these are much larger than the smallest seven consecutive composites: $90, 91, 92, \ldots, 96$.

10.2.2 Twin Primes

Theorem 10.2.4 shows that there are arbitrary long runs of integers containing no primes. On the other hand, primes may often close together. Note that the only consecutive primes are 2 and 3. However, there may exist pairs of primes which differ by 2. These pairs of primes are known as twin primes. The pairs $(3, 5)$, $(5, 7)$, $(11, 13)$, $(17, 19)$, $(29, 31)$, $(41, 43)$, $(59, 61)$, $(71, 73)$, $(101, 103)$, $(107, 109)$,

(137,139), (149,151), (179,181), (191,193), (197,199), (227,229) are some of the examples of twin prime pairs.

The twin primes are used to define a very special type of constant, known as Brun's constant. In 1919, Viggo Brun proved that the sum of the reciprocals of all the twin primes, i.e., $(\frac{1}{3} + \frac{1}{5}) + (\frac{1}{5} + \frac{1}{7}) + (\frac{1}{11} + \frac{1}{13}) + \cdots$, converges to a finite value, known as Brun's constant. In contrast, the series of all prime reciprocals diverges to infinity.

Despite this proof by Viggo Brun regarding Brun's constant, the question of whether there exist infinitely many twin primes has been one of the great open questions in number theory for many years. The famous twin prime conjecture states that there are infinitely many primes p such that $p + 2$ is also prime. In 1849 de Polognac made the more general conjecture that for every natural number k, there are infinitely many prime pairs p and p' such that $p' - p = 2k$. When $k = 1$, we get the twin prime conjecture.

10.3 Multiplicative Functions

In number theory, multiplicative functions play an important role. In this section, we first study some general properties of this multiplicative function. Then we shall discuss some special multiplicative functions, such as, Euler phi-function, the number of positive divisor function and the sum of divisor function. Let us start with some definitions.

Definition 10.3.1 A function that is defined for all positive integers is called an arithmetic function.

We are interested in a particular type of arithmetic function which leads to the following definition.

Definition 10.3.2 An arithmetic function f is said to be a multiplicative function iff $\forall a, b \in \mathbf{N}^+$ with $\gcd(a, b) = 1$ we have $f(ab) = f(a)f(b)$.

Example 10.3.1 The functions $f, g : \mathbf{N}^+ \to \mathbf{N}^+$ defined by $f(n) = 1$ and $g(n) = n$, $\forall n \in \mathbf{N}^+$ are examples of multiplicative functions.

Given the prime factorization of a positive integer, a simple formulation for a multiplicative function can be found through the following theorem.

Theorem 10.3.1 *Let f be a multiplicative function and for the natural number n, let $n = p_1^{\alpha_1} p_2^{\alpha_2} \cdots p_r^{\alpha_r}$ be the prime factorization of n, where p_i's are distinct primes and $\alpha_i \geq 1$, for $i = 1, 2, \ldots, r$. Then $f(p_1^{\alpha_1} p_2^{\alpha_2} \cdots p_r^{\alpha_r}) = f(p_1^{\alpha_1})f(p_2^{\alpha_2}) \cdots f(p_r^{\alpha_r})$.*

Proof We shall prove the result by using the principle of mathematical induction on r. For $r = 1$, the result is obvious. So, we assume that the result is true for some $k > 1$, i.e., $f(p_1^{\alpha_1} p_2^{\alpha_2} \cdots p_k^{\alpha_k}) = f(p_1^{\alpha_1}) f(p_2^{\alpha_2}) \cdots f(p_k^{\alpha_k})$. Now, as $\gcd(p_1^{\alpha_1} p_2^{\alpha_2} \cdots p_k^{\alpha_k}, p_{k+1}^{\alpha_{k+1}}) = 1$, $f(p_1^{\alpha_1} p_2^{\alpha_2} \cdots p_k^{\alpha_k} p_{k+1}^{\alpha_{k+1}}) = f(p_1^{\alpha_1} p_2^{\alpha_2} \cdots p_k^{\alpha_k})$ $f(p_{k+1}^{\alpha_{k+1}}) = f(p_1^{\alpha_1}) f(p_2^{\alpha_2}) \cdots f(p_k^{\alpha_k}) f(p_{k+1}^{\alpha_{k+1}})$, proving the result for $k + 1$. Hence by the principle of mathematical induction, we have the required result. □

Next we shall prove another important result that will be very useful in proving other results on multiplicative functions of a special type.

Theorem 10.3.2 *Let f be a multiplicative function. Then the arithmetic function F defined by $F(n) = \sum_{d \mid n} f(d)$ is also a multiplicative function.*

Proof Let $\gcd(a, b) = 1$. Note that for each pair of divisors d_1 of a and d_2 of b, there corresponds a divisor $d = d_1 d_2$ of ab. On the other hand each positive divisor of ab can be written uniquely as the product of relatively prime divisors d_1 of a and d_2 of b. Thus

$$F(ab) = \sum_{d \mid ab} f(d) = \sum_{\substack{d_1 \mid a \\ d_2 \mid b}} f(d_1 d_2) = \sum_{\substack{d_1 \mid a \\ d_2 \mid b}} f(d_1) f(d_2)$$

$$= \sum_{d_1 \mid a} f(d_1) \sum_{d_2 \mid b} f(d_2) = F(a) F(b),$$

proving that F is a multiplicative function. □

Next we discuss a special type of multiplicative function, known as Euler phi-function, which is not only very useful in number theory but also plays an important role in constructing many important cryptographic schemes.

10.3.1 Euler phi-Function

Among all the multiplicative functions, Euler phi-function is the most important function in number theory. This function, named after the great Swiss mathematician Leonhard Euler (1707–1783), is defined as follows.

Definition 10.3.3 For a positive integer n, the Euler phi-function, denoted by $\phi(n)$, is defined to be the number of positive integers not exceeding n and relatively prime to n.

Euler phi-function and the ring $\mathbf{Z}_n$, for $n > 1$, have a nice connection. To deduce the relation, let us first prove the following lemma.

Lemma 10.3.1 *An equivalence class (a) in $\mathbf{Z}_n$ is a unit in $\mathbf{Z}_n$ if and only if $\gcd(a, n) = 1$.*

Proof Let (a) in $\mathbf{Z}_n$ be a unit in $\mathbf{Z}_n$. Then there exists an equivalence class (b) in $\mathbf{Z}_n$ such that $(a)(b) = (1)$. Thus $ab \equiv 1 \pmod{n}$, i.e., $ab = 1 + kn$ for some integer k. This implies $ab + n(-k) = 1$, with the result that $\gcd(a, n) = 1$, by the Corollary of Theorem 1.4.5.

Conversely, let $\gcd(a, n) = 1$. Then there exist integers u and v such that $au + nv = 1$ which implies $au \equiv 1 \pmod{n}$, with the result $(a)(u) = (1)$. So, (a) is a unit in $\mathbf{Z}_n$. $\qquad\square$

The immediate consequence of the above lemma is the following theorem, which provides a nice connection between number theory and ring theory.

Theorem 10.3.3 *For $n > 1$, the number of units of the ring $\mathbf{Z}_n$ is precisely $\phi(n)$.*

Proof The proof follows directly from Lemma 10.3.1. $\qquad\square$

Now we shall study some of the basic properties of the Euler phi-function. We start by showing that the Euler phi-function is a multiplicative function. To prove this we need the following lemmas.

Lemma 10.3.2 *Let n and m be two positive integers such that $\gcd(n, m) = 1$. Then $((a), (b))$ is a generator of $\mathbf{Z}_n \times \mathbf{Z}_m$ if and only if (a) is a generator of $\mathbf{Z}_n$ and (b) is a generator of $\mathbf{Z}_m$.*

Proof Let $((a), (b)) \in \mathbf{Z}_n \times \mathbf{Z}_m$ be a generator of $\mathbf{Z}_n \times \mathbf{Z}_m$. Let $(f) \in \mathbf{Z}_n$ and $(g) \in \mathbf{Z}_m$. Then $((f), (g)) \in \mathbf{Z}_n \times \mathbf{Z}_m$. Hence there exists an integer x such that $((f), (g)) = x((a), (b))$. This implies that $(f) = x(a)$ and $(g) = x(b)$, implying that $\mathbf{Z}_n = \langle(a)\rangle$ and $\mathbf{Z}_m = \langle(b)\rangle$, proving that (a) is a generator of $\mathbf{Z}_n$ and (b) is a generator of $\mathbf{Z}_m$.

Conversely, let (a) be a generator of $\mathbf{Z}_n$ and (b) be a generator of $\mathbf{Z}_m$. The order of (a) and (b) are, respectively, n and m. Now $nm((a), (b)) = ((0)_{\mathbf{Z}_n}, (0)_{\mathbf{Z}_m}) = 0_{\mathbf{Z}_n \times \mathbf{Z}_m}$. Further let $\lambda((a), (b)) = ((0)_{\mathbf{Z}_n}, (0)_{\mathbf{Z}_m})$. This implies n divides λ and m divides λ. As m and n are co-prime, nm divides λ, implying that the order of $((a), (b))$ is mn, proving that $((a), (b))$ is a generator of $\mathbf{Z}_n \times \mathbf{Z}_m$. $\qquad\square$

The following lemma again provides a nice connection between number theory and cyclic group.

Lemma 10.3.3 *The number of generators of a finite cyclic group of order n, $n \geq 2$, is $\phi(n)$.*

Proof Let G be a finite cyclic group of order n. Then $G \simeq \mathbf{Z}_n$. Let (g) be a generator of $\mathbf{Z}_n$. As $(1) \in \mathbf{Z}_n$, $(1) = (m)(g)$, for some integer $m \in \mathbf{Z}$. Thus there exists some integer k such that $1 - mg = kn$, i.e., $kn + mg = 1$. Hence by the Corollary of Theorem 1.4.5, we have $\gcd(n, g) = 1$. Thus the number of generators of $\mathbf{Z}_n$ is $\phi(n)$. $\square$

Now we are in a position to prove the following theorem.

Theorem 10.3.4   *Euler phi-function is a multiplicative function.*

Proof Let n and m be two positive integers such that $\gcd(n, m) = 1$. Let us consider the four groups $\mathbf{Z}_n$, $\mathbf{Z}_m$, $\mathbf{Z}_{nm}$ and $\mathbf{Z}_n \times \mathbf{Z}_m$. As $(n, m) = 1$, $\mathbf{Z}_n \times \mathbf{Z}_m$ is a cyclic group which is isomorphic to $\mathbf{Z}_{nm}$. By Lemma 10.3.3, it follows that the number of generators of $\mathbf{Z}_n$, $\mathbf{Z}_m$ and $\mathbf{Z}_{nm}$ are, respectively, $\phi(n)$, $\phi(m)$ and $\phi(nm)$. Now by Lemma 10.3.2, it follows that the number of generators of $\mathbf{Z}_n \times \mathbf{Z}_m$ is $\phi(n) \cdot \phi(m)$. As $\mathbf{Z}_{nm}$ and $\mathbf{Z}_n \times \mathbf{Z}_m$ are isomorphic groups, they must have the same number of generators, proving the fact that $\phi(nm) = \phi(n) \cdot \phi(m)$. Consequently, ϕ is a multiplicative function. $\square$

Now we shall show through the following theorems how to evaluate $\phi(n)$ for a given positive integer n, provided the factorization of n is known.

Theorem 10.3.5   *Let p be any prime integer. Then for any positive integer n,* $\phi(p^n) = p^n - p^{n-1}$.

Proof The only positive integers not exceeding p^n and not relatively prime to p^n are λp, where λ is an integer such that $1 \leq \lambda \leq p^{n-1}$. Thus the number of positive integers not exceeding p^n and relatively prime to p^n is $p^n - p^{n-1}$. $\square$

The next theorem is for general n whose prime factorization is known.

Theorem 10.3.6   *Let $n = p_1^{\alpha_1} p_2^{\alpha_2} \cdots p_r^{\alpha_r}$ be the prime factorization of n, where p_i's are distinct primes. Then $\phi(n) = n(1 - \frac{1}{p_1})(1 - \frac{1}{p_2}) \cdots (1 - \frac{1}{p_r})$.*

Proof The result follows from Theorems 10.3.4, 10.3.1, and 10.3.5. $\square$

10.3.2   Sum of Divisor Functions and Number of Divisor Functions

Beside the Euler phi-function, there are two more important multiplicative functions in number theory, namely the sum of divisor function and number of divisor function as defined below.

Definition 10.3.4   For a positive integer n, the sum of divisor function, denoted by σ, is defined by setting $\sigma(n)$ to be equal to the sum of all positive divisors of n.

Example 10.3.2 For $n = 6$, $\sigma(6) = 1 + 2 + 3 + 6 = 12$.

Definition 10.3.5   For a positive integer n, the number of divisor function, denoted by τ, is defined by setting $\tau(n)$ to be equal to the number of all positive divisors of n.

Example 10.3.3 For $n = 6$, $\tau(6) = 4$.

We now show that both σ and τ are multiplicative functions.

Theorem 10.3.7 σ *and* τ *are multiplicative functions.*

Proof Note that $f(n) = n$ and $g(n) = 1$ are multiplicative functions. The rest of the theorem follows from Theorem 10.3.2 and the fact that $\sigma(n) = \sum_{d|n} f(d)$ and $\tau(n) = \sum_{d|n} g(d)$. $\qquad\qquad\qquad\qquad\qquad\qquad\qquad\qquad\qquad\qquad\qquad\square$

Theorem 10.3.8 *Let* p *be a prime integer and* n *be a positive integer. Then* $\sigma(p^n) = \frac{p^{n+1}-1}{p-1}$ *and* $\tau(p^n) = n + 1$.

Proof The divisors of p^n are $1, p, p^2, p^3, \ldots, p^n$. Thus p^n has exactly $n + 1$ positive divisors, with the result $\tau(p^n) = n + 1$ and $\sigma(n) = 1 + p + p^2 + \cdots + p^n = \frac{p^{n+1}-1}{p-1}$. $\qquad\qquad\qquad\qquad\qquad\qquad\qquad\qquad\qquad\qquad\qquad\qquad\qquad\square$

The above result may be further generalized for any positive integer n.

Theorem 10.3.9 *Let* $n = p_1^{\alpha_1} p_2^{\alpha_2} \cdots p_r^{\alpha_r}$ *be the prime factorization of* n, *where* p_i's *are distinct primes. Then* $\sigma(n) = \frac{p_1^{\alpha_1+1}-1}{p_1-1} \frac{p_2^{\alpha_2+1}-1}{p_2-1} \cdots \frac{p_r^{\alpha_r+1}-1}{p_r-1}$ *and* $\tau(n) = (\alpha_1 + 1)(\alpha_2 + 1) \cdots (\alpha_r + 1)$.

Proof The result follows from Theorems 10.3.7, 10.3.1, and 10.3.8. $\qquad\qquad\square$

10.4 Group of Units

The group of units play an important role in number theory and it has tremendous applications in the field of cryptology and coding theory. While studying ring theory, we have seen that the ring $(\mathbf{Z}_n, +, \cdot)$ contains zero divisors if and only if n is composite. For example, $\mathbf{Z}_6$ contains (2) and (3) such that $(2)(3) = (3)(2) = (0)$. This fact implies that $\mathbf{Z}_n$ is not a field under usual addition and multiplication modulo n if n is composite. Therefore, $(\mathbf{Z}_n \setminus \{(0)\}, \cdot)$ may not be a group in general. However, $(\mathbf{Z}_n, +, \cdot)$ forms a commutative ring. Now we may look at the units of $\mathbf{Z}_n$ and may try to give some algebraic structures on the set of all units of $\mathbf{Z}_n$.

Definition 10.4.1 An equivalence class (a) in $\mathbf{Z}_n$, $n \geq 2$, is called a unit in $\mathbf{Z}_n$, iff there exists an equivalence class (b) in $\mathbf{Z}_n$ such that $(a)(b) = (b)(a) = (1)$.

Definition 10.4.2 The collection of all units in $\mathbf{Z}_n$, $n \geq 2$, is called the set of units of $\mathbf{Z}_n$, and is denoted by $\mathbf{Z}_n^*$.

We are now trying to provide some algebraic structures on $\mathbf{Z}_n^*$. The subsequent discussions will reveal that $\mathbf{Z}_n^*$ has indeed some rich algebraic structures. The following theorem shows that $\mathbf{Z}_n^*$ forms a group, known as group of units, under the usual multiplication modulo n.

Theorem 10.4.1   *For each integer $n \geq 2$, the set $\mathbf{Z}_n^*$ forms a commutative group.*

Proof As $(1) \in \mathbf{Z}_n^*$, $\mathbf{Z}_n^* \neq \emptyset$. So, the set $\mathbf{Z}_n^*$ is a not empty subset of the commutative ring $(\mathbf{Z}_n, +, \cdot)$. Let $(a), (b) \in \mathbf{Z}_n^*$. Then by definition of $\mathbf{Z}_n^*$, there exist (u), $(v) \in \mathbf{Z}_n^*$ such that $(a)(u) = (1)$ and $(b)(v) = (1)$, with the result $(ab)(uv) = (1)$. Associativity and the commutativity of $\mathbf{Z}_n^*$ follows from the hereditary property of $(\mathbf{Z}_n, +, \cdot)$. Finally, for each $(a) \in \mathbf{Z}_n^*$, there exists $(b) \in \mathbf{Z}_n^*$ such that $(a)(b) = (1)$ follows from the definition of $\mathbf{Z}_n^*$. Combining all the arguments, it follows that $\mathbf{Z}_n^*$ forms a commutative group.   □

The following proposition provides the order of the group $\mathbf{Z}_n^*$.

Proposition 10.4.1 $|\mathbf{Z}_n^*| = \phi(n)$, *where $|\mathbf{Z}_n^*|$ denotes the order of the group $\mathbf{Z}_n^*$.*

Proof Follows from Lemma 10.3.1.   □

Using the properties of the group $(\mathbf{Z}_n^*, \cdot)$, we deduce the following important theorems of number theory.

Theorem 10.4.2 (Fermat's Little Theorem)   *Let p be a prime and a be an integer such that $\gcd(a, p) = 1$. Then $a^{p-1} \equiv 1 \pmod{p}$.*

Proof Consider the group $\mathbf{Z}_p^*$ with $|\mathbf{Z}_p^*| = p - 1$. Let $(a) \in \mathbf{Z}_p^*$. Then by the definition of $\mathbf{Z}_p^*$, $\gcd(a, p) = 1$. Let the order of (a) in $\mathbf{Z}_p^*$ be n. Then, $(a)^n = (1)$ and the order of the subgroup generated by (a) is n. Thus by Lagrange's Theorem, $n \mid (p - 1)$. Then there exists some integer k such that $p - 1 = kn$. Thus $(a)^{p-1} = (a)^{kn} = ((a)^n)^k = (1)$, with the result $a^{p-1} \equiv 1 \pmod{p}$.   □

Corollary   *Let p be a prime and a be any integer. Then $a^p \equiv a \pmod{p}$.*

Proof We divide the proof into two cases. In the first case, let us assume that $\gcd(a, p) = 1$. Then the result, follows from Theorem 10.4.2. For the second case, let us assume that $\gcd(a, p) \neq 1$. In that case, p must divide a. Thus $a \equiv 0 \pmod{p}$, with the result $a^p \equiv 0 \equiv a \pmod{p}$.   □

Theorem 10.4.3 (Euler's Theorem)   *Let n be a positive integer and a be an integer such that $\gcd(a, n) = 1$. Then $a^{\phi(n)} \equiv 1 \pmod{n}$.*

Proof Let us consider the group $\mathbf{Z}_n^*$. Then by Proposition 10.4.1, the order of the group is $\phi(n)$. As $\gcd(a, n) = 1$, $(a) \in \mathbf{Z}_n^*$. Let the order of the element (a) in

$\mathbf{Z}_n^*$ be k. Then $(a)^k = (1)$ and the order of the subgroup generated by (a) is also k. Thus by Lagrange's Theorem, k divides $\phi(n)$. Let $\phi(n) = kt$ for some integer t. Then $(a)^{\phi(n)} = (a)^{kt} = ((a)^k)^t = (1)$ implies that $a^{\phi(n)} \equiv 1 \pmod{n}$.   □

Theorem 10.4.4 (Wilson's Theorem)   *For any prime integer p, $(p - 1)! \equiv -1 \pmod{p}$.*

Proof The non-zero elements of $\mathbf{Z}_p$ i.e., the elements of $\mathbf{Z}_p^*$ form a multiplicative group of order $p - 1$ and the self inverse elements of $\mathbf{Z}_p^*$ are only (1) and $(-1) = (p-1)$. The other non-null elements of $\mathbf{Z}_p^*$ can be paired by inverse elements so that the product of each pair is the identity (1). Consequently, the product of all the non-zero elements is $(1)(2)\ldots(p-1) = (-1)$ in $\mathbf{Z}_p^*$. This shows that $([p-1]!) = (-1)$ and hence $(p - 1)! \equiv -1 \pmod{p}$.   □

Definition 10.4.3   If $\mathbf{Z}_n^*$ is cyclic, then any generator (g) of $\mathbf{Z}_n^*$ is called a primitive root for $\mathbf{Z}_n^*$.

Example 10.4.1 (2) is a *primitive root* for $\mathbf{Z}_5^*$, whereas there are no primitive roots for $\mathbf{Z}_8^*$ as $\mathbf{Z}_8^*$ is not cyclic.

From the above example, it is clear that primitive root may or may not exist. Even if it exits, finding primitive roots in $\mathbf{Z}_n^*$ is a non-trivial problem. Till date, no efficient algorithm is known which can efficiently find primitive roots. One of the obvious but tedious methods is to try each of the $\phi(n)$ units $(u) \in \mathbf{Z}_n^*$ and check the order of (u) in $\mathbf{Z}_n^*$. If we find some element of order $\phi(n)$, we say that the element must be a primitive root. But a slightly better result for a more efficient test for primitive roots is as follows:

Proposition 10.4.2   *An element (g) is a primitive root for $\mathbf{Z}_n^*$ if and only if $(g)^{\phi(n)/p} \neq (1)$ in $\mathbf{Z}_n^*$ for each prime p dividing $\phi(n)$.*

Proof Let (g) be a primitive root for $\mathbf{Z}_n^*$. Then the order of (g) is $\phi(n)$. The proof follows from the definition of the order of (g), i.e., $(g)^i \neq (1)$ for all i such that $1 \leq i < \phi(n)$.

Conversely, let $(g)^{\phi(n)/p} \neq (1)$ in $\mathbf{Z}_n^*$ for each prime p dividing $\phi(n)$. If possible, let (g) not be a primitive root for $\mathbf{Z}_n^*$. Then its order, say, m must be a proper factor of $\phi(n)$, with the result $\phi(n)/m > 1$. If p is any prime factor of $\phi(n)/m$, then m must divide $\phi(n)/p$, with the result $(g)^{\phi(n)/p} = (1)$ in $\mathbf{Z}_n^*$, contradicting the hypothesis. Thus (g) must be a primitive root for $\mathbf{Z}_n^*$.   □

Next we shall prove a very important theorem, which plays an important role in cryptography. We know that $(\mathbf{Z}_p, +, \cdot)$ forms a field, if p is a prime number. As a result, $(\mathbf{Z}_p \setminus \{(0)\}, \cdot) = \mathbf{Z}_p^*$ is a commutative group. We now prove that $(\mathbf{Z}_p^*, \cdot)$ is not only a commutative group, but a cyclic group of order $p - 1$. In fact, we shall prove a much general result for finite fields. To prove that first prove the following lemma.

Lemma 10.4.1 *Let G be a commutative group and $a, b \in G$ be such that $o(a) = m$, $o(b) = n$ and $\gcd(m, n) = 1$. Then $o(ab) = mn$.*

Proof Let $o(ab) = k$. As $(ab)^{mn} = e_G$, the identity element of G, k divides mn. On the other hand, as G is a commutative group $(ab)^k = e_G$ implies $a^{kn} = b^{-kn}$, with the result $a^{kn} = e_G$. Thus m divides kn. As $\gcd(m, n) = 1$, m divides kn implies m divides k. Similarly, it can be shown that n divides k, with the result that mn divides k. Hence we have $mn = k$, with the result $o(ab) = mn$. $\qquad \square$

Now we shall prove the following lemma for finite fields.

Lemma 10.4.2 *For any finite field F_q having q elements, the multiplicative group $(F_q \setminus \{0\} = F_q^*, \cdot)$ is a cyclic group of order $q - 1$.*

Proof If $q = 2$, the result is trivial. So we assume that $q \geq 3$. Then $|F_q^*| = q - 1 = h$, say. Let $h = p_1^{\alpha_1} p_2^{\alpha_2} \cdots p_m^{\alpha_m}$ be the prime factorization of h. Now every polynomial $x^{h/p_i} - 1 \in F_q[x]$ has at most h/p_i roots in F_q. Since $h/p_i < h$, there are non-zero elements in F_q which are not roots of $x^{h/p_i} - 1$. Let a_i be such an element. So, $a_i^{h/p_i} \neq 1$, the identity element of the field F_q. Further, let $b_i = a_i^{h/p_i^{\alpha_i}}$. Then $b_i^{p_i^{\alpha_i}} = a_i^h = 1$ and $b_i^{p_i^{\alpha_i - 1}} = a_i^{h/p_i} \neq 1$. Hence by Lemma 10.10.10 (used in the setting of multiplicative group), $o(b_i) = p_i^{\alpha_i}$. Now by applying the result of Lemma 10.4.1 inductively, we get $o(b_1 b_2 \cdots b_m) = o(b_1) o(b_2) \cdots o(b_m) = p_1^{\alpha_1} p_2^{\alpha_2} \cdots p_m^{\alpha_m} = q - 1$. Thus F_q^* contains an element $b_1 b_2 \cdots b_m$ of order $q - 1$ which is same as the order of the group F_q^*, with the result that F_q^* is a cyclic group. $\qquad \square$

Theorem 10.4.5 $\mathbf{Z}_p^*$ *is cyclic if p is a prime integer.*

Proof The theorem follows directly from Lemma 10.4.2. $\qquad \square$

Remark For any positive integer $n \geq 2$, the group $\mathbf{Z}_n^*$ may not always be cyclic. The cyclic property of the group $\mathbf{Z}_n^*$ is characterized in Theorem 10.4.8.

We now deal with the case for $\mathbf{Z}_n^*$ in which n is an odd prime power. We shall prove that if p is an odd prime, $\mathbf{Z}_{p^e}^*$ is a cyclic group. Before proving the main result, we shall prove the following lemmas.

Lemma 10.4.3 *Let (g) be a primitive root for $\mathbf{Z}_p^*$, where p is an odd prime integer. Then (g) or $(g + p)$ is a primitive root for $\mathbf{Z}_{p^2}^*$.*

Proof Theorem 10.4.5 ensures the existence of a primitive root, say (g), for $\mathbf{Z}_p^*$. Thus $g^{p-1} \equiv 1 \pmod{p}$, but $g^i \not\equiv 1 \mod(p)$, for $1 \leq i < p - 1$. Note that $(g) \in \mathbf{Z}_{p^2}^*$ as $\gcd(g, p^2) = 1$. Let the order of (g) in $\mathbf{Z}_{p^2}^*$ be d. Then d divides $\phi(p^2) = p(p - 1)$. Again, from the definition of the order of (g), it follows that

$g^d \equiv 1 \pmod{p^2}$. So $g^d \equiv 1 \pmod{p}$. Also, (g) has order $p - 1$ in $\mathbf{Z}_p^*$. Thus $p - 1$ divides d. These two facts imply that either $d = p(p - 1)$ or $d = p - 1$. If $d = p(p - 1)$, then (g) is a primitive root for $\mathbf{Z}_{p^2}^*$ and hence we are done in that case. So assume that $d = p - 1$. Let $h = g + p$. Since $h \equiv g \pmod{p}$, (h) is also a primitive root for $\mathbf{Z}_p^*$. Thus arguing as before we see that (h) has order $p(p - 1)$ or $p - 1$ in $\mathbf{Z}_{p^2}^*$. Since $g^d = g^{p-1} \equiv 1 \bmod(p^2)$ and $\mathbf{Z}_{p^2}^*$ is a commutative group, it follows that

$$h^{p-1} = (g + p)^{p-1} = g^{p-1} + (p-1)g^{p-2}p + \cdots + p^{p-1} \equiv 1 - pg^{p-2} \pmod{p^2},$$

where the dots represent terms divisible by p^2. Since g is relatively prime to p, we have $pg^{p-2} \not\equiv 0 \bmod(p^2)$ and hence $h^{p-1} \not\equiv 1 \bmod(p^2)$. Thus (h) is not of order $p - 1$ in $\mathbf{Z}_{p^2}^*$, so it must be of order $p(p - 1)$ and is therefore a primitive root for $\mathbf{Z}_{p^2}^*$. Thus there exists a primitive root modulo p^2. $\square$

Lemma 10.4.4 *Let (h) be a primitive root for $\mathbf{Z}_{p^2}^*$ for an odd prime p. Then (h) is also a primitive root for $\mathbf{Z}_{p^e}^*$, for all integers $e \geq 2$.*

Proof We shall prove the result by the principle of mathematical induction on e. The result is trivially true for $e = 2$. Let (h) be a primitive root modulo p^e for some $e > 2$ and d be the order of (h) modulo p^{e+1}. An argument similar to that as described in Lemma 10.4.3 shows that d divides $\phi(p^{e+1}) = p^e(p - 1)$ and is divisible by $\phi(p^e) = p^{e-1}(p - 1)$. Thus either $d = p^e(p - 1)$ or $d = p^{e-1}(p - 1)$. In the first case, (h) is a primitive root modulo p^{e+1}, as required. Thus it is sufficient to eliminate the second case by showing that $h^{p^{e-1}(p-1)} \not\equiv 1 \pmod{p^{e+1}}$. Since (h) is a primitive root modulo p^e, it has order $\phi(p^e) = p^{e-1}(p - 1)$ in $\mathbf{Z}_{p^e}^*$. Thus $h^{p^{e-2}(p-1)} \not\equiv 1 \pmod{p^e}$. However, $p^{e-2}(p - 1) = \phi(p^{e-1})$, so, $h^{p^{e-2}(p-1)} \equiv 1 \pmod{p^{e-1}}$ by Euler's Theorem. Combining these two results, we see that $h^{p^{e-2}(p-1)} = 1 + kp^{e-1}$, where k is relatively prime to p. Again as before,

$$h^{p^{e-1}(p-1)} = \left(1 + kp^{e-1}\right)^p$$

$$= 1 + \binom{p}{1}kp^{e-1} + \binom{p}{2}\left(kp^{e-1}\right)^2 + \cdots + \left(kp^{e-1}\right)^p$$

$$= 1 + kp^e + \frac{1}{2}k^2p^{2e-1}(p - 1) + \cdots + \left(kp^{e-1}\right)^p,$$

where the dots represent terms divisible by $(p^{e-1})^3$ and hence by p^{e+1}, since $3(e - 1) \geq (e + 1)$ for $e \geq 2$. Thus

$$h^{p^{e-1}(p-1)} \equiv 1 + kp^e + \frac{1}{2}k^2p^{2e-1}(p - 1) \pmod{p^{e+1}}.$$

Now as p is odd, the third term $k^2p^{2e-1}(p - 1)/2$ is also divisible by p^{e+1}, since $2e - 1 \geq e + 1$ for $e \geq 2$. Thus $h^{p^{e-1}(p-1)} \equiv 1 + kp^e \pmod{p^{e+1}}$.

Since p does not divide k, we must have $h^{p^{e-1}(p-1)} \not\equiv 1 \pmod{p^{e+1}}$, as required. Thus the lemma follows from the principle of mathematical induction. $\qquad\square$

Theorem 10.4.6 *For all* $e > 0$, $\mathbf{Z}_{p^e}^*$ *is cyclic if* p *is an odd prime integer.*

Proof The case with $e = 1$ follows from Theorem 10.4.5. The rest of the theorem follows from Lemmas 10.4.3 and 10.4.4. $\qquad\square$

Remark Note that we need p to be odd, because if $p = 2$, then the third term $k^2 p^{2e-1}(p-1)/2 = k^2 2^{2e-2}$ is not divisible by 2^{e+1} when $e = 2$, so the first step of the induction argument fails.

Theorem 10.4.7 *The group* $\mathbf{Z}_{2^e}^*$ *is cyclic if and only if* $e = 1$ *or* $e = 2$.

Proof The groups $\mathbf{Z}_2^* = \{(1)\}$ and $\mathbf{Z}_4^* = \{(1), (3)\}$ are cyclic, generated by (1) and by (3), respectively. So it is sufficient to prove that $\mathbf{Z}_{2^e}^*$ is not cyclic for $e \geq 3$. We shall prove that $\mathbf{Z}_{2^e}^*$ contains no element of order $\phi(2^e) = 2^{e-1}$ by showing that $a^{2^{e-2}} \equiv 1 \pmod{2^e}$ for all odd a. We shall prove this by the method of principle of mathematical induction on e. Let us first show that the relation holds for the base value when $e = 3$, that is, we need to show $a^2 \equiv 1 \pmod 8$ for all odd a. This is true, since if $a = 2b + 1$ then $a^2 = 4b(b+1) + 1 \equiv 1 \pmod 8$. Now we assume that for some exponent $k \geq 3$, $a^{2^{k-2}} \equiv 1 \pmod{2^k}$ for all odd a. Then for each odd a we have $a^{2^{k-2}} = 1 + 2^k t$ for some integer t. Squaring both sides we get $a^{2^{(k+1)-2}} = (1 + 2^k t)^2 = 1 + 2^{(k+1)} t + 2^{2k} t^2 = 1 + 2^{k+1}(t + 2^{k-1} t^2) \equiv 1 \pmod{2^{k+1}}$, proving that the result is true for $k + 1$. Hence by the principle of mathematical induction, the result is true for all integers $e \geq 3$. This completes the proof. $\qquad\square$

We are going to prove the following lemma that will be useful in characterizing all cyclic groups of the form $\mathbf{Z}_n^*$.

Lemma 10.4.5 *If* $n = ab$ *where* a *and* b *are relatively prime and are both greater than* 2, *then* $\mathbf{Z}_n^*$ *is not cyclic.*

Proof Since $\gcd(a, b) = 1$, we have $\phi(n) = \phi(a)\phi(b)$. Moreover, as both a and b are greater than 2, both $\phi(a)$ and $\phi(b)$ are even, with the result that $\phi(n)$ to be an integer divisible by 4. Further, the integer $e = \phi(n)/2$ is divisible by both $\phi(a)$ and $\phi(b)$. Note that if (x) is a unit in $\mathbf{Z}_n$, then (x) is also a unit in $\mathbf{Z}_a$ as well as in $\mathbf{Z}_b$. Thus $x^{\phi(a)} \equiv 1 \pmod a$ and $x^{\phi(b)} \equiv 1 \pmod b$. Since both $\phi(a)$ and $\phi(b)$ divide e, we therefore have $x^e \equiv 1 \bmod(a)$ and $x^e \equiv 1 \bmod(b)$. Since a and b are co-prime, this implies that $x^e \equiv 1 \bmod(ab)$, that is, $x^e \equiv 1 \bmod(n)$. Thus every elements of $\mathbf{Z}_n^*$ has order dividing e, and since $e < \phi(n)$, this means that there is no primitive root $\mathbf{Z}_n^*$. $\qquad\square$

Now we are in a position to provide a necessary and sufficient condition for $\mathbf{Z}_n^*$ to be cyclic.

Theorem 10.4.8 *The group $\mathbf{Z}_n^*$ is cyclic if and only if*

$$n = 2, 4, p^e \text{ or } 2p^e, \text{ where } p \text{ is an odd prime integer}.$$

Proof ($\Leftarrow$) Clearly $\mathbf{Z}_2^*$ and $\mathbf{Z}_4^*$ are cyclic groups generated by (1) and (3), respectively. Further, Theorem 10.4.6 ensures the case with odd prime powers. So, we may assume that $n = 2p^e$, where p is an odd prime integer. Now by Theorem 10.3.4, we have $\phi(n) = \phi(2)\phi(p^e) = \phi(p^e)$. Also Theorem 10.4.6 ensures the existence of a primitive root (g) in $\mathbf{Z}_{p^e}^*$. Then $(g + p^e)$ is also a primitive root modulo p^e and one of the g or $g + p^e$ must be odd, resulting in the existence of an odd primitive root, say (h) modulo p^e. We will show that (h) is also a primitive root modulo $2p^e$. By its construction, h is co-prime to both 2 and p^e, so (h) is a unit in $\mathbf{Z}_{2p^e}$ i.e., $(h) \in \mathbf{Z}_{2p^e}^*$. If $h^i \equiv 1 \pmod{2p^e}$, then certainly $h^i \equiv 1 \pmod{p^e}$. Again, since (h) is a primitive root modulo p^e, $\phi(p^e)$ divides i. Since $\phi(p^e) = \phi(2p^e)$, it follows that $\phi(2p^e)$ divides i, so (h) has order $\phi(2p^e)$ in $\mathbf{Z}_{2p^e}^*$ and is therefore a primitive root, with the result that $\mathbf{Z}_{2p^e}^*$ is a cyclic group.

($\Rightarrow$) Let us now prove the converse part. If $n \neq 2, 4$, p^e or $2p^e$, where p is an odd prime integer, then we have either

(a) $n = 2^e$ where $e \geq 3$, or
(b) $n = 2^e p^f$ where $e \geq 2$, $f \geq 1$ and p is odd prime, or
(c) n is divisible by at least two odd primes.

We shall show that in all the cases, $\mathbf{Z}_n^*$ is not cyclic. The case (a) follows from Theorem 10.4.7. For the case (b), we can take $a = 2^e$ and $b = p^f$, while for case (c) we can take a to be of the form p^e for some odd prime p dividing n such that p^e divides n but p^{e+1} does not divide n, and $b = n/a$. In either case, $n = ab$ where a and b are co-prime and greater than 2. Thus Lemma 10.4.5 ensures that $\mathbf{Z}_n^*$ is not cyclic. This completes the proof of the theorem. $\qquad\square$

10.5 Quadratic Residues and Quadratic Reciprocity

Let p be an odd prime and a an integer such that $\gcd(a, p) = 1$, i.e., a and p are relatively prime. In this section, we shall deal with the question: whether a is a perfect square modulo p? Towards finding the answer, let us start with the following definition.

Definition 10.5.1 If n is a positive integer, we say that the integer a is a quadratic residue modulo n iff $\gcd(a, n) = 1$ and the congruence $x^2 \equiv a \pmod{n}$ has a solution in the set of integers. If the congruence $x^2 \equiv a \pmod{n}$ has no solution in the set of integers, we say that a is a quadratic non-residue of n.

Remark 1 While defining an integer a to be a quadratic residue modulo a positive integer n, instead of considering a to be an integer, we may very well consider it to be an equivalence class $(a) \in \mathbf{Z}_n^*$. Thus the Definition 10.5.1 my be restated as follows:

If n is a positive integer, we say that the integer a is a quadratic residue modulo n iff $\gcd(a, n) = 1$ and there exists an equivalence class $(x) \in \mathbf{Z}_n^*$ such that $(a) = (x)^2$.

Remark 2 The above definition can be extended for any group. Given a group G, an element $a \in G$ is a quadratic residue if there exists an element $x \in G$ such that $x^2 = a$. In this case, we call x as a square-root of a. An element that is not a quadratic residue is called a quadratic non-residue.

Now we are going to explore some algebraic properties of the set of quadratic residues.

Proposition 10.5.1 *In an abelian group G, the set of all quadratic residues, denoted by $\mathcal{QR}$, forms a subgroup of G.*

Proof The identity element of G is always a quadratic residue. Thus $\mathcal{QR} \neq \emptyset$. Let $x, y \in \mathcal{QR}$. Then there exist some $z, w \in G$ such that $x = z^2$ and $y = w^2$. This implies, $xy^{-1} = (zw^{-1})^2$, with the result $xy^{-1} \in \mathcal{QR}$. As, $\mathcal{QR} \subseteq G$, it follows that $\mathcal{QR}$ is a subgroup of G. $\square$

Example 10.5.1 Let us consider $n = 11$. We are going to find all integers a modulo 11 such that $x^2 \equiv a \pmod{11}$ has a solution in $\mathbf{Z}$. To find all such a's let us start from other direction by calculating the following: $1^2 \equiv 10^2 \equiv 1 \pmod{11}$, $2^2 \equiv 9^2 \equiv 4 \pmod{11}$, $3^2 \equiv 8^2 \equiv 9 \pmod{11}$, $4^2 \equiv 7^2 \equiv 5 \pmod{11}$, $5^2 \equiv 6^2 \equiv 3 \pmod{11}$. These calculations imply that in the group $\mathbf{Z}_{11}^*$, the conjugacy classes (1), (4), (9), (5), (3) modulo 11 are the only elements which are the quadratic residues modulo 11 while (2), (6), (7), (8), (10) are the only elements in $\mathbf{Z}_{11}^*$ which are the quadratic non-residues modulo 11. Also note that the set of all quadratic residues modulo 11, denoted by $\mathcal{QR}_{11}$, forms a subgroup of $\mathbf{Z}_{11}^*$.

The above example demonstrates that not all elements of $\mathbf{Z}_{11}^*$ are quadratic residues modulo 11. Also, the congruence $x^2 \equiv a \pmod{11}$ has either no solution or exactly two incongruent solutions modulo 11. This observation leads to the following theorem.

Theorem 10.5.1 *Let p be an odd prime and a be an integer such that $\gcd(a, p) = 1$. Then the congruence $x^2 \equiv a \pmod{p}$ has either no solution or exactly two incongruent solutions modulo p.*

Proof Let the congruence $x^2 \equiv a \pmod{p}$ have a solution in $\mathbf{Z}$, say y_1. Then $y_1^2 \equiv a \pmod{p}$. This implies $(-y_1)^2 \equiv a \pmod{p}$ which implies $(-y_1)$ is also a solution of the congruence $x^2 \equiv a \pmod{p}$. Now we shall show that y_1 and $-y_1$ are incongruent modulo p. If not, then $y_1 \equiv (-y_1) \pmod{p}$, i.e., p divides $2y_1$. As p is an odd prime, p does not divide 2. Thus $p|y_1$. So, $y_1^2 \equiv 0 \pmod{p}$. Also, $y_1^2 \equiv a \pmod{p}$ implies that $a \equiv 0 \pmod{p}$, i.e., $p|a$, contradicting the fact that

$\gcd(a, p) = 1$. Hence, $y_1 \not\equiv -y_1 \pmod{p}$. Now we shall prove that there exist exactly two incongruent solutions modulo p. Let z be any solution of $x^2 \equiv a \pmod{p}$. Then $z^2 \equiv a \pmod{p}$. As $y_1^2 \equiv a \pmod{p}$, $y_1^2 \equiv z^2 \pmod{p}$ which implies that $p|(y_1 + z)(y_1 - z)$ which implies either $p|(y_1 + z)$ or $p|(y_1 - z)$, with the result that either $z \equiv y_1 \pmod{p}$ or $z \equiv -y_1 \pmod{p}$. Hence the congruence $x^2 \equiv a \pmod{p}$ has either no solution or exactly two incongruent solutions modulo p.   □

Corollary *If p is an odd prime integer, then the equation $x^2 \equiv 1 \pmod{p}$ has exactly two incongruent solutions, namely 1 and -1.*

Proof For $p = 2$, the result is trivial as $1 \equiv -1 \pmod{2}$. The rest of the corollary follows from the above Theorem 10.5.1.   □

The following result demonstrates that if p is an odd prime, then there are exactly as many quadratic residues as quadratic non-residues in $\mathbf{Z}_p^*$.

Proposition 10.5.2 *Let p be an odd prime integer. Then exactly half the elements of $\mathbf{Z}_p^*$ are quadratic residues.*

Proof Define a map $\mathrm{sq}_p : \mathbf{Z}_p^* \to \mathbf{Z}_p^*$, defined by $\mathrm{sq}_p(x) = x^2$, for all $x \in \mathbf{Z}_p^*$. It follows from Theorem 10.5.1 that sq_p is a two-to-one function for any odd prime p. This immediately implies that exactly half the elements of $\mathbf{Z}_p^*$ are quadratic residues. We denote the set of quadratic residues modulo p by $\mathcal{QR}_p$, and the set of quadratic non-residues by $\mathcal{QNR}_p$. Then

$$|\mathcal{QR}_p| = |\mathcal{QNR}_p| = \frac{|\mathbf{Z}_p^*|}{2} = \frac{p-1}{2}.$$ □

A special notation, known as Legendre symbol, associated with quadratic residues is defined as follows.

Definition 10.5.2 Let p be an odd prime and a an integer such that $\gcd(a, p) = 1$. Then the Legendre symbol $\left(\frac{a}{p}\right)$ is defined by

$$\left(\frac{a}{p}\right) = \begin{cases} 1, & \text{if } a \text{ is a quadratic residue modulo } p, \\ -1, & \text{if } a \text{ is not a quadratic residue modulo } p. \end{cases}$$

Example 10.5.2 From Example 10.5.1, we find that

$$\left(\frac{a}{11}\right) = \begin{cases} 1, & \text{for } a = 1, 3, 4, 5, 9, \\ -1, & \text{for } a = 2, 6, 7, 8, 10. \end{cases}$$

We now characterize the quadratic residues in $\mathbf{Z}_p^*$ for odd prime p. We may recall the fact that $\mathbf{Z}_p^*$ is a cyclic group of order $p - 1$ (see Theorem 10.4.5). Let g be a generator of $\mathbf{Z}_p^*$. As p is an odd prime, $\mathbf{Z}_p^*$ may be written as, $\mathbf{Z}_p^* =$

$\{g^0, g^1, g^2, \ldots, g^{\frac{p-1}{2}-1}, g^{\frac{p-1}{2}}, g^{\frac{p-1}{2}+1}, \ldots, g^{p-2}\}$. As in $\mathbf{Z}_p^*$, $g^i = g^{i \ (\mathrm{mod}\, p-1)}$, for all $i \in \mathbf{Z}$, squaring each element in this list and reducing modulo $p-1$ in the exponent yields a multi-set $\{g^0, g^2, g^4, \ldots, g^{p-3}, g^0, g^2, g^4, \ldots, g^{p-3}\}$, containing the list of all the quadratic residues in $\mathbf{Z}_p^*$. Note that in the multi-set, each quadratic residue appears exactly twice. Thus $\mathcal{QR}_p = \{g^0, g^2, g^4, \ldots, g^{p-3}\}$. We see that the quadratic residues in $\mathbf{Z}_p^*$ are exactly those elements that can be written as g^i with $i \in \{0, \ldots, p-3\}$ an even integer. This leads to the following proposition.

Proposition 10.5.3 *If g is a generator of $\mathbf{Z}_p^*$, then*

$$g^n \text{ is a} \begin{cases} \text{quadratic residue,} & \text{if } n \text{ is even,} \\ \text{quadratic non-residue,} & \text{if } n \text{ is odd.} \end{cases}$$

The above characterization leads to a simple way to compute the Legendre symbol and this tells us whether a given element $x \in \mathbf{Z}_p^*$ is a quadratic residue or not by using the following proposition.

Proposition 10.5.4 *Let p be an odd prime and a be an integer such that $\gcd(a, p) = 1$. Then*

$$\left(\frac{a}{p} \right) \equiv a^{\frac{p-1}{2}} \ (\mathrm{mod}\, p).$$

Proof Let a be a quadratic residue modulo p. Then $\left(\frac{a}{p} \right) = 1$. Also, let (g) be a generator of $\mathbf{Z}_p^*$. Then by Proposition 10.5.3, $(a) = (g)^{2i}$, for some integer i. Thus $a^{\frac{p-1}{2}} = (g^{2i})^{\frac{p-1}{2}} = (g^{p-1})^i \equiv 1 \ (\mathrm{mod}\, p)$, by Fermat's Little Theorem 10.4.2. So, $\left(\frac{a}{p} \right) \equiv 1 \ (\mathrm{mod}\, p) \equiv a^{\frac{p-1}{2}} \ (\mathrm{mod}\, p)$, as claimed.

On the other hand, if a is not a quadratic residue, then by Proposition 10.5.3, $(a) = (g)^{2i+1}$, for some integer i. Thus $a^{\frac{p-1}{2}} = (g^{2i+1})^{\frac{p-1}{2}} = (g^{2i})^{\frac{p-1}{2}} g^{\frac{p-1}{2}} = (g^{p-1})^i g^{\frac{p-1}{2}} \equiv g^{\frac{p-1}{2}} \ (\mathrm{mod}\, p)$. Now, $(g^{\frac{p-1}{2}})^2 = (g)^{p-1} \equiv 1 \ (\mathrm{mod}\, p)$. Thus $(g)^{\frac{p-1}{2}}$ is either $+(1)$ or $-(1)$ by the Corollary of Theorem 10.5.1. As, (g) is a generator of $\mathbf{Z}_p^*$, the order of (g) is $p-1$, and hence $(g)^{\frac{p-1}{2}}$ cannot be $+(1)$. Thus $a^{\frac{p-1}{2}} \equiv -1 \ (\mathrm{mod}\, p) \equiv \left(\frac{a}{p} \right)$. $\square$

Using the above Proposition 10.5.4, we state the following proposition.

Proposition 10.5.5 *Let p be an odd prime and a, b be two integers such that $\gcd(a, p) = 1$ and $\gcd(b, p) = 1$. Then*

(i) $\left(\frac{ab}{p} \right) = \left(\frac{a}{p} \right)\left(\frac{b}{p} \right)$;

(ii) $a \equiv b \ (\mathrm{mod}\, p)$ *implies* $\left(\frac{a}{p} \right) = \left(\frac{b}{p} \right)$;

(iii) $\left(\frac{a^2}{p} \right) = 1, \left(\frac{1}{p} \right) = 1, \left(\frac{-1}{p} \right) = (-1)^{\frac{p-1}{2}}$.

The following proposition is an immediate consequence of Proposition 10.5.5.

Proposition 10.5.6 *Let $\mathcal{QR}_p$ and $\mathcal{QNR}_p$ denote, respectively, the set of all quadratic residues and the set of all quadratic non-residues modulo an odd prime p. Let $a, a' \in \mathcal{QR}_p$ and $b, b' \in \mathcal{QNR}_p$. Then*

(i) $aa' \in \mathcal{QR}_p$;
(ii) $bb' \in \mathcal{QR}_p$;
(iii) $ab \in \mathcal{QNR}_p$.

The following proposition leads an alternative proof of Proposition 10.5.2.

Proposition 10.5.7 *Let p be an odd prime and a be an integer such that $1 \le a \le p - 1$. Then*

$$\sum_{a=1}^{p-1} \left(\frac{a}{p}\right) = 0.$$

Proof Let (g) be a generator of $\mathbf{Z}_p^*$. As $(a) \in \mathbf{Z}_p^*$, there exists unique i, $1 \le i \le p - 1$, such that $(a) = (g)^i$. Now, we claim that $\left(\frac{g}{p}\right) = -1$. If not, then $\left(\frac{g}{p}\right) = 1$. This implies that $g^{\frac{p-1}{2}} \equiv 1 \pmod{p}$, contradicting the fact that order of g is $p - 1$. Thus $\left(\frac{g}{p}\right) = -1$. Now,

$$\left(\frac{a}{p}\right) = \left(\frac{g^i}{p}\right) = \left(\frac{g}{p}\right)\left(\frac{g}{p}\right) \cdots \left(\frac{g}{p}\right) = (-1)^i.$$

So,

$$\sum_{a=1}^{p-1} \left(\frac{a}{p}\right) = \sum_{i=1}^{p-1} (-1)^i = 0. \qquad \square$$

The following remark is an immediate consequence of the above proposition and provides an alternative proof of Proposition 10.5.2.

Remark Let p be an odd prime integer. Then there are precisely $\frac{p-1}{2}$ quadratic residues and $\frac{p-1}{2}$ quadratic non-residues modulo p.

Gauss provided another elegant criterion to determine whether an integer a relatively prime to p is a quadratic residue modulo an odd prime p.

Theorem 10.5.2 (Gauss Lemma) *Let p be an odd prime and a be an integer such that $\gcd(a, p) = 1$. Consider the integers $a, 2a, 3a, \ldots, \frac{p-1}{2}a$ and their least non-negative residues modulo p that exceeds $\frac{p}{2}$. Let μ denote the number of these residues modulo p that exceeds $\frac{p}{2}$. Then*

$$\left(\frac{a}{p}\right) = (-1)^\mu.$$

Proof Let $\mathcal{R} = \{a, 2a, 3a, \ldots, \frac{p-1}{2}a\}$. As $\gcd(a, p) = 1$, none of these $\frac{p-1}{2}$ integers in $\mathcal{R}$ is congruent to 0 modulo p and no two of them are congruent to each other modulo p. Let $r_1, r_2, \ldots, r_\mu$ denote the least positive residues of these elements of $\mathcal{R}$ that exceed $\frac{p}{2}$ and let $s_1, s_2, \ldots, s_t$ denote the remaining least positive residues of the elements of $\mathcal{R}$. Note that $\mu + t = \frac{p-1}{2}$ and $0 < p - r_i < \frac{p}{2}$ for $1 \leq i \leq \mu$. So the integers $p - r_1, p - r_2, \ldots, p - r_\mu, s_1, s_2, \ldots, s_t$ are all positive integers and less than $\frac{p}{2}$. We shall prove that these $\frac{p-1}{2}$ integers are all distinct. To prove that it is sufficient to prove that no $p - r_i$ is equal to any s_j. If possible, let for some i and j, $p - r_i = s_j$, i.e., $r_i + s_j = p$. Then there exist integers x and y such that $r_i \equiv ax \pmod{p}$ and $s_j \equiv ay \pmod{p}$ such that $1 \leq x, y \leq \frac{p-1}{2}$. Thus $(x + y)a \equiv r_i + s_j = p \equiv 0 \pmod{p}$. As $\gcd(a, p) = 1$, $p|(x + y)$, contradicting the fact that $1 \leq x, y \leq \frac{p-1}{2}$. Hence $p - r_1, p - r_2, \ldots, p - r_\mu, s_1, s_2, \ldots, s_t$ are just all the integers $1, 2, \ldots, \frac{p-1}{2}$, in some order. Hence their product is simply $(\frac{p-1}{2})!$. Thus we have

$$\left(\frac{p-1}{2}\right)! = (p - r_1) \cdot (p - r_2) \cdots (p - r_\mu) \cdot s_1 \cdot s_2 \cdots s_t$$

$$\equiv (-1)^\mu r_1 \cdot r_2 \cdots r_\mu \cdot s_1 \cdot s_2 \cdot s_t \pmod{p}.$$

As $r_1, r_2, \ldots, r_\mu, s_1, s_2, \ldots, s_t$ are least positive residues modulo p of $a, 2a, \ldots, \frac{p-1}{2}$, in some order, we have

$$\left(\frac{p-1}{2}\right)! \equiv (-1)^\mu a \cdot 2a \cdots \left(\frac{p-1}{2}\right)a \pmod{p}$$

$$\equiv (-1)^\mu a^{\frac{p-1}{2}} \left(\frac{p-1}{2}\right)! \pmod{p}.$$

As $\gcd((\frac{p-1}{2})!, p) = 1$, $(\frac{p-1}{2})!$ is canceled out from both sides, resulting in

$$1 \equiv (-1)^\mu a^{\frac{p-1}{2}} \pmod{p}.$$

Multiplying both sides by $(-1)^\mu$, we have

$$a^{\frac{p-1}{2}} \equiv (-1)^\mu \bmod p.$$

Now, from Proposition 10.5.4, it follows that $\left(\frac{a}{p}\right) \equiv (-1)^\mu \bmod p$ and hence $\left(\frac{a}{p}\right) = (-1)^\mu$. $\qquad \square$

Example 10.5.3   To check whether the congruence $x^2 \equiv 5 \pmod{13}$ has solutions in the set of integers, we may use the Gauss Lemma. Here $a = 5$, $p = 13$ and the set $\mathcal{R} = \{5, 10, 15, 20, 25, 30\}$. Now the least positive residues modulo 13 of these elements of $\mathcal{R}$ are 5, 10, 2, 7, 12, 4, respectively. Out of these elements, only 10, 7, and 12 exceed $p/2 = 6\cdot5$. Thus here $\mu = 3$. Hence by the Gauss Lemma, $\left(\frac{5}{13}\right) =$

$(-1)^\mu = (-1)^3 = -1$, with the result that the congruence $x^2 \equiv 5 \pmod{13}$ has no solution in the set of integers.

Using the Gauss Lemma, we now characterize all primes that have 2 as a quadratic residue modulo an odd prime p.

Theorem 10.5.3 *Let p be an odd prime integer. Then*

$$\left(\frac{2}{p}\right) = (-1)^{\frac{p^2-1}{8}}.$$

Proof To prove the theorem, it is sufficient to prove that 2 is a quadratic residue modulo p if and only if $p \equiv 1 \pmod 8$ or $p \equiv 7 \pmod 8$. Let us consider the set $S = \{2, 4, \ldots, p-1\}$. Then the number μ of integers from the set S that exceeds $p/2$ is same as the number of integers that exceeds $p/4$ in the set $\{1, 2, \ldots, \frac{p-1}{2}\}$. Thus $\mu = \frac{p-1}{2} - \lfloor \frac{p}{4} \rfloor$.

Now as p is an odd prime, depending on the remainder modulo 8, there are the following four cases:

Case 1: If $p = 8k + 1$, for some $k \geq 1$, then $\mu = 4k - 2k = 2k$, an even integer.

Case 2: If $p = 8k + 3$, for some $k \geq 0$, then $\mu = 4k + 1 - 2k = 2k + 1$, an odd integer.

Case 3: If $p = 8k + 5$, for some $k \geq 0$, then $\mu = 4k + 2 - 2k - 1 = 2k + 1$, an odd integer.

Case 4: If $p = 8k + 7$, for some $k \geq 0$, then $\mu = 4k + 3 - 2k - 1 = 2k + 2$, an even integer.

The theorem now follows from Theorem 10.5.2 (Gauss Lemma).   □

Remark The integer 2 is a quadratic residue of all primes $p \equiv \pm 1 \pmod 8$ and a quadratic non-residue of all primes $p \equiv \pm 3 \pmod 8$.

We are now going to prove the following lemma that facilitates a passage from the Gauss Lemma to the proof of one of the celebrated theorems in number theory, known as, Law of Quadratic Reciprocity.

Lemma 10.5.1 *Let p be an odd prime and a be an odd integer with $\gcd(a, p) = 1$. Then*

$$\left(\frac{a}{p}\right) = (-1)^{\sum_{i=1}^{(p-1)/2}[ia/p]},$$

where $[ia/p]$ denotes the greatest integer not exceeding ia/p.

Proof Let us consider the integers $a, 2a, \ldots, [\frac{p-1}{2}]a$. For $i = 1, 2, \ldots, [\frac{p-1}{2}]$, each ia can be uniquely represented by $ia = q_i p + u_i$ where $1 \leq u_i < p$. Thus $ia/p =$

$q_i + u_i/p$. So $[ia/p] = q_i$. Hence, for $i = 1, 2, \ldots, [\frac{p-1}{2}]$,

$$ia = \left[\frac{ia}{p}\right]p + u_i. \tag{10.1}$$

As in the Gauss Lemma (Theorem 10.5.2), let us use the notations for r_i and s_j, i.e., if the remainder $u_i > p/2$, then it is one of the integers $r_1, r_2, \ldots, r_\mu$, while if $u_i < p/2$, then it is one of the integers $s_1, s_2, \ldots, s_t$. Now from (10.1), we have

$$\sum_{i=1}^{(p-1)/2} ia = \sum_{i=1}^{(p-1)/2}\left[\frac{ia}{p}\right]p + \sum_{i=1}^{\mu} r_i + \sum_{i=1}^{t} s_i. \tag{10.2}$$

As it was argued while proving the Gauss Lemma that the $\frac{p-1}{2}$ integers $p - r_1, p - r_2, \ldots, p - r_\mu, s_1, s_2, \ldots, s_t$ are just all the integers $1, 2, \ldots, \frac{p-1}{2}$, in some order, we have

$$\sum_{i=1}^{(p-1)/2} i = \sum_{i=1}^{\mu}(p - r_i) + \sum_{i=1}^{t} s_i = p\mu - \sum_{i=1}^{\mu} r_i + \sum_{i=1}^{t} s_i. \tag{10.3}$$

Subtracting (10.3) from (10.2), we get

$$(a - 1)\sum_{i=1}^{(p-1)/2} i = p\left(\sum_{i=1}^{(p-1)/2}\left[\frac{ia}{p}\right] - \mu\right) + 2\sum_{i=1}^{\mu} r_i. \tag{10.4}$$

As a and p are both odd, $a \equiv p \equiv 1 \pmod 2$. Thus equation (10.4) becomes

$$0 \cdot \sum_{i=1}^{(p-1)/2} i \equiv 1 \cdot \left(\sum_{i=1}^{(p-1)/2}\left[\frac{ia}{p}\right] - \mu\right) \pmod 2,$$

resulting in

$$\mu = \sum_{i=1}^{(p-1)/2}\left[\frac{ia}{p}\right] \pmod 2.$$

Thus from the Gauss Lemma, we have

$$\left(\frac{a}{p}\right) = (-1)^\mu = (-1)^{\sum_{i=1}^{(p-1)/2}[\frac{ia}{p}] \pmod 2} = (-1)^{\sum_{i=1}^{(p-1)/2}[\frac{ia}{p}]},$$

as required.   □

We are now in a position to prove the Law of Quadratic Reciprocity, a celebrated theorem in number theory. Suppose p and q are distinct odd primes and we know whether q is a quadratic residue of p. Then the question is: whether p is a quadratic residue modulo q? In mid-1700s, Euler found the answer of the question through examining numerical evidence, but failed to prove the result. Later, in 1785, Legen-

Fig. 10.1 Lattice points
within the rectangle

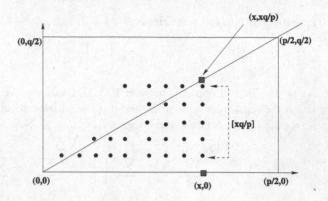

dre re-formulated Euler's answer, in its modern form. Though Legendre proposed
several proofs of the theorem, each of his proofs contained a serious gap. The first
correct proof was proposed by Gauss, who claimed to have rediscovered this result
when he was 18 years old. Gauss devised seven more proofs, each based on a different approach. The proof presented below, a variant of one of Gauss' own arguments,
is due to his student Ferdinand Eisenstein.

Theorem 10.5.4 (Law of Quadratic Reciprocity) *Let p and q be two distinct odd
primes. Then*

$$\left(\frac{p}{q}\right)\left(\frac{q}{p}\right) = (-1)^{\frac{p-1}{2}\cdot\frac{q-1}{2}}.$$

Proof Let us consider a rectangle in the xy coordinate plane, whose vertices are
$(0, 0)$, $(p/2, 0)$, $(p/2, q/2)$ and $(0, q/2)$ as shown in Fig. 10.1.

We consider the points, whose coordinates are integers, within this rectangle i.e.,
not including any points on the bounding lines i.e., the points on the x axis, y axis,
and the lines $x = p/2$ and $y = q/2$ are excluded. The points within the rectangle
are called lattice points within the rectangle. We are now going to count the number
of these lattice points in two different ways. As p and q are both odd integers, the
lattice points within the rectangle consists of all points (x, y) such that $1 \le x \le \frac{p-1}{2}$
and $1 \le y \le \frac{q-1}{2}$. Thus the number of such points is $\frac{p-1}{2} \cdot \frac{q-1}{2}$.

Now we are going to count the number of these points in a different way. Let us
consider the diagonal D passing through the points $(0, 0)$ and $(p/2, q/2)$. Then the
equation of the diagonal D is given by $y = (q/p)x$. As $\gcd(p, q) = 1$, we claim
that none of the lattice points within the rectangle lies on the diagonal D. If possible
let there exist some lattice point on the diagonal, say (a, b), within the rectangle.
Then, we have $b = (q/p)a$. That implies p divides a and q divides b, contradicting
the fact that $1 \le a \le \frac{p-1}{2}$ and $1 \le b \le \frac{q-1}{2}$. Hence, none of the lattice points within
the rectangle lie on the diagonal D. As, the diagonal D divides the rectangle into
two parts, let B denote the portion of the rectangle below the diagonal D and let U
denote the portion above. To count the total number of points within the rectangle, it
is sufficient to count the total number of points inside each of the portions B and U.

First let us count the number of lattice points within B. Let us fix any point $(x, 0)$ on the x-axis such that x is an integer with $1 \le x \le p/2$. Now if we draw a vertical line through the point $(x, 0)$, then it cuts the diagonal at $(x, qx/p)$. As, the lattice points within the rectangle neither lie on the x-axis nor on the diagonal, there are precisely $[\frac{xq}{p}]$ lattice points in B just above the point $(x, 0)$ and below D. Now, if we range x from 1 to $(p-1)/2$, the total number of lattice points contained in B is $\sum_{x=1}^{(p-1)/2}[\frac{xq}{p}]$. By similar argument, with the roles of p and q interchanged, it can also be shown that the total number of lattice points within U is $\sum_{y=1}^{(q-1)/2}[\frac{yp}{q}]$. Thus equating the two different types of counting, we have

$$\frac{p-1}{2} \cdot \frac{q-1}{2} = \sum_{x=1}^{(p-1)/2}\left[\frac{xq}{p}\right] + \sum_{y=1}^{(q-1)/2}\left[\frac{yp}{q}\right].$$

Now by Lemma 10.5.1, we have

$$\begin{aligned}\left(\frac{p}{q}\right)\left(\frac{q}{p}\right) &= (-1)^{\sum_{y=1}^{(q-1)/2}[\frac{yp}{q}]} \cdot (-1)^{\sum_{x=1}^{(p-1)/2}[\frac{xq}{p}]} \\ &= (-1)^{\sum_{x=1}^{(p-1)/2}[\frac{xq}{p}]+\sum_{y=1}^{(q-1)/2}[\frac{yp}{q}]} \\ &= (-1)^{\frac{p-1}{2}\frac{q-1}{2}}.\end{aligned}$$

This completes the proof of the law of quadratic reciprocity.   □

As a consequence, the following corollary is immediate.

Corollary *Let p and q be two distinct odd primes. Then*

$$\left(\frac{p}{q}\right) = \begin{cases} \left(\frac{q}{p}\right) & \text{if } p \equiv 1 \ (\text{mod } 4) \text{ or } q \equiv 1 \ (\text{mod } 4) \\ -\left(\frac{q}{p}\right) & \text{if } p \equiv q \equiv 3 \ (\text{mod } 4). \end{cases}$$

Remark The law of quadratic reciprocity is some times very useful in calculating $\left(\frac{a}{p}\right)$ with large prime p and comparatively small a, where the factorization of a is known. For example, if $a = p_1^{\alpha_1} p_2^{\alpha_2} \cdots p_k^{\alpha_k}$ is the prime factorization of a, then $\left(\frac{a}{p}\right) = \prod_{i=1}^{k}\left(\frac{p_i}{p}\right)^{\alpha_i}$. Now each of $\left(\frac{p_i}{p}\right)$ may be calculated as $\left(\frac{p_i}{p}\right) = \left(\frac{p \ (\text{mod } p_i)}{p_i}\right)$, if $p \equiv 1 \ (\text{mod } 4)$ or $p_i \equiv 1 \ (\text{mod } 4)$ and $\left(\frac{p_i}{p}\right) = -\left(\frac{p \ (\text{mod } p_i)}{p_i}\right)$, if $p \equiv p_i \equiv 3$ (mod 4). So instead of calculating in modulo p, we can calculate in modulo the smaller prime p_i.

We are now going to define Jacobi Symbol, named after the German mathematician Carl Jacobi (1804–1851) who introduced the symbol which is a generalization of the Legendre symbol. Jacobi symbol plays an important role in primality testing to be discussed latter.

Definition 10.5.3 Let n be an odd positive integer with prime factorization $n = p_1^{\alpha_1} p_2^{\alpha_2} \cdots p_k^{\alpha_k}$ and let a be an integer such that $\gcd(a, n) = 1$. Then, the Jacobi symbol $\binom{a}{n}$ is defined by

$$\left(\frac{a}{n}\right) = \left(\frac{a}{p_1}\right)^{\alpha_1} \left(\frac{a}{p_2}\right)^{\alpha_2} \cdots \left(\frac{a}{p_k}\right)^{\alpha_k},$$

where the symbols on the right hand side of the equality are the Legendre symbols.

Proposition 10.5.8 *Let n be an odd positive integer and a be an integer such that* $\gcd(a, n) = 1$. *If the congruence* $x^2 \equiv a \pmod{n}$ *has a solution in* **Z**, *then* $\left(\frac{a}{n}\right) = 1$.

Proof Let p be any prime factor of n. As the congruence $x^2 \equiv a \pmod{n}$ has a solution in **Z**, the congruence $x^2 \equiv a \pmod{p}$ has also a solution in **Z**. Thus $\left(\frac{a}{p}\right) = 1$. Consequently,

$$\left(\frac{a}{n}\right) = \prod_{i=1}^{k} \left(\frac{a}{p_i}\right)^{\alpha_i} = 1, \quad \text{where } n = p_1^{\alpha_1} p_2^{\alpha_2} \cdots p_k^{\alpha_k}.$$

$\square$

Remark The converse of the above proposition may not be true. For example, $\left(\frac{2}{15}\right) = \left(\frac{2}{5}\right) \cdot \left(\frac{2}{3}\right) = (-1) \cdot (-1) = 1$. However, there are no solutions to the congruence $x^2 \equiv 2 \pmod{15}$ in **Z** as the congruences $x^2 \equiv 2 \pmod{3}$ and $x^2 \equiv 2 \pmod{5}$ have no solutions in **Z**.

Now we shall show that the Jacobi symbol enjoys some of the properties similar to those of Legendre symbol.

Proposition 10.5.9 *Let n be an odd positive integer and let a and b be integers such that* $\gcd(a, n) = 1$ *and* $\gcd(b, n) = 1$. *Then*

(i) *if,* $a \equiv b \pmod{n}$ *then* $\left(\frac{a}{n}\right) = \left(\frac{b}{n}\right)$;

(ii) $\left(\frac{ab}{n}\right) = \left(\frac{a}{n}\right) \cdot \left(\frac{b}{n}\right)$;

(iii) $\left(\frac{-1}{n}\right) = (-1)^{(n-1)/2}$;

(iv) $\left(\frac{2}{n}\right) = (-1)^{(n^2-1)/8}$.

Proof Let $n = p_1^{\alpha_1} p_2^{\alpha_2} \cdots p_k^{\alpha_k}$ be the prime factorization of n. Now we prove the four parts one by one.

Proof of (i). Let p be any prime factor of n. As $a \equiv b \pmod{n}$, $a \equiv b \pmod{p}$. Thus by Proposition 10.5.5(ii), it follows that $\left(\frac{a}{p}\right) = \left(\frac{b}{p}\right)$. Now

$$\left(\frac{a}{n}\right) = \left(\frac{a}{p_1}\right)^{\alpha_1} \cdot \left(\frac{a}{p_2}\right)^{\alpha_2} \cdots \left(\frac{a}{p_k}\right)^{\alpha_k}$$

$$= \left(\frac{b}{p_1}\right)^{\alpha_1} \cdot \left(\frac{b}{p_2}\right)^{\alpha_2} \cdots \left(\frac{b}{p_k}\right)^{\alpha_k}$$

$$= \left(\frac{b}{n}\right).$$

Proof of (ii). The result follows directly from Proposition 10.5.5(i).

Proof of (iii). Proposition 10.5.5(iii) asserts that for a prime factor p of n, $\left(\frac{-1}{p}\right) = (-1)^{(p-1)/2}$. Consequently,

$$\left(\frac{-1}{n}\right) = \left(\frac{-1}{p_1}\right)^{\alpha_1} \cdot \left(\frac{-1}{p_2}\right)^{\alpha_2} \cdots \left(\frac{-1}{p_k}\right)^{\alpha_k}$$

$$= (-1)^{\alpha_1(p_1-1)/2 + \alpha_2(p_2-1)/2 + \cdots + \alpha_k(p_k-1)/2}. \qquad (10.5)$$

Now from the prime factorization of n, we have

$$n = \left(1 + (p_1 - 1)\right)^{\alpha_1} \cdot \left(1 + (p_2 - 1)\right)^{\alpha_2} \cdots \left(1 + (p_k - 1)\right)^{\alpha_k}.$$

As each p_i is odd, we have

$$\left(1 + (p_i - 1)\right)^{\alpha_i} \equiv \left(1 + \alpha_i(p_i - 1)\right) \pmod 4$$

and

$$\left(1 + \alpha_i(p_i - 1)\right) \cdot \left(1 + \alpha_j(p_j - 1)\right)$$
$$\equiv \left(1 + \alpha_i(p_i - 1) + \alpha_j(p_j - 1)\right) \pmod 4, \quad \text{for } i \neq j.$$

Therefore,

$$n \equiv 1 + \alpha_1(p_1 - 1) + \alpha_2(p_2 - 1) + \cdots + \alpha_k(p_k - 1) \pmod 4.$$

Consequently,

$$\frac{n-1}{2} \equiv \left(\alpha_1(p_1 - 1)/2 + \alpha_2(p_2 - 1)/2 + \cdots + \alpha_k(p_k - 1)/2\right) \pmod 2. \quad (10.6)$$

Thus combining (10.5) and (10.6), we get $\left(\frac{-1}{n}\right) = (-1)^{(n-1)/2}$, as required.

Proof of (iv). From Theorem 10.5.3, we have $\left(\frac{2}{p}\right) = (-1)^{(p^2-1)/8}$, for any odd prime p. Hence,

$$\left(\frac{2}{n}\right) = \left(\frac{2}{p_1}\right)^{\alpha_1} \cdot \left(\frac{2}{p_2}\right)^{\alpha_2} \cdots \left(\frac{2}{p_k}\right)^{\alpha_k}$$

$$= (-1)^{\alpha_1(p_1^2-1)/8 + \alpha_2(p_2^2-1)/8 + \cdots + \alpha_k(p_k^2-1)/8}. \qquad (10.7)$$

Now from the prime factorization of n, we have

$$n^2 = \left(1 + (p_1^2 - 1)\right)^{\alpha_1} \cdot \left(1 + (p_2^2 - 1)\right)^{\alpha_2} \cdots \left(1 + (p_k^2 - 1)\right)^{\alpha_k}.$$

As $p_i^2 - 1 \equiv 0 \pmod 8$, we have

$$\left(1 + (p_i^2 - 1)\right)^{\alpha_i} \equiv 1 + \alpha_i(p_i^2 - 1) \pmod{64}$$

and

$$(1 + \alpha_i (p_i^2 - 1)) \cdot (1 + \alpha_j (p_j^2 - 1))$$
$$\equiv 1 + \alpha_i (p_i^2 - 1) + \alpha_j (p_j^2 - 1) \pmod{64}, \quad \text{for } i \neq j.$$

Therefore,

$$n^2 \equiv 1 + \alpha_1 (p_1^2 - 1) + \alpha_2 (p_2^2 - 1) + \cdots + \alpha_k (p_k^2 - 1) \pmod{64}.$$

Consequently,

$$\frac{n^2 - 1}{8} \equiv \alpha_1 (p_1^2 - 1)/8 + \alpha_2 (p_2^2 - 1)/8 + \cdots + \alpha_k (p_k^2 - 1)/8 \pmod{8}. \quad (10.8)$$

Thus combining (10.7) and (10.8), we have $\left(\frac{2}{n}\right) = (-1)^{(n^2-1)/8}$, as required. $\qquad \square$

Theorem 10.5.5 *Let n and m be two odd positive integers such that $\gcd(n, m) = 1$. Then*

$$\binom{n}{m}\binom{m}{n} = (-1)^{\frac{m-1}{2}\frac{n-1}{2}}.$$

Proof Let $n = p_1^{\alpha_1} p_2^{\alpha_2} \cdots p_k^{\alpha_k}$ and $m = q_1^{\beta_1} q_2^{\beta_2} \cdots q_t^{\beta_t}$ be the prime factorizations of n and m, respectively. Now

$$\binom{n}{m} = \prod_{j=1}^{t} \binom{n}{q_j}^{\beta_j} = \prod_{j=1}^{t} \prod_{i=1}^{k} \binom{p_i}{q_j}^{\beta_j \alpha_i}$$

and

$$\binom{m}{n} = \prod_{i=1}^{k} \binom{m}{p_i}^{\alpha_i} = \prod_{i=1}^{k} \prod_{j=1}^{t} \binom{q_j}{p_i}^{\alpha_i \beta_j}.$$

Hence,

$$\binom{n}{m}\binom{m}{n} = \prod_{i=1}^{k} \prod_{j=1}^{t} \left(\binom{p_i}{q_j}\binom{q_j}{p_i}\right)^{\alpha_i \beta_j}.$$

Using the law of quadratic reciprocity for the primes p_i and q_j, we have

$$\binom{n}{m}\binom{m}{n} = \prod_{i=1}^{k} \prod_{j=1}^{t} (-1)^{\alpha_i \left(\frac{p_i-1}{2}\right)\beta_j \left(\frac{q_j-1}{2}\right)}$$

$$= (-1)^{\sum_{i=1}^{k} \sum_{j=1}^{t} \alpha_i \left(\frac{p_i-1}{2}\right)\beta_j \left(\frac{q_j-1}{2}\right)}$$

$$= (-1)^{\sum_{i=1}^{k} \alpha_i \left(\frac{p_i-1}{2}\right) \sum_{j=1}^{t} \beta_j \left(\frac{q_j-1}{2}\right)} \qquad (10.9)$$

By using the similar argument as in the Proof of (iii) in Proposition 10.5.9, we have

$$\sum_{i=1}^{k} \alpha_i \left(\frac{p_i - 1}{2} \right) \equiv \frac{n-1}{2} \pmod{2}$$

and

$$\sum_{j=1}^{t} \beta_j \left(\frac{q_j - 1}{2} \right) \equiv \frac{m-1}{2} \pmod{2}.$$

Thus from (10.9), we have

$$\binom{n}{m}\binom{m}{n} = (-1)^{\frac{n-1}{2} \frac{m-1}{2}},$$

as required.   □

10.6 Fermat Numbers

In this section, we shall look at a special class of numbers, known as Fermat number. In number theory, this special class of numbers plays an important role.

Definition 10.6.1 A Fermat number is an integer F_n of the form $F_n = 2^{2^n} + 1$, $n \geq 0$. If F_n is prime, then it is called Fermat prime.

Fermat, having great mathematical intuition, observed that $F_0 = 3$, $F_1 = 5$, $F_2 = 17$, $F_3 = 257$, $F_4 = 65537$ are all primes. He had a belief that F_n is prime for each value of n and expressed his confidence while communicating his idea to another great mathematician Mersenne. In 1732, the belief of Fermat was proven to be wrong by Euler showing that 641 divides $F_5 = 4294967297$. G. Bennett gave the following alternative elementary proof that 641 divides F_5.

Proposition 10.6.1 641 *divides* F_5.

Proof Observe that $641 = 2^7 \cdot 5 + 1$. Let $a = 2^7$ and $b = 5$. Now $1 + ab - b^4 = 1 + b(a - b^3) = 1 + b(2^7 - 5^3) = 1 + 3b = 2^4$. But this implies, $F_5 = 2^{2^5} + 1 = 2^4(2^7)^4 + 1 = (1 + ab - b^4)a^4 + 1 = (1 + ab)a^4 - a^4 b^4 + 1 = (1 + ab)a^4 + (1 + a^2 b^2)(1 - a^2 b^2) = (1 + ab)a^4 + (1 + ab)(1 - ab)(1 + a^2 b^2) = (1 + ab)(a^4 + (1 - ab)(1 + a^2 b^2))$. This implies that $641 = ab + 1$ divides F_5.   □

The following proposition gives a nice form of the prime factors of F_n.

Proposition 10.6.2 *Every prime factor of* F_n ($n \geq 2$) *must be of the form* $2^{n+2}k+1$, *for some positive integer* k.

Proof Let p be a prime factor of $F_n = 2^{2^n} + 1$, where $n \geq 2$. Then $2^{2^n} \equiv -1 \pmod{p}$. This implies that $2^{2^n} \cdot 2^{2^n} \equiv 1 \pmod{p}$, with the result $2^{2^{n+1}} \equiv 1 \pmod{p}$. We claim that the order of 2 $\pmod{p}$ in $\mathbf{Z}_p^*$ is 2^{n+1}. If possible, let the order of 2 $\pmod{p}$ in $\mathbf{Z}_p^*$ be $t < 2^{n+1}$. Then t divides 2^{n+1}. Thus t must be of the form 2^k, $k = 1, 2, \ldots, n$. Thus $2^{2^k} \equiv 1 \pmod{p}$, contradicting the fact that $2^{2^n} \equiv -1 \pmod{p}$. Hence the order of 2 $\pmod{p}$ in $\mathbf{Z}_p^*$ is 2^{n+1}. Thus by Lagrange's Theorem, 2^{n+1} divides $p - 1$. In particular, as $n = 2$, $8 = 2^3$ divides $p - 1$. So p must be of the form $8k + 1$. Hence by Theorem 10.5.3 and Proposition 10.5.4, $\left(\frac{2}{p}\right) = 1 \equiv 2^{\frac{p-1}{2}} \pmod{p}$. Thus 2^{n+1} divides $\frac{p-1}{2}$ which implies $p = k2^{n+2} + 1$, for some positive integer k. $\qquad\square$

Remark Proposition 10.6.2 may be used to provide a simple alternative proof of the fact that $641 = 2^{5+2} \cdot 5 + 1$ divides F_5 (Proposition 10.6.1).

Since the number F_n increases very rapidly with n, it is not very easy to check the primality for F_n (i.e., to check whether F_n is prime or not). But in 1877, Pepin first introduced the following primality test for F_n.

Proposition 10.6.3 *Let F_n denote the nth Fermat number with $n \geq 2$. For the integer $k \geq 2$, the following conditions are equivalent.*

1. F_n *is prime and the Legendre symbol* $\left(\frac{k}{F_n}\right) = -1$.
2. $k^{\frac{F_n-1}{2}} \equiv -1 \pmod{F_n}$.

Proof $(1) \Rightarrow (2)$ Let F_n be prime and $\left(\frac{k}{F_n}\right) = -1$. Then by the definition of Legendre symbol, $k^{\frac{F_n-1}{2}} \equiv -1 \pmod{F_n}$.

$(2) \Rightarrow (1)$ Let $1 \leq a < F_n$ be such that $a \equiv k \pmod{F_n}$. Then $a^{\frac{F_n-1}{2}} \equiv -1 \pmod{F_n}$. Thus $a^{F_n-1} \equiv 1 \pmod{F_n}$. As $a^{\frac{F_n-1}{2}} \equiv -1 \pmod{F_n}$ and 2 is the only prime factor of $F_n - 1$, by Proposition 10.9.3, F_n is a prime integer. Finally, the rest of the proposition follows from Proposition 10.5.4. $\qquad\square$

Remark Till date it is not known whether there exist infinitely many Fermat primes. Researches are going on in that direction. The Fermat number have been studied intensively, often with the aid of computers. Fermat primes play an important role in geometry as showed by Gauss in 1801 that a regular polygon with k sides can be constructed by ruler and compass methods if and only if $k = 2^e p_1 p_2 \cdots p_r$, where $p_1, p_2, \ldots, p_r$ are distinct Fermat primes.

We now study some more properties of Fermat numbers.

Proposition 10.6.4 *Let F_n denote the nth Fermat number. Then for all positive integers n, $F_0 F_1 F_2 \cdots F_{n-1} = F_n - 2$.*

Proof We prove the result by the principle of mathematical induction on n. For $n = 1$, $F_0 = 3$ and $F_1 = 5$. So the condition $F_0 = F_1 - 2$ holds for $n = 1$. Let us assume that the condition holds for some $k > 1$, i.e., $F_0 F_1 F_2 \cdots F_{k-1} = F_k - 2$. Now, $F_0 F_1 F_2 \cdots F_{k-1} F_k = (F_0 F_1 F_2 \cdots F_{k-1}) F_k = (F_k - 2) F_k = (2^{2^k} + 1 - 2)(2^{2^k} + 1) = 2^{2^{k+1}} - 1 = F_{k+1} - 2$, showing that the result is true for $k + 1$. Hence the result follows from the principle of mathematical induction. $\square$

We are going to show another interesting property of Fermat numbers which leads to an alternative proof of Theorem 10.2.2.

Proposition 10.6.5 *Distinct Fermat numbers are co-prime.*

Proof Let d be the gcd of F_n and F_{n+m}, where $m \in \mathbf{N}^+$, i.e., $d = \gcd(F_n, F_{n+m})$. Note that $x = -1$ is a root of the polynomial $x^{2^m} - 1$. Thus $x + 1$ divides $x^{2^m} - 1$. Now substituting $x = 2^{2^n}$, $F_n = 2^{2^n} + 1$ divides $2^{2^{n+m}} - 1 = F_{n+m} - 2$, with the result that d divides 2. As both F_n and F_{n+m} are odd, d must be 1. $\square$

Alternative Proof of Theorem 10.2.2 From the above theorem, it follows that for any given positive integer n, each of the Fermat numbers F_i is divisible by an odd prime which does not divide any of the other Fermat number F_j, for $i \neq j$, $i, j \in \{1, 2, \ldots, n\}$. So for any given positive integer n, there are at least n distinct odd primes not exceeding F_n, proving that the number of primes is infinite. $\square$

Remark Note that the $n + 1$th prime $p_{n+1} \leq F_n = 2^{2^n} + 1$. This inequality is slightly stronger than the inequality as mentioned in Corollary 1 of Theorem 10.2.2.

10.7 Perfect Numbers and Mersenne Numbers

Due to some mystical beliefs, the ancient Greeks were interested on a special type of integers, known as perfect numbers that are equal to the sum of all their proper positive divisors. Surprisingly, the ancient Greeks knew how to find even perfect numbers.

Definition 10.7.1 A positive integer n is said to be a perfect number if $\sigma(n) = 2n$, i.e., the sum of all the proper positive divisors of n is equal to n itself.

Example 10.7.1 Since $\sigma(6) = 1 + 2 + 3 + 6 = 12$ and $\sigma(28) = 1 + 2 + 4 + 7 + 14 + 28 = 56$, 6 and 28 are examples of perfect numbers. The next perfect number is 496.

Next we shall characterize the even perfect numbers through the following theorem.

Theorem 10.7.1 *An even integer $n > 0$ is a perfect number if and only if $n = 2^{k-1}(2^k - 1)$, where $k \geq 2$ and $2^k - 1$ is a prime integer.*

Proof First let n be an even perfect number. Let $n = 2^t m$, where $t, m \geq 1$ and m is odd. As σ is a multiplicative function, $\sigma(n) = \sigma(2^t m) = \sigma(2^t)\sigma(m) = (2^{t+1} - 1)\sigma(m)$. Again as n is a perfect number, $\sigma(n) = 2n = 2^{t+1}m$. Thus we have

$$2^{t+1}m = (2^{t+1} - 1)\sigma(m). \tag{10.10}$$

Since 2^{t+1} and $2^{t+1} - 1$ are relatively prime, 2^{t+1} divides $\sigma(m)$. Thus there exists some integer q such that

$$\sigma(m) = 2^{t+1}q. \tag{10.11}$$

Consequently, from (10.10), we have

$$m = (2^{t+1} - 1)q. \tag{10.12}$$

From (10.12), q divides m. We claim that $q \neq m$. If $q = m$, then from (10.12), we have $1 = 2^{t+1} - 1$, implying $t = 0$, a contradiction as $t \geq 1$, with the result that $q \neq m$. Now from (10.11) and (10.12), we see that

$$m + q = (2^{t+1} - 1)q + q = 2^{t+1}q = \sigma(m). \tag{10.13}$$

Now we claim that $q = 1$. If possible, let $q \neq 1$. As $q \neq m$, there are at least three positive divisors of m, namely $1, q$, and m, implying $\sigma(m) \geq 1 + q + m > q + m = \sigma(m)$, by (10.13), a contradiction. Hence $q = 1$. Thus from (10.12) we have

$$m = 2^{t+1} - 1. \tag{10.14}$$

Also from (10.13), we have $\sigma(m) = m + 1$, which implies m must be a prime integer. Hence from (10.14), we can conclude that $n = 2^t m = 2^t(2^{t+1} - 1)$, where $t \geq 1$ and $2^{t+1} - 1$ is a prime integer.

Conversely, we assume that $n = 2^{k-1}(2^k - 1)$, where $k \geq 2$ and $2^k - 1$ is a prime integer. We have to show that $\sigma(n) = 2n$. Since $2^k - 1$ is odd, 2^{k-1} and $2^k - 1$ are relatively prime. Thus $\sigma(n) = \sigma(2^{k-1})\sigma(2^k - 1) = (1 + 2 + 2^2 + \cdots + 2^{k-1})(1 + 2^k - 1)$, as $2^k - 1$ is a prime integer. This implies that $\sigma(n) = 2(2^{k-1}(2^k - 1)) = 2n$, proving that n is a perfect number. $\square$

Theorem 10.7.1 asserts that to get an even perfect number we need to have a prime number of the form $2^k - 1$, where $k \geq 2$. Then the following theorem will help us in finding such primes.

Theorem 10.7.2 *For some positive integers a and $n > 1$, if $a^n - 1$ is a prime integer then $a = 2$ and n is a prime integer.*

Proof First we prove that $a = 2$. As $a^n - 1 = (a - 1)(a^{n-1} + a^{n-2} + \cdots + 1)$ is prime and a is a positive integer great than 1, $a - 1$ must be equal to 1, implying $a = 2$. Now we claim that n is a prime integer. If not, then let $n = rs, 1 < r, s < n$. Thus $(a^n - 1) = a^{rs} - 1 = (a^r - 1)((a^r)^{s-1} + (a^r)^{s-2} + \cdots + 1)$, a contradiction as both $(a^r - 1)$ and $((a^r)^{s-1} + (a^r)^{s-2} + \cdots + 1)$ are positive integers greater than 1, with the result that n to be a prime integer. $\square$

Theorem 10.7.2 asserts that for primes of the form $2^n - 1$, we need to consider only integers n that are prime. Motivated by the study of perfect numbers, these special type of integers have been studied in great depth by a great French monk of the 17th century, Mersenne. These special type of integers are known as Mersenne numbers, which are defined as follows.

Definition 10.7.2   If n is a positive integer, then $M_n = 2^n - 1$ is called the nth Mersenne number. Further, if $M_p = 2^p - 1$ is a prime integer, then M_p is called the Mersenne prime.

Example 10.7.2   Already at the time of Mersenne, it was known that some Mersenne numbers, such as, $M_2 = 3$, $M_3 = 7$, $M_5 = 31$, $M_7 = 127$ are primes, while $M_{11} = 2047 = 23 \times 89$ and $M_{37} = 2^{37} - 1 = 137438953471 = 223 \times 616318177$ are composite. In 1640, Mersenne stated that M_p is also a prime for $p = 13, 17, 19, 31, 67, 127, 257$. Unfortunately, he was wrong about 67 and 257, and he did not include 61, 89, 107 (in the list among those less than 257), which also produce Mersenne primes. However, we must appreciate that his statement was quite astonishing, in view of the size of the numbers involved.

Since M_n increases exponentially, it is very difficult to check the primality of M_n for even moderate n. But the following theorem plays an important role in checking the primality for M_n. To prove this let us first prove the following lemma.

Lemma 10.7.1   *For any two positive integers a and b, $\gcd(2^a - 1, 2^b - 1) = 2^{\gcd(a,b)} - 1$.*

Proof Let $c = \gcd(a, b)$. Then $2^c - 1$ divides both $2^a - 1$ and $2^b - 1$. Let d divide both $2^a - 1$ and $2^b - 1$. We show that d also divides $2^c - 1$. If possible, suppose d does not divide $2^c - 1$. Then there exists a prime p and a positive integer n such that p^n divides d but p^n does not divide $2^c - 1$. Now p^n divides d implies p^n divides both $2^a - 1$ and $2^b - 1$. Note that as both $2^a - 1$ and $2^b - 1$ are odd, p must be an odd prime integer. Also as $c = \gcd(a, b)$, there exist integers x and y such that $c = ax + by$. Further, $2^a \equiv 1 \pmod{p^n}$ and $2^b \equiv 1 \pmod{p^n}$ imply that $2^c = 2^{ax+by} \equiv 1 \pmod{p^n}$, with the result that p^n divides $2^c - 1$, a contradiction. Thus d divides $2^c - 1$. Hence $\gcd(2^a - 1, 2^b - 1) = 2^{\gcd(a,b)} - 1$. This completes the proof. $\square$

Using the above lemma, we prove the following theorem.

Theorem 10.7.3 *If p is an odd prime, then every divisor of the pth Mersenne number $M_p = 2^p - 1$ is of the form $2kp + 1$, where k is a positive integer.*

Proof To show the result, it is sufficient to prove that any prime q dividing M_p is of the form $2kp + 1$, for some positive integer k. By Fermat's Little Theorem, q divides $2^{q-1} - 1$. Now by Lemma 10.7.1, we have $\gcd(2^p - 1, 2^{q-1} - 1) = 2^{\gcd(p, q-1)} - 1$. As q is a common divisor of $2^p - 1$ and $2^{q-1} - 1$, $\gcd(2^p - 1, 2^{q-1} - 1) > 1$. We claim that $\gcd(p, q - 1) \neq 1$. If possible, let $\gcd(p, q - 1) = 1$. Then $\gcd(2^p - 1, 2^{q-1} - 1) = 1$, a contradiction. Therefore $\gcd(p, q - 1) = p$. Hence p divides $q - 1$. Thus there exists a positive integer t such that $q - 1 = tp$. As p is odd and $q - 1$ is even, t must be even, say $t = 2k$, for some positive integer k. Hence $q = 2kp + 1$, as required. $\square$

10.8 Analysis of the Complexity of Algorithms

In this section, we study how to analyze an algorithm and how the computational complexity of an algorithm is computed. This plays an important role in the theory of computational number theory as the analysis enables us to judge whether an algorithm is good/efficient or not. Computational complexity theory is an important branch in theoretical computer science and mathematics. Complexity theory, or more precisely, computational complexity theory, deals with the resources required during some computation to solve a given problem. The main objective of the computational complexity theory is to classify the computational problems according to their inherent difficulties. Again in computational complexity theory, complexity analysis of an algorithm plays an important role. Complexity analysis is a tool that allows us to explain how an algorithm behaves as the input grows larger i.e., if we feed an algorithm a different input, the question is how will the algorithm behave? More specifically, if one algorithm takes 1 second to run for an input of size 1000, how will it behave if we double the input size? Will it run just as fast, half as fast, or four times slower? In practical programming, this is important as it allows us to predict how our algorithm will behave when the input data becomes larger. The process of computing involves the consumption of different resources like time taken to perform the computation, amount of memory used, power consumed by the system performing the computation, etc. To measure the amount of resources utilized by different algorithms, we need the following notion of asymptotic notations.

10.8.1 Asymptotic Notation

Here we discuss some standard notations related to the rate of growth of functions. This concept will be useful not only in discussing the running time of algorithms but also in a number of other contexts. In algorithm analysis, it is convenient to have a notation for running time which is not clogged up by too much details, because often it is not possible and often it would even be undesirable to have an exact formula for the number of operations. For this, "O-notation" (read it as "big-Oh"-notation) is commonly used. We give the definitions and basic rules for manipulating bounds for any given algorithm.

Let f and g be real-valued functions, both defined either on the set of non-negative integers ($\mathbf{N}$), or on the set of non-negative reals $\mathbf{R}^*$. Here we are interested about the behavior of both $f(x)$ and $g(x)$, for large x, mostly for $x \to \infty$. For this reason, we only require that $f(x)$ and $g(x)$ are defined for all sufficiently large x. We further assume that $g(x) > 0$ for all sufficiently large x.

Definition 10.8.1 For the function $g : \mathbf{N} \to \mathbf{R}^*$, let us define

- $O(g(n)) = \{f(n) | f : \mathbf{N} \to \mathbf{R}^* \text{ and } \exists \text{ positive real constant } c \text{ and } n_0 \in \mathbf{N} \text{ such that } 0 \le f(n) \le cg(n), \forall n \ge n_0\}$.

  Remark Instead of writing $f(n) \in O(g(n))$, we write $f(n) = O(g(n))$. This means that $g(n)$ is an asymptotic upper bound for $f(n)$. For example, if $f(n) = n^2 + 7n + 10000$, then $\forall n \ge 10000$, $n^2 + 7n + 10000 \le n^2 + n^2 + n^2 = 3n^2$. Thus in this case, $n_0 = 10000$ and $c = 3$. So, $n^2 + 7n + 10000 = O(n^2)$. Thus O-notation classifies functions by simple representative growth functions even if the exact values of the functions are unknown. Note that since we demand a certain behavior of a function only for $n \ge n_0$, it does not matter if the functions that we consider are not defined or not specified for $n < n_0$.

- $\Theta(g) = \{f | f : \mathbf{N} \to \mathbf{R}^* \text{ and } \exists \text{ positive real constants } c_1, c_2 \text{ and } n_0 \in \mathbf{N} \text{ such that } 0 \le c_1 g(n) \le f(n) \le c_2 g(n), \forall n \ge n_0\}$.

  Remark Instead of writing $f(n) \in \Theta(g(n))$, we write $f(n) = \Theta(g(n))$. Informally, Θ notation stands for equality in asymptotic sense. For example, if $f(n) = n^2/2 - 2n$, then $\forall n \ge 8$, $n^2/4 \le n^2/2 - 2n \le n^2/2$. So here, $c_1 = 1/4$, $c_2 = 1/2$ and $n_0 = 8$. Thus $f(n) = \Theta(n^2)$. Also note that $6n^3 \ne \Theta(n^2)$, as if $6n^3 = \Theta(n^2)$, there exist $c_2 \in \mathbf{R}^+$ and $n_0 \in \mathbf{N}$ such that $\forall n \ge n_0$, $6n^3 \le c_2 n^2$ which implies $n \le c_2/6$, a contradiction, proving that $6n^3 \ne \Theta(n^2)$. The notation $\Theta(1)$ is used either for a constant or for a constant function with respect to some variable. Finally note that $\Theta(g(n)) \subseteq O(g(n))$, as $f(n) = \Theta(g(n))$ implies $f(n) \in O(g(n))$.

- $\Omega(g(n)) = \{f(n) | f : \mathbf{N} \to \mathbf{R}^* \text{ and } \exists \text{ positive real constant } c \text{ and } n_0 \in \mathbf{N} \text{ such that } 0 \le cg(n) \le f(n), \forall n \ge n_0\}$.

  Remark Instead of writing $f(n) \in \Omega(g(n))$, we write $f(n) = \Omega(g(n))$. This means that $g(n)$ is an asymptotic lower bound for $f(n)$. For example, if $f(n) = n^2 + 7n + 10000$, then $\forall n \ge 1$, $n^2 \le n^2 + 7n + 10000$. Thus in this case, $n_0 = 1$ and $c = 1$. So, $n^2 + 7n + 10000 = \Omega(n^2)$. Note that for any two functions f and g, $f(n) = \Theta(g(n))$ if and only if $f(n) = O(g(n))$ and $f(n) = \Omega(g(n))$.

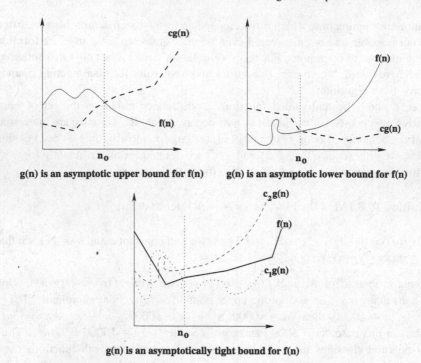

Fig. 10.2 Asymptotic bounds

- $o(g(n)) = \{f(n) \mid f : \mathbf{N} \to \mathbf{R}^* \text{ and } \forall \text{ positive real constant } c \text{ and } n_0 \in \mathbf{N} \text{ such that } 0 \le f(n) < cg(n), \forall n \ge n_0\}$.

Remark Instead of writing $f(n) \in o(g(n))$, we write $f(n) = o(g(n))$. Here, the bound is not an asymptotically tight upper bound for $f(n)$. For example, $2n = o(n^2)$ but $2n \ne o(n)$. The main difference between O and o is that in $f(n) = O(g(n))$, the bound $0 \le f(n) \le cg(n)$ holds for some constant $c \in \mathbf{R}^+$, but in $f(n) = o(g(n))$, the bound $0 \le f(n) < cg(n)$ holds for all constant $c \in \mathbf{R}^+$. Intuitively, in o-notation, the function $f(n)$ becomes insignificant relative to $g(n)$ as n approaches infinity, i.e., $\lim_{n \to \infty} \frac{f(n)}{g(n)} = 0$.

Note In Fig. 10.2, the above asymptotic bounds are described.

- $\omega(g(n)) = \{f(n) \mid f : \mathbf{N} \to \mathbf{R}^* \text{ and } \forall \text{ positive real constant } c \text{ and } n_0 \in \mathbf{N} \text{ such that } 0 \le cg(n) < f(n), \forall n \ge n_0\}$.

Remark Instead of writing $f(n) \in \omega(g(n))$, we write $f(n) = \omega(g(n))$. Here, the bound is not an asymptotically tight lower bound for $f(n)$. For example, $2n^2 = \omega(n)$ but $2n^2 \ne \omega(n^2)$. An alternative way of defining the notation ω is $f(n) \in \omega(g(n))$ if and only if $\lim_{n \to \infty} \frac{f(n)}{g(n)} = \infty$. Finally, note that $f(n) \in \omega(g(n))$ if and only if $g(n) \in o(f(n))$.

10.8.2 Relational Properties Among the Asymptotic Notations

Here we shall discuss some of the relational properties among the asymptotic notations.

Transitivity: $f(n) = \Theta(g(n))$ and $g(n) = \Theta(h(n))$ imply $f(n) = \Theta(h(n))$. Similar result holds for O, Ω, o, ω.

Reflexivity: $f(n) = \Theta(f(n))$. The same result holds good for O and Ω. But reflexivity does not hold for o and ω.

Symmetry: $f(n) = \Theta(g(n))$ if and only if $g(n) = \Theta(f(n))$. The same result does not hold for O, Ω, o, ω.

Transpose Symmetry: $f(n) = O(g(n))$ if and only if $g(n) = \Omega(f(n))$. Also, $f(n) = o(g(n))$ if and only of $g(n) = \omega(f(n))$.

10.8.3 Poly-time and Exponential-Time Algorithms

In this section, we shall introduce informally some of the important concepts of complexity theory. Instead of using a fully abstract model of computation, such as, Turing machines, in this section, we consider all algorithms running on a digital computer with a typical instruction set, an infinite number of bits of memory and constant-time memory access. This model may be thought of as the random access machine (or register machine) model.

A computational problem is specified by an input (of a certain form) and an output (satisfying certain properties relative to the input). An instance of a computational problem is a specific input. The input size of an instance of a computational problem is the number of bits required to represent the instance. The output size of an instance of a computational problem is the number of bits necessary to represent the output. A decision problem is a computational problem where the output is either "yes" or "no".

An algorithm to solve a computational problem is called deterministic if it does not make use of any randomness. We will study the asymptotic complexity of deterministic algorithms by counting the number of bit operations performed by the algorithm expressed as a function of the input size. Upper bounds on the complexity are presented using "big O" notation. When giving complexity estimates using big O notation we implicitly assume that there is a countably infinite number of possible inputs to the algorithm.

Order notation allows us to define several fundamental concepts that are used to get a rough bound on the computational complexity of mathematical problems. Suppose that we are trying to solve a certain type of mathematical problem, where the input to the problem is a number whose size may vary. As an example, consider the Integer Factorization Problem, whose input is a number N and whose output is a prime factor of N. We are interested in knowing how long it takes to solve the problem in terms of the size of the input. Typically, one measures the size of the input by its number of bits, i.e., how much storage it takes to record the input.

Definition 10.8.2 A problem is said to be solvable in polynomial time if there is a constant $c \geq 0$, independent of the size of the input, such that for inputs of size $O(k)$ bits, there is an algorithm to solve the problem in $O(k^c)$ steps.

Remark If we take c to be 1 in Definition 10.8.2, then the problem is solvable in linear time, while if we take c to be 2, then we say that the problem is solvable in quadratic time. In general, polynomial-time algorithms are considered to be fast algorithms.

Definition 10.8.3 A problem is said to be solvable in exponential time if there is a constant $c > 0$ such that for inputs of size $O(k)$ bits, there is an algorithm to solve the problem in $O(e^{ck})$ steps.

Remark Exponential-time algorithms are considered to be slow algorithms.

Remark In the theory of complexity, problems solvable in polynomial time are considered to be "easy" while problems that require exponential time are considered to be "hard".

As we are going to analysis algorithms, we need to know what do we mean by an algorithm. Before that the notations related to algorithms as discussed in the next section.

10.8.4 Notations for Algorithms

The notion of an algorithm is one of the basic concepts in theoretical computer science. Informally, an algorithm is a sequence of well-defined computational procedure that takes as input some value, or a set of values and produces some value, or set of values, as output. The notion of an algorithm has been formalized as a program for a particular theoretical machine model known as the Turing machine model in the theory of algorithms and computational complexity. Let us first start with a simple notion and notations related to algorithms for integers.

- An integer variable contains an integer. Variables are denoted by typewriter type names like a, b, k, l, and so on.
- An array corresponds to a sequence of similar type of variables, indexed by a segment of the natural numbers. For example, int $a[10]$ denotes an array with 10 entities of integer type, while char a[10] denotes an array with 10 entities of character type. An array element is given by the name with the index in brackets. For example, $a[0]$ denotes the first entity, while $a[5]$ denotes the sixth entity of this array.
- If x is some integral value obtained by evaluating some expression and v is an integer type variable, then $v \leftarrow x$ denotes an instruction that causes the value x to be put into the variable v.

- **if-then** statement and **if-then-else** statement are used with the obvious semantics. Let **Bool-exp** denote an expression that evaluates to a boolean value and **stat** denote a (simple or composite) statement. Then the statement

$$\textbf{if } \text{Bool-exp } \textbf{then } \text{stat}$$

is executed as follows: first the boolean expression **Bool-exp** is evaluated to either **true** or **false**. If the result is true, the (simple or composite) statement **stat** is carried out. Similarly, the statement

$$\textbf{if } \text{Bool-exp } \textbf{then } \text{stat1 } \textbf{else } \text{stat2}$$

is executed as follows: if the value of the Boolean expression **Bool-exp** is true, then **stat1** is carried out, otherwise **stat2** is carried out.
- In an algorithm, it is some times required to perform a set of instructions repeatedly. This involves repeating some part of the algorithm either a specified number of times or until a particular condition is being satisfied. This repeated operations may be done using loop control instructions through the **while** statement. The statement

$$\textbf{while } \text{Bool-exp } \textbf{do } \text{stat}$$

may be executed as follows: first the boolean expression bool-exp is evaluated. If the outcome is true, the body "stat" is carried out once, and we start again carrying out the whole while statement. Otherwise the execution of the statement is finished.
- The loop body may contain a special instruction, known as **break**, which, when executed, immediately terminates the execution of the inner loop in which the **break** statement is executed.
- The **return** statement is used to immediately finish the execution of the algorithm.

10.8.5 Analysis of Few Important Number Theoretic Algorithms

Efficiency of an algorithm can be measured in terms of execution time (time complexity) and the amount of memory required (space complexity). Now the question is: which measure is more important? Answer often depends on the limitations of the technology available at time of analysis. However, in this section, we are going to deal only with the time complexity but not the space complexity. Note that time complexity analysis for an algorithm must be independent of the programming language and the machine used. The major objectives of time complexity analysis are to determine the feasibility of an algorithm by estimating an upper bound on the amount of work performed by the machine and to compare different algorithms before deciding on which one to implement. To simplify the analysis, we sometimes ignore the work that takes a constant amount of time.

In this section, as we will be analyzing some number theoretic algorithms, let us first look at how a non-negative integer n is written to the base b. Let $n = c_{k-1}b^{k-1} + c_{k-2}b^{k-2} + \cdots + c_1 b + c_0$, where c_i's are integer between 0 and $b - 1$. Then we say that $(c_{k-1}c_{k-2} \cdots c_1 c_0)_b$ is the representation of n to the base b. If c_{k-1} is non-zero, then we call n a k-digit base b number. Note that any integer in the interval $[b^{k-1}, b^k - 1]$ is a k-digit number to the base b.

Theorem 10.8.1 *Let n be a positive integer. Then the number of digits in the representation of n to the base b is $[\log_b n] + 1$, where $[x]$ represents the greatest integer not exceeding x.*

Proof Note that any integer satisfying $b^{k-1} \leq n < b^k$ has k digits to the base b. Then $\log_b b^{k-1} \leq \log_b n < \log_b b^k$ implies $k - 1 \leq \log_b n < k$ that implies $[\log_b n] = k - 1$, with the result $k = [\log_b n] + 1$.   □

Bit Operations   While analyzing any program or algorithm to be run by a computer, we are interested in calculating the total time taken by the computer to run the algorithm. For example, suppose by using two different logics we write two different programs, say P_1 and P_2, that can check whether a given integer is prime or not. On the same input in a same machine, suppose P_1 takes less time than P_2 to get a correct output. Then surely we will call P_1 a better algorithm than P_2. So, it seems that we can measure the goodness of an algorithm by its run time. But if we consider only the run time, we may lead to a wrong conclusion. Suppose we are running the program P_1 in a very old machine, say 15 years old machine while running the program P_2 in a latest machine having latest hardware specification. It may happen that P_2 may take much less time than P_1 to give correct answer on the same input. So run time of a program should not be a measure of goodness of an algorithm. We need something more basic that should be machine independent. One of the best measures is to calculate the number of bit operations required to complete the algorithm. The amount of time a computer takes to perform a task is proportional to the number of bit operations. So now on, when we speak of estimating the time required by a computer to perform a particular task, we mean finding an estimation of the number of bit operations required by it. Let us explain through the following example what we exactly mean by the bit operations.

Example 10.8.1 (Addition of two k bits long binary strings) Suppose we want to add two binary strings, say a and b. If one of the integers has fewer bits than the other, we fill in zeros to the left of the integer to make both integers of the same length, say k. Let us explain the addition operation through the following example:

	1	1	1	1	1		← carry bits
	1	0	0	1	1	1	← the 6 bit long binary string a
+		1	1	1	0	1	← the 5 bit long binary string b
1	0	0	0	1	0	0	← the output binary string $a + b$

Let us analyze the addition algorithm in detail. We need to repeat the following steps k times starting from right to left:

1. Look at the top and the bottom bit and also at whether there is any carry above the top bit. Note that for the right-most bits, there will be no carry bit. For example, in the above case, the right-most bits of a and b are 1 and 1, respectively, without any carry bit above the right-most bit of a.
2. There will be no carry, if both bits are 0. Put down 0 and move on to the left, if there are more bits left.
3. If either (i) both bits are 0 and there is a carry, or (ii) one of the bits is 0, the other is 1, and there is no carry, then put down 1 and move to the left, if more bits are left.
4. If either (i) one of the bits is 0, the other is 1, and there is a carry, or else (ii) both bits are 1 and there is no carry, then put down 0 along with a carry in the next column, and move on to the left, if more bits are left.
5. If both bits are 1 and there is a carry, then put down 1, put a carry in the next column and move on to the left, if more bits are left.

Doing this procedure once is called a bit operation. So, the addition of two k bit binary strings requires $O(k)$ bit operations.

Example 10.8.2 (Multiplication of a k-bit long binary string with an l-bit long binary string) Suppose, we want to multiply a k-bit integer n by an l-bit integer m. Let us explain the method through an example by taking the binary strings $n = 10110$ and $m = 1011$. So here $k = 5$ and $l = 4$. The multiplication may be done as follows:

				1	0	1	1	0	← binary representation of n
			×	1	0	1	1		← binary representation of m
				1	0	1	1	0	← Row 1:
		1	0	1	1	0	$\underline{0}$		← Row 2:
									← Row 3:
1	0	1	1	0	$\underline{0}$	$\underline{0}$	$\underline{0}$		← Row 4:
1	1	1	1	0	0	1	0		← Row 5: output of the multiplication

As we are multiplying the number n by an l-bit integer m, we obtain at most l rows as shown above, where each row consists of a copy of n, shifted to the left a certain number of times, i.e., with $\underline{0}$'s put on at the end. Note that, corresponding to each occurrence of '0' in m, a row is reduced. For example, in the above example, due to the occurrence of a single '0' in m, the third row in the summation does not appear. Suppose there are l' rows appearing in the summation part (Row 1 to Row 4). Clearly, $l' \le l$. As we are interested in counting the number of bit operations, we cannot simultaneously add all the rows, if $l' > 2$. Instead we shall first add the first two rows and then add the resulting to the third row and so on. The details of the addition process is as follows:

1. At each stage, we count the number of places to the left the number n has been shifted to form the new row. For example, in the above multiplication, the second

row of the summation part is a single shift of n to the left followed by a $\underline{0}$ put on to the end, i.e., the second row becomes $10110\underline{0}$.

2. We copy down that many right-most bits of the partial sum, and then add to n the integer formed from the rest of the partial sum. As explained in Example 10.8.1, it takes $O(k)$ bit operations. For example, in the above sum, while adding Row 1 and Row 2, we first count the number of places, the number n has been shifted to the left to form the Row 2. We see that, in Row 2, the number n has been shifted one position to the left. So, we cut that many bits, here only one bit, from the right of the first row 10110 to form a new binary string 1011 and add this newly formed number with the Row 2 to get $s = 100001$. The partial sum of the Row 1 and Row 2 is obtained by appending to the right of $s = 100001$, the last bit that was cut from the right of the first row, it resulting that the partial sum is $p = 1000010$. Next we are going to add this partial sum with the Row 3. Due to the occurrence of 0 in m, Row 3 is ignored or filled up with zeros. So, the partial sum of Row 1 to Row 3 is same as that of Row 1 and Row 2. So, we look at the Row 4 and count the number of left shifts. We find that the number of left shifts are 3. So, we cut three bits from the right of the partial sum $p = 100010$ and get the binary string 1000. We add this newly formed string with $n = 10110$ to get 11110. Finally we append the three bits 010 that was cut from the right of the partial sum $p = 100010$ to the right of 11110 to get the final answer as 11110010.

3. The example shows that the multiplication task can be broken down into $l' - 1$ additions, each taking k bit operations. As $l' \leq l$, the number of bit operation required is bounded by $O(kl)$.

Remark In the above example, we only count the number of bit operations and neglect the time that a machine takes to shift the bits in n a few places to the left, or the time it takes to copy or to cut suitable number of bits of the partial sum corresponding to the places through which n has been shifted to the left in the new row. For the machines with latest specifications, the shifting, copying, and cutting operations are very fast compared to the large number of bit operations. As a result, we can safely ignore them. So, the time estimation for an arithmetic task may be defined as an upper bound for the number of bit operations, without including any consideration of shift operations, changing registers (copying), memory access etc.

10.8.6 Euclidean and Extended Euclidean Algorithms

Given two integers x and y, not both zero, the greatest common divisor of x and y, denoted by $\gcd(x, y)$ is defined to be the largest integer g dividing both x and y. If by some means, we know the prime factorizations of both x and y, then finding the $\gcd(x, y)$ is very easy. The gcd is simply the product of all primes which occur in both factorizations raised to the minimum of the two exponents. For example, let $x = 868268816108642945649520325611107 = 53^5 \cdot 149^3 \cdot 251^7$ and $y = 219196895380972729174851634933 = 53^4 \cdot 149^2 \cdot 277^7$. Then the $\gcd(x, y) = 53^4 \cdot$

$149^2 = 175176568681$. Now if we deal with large numbers like $x = 114555615673$
$899844817675135734699353962887022183538809518439182723858423697776$
$654969773636025351593611615831027792054723291491195805742582798507756943658064635$ $24514043006302906288887814886595167524820823509595753536314246808029571192806848851$ 426825458311 having 252 digits and $y = 1320$
$736327839163115880849462291291516297119070194585300759588080043821791918735106562803348580741497531553011157325248505169907528308065250382082906933108203122745125366467088710109313205616010071002485676467559886188923580532087187320097935537658916537250176345966729$
71 having 270 digits, then finding $\gcd(x, y)$ using the prime factorization method
is not an effective one as it is likely that the prime factorizations of x and y are not
known. In fact, in the theory of numbers, an important area of research is the search
for quicker methods of factoring large integers. In that respect, we are fortunate
enough to have a quick way to find gcd of two integers x and y, even when we do
not have any idea of the prime factorizations of x and y.

In Chap. 1, we have discussed about the greatest common divisor (gcd) of two
positive integers. In this chapter, we shall discuss how this can be efficiently computed by the Euclidean algorithm. We further analyze the efficiency of the algorithm.
We shall show that this algorithm is very efficient in the sense that the algorithm runs
in polynomial time in the size of inputs. For that let us first describe the algorithm.

Euclidean Algorithm Let $r_0 = a$ and $r_1 = b$ be two integers with $a \geq b > 0$. If the
division algorithm is successively applied to obtain $r_j = r_{j+1}q_{j+1} + r_{j+2}$, where
$0 < r_{j+2} < r_{j+1}$, for $j = 0, 1, \ldots, n - 2$ and $r_{n+1} = 0$, then $\gcd(a, b) = r_n$, the last
non-zero remainder.

Let us try to illustrate the algorithm through the following example.

Example 10.8.3 Let us try to compute $\gcd(55, 34)$ using Euclidean algorithm. The
steps are as follows:

$$55 = 34 \cdot 1 + 21,$$
$$34 = 21 \cdot 1 + 13,$$
$$21 = 13 \cdot 1 + 8,$$
$$13 = 8 \cdot 1 + 5,$$
$$8 = 5 \cdot 1 + 3,$$
$$5 = 3 \cdot 1 + 2,$$
$$3 = 2 \cdot 1 + \mathbf{1} \leftarrow r_n,$$
$$2 = 1 \cdot 2 + \mathbf{0} \leftarrow r_{n+1}.$$

Thus the $\gcd(55, 34) = 1$.

A pseudocode for the Euclidean algorithm is presented in Table 10.1.

Table 10.1 Euclidean
algorithm

Input: Two positive integers a and b.
Method:

1. $r_0 \leftarrow a$
2. $r_1 \leftarrow b$
3. $i \leftarrow 1$
4. **while** $r_i \neq 0$
5. **do** $\begin{cases} q_i \leftarrow \lceil \frac{r_{i-1}}{r_i} \rceil \\ r_{i+1} \leftarrow r_{i-1} - q_i r_i \\ i \leftarrow i+1 \end{cases}$
6. $i \leftarrow i - 1$
7. **return** $\gcd(a, b) = r_i$

Now we shall show that Euclidean algorithm always gives the greatest common divisor of two positive integers in a finite number of steps. For that the following lemma plays an important role.

Lemma 10.8.1 *If a and b are two positive integers and $a = bq + r$, where q and r are integers such that $0 \leq r < b$, then $\gcd(a, b) = \gcd(b, r)$.*

Proof Let d be a common divisor of a and b. As $r = a - bq$, d divides r. Thus if d is a common divisor of b and r, then since $a = bq + r$, d is also a divisor of a. Since the common divisors of a and b are the same as the common divisors of b and r, we get $\gcd(a, b) = \gcd(b, r)$. $\square$

Now we are going to prove the correctness of the Euclidean algorithm.

Proof Given that $r_0 = a$ and $r_1 = b$. Now by successively applying the division algorithm, we have

$$r_0 = r_1 q_1 + r_2, \qquad 0 < r_2 < r_1$$
$$r_1 = r_2 q_2 + r_3, \qquad 0 < r_3 < r_2$$
$$\vdots$$
$$r_{j-2} = r_{j-1} q_{j-1} + r_j, \qquad 0 < r_j < r_{j-1}$$
$$\vdots$$
$$r_{n-3} = r_{n-2} q_{n-2} + r_{n-1}, \quad 0 < r_{n-1} < r_{n-2}$$
$$r_{n-2} = r_{n-1} q_{n-1} + r_n, \qquad 0 < r_n < r_{n-1}$$
$$r_{n-1} = r_n q_n.$$

Since the sequence $a = r_0 \geq r_1 > r_2 > r_3 > \cdots \geq 0$ is a strictly decreasing sequence of non-negative integers, we eventually obtain a remainder zero. Thus by Lemma 10.8.1, we have $\gcd(a, b) = \gcd(r_0, r_1) = \gcd(r_1, r_2) = \gcd(r_2, r_3) = \cdots = \gcd(r_{n-2}, r_{n-1}) = \gcd(r_{n-1}, r_n) = \gcd(r_n, 0) = r_n$, the last non-zero remainder. This completes the proof. $\square$

Remark The Euclidean algorithm may be used to determine whether a positive integer a, less than a given positive integer n, has a multiplicative inverse modulo n. If the $\gcd(a, n) = 1$, we can say that the multiplicative inverse of a modulo n exists. However, the Euclidean algorithm does not directly provide the multiplicative inverse modulo n, if it exists.

Now to analyze the time complexity of the Euclidean algorithm, we need the following lemma.

Lemma 10.8.2   *In the above mentioned Euclidean algorithm, $r_{j+2} < \frac{1}{2}r_j$, for all $j = 0, 1, \ldots, n - 2$.*

Proof If $r_{j+1} \leq \frac{1}{2}r_j$, then we have $r_{j+2} < r_{j+1} \leq \frac{1}{2}r_j$. So, in this case, we are done. Now suppose, $r_{j+1} > \frac{1}{2}r_j$. Note that $\frac{1}{2}r_j < r_{j+1}$ and $r_{j+1} < r_j$, imply $r_{j+1} < r_j < 2r_{j+1}$. Thus in the expression $r_j = r_{j+1}q_{j+1} + r_{j+2}$, we must have $q_{j+1} = 1$, i.e., $r_{j+2} = r_j - r_{j+1} < r_j - \frac{1}{2}r_j = \frac{1}{2}r_j$.   □

Now we are in a position to analyze the Euclidean algorithm through the following theorem.

Theorem 10.8.2   *The time complexity for computing the Euclidean algorithm is $O((\log_2 a)^3)$.*

Proof Lemma 10.8.2 asserts that every two steps in the Euclidean algorithm must result in cutting the size of the remainder at least in half and as the remainder never goes below 0, it follows that there are at most $2\lceil \log_2 a \rceil$ many above steps. Again, as each step involves division and each division involves numbers not larger than a, it takes altogether $O((\log_2 a)^2)$ bit operations. Thus total time required is $O(\log_2 a) \cdot O((\log_2 a)^2) = O((\log_2 a)^3)$.   □

Remark A more careful analysis of the number of bit operations required for the computation of Euclidean algorithm, taking into account of the decreasing size of the numbers in the successive divisions, can improve the time complexity of the Euclidean algorithm to $O((\log_2 a)^2)$ bit operations.

As an extension of the Euclidean algorithm, we state the following algorithm known as Extended Euclidean Algorithm. In Chap. 1, we have already seen that if d is the greatest common divisor of a and b, then there exist integers u and v such that $d = au + bv$. The Extended Euclidean algorithm actually provides a method to find such u and v. To understand the Extended Euclidean algorithm, let us first define two sequences of integers $u_0, u_1, u_2, \ldots, u_n$ and $v_0, v_1, v_2, \ldots, v_n$ according to the following recurrences in which the q_i's are defined as in Table 10.1:

$$u_i = \begin{cases} 0 & \text{if } i = 0 \\ 1 & \text{if } i = 1 \\ u_{i-2} - q_{i-1}u_{i-1} & \text{if } i \geq 2 \end{cases}$$

and

$$v_i = \begin{cases} 1 & \text{if } i = 0 \\ 0 & \text{if } i = 1 \\ v_{i-2} - q_{i-1}v_{i-1} & \text{if } i \geq 2. \end{cases}$$

Now we are going to prove the following theorem which provides a method to find u and v, for given two positive integers $a = r_0$ and $b = r_1$, such that $\gcd(a, b) = au + bv$.

Theorem 10.8.3 *For $0 \leq i \leq n$, let us consider r_i as defined in the Table* 10.1 *and u_i and v_i as defined above. Then for $0 \leq i \leq n$, $r_i = r_0 v_i + r_1 u_i$.*

Proof We shall prove the theorem by using the principle of mathematical induction on i. The result is trivially true for $i = 0$ and $i = 1$. Let us assume that the result is true for all $i < n$, where $n \geq 2$. Now we shall prove the result for $i = n$. By the induction hypothesis, we have

$$r_{n-2} = v_{n-2}r_0 + u_{n-2}r_1$$

and

$$r_{n-1} = v_{n-1}r_0 + u_{n-1}r_1.$$

Now,

$$\begin{aligned} r_n &= r_{n-2} - q_{n-1}r_{n-1} \\ &= v_{n-2}r_0 + u_{n-2}r_1 - q_{n-1}(v_{n-1}r_0 + u_{n-1}r_1) \\ &= (v_{n-2} - q_{n-1}v_{n-1})r_0 + (u_{n-2} - q_{n-1}u_{n-1})r_1 \\ &= r_0 v_n + r_1 u_n. \end{aligned}$$

Hence the result is true for $i = n$. Consequently, by the principle of mathematical induction, the result is true for all $i \geq 0$. This completes the proof of the theorem. $\square$

A pseudocode for the Extended Euclidean algorithm is presented in Table 10.2.

Remark By a similar argument as in the case of the Euclidean algorithm, it can also be shown that the time required to execute the Extended Euclidean Algorithm is $O((\log_2 a)^3)$ bit operations.

Remark The Extended Euclidean algorithm may be used to determine the multiplicative inverse of a given positive integer a modulo n, if it exists. Let $\gcd(a, n) = 1$. Then using the Extended Euclidean algorithm we can find u and v such that $1 = au + nv$. Then $1 \equiv au \pmod{n}$, with the result that (u) is the multiplicative inverse of (a) in $\mathbf{Z}_n^*$.

Table 10.2 Extended Euclidean algorithm: given two positive integers a and b, this algorithm returns r, u, and v such that $r = \gcd(a, b) = au + bv$

Input: Two positive integers a and b.
Method:

1. $a_0 \leftarrow a$
2. $b_0 \leftarrow b$
3. $v_0 \leftarrow 0$
4. $v \leftarrow 1$
5. $u_0 \leftarrow 1$
6. $u \leftarrow 0$
7. $q \leftarrow \lceil \frac{a_0}{b_0} \rceil$
8. $r \leftarrow a_0 - q b_0$
9. **while** $r > 0$
10. **do** $\begin{cases} temp \leftarrow v_0 - qv \\ v_0 \leftarrow v \\ v \leftarrow temp \\ temp \leftarrow v_0 - qv \\ u_0 \leftarrow u \\ u \leftarrow temp \\ a_0 \leftarrow b_0 \\ b_0 \leftarrow r \\ q \leftarrow \lceil \frac{a_0}{b_0} \rceil \\ r \leftarrow a_0 - q b_0 \end{cases}$
11. $r \leftarrow b_0$
12. **return** r, u, and v

10.8.7 Modular Arithmetic

If we can find efficient algorithms for the basic arithmetic operations over the set of integers, then immediately we can find efficient algorithms for the corresponding operations modulo some positive integer n. We note the following:

Suppose n is a k-bit integer and $0 \leq m_1, m_2 \leq n - 1$. Then

- the computation of $(m_1 + m_2) \pmod{n}$ can be done in time $O(k)$;
- the computation of $(m_1 - m_2) \pmod{n}$ can be done in time $O(k)$;
- the computation of $(m_1 \cdot m_2) \pmod{n}$ can be done in time $O(k^2)$;
- the computation of $(m_1^{-1}) \pmod{n}$ can be done in time $O(k^3)$.

10.8.8 Square and Multiply Algorithm

We are now going to compute a function of the form $x^c \pmod{n}$ for given non-negative integer c and positive integer n. Computation of $x^c \pmod{n}$ may be done using $c - 1$ modular multiplications. Note that c might be as big as $\phi(n)$ which is almost as big as n and exponentially large compared to the size of n. So, this

Table 10.3 Square and
multiply algorithm to
compute x^c $(\bmod\, n)$

Input: Three positive integers $x, c,$ and n.
Method:

1. $a \leftarrow 1$
2. $i \leftarrow l - 1$
3. **while** $i \geq 0$
4. **do** $\begin{cases} a \leftarrow a^2\,(\bmod\, n) \\ \textbf{if } c_i = 1 \\ \quad \textbf{then } a \leftarrow (a * x)\,(\bmod\, n) \\ i \leftarrow i - 1 \end{cases}$
5. **return** a

Table 10.4 Steps to calculate
2013^{210} $(\bmod\, 3197)$ using
square and multiply algorithm

i	c_i	a
7	1	$41^2 * 2013\,(\bmod\, 3197) = 2013$
6	1	$2013^2 * 2013\,(\bmod\, 3197) = 1774$
5	0	$1774^2\,(\bmod\, 3197) = 1228$
4	1	$1228^2 * 2013\,(\bmod\, 3197) = 1110$
3	0	$1110^2\,(\bmod\, 3197) = 1255$
2	0	$1255^2\,(\bmod\, 3197) = 2101$
1	1	$2101^2 * 2013\,(\bmod\, 3197) = 55$
0	0	$55^2\,(\bmod\, 3197) = 3025$

method is very inefficient, if c is very large. However, there exists an efficient algorithm, known as square and multiply algorithm, which reduces the number of modular multiplications required to compute x^c $(\bmod\, n)$ to at most $2l$, where l is the number of bits in the binary representation of c. Let the binary representation of c be $(c_{l-1}c_{l-2} \cdots c_2 c_1)_2$, where $c = \sum_{i=0}^{l-1} c_i 2^i$ and $c_i \in \{0, 1\}$. Let us first explain the algorithm to compute x^c $(\bmod\, n)$ in Table 10.3.

Let us now illustrate the square and multiply algorithm through the following example.

Example 10.8.4 Using square and multiply algorithm, let us compute 2013^{210} $(\bmod\, 3197)$. Note that the binary representation of 210 is $(11010010)_2$. Thus here $l = 8$. The steps are explained in Table 10.4. Thus 2013^{210} $(\bmod\, 3197) = 3025$.

Remark Note that in the square and multiply algorithm, the number of squaring performed is l. The number of modular multiplications of type $a \leftarrow a * x$ $(\bmod\, n)$ is equal to the number of 1 present in the binary representation of c. Thus, $O(\log_2 c(\log_2 l)^2)$ bit operations are required to execute square and multiply algorithm to compute x^c $(\bmod\, n)$.

Next we deal with one of the most important problems in number theory, known as "primality testing". Algebraic techniques are very much useful in primality testing. In the next section, we show how algebra can be an important tool in primality testing.

10.9 Primality Testing

Due to the great invention of public key cryptography at the end of the 20th century, the interest in primality testing i.e., checking whether a number is prime or not, has grown rapidly in the past three decades. The security of this type of cryptographic schemes primarily relies on the difficulty involved in factoring the product of very large primes. Integer factorization poses many problems, a key one being the testing of numbers for primality. A reliable and fast test for primality would help us in constructing efficient and secure cryptosystem. Therefore, the mathematics and computer science communities have begun to address the problem of primality testing. A primality test is simply a function that determines if a given integer greater than 1 is prime or composite. The following theorem plays an important role in primality testing.

Theorem 10.9.1 *If n is a positive composite integer, then n has a prime divisor not exceeding $\sqrt{n}$.*

Proof As n is composite, we can write $n = ab$, where $1 < a \le b < n$. We claim that $a \le \sqrt{n}$. If not, then $b \ge a > \sqrt{n}$ which implies $n = ab > \sqrt{n}\sqrt{n} = n$, a contradiction. Hence, $a \le \sqrt{n}$. Again, as $a > 1$, by Theorem 10.2.1, a must have a prime divisor, say p which is clearly less than or equal to $\sqrt{n}$. $\qquad\square$

Though the recent interest in primality testing grew at the end of the 20th century, the quest for discovering a good test is by no means a new one, and very likely one of the oldest issues in mathematics. The ancient priests in Uruk the Sheepfold, circa 2500 B.C., are known to have inscribed long lists of prime numbers. Both the ancient Greeks and the ancient Chinese independently developed primality testing.

10.9.1 Deterministic Primality Testing

One of the simplest and most famous primality test is the Sieve of Eratosthenes who lived in Greece at around circa 200 B.C. His method for determining primality is as follows. Suppose we want to determine if n is prime or not. First, we make a list of all integers $2, 3, \ldots, \lceil\sqrt{n}\rceil$. Next, we circle 2 and cross off from the list all multiples of two. Then we circle 3 and cross off its multiples. We now continue through the list, each time advancing to the least integer that is not crossed off, circling that integer and crossing off all its multiples. We then test to see if any of the circled numbers divide n. If the list of circled numbers is exhausted and no divisor is found, then n is prime. This algorithm is based on the simple observation that, if n is composite, then n has a prime factor less than or equal to $\sqrt{n}$. Though the algorithm itself is fairly straightforward and easy to implement, it is by no means efficient.

Next we are going to prove some of the results that will lead to the notions of deterministic primality testing. Among these results, Fermat's Little Theorem plays an important role. Recall that this theorem states that if p is a prime and a is any integer such that p does not divide a, then $a^{p-1} \equiv 1 \pmod{p}$. However, the converse of this theorem is not true. There exist composite integers N and an integer a such that $a^{N-1} \equiv 1 \pmod{N}$. These numbers play an important role in primality questions. Nevertheless, a partial converse of Fermat's Little Theorem was discovered by Lucas in 1876. The following proposition may be used for primality test.

Proposition 10.9.1 *Let $N > 1$. Assume that there exists an integer $a > 1$ such that*

1. $a^{N-1} \equiv 1 \pmod{N}$;
2. $a^m \not\equiv 1 \pmod{N}$, *for $m = 1, 2, \ldots, N - 2$.*

Then N is a prime integer.

Proof Let us consider the group of units $(\mathbf{Z}_N^*, \cdot)$. Then the order of $\mathbf{Z}_N^*$ is $\phi(N)$. As $a^{N-1} \equiv a^{N-2} \cdot a \equiv 1 \pmod{N}$, a is a unit in the ring $(\mathbf{Z}_N, +, \cdot)$. Thus $(a) \in \mathbf{Z}_N^*$. To show that N is prime, it is sufficient to show that $\phi(N) = N - 1$, where $\phi(n)$ is the Euler ϕ function. By hypothesis, order of (a) in the group $\mathbf{Z}_N^*$ is $N - 1$. As the order of an element in a finite group divides the order of the group, $N - 1$ divides $\phi(N)$. As $N - 1 \leq \phi(N)$, we have $\phi(N) = N - 1$. Hence N is prime. $\square$

Remark Though it might seem perfect, it requires $N - 2$ successive multiplications by a, and finding residues modulo N. So, for large N, the problem remains.

In 1891, Lucas gave the following test for primality:

Proposition 10.9.2 *Let $N > 1$. Assume that there exists an integer $a > 1$, relatively prime to N such that*

1. $a^{N-1} \equiv 1 \pmod{N}$;
2. $a^m \not\equiv 1 \pmod{N}$, *for every $m < N - 1$ such that m divides $N - 1$.*

Then N is a prime integer.

Proof The proof follows directly from Proposition 10.9.1. $\square$

Remark To apply the proposition, it requires to know all factors of $N - 1$. So, this test may only be easily applicable when $N - 1$ can be factored. For example, $N = 2^n + 1$ or $N = 3 \cdot 2^n + 1$ etc.

Brillhart, Lehmer, and Selfridge came up with the following test in 1975:

Proposition 10.9.3 *Let $N > 1$. Assume that for every prime divisor q of $N - 1$, there exists an integer $a > 1$ such that*

1. $a^{N-1} \equiv 1 \pmod{N}$;
2. $a^{(N-1)/q} \not\equiv 1 \pmod{N}$.

Then N is prime.

Proof As before $(a) \in \mathbf{Z}_N^*$. To show that N is prime, it is sufficient to show that $N - 1$ divides $\phi(N)$. If possible, suppose $N - 1$ does not divide $\phi(N)$. Then there exist a prime factor p of $N - 1$ and a positive integer n such that p^n divides $N - 1$ but does not divide $\phi(N)$. Let $N - 1 = X_1 p^n$, for some integer X_1. Further let the order of (a) in $\mathbf{Z}_N^*$ be e. As by hypothesis, $(a)^{N-1} = (1)$, e divides $N - 1 = X_1 p^n$ and e does not divide $(N - 1)/p = X_1 p^{n-1}$. Let $N - 1 = X_1 p^n = e X_2$, for some integer X_2. Thus $X_1 p^n = X_1 p^{n-1} p = e X_2$ implies p divides $e X_2$. As p is prime, either p divides X_2 or p divides e. Let p divide X_2. Then $X_2 = p X_3$ for some integer X_3. Thus $X_1 p^n = X_1 p^{n-1} p = X_2 e$ implies $X_1 p^{n-1} = e X_3$ which implies e divides $X_1 p^{n-1} = (N - 1)/p$, a contradiction. Thus p does not divide X_2. As p^n divides $e X_2$, and p does not divide X_2, p^n must divide e. Also as $(a)^{\phi(N)} = (1)$ and order of (a) in $\mathbf{Z}_N^*$ is e, e divides $\phi(N)$, with the result that p^n divides $\phi(N)$, a contradiction. Hence the proposition follows. □

Remark To apply the proposition, once again, it requires the knowledge of all factors of $N - 1$. But here fewer congruences have to be satisfied than Proposition 10.9.2.

10.9.2 AKS Algorithm

The AKS Algorithm is the first deterministic polynomial-time primality test named after its authors M. Agrawal, N. Kayal, and N. Saxena. In August 2002, this algorithm was presented in the paper "PRIMES is in P" Agrawal et al. (2002). This solved the longstanding problem of solving the primality testing deterministically of an integer N in polynomial time. The running time (with fast multiplication), was originally evaluated as essentially $O((\log N)^{12})$ and lately lowered to $O((\log N)^{7.5})$.

The main idea in the new primality testing algorithm is the following identity characterizing primes:

$$N \text{ is prime} \quad \text{if and only if} \quad (1 - X)^N \equiv 1 - X^N \pmod{N}.$$

For more detailed result, the readers may refer the book Dietzfelbinger (2004).

10.10 Probabilistic or Randomized Primality Testing

The main aim of this section is to bring out some of the efficient but simple probabilistic tests for primality by using group theoretic results. The first question that comes to our mind is: *"What do we mean by a probabilistic test?"* The next immediate question may be: *"Is there any advantage of using probabilistic or randomized algorithm over deterministic algorithm?"*

First let us try to explore some of the advantages of the use of probabilistic or randomized algorithms over the deterministic algorithms in the context of primality testing. Suppose we are given an integer $n > 1$ and we want to determine whether the given integer n is prime or composite. If n is composite, Theorem 10.9.1 ensures the existence of a prime factor of n within the range from 2 to $\lfloor \sqrt{n} \rfloor$, the greatest integer not exceeding $\sqrt{n}$. So, we simply divide n by 2, 3, up to $\lfloor \sqrt{n} \rfloor$, testing if any of these numbers divide n. Note that this deterministic algorithm does, not only the primality testing for n, but also it produces a non-trivial factor of n, whenever n is composite. Of course, the major drawback of this deterministic algorithm is that it is terribly inefficient as it requires $O(\sqrt{n})$ arithmetic operations, which is exponential in the binary length of n. Thus for practical purposes, this algorithm is limited only to small values of n. Let us try to see what happens when this algorithm is used to test the primality of an integer n having 100 decimal digits in a computer that can perform 1 billion divisions per second. Then to perform $\sqrt{n}$ divisions, it would take on the order of 10^{33} years which is quite impractical. So the question that comes to our mind is "does there exist a deterministic primality testing algorithm which is efficient?" In 2002, Agrawal et al. (2002) first came up with a path breaking work which provides a deterministic algorithm to check whether a given number is prime or composite in polynomial time. However, we must admit that till date no deterministic primality testing algorithm is known which is as efficient as the probabilistic primality testing algorithms, such as Solovay–Strassen primality testing or Miller–Rabin primality testing. In this section, we shall develop those probabilistic primality tests that allow 100 decimal digit numbers to be tested for primality in less than a second. One important thing we should note that these algorithms are probabilistic, and may make mistakes. However, the probability that they commit a mistake can be made so small as to be irrelevant for all practical purposes. For example, we can easily make the probability of error as small as 2^{-100}: should one really care about an event that happens with such a negligible probability? Let us now try to answer the question *"what do we mean by a probabilistic or randomized test?"* For better understanding we need to know two notions, namely, decision problem and randomized algorithm. A decision problem is a problem in which a question is to be answered "yes" or "no" while a randomized algorithm is any algorithm that uses random numbers, in contrast, an algorithm that does not use random numbers is called a deterministic algorithm. For the randomized algorithms, the following definition plays an important role.

Definition 10.10.1 For a decision problem, a yes-biased Monte Carlo algorithm is a randomized algorithm in which a "yes" answer is always correct but a "no"

Table 10.5 The decisional
problem: Is-Composite

Problem: Is-Composite.

Instance: A positive integer $n \geq 2$.

Question: Is n composite?

Table 10.6
Solovay–Strassen primality
test

Input: Odd integer $n \geq 3$.
Method:

1. Choose an integer a randomly from $\{1, 2, 3, \ldots, n - 1\}$.
2. if $\gcd(a, n) \neq 1$
3.   return ("Yes: The number is a composite integer").
4. $x = \left(\frac{a}{n}\right)$, the Jacobi symbol of a modulo n
5. $y \equiv a^{(n-1)/2} \pmod{n}$
6. if $x \equiv y \pmod{n}$
7.   return ("No: The number is a prime integer").
8. else
9.   return ("Yes: The number is a composite integer").

answer may not be correct. Similarly, for a decision problem, a no-biased Monte Carlo algorithm is a randomized algorithm in which a "no" answer is always correct but a "yes" answer may not be correct.

As we see that for a yes-biased or no-biased Monte Carlo algorithm, there is a chance of getting error in the output, the following definition is very important for any Monte Carlo algorithm.

Definition 10.10.2 For a decision problem, a yes-biased Monte Carlo algorithm has an error probability ϵ if for any instance the answer is "yes", but the algorithm will give an incorrect answer "no" with probability almost ϵ, where the probability is taken over all random choices made by the algorithm when it is run with a given input.

One of the very important decision problems, known as "Is-Composite" is described in Table 10.5.

As the answer to the problem "Is-Composite" is either "yes" or "no", the problem is a decisional problem. In the next section, we are going to provide some efficient yes-biased Monte Carlo algorithms that will provide the answer in an efficient way.

10.10.1 Solovay–Strassen Primality Testing

The primality test of Solovay and Strassen is a randomized algorithm based on the theory of quadratic reciprocity. This test is capable of recognizing composite numbers with a probability of at least $\frac{1}{2}$. Let us first explain the algorithm in Table 10.6.

First note that as the gcd algorithm, the square and multiply algorithm and the algorithm to compute the Jacobi symbol can be done in time $O((\log n)^3)$, the Solovay–Strassen Primality testing algorithm can run in time $O((\log n)^3)$. Now, we shall show that Solovay–Strassen Primality testing algorithm is a yes-biased Monte Carlo Algorithm with error probability at most $\frac{1}{2}$. To prove this we shall use the following proposition and lemma.

Proposition 10.10.1 *Let n be an odd integer greater than 1. Let $SS(n) = \{a \in \mathbf{Z}_n^* | \binom{a}{n} \equiv a^{(n-1)/2} \pmod{n}\}$. Then $SS(n)$ is a subgroup of $\mathbf{Z}_n^*$.*

Proof As $(1) \in SS(n)$, $SS(n) \neq \emptyset$. Let $a, b \in SS(n)$. Then $\binom{a}{n} \equiv a^{(n-1)/2} \pmod{n}$ and $\binom{b}{n} \equiv b^{(n-1)/2} \pmod{n}$. Now, $\binom{ab}{n} = \binom{a}{n}\binom{b}{n} \equiv a^{(n-1)/2} \pmod{n} b^{(n-1)/2}$ $\pmod{n} \equiv (ab)^{(n-1)/2} \pmod{n}$. Thus $ab \in SS(n)$. So, $SS(n)$ is closed under composition modulo n. As $SS(n)$ is a finite subset of $\mathbf{Z}_n^*$, $SS(n)$ is a subgroup of $\mathbf{Z}_n^*$. $\qquad\square$

Our next aim is to show that $SS(n)$ is a proper subgroup of $\mathbf{Z}_n^*$. To prove that we shall take the help of the following lemmas.

Lemma 10.10.1 *Let $n = p^k m$ be an odd integer greater than 1 with p an odd prime, $\gcd(p, m) = 1$ and $k \geq 2$. Then there exists an element $(g) \in \mathbf{Z}_n^*$ such that $\binom{g}{n} \not\equiv g^{(n-1)/2} \pmod{n}$.*

Proof As $\phi(n) = p^{k-1}(p-1)\phi(m)$, p divides $\phi(n)$ which is the order of the group $\mathbf{Z}_n^*$. Thus by Cauchy's Theorem, there exists an element, say $(g) \in \mathbf{Z}_n^*$ such that order of (g) is p. We claim that $\binom{g}{n} \not\equiv g^{(n-1)/2} \pmod{n}$. If possible, let $\binom{g}{n} \equiv g^{(n-1)/2} \pmod{n}$. As $(g) \in \mathbf{Z}_n^*$, $\gcd(g, n) = 1$. Now the value of $\binom{g}{n}$ is either $+1$ or -1. Thus $g^{(n-1)/2} \equiv \pm 1 \pmod{n}$ implies $g^{n-1} \equiv 1 \pmod{n}$. Hence order of (g) divides $n - 1$, i.e., p divides $n - 1$. Again, as $n = p^k m$, p divides n, a contradiction. Thus $\binom{g}{n} \not\equiv g^{(n-1)/2} \pmod{n}$. This completes the proof. $\qquad\square$

Lemma 10.10.2 *Let $n = p_1 p_2 \ldots p_k$, where p_i's are distinct odd primes. Suppose $a \equiv u \pmod{p_1}$ and $a \equiv 1 \pmod{p_2 p_3 \ldots p_k}$, where u is a quadratic non-residue modulo p_1. Then $\binom{a}{n} \not\equiv a^{(n-1)/2} \pmod{n}$.*

Proof Note that $\binom{a}{n} \equiv \binom{a}{p_1}\binom{a}{p_2 \cdots p_k} \pmod{n} \equiv \binom{u}{p_1}\binom{1}{p_2 \cdots p_k} \pmod{n} \equiv (-1) \pmod{n}$. We claim that $a^{\frac{n-1}{2}} \not\equiv \binom{a}{n} \pmod{n}$. If possible let $a^{\frac{n-1}{2}} \equiv \binom{a}{n} \pmod{n}$. Then $a^{\frac{n-1}{2}} \equiv (-1) \pmod{n} \equiv (-1) \pmod{p_1 p_2 \cdots p_k}$, contradicting the fact that $a^{\frac{n-1}{2}} \equiv 1 \pmod{p_2 \cdots p_k}$. Thus $a^{\frac{n-1}{2}} \not\equiv \binom{a}{n} \pmod{n}$. This completes the proof. $\square$

We are now in a position to prove the following theorem.

Theorem 10.10.1 *The Solovay–Strassen Primality testing algorithm is a yes-biased Monte Carlo Algorithm with error probability at most $\frac{1}{2}$.*

Proof Clearly, the Solovay–Strassen primality testing algorithm is a yes-biased Monte-Carlo algorithm. Note that the error in the algorithm may occur when n is a composite integer but the algorithm returns "no", i.e., when $a \in SS(n)$. So we need to find a bound on the size of $SS(n)$. To get that we first show that $SS(n)$ is a proper subgroup of $\mathbf{Z}_n^*$. As n is an odd positive integer, the form of n is either $n = p^k m$ with p an odd prime, $\gcd(p, m) = 1$ and $k \geq 2$ or $n = p_1 p_2 \ldots p_k$, where p_i's are distinct odd primes. In both the cases, from Lemmas 10.10.1 and 10.10.2, we see that there exists an element, say (g), such that $(g) \in \mathbf{Z}_n^*$ but $(g) \notin SS(n)$, it resulting that $SS(n)$ is a proper subgroup of $\mathbf{Z}_n^*$. As order of $SS(n)$ divides the order of $\mathbf{Z}_n^*$, let $\frac{|\mathbf{Z}_n^*|}{|SS(n)|} = t$. Then $t \geq 2$. Thus $|SS(n)| = \frac{|\mathbf{Z}_n^*|}{t} \leq \frac{|\mathbf{Z}_n^*|}{2} \leq \frac{n-1}{2}$. Consequently, the error probability of the algorithm is at most $\frac{n-1}{2(n-1)} = \frac{1}{2}$. $\qquad\square$

10.10.2 Pseudo-primes and Primality Testing Based on Fermat's Little Theorem

Fermat's Little Theorem 10.4.2 ensures us that if p is a prime and a is any integer such that $1 \leq a < n$, then $a^{p-1} \equiv 1 \pmod{p}$. Consequently, for a given n, if we can find an integer b such that $1 \leq b < n$ and $b^{n-1} \not\equiv 1 \pmod{n}$ then we know that n must be composite. We call this number b, $1 \leq b < n$, as an F-witness for n. So, if n has an F-witness, n must be a composite number, i.e., the F-witness b for n provides a certificate for the compositeness of n. It can be shown that 2 is an F-witness for all composite numbers not exceeding 340. However, $2^{340} \equiv 1 \pmod{341}$, even though $341 = 11 \cdot 31$ is a composite number. So 2 is not an F-witness for 341. We call 2 as F-liar for 341. So in general, for an odd composite number n, an integer a, $1 \leq a < n$, is called an F-liar if $a^{n-1} \equiv 1 \pmod{n}$. Thus 2 is a F-liar for 341 while 3 is an F-witness for 341. So we see that there exist some composite integers which pass Fermat's test for some base. This observation leads to the definition of a special type of numbers, known as pseudo-primes. Let us first define what is a pseudo-prime to some base.

Definition 10.10.3 Let n be a positive composite integer and a be a positive integer. Then the integer n is said to be a pseudo-prime to the base a if $a^n \equiv a \pmod{n}$.

Remark 1 Thus a positive odd composite integer n is a pseudo-prime to some base a, if a is an F-liar for n.

Remark 2 If $\gcd(a, n) = 1$, then the congruence $a^n \equiv a \pmod{n}$ is equivalent to the congruence $a^{n-1} \equiv 1 \pmod{n}$.

Example 10.10.1 The integers $341 = 11 \times 31$, $561 = 3 \times 11 \times 17$, $645 = 3 \times 5 \times 43$ are examples of pseudo-primes to the base 2.

It has been found that the pseudo-primes to some fixed base a is much rarer than prime numbers. In particular, in the interval $(2, 10^{10})$, there are 455052512 primes while there are only 14884 pseudo-primes to the base 2. Although pseudo-primes to any given base are rare, there are, nevertheless, infinitely many pseudo-primes to any given base. Here we shall prove this for the base 2 only. The following lemmas are useful to prove the result.

Lemma 10.10.3   *If k and n are positive integers such that k divides n, then $2^k - 1$ divides $2^n - 1$.*

Proof Let $n = kq$ for some positive integer q. Then $2^n - 1 = 2^{kq} - 1 = (2^k - 1)((2^k)^{q-1} + (2^k)^{q-2} + \cdots + 1)$ which implies that $2^k - 1$ divides $2^n - 1$. $\square$

Lemma 10.10.4   *If an odd positive integer n is a pseudo-prime to the base 2, then $2^n - 1$ is also a pseudo-prime to the base 2.*

Proof Lemma 10.10.3 ensures that $N = 2^n - 1$ is a composite integer. As $\gcd(2, N) = 1$, to show that N is a pseudo-prime, we only need to show that $2^{N-1} \equiv 1 \pmod{N}$. Since by hypothesis, n is a pseudo-prime to the base 2, $2^{n-1} \equiv 1 \pmod{n}$. Thus there exists some integer k such that $2^{n-1} - 1 = kn$. Now $2^{N-1} = 2^{2^n - 2} = 2^{2kn}$. As n divides $2kn$, by Lemma 10.10.3, $N = 2^n - 1$ divides $2^{2kn} - 1 = 2^{N-1} - 1$, with the result $2^{N-1} \equiv 1 \pmod{N}$. Hence $N = 2^n - 1$ is a pseudo-prime to the base 2. $\square$

Now we are in a position to prove the existence of infinitely many pseudo-primes to the base 2.

Theorem 10.10.2   *There are infinitely many pseudo-primes to the base 2.*

Proof Note that 341 is an odd pseudo-prime to the base 2. Now by Lemma 10.10.4, we will be able to construct infinitely many odd pseudo-primes to the base 2 by taking $ps_1 = 341$, $ps_2 = 2^{ps_1} - 1$, $ps_3 = 2^{ps_2} - 1, \ldots$. Note that these odd integers are all distinct since $ps_1 < ps_2 < ps_3 < \cdots$. Therefore, the infinite sequence $ps_1, ps_2, \ldots$ shows the existence of infinitely many pseudo-primes to the base 2. $\square$

Thus we see that while testing the primality of an integer n, if we find that $2^{n-1} \equiv 1 \pmod{n}$, then we know that n is either prime or a pseudo-prime to the base 2. One follow-up approach may be taken by testing whether $a^{n-1} \equiv 1 \pmod{n}$ for various positive integers a. If we find any value b, $1 \le b < n$ with $\gcd(b, n) = 1$ and $b^{n-1} \not\equiv 1 \pmod{n}$, then we can readily say that n is composite. For example, we have seen that 341 is a pseudo-prime to the base 2. But, $3^{340} \equiv 56 \not\equiv 1 \pmod{341}$, with the result that 341 is a composite integer.

Now the question that comes to our mind is: "exploiting the above fact, can we develop a yes-biased Monte Carlo algorithm for the problem Is-Composite?"

For that let us first prove the following useful result which is some kind of inverse of Fermat's Little Theorem.

Table 10.7 Fermat's test

> **Input:** Odd integer $n \geq 3$.
> **Method:**
>
> 1. Choose an integer a randomly from $\{2, 3, \ldots, n-2\}$.
> 2. if $(a^{n-1} \not\equiv 1 \pmod{n})$
> 3.   return ("Yes: The number is a composite integer").
> 4. else
> 5.   return ("No: The number is a prime integer").

Theorem 10.10.3 *For an integer $n \geq 2$, if $a^{n-1} \equiv 1 \pmod{n}$ for all a, $1 \leq a < n$, then n must be a prime number.*

Proof As $n \geq 2$, for all a, $1 \leq a < n$, $a \cdot a^{n-2} = a^{n-1} \equiv 1 \pmod{n}$ implies that (a) has an inverse in the ring $\mathbf{Z}_n$, i.e., $(a) \in \mathbf{Z}_n^*$. Thus a and n are relatively prime for all a, $1 \leq a < n$. Hence n must be a prime integer. $\square$

Theorem 10.10.3 ensures that if n is a composite integer, then there must exist at least one F-witness for n. Further note that as $(1)^{n-1} \equiv 1 \pmod{n}$ and $(n-1)^{n-1} \equiv (-1)^{n-1} \pmod{n} \equiv 1 \pmod{n}$ for all odd integer $n \geq 3$, 1 and $n - 1$ are always F-liars for n, called the trivial F-liars for n. The above observation leads to the randomized primality testing algorithm known as Fermat's Test as described in Table 10.7.

Now we shall show that the algorithm is a yes-biased Monte Carlo algorithm for n, provided there is at least one F-witness a for n such that $\gcd(a, n) = 1$. To prove that let us first prove the following lemma.

Lemma 10.10.5 *For an odd composite integer $n \geq 3$, if there exists at least one F-witness c for n such that $\gcd(c, n) = 1$, then the set of all F-liars for n, i.e., $\mathcal{L}_n^F = \{(a) \in \mathbf{Z}_n : 1 \leq a < n \text{ and } a^{n-1} \equiv 1 \pmod{n}\}$ is a proper subgroup of $\mathbf{Z}_n^*$.*

Proof Clearly, $(1) \in \mathcal{L}_n^F$ and hence $\mathcal{L}_n^F$ is a non-empty subset of $\mathbf{Z}_n^*$. Now let $(a), (b) \in \mathcal{L}_n^F$. Then $a^{n-1} \equiv 1 \pmod{n}$ and $b^{n-1} \equiv 1 \pmod{n}$. Thus $(ab)^{n-1} \equiv a^{n-1}b^{n-1} \equiv 1 \pmod{n}$, resulting in $(ab) \in \mathcal{L}_n^F$. Hence from the property of finite group, $\mathcal{L}_n^F$ is a subgroup of $\mathbf{Z}_n^*$. Further, as there exists at least one F-witness, say c for n such that $(c) \in \mathbf{Z}_n^*$ and $c^{n-1} \not\equiv 1 \pmod{n}$, it follows that $(c) \notin \mathcal{L}_n^F$. Hence $\mathcal{L}_n^F \subsetneq \mathbf{Z}_n^*$. $\square$

Now we shall prove the following theorem.

Theorem 10.10.4 *If $n \geq 3$ is an odd composite integer such that there exists at least one F-witness a for n such that $(a) \in \mathbf{Z}_n^*$, then the algorithm as described in Table 10.7 is a yes-biased Monte Carlo algorithm with error probability at most $\frac{1}{2}$.*

Proof Clearly the algorithm as described in Table 10.7 is a yes-biased Monte Carlo Algorithm. Now the error occurs when $(a) \in \mathcal{L}_n^F \setminus \{(1), (n-1)\}$. As $\mathcal{L}_n^F \subsetneq \mathbf{Z}_n^*$,

Table 10.8 Repeated
Fermat's test

Input: Odd integer $n \geq 3$ and an integer $k \geq 1$.
Method:
1. repeat k times
2.   Choose an integer a randomly from $\{2, 3, \ldots, n-2\}$
3.   if $(a^{n-1} \not\equiv 1 \pmod{n})$
4.   return ("Yes: The number is a composite integer").
5. return ("No: The number is a prime integer").

$|\mathcal{L}_n^F|$ is a proper divisor of $|\mathbf{Z}_n^*| = \phi(n) < n - 1$. Thus $|\mathcal{L}_n^F| \leq \frac{n-1}{2}$. As a result, the probability that an a randomly chosen from $\{2, 3, \ldots, n-2\}$ is in $\mathcal{L}_n^F \setminus \{(1), (n-1)\}$ is at most $\frac{\frac{n-1}{2} - 2}{n-3} = \frac{n-5}{2(n-3)} < \frac{1}{2}$. $\square$

Remark To increase the confidence level of the primality testing as described in Table 10.7, we may repeat the process as described in Table 10.8.

Remark Note that if the algorithm as described in Table 10.8 outputs "Yes: The number is a composite integer", then the algorithm finds an F-witness for n and hence n must be a composite integer. On the other hand, if n is composite and there exists at least one F-witness for n in $\mathbf{Z}_n^*$, then the error probability in all k attempts is at most $(\frac{1}{2})^k$. Thus by choosing k large enough, the error probability can be made as small as desired.

Unfortunately, there are composite integers that cannot be shown to be composite using the above approach, because there are integers which are pseudo-primes to every relatively prime base, that is, there are composite integers n such that $b^{n-1} \equiv 1 \pmod{n}$, for all b relatively prime with n. This leads to the following important definition.

Definition 10.10.4 A composite integer n which satisfies $a^{n-1} \equiv 1 \pmod{n}$ for all positive integer a with $\gcd(a, n) = 1$ is called a Carmichael number named after R Carmichael who studied them in the early part of the 20th century.

Example 10.10.2 The integer $n = 561$ is the smallest Carmichael number.

In 1912, Carmichael conjectured that there exist infinitely many Carmichael numbers. After 8 years, Alford, Granville and Pomerance showed the correctness of the conjecture. As the proof of the conjecture is out of the scope of this book, we are not discussing that here. However, the following theorems provide some useful properties of Carmichael number.

Theorem 10.10.5 *Let* $n = p_1 p_2 \cdots p_k$, *where* p_i's *are distinct primes such that* $p_i - 1$ *divides* $n - 1$ *for all* $i \in \{1, 2, \ldots, k\}$. *Then* n *is a Carmichael number.*

Proof Let a be a positive integer such that $\gcd(a, n) = 1$. Then $\gcd(a, p_i) = 1$ for all $i \in \{1, 2, \ldots, k\}$. Hence by Fermat's Little Theorem 10.4.2, $a^{p_i-1} \equiv 1 \pmod{p_i}$ for all i which implies $a^{n-1} \equiv 1 \pmod{p_i}$ for all i as $p_i - 1$ divides $n - 1$, by hypothesis. The rest of the theorem follows from Chinese Remainder Theorem. $\Box$

Example 10.10.3 Using Theorem 10.10.5, we can conclude that $n = 6601 = 7 \cdot 23 \cdot 41$ is a Carmichael number.

Now we shall show that the converse of the Theorem 10.10.5 is also true. To prove that we need some concepts from group theory. For an abelian group $(G, \cdot)$, we say that an integer k kills the group G iff $G^k = \{e\}$, where e is the identity element of G. Let $\mathcal{K}_G = \{k \in \mathbf{Z} : k \text{ kills } G\}$. Then clearly $\mathcal{K}_G$ is a subgroup of the group $(\mathbf{Z}, +)$ and hence of the form $n\mathbf{Z}$ for a uniquely determined non-negative integer n. This integer n is called exponent of G. Note that if $n \neq 0$, then this n is the least positive integer that kills G. Further note that for any integer k such that $G^k = \{e\}$, the exponent n of G divides k. Also, if G is a finite cyclic group, then the exponent of G is same as the order of the group G. Using the concept of exponent of the group, we prove the following theorem.

Theorem 10.10.6 *The converse of the Theorem 10.10.5 is also true.*

Proof Let n be a Carmichael number. Then n is a composite integer and $a^{n-1} \equiv 1 \pmod{n}$, for all $(a) \in \mathbf{Z}_n^*$. Hence $n - 1$ kills the group $\mathbf{Z}_n^*$. Let $n = p_1^{\alpha_1} p_2^{\alpha_2} \cdots p_k^{\alpha_k}$ be the prime factorization of n, where p_i's are distinct primes and $\alpha_i \geq 1$, for all $i = 1, 2, \ldots, k$. Now by Chinese Remainder Theorem, we have $\mathbf{Z}_n^*$ is isomorphic to the group $\mathbf{Z}^*_{p_1^{\alpha_1}} \times \mathbf{Z}^*_{p_2^{\alpha_2}} \times \cdots \times \mathbf{Z}^*_{p_k^{\alpha_k}}$, where each of the $\mathbf{Z}^*_{p_i^{\alpha_i}}$ is a cyclic group of order $p_i^{\alpha_i-1}(p_i - 1)$. Thus $n - 1$ kills the group $\mathbf{Z}_n^*$ iff $n - 1$ kills each the groups $\mathbf{Z}^*_{p_i^{\alpha_i}}$, i.e., iff $p_i^{\alpha_i-1}(p_i - 1)$ divides $n - 1$, for all $i = 1, 2, \ldots, k$. Now we claim that $\alpha_i = 1$, for all $i = 1, 2, \ldots, k$. If not, then there exists p_i, for some i such that p_i will divide both $n - 1$ and n, a contradiction. Thus $\alpha_i = 1$ for all $i = 1, 2, \ldots, k$ and hence $n = p_1 p_2 \cdots p_k$ and $p_i - 1$ divides $n - 1$, for all $i = 1, 2, \ldots, k$. $\Box$

The following theorem also provides useful information about Carmichael numbers.

Theorem 10.10.7 *A Carmichael number must have at least three distinct odd prime factors.*

Proof From Theorem 10.10.6, it follows that if an integer $n > 2$ is a Carmichael number, then $n = p_1 p_2 \cdots p_k$, where p_i's are distinct odd primes such that $p_i - 1$ divides $n - 1$ for all $i \in \{1, 2, \ldots, k\}$. We need to show that $k \geq 3$. As n is a positive composite integer, $k > 1$. If possible, let $k = 2$, i.e., let $n = pq$, where p and q are distinct primes. Without loss of generality, let us assume that $p > q$. Now $p - 1 | n -$

$1 = pq - 1 = q(p-1) + (q-1)$. Thus $p - 1 | q - 1$, a contradiction. So, n must be a product of at least three distinct odd primes. $\square$

In the next section we are going to introduce another probabilistic primality testing based on finding non-trivial square root of 1 modulo n.

10.10.3 Miller–Rabin Primality Testing

If n is a Carmichael number, then the probability that Fermat's test for "Is-Composite" returns a wrong answer is $\frac{\phi(n)-2}{n-3} > \frac{\phi(n)}{n} = \prod_{p|n}(1 - \frac{1}{p})$ which is very close to 1, if n has only few and large prime factors. As a result, the repetition trick for reducing the error probability does not hold. So we need to search for other techniques.

The next primality testing algorithm is based on finding non-trivial square root of 1 modulo n. First let us try to explain what do we mean by a square root of 1 modulo n. An integer a, $1 \leq a < n$ is called a square root of 1 modulo n if $a^2 \equiv 1 \pmod{n}$. Note that, 1 and $n - 1$ are always square roots of 1 modulo n. Thus these numbers are called trivial square roots of 1 modulo n. Recall that the Corollary of Theorem 10.5.1 ensures that if p is an odd prime number then 1 has no non-trivial square root modulo p. Thus to develop a primality testing algorithm, we may use the fact "If there exists a non-trivial square root of 1 modulo n, then n must be a composite number." Before developing a "yes-biased" Monte Carlo algorithm for the problem "Is-Composite", let us first look at the Fermat's primality testing little closely. As we are interested in odd n, let us assume that $n = 2^h \cdot t + 1$, where $\gcd(2, t) = 1$ and $h \geq 1$. Thus $a^{n-1} \equiv ((a^t \pmod{n})^{2^h}) \pmod{n}$ implies that $a^{n-1} \pmod{n}$ may be calculated in $h + 1$ intermediate steps if we let $b_0 \equiv a^t \pmod{n}$; $b_i \equiv b_{i-1}^2 \pmod{n}$, for $i = 1, 2, \ldots, h$. Thus $b_h \equiv a^{n-1} \pmod{n}$. Now let us try to explain through an example how this observation may be helpful in developing a probabilistic primality testing.

Let $n = 325$. Then $n - 1 = 81 \cdot 2^2$. In the Table 10.9, we calculate the powers $b_0 = a^{81} \pmod{n}$, $b_1 = (a^{81})^2 \equiv a^{162} \pmod{n}$, $b_2 \equiv a^{324} \pmod{n}$ for a list of different values of a, $2 \leq a < n - 1$ and show that how this table plays an important role for primality testing.

Note that 2 is an F-witness for 325 having $\gcd(2, 325) = 1$ while 130 is also an F-witness with $\gcd(130, 325) = 65 > 1$. Note that 7, 18, 32, 118, 126, 199, 224 and 251 are all F-liars for 325. So, these numbers will not help us in the primality testing using Fermat's test. But, if we start calculation with 118, 224 and 251, we see that 274, 274 and 51 are non-trivial square roots of 1 modulo 325. This observation directly implies that 325 is a composite number. On the other hand, the calculation with 7, 18, 32 and 199 provide no information regarding primality testing based on non-trivial square root modulo 1 as $-1 \equiv 324 \pmod{325}$ is a trivial square root modulo 325 and $7^{162} \equiv -1 \pmod{325}$, $18^{162} \equiv -1 \pmod{325}$, $32^{162} \equiv -1 \pmod{325}$ and $199^{81} \equiv -1 \pmod{325}$. Likewise, calculation with 126 does not provide any

Table 10.9 Calculation of a^{324} (mod 325) for different base values of a

a	$b_0 = a^{81}$ (mod 325)	$b_1 = (a^{162})$ (mod 325)	$b_2 = a^{324}$ (mod 325)
2	252	129	66
7	307	324	1
18	18	324	1
32	57	324	1
118	118	274	1
126	1	1	1
130	0	0	0
199	324	1	1
224	274	1	1
251	51	1	1

Table 10.10 General form of the sequence of b_i's, where "$*$" denotes an arbitrary element from $\{2, 3, \ldots, n-3\}$

SN.	b_0	b_1	...				...	b_{h-1}	b_h	Information
1	1	1	...	1	1	1	...	1	1	No information
2	-1	1	...	1	1	1	...	1	1	No information
3	$*$	1	...	1	1	1	...	1	1	n is composite
4	$*$	-1	...	1	1	1	...	1	1	No information
5	$*$	$*$	...	1	1	1	...	1	1	n is composite
6	$*$	$*$	...	-1	1	1	...	1	1	No information
7	$*$	$*$	...	$*$	1	1	...	1	1	n is composite
8	$*$	$*$	...	$*$	-1	1	...	1	1	No information
9	$*$	$*$	...	$*$	$*$	1	...	1	1	n is composite
10	$*$	$*$	...	$*$	$*$	-1	...	1	1	No information
11	$*$	$*$	...	$*$	$*$	$*$	...	1	1	n is composite
12	$*$	$*$	...	$*$	$*$	$*$	...	-1	1	No information
13	$*$	$*$	...	$*$	$*$	$*$	...	$*$	1	n is composite
14	$*$	$*$	...	$*$	$*$	$*$	...	$*$	-1	n is composite
15	$*$	$*$	...	$*$	$*$	$*$	...	$*$	$*$	n is composite

information on primality testing as $126^{81} \equiv 1$ (mod 325) is also a trivial square root modulo 325. So depending on different values of a, we get useful information from the sequence of b_i's. Thus we need to know a general form of the sequence of b_i's, $i = 0, 1, 2, \ldots, h$ which is described in Table 10.10.

As Row 1, Row 2, Row 4, Row 6, Row 8, Row 10, and Row 12 of Table 10.10 do not ensure the existence of any non-trivial square root of 1 modulo n, we get no information regarding the primality testing. On the other hand, Row 3, Row 5, Row 7, Row 9, Row 11 and Row 13 ensure the existence of non-trivial square root of 1 modulo n, thereby provides the information that n is a composite integer. For Row 14 and Row 15, as $b_h \not\equiv 1 \pmod{n}$, the corresponding a is an F-witness for n. The above observation leads to the following definition.

Definition 10.10.5 Let $n = 2^h \cdot t + 1 \geq 3$ be an odd integer, where $\gcd(2, t) = 1$ and $h \geq 1$. An integer a, $1 \leq a < n$, is called an A-witness for n if $a^t \not\equiv 1 \pmod{n}$ and $a^{t2^i} \not\equiv -1 \pmod{n}$ for all i, $0 \leq i < h$. On the other hand, if n is composite and a is not an A-witness for n, that is, either $a^t \equiv 1 \pmod{n}$ or $a^{t2^i} \equiv -1 \pmod{n}$ for some i, $0 \leq i < h$, then a is said to be an A-liar for n.

Example 10.10.4 For $n = 325$, $a = 2, 118, 130, 224, 251$ are examples of A-witness while $a = 7, 18, 32, 126$ and 199 are examples of A-liars.

Now we are going to prove the following lemma which will help us in developing a probabilistic primality testing.

Lemma 10.10.6 *If a is an A-witness for n, then n must be a composite integer.*

Proof Let a be an A-witness for n. Let us consider the sequence $b_0, b_1, b_2, \ldots, b_{h-1}$ as in the Table 10.10. Now we may have two cases. First let $b_h \not\equiv 1 \pmod{n}$. Then a must be an F-witness for n, implying that n is a composite integer. Now let $b_h \equiv 1 \pmod{n}$. As a is an A-witness, $b_0 \not\equiv 1 \pmod{n}$ and $n - 1$ does not occur in the sequence $b_0, b_1, b_2, \ldots, b_{h-1}$. Let i be the minimum $i \geq 1$ such that $b_i \equiv 1 \pmod{n}$. Such i always exists as $b_h \equiv 1 \pmod{n}$. As $b_{i-1} \notin \{1, n - 1\}$, b_{i-1} is a non-trivial square root of 1 modulo n and thereby n is a composite integer. $\qquad\square$

Based on the above lemma, very soon we are going to develop a primality testing. Before that let us try to find some properties of A-liars. For that the following definition plays an important role.

Definition 10.10.6 A positive integer $n \geq 3$ of the form $2^h t + 1$, where $\gcd(2, t) = 1$, passes Miller's test for a base a, $1 \leq a < n$, if either $a^t \equiv 1 \pmod{n}$ or $a^{2^i t} \equiv -1 \pmod{n}$, for some i with $0 \leq i \leq h - 1$. In other words, n passes Miller's test for a base a, $1 \leq a < n$, if a is an A-liar for n.

Through the following theorems, we try to find some of the properties of those odd integers which pass the Miller's test for some base a. These properties will lead to the primality testing algorithm.

Theorem 10.10.8 *If n is an odd prime and a is a positive integer such that $1 \leq a < n$, then n passes Miller's test for the base a.*

Proof Let $a_k = a^{(n-1)/2^k}$, $k = 0, 1, \ldots, h$. By Fermat's Little Theorem, we have $a_0 = a^{2^h t} = a^{n-1} \equiv 1 \pmod{n}$. Again as $a_1^2 = (a^{2^{h-1}t})^2 = a^{n-1} = a_0 \equiv 1 \pmod{n}$, by the Corollary of Theorem 10.5.1, it follows that either $a_1 \equiv 1 \pmod{n}$ or $a_1 \equiv -1 \pmod{n}$. So, in general, if we have found that $a_0 \equiv a_1 \equiv \cdots \equiv a_k \equiv 1 \pmod{n}$ with $k < h$, then since $a_{k+1}^2 \equiv a_k \equiv 1 \pmod{n}$, we have either $a_{k+1} \equiv 1 \pmod{n}$ or $a_{k+1} \equiv -1 \pmod{n}$. Thus continuing this process, we find that either $a_k \equiv 1 \pmod{n}$ for $k = 0, 1, 2, \ldots, h$ or $a_k \equiv -1 \pmod{n}$ for some integer k. Hence, n passes Miller's test for the base a. $\qquad \square$

Theorem 10.10.9 *If a composite integer n passes Miller's test for a base a, then n is a pseudo-prime to the base a.*

Proof As n passes Miller's test for the base a, either $a^t \equiv 1 \pmod{n}$ or $a^{2^i t} \equiv -1 \pmod{n}$, for some i with $0 \le i \le h - 1$, where $n = 2^h t + 1$, $h \ge 1$ and $\gcd(2, t) = 1$. Thus $a^n = a^{n-1} a = (a^{2^i t})^{2^{h-i}} a = (\pm 1)^{2^{h-i}} a \equiv a \pmod{n}$, with the result that n is a pseudo-prime to the base a. $\qquad \square$

The above theorem leads to the following important definition.

Definition 10.10.7 If n is a composite integer that passes Miller's test for a base a, then we say that n is a strong pseudo-prime to the base a.

Example 10.10.5 Let $n = 2047 = 23 \cdot 89$. Then $n - 1 = 2046 = 2^1 \cdot 1023$. As both $2^{2046} \equiv 1 \pmod{2047}$ and $2^{1023} \equiv 1 \pmod{2047}$, 2047 is a strong pseudo-prime to the base 2.

The following theorem shows the existence of infinitely many strong pseudo-primes to the base 2. To prove this let us first prove the following lemma.

Lemma 10.10.7 *Let n be an odd integer which is a pseudo-prime to the base 2. Then $N = 2^n - 1$ is a strong pseudo-prime to the base 2.*

Proof As n is a composite integer, Lemma 10.10.3 ensures that $N = 2^n - 1$ is also a composite integer. Also as $2^{n-1} \equiv 1 \pmod{n}$, there exists some odd integer k such that $2^{n-1} = kn + 1$. Now, $N - 1 = 2^n - 2 = 2(2^{n-1} - 1) = 2nk$ and $2^n = N + 1 \equiv 1 \pmod{N}$. Thus $2^{\frac{N-1}{2}} = 2^{nk} = (2^n)^k \equiv 1 \pmod{N}$, which implies that N passes Miller's test for the base 2. Consequently, N is a strong pseudo-prime to the base 2. $\qquad \square$

We are now in a position to prove the existence of infinitely many strong pseudo-primes to the base 2.

Theorem 10.10.10 *There are infinitely many strong pseudo-primes to the base 2.*

Proof Lemma 10.10.7 ensures that every odd pseudo-prime to the base 2 yields a strong pseudo-prime $2^n - 1$ to the base 2. The rest of the theorem follows from Theorem 10.10.2. $\qquad \square$

Table 10.11 Miller–Rabin primality test

> **Input:** Odd integer $n \geq 3$.
> **Method:**
>
> 1. write $n - 1 = 2^h t$, where $h \geq 1$ and $\gcd(2, t) = 1$.
> 2. choose a random integer a such that $2 \leq a \leq n - 2$.
> 3. $b \leftarrow a^t \pmod{n}$
> 4. **if** $b \equiv 1 \pmod{n}$
> 5.   return ("No: the number is a prime integer").
> 6. **for** $i \leftarrow 0$ **to** $h - 1$
> 7. **do** $\begin{cases} \text{if } b \equiv -1 \pmod{n} \\ \quad \text{return ("No: the number is a prime integer").} \\ b \leftarrow b^2 \pmod{n} \end{cases}$
> 8. return ("Yes: the number is a composite integer").

Remark The smallest odd strong pseudo-prime to the base 2 is 2047. So, if n is an odd integer less than 2047 and n passes Miller's test to the base 2, then n must be a prime integer. Using the concept of strong pseudo-prime, we are now going to present another yes-biased Monte Carlo algorithm for Composites, known as the Miller–Rabin primality test (also known as "strong pseudo-prime test"). Let us first describe the algorithm in Table 10.11, where n is an odd integer greater than 1.

Now we shall show that the algorithm as described in Table 10.11 is a yes-biased Monte Carlo algorithm.

Theorem 10.10.11 *The Miller–Rabin primality testing algorithm is a yes-biased Monte Carlo algorithm.*

Proof To show that the Miller–Rabin primality testing for an odd integer $n \geq 3$ is a yes-biased Monte Carlo algorithm, we need to show that if the algorithm returns "Yes: the number is a composite integer", then n must be composite. We shall prove the result by the method of contradiction. Suppose, the algorithm returns "Yes: the number is a composite integer" for some prime n. Since it returns "Yes: the number is a composite integer", it must be the case that

$$a^t \not\equiv 1 \pmod{n}. \tag{10.15}$$

Now we consider the sequence of values $b_0 = a^t, b_1 = b_0^2 = a^{2t}, b_2 = b_1^2 = b_0^{2^2} = a^{2^2 t}, \ldots, b_{h-1} = b_0^{2^{h-1}} = a^{2^{h-1} t}$. As the algorithm returns "Yes: the number is a composite integer", we must say that $a^{2^i t} \not\equiv -1 \pmod{n}, 0 \leq i \leq h - 1$. On the other hand, as n is prime and $2 \leq a \leq n - 2$, by Fermat's Little Theorem 10.4.2, we have $a^{n-1} \equiv 1 \pmod{n}$, i.e., $a^{2^h t} \equiv 1 \pmod{n}$. Thus by Corollary of Theorem 10.5.1, $a^{2^{h-1} t} \equiv \pm 1 \pmod{n}$. As $a^{2^{h-1} t} \not\equiv -1 \pmod{n}$, we have $a^{2^{h-1} t} \equiv 1 \pmod{n}$. Then $a^{2^{h-2} t}$ must be a square root of 1. Thus continuing similar argument, we have $a^t \equiv 1 \pmod{n}$, contradicting (10.15), proving that the Miller–Rabin primality testing algorithm is a yes-biased Monte Carlo algorithm. $\square$

To obtain an error bound of the Miller–Rabin algorithm for a given odd integer $n \geq 3$, we need to consider the set of A-liars for n. But unfortunately, unlike the set of all F-liars for n, the set of all A-liars for n does not have a good algebraic structure like group structure. To explain that let us revisit the Table 10.9. Note that 7 and 32 are A-liars for 325 while their product 224 is an A-witness for 325. Thus to obtain an error bound, we consider the set $MR(n)$, for a given odd positive integer n, defined by $MR(n) = \{a \in \mathbf{Z}_n \setminus \{(0)\} : a^{n-1} = (1) \text{ and for } j = 0, 1, \ldots, h - 1, a^{t2^{j+1}} = (1) \text{ implies } a^{t2^j} = \pm(1)\}$, where $n = 2^h t + 1$, $h \geq 1$ and $\gcd(2, t) = 1$. We shall prove that the set of all A-liars is contained in $MR(n)$. Further we shall prove the main result that if n is a composite integer, then $|MR(n)| \leq \frac{n-1}{4}$. To prove the main result, let us first prove the following lemmas.

Lemma 10.10.8 *If n passes Miller's test for a base a, i.e., if a is an A-liar for n, then $a \in MR(n)$.*

Proof The lemma follows from Theorem 10.10.9, Table 10.10 and the Definition of $MR(n)$. □

Lemma 10.10.9 *Let $f : \mathbf{Z}_n^* \to \mathbf{Z}_n^*$ be a group homomorphism defined by $f(x) = x^{n-1}, \forall x \in \mathbf{Z}_n^*$. Then $|\ker f| = \gcd(\phi(n), n-1)$, where $n = p^e$, for some odd prime p with $e \geq 1$.*

Proof As p is an odd prime and $e \geq 1$, by Theorem 10.4.6, $\mathbf{Z}_n^*$ is a cyclic group of order $\phi(n)$. So, the group $(\mathbf{Z}_n^*, .)$ is isomorphic to the group $(\mathbf{Z}_{\phi(n)}, +)$. Thus the homomorphism $x \mapsto x^{n-1}$ in $\mathbf{Z}_n^*$ induces a homomorphism $g : \mathbf{Z}_{\phi(n)} \to \mathbf{Z}_{\phi(n)}$ in $\mathbf{Z}_{\phi(n)}$ defined by $g(x) = (n-1)x, \forall x \in \mathbf{Z}_{\phi(n)}$. Hence $|\ker f| = |\ker g|$. Now we are going to calculate $|\ker g|$. Note that $\ker g = \{x \in \mathbf{Z}_{\phi(n)} : (n-1)x \equiv 0 \pmod{\phi(n)}\}$. Let $d = \gcd(\phi(n), n-1)$. Thus $y \in \ker g$ if and only if $(n-1)y \equiv 0 \pmod{\phi(n)}$ if and only if $\phi(n)$ divides $(n-1)y$ if and only if $\frac{\phi(n)}{d}$ divides $((n-1)/d)y$ if and only if $\frac{\phi(n)}{d}$ divides y if and only if $y \equiv 0 \pmod{\frac{\phi(n)}{d}}$. Thus kernel of g contains precisely the d residue classes, namely, $(\frac{i\phi(n)}{d})$, where $i = 0, 1, \ldots, d - 1$, with the result $|\ker f| = |\ker g| = \gcd(\phi(n), n-1)$. □

Lemma 10.10.10 *Suppose that a is an element of an additive abelian group. Suppose further that for some prime p and an integer $e \geq 1$, $p^e a = (0)$ and $p^{e-1}a \neq (0)$. Then the order of the element a is p^e.*

Proof Let m be the order of the element a. Then m divides p^e. As p is a prime integer, m must be of the form p^i for some $i \in \{0, 1, 2, \ldots, e\}$. We claim that $i = e$. If $i < e$, then $ma = p^i a = (0)$ which implies that $p^{e-1}a = (0)$, contradicting the fact that $p^{e-1}a \neq (0)$. Hence i must be equal to e, with the result $m = p^e$. □

Now we are in a position to prove the main theorem. But before proving that let us first discuss some notions of random variable and uniform distribution from probability theory. Here, we introduce the concepts from probability theory, starting with the basic notions of probability distributions on finite sample spaces as defined below.

Definition 10.10.8 Let Ω be a finite non-empty set. A finite probability distribution on Ω is a function $P : \Omega \to [0, 1]$ that satisfies $\sum_{\omega \in \Omega} P(\omega) = 1$. The set Ω is called the sample space of P.

Intuitively, the elements of Ω represent the possible outcomes of a random experiment, where the probability of outcome $\omega \in \Omega$ is $P(\omega)$.

Example 10.10.6 If we think of tossing a fair coin, then setting $\Omega = \{H, T\}$ and $P(\omega) = \frac{1}{2}$ for all $\omega \in \Omega$ gives a probability distribution that naturally describes the possible outcomes of the experiment.

The above example leads to a very important distribution in the theory of probability, known as uniform distribution.

Definition 10.10.9 Let Ω be a finite non-empty set. If $P(w) = \frac{1}{|\Omega|}$ for all $\omega \in \Omega$, then P is called the uniform distribution on Ω.

Now let us define another important term in the theory of probability, an event.

Definition 10.10.10 An event is a subset A of Ω and the probability of A is defined to be

$$P[A] = \sum_{\omega \in A} P(\omega).$$

It is sometimes convenient to associate a real number, or other mathematical object, with each outcome of a random experiment. This idea has been formalized by the notion of a random variable as defined below.

Definition 10.10.11 Let P be a probability distribution on a sample space Ω. A random variable X is a function $X : \Omega \to S$, where S is some set. We say that X takes values in S. In particular, if $S = \mathbf{R}$, i.e., the values taken by X are real numbers, then we say that X is real valued.

For $s \in S$, "$X = s$" denotes the event $\{\omega \in \Omega : X(\omega) = s\}$. Thus $P[X = s] = \sum_{\omega \in X^{-1}(\{s\})} P(\omega)$. It is further possible to combine random variables to define new random variables. Suppose, $X_1, X_2, \ldots, X_m$ are random variables, where $X_i : \Omega \to S_i, i = 1, 2, \ldots, m$. From these m random variables, we can get a new random variable $(X_1, X_2, \ldots, X_m)$ that maps $\omega \in \Omega$ to $(X_1(\omega), X_2(\omega), \ldots, X_m(\omega)) \in S_1 \times S_2 \times \cdots \times S_m$. More generally, if $g : S_1 \times S_2 \times \cdots \times S_m \to T$ is a function, then $g((X_1, X_2, \ldots, X_m))$ denotes the random variable that maps ω to $g((X_1(\omega), X_2(\omega), \ldots, X_m(\omega)))$. Next we are going to define the distribution of a random variable X.

Definition 10.10.12 Let $X : \Omega \to S$ be a random variable. The variable determines a probability distribution $P_X : S \to [0, 1]$ on the set S, where P_X is defined by $P_X(a) = P[X = a]$, for all $a \in S$. P_X is called the distribution of X. In particular, if P_X is the uniform distribution on S, then we say that the random variable X is uniformly distributed over S.

Uniform distributions have many very nice and simple properties. To understand the nature of uniform distribution, we should know some simple criteria that will ensure that certain random variables have uniform distributions. For that we need the following definition.

Definition 10.10.13 Let T and S be finite sets. We say that a function $f : S \to T$ is a regular function if every element in the image of f has the same number of pre-images under f.

Based on the definition of regular function, let us prove the following theorem.

Theorem 10.10.12 *Let $f : S \to T$ be a surjective, regular function and X be a random variable that is uniformly distributed over S. Then $f(X)$ is also uniformly distributed over T.*

Proof As f is surjective and regular, for every $x \in T$, the order of the set $S_x = f^{-1}(\{x\})$ will be $\frac{|S|}{|T|}$. Thus for each $x \in T$, we have

$$P\big[f(X) = x\big] = \sum_{\omega \in X^{-1}(S_x)} P(\omega) = \sum_{s \in S_x} \sum_{\omega \in X^{-1}(\{s\})} P(\omega)$$

$$= \sum_{s \in S_x} P[X = s] = \sum_{s \in S_x} \frac{1}{|S|} = \frac{|S|}{|T|} \frac{1}{|S|} = \frac{1}{|T|}.$$

Hence the result.   □

Based on this theorem, we prove the following theorem, which will help us in proving an error bound for Miller–Rabin primality testing.

Theorem 10.10.13 *Let $f : G \to G'$ be a group epimorphism from the abelian group G onto the abelian group G' and X be a random variable that is uniformly distributed over G. Then $f(G)$ is also uniformly distributed over G'.*

Proof To prove the theorem, it is sufficient to prove that f is regular. For that let us consider ker f. Then ker f is a subgroup of G and for every $g' \in G'$, the set $f^{-1}(\{g'\})$ is a coset of ker f. The theorem follows from the fact that every coset of ker f has the same size.   □

Now we are in a position to prove the main result in this section.

Theorem 10.10.14 *If n is an odd prime, then $MR(n) = \mathbf{Z}_n^*$. If n is a positive composite integer, then $|MR(n)| \leq \frac{n-1}{4}$.*

Proof First let n be an odd prime. Then by Theorem 10.4.5, $(\mathbf{Z}_n^*, \cdot)$ is a cyclic group of order $n - 1$. As $MR(n) \subseteq \mathbf{Z}_n^*$, we need to show that $\mathbf{Z}_n^* \subseteq MR(n)$. Let $a \in \mathbf{Z}_n^*$. As n is a prime integer, $a^{n-1} = (1)$. Now consider any $j = 0, 1, \ldots, h - 1$ such that $a^{2^{j+1}t} = (1)$. Such a j exists, as for $j = h - 1$, $a^{n-1} = (1)$. Let $b = a^{2^j t}$. Then $b^2 = a^{2^{j+1}t} = (1)$. As n is prime, then by the Corollary of Theorem 10.5.1, the only possible choices for b are $\pm(1)$. Thus $a^{2^{j+1}t} = (1)$ implies $a^{2^j t} = \pm(1)$, with the result that $a \in MR(n)$. Hence $MR(n) = \mathbf{Z}_n^*$.

Now we assume that n is composite. We divide the proof into two cases.

Case 1: Let $n = p^e$, where p is an odd prime and $e > 1$. Let us consider the map $\psi : \mathbf{Z}_n^* \to \mathbf{Z}_n^*$ defined by $\psi(x) = x^{n-1}, \forall x \in \mathbf{Z}_n^*$. Then clearly ψ is a homomorphism and $MR(n) \subseteq \ker \psi$. Now by Lemma 10.10.9, we have $|\ker \psi| = \gcd(\phi(n), n - 1)$. Thus $|MR(n)| \leq |\ker \psi| = \gcd(p^{e-1}(p - 1), p^e - 1) = p - 1 = \frac{p^e - 1}{p^{e-1} + p^{e-2} + \cdots + 1} \leq \frac{n-1}{4}$.

Case 2: Let $n = p_1^{\alpha_1} p_2^{\alpha_2} \ldots p_k^{\alpha_k}$ be the prime factorization of n with $k > 1$. Let $n - 1 = t2^h$, where $h \geq 1$ and $\gcd(t, 2) = 1$. Let $\psi : \mathbf{Z}_{p_1^{\alpha_1}}^* \times \mathbf{Z}_{p_2^{\alpha_2}}^* \times \cdots \times \mathbf{Z}_{p_k^{\alpha_k}}^* \to \mathbf{Z}_n^*$ be the ring isomorphism as defined in the Chinese Remainder Theorem. Let $\phi(p_i^{\alpha_i}) = t_i 2^{h_i}$, where ϕ is the Euler ϕ-function, $h_i \geq 1$ and $\gcd(t_i, 2) = 1$, for $i = 1, 2, \ldots, k$. Let $l = \min\{h, h_1, h_2, \ldots, h_k\}$. Then $l \geq 1$. Also by Theorem 10.4.6, each of $\mathbf{Z}_{p_i^{\alpha_i}}^*$ is a cyclic group of order $t_i 2^{h_i}$.

We claim that for $a \in MR(n)$, $a^{t2^l} = (1)$. If $l = h$, then from the definition of $MR(n)$, $a^{t2^h} = (1)$. So we assume that $l < h$. We shall prove that in this case also $a^{t2^l} = (1)$. We shall prove this by the method of contradiction. If possible let $a^{t2^l} \neq (1)$ and let j be the smallest index in the range $l, l+1, \ldots, h - 1$ such that $a^{t2^{j+1}} = (1)$. Then by the definition of $MR(n)$, we have $a^{t2^j} = -(1)$. Since $l < h$, we have $l = h_i$ for some $i \in \{1, 2, \ldots, k\}$. Let $a = \psi(a_1, a_2, \ldots, a_k)$. Then we have $a_i^{t2^j} = -(1)$. Then by Lemma 10.10.10, the order of a_i^t in $\mathbf{Z}_{p_i^{\alpha_i}}^*$ is 2^{j+1}. Now by Lagrange's Theorem, 2^{j+1} must divide $t_i 2^{h_i}$, a contradiction as $j + 1 > j \geq l = h_i$ and $\gcd(2, t_i) = 1$. Thus $a^{t2^l} = (1)$.

From the claim in the previous paragraph and the definition of $MR(n)$, it follows that for any $a \in MR(n)$, $a^{t2^{l-1}} = \pm(1)$. Now we are going to use some results from the probability theory. We consider an experiment in which a is chosen at random (i.e., with a uniform distribution) from $\mathbf{Z}_n^*$. To prove the main result, it suffices to prove that $\Pr[a^{t2^{l-1}} = \pm(1)] \leq \frac{1}{4}$. Let $a = \psi(a_1, a_2, \ldots, a_k)$. As a is uniformly distributed over $\mathbf{Z}_n^*$, each $a_i \in \mathbf{Z}_{p_i^{\alpha_i}}^*$ is also uniformly distributed over $\mathbf{Z}_{p_i^{\alpha_i}}^*$ and the collection of all the a_i's is a mutually independent collection of random variables. Let us consider the group homomorphism $f_j^i : \mathbf{Z}_{p_i^{\alpha_i}}^* \to \mathbf{Z}_{p_i^{\alpha_i}}^*$ defined by $f_j^i(a_i) = a_i^{t2^j}$, for $i = 1, 2, \ldots, k$ and $j = 0, 1, 2, \ldots, h$. Then by Lemma 10.10.9

we have $|\ker f_j^i| = \gcd(t_i 2^{h_i}, t2^j)$. Also, by using first isomorphism theorem we have $|\operatorname{Im}(f_j^i)| = \frac{t_i 2^{h_i}}{\gcd(t_i 2^{h_i}, t2^j)}$. Further, as $l \leq h$ and $l \leq h_i$, we have $\gcd(t_i 2^{h_i}, t2^l)$ divides $\gcd(t_i 2^{h_i}, t2^h)$ and $\gcd(t_i 2^{h_i}, t2^l)$ divides $\gcd(t_i 2^{h_i}, t2^{h_i})$, it follows that $|\operatorname{Im}(f_h^i)|$ divides $|\operatorname{Im}(f_l^i)|$ and $|\operatorname{Im}(f_l^i)|$ divides $|\operatorname{Im}(f_{l-1}^i)|$. Thus $|\operatorname{Im}(f_{l-1}^i)|$ is even and contains at least $2|\operatorname{Im}(f_h^i)|$ elements. As $\mathbf{Z}^*_{p_i^{\alpha_i}}$ is cyclic and $\operatorname{Im}(f_{l-1}^i)$ is a subgroup of $\mathbf{Z}^*_{p_i^{\alpha_i}}$, $\operatorname{Im}(f_{l-1}^i)$ is also cyclic. So, in particular, $\operatorname{Im}(f_{l-1}^i)$ contains a unique subgroup of order 2, namely $\{(1), -(1)\}$. Thus $-(1) \in \operatorname{Im}(f_{l-1}^i)$. Now let us consider two events E_1 and E_2 as follows:

Note that the event $a^{t2^{l-1}} = \pm(1)$ occurs if and only if either

1. E_1: $a_i^{t2^{l-1}} = +(1)$ for $i = 1, 2, \ldots, k$ or
2. E_2: $a_i^{t2^{l-1}} = -(1)$ for $i = 1, 2, \ldots, k$.

Further note that the events E_1 and E_2 are disjoint, and since the values $a_i^{t2^{l-1}}$ are mutually independent, with each value $a_i^{t2^{l-1}}$ uniformly distributed over $\operatorname{Im}(f_{l-1}^i)$ by Theorem 10.10.13, and since $\operatorname{Im}(f_{l-1}^i)$ contains $\pm(1)$, we have

$$\Pr\left[a^{t2^{l-1}} = \pm(1)\right] = \Pr[E_1] + \Pr[E_2] = 2\prod_{i=1}^{k} \frac{1}{|\operatorname{Im}(f_{l-1}^i)|},$$

and since $|\operatorname{Im} f_{l-1}^i| \geq 2|(\operatorname{Im} f_h^i)|$, we have

$$\Pr\left[a^{t2^{l-1}} = \pm(1)\right] \leq 2^{-k+1} \prod_{i=1}^{k} \frac{1}{|\operatorname{Im}(f_h^i)|}. \tag{10.16}$$

If $k \geq 3$, then from (10.16) we can directly say that $\Pr[a^{t2^{l-1}} = \pm(1)] \leq \frac{1}{4}$, and in that case, we are done. Now let us suppose that $k = 2$. In this case, Theorem 10.10.7 implies that n is not a Carmichael number, which implies that for some $i = 1, \ldots, l$, we must have $\operatorname{Im}(f_h^i) \neq \{(1)\}$, with the result $|\operatorname{Im}(f_h^i)| \geq 2$, and (10.16) finally implies that $\Pr[a^{t2^{l-1}} = \pm(1)] \leq \frac{1}{4}$, as required. $\qquad\square$

10.11 Exercise

Exercises-I

1. Show that the last digit in the decimal expression of $F_n = 2^{2^n} + 1$ is 7 if $n \geq 2$.
2. Use the fact that every prime divisor of $F_4 = 65537$ is of the form $2^6 k + 1$ to verify that F_4 is a prime integer.
3. Estimate the number of decimal digits in the mth Fermat number F_m.
4. Show that 91 is a pseudo-prime to the base 3.

5. Show that every odd composite integer is a pseudo-prime to the base 1 and -1.

6. Show that if p is a prime and $2^p - 1$ is composite, then $2^p - 1$ is a pseudo-prime to the base 2.

7. Show that if n is a pseudo-prime to the bases a and b, then n is also a pseudo-prime to the base ab.

8. Show that if a and n are positive integers with $\gcd(a, n) = \gcd(a - 1, n) = 1$, then $1 + a + a^2 + \cdots + a^{\phi(n)-1} \equiv 0 \pmod{m}$.

9. Show that $a^{\phi(b)} + b^{\phi(a)} \equiv 1 \pmod{ab}$, if a and b are relatively prime positive integers.

10. Use Euler's Theorem to find the last digit in the decimal representation of 7^{1000}.

11. Find the last digit in the decimal representation of $17^{17^{17}}$.

12. Prove the following:

(a) If n is an integer greater than 2 then $\phi(n)$ is even.

(b) $\phi(3n) = 3\phi(n)$ if and only if 3 divides n.

(c) $\phi(3n) = 2\phi(n)$ if and only if 3 does not divide n.

(d) If n is odd then $\phi(2n) = \phi(n)$.

(e) If n is even then $\phi(2n) = 2\phi(n)$.

(f) If m divides n then $\phi(m)$ divides $\phi(n)$.

(g) If n is a composite positive integer and $\phi(n)$ divides $n - 1$, then n is a square free integer and is the product of at least three distinct primes.

13. For all positive integer n and for all positive integer $a \geq 2$, prove that n divides $\phi(a^n - 1)$.

14. Which of the following congruences have solutions and how many: $x^2 \equiv 2 \pmod{11}$ and $x^2 \equiv -2 \pmod{59}$.

15. Prove that $\sigma(n)$ is an odd integer if and only if n is a perfect square or a double of a perfect square.

16. Prove that there exist infinitely many primes of the form $8k + 1$.

17. Prove that there exist infinitely many primes of the form $8k - 1$.

18. Prove that if p and q are primes of the form $4k + 3$ and if $x^2 \equiv p \pmod{q}$ has no solutions, then prove that $x^2 \equiv q \pmod{p}$ has two solutions.

19. Find all solutions of $x^2 \equiv 1 \pmod{15}$.

20. Find all primes p such that $x^2 \equiv -1 \pmod{p}$.

21. Write down the last two digits of 9^{1500}.

Exercises-II   Identify the correct alternative(s) (there may be more than one) from the following list:

1. Which of the following numbers is the largest?
   $$2^{3^4}, 2^{4^3}, 3^{2^4}, 3^{4^2}, 4^{2^3}, 4^{3^2}.$$

        (a) 4^{3^2} (b) 3^{4^2} (c) 2^{3^4} (d) 4^{2^3}.

2. Suppose the sum of the seven positive numbers is 21. What is the minimum possible value of the average of the squares of these numbers?

$$\text{(a) } 9 \qquad \text{(b) } 21 \qquad \text{(c) } 63 \qquad \text{(d) } 7.$$

3. The number of elements in the set $\{n : 1 \leq n \leq 1000, n$ and 1000 are relatively prime$\}$ is

$$\text{(a) } 300 \qquad \text{(b) } 250 \qquad \text{(c) } 100 \qquad \text{(d) } 400.$$

4. The number of 4 digit numbers with no two digits common is

$$\text{(a) } 5040 \qquad \text{(b) } 3024 \qquad \text{(c) } 4536 \qquad \text{(d) } 4823.$$

5. The unit digit of 2^{100} is

$$\text{(a) } 2 \qquad \text{(b) } 8 \qquad \text{(c) } 6 \qquad \text{(d) } 4.$$

6. The number of multiples of 10^{44} that divide 10^{55} is

$$\text{(a) } 121 \qquad \text{(b) } 12 \qquad \text{(c) } 11 \qquad \text{(d) } 144.$$

7. The number of positive divisors of $50,000$ is

$$\text{(a) } 40 \qquad \text{(b) } 30 \qquad \text{(c) } 20 \qquad \text{(d) } 50.$$

8. The last digit of $(38)^{2011}$ is

$$\text{(a) } 4 \qquad \text{(b) } 6 \qquad \text{(b) } 2 \qquad \text{(d) } 8.$$

9. The last two digits of 7^{81} are

$$\text{(a) } 37 \qquad \text{(b) } 17 \qquad \text{(c) } 07 \qquad \text{(d) } 47.$$

10. Given a positive integer n, let $\phi(n)$ denote the number of integers k such that $1 \leq k \leq n$ and $\gcd(k, n) = 1$. Then identify the correct statement(s):

(a) $\phi(m)$ divides m for every positive integer m;
(b) a divides $\phi(a^m - 1)$ for all positive integers a and m such that $\gcd(a, m) = 1$;
(c) m divides $\phi(a^m - 1)$ for all positive integers a and m such that $\gcd(a, m) = 1$;
(d) m divides $\phi(a^m - 1)$ for all positive integers a and m.

10.12 Additional Reading

We refer the reader to the books (Adhikari and Adhikari 2003; Burton 1989; Hoffstein et al. 2008; Ireland and Rosen 1990; Jones and Jones 1998; Katz and Lindell 2007; Koblitz 1994; Rosen 1992; Mollin 2009; Ribenboim 2004; Stinson 2002) for further details.

References

Adhikari, M.R., Adhikari, A.: Groups, Rings and Modules with Applications, 2nd edn. Universities Press, Hyderabad (2003)

Agrawal, M., Kayal, N., Saxena, N.: Primes is in P. Preprint, IIT Kanpur. http://www.cse.iitk.ac.in/news/primality.pdf (August 2002)

Burton, D.M.: Elementary Number Theory. Brown, Dulreque (1989)

Dietzfelbinger, M.: Primality Testing in Polynomial Time from Randomized Algorithms to "PRIMES is in P". LNCS, vol. 3000, Tutorial. Springer, Berlin (2004)

Hoffstein, J., Pipher, J., Silverman, J.H.: An Introduction to Mathematical Cryptography. Springer, Berlin (2008)

Ireland, K., Rosen, M.: A Classical Introduction to Modern Number Theory, 2nd edn. Springer, Berlin (1990)

Jones, G.A., Jones, J.M.: Elementary Number Theory. Springer, London (1998)

Katz, J., Lindell, Y.: Introduction to Modern Cryptography. Chapman & Hall/CRC, London/Boca Raton (2007)

Koblitz, N.: A Course in Number Theory and Cryptography, 2nd edn. Springer, Berlin (1994)

Rosen, Kenneth.H.: Elementary Number Theory & Its Applications, 3rd edn. Addition-Wesley, Reading (1992)

Mollin, R.A.: Advanced Number Theory with Applications. CRC Press/Chapman & Hall, Boca Raton/London (2009)

Ribenboim, P.: The Little Book of Bigger Primes. Springer, New York (2004)

Stinson, D.R.: Cryptography, Theory & Practice. CRC Press Company, Boca Raton (2002)

Chapter 11
Algebraic Numbers

Algebraic number theory arose through the attempts of mathematicians to prove Fermat's Last Theorem. One of the continuous themes since early 20th century which motivated algebraic number theory was to establish analogy between algebraic number fields and algebraic function fields. The study of algebraic number theory was initiated by many mathematicians. This list includes the names of Kronecker, Kummer, Dedekind, Dirichlet, Gauss, and many others. Gauss called algebraic number theory 'Queen of Mathematics'. Andrew Wiles established Fermat's Last Theorem a few years back. An algebraic number is a complex number which is algebraic over the field $\mathbf{Q}$ of rational numbers. An algebraic number field is a subfield of the field $\mathbf{C}$ of complex numbers, which is a finite field extension of the field $\mathbf{Q}$ and is obtained from $\mathbf{Q}$ by adjoining a finite number of algebraic elements. The concepts of algebraic numbers, algebraic integers, Gaussian integers, algebraic number fields and quadratic fields are introduced in this chapter after a short discussion on general properties of field extensions and finite fields. Moreover, the countability of algebraic numbers, existence of transcendental numbers, impossibility of duplication of general cube and impossibility of trisection of a general angle by straight edge and compass are shown. The celebrated theorem known as Fundamental Theorem of Algebra is also proved in this chapter by using the tools from homotopy theory as discussed in Chap. 2. This theorem proves the algebraic completeness of the the field of complex numbers. Like the field of complex numbers, we further prove that the field of algebraic numbers is also algebraically closed.

11.1 Field Extension

We begin with a review of basic concepts of field extension followed by finite fields. While studying field extension, we mainly consider a pair of fields F and K such that F is a subfield of K. Taking F as the basic field, an extension field K of F is a field K which contains F as a subfield. The basic results needed for the study of field extensions are discussed first followed by a discussion on simple extensions.

M.R. Adhikari, A. Adhikari, *Basic Modern Algebra with Applications*,
DOI 10.1007/978-81-322-1599-8_11, © Springer India 2014

Definition 11.1.1 Let K be a field and F be a subfield of K. Then K is said to be a field extension of F, written as K/F.

Example 11.1.1 (i) $\mathbf{C}$ is a field extension of $\mathbf{R}$, denoted by $\mathbf{C}/\mathbf{R}$.

(ii) $\mathbf{Q}(\sqrt{2}) = \{a + b\sqrt{2} : a, b \in \mathbf{Q}\}$ is a field extension of $\mathbf{Q}$, denoted by $\mathbf{Q}(\sqrt{2})/\mathbf{Q}$.

(iii) $\mathbf{R}$ is a field extension of $\mathbf{Q}$, denoted by $\mathbf{R}/\mathbf{Q}$.

(iv) Let K be a field. If char K is 0, then K contains a subfield F isomorphic to $\mathbf{Q}$. If char $K = p > 0$, for some prime p, then K contains a subfield F isomorphic to $\mathbf{Z}_p$. Thus K can be viewed as a field extension of F, where F is a field isomorphic to $\mathbf{Q}$ or $\mathbf{Z}_p$ according as char K is 0 or p.

(v) Let F be a field and $K = F(x)$, the field of rational functions in x over F, which is the quotient field of the integral domain $F[x]$. Then K is a field extension of F.

Definition 11.1.2 Let K be a field extension of F and $\alpha_1, \alpha_2, \ldots, \alpha_n$ be in K. Then the smallest field extension of F containing both F and $\{\alpha_1, \alpha_2, \ldots, \alpha_n\}$ is called the field generated by F and $\{\alpha_1, \alpha_2, \ldots, \alpha_n\}$ and is denoted by $F(\alpha_1, \alpha_2, \ldots, \alpha_n)$.

We now describe the elements of $F(\alpha_1, \alpha_2, \ldots, \alpha_n)$.

Theorem 11.1.1 *Let K be a field extension of F and $\alpha_1, \alpha_2, \ldots, \alpha_n$ be in K. Then*

$$F(\alpha_1, \alpha_2, \ldots, \alpha_n)$$
$$= \{f(\alpha_1, \alpha_2, \ldots, \alpha_n)/g(\alpha_1, \alpha_2, \ldots, \alpha_n) : f, g \in F[x_1, x_2, \ldots, x_n] \text{ and}$$
$$g(\alpha_1, \alpha_2, \ldots, \alpha_n) \neq 0\}.$$

Proof Let S be the set defined by $S = \{f(\alpha_1, \alpha_2, \ldots, \alpha_n)/g(\alpha_1, \alpha_2, \ldots, \alpha_n) : f, g \in F[x_1, x_2, \ldots, x_n] \text{ and } g(\alpha_1, \alpha_2, \ldots, \alpha_n) \neq 0\}$. Then S is a subfield of K such that $F \subseteq S$, since $a = a/1 \in S$ for every $a \in F$. Let L be a subfield of K containing F and $\{\alpha_1, \alpha_2, \ldots, \alpha_n\}$. Then $L \supseteq S$, because L contains $f(\alpha_1, \alpha_2, \ldots, \alpha_n)$ and also $f(\alpha_1, \alpha_2, \ldots, \alpha_n)/g(\alpha_1, \alpha_2, \ldots, \alpha_n)$, if $g(\alpha_1, \alpha_2, \ldots, \alpha_n) \neq 0$. This shows that S is the smallest subfield of K containing F and $\{\alpha_1, \alpha_2, \ldots, \alpha_n\}$. Hence $F(\alpha_1, \alpha_2, \ldots, \alpha_n) = S$. $\square$

Corollary *Let K be a field extension of F and $\alpha \in K$. Then $F(\alpha) = \{f(\alpha)/g(\alpha)$, where $f(x), g(x) \in F[x]$ and $g(\alpha) \neq 0\}$ is the quotient field of $F[\alpha]$.*

Definition 11.1.3 Let K be a field extension of F. The field K is said to be a simple extension of F iff there exists an element α in K such that $K = F(\alpha)$. The element $\alpha \in K$ is said to be a primitive element of the extension and $F(\alpha)$ is said to be generated by F and α.

Example 11.1.2   The field extension $\mathbf{Q}(\sqrt{5})$ of $\mathbf{Q}$ is a simple extension with $\sqrt{5}$ as its primitive element. The field $\mathbf{Q}(\sqrt{5})$ is generated by $\mathbf{Q}$ and a root $\sqrt{5}$ of the equation $x^2 - 5 = 0$ and consists of all real numbers $a + b\sqrt{5}$ with rational coefficients a and b.

Definition 11.1.4   A field F is said to be a prime field iff F has no proper subfield.

Example 11.1.3   (i) $\mathbf{Q}$ is a prime field, because $\mathbf{Q}$ has no proper subfield.
  (ii) $\mathbf{Z}_p$ is a prime field for every prime integer p.

Definition 11.1.5   Let K be a field extension of the field F. Then an element α of K is said to be

(a) a root of a polynomial $f(x) = a_0 + a_1x + \cdots + a_nx^n$ in $F[x]$ iff $f(\alpha) = a_0 + a_1\alpha + \cdots + a_n\alpha^n = 0$;
(b) algebraic over F iff α is a root of some non-null polynomial $f(x)$ in $F[x]$;
(c) transcendental over F iff α is not a root of any non-null polynomial in $F[x]$.

Remark   Every element a of a field F is algebraic over F.

Definition 11.1.6   Let K be a field extension of the field F. Then K is said to be an algebraic extension over F iff every element of K is algebraic over F. Otherwise, K is called a transcendental extension over F. A simple extension $F(\alpha)$ is said to be an algebraic or transcendental over F according to whether the element α is algebraic or transcendental over F.

Example 11.1.4

(a)   (i) For the field extension $\mathbf{Q}(\sqrt{2})$ of $\mathbf{Q}$, every element $\alpha = a + b\sqrt{2} \in \mathbf{Q}(\sqrt{2})$ is algebraic over $\mathbf{Q}$.
   (ii) The element $i \in \mathbf{C}$ is algebraic over $\mathbf{R}$.
   (iii) The element π in $\mathbf{R}$ is transcendental over $\mathbf{Q}(\sqrt{2})$.
   (iv) The element $2\pi i \in \mathbf{C}$ is algebraic over $\mathbf{R}$ but transcendental over $\mathbf{Q}$.
(b) For the field extension $\mathbf{R}$ of $\mathbf{Q}$,

   (i) e and π are both transcendental over $\mathbf{Q}$;
        For proof see [Hardy and Wright (2008, pp. 218–227)];
   (ii) $\sqrt{\pi}$ is transcendental over $\mathbf{Q}$;
   (iii) π^2 is transcendental over $\mathbf{Q}$.

Remark   Example 11.1.4(a)(iv) shows that the two properties algebraic and transcendental vary depending on the base field.

Theorem 11.1.2   *Let F be a field and $F(x)$ be the field of rational functions in x over F. Then the element x of $F(x)$ is transcendental over F.*

Proof $F(x)$ is clearly a field extension of the field F. If x is not transcendental over F, then there is a non-null polynomial $f(y) = a_0 + a_1 y + \cdots + a_n y^n$ over F, of which x is a root. This shows that $a_0 + a_1 x + \cdots + a_n x^n = 0$. Hence each $a_i = 0$ implies that $f(y)$ is a null polynomial. This gives a contradiction. $\square$

Theorem 11.1.3 *Let K be a field extension of F and an element α of K be algebraic over F. Then $F(\alpha)$ and $F[x]/\langle f(x) \rangle$ are isomorphic for some monic irreducible polynomial $f(x)$ in $F[x]$, of which α is a root.*

Proof Define the mapping $\mu : F[x] \to F[\alpha]$, $f(x) \mapsto f(\alpha)$.

Then μ is an epimorphism and hence by the First Isomorphism Theorem, $F[x]/\ker \mu \cong F[\alpha]$. But α is algebraic over $F \Rightarrow \ker \mu \neq \{0\}$. Again $\ker \mu$ is an ideal of the Euclidean domain $F[x]$. But $F[x]$ is a PID $\Rightarrow \ker \mu$ is a principal ideal of $F[x]$. Hence $\ker \mu = \langle g(x) \rangle$ for some $g(x) \in F[x]$. If a is the leading coefficient of $g(x)$, then $f(x) = a^{-1} g(x)$ is a monic polynomial of $F[x]$ and $\ker \mu = \langle g(x) \rangle = \langle f(x) \rangle$. Clearly, $f(x)$ is irreducible in $F[x]$ and hence $\langle f(x) \rangle$ is a maximal ideal in $F[x]$. Consequently, $F[x]/\langle f(x) \rangle$ is a field. Thus $F(\alpha) = Q(F[\alpha]) \cong Q(F[x]/\langle f(x) \rangle) = F[x]/\langle f(x) \rangle$. Hence the theorem follows. $\square$

Corollary *Let K be a field extension of the field F. If an element α of K is algebraic over F, then there exists a unique monic irreducible polynomial in $F[x]$ of which α is a root.*

Proof Using Theorem 11.1.3, there exists a monic irreducible polynomial $f(x)$ in $F[x]$ such α is a root of $f(x)$. If $g(x)$ is a monic irreducible polynomial in $F[x]$ such that α is a root of $g(x)$, then by the above definition of μ, $g(x) \in \ker \mu = \langle f(x) \rangle$. Consequently, $f(x)$ divides $g(x)$ in $F[x]$. Suppose $g(x) = q(x) f(x)$ for some $q(x) \in F[x]$. Since $g(x)$ is irreducible in $F[x]$, it follows that either $q(x)$ or $f(x)$ is a unit in $F[x]$. As $f(x)$ is not a unit in $F[x]$, $q(x)$ is a unit and thus $q(x) = c$ for some non-zero element of F. Again, since $f(x)$ and $g(x)$ are both monic, $c = 1$ and hence $f(x) = g(x)$. $\square$

This corollary leads to Definition 11.1.7.

Definition 11.1.7 Let F be a field and K be a field extension of F. Then the unique monic irreducible polynomial $f(x)$ in $F[x]$ having α as a root is called the *minimal polynomial* of α over F and the degree of $f(x)$ is called the *degree* of α over F.

We now show that, under a suitable condition, $F(\alpha) = F[\alpha]$.

Proposition 11.1.1 *Let K be a field extension of the field F. Then $F(\alpha) = F[\alpha]$ iff α is algebraic over F.*

Proof Suppose α is algebraic over F. Then as proved in Theorem 11.1.3, $F[\alpha] \cong F[x]/\langle f(x) \rangle$, where $f(x)$ is a monic irreducible polynomial in $F[x]$, of which α is a root. Since $F[x]/\langle f(x) \rangle$ is a field, $F[\alpha]$ is a field and hence $F(\alpha) = F[\alpha]$.

Conversely, suppose $F(\alpha) = F[\alpha]$. If $\alpha = 0$, then α is the root of the polynomial $x \in F[x]$. If $\alpha \neq 0$, then $\alpha^{-1} \in F(\alpha)$ and hence $\alpha^{-1} = a_0 + a_1\alpha + \cdots + a_n\alpha^n$ for some $a_i \in F$, $i = 0, 1, \ldots, n$, $n \in \mathbf{N}$. Consequently, $0 = -1 + a_0\alpha + a_1\alpha^2 + \cdots + a_n\alpha^{n+1} \Rightarrow \alpha$ is a root of the non-null polynomial $f(x) = -1 + a_0x + a_1x^2 + \cdots + a_nx^{n+1} \in F[x] \Rightarrow \alpha$ is algebraic over F. $\qquad\square$

Example 11.1.5 (i) The element i of $\mathbf{C}$ is algebraic over $\mathbf{R}$ and $x^2 + 1$ is the minimal polynomial of i over $\mathbf{R}$. Again $x - i$ is the minimal polynomial of i over $\mathbf{C}$.

(ii) The element $\sqrt{5}$ of $\mathbf{Q}(\sqrt{5})$ is algebraic over $\mathbf{Q}$ and $x^2 - 5$ is the minimal polynomial of $\sqrt{5}$ over $\mathbf{Q}$. The degree of $\sqrt{5}$ over $\mathbf{Q}$ is 2.

Theorem 11.1.4 *Let K be a field extension of the field F and the element α of K is algebraic over F. Suppose $f(x)$ is the minimal polynomial of α over F.*

(a) *If a polynomial $g(x)$ of $F[x]$ has a root α, then $f(x)$ divides $g(x)$ in $F[x]$.*
(b) *$f(x)$ is the monic polynomial of smallest degree in $F[x]$ such that α is a root of $f(x)$.*

Proof (a) follows from the Corollary to Theorem 11.1.3.

(b) If $f(x)$ is not of the smallest degree polynomial in $F[x]$, of which α is a root, then there exists a monic polynomial $g(x)$ in $F[x]$ such that α is a root of $g(x)$ and $\deg g(x) < \deg f(x)$. This gives a contradiction. $\qquad\square$

Definition 11.1.8 Let L and K be two field extensions of the same field F. An isomorphism $\psi : L \to K$ of fields, whose restriction to F is the identity homomorphism, is called an isomorphism of field extensions or an F-isomorphism. Two field extensions L and K of the same field F are said to be isomorphic field extensions iff there exists an F-isomorphism $\psi : L \to K$.

Proposition 11.1.2 *Let L and K be two field extensions of the same field F and $\psi : L \to K$ be an F-isomorphism. Suppose $f(x) \in F[x]$ and α is a root of $f(x)$ in L. If $\beta = \psi(\alpha) \in K$, then β is also a root of $f(x)$.*

Proof If $f(x) = a_0 + a_1x + \cdots + a_nx^n \in F[x]$, then $\psi(a_i) = a_i$ and $\psi(\alpha) = \beta$. Again $f(\alpha) = 0$ gives $0 = \psi(0) = \psi(f(\alpha)) = \psi(a_0 + a_1\alpha + \cdots + a_n\alpha^n) = \psi(a_0) + \psi(a_1)\psi(\alpha) + \cdots + \psi(a_n)(\psi(\alpha))^n$, since ψ is a homomorphism. This shows that $0 = a_0 + a_1\beta + \cdots + a_n\beta^n$. Hence $\beta = \psi(\alpha)$ is a root of $f(x)$. $\qquad\square$

Theorem 11.1.5 *Let F be a field and L, K be two field extensions of F. If $\alpha \in L$ and $\beta \in K$ are both algebraic elements over F, then there is an isomorphism $\psi : F(\alpha) \to F(\beta)$ of fields, such that $\psi(\alpha) = \beta$ and $\psi|_F = $ identity, iff α and β have the same minimal polynomial over F.*

Proof Suppose that $f(x)$ is the minimal polynomial of both α and β over F. Then by Theorem 11.1.3, there exist two isomorphisms $\mu : F[x]/\langle f(x)\rangle \to F(\alpha)$ and $\lambda :$

$F[x]/\langle f(x)\rangle \to F(\beta)$. Hence the composite isomorphism $\psi = \lambda \circ \mu^{-1} : F(\alpha) \overset{\sim}{\to} F(\beta)$ shows that $F(\alpha)$ and $F(\beta)$ are isomorphic under ψ.

Conversely, suppose $\psi : F(\alpha) \to F(\beta)$ is an isomorphism such that $\psi(\alpha) = \beta$ and $\psi|_F = $ identity. If $f(x) \in F[x]$ is a polynomial such that $f(\alpha) = 0$, then $\psi(\alpha) = \beta$ is also a root of $f(x)$ by Proposition 11.1.2. Hence α and β have the same minimal polynomial. $\square$

Example 11.1.6 If $\omega = e^{2\pi/3}$ is a complex cube root of 1 and α is a real root of the polynomial $f(x) = x^3 - 3$ in $\mathbf{Q}[x]$, then $f(x)$ is the minimal polynomial of both α and $\omega\alpha$ over $\mathbf{Q}$. Hence by Theorem 11.1.5 there exists a $\mathbf{Q}$-isomorphism $\psi : \mathbf{Q}(\alpha) \to \mathbf{Q}(\omega\alpha)$, such that $\psi(\alpha) = \omega\alpha$. Note that the elements of $\mathbf{Q}(\alpha)$ are real numbers, but the elements of $\mathbf{Q}(\omega\alpha)$ are not.

Let F be a field and K be a field extension of F. Then we may consider K as a vector space over F by using the field operations.

Definition 11.1.9 Let F be a field and K a field extension of F. Then the dimension of the vector space K over F is called the degree or dimension of K over F and is denoted by $[K : F]$. If $[K : F]$ is finite, then K is called a finite extension over F (or K is said to be finite over F) or K/F is called a finite extension. Otherwise, K is said to be an infinite extension of F.

Example 11.1.7 (i) $\mathbf{C}$ is a finite extension over $\mathbf{R}$. This is so because $\{1, i\}$ forms a basis of $\mathbf{C}$ over $\mathbf{R}$ and hence $[\mathbf{C} : \mathbf{R}] = 2$.

(ii) Let $K = \mathbf{Q}(\{\sqrt{q} : q$ is a prime integer$\}) \subset \mathbf{R}$. Then $[K : \mathbf{Q}]$ is not finite.

[*Hint.* As q is a prime integer, $\sqrt{q}$ is not an element of $\mathbf{Q}$. Let us assume that $p_1, p_2, \ldots, p_n$ be distinct prime integers (each is different from q) such that $\sqrt{q} \notin \mathbf{Q}(\sqrt{p_1}, \sqrt{p_2}, \ldots, \sqrt{p_n})$, which is our induction hypothesis. We claim that $\sqrt{q} \notin \mathbf{Q}(\sqrt{p_1}, \sqrt{p_2}, \ldots, \sqrt{p_n}, \sqrt{p_{n+1}})$, where $p_1, p_2, \ldots, p_n, p_{n+1}$ are distinct prime integers such that each is different from q. As $\sqrt{q} \notin \mathbf{Q}$, the induction hypothesis is true for $n = 0$. If $\sqrt{q} \in \mathbf{Q}(\sqrt{p_1}, \sqrt{p_2}, \ldots, \sqrt{p_n}, \sqrt{p_{n+1}})$, then there exist elements $x, y \in \mathbf{Q}(\sqrt{p_1}, \ldots, \sqrt{p_n})$ such that $\sqrt{q} = x + y\sqrt{p_{n+1}}$. As $y = 0$ contradicts the induction hypothesis, it follows that $y \neq 0$. If $x = 0$, then $q = y^2 p_{n+1}$ gives also a contradiction, since q and p_{n+1} are distinct primes. Again for $x \neq 0$ and $y \neq 0$, $q = x^2 + 2xy\sqrt{p_{n+1}} + p_{n+1}y^2 \Rightarrow \sqrt{p_{n+1}} = (q - x^2 - p_{n+1}y^2)/2xy \in \mathbf{Q}(\sqrt{p_1}, \sqrt{p_2}, \ldots, \sqrt{p_n}) \Rightarrow \sqrt{q} \in \mathbf{Q}(\sqrt{p_1}, \sqrt{p_2}, \ldots, \sqrt{p_n})$. This gives a contradiction of hypothesis again. Thus by the induction hypothesis, we find that for any positive integer r, if $p_1, p_2, \ldots, p_r, q$ are distinct prime integers, then $\sqrt{q} \notin \mathbf{Q}(\sqrt{p_1}, \sqrt{p_2}, \ldots, \sqrt{p_r})$. Consequently, we get a strictly infinite ascending chain of fields such that $\mathbf{Q} \subsetneq \mathbf{Q}(\sqrt{2}) \subsetneq \mathbf{Q}(\sqrt{2}, \sqrt{3}) \subsetneq \cdots$. This shows that $[K : \mathbf{Q}]$ is not finite. Clearly, $[\mathbf{R} : \mathbf{Q}]$ is not finite.]

Theorem 11.1.6 *Let K be a finite field extension of the field F. If $[K : F] = n$, then*

(a) *there exists a basis of K over F, consisting of n elements of K and*
(b) *any set of $(n + 1)$ (or more) elements of K is always linearly dependent over F.*

Proof As K is an n-dimensional vector space over F, the theorem follows from the properties of finite dimensional vector spaces. $\qquad\square$

Corollary 1 *The degree of an algebraic element α over a field F is equal to the dimension of the extension field $F(\alpha)$, regarded as a vector space over F, with a basis $B = \{1, \alpha, \alpha^2, \ldots, \alpha^{n-1}\}$ if $[F(\alpha) : F] = n$.*

Corollary 2 *Let α and β be two algebraic elements over the same field F such that $F(\alpha) = F(\beta)$. Then α and β have the same degree over F.*

Theorem 11.1.7 *Let K be a finite field extension of the field F and an element α of K be algebraic over F. If n is the degree of the minimal polynomial $f(x)$ of α over F, then $[F(\alpha) : F] = n$.*

Proof It is sufficient to prove that the set $S = \{1, \alpha, \ldots, \alpha^{n-1}\}$ forms a basis of the vector space $F(\alpha)$ over F. Suppose $a_0 + a_1\alpha + \cdots + a_{n-1}\alpha^{n-1} = 0$, where each $a_i \in F$. Then α is a root of the polynomial $g(x) = a_0 + a_1x + \cdots + a_{n-1}x^{n-1} \in F[x]$. Hence $f(x)$ divides $g(x)$ in $F[x]$ by Theorem 11.1.4(a). But this is possible only when $a_0 = a_1 = \cdots = a_{n-1} = 0$, since $\deg g(x) < \deg f(x) = n$. This concludes that S is linearly independent over F. We now claim that the set S generates the vector space $F(\alpha)$. Let $h(\alpha) \in F(\alpha)$. As α is algebraic over F, $F(\alpha) = F[\alpha]$ (see Proposition 11.1.1). Hence $h(\alpha) \in F[\alpha]$. Let $h(x)$ be the polynomial in $F[x]$ corresponding to $h(\alpha)$ in $F[\alpha]$. Again F is a field $\Rightarrow F[x]$ is a Euclidean domain $\Rightarrow$ there exist polynomials $q(x)$ and $r(x)$ in $F[x]$ such that $h(x) = f(x)q(x) + r(x)$, where $r(x) = 0$ or $\deg r(x) < \deg f(x)$. Hence for $h(\alpha) = f(\alpha)g(\alpha) + r(\alpha)$, either $h(\alpha) = 0$ or $h(\alpha) = r(\alpha)$ is a linear combination of elements of S, since $\deg r(x) < n$. Thus S forms a basis of $F(\alpha)$ over K and hence $[F(\alpha) : F] = n$. $\qquad\square$

Theorem 11.1.8 *Let K be a finite field extension of the field F. If $S = \{\alpha_1, \alpha_2, \ldots, \alpha_n\} \subset K$ forms a basis of the vector space K over F, then $K = F(\alpha_1, \alpha_2, \ldots, \alpha_n)$.*

Proof Clearly, $F(\alpha_1, \alpha_2, \ldots, \alpha_n) \subseteq K$, since $F(\alpha_1, \alpha_2, \ldots, \alpha_n)$ is the smallest field containing F and S. For the reverse inclusion, let $\alpha \in K$. As S forms a basis of K, there exist elements $a_1, a_2, \ldots, a_n$ in F such that $\alpha = a_1\alpha_1 + a_2\alpha_2 + \cdots + a_n\alpha_n \in F(\alpha_1, \alpha_2, \ldots, \alpha_n)$. Thus $K \subseteq F(\alpha_1, \alpha_2, \ldots, \alpha_n)$ and hence $K = F(\alpha_1, \alpha_2, \ldots, \alpha_n)$. $\qquad\square$

We now prove a relation between a finite field extension and an algebraic extension.

Theorem 11.1.9 *Every finite field extension of F is an algebraic extension.*

Proof Let K be a finite field extension of the field F. Suppose $[K : F] = n$ and α is a non-zero element of K. Then the $(n + 1)$ elements $1, \alpha, \alpha^2, \ldots, \alpha^n$ of the vector space K over F are linearly dependent. Hence there exist elements $a_0, a_1, \ldots, a_n$ (not all zero) in F such that $a_0 + a_1\alpha + \cdots + a_n\alpha^n = 0$. This shows that α is a root of the non-null polynomial $a_0 + a_1x + \cdots + a_nx^n \in F[x]$. This implies α and hence K is algebraic over F.   □

Corollary *If K is a finite field extension of F, then every element of K is algebraic over F.*

Remark The converse of Theorem 11.1.9 is not true in general. For example, the field K defined in Example 11.1.7(ii) is an algebraic field extension of $\mathbf{Q}$ but K is not a finite field extension of $\mathbf{Q}$. This is so because for $\alpha \in K$, there exist prime integers $p_1, p_2, \ldots, p_n$ such that $\alpha \in \mathbf{Q}(\sqrt{p_1}, \sqrt{p_2}, \ldots, \sqrt{p_n})$. But $\mathbf{Q}(\sqrt{p_1}, \sqrt{p_2}, \ldots, \sqrt{p_n})$ is a finite extension of $\mathbf{Q}$ implies that α is algebraic over $\mathbf{Q}$. Thus K is algebraic over $\mathbf{Q}$ but $[K : \mathbf{Q}]$ is not finite. The partial converse of Theorem 11.1.9 is true for the following particular case.

Theorem 11.1.10 *The field extension $K(\alpha)$ over K is finite iff α is algebraic over K.*

Proof If $K(\alpha)$ is a finite field extension over K, then by the Corollary to Theorem 11.1.9, it follows that α is algebraic over K. Conversely, let α be algebraic over K. Then $K(\alpha)$ is an extension over K implies that $[K(\alpha) : K] = n$, where n is the degree of the minimal polynomial $f(x)$ of α over K. This proves that $K(\alpha)$ is a finite extension over K.   □

Corollary *Every element of a simple algebraic extension $F(\alpha)$ is algebraic over F.*

Remark A transcendental element cannot appear in a simple algebraic extension.

Definition 11.1.10 Let K be a field extension of the field F. A subfield L of K is said to be an intermediate field of K/F iff $F \subseteq L \subseteq K$ holds and this intermediate field L is said to be proper iff $L \neq K$ and $L \neq F$.

Example 11.1.8 Let K be a field extension of a field F and α be an element of K. Then the field $F(\alpha)$ generated by F and α is an intermediate field, because $F \subseteq F(\alpha) \subseteq K$.

Remark Let L be an intermediate field of K/F. Then

(a) K is a vector space over F and L is a subspace of K;
(b) K/F is an algebraic extension iff K/L and L/F are both algebraic extensions.

Theorem 11.1.11 *Let K be a finite field extension of the field F and L be an intermediate field of K/F. Then $[K : F] = [K : L][L : F]$.*

Proof Let V be a basis of the vector space K over L and U be a basis of the vector space L over F. Suppose $[L : F] = m$ and $[K : L] = n$. Let $V = \{v_1, v_2, \ldots, v_n\}$ and $U = \{u_1, u_2, \ldots, u_m\}$. Construct $W = \{uv : u \in U, v \in V\}$. Then W forms a basis of K over F. Finally, card $W = nm$ proves the theorem. $\square$

Remark If either card V or card U is infinite, then card W is also infinite.

Corollary 1 *If K is a field extension of the field F and each of the elements $\alpha_1, \alpha_2, \ldots, \alpha_n$ of K are algebraic over F, then $K(\alpha_1, \alpha_2, \ldots, \alpha_n)$ is a finite extension of F.*

Proof The corollary follows by induction on n. $\square$

Corollary 2 *If $V = \{v_1, v_2, \ldots, v_n\}$ is a basis of a finite field extension K of L and $U = \{u_1, u_2, \ldots, u_m\}$ is a basis of a finite field extension L of F, then $W = \{uv : u \in U, v \in V\}$, consisting of mn elements, forms a basis of K over F.*

Corollary 3 *Let F be a field and K be a field extension of F such that $[K : F] = n$. If an element α of K is algebraic over F, then the degree of α over F divides n.*

Proof Consider the tower of fields: $F \subset F(\alpha) \subset K$. Then $n = [K : F] = [K : F(\alpha)][F(\alpha) : F]$. As α is algebraic over F, then $[F(\alpha) : F]$ is the degree of α over F. Hence the corollary follows. $\square$

Corollary 4 *Let F be a field and K be a field extension of F of prime degree p. If α is an element of K but not in F, then α has the degree p over F and $K = F(\alpha)$.*

Proof $[K : F] = p \Rightarrow [K : F(\alpha)][F(\alpha) : F] = p$. Again $\alpha \notin F \Rightarrow [F(\alpha) : F] \neq 1 \Rightarrow [K : F(\alpha)] = 1$ and $[F(\alpha) : F] = p$. The corollary follows from Ex. 1 of Exercise-I. $\square$

Corollary 5 *Let $F \subset L \subset K$ be a tower of fields such that K is algebraic over L and L is algebraic over F. Then K is algebraic over F.*

Proof It is sufficient to show that every element $\alpha \in K$ is algebraic over F. By hypothesis, K is algebraic over $L \Rightarrow \exists a_0, a_1, \ldots, a_{n-1}$ (not all zero) $\in L$ such that $\alpha^n + a_{n-1}\alpha^{n-1} + \cdots + a_1\alpha + a_0 = 0 \Rightarrow \alpha$ is algebraic over the field $F(a_0, a_1, \ldots, a_{n-1})$. Consider the tower of fields: $F \subset F(a_0) \subset F(a_0, a_1) \subset \cdots \subset F(a_0, a_1, \ldots, a_{n-1}) \subset F(a_0, a_1, \ldots, a_{n-1}, \alpha)$. Clearly, each extension of the above tower is finite. Hence by Theorem 11.1.11, the degree of $F(a_0, a_1, \ldots, a_{n-1}, \alpha)$ over F is finite. Then it follows that α is algebraic over F. Hence K is algebraic over F. $\square$

The structure of a simple transcendental extension is given in the next theorem.

Theorem 11.1.12 *If α is transcendental over a field F, then the field $F(\alpha)$ generated by F and α is isomorphic to the field $F(x)$ of all rational functions in x over F.*

Proof The extension $F(\alpha)$ is given by $F(\alpha) = \{f(\alpha)/g(\alpha) : f(x), g(x) \in F[x]$ and $g(\alpha) \neq 0\}$, using Corollary to Theorem 11.1.1. If two polynomial expressions $f_1(\alpha)$ and $f_2(\alpha)$ are equal in $F(\alpha)$, their coefficients must be equal term by term, otherwise, the difference $f_1(\alpha) - f_2(\alpha)$ will give a polynomial equation in α with coefficients, not all zero. This contradicts our assumption that α is transcendental over F. This shows that the map $\psi : F[\alpha] \to F[x]$, $f(\alpha) \mapsto f(x)$ is a bijection and hence it is an isomorphism under usual operations of polynomials. This isomorphism gives rise to an isomorphism $F(\alpha) \to F(x)$, $f(\alpha)/g(\alpha) \mapsto f(x)/g(x)$. $\qquad \square$

11.1.1 Exercises

Exercises-I

1. Let K be a field extension of the field F. Prove that $[K : F] = 1$ iff $K = F$.

   [*Hint.* $[K : F] = 1 \Rightarrow K$ is a vector space over F of dimension 1 with $\{1\}$ as a basis. Now $x \in K \Rightarrow x = a.1$ for some $a \in F \Rightarrow K \subseteq F$. Then $F \subseteq K$ and $K \subseteq F \Rightarrow K = F$. Conversely, if $K = F$, then $\{1\}$ is a basis of the vector space K over F. Hence $[K : F] = 1$.]

2. Let K be a field extension of the field F such that $[K : F]$ is finite. If $f(x)$ is an irreducible polynomial in $F[x]$ such that $f(\alpha) = 0$ for some $\alpha \in K$, then prove that $\deg f(x)$ divides $[K : F]$.

   [*Hint.* Suppose $\deg f(x) = n$. Then $[F(\alpha) : F] = n \Rightarrow [K : F] = [K : F(\alpha)][F(\alpha) : F] = [K : F(\alpha)] \cdot n$.]

3. Find $[\mathbf{Q}(\sqrt[3]{5}) : \mathbf{Q}]$.

   [*Hint.* $\sqrt[3]{5}$ has the minimal polynomial $x^3 - 5$ over $\mathbf{Q}$. Hence $[\mathbf{Q}(\sqrt[3]{5}) : \mathbf{Q}] = 3$.]

4. Let K be a field extension of the field F and $a, b \in K$ be algebraic over F. If a has degree m over F and $b \neq 0$ has degree n over F, then show that the elements $a + b$, ab, $a - b$ and ab^{-1} are algebraic over F and each has at most degree mn over F.

5. Let K be a field extension of F such that $[K : F] = p$, where p is a prime integer. Then show that K/F has no proper intermediate field.

   [*Hint.* Suppose K/F has an intermediate field L. Then $[K : F] = [K : L][L : F] \Rightarrow [L : F]$ divides $p \Rightarrow$ either $[L : F] = 1$ or $[L : F] = p \Rightarrow$ either $L = F$ or $L = K$.]

6. Let K be a field extension of F and L be the set of all elements of K which are algebraic over F. Then show that L is an intermediate field of K/F.

7. Let K be a field extension of F and L be an intermediate field of K/F. Show that L is an algebraic extension iff both K/F and L/F are algebraic extensions.

8. Let K be a field extension of the field F and R be a ring such that $F \subseteq R \subseteq K$. If every element of R is algebraic over F, then prove that R is a field.

   [*Hint.* Let a be a non-zero element of R and $f(x) = x^n + \cdots + a_1 x + a_0$ be its minimal polynomial over F. Then $a^n + \cdots + a_1 a + a_0 = 0$. If $a_0 = 0$, we have a contradiction, since the minimal polynomial of α over F has degree n. Again if $a_0 \neq 0$, then a^{-1} exists.]

9. Show that $\mathbf{Q}(\sqrt{5}, -\sqrt{5}) = \mathbf{Q}(\sqrt{5})$.

   [*Hint.* $\mathbf{Q}(\sqrt{5}, -\sqrt{5})$ is the smallest field containing $\sqrt{5}, -\sqrt{5}$ and $\mathbf{Q}$. Again $\sqrt{5} \in \mathbf{Q}(\sqrt{5}) \Rightarrow -\sqrt{5} \in \mathbf{Q}(\sqrt{5})$. Moreover, $\mathbf{Q} \subset \mathbf{Q}(\sqrt{5}), \mathbf{Q}(\sqrt{5}, -\sqrt{5}) \subseteq \mathbf{Q}(\sqrt{5})$ and $\sqrt{5} \in \mathbf{Q}(\sqrt{5}, -\sqrt{5}) \Rightarrow \mathbf{Q}(\sqrt{5})$ is the smallest field containing $\mathbf{Q}, \sqrt{5}$ and $-\sqrt{5}$.]

10. Show that $x^2 - 5$ is irreducible over $\mathbf{Q}(\sqrt{2})$.

    [*Hint.* Suppose $x^2 - 5$ is not irreducible over $\mathbf{Q}(\sqrt{2})$. Then $x^2 - 5 = (x - a)(x - b)$ for some $a, b \in \mathbf{Q}(\sqrt{2})$. Hence $a + b = 0$ and $ab = -5 \Rightarrow \sqrt{5} \in \mathbf{Q}(\sqrt{2})$. This is a contradiction.]

11. Let K be finite extension of the field F. If $\alpha, \beta \in K$ are algebraic over F, then show that $\alpha + \beta, \alpha\beta$ and α^{-1} ($\alpha \neq 0$) are also algebraic over F.

    [*Hint.* Clearly, $F(\alpha, \beta)$ is a finite extension of F shows that $F(\alpha, \beta)$ is an algebraic extension of F.]

12. (a) If α is transcendental over a field F, then show that

    (i) the map $\mu : F[x] \to F[\alpha], f(x) \mapsto f(\alpha)$ is an isomorphism;
    (ii) $F(\alpha)$ is isomorphic to the field $F(x)$ of rational functions over F in the indeterminate x.

    (b) Show that the field $\mathbf{Q}(\pi)$ is isomorphic to the field $\mathbf{Q}(x)$ of rational functions over $\mathbf{Q}$.

    [*Hint.* See Theorem 11.1.12.]

13. For the field extension $\mathbf{R}$ of $\mathbf{Q}$, show that π is transcendental over $\mathbf{Q}(\sqrt{2})$.

14. Let K be a field extension of F and L be also a field extension of F. If $\alpha \in K$ and $\beta \in L$ are both algebraic over F, then show that there is an isomorphism of fields: $\mu : F(\alpha) \to F(\beta)$, which is the identity on the subfield F and which maps $\alpha \mapsto \beta$ iff the monic irreducible polynomials for α and β over F are the same.

15. If $f(x)$ is an irreducible polynomial of degree n over a field F, then show that there is an extension K of F such that $[K : F] = n$ and $f(x)$ has a root in K.

16. Let K be a field extension of the field F and α, β be two algebraic elements of K over F having degrees m and n respectively. If $\gcd(m, n) = 1$, show that $[F(\alpha, \beta) : F] = mn$.

17. Show that every irreducible polynomial in $\mathbf{R}[x]$ has degree 1 or 2.

    [*Hint.* Let $f(x)$ be an irreducible polynomial in $\mathbf{R}[x]$. Then $f(x)$ has a root in $\mathbf{C}$ by s of Algebra. Since $[\mathbf{C} : \mathbf{R}] = 2$, the degree of α over $\mathbf{R}$ divides 2.]

11.2 Finite Fields

Given a prime integer p and a positive integer n, we have shown in Chap. 6 the existence of a finite field with p^n elements.

Finite fields form an important class of fields. A field having finitely many elements is called a finite field. If F is a finite field, the kernel of the homomorphism $\psi : \mathbf{Z} \to F, n \mapsto n 1_F$ is a prime ideal. Then $\ker \psi \neq \{0\}$, because F is finite but $\mathbf{Z}$ is infinite. Thus $\ker \psi = \langle p \rangle$, generated by a prime integer p and $\psi(\mathbf{Z})$ is isomorphic to the quotient field $\mathbf{Z}/\langle p \rangle = \mathbf{Z}_p$. So F contains a subfield K isomorphic to the prime field $\mathbf{Z}_p$. Hence F can be considered as an extension of $\mathbf{Z}_p$.

In this section we study finite fields. Throughout the section p denotes a prime integer.

Definition 11.2.1 A field having finitely many elements is called a finite field or Galois field.

Let K be a finite field. If char $K = p$, then by Example 11.1.1(iv), K may be considered as a finite dimensional vector space over $\mathbf{Z}_p$. Then the dimension of K over $\mathbf{Z}_p$ is finite and it is denoted by $[K : \mathbf{Z}_p]$.

Theorem 11.2.1 *Let K be a finite field of characteristic p and $[K : \mathbf{Z}_p] = n$. Then K contains exactly p^n elements.*

Proof Let $B = \{x_1, x_2, \ldots, x_n\}$ be a basis of K over $\mathbf{Z}_p$ and $x \in K$. Then x can be expressed as $x = a_1 x_1 + a_2 x_2 + \cdots + a_n x_n$, where $a_i \in \mathbf{Z}_p$. As $\mathbf{Z}_p$ has p elements, K has at most p^n elements. Since B is a basis of K, the elements $a_1 x_1 + a_2 x_2 + \cdots + a_n x_n$ are all distinct for every distinct choice of elements $a_1, a_2, \ldots, a_n$ of $\mathbf{Z}_p$. Thus K has exactly p^n elements. $\square$

Theorem 11.2.2 *Let K be a finite field of characteristic p. Then every element of K is a root of the polynomial $x^{p^n} - x$ over $\mathbf{Z}_p$.*

Proof Let $[K : \mathbf{Z}_p] = n$. Then K is a finite field of p^n elements. Hence the multiplicative group $K \setminus \{0\}$ is of order $p^n - 1$. If y is a non-zero element of K, then $y^{p^n - 1} = 1$ and hence $y^{p^n} = y$. Moreover, $0^{p^n} = 0$. Thus every element of K is a root of the polynomial $x^{p^n} - x$ over $\mathbf{Z}_p$. $\square$

To describe finite fields, we need the concept of splitting fields.

Definition 11.2.2 Let F be a field. A polynomial $f(x)$ in $F[x]$ is said to split over a field K containing F iff $f(x)$ can be factored as a product of linear factors in $K[x]$. A field K containing F is said to be a splitting field for $f(x)$ over F iff $f(x)$ splits over K and there is no proper intermediate field L of K/F (i.e., if L is a subfield of K such that $F \subseteq L \subseteq K$, then either $L = F$ or $L = K$).

Remark 1 Let F be a field and $f(x)$ be a polynomial in $F[x]$ of positive degree n. Then an extension K of F is a splitting field of $f(x)$ such that

(i) $f(x)$ factors into linear factors in $K[x]$ as $f(x) = a(x-\alpha_1)(x-\alpha_2)\cdots(x-\alpha_n)$, with $\alpha_i \in K$ and $a \in F$;
(ii) K is generated by F and the roots of $f(x)$, i.e., $K = F(\alpha_1, \alpha_2, \ldots, \alpha_n)$.

The second condition shows that K is the smallest extension of F which contains all the roots of $f(x)$.

Remark 2 If F is a field, then every polynomial $f(x)$ in $F[x]$ of positive degree has a splitting field over F.

Remark 3 Let F be a field and $f(x) \in F[x]$ is of positive degree. Then any two splitting fields of $f(x)$ are isomorphic and hence the splitting field of $f(x)$ is unique up to isomorphism.

Example 11.2.1 (i) $\mathbf{C}$ is the splitting field of the polynomial $x^2 + 1$ over $\mathbf{R}$ but $\mathbf{R}$ is not splitting field of $x^2 + 1$ over $\mathbf{Q}$.

(ii) Let K be a finite field of characteristic p. Then K is the splitting field of $x^{p^n} - x$ over $\mathbf{Z}_p$. This is so because if S is the splitting field of $x^{p^n} - x$ over $\mathbf{Z}_p$, then S contains all the roots of the polynomial $x^{p^n} - x$ over $\mathbf{Z}_p$. Hence $S \subseteq K$. Again by Theorem 11.2.2, $K \subseteq S$. Consequently $K = S$.

11.2.1 Exercises

Exercises-II

1. Show that any two finite fields with same number of elements are isomorphic.
   [*Hint.* Let K and S be two finite fields containing p^n elements, where p is prime and n is a positive integer. Then K and S are both the splitting fields of the same polynomial $x^{p^n} - x$ over $\mathbf{Z}_p$ and hence they are isomorphic.]
2. Corresponding to a given prime integer p, and a positive integer n, show that there exists a finite field consisting of $p^n = q$ roots of $x^q - x$ over $\mathbf{Z}_p$, which is determined uniquely up to an isomorphism and is denoted by $\mathbf{F}_{p^n}$. This field is sometimes called the Galois field $GF(p^n)$.
3. Prove that $\mathbf{F}_{p^n}$ is a subfield of $\mathbf{F}_{p^m}$ iff n is a divisor of m.
   [*Hint.* Let H be a subfield of $\mathbf{F}_{p^m}$. Then H is a finite extension of $\mathbf{Z}_p$ and $H = \mathbf{F}_{p^n}$, where $n = [H : \mathbf{Z}_p]$. Thus $m = [\mathbf{F}_{p^m} : \mathbf{Z}_p] = [\mathbf{F}_{p^m} : H][H : \mathbf{Z}_p]$. Conversely, if n is a divisor of m, then $(p^n - 1)|(p^m - 1)$ and hence $(x^{p^n} - x)|(x^{p^m} - x)$. Consequently, all the roots of $x^{p^n} - x$, over $\mathbf{Z}_p$, are contained among the roots of $x^{p^m} - x$ over $\mathbf{Z}_p$.]

4. Let F be a field and G be a finite multiplicative subgroup of $F^* = F \setminus \{0\}$. Then show that G is cyclic and G consists of all the nth roots of unity in F, where $|G| = n$. (A generator of G is called a primitive element for F.)

   [*Hint*. Use the Structure Theorem for abelian groups or see Chap. 10.]

5. Let F be a finite field. For the finite extension $F(a, b)$ of F with a, b algebraic over F, show that there exists an element c in $F(a, b)$ such that $F(a, b) = F(c)$, i.e., $F(a, b)$ is a simple extension of F.

6. Let F be a field of p^n elements, where p is prime and n is a positive integer. Then show that $[F : \mathbf{Z}_p] = n$.

   [*Hint*. Suppose $[F : \mathbf{Z}_p] = m$. Then by Theorem 11.2.1, F has exactly p^m elements. Hence $p^m = p^n \Rightarrow m = n$.]

7. Let F be a finite field of characteristic p. Then prove that

   (a) if c is a primitive element for F, then c^p is also;
   (b) if $n = [F : \mathbf{Z}_p]$, then there exists an element c in F such that c is algebraic over $\mathbf{Z}_p$, of degree n and $F = \mathbf{Z}_p(c)$.

8. Every finite field F of characteristic p has an automorphism $\psi : F \to F$, $x \mapsto x^p$. Find the order of ψ in the automorphism group of F. If F is not finite, examine the validity of the above result.

   [*Hint*. For the field $F = \mathbf{Z}_p(x)$, the field of rational functions, the map $\psi : F \to F, x \mapsto x^p$ is not an automorphism.]

9. Let F be a finite field and K a finite field extension of F. Then prove that the number of elements of K is some power of the number of elements of F.

   [*Hint*. If char $F = p > 0$, then F has $p^m = q$ (say) elements. Again if $[K : F] = n$, then K has q^n elements.]

11.3 Algebraic Numbers

An algebraic number is a complex number which is algebraic over the field $\mathbf{Q}$ of rational numbers. In this section we shall discuss the basic properties of algebraic numbers. The theory of algebraic numbers was born and developed through the attempts to prove Fermat's Last Theorem. In this section we study the properties of algebraic numbers, such as the set $\mathcal{A}$ of algebraic numbers is countable and forms a field, called the algebraic number field which is algebraically closed. We prove the fundamental theorem of algebra and show the existence of real transcendental numbers.

Definition 11.3.1 A complex number α which is algebraic over $\mathbf{Q}$ (i.e., α is a root of some non-null polynomial over $\mathbf{Q}$) is called an algebraic number. Otherwise, it is said to be transcendental.

Example 11.3.1 (i) Every rational number q is an algebraic number. This is so because q satisfies the equation $x - q = 0$ over $\mathbf{Q}$.

(ii) $1/\sqrt{5}$ is an algebraic number. This is so because it satisfies the equation $x^2 - 1/5 = 0$ over $\mathbf{Q}$.

(iii) e and π are transcendental numbers [see Hardy and Wright (2008, pp. 218–227)].

Theorem 11.3.1 *An algebraic number α satisfies a unique monic irreducible polynomial equation $f(x) = 0$ over $\mathbf{Q}$. Moreover, every polynomial equation $g(x) = 0$ over $\mathbf{Q}$ satisfied by α is divisible by $f(x)$ in $\mathbf{Q}[x]$.*

Proof Let $\mathbf{S}$ be the set of all polynomial equations over $\mathbf{Q}$ satisfied by the given algebraic number α and $h(x) = 0$ be one of the lowest degree polynomial equations in $\mathbf{S}$. If the leading coefficient of $h(x)$ is q, define $f(x)$ by $f(x) = q^{-1}h(x)$. Hence $f(\alpha) = 0$ and $f(x)$ is monic. We claim that $f(x)$ is irreducible in $\mathbf{Q}[x]$. Suppose $f(x) = f_1(x)f_2(x)$ in $\mathbf{Q}[x]$. Then at least one of $f_1(\alpha) = 0$ and $f_2(\alpha) = 0$ holds. This implies a contradiction, because $f(x) = 0$ is a polynomial equation of lowest degree in $\mathbf{S}$ having α as a root. Again, since $\mathbf{Q}$ is a field, $\mathbf{Q}[x]$ is Euclidean domain. Hence for the polynomials $f(x)$ and $g(x)$, there exist polynomials $q(x)$ and $r(x)$ in $\mathbf{Q}[x]$ such that $g(x) = f(x)q(x) + r(x)$, where $r(x) = 0$ or $\deg r(x) < \deg f(x)$. Then $r(x)$ must be identically zero, otherwise, the degree of $r(x)$ would be less than the degree of $f(x)$ and α would be a root of the polynomial equation $r(x) = 0$. Consequently, $f(x)$ divides $g(x)$ in $\mathbf{Q}[x]$. To prove the uniqueness of $f(x)$, suppose $m(x)$ is an irreducible monic polynomial over $\mathbf{Q}$ such that $m(\alpha) = 0$. Then $f(x)$ divides $m(x)$. Hence there exists some polynomial $n(x)$ in $\mathbf{Q}[x]$ such that $m(x) = f(x)n(x)$. Since $m(x)$ is irreducible in $\mathbf{Q}[x]$, $n(x)$ must be constant in $\mathbf{Q}[x]$. Again, since $f(x)$ and $m(x)$ are both monic polynomials, it follows that $n(x) = 1$. This proves the uniqueness of $f(x)$. □

This theorem leads to the following definition.

Definition 11.3.2 Let α be an algebraic number. Then the irreducible monic polynomial $f(x)$ over $\mathbf{Q}$ having α as a root is called the minimal polynomial of α and the degree of $f(x)$ is called the degree of α.

Example 11.3.2 $\sqrt{3}$ is an algebraic number and $f(x) = x^2 - 3$ is the minimal polynomial of $\sqrt{3}$ over $\mathbf{Q}$. The field extension $\mathbf{Q}(\sqrt{3})$ of $\mathbf{Q}$ is of degree 2 with a basis $B = \{1, \sqrt{3}\}$.

Theorem 11.3.2 *Let F be a finite field extension of $\mathbf{Q}$ of degree n. Then every element α of F is an algebraic number and is a root of an irreducible polynomial over $\mathbf{Q}$, of degree $\leq n$.*

Proof Clearly, the set $S = \{1, \alpha, \alpha^2, \ldots, \alpha^{n-1}, \alpha^n\}$, consisting of $n + 1$ elements of the n-dimensional vector space F over $\mathbf{Q}$, is linearly dependent over $\mathbf{Q}$. This implies that α is algebraic over $\mathbf{Q}$ and hence α is an algebraic number. □

Corollary *Every element of a simple algebraic extension $\mathbf{Q}(\alpha)$ is an algebraic number.*

Definition 11.3.3 A field F is said to be algebraically closed (or complete) iff every polynomial in $F[x]$, with degree ≥ 1, has a root in F.

Two important algebraically closed fields, such as the field $\mathbf{C}$ of complex numbers and the field of algebraic numbers are discussed here.

First we show that $\mathbf{C}$ is algebraically closed by the Fundamental Theorem of Algebra. There are several proofs of the Fundamental Theorem of Algebra. We present one of them by using the concept of homotopy as discussed in Sect. 2.10 of Chap. 2.

Theorem 11.3.3 (Fundamental Theorem of Algebra) *Every non-constant polynomial with complex coefficients has a complex root.*

Proof To prove the theorem it is sufficient to prove that a non-constant polynomial

$$f(z) = z^n + a_{n-1}z^{n-1} + \cdots + a_1z + a_0, \quad a_i \in \mathbf{C} \tag{11.1}$$

has a root in $\mathbf{C}$. If $a_0 = 0$, then 0 is a root of $f(z)$. So we assume that $a_0 \neq 0$. We may consider f as a continuous function $f : \mathbf{C} \to \mathbf{C}$, defined by (11.1). Suppose $f(z)$ has no root in $\mathbf{C}$. Then $f(z)$ is never zero. Consider the unit circle $S^1 = \{z \in \mathbf{C} : |z| = 1\}$ in the complex plane. Define a continuous family of maps $f_t : S^1 \to S^1, z \mapsto f(z)/|f(tz)|$ for non-negative real numbers t. Then any two maps of this family are homotopic. For $t = 0$, f_0 is a constant map. On the other hand, for sufficiently large t, f_t is homotopic to g. where $g(z) = z^n$, because z^n is dominant in the expression of $f(z)$ for sufficiently large t. Hence f_0 cannot be homotopic to g, because their degrees are different. This contradiction concludes that $f(z)$ has a root in $\mathbf{C}$. $\square$

Corollary 1 *The field $\mathbf{C}$ of complex numbers is algebraically closed.*

Remark This corollary proves the algebraic completeness of the field of complex numbers.

Corollary 2 *The field $\mathbf{R}$ of real numbers is embedded in the algebraically closed field $\mathbf{C}$ of complex numbers.*

We now show that the set of algebraic numbers forms an algebraically closed field.

Theorem 11.3.4 *(a) The set $\mathcal{A}$ of algebraic numbers is a field, called the algebraic number field.*
(b) The field $\mathcal{A}$ is algebraically closed.

Proof (a) It is sufficient to show that the sum, product, difference and quotient of any two elements α and $\beta \neq 0$ in $\mathcal{A}$ are also in $\mathcal{A}$. Clearly, $\alpha + \beta$, $\alpha - \beta$, $\alpha\beta$ and $\alpha\beta^{-1}$ are in the subfield $\mathbf{Q}(\alpha, \beta)$ of $\mathbf{C}$, which is generated by $\mathbf{Q}$ and two elements

α and β of $\mathcal{A}$. Again α is algebraic over $\mathbf{Q} \Rightarrow \mathbf{Q}(\alpha)$ is a finite extension of $\mathbf{Q}$ and similarly β is algebraic over $\mathbf{Q}(\alpha) \Rightarrow \mathbf{Q}(\alpha, \beta)$ is a finite extension over $\mathbf{Q}(\alpha)$. Hence $\mathbf{Q}(\alpha, \beta)$ is a finite extension of $\mathbf{Q} \Rightarrow$ each element of $\mathbf{Q}(\alpha, \beta)$ is an algebraic number. This shows that $\mathcal{A}$ is a field.

(b) Let $f(x) = x^n + a_{n-1}x^{n-1} + \cdots + a_0$ be a polynomial in $\mathcal{A}[x]$. Then the coefficients $a_0, a_1, \ldots, a_{n-1}$ generate an extension $F = \mathbf{Q}(a_0, a_1, \ldots, a_{n-1})$. Clearly, F is a finite extension of the field $\mathbf{Q}$. Any complex root α of $f(x)$ is algebraic over the field F and hence $F(\alpha)$ is a finite extension of F. Consequently, $F(\alpha)$ is also a finite extension of $\mathbf{Q}$. Thus the element α of this extension is an algebraic number, in the field $\mathcal{A}$. This leads us to conclude that the field $\mathcal{A}$ is algebraically closed. $\square$

Corollary *The field* $\mathbf{Q}$ *of rational numbers is embedded in the field* $\mathcal{A}$ *of algebraically closed field* $\mathcal{A}$.

We now show the existence of transcendental numbers. It does not follow immediately that there are transcendental numbers, though almost all real numbers are transcendental. So we need an alternative definition of an algebraic number.

An algebraic number is a complex number α which satisfies an algebraic equation of the form $a_0x^n + a_1x^{n-1} + \cdots + a_n = 0$, where $a_0, a_1, \ldots, a_n$ are all integers, not all zero. A number which is not an algebraic number is called a transcendental number.

Theorem 11.3.5 *The set of all algebraic numbers is countable.*

Proof To prove this theorem we define the rank r of an equation $a_0x^n + a_1^{n-1} + \cdots + a_n = 0$, $a_i \in \mathbf{Z}$ and $a_0 \neq 0$ as $r = n + |a_0| + |a_1| + \cdots + |a_n|$. The minimum value of rank r is 2. As there exist only a finite number of such equations of rank r, we may write them as:

$$A_{r,1}, A_{r,2}, \ldots, A_{r,m_r}.$$

Arranging the equations in the sequence:

$$A_{2,1}, A_{2,2}, \ldots, A_{2,r_2}, A_{3,1}, A_{3,2}, \ldots, A_{3,r_3}, A_{4,1}, \ldots,$$

we can assign to them the integers $1, 2, 3, \ldots$. This shows that the aggregate of these equations is countable. Clearly, every algebraic number corresponds to at least one of these equations, and the number of algebraic numbers corresponding to at least one of these equations is finite. Hence the theorem follows. $\square$

Corollary *The set of all real algebraic numbers is countable and has measure zero.*

Existence of Transcendental Numbers The set of all real numbers is not countable. On the other hand, the set of all real algebraic numbers is countable. Hence there exist real numbers which are not algebraic. This shows the existence of real transcendental numbers. This leads us to conclude the following.

Theorem 11.3.6 *Almost all real numbers are transcendental.*

11.4 Exercises

Exercises-III

1. Show that the degree of an algebraic number α is equal to the dimension $n = [\mathbf{Q}(\alpha) : \mathbf{Q}]$, with a basis $B = \{1, \alpha, \ldots, \alpha^{n-1}\}$. (Here $\mathbf{Q}(\alpha)$ is regarded as a vector space over $\mathbf{Q}$.)
2. If two algebraic numbers α and β generate the same extension fields $\mathbf{Q}(\alpha)$ and $\mathbf{Q}(\beta)$, i.e., $\mathbf{Q}(\alpha) = \mathbf{Q}(\beta)$, then show that α and β have the same degree.
3. Let F be a field. Show that the following statements are equivalent:

    (a) F is algebraically closed;
    (b) every irreducible polynomial in $F[x]$ is of degree 1;
    (c) if a polynomial $f(x) \in F[x]$ is of degree ≥ 1, then $f(x)$ can be expressed as a product of linear factors in $F[x]$;
    (d) if K is an algebraic field extension of F, then $F = K$.

4. Let F be a finite field. Then show that F cannot be algebraically closed.

    We now introduce the concept of algebraic integers.

11.5 Algebraic Integers

The concept of algebraic integers is one of the most important discoveries in number theory. The ring of algebraic integers shares some properties of the ring of integers but it differs as regards many other properties.

Definition 11.5.1 A complex number α is called an algebraic integer iff α is a root of a monic irreducible polynomial $f(x)$ in $\mathbf{Q}[x]$ with integral coefficients of the form

$$f(x) = x^n + a_{n-1}x^{n-1} + \cdots + a_0.$$

Remark 1 An algebraic number α is an algebraic integer iff α satisfies some monic polynomial equation $f(x) = 0$ with integral coefficients.

Remark 2 While defining algebraic integers, some authors do not insert the condition of irreducibility of $f(x)$ in $\mathbf{Q}[x]$, because of Theorem 11.5.2. The irreducible equation satisfied by a rational number p/q is just the linear equation $x - p/q = 0$. Thus a rational number is an algebraic integer iff it is an integer in the ordinary sense. Such an integer in $\mathbf{Z}$ is called a rational integer to distinguish it from other algebraic integers.

Theorem 11.5.1 *Every algebraic number is of the form α/β, where α is an algebraic integer and β is a non-zero integer in $\mathbf{Z}$.*

Proof Let γ be an algebraic number. Then γ is a root of a monic irreducible polynomial $f(x) = x^n + a_{n-1}x^{n-1} + \cdots + a_1 x + a_0 \in \mathbf{Q}[x]$. If β is the lcm of the denominators of $a_0, a_1, \ldots, a_{n-1}$, then β is a positive integer and each βa_i is an integer, for $i = 1, 2, \ldots, n - 1$. Then $\beta\gamma$ is a root of a monic polynomial in $\mathbf{Z}[x]$. Hence $\beta\gamma$ is an algebraic integer, α (say). Thus $\gamma = \alpha/\beta$, where α is an algebraic integer and β is a non-zero integer. $\qquad\square$

Definition 11.5.2 An algebraic element $\alpha \neq 0$ is called a unit iff both α and α^{-1} are algebraic integers.

Example 11.5.1 (i) $\sqrt{3}$ is an algebraic integer, because it satisfies the equation $x^2 - 3 = 0$ over $\mathbf{Z}$.

(ii) $(1 + \sqrt{13})/2$ is an algebraic integer, because it satisfies the equation $x^2 - x - 3 = 0$ over $\mathbf{Z}$.

(iii) $\omega = \frac{1}{2}(-1 + \sqrt{3}i)$ is an algebraic integer, because it satisfies the equation $x^2 + x + 1 = 0$ over $\mathbf{Z}$. In general, every root of unity is an algebraic integer, because it satisfies the monic polynomial equation $x^n - 1 = 0$ over $\mathbf{Z}$ for some positive integer n.

(iv) $0, \pm 1, \pm 2, \ldots$ are the only algebraic integers in $\mathbf{Q}$ (referred to as "rational integers").

[*Hint.* Every integer n is an algebraic integer, because n is a root of $(x - n) \in \mathbf{Z}[x]$. On the other hand, suppose a rational number m/q with $\gcd(m, q) = 1$ is an algebraic integer. Then there exists a polynomial $f(x) = x^n + a_{n-1}x^{n-1} + \cdots + a_0 \in \mathbf{Q}[x]$ such that $f(m/q) = 0$ Consequently, $m^n + a_{n-1}qm^{n-1} + \cdots + a_0 q^n = 0 \Rightarrow q | m^n \Rightarrow q = \pm 1$, since $\gcd(m, q) = 1 \Rightarrow m/q$ is an integer.]

(v) $1/\sqrt{2}$ is an algebraic number but it is not an algebraic integer.

[*Hint.* As $1/\sqrt{2}$ satisfies the equation $x^2 - 1/2$ over $\mathbf{Q}$. So it is an algebraic number. If possible, let $1/\sqrt{2}$ be an algebraic integer. Then it satisfies a monic polynomial $f(x)$ over $\mathbf{Z}$ such that $f(1/\sqrt{2}) = (1/\sqrt{2})^n + a_{n-1}(1/\sqrt{2})^{n-1} + a_{n-2}(1/\sqrt{2})^{n-2} + \cdots + a_0 = 0$. Then $(1 + 2a_{n-2} + 4a_{n-4} + \cdots) + \sqrt{2}(a_{n-1} + 2a_{n-3} + \cdots) = 0$. If $a = 1 + 2a_{n-2} + 4a_{n-4} + \cdots$, and $b = a_{n-1} + 2a_{n-3} + \cdots$, then $a + b\sqrt{2} = 0$. Again if $b \neq 0$, then $\sqrt{2} = -a/b$ is the quotient of two integers. This gives a contradiction. Hence $b = 0$ and thus $a + b\sqrt{2} = 0$ gives $a = 0$. This also gives a contradiction, since a is odd.]

Remark In examining whether a given algebraic number is an algebraic integer, it is not necessary to appeal to an irreducible polynomial equation. This follows from Theorem 11.5.2.

Theorem 11.5.2 *A complex number α is an algebraic integer iff it satisfies over $\mathbf{Q}$ a monic polynomial equation with integral coefficients.*

Proof If α is an algebraic integer, then clearly α satisfies a monic polynomial $p(x)$ over $\mathbf{Q}$ with integral coefficients. Conversely, let a complex number α satisfy a monic polynomial equation over $\mathbf{Q}$ with integral coefficients. Then α also satisfies an irreducible polynomial, say $f(x)$ over $\mathbf{Q}$ with integral coefficients. Note that any common divisor of these integral coefficients may be removed. So without loss of generality, we may assume that the gcd of these integral coefficient is 1, resulting $f(x)$ to be a primitive polynomial in $\mathbf{Z}[x]$. The given polynomial $p(x)$ is monic and hence also primitive in $\mathbf{Z}[x]$. Now, the polynomial $p(x)$ is divisible in $\mathbf{Q}[x]$ by the irreducible polynomial $f(x)$. Let $p(x) = f(x)g(x)$, where $g(x) \in \mathbf{Q}[x]$. Since both $p(x)$ and $f(x)$ are primitive polynomials in $\mathbf{Z}[x]$, it follows that $g(x) \in \mathbf{Z}[x]$. Thus the leading coefficient 1 in $p(x)$ is the product of the leading coefficients in $f(x)$ and $g(x)$. Hence, $\pm f(x)$ is monic, resulting that α is an algebraic integer by definition.   □

11.6 Gaussian Integers

We recall that $\mathbf{Z}[i] = \{a + bi : a, b \in \mathbf{Z}\}$ forms a ring, called Gaussian ring, named after C.F. Gauss. The elements of $\mathbf{Z}[i]$, called Gaussian integers are the points of a square lattice in the complex plane. In this section we define Gaussian integers and Gaussian prime integers. Moreover, we study their properties with an eye to the corresponding properties of integers.

Definition 11.6.1 A complex number of the form $a + bi$, $a, b \in \mathbf{Z}$ (i.e., an element of $\mathbf{Z}[i]$), is called a Gaussian integer.

The set of all Gaussian integers is a subring of $\mathbf{C}$. C.F. Gauss developed the properties of Gaussian integers in his work on biquadratic reciprocity. $\mathbf{Z}[i]$ is an Euclidean domain with norm function defined by $N(\alpha) = \alpha\bar{\alpha} = m^2 + n^2$, for $\alpha = m + ni \in \mathbf{Z}[i]$. Clearly, $N(\alpha)$ has the following properties in $\mathbf{Z}[i]$:

(a) $N(\alpha\beta) = N(\alpha)N(\beta)$ for all $\alpha, \beta \in \mathbf{Z}[i] \setminus \{0\}$;
(b) $N(\alpha) = 1$ iff α is a unit;
(c) $1, -1, i$ and $-i$ are the only units in $\mathbf{Z}[i]$;
(d) $N(\alpha) = \begin{cases} 1 & \text{if } \alpha \text{ is a unit} \\ >1 & \text{if } \alpha \notin \{0, 1, -1, i, -i\}. \end{cases}$

Definition 11.6.2 A prime element in the Gaussian ring $\mathbf{Z}[i]$ is called a Gaussian prime integer.

Remark Every Gaussian prime is non-zero and non-unit.

Example 11.6.1 (i) 3 is a Gaussian prime integer.
  (ii) 5 is not a Gaussian prime integer.
  [*Hint*. As there are four units in $\mathbf{Z}[i]$, there are four factorizations of the integer 5 in $\mathbf{Z}[i]$, which are considered equivalent: $5 = (2 + i)(2 - i)$. This is the prime factorization of 5 in $\mathbf{Z}[i]$. Every non-zero element $\alpha \in \mathbf{Z}[i]$ has four associates, viz., $\pm\alpha, \pm i\alpha$. For example, the associates of $2 + i$ are $2 + i, -2 - i, -1 + 2i, 1 - 2i$.]

Example 11.6.2 $3 + 2i$ is an algebraic integer, because it is a root of $x^2 - 6x + 13 = 0$.

Proposition 11.6.1 *For an element α in $\mathbf{Z}[i]$, if $N(\alpha) = p$, a prime integer, then α is prime in $\mathbf{Z}[i]$.*

Proof Clearly, α is a non-zero non-unit, since $p \neq 0$, $p \neq 1$ and $1, -1, i$ and $-i$ are the only units in $\mathbf{Z}[i]$. If possible, let $\alpha = \beta\gamma$ for some $\beta, \gamma \in \mathbf{Z}[i]$. Then $N(\alpha) = N(\beta)N(\gamma) \Rightarrow p = N(\beta)N(\gamma) \Rightarrow$ either $N(\beta) = 1$ or $N(\gamma) = 1$, since p is prime and $N(\beta), N(\gamma)$ are both positive integers. Hence either γ or β is a unit in $\mathbf{Z}[i]$. This shows that α is irreducible in $\mathbf{Z}[i]$. As $\mathbf{Z}[i]$ is an Euclidean domain, $\mathbf{Z}[i]$ is also a principal ideal domain. Hence α is a prime element in $\mathbf{Z}[i]$. $\square$

Remark $N(3) = 9 \Rightarrow$ norm of a Gaussian prime integer may not be a prime integer.

Example 11.6.3 The Gaussian integer $1 + i$ is Gaussian prime, since $N(1 + i) = 2$ is a prime integer.

Remark All prime integers are not Gaussian primes. Let p be a prime integer. Then p is a Gaussian prime iff the ring $\mathbf{Z}[i]/\langle p \rangle$ is a field.

We now describe all Gaussian primes.

Theorem 11.6.1 *The Gaussian primes are precisely of the following types*:

(a) *all prime integers of the form $4n + 3$ and their associates in $\mathbf{Z}[i]$*;
(b) *the number $1 + i$ and its associates in $\mathbf{Z}[i]$*;
(c) *all Gaussian integers α associated with either $a + bi$ or $a - bi$, where $a > 0$, $b > 0$, one of a or b is even and $N(\alpha) = a^2 + b^2$ is a prime integer of the form $4m + 1$*.

Proof Let $\alpha = a + bi$ be a prime element in $\mathbf{Z}[i]$. Then $N(\alpha) = a^2 + b^2$ is an integer >1. Suppose $N(\alpha) = \alpha\bar{\alpha} = p_1^{\alpha_1} p_2^{\alpha_2} \cdots p_m^{\alpha_m}$, where $p_1, p_2, \ldots, p_m$ are distinct prime integers. Then $\alpha | p_1^{\alpha_1} p_2^{\alpha_2} \cdots p_m^{\alpha_m}$ in $\mathbf{Z}[i]$. We claim that α divides only one of the integers $p_1, p_2, \ldots, p_m$. If possible, let α divide two distinct primes p and q in the above expression. Now $\gcd(p, q) = 1 \Rightarrow 1 = px + qy$ for some $x, y \in \mathbf{Z} \Rightarrow \alpha | 1 \Rightarrow$ a contradiction, since α is a prime element in $\mathbf{Z}[i]$. This shows that α divides only one prime p (say) in $\mathbf{Z}[i]$. On dividing p by 4, we have either $p \equiv 1 \pmod 4$ or $p \equiv 2 \pmod 4$ or $p \equiv 3 \pmod 4$.

(a) Suppose $p \equiv 3 \pmod 4$. Now $\alpha | p$ in $\mathbf{Z}[i] \Rightarrow N(\alpha) | N(p) = p^2 \Rightarrow N(\alpha) = p$ or p^2, since $N(\alpha) > 1$. If $N(\alpha) = p$, then $a^2 + b^2 = p$. Since p is an odd prime, either a is even and b is odd or a is odd and b is even which imply $p \equiv 1 \pmod 4 \Rightarrow$ a contradiction. Thus, $N(\alpha) = p^2$. Again $\alpha | p$ in $\mathbf{Z}[i] \Rightarrow \alpha\beta = p$ for some $\beta \in \mathbf{Z}[i] \Rightarrow N(\alpha)N(\beta) = N(p) \Rightarrow p^2 N(\beta) = p^2 \Rightarrow N(\beta) = 1 \Rightarrow \beta$ is a unit in $\mathbf{Z}[i] \Rightarrow \alpha$ and p are associates in $\mathbf{Z}[i]$. Thus p is also a prime element in $\mathbf{Z}[i]$, as required in (a).

(b) Suppose $p \equiv 2 \pmod 4$. Then $p = 2 = (1+i)(1-i)$. Now $N(1+i) = 2 \Rightarrow 1+i$ is prime in $\mathbf{Z}[i]$. Again $1 - i = -i(1 + i)$ shows that $1 - i$ is an associate of $(1 + i)$. Thus $1 + i$ and its associates are prime elements in $\mathbf{Z}[i]$, as required in (b).

(c) Suppose $p \equiv 1 \pmod 4$. Now $\alpha|p \Rightarrow N(\alpha)|N(p) \Rightarrow N(\alpha) = p$ or p^2. Suppose $N(\alpha) = p^2$. As $p \equiv 1 \pmod 4$, suppose $p = 4k + 1$, $k \in \mathbf{Z}$. Then the quadratic reciprocity of -1 modulo p, i.e., $\left(\frac{-1}{p}\right) = (-1)^{\frac{p-1}{2}} \pmod p = (-1)^{2k} = 1$. Hence there exists an integer n such that $n^2 \equiv -1 \pmod p \Rightarrow p|(n^2 + 1)$ in $\mathbf{Z} \Rightarrow p|(n^2 + 1)$ in $\mathbf{Z}[i] \Rightarrow p|(n+i)(n-i)$ in $\mathbf{Z}[i]$. But $\alpha|p \Rightarrow \alpha|(n+i)(n-i) \Rightarrow \alpha|(n-i)$ or $\alpha|(n-i)$, since α is a Gaussian prime. We claim that p is not an associate of α. Suppose p is an associate of α. Then $p|\alpha$ in $\mathbf{Z}[i] \Rightarrow p|(n+i)$ or $p|(n-i)$ in $\mathbf{Z}[i] \Rightarrow$ either $\frac{n}{p} + \frac{1}{p}i$ or $\frac{n}{p} - \frac{1}{p}i$ is a Gaussian integer $\Rightarrow$ a contradiction, since $1/p$ is not an integer $\Rightarrow p$ is not an associate of α. Suppose $N(\alpha) = N(p) = p^2$. Then $\alpha|p \Rightarrow p = \alpha\beta$ for some $\beta \in \mathbf{Z}[i]$. Hence $N(p) = N(\alpha)N(\beta) \Rightarrow N(\beta) = 1 \Rightarrow \beta$ is a unit in $\mathbf{Z}[i] \Rightarrow p$ is an associate of $\alpha \Rightarrow$ a contradiction $\Rightarrow N(\alpha) \neq p^2 \Rightarrow N(\alpha) = p$ (which is the other possibility) $\Rightarrow N(\bar{\alpha}) = p \Rightarrow a - bi$ is also a Gaussian prime integer. Suppose $a - bi$ is an associate of $a + bi$. Then $a - bi = u(a + bi)$, where $u \in \{1, -1, i, -i\}$. If $u = 1$, then $b = 0$. Hence $a^2 = N(\alpha) = p$ is not possible. Thus $u \neq 1$. Similarly, $u \neq -1, i, -i$. Consequently, $a + bi$ is not an associate of $a - bi$. Again $N(\alpha) = p \Rightarrow a^2 + b^2 = p \Rightarrow$ one of a and b must be even and the other must be odd. This proves (c). $\square$

Like division algorithm for ordinary integers and for polynomials, a division algorithm can be developed for Gaussian integers.

Theorem 11.6.2 *For given Gaussian integers α and $\beta \neq 0$, there exist Gaussian integers γ and λ in $\mathbf{Z}[i]$ such that*

$$\alpha = \beta\gamma + \lambda, \quad \text{where either } \lambda = 0 \text{ or } \quad N(\lambda) < N(\beta). \quad (11.2)$$

Proof Suppose $\alpha/\beta = r + ti$ and choose integers r' and t' as close as possible to the rational numbers r and t, respectively. Then $\alpha/\beta = (r'+t'i)+[(r-r')+(t-t')i] = \gamma + \mu$ (say), where $|r - r'| \leq 1/2$, $|t - t'| \leq 1/2$. Now $\alpha = \beta\gamma + \beta\mu$. As α, β and $\gamma \in \mathbf{Z}[i]$, $\beta\mu = \alpha - \beta\gamma \in \mathbf{Z}[i]$. If $r = r'$ and $t = t'$, then $\mu = 0$. If not, then $\mu \neq 0$. Let $\beta = a + bi \in \mathbf{Z}[i]$. Then $N(\beta\mu) = (a^2+b^2)[(r-r')^2+(t-t')^2] < (a^2+b^2) = N(\beta)$. Thus the proposition follows by assuming $\lambda = \beta\mu$. $\square$

Proposition 11.6.2 *Two Gaussian integers α and β, not both zero, have a greatest common divisor δ in $\mathbf{Z}[i]$, which is a Gaussian integer given by $\delta = \lambda\alpha + \mu\beta$, where λ and μ are Gaussian integers.*

Proof Let $J = \langle \alpha, \beta \rangle$ be the ideal generated by α and β in the ring $\mathbf{Z}[i]$. Suppose δ is one of the non-zero elements of J such that $N(\delta)$ is of minimum non-zero norm and $\alpha = \delta\gamma + \psi$ and $\beta = \delta\gamma_1 + \psi_1$ as in Theorem 11.6.2. We claim that ψ and ψ_1

are both zero. If not, then the remainders ψ and ψ_1 are in J with $N(\psi) < N(\delta)$ and $N(\psi_1) < N(\delta)$, which implies a contradiction. Hence $\psi = 0 = \psi_1$ shows that $\alpha = \delta\gamma$ and $\beta = \delta\gamma_1$. Consequently, δ is a common divisor of α and β. Again, since $\delta \in J$, it has the form $\delta = \lambda\alpha + \mu\beta$, $\lambda, \mu \in \mathbf{Z}[i]$. Thus δ is a multiple of every common divisor of α and β. This proves that δ is the greatest common divisor of α and β. $\qquad\square$

The rest of the method of decomposition of Gaussian integers can be proved by the method of rational integers.

Proposition 11.6.3 (a) *If λ is a non-zero non unit Gaussian integer, then there exists a Gaussian prime p such that p divides λ in $\mathbf{Z}[i]$.*

(b) *If λ is Gaussian prime and α, β are Gaussian integers such that $\lambda|\alpha\beta$, then $\lambda|\alpha$ or $\lambda|\beta$ in $\mathbf{Z}[i]$.*

Proof Left as an exercise. $\qquad\square$

Theorem 11.6.3 *Every non-zero, non unit Gaussian integer α can be expressed uniquely as a product $\alpha = \lambda_1\lambda_2\ldots\lambda_n$ of prime Gaussian integers, such that any other decomposition of α in $\mathbf{Z}[i]$ into prime Gaussian integers has the same number of factors and can be arranged in such a way that the corresponding placed factors are associates.*

Proof Left as an exercise. $\qquad\square$

Theorem 11.6.4 *A complex number α in the field $\mathbf{Q}(i)$ is a Gaussian integer iff the monic irreducible polynomial equation satisfied by α over $\mathbf{Q}$ has integral coefficients.*

Proof Let $\alpha = a + bi$ be a Gaussian integer which is not a rational integer. Then $b \neq 0$ and α satisfies a monic irreducible quadratic equation: $x^2 - 2ax + (a^2 + b^2) = 0$ over $\mathbf{Q}$ with rational integers as coefficients. Conversely, let a number $\alpha = p + qi \in \mathbf{Q}(i)$ satisfy a monic irreducible polynomial $f(x)$ in $\mathbf{Q}[x]$ with integral coefficients. Then it follows that α is a Gaussian integer. $\qquad\square$

11.7 Algebraic Number Fields

An algebraic number field is a field which is obtained from the field $\mathbf{Q}$ by adjoining a finite number of algebraic numbers. In this section we shall give some basic properties of algebraic number fields.

Definition 11.7.1 An algebraic number field K is a subfield of $\mathbf{C}$ of the form $K = \mathbf{Q}(\alpha_1, \alpha_2, \ldots, \alpha_n)$, where $\alpha_1, \alpha_2, \ldots, \alpha_n$ are algebraic numbers.

Example 11.7.1 (i) $\mathbf{Q}(\sqrt{2}, \sqrt{7}, \sqrt{11}, \sqrt{13})$ is an algebraic number field.

   (ii) $\mathbf{Q}(\sqrt[17]{5}, \frac{1}{\sqrt{5}})$ is an algebraic number field.

   (iii) $\mathbf{Q}(\pi^2)$ is not an algebraic number field.

Proposition 11.7.1 *The roots of unity which lie in any given algebraic number field form a cyclic group.*

Proof Left as an exercise.   □

Remark For more results see Exercises-IV.

11.8 Quadratic Fields

Explicit computation with algebraic integers is difficult. Instead of developing a general theory, we study algebraic integers in a quadratic field. More precisely, in this section we describe algebraic integers in a field $\mathbf{Q}(\alpha)$, where the complex number α is root of an irreducible quadratic polynomial in $\mathbf{Q}[x]$.

Definition 11.8.1 A subfield F of $\mathbf{C}$ is called a quadratic field iff there exists a complex number α such that $F = \mathbf{Q}(\alpha)$, where α is a root of an irreducible quadratic polynomial in $\mathbf{Q}[x]$.

Remark $\mathbf{Q}(\alpha) = \{a + b\alpha : a, b \in \mathbf{Q}\}$ and $[\mathbf{Q}(\alpha) : \mathbf{Q}] = 2$.

Example 11.8.1 $\mathbf{Q}(\sqrt{7})$ is a quadratic field. This is so because $x^2 - 7$ is irreducible in $\mathbf{Q}[x]$.

Theorem 11.8.1 *Let F be a quadratic field. Then there exists a unique square free integer n such that $F = \mathbf{Q}(\sqrt{n})$.*

Proof Let $F = \mathbf{Q}(\alpha)$, where α is a root of the irreducible polynomial $x^2 + bx + c$ over $\mathbf{Q}$. If α_1 and α_2 are the two values of α, then $\alpha_1 = \frac{-b+\sqrt{b^2-4c}}{2}$ and $\alpha_2 = \frac{-b-\sqrt{b^2-4c}}{2}$. Hence $\alpha_1 + \alpha_2 = -b \in \mathbf{Q} \Rightarrow \mathbf{Q}(\alpha_1) = \mathbf{Q}(\alpha_2) \Rightarrow F = \mathbf{Q}(\alpha) = \mathbf{Q}(\alpha_1) = \mathbf{Q}(\frac{-b+\sqrt{b^2-4c}}{2}) = \mathbf{Q}(\sqrt{a})$, where $a = b^2 - 4c \in \mathbf{Q}$. But this a is not the square of a rational number, since $x^2 + bx + c$ is irreducible in $\mathbf{Q}[x]$ by hypothesis. Suppose $a = p/q$, where p, q are integers, $q > 0$ and $\gcd(p, q) = 1$. Let l^2 be the largest square dividing pq. Then $pq = l^2 n$, where n $(\neq 1)$ is a square free integer and $F = \mathbf{Q}(\sqrt{a}) = \mathbf{Q}(\sqrt{p/q}) = \mathbf{Q}(\sqrt{pq}) = \mathbf{Q}(l\sqrt{n}) = \mathbf{Q}(\sqrt{n})$. To show the uniqueness of n, let m be another square free integer such that $F = \mathbf{Q}(\sqrt{m})$. Then $\mathbf{Q}(\sqrt{n}) = \mathbf{Q}(\sqrt{m}) \Rightarrow \sqrt{n} = x + y\sqrt{m}$ for some $x, y \in \mathbf{Q} \Rightarrow n = x^2 + my^2 + 2xy\sqrt{m}$. If $xy \neq 0$, then $\sqrt{m} = (n - x^2 - my^2)/2xy$. This is a contradiction, since $\sqrt{m} \notin \mathbf{Q}$, as m is square free. Hence $xy = 0$. Again if $y = 0$, then $x = \sqrt{n}$. This is again a contradiction, since $\sqrt{n} \notin \mathbf{Q}$ as n is square free. Consequently, $x = 0$ and $n = my^2$. Again n is square free and hence $y^2 = 1$. Consequently, $m = n$. $\qquad\qquad\Box$

We now describe the form of the algebraic integers α in the quadratic field $F = \mathbf{Q}(\sqrt{n})$, where n is a square free integer.

Let $\alpha = \frac{l + m\sqrt{n}}{d}$, where l, m, n are integers such that $d > 0$ and $\gcd(l, m, d) = 1$. If $m = 0$, then $\alpha = l/d \Rightarrow \alpha$ is a quadratic integer iff $d = 1$. Next suppose $m \neq 0$. Now α is a root of a polynomial $x^2 + bx + c \in \mathbf{Z}[x] \Rightarrow (\frac{l + m\sqrt{n}}{d})^2 + b(\frac{l + m\sqrt{n}}{d}) + c = 0 \Rightarrow l^2 + m^2 n + bld + cd^2 = 0$ and $2l + bd = 0$, since $m \neq 0$. Let $\gcd(l, d) = t > 1$. Then there exists a prime factor p of t and hence $p|l$ and $p|d \Rightarrow p^2|m^2 n \Rightarrow p^2|m^2$, since n is square free $\Rightarrow p|m$. This is a contradiction, since $\gcd(l, m, d) = 1$ and hence $t = 1$. Thus $\gcd(l, d) = 1 \Rightarrow d|2$, since $2l + bd = 0 \Rightarrow d = 1$ or $d = 2$. If $d = 1$, then $\alpha = l + m\sqrt{n}$ is a quadratic integer. If $d = 2$, then $\alpha = \frac{l + m\sqrt{n}}{2} \Rightarrow (\alpha - (l/2))^2 = \frac{m^2}{4}n \Rightarrow \alpha$ is a root of the quadratic equation $x^2 - lx + \frac{l^2 - m^2 n}{4} = 0$ over $\mathbf{Z} \Rightarrow l^2 \equiv m^2 n \pmod{4}$, since the above coefficients are all integers. Again $\gcd(l, d) = 1 \Rightarrow \gcd(l, 2) = 1 \Rightarrow l$ is odd $\Rightarrow 1 \equiv m^2 n \pmod{4}$, since $l^2 = m^2 n \pmod{4}$. We now consider the possible cases: $n \equiv 1 \pmod{4}$ or $n \equiv 2 \pmod{4}$ or $n \equiv 3 \pmod{4}$.

Case 1 If $n \equiv 1 \pmod{4}$, then $1 \equiv m^2 \pmod{4}$. This is true iff m is odd.

Case 2 If $n \equiv 2 \pmod{4}$, then $1 \equiv 2m^2 \pmod{4}$. But there is no integer m such that $1 \equiv 2m^2 \pmod{4}$.

Case 3 If $n \equiv 3 \pmod{4}$, then $1 \equiv 3m^2 \pmod{4}$. But there is no integer m such that $1 \equiv 3m^2 \pmod{4}$.

From the above discussion there follows Theorem 11.8.2.

Theorem 11.8.2 *Let $\mathcal{A}$ be the set of all algebraic integers in the quadratic field $F = \mathbf{Q}(\sqrt{n}) = \{a + b\sqrt{n} : a, b \in \mathbf{Q}\}$, where n is a square free integer. Then the set $\mathcal{A}$ is given by*

$$\mathcal{A} = \begin{cases} l + m\sqrt{n}, & \text{where } l, m \text{ are integers and } n \not\equiv 1 \pmod{4} \\ \frac{l + m\sqrt{n}}{2}, & \text{where } l, m \text{ are odd integers and } n \equiv 1 \pmod{4}. \end{cases}$$

11.9 Exercises

Exercises-IV

1. Show that a simple transcendental extension of a field F is isomorphic to the field of rational functions over F and is of infinite degree over F; moreover, any two simple transcendental extensions of F are isomorphic.

2. Let $F(\alpha)$ be a simple algebraic extension of a field F. Prove that

   (a) $F(\alpha) \cong F[x]/\langle f(x) \rangle$, where $f(x)$ is the minimal polynomial of α over F;
   (b) If n is the degree of $f(x)$, then $S = \{1, \alpha, \alpha^2, \ldots, \alpha^{n-1}\}$ forms a basis of $F(\alpha)$ over F;
   (c) If β is an algebraic element of $F(\alpha)$, then the degree of β over F is a divisor of degree of α.

3. (a) Let F be a subfield of $\mathbf{C}$. If $\alpha \in \mathbf{C}$ and $\beta \in \mathbf{C}$ are algebraic over F, show that there exists an element $\gamma \in \mathbf{C}$ such that $F(\alpha, \beta) = F(\gamma)$.
   (b) Let F be a subfield of $\mathbf{C}$. If $\alpha_1, \alpha_2, \ldots, \alpha_n \in \mathbf{C}$ are algebraic over F, show that there exists an element $\gamma \in \mathbf{C}$ such that $F(\alpha_1, \alpha_2, \ldots, \alpha_n) = F(\gamma)$.
   (c) If F is an algebraic number field, prove that there exists an algebraic integer γ such that $F = \mathbf{Q}(\gamma)$.
       [*Hint.* Use 3(b).]
   (d) Let $F = \mathbf{Q}(\gamma)$ be an algebraic number field, where γ is an algebraic integer. If the degree of the minimal polynomial of γ over $\mathbf{Q}$ is n, then prove that every element of F can be expressed uniquely in the form $a_0 + a_1\gamma + \cdots + a_{n-1}\gamma^{n-1}$, where a_i are rational numbers.
       [*Hint.* F is vector space over Q of dimension n.]
   (e) Let F be an algebraic number field and K be the set of all algebraic integers. Then show that the set $H = K \cap F$ is an integral domain (called the ring of integers of the algebraic number field K) and the quotient field of H is F.
   (f) Let F be a field of characteristic 0.

      (i) If α, β are algebraic over F, show that there exists an element $\gamma \in F(\alpha, \beta)$ such that $F(\alpha, \beta) = F(\gamma)$.
     (ii) Prove that any finite field extension of the field F is a simple extension.
          [*Hint.* Use induction argument to case (a).]

4. Let F be field and K be a field extension of F of prime dimension p. If α is an element of K but not in F, show that α has degree p and $K = F(\alpha)$.
   [*Hint.* $[K : F] = p \Rightarrow [K : F(\alpha)][F(\alpha) : F] = p \Rightarrow [K : F(\alpha)] = 1$, since $[F(\alpha) : F] \neq 1$, as $\alpha \notin F$. Hence $[F(\alpha) : F] = p$ and $K = F(\alpha)$.]
5. Let F be a subfield of the field $\mathbf{C}$. If an element α of $\mathbf{C}$ is algebraic over F, then prove that

   (a) every element β of $F(\alpha)$ is algebraic over F;
   (b) degree of β over $F \leq$ degree of α over F.

6. Let $F \subset L \subset K$ be a tower of fields such that K is algebraic over L and L is algebraic over F. Prove that K is algebraic over F.
7. Let F be field. Prove that F is algebraically closed iff every irreducible polynomial in $F[x]$ is of degree 1.
8. (a) Let F be field. Show that there exists an algebraic field extension K of F such that K is algebraically closed (called an algebraic closure of F).

(b) If K and L are two algebraic closures of F, then show that there is an isomorphism $\psi : K \to L$ such that $\psi(a) = a$, $\forall a \in F$.

Remark Given a field F, its algebraic closure is unique up to isomorphisms.

9. (a) Prove that a number α is an algebraic integer iff the additive group generated by all the powers $1, \alpha, \alpha^2, \alpha^3, \ldots$ of α can be generated by a finite number of elements.
   (b) If all the positive powers of an algebraic number α lie in an additive group generated by a finite set of numbers $\alpha_1, \alpha_2, \ldots, \alpha_n$, then show that α is an algebraic integer.
   (c) Prove that the set of all algebraic integers is an integral domain.
   (d) In any field of algebraic numbers, prove that the set of algebraic integers is also an integral domain.
10. Show that the minimal polynomial of an algebraic integer is monic with integral coefficients.
11. (a) Let K be finite extension over F such that $[K : F] = n$. Prove that every element α of K has a degree m over F such that m is a divisor of n.
    (b) Let $f(x)$ be an irreducible cubic polynomial over a field F. If K is an extension of F of degree 2^n, then show that $f(x)$ is irreducible in $K[x]$.
    (c) Using (b) as the algebraic basis show that by straight edge and compass alone, it is impossible to

    (i) duplicate a general cube (duplication of a cube);
    (ii) trisect a general angle (trisection of an angle).

       If a real number can be constructed by straight edge and compass alone, then the number is said to be constructible.

[*Hint.* (a) The element α generates a simple extension $F(\alpha)$ of F. Hence $[K : F(\alpha)][F(\alpha) : F] = n$.

(b) If $f(x)$ is a reducible polynomial over K of degree 2^n, then the cubic polynomial $f(x)$ must have at least one linear factor in $K[x]$. This shows that K contains a root α of $f(x)$. But such an element α of degree 3 over F cannot lie in a field K of degree 2^n over F by (a).

(c) (i) If it possible to construct another cube of double the volume of a unit cube, then the side x of the new cube must satisfy the equation $x^3 - 2 = 0$. But by the Eisenstein criterion, the polynomial $x^3 - 2$ is irreducible in $\mathbf{Q}[x]$. Over any field F corresponding to the straight edge and compass construction, the polynomial $x^3 - 2$ is also irreducible by (b).

(ii) A $60°$ angle is constructible, because $\cos 60°$ is half. But if an angle of $60°$ is trisected by straight edge and compass, then $\cos 60° = 4\cos^3 20° - 3\cos 20°$ shows that $x = \cos 20°$ must satisfy the cubic equation $8x^3 - 6x - 1 = 0$ in $\mathbf{Q}[x]$. But $8x^3 - 6x - 1$ is irreducible in $\mathbf{Q}[x]$.]

12. An extension K of a field F is a root field of a polynomial $f(x)$ of degree ≥ 1, with coefficients in F iff

(i) $f(x)$ can be factored into linear factors $f(x) = a(x - \alpha_1) \cdots (x - \alpha_n)$ in $K[x]$; and

(ii) $K = F(\alpha_1, \alpha_2, \ldots, \alpha_n)$.
      Prove that

   (a) any polynomial over any field has a root field (existence).
   (b) all root fields of a given polynomial over a given field F are isomorphic
        (uniqueness).
   (c) for any prime integer p and any positive integer n, there exists a finite
        field with $q = p^n$ elements, which is the root field of $x^q - x$ over $\mathbf{Z}_p$.

Exercises-V   Identify the correct alternative(s) (there may be more than one) from
the following list:

1. Let R be the quotient ring $\mathbf{Z}[i]/n\mathbf{Z}[i]$. Then R is an integral domain if n is

      (a) 19   (b) 7   (c) 13   (d) 1.

2. The field extension $\mathbf{Q}(\sqrt{2} + \sqrt[3]{2})$ over the field $\mathbf{Q}(\sqrt{2})$ has degree

      (a) 1   (b) 3   (c) 2   (d) 6.

3. If the polynomial $x^4 + x + 6$ has a root of multiplicity greater than 1 over a field
    of characteristic p (prime), then p is

      (a) $p = 5$ (b) $p = 3$ (c) $p = 2$ (d) $p = 7$.

4. Let F be a field of eight elements and B be a subset of F such that $B = \{x \in F : x^7 = 1$ and $x^n \neq 1$ for all positive integers n less than 7$\}$. Then the number
    of elements in B is

      (a) 3   (b) 2   (c) 1   (d) 6.

5. Let $f(x) = 2x^2 + x + 2$ and $g(x) = x^3 + 2x^2 + 1$ be two polynomials over the
    field $\mathbf{Z}_3$. Then

   (a) $f(x)$ and $g(x)$ are both irreducible;
   (b) neither $f(x)$ nor $g(x)$ is irreducible;
   (c) $f(x)$ is irreducible, but $g(x)$ is not;
   (d) $g(x)$ is irreducible, but $f(x)$ is not.

6. Let F be a field and the polynomial $f(x) = x^3 - 312312x + 123123$ be irre-
    ducible in $F[x]$. Then

   (a) F is a finite field with 7 elements;
   (b) F is a finite field with 13 elements;
   (c) F is a finite field with 3 elements;
   (d) $F = \mathbf{Q}$ of rational numbers.

7. Let ω be an imaginary cube root of unity i.e., ω is a complex number such that
    $\omega^3 = 1$ and $\omega \neq 1$. If K is the field $\mathbf{Q}(\sqrt[3]{2}, \omega)$ generated by $\sqrt[3]{2}$ and ω over the

field $\mathbf{Q}$ of rational numbers and n is the number of subfields L of K such that $\mathbf{Q} \subsetneq L \subsetneq K$, then n is

(a) 4   (b) 3   (c) 5   (d) 2.

8. Let F_{5^n} be the finite field with 5^n elements. If F_{5^n} contains a non-trivial root of unity, then n is

(a) 15   (b) 6   (c) 92   (d) 30.

9. Which of the following statements is (are) correct?

(a) $\sin 7°$ is algebraic over $\mathbf{Q}$;
(b) $\sin^{-1} 1$ is algebraic over $\mathbf{Q}$;
(c) $\cos(\pi/17)$ is algebraic over $\mathbf{Q}$;
(d) $\sqrt{2} + \sqrt{\pi}$ is algebraic over $\mathbf{Q}(\pi)$.

10. Which of the following statements is (are) correct?

(a) There exists a finite field in which the additive group is not cyclic;
(b) Every infinite cyclic group is isomorphic to the additive group of integers;
(c) If F is a finite field, there exists a polynomial p over F such that $p(x) = 0$ for all $x \in F$, where 0 denotes the zero in F;
(d) Every finite field is isomorphic to a subfield of the field of complex numbers.

11. Consider the ring $\mathbf{Z}_n$ for $n \geq 2$. Which of the following statements is (are) correct?

(a) If $\mathbf{Z}_n$ is a field, then n is a composite integer;
(b) If $\mathbf{Z}_n$ is a field iff n is a prime integer;
(c) If $\mathbf{Z}_n$ is an integral domain, then n is prime integer;
(d) If there is an injective ring homomorphism of $\mathbf{Z}_5$ to $\mathbf{Z}_n$, then n is prime.

12. Let F_p be the field $\mathbf{Z}_p$, where p is a prime. Let $F_p[x]$ be the associated polynomial ring. Which of the following quotient rings is (are) fields?

(a) $F_5[x]/\langle x^2 + x + 1 \rangle$;   (b) $F_2[x]/\langle x^3 + x + 1 \rangle$;

(c) $F_3[x]/\langle x^3 + x + 1 \rangle$;   (d) $F_7[x]/\langle x^2 + 1 \rangle$.

11.10 Additional Reading

We refer the reader to the books (Adhikari and Adhikari 2003, 2004; Alaca and Williams 2004; Artin 1991; Birkhoff and Mac Lane 2003; Hardy and Wright 2008; Hungerford 1974; Rotman 1988) for further details.

References

Adhikari, M.R., Adhikari, A.: Groups, Rings and Modules with Applications, 2nd edn. Universities Press, Hyderabad (2003)

Adhikari, M.R., Adhikari, A.: Text Book of Linear Algebra: An Introduction to Modern Algebra. Allied Publishers, New Delhi (2004)

Alaca, S., Williams, K.S.: Introductory Algebraic Number Theory. Cambridge University Press, Cambridge (2004)

Artin, M.: Algebra. Prentice-Hall, Englewood Cliffs (1991)

Birkhoff, G., Mac Lane, S.: A Survey of Modern Algebra. Universities Press, Hyderabad (2003)

Hardy, G.H., Wright, E.M.: An Introduction to the Theory of Numbers, 6th edn. Oxford University Press, London (2008)

Hungerford, T.W.: Algebra. Springer, New York (1974)

Rotman, J.J.: An Introduction to Algebraic Topology. Springer, New York (1988)

Chapter 12
Introduction to Mathematical Cryptography

The aim of this chapter is to introduce various cryptographic notions starting from historical ciphers to modern cryptographic notions by using mathematical tools mainly based on number theory, modern algebra, probability theory and information theory. In the modern busy digital world, the word "cryptography" is well known to many of us. Everyday, knowingly or unknowingly, in many places we use different techniques of cryptography. Starting from the logging on a PC, sending e-mails, withdrawing money from ATM by using a PIN code, operating the locker at a bank with the help of a designated person from the bank, sending message by using a mobile phone, buying things through Internet by using a credit card, transferring money digitally from one account to another over the Internet, we are applying cryptography everywhere. If we observe carefully, we see that in every case we need to hide some information or it is necessary to transfer information secretly. So intuitively we can guess that cryptography has something to do with security. In this chapter, we introduce cryptography and provide a brief overview of the subject and discuss the basic goals of cryptography and understand the subject, both intuitively and mathematically. More precisely various cryptographic notions starting from the historical ciphers to modern cryptographic notions like public-key encryption schemes, signature schemes, oblivious transfer, secret sharing schemes and visual cryptography by using mathematical tools mainly based on modern algebra are explained. Finally, the implementation issues of three public-key cryptographic schemes, namely RSA, ElGamal, and Rabin using the open-source software SAGE are discussed.

12.1 Introduction to Cryptography

Cryptography has been used almost since the time when writing concept was invented. For the larger part of its history, cryptography remained an art, a game of ad hoc designs and attacks. Historically, the notion of cryptography arose as a means to enable parties to maintain privacy of the information they sent to each other, even in the presence of an enemy who even had an access to the communicated message.

M.R. Adhikari, A. Adhikari, *Basic Modern Algebra with Applications*,
DOI 10.1007/978-81-322-1599-8_12, © Springer India 2014

The name cryptography comes from the Greek words "kruptos" (means hidden) and "graphia" (means writing). In short, cryptography means the art of writing secrets message. In general, the aim of cryptography is to construct efficient schemes achieving some desired functionality, even in an adversarial environment. For example, the most basic question in cryptography is that of secure communication across an insecure channel. Before we start describing the major objectives of cryptography, let us first introduce the basic model of cryptography. Under this model, two parties, conventionally named Alice (sender) and Bob (receiver), wish to communicate with each other over an insecure channel (think of the Internet or mobile network) in which Alice wishes to send a secret message to Bob. Now the first question is: what do we mean by an insecure channel? Insecure channel means a channel through which the message will be sent to Bob is controlled by a whole lot of adversaries (enemies) who are trying to get hold of the message of Alice or to modify it and trick Bob into believing that some other message was sent by Alice. In cryptography, we assume that the adversary has practically unlimited computation power. He cannot only eavesdrop over any message, but also even take the message and change some of its bits. Traditionally, the two basic goals of cryptography have been the secrecy and authenticity (or data-integrity). It is necessary in secrecy that the message should be guarded from the adversary, that is, the message that has been sent by Alice to Bob should be hidden from the adversary and only the intended recipient, i.e., Bob can see the message. On the other hand, authenticity ensures that Bob should not be tricked into accepting a message that did not originate from Alice, that is, the message received by Bob must be the same as the message that was sent by Alice. To achieve such goals, we must note that there must be something that Bob knows but the adversary does not know, otherwise the adversary could simply apply the same method that Bob does and thus be able to read the messages sent by Alice.

Under these circumstances in presence of adversaries, achieving these security goals such as privacy or authenticity is provided by cryptography with the help of a weapon, known as a protocol to Alice and Bob. A protocol is just a collection of programs or algorithms. A protocol for Alice will guide her to package or encapsulate her data for transmission, on the other hand, the protocol for Bob will guide him to decapsulate the received package from Alice to recover the data together possibly with associated information telling him whether or not to regard it as authentic.

To understand the basic cryptography, we first need to understand the basic terminologies associated with cryptography. The original secret message that Alice wants to send to Bob is called plain text, while the disguised message that is sent by Alice to Bob is called cipher text. The procedure of converting a plain text into a cipher text is known as encryption, while the procedure to convert a cipher text into the plain text is known as decryption.

Let us now define formally what do we mean by a cryptosystem.

Definition 12.1.1 A cryptosystem is a five-tuple $(\mathcal{P}, \mathcal{C}, \mathcal{K}, \mathcal{E}, \mathcal{D})$, where the following conditions are satisfied:

- $\mathcal{P}$ is a finite set of possible plain texts;
- $\mathcal{C}$ is a finite set of possible cipher texts;
- $\mathcal{K}$, the key space, is a finite set of possible keys;
- For each $K \in \mathcal{K}$, there is an encryption rule $e_K \in \mathcal{E}$ and a corresponding decryption rule $d_K \in \mathcal{D}$. Each $e_K : \mathcal{P} \to \mathcal{C}$ and $d_K : \mathcal{C} \to \mathcal{P}$ are functions such that $d_K(e_K(x)) = x$ for every plain text element $x \in \mathcal{P}$.

12.2 Kerckhoffs' Law

In cryptography, Kerckhoffs' law (also called Kerckhoffs' assumption or Kerckhoffs' principle) was stated by Auguste Kerckhoffs (1835–1903), a professor of languages at the School of Higher Commercial Studies, Paris, in the 19th century. It states that a cryptosystem should be secure even if everything about the system, except the key, is of public knowledge. It was reformulated (perhaps independently) by Claude Shannon as "the enemy knows the system". The idea is that if any part of a cryptosystem (except the key) has to be kept secret then the cryptosystem is not secure. So in cryptography it is usually assumed that the enemy i.e., the adversary knows the cryptosystem that is being used. Of course, if the adversary does not know the cryptosystem that is being used, then it will make the task of the adversary more difficult. Thus the goal of a cryptographer is to design a secure cryptosystem which satisfies Kerckhoffs' principle.

Broadly speaking, a cryptographer's work may be divided into two parts: namely, the constructive part (cryptography) and the destructive part (cryptanalysis). In the constructive part, the aim of a cryptographer is to construct a secure cryptosystem, on the other hand, in the destructive part, the aim of a cryptographer is to find weaknesses of some existing cryptosystem to break the system. For example, the cryptographers associated with the banking sectors try to make their systems secure so that all the transactions over the Internet can be made safely, whereas the cryptographers of the defense or intelligence organizations try to break the code that had been transmitted between two suspects. As a whole, both cryptography and cryptanalysis are included in the subject, called "CRYPTOLOGY". There is no doubt that cryptography and cryptanalysis are complementing each other and are being developed side by side. To construct a good and secure cryptographic scheme one has to have a good knowledge in the cryptanalysis techniques. On the other hand, to attack a scheme, one has to have a good knowledge in the construction of the scheme. Thus unless one has a good knowledge in cryptanalysis, it is not possible for him/her to construct a good cryptographic scheme and vice versa.

12.3 Cryptanalysis

Cryptanalysis is the study of a cryptographic system for the purpose of finding weaknesses in the system and breaking the code used to encrypt the data without accessing the secret information which is normally required to do so. Typically, this involves finding the secret key. Although the actual word "cryptanalysis" is relatively recent (it was coined by William Friedman in 1920), methods for breaking codes and ciphers are much older. The first known recorded explanation of cryptanalysis was given in 9th Century by Arabic polymath Abu Yusuf Yaqub ibn Ishaq al-Sabbah Al-Kindi in A Manuscript on Deciphering Cryptographic Messages. This manuscript includes a description of the method of frequency analysis.

Even though the goal has been the same, the methods and techniques of cryptanalysis have changed drastically through the evolution of cryptography, ranging from the pen-and-paper methods of the past, through machines like Enigma in World War II, to the computer-based schemes of the present. Even the results of cryptanalysis have changed, it is no longer possible to have unlimited success in code breaking. In the mid-1970s, a new class of cryptography, known as asymmetric cryptography, was introduced in the literature of cryptology. Methods for breaking these cryptosystems are different from before and usually involve solving a carefully constructed problem in pure mathematics, the most well known being integer factorization.

Now we will discuss different attack models on cryptography. The most common types of attacks are given below.

• *Cipher text only attack*: In cryptography, a cipher text only attack is a form of cryptanalysis, where the attacker is assumed to have access only to a set of cipher texts. In the history of cryptography, early ciphers, implemented using pen-and-paper, were routinely broken using cipher texts alone. Cryptographers developed a variety of statistical techniques for attacking cipher text, such as, frequency analysis. Mechanical encryption devices, such as, Enigma made these attacks much more difficult. This cipher machine was invented by a German engineer, Arthur Scherbius. The Enigma machine was an ingenious advance in technology. It is an electromechanical machine, very similar to a typewriter, with a plugboard to swap letters, rotors to further scramble the alphabet and a lamp panel to display the result. Most models of Enigma used 3 or 4 rotors with a reflector to allow the same settings to be used for enciphering and deciphering. The German Navy and Army adopted the Enigma in 1926 and 1928, respectively, but only added the plugboard in 1930. The history of breaking the code for Enigma by Polish mathematicians is very exciting. The Polish were understandably nervous about German aggression and on September 1, 1932 the Polish Cipher Bureau hired a 27-year-old Polish mathematician, Marian Rejewski, along with two fellow Poznan University mathematics graduates, Henryk Zygalski and Jerzy Rozycki, to try to break the code of the Enigma machine. This was an early insight into the role of mathematics in code breaking. The three Polish code breakers had access to an Enigma machine, but did not know the rotor wiring. Through a German spy, the French gained access to two months of Enigma key settings. But without the rotor wire, they were unable to make use of this information. They passed this information to their British and Polish colleagues and the Polish were able to quickly solve the Enigma puzzle, recreating the three rotors then in use to mount a successful cipher text-only cryptanalysis to the Enigma. This was in March 1933 and they continued to break the code until the Nazis invaded Poland on September 1, 1939, marking the start of WW2.

- *Known plain text attack*: In cryptography, a known plain text attack is a form of cryptanalysis, where the attacker has a cipher text corresponding to a plain text of an arbitrary message not of his choice.
- *Chosen plain text attack*: In cryptography, a chosen plain text attack is a form of cryptanalysis, where the attacker has the capability to compute the cipher text corresponding to an arbitrary plain text message of his choice.
- *Chosen cipher text attack*: In cryptography, a chosen cipher text attack is a form of cryptanalysis, where the attacker has the temporary access to the decryption machinery to find the plain text corresponding to a cipher text message of his choice.

In each case the objective to the attacker is to determine the key that was used. We observe that chosen cipher text attack is most relevant to a special type of cryptosystem known as asymmetric or public-key cryptosystem.

12.3.1 Brute-Force Search

Now we will discuss informally an important term in cryptanalysis known as brute-force search. Suppose a cryptanalyst has found a plain text and a corresponding cipher text but he does not know the key. He can simply try encrypting the plain text using each possible key, until the cipher text matches or try decrypting the cipher text to match the plain text. This is called brute-force search. So every cryptosystem can be broken using brute-force search attack. However, every well-designed cryptosystem has such a large key space that this brute-force search is not practical.

In academic cryptography, a weakness or a break in a scheme is usually defined quite conservatively. Bruce Schneier sums up this approach: "Breaking a cipher simply means finding a weakness in the cipher that can be exploited with a complexity less than brute-force. Never mind that brute-force might require 2^{128} encryptions; an attack requiring 2^{110} encryptions would be considered a break...simply put, a break can just be a certificational weakness: evidence that the cipher does not perform as advertised."

12.4   Some Classical Cryptographic Schemes

Cryptography is as old as writing itself and has been used for thousands of years to safeguard military and diplomatic communications. It has a long fascinating history. Kahn's (1967) *The Codebreakers* is the most complete non-technical account of the subject. This book traces cryptography from its initial and limited use by Egyptians some 4000 years ago, to the twentieth century where it played a critical role in the outcome of both the world wars. Before the 1980s, cryptography was used primarily for military and diplomatic communications and in fairly limited contexts. But now cryptography is the only known practical method for protecting information transmitted through communication networks that use land lines, communication satellites and microwave facilities. In some instances it can be the most economical way to protect stored data.

Let us start with some classical examples of cryptographic schemes in which both the sender and the receiver agree upon a common key secretly before the actual communication starts. We shall explain later that this concept leads to a very important notion of cryptography known as symmetric key cryptosystem.

12.4.1 Caesarian Cipher or Shift Cipher

This cryptographic scheme was discovered by Julius Caesar, used around 2000 years ago, much earlier than the invention of group theory. Caesar used his scheme to send instructions or commands through messengers to his generals and allies who stayed at the war front. Now if the message was sent in a simple text form (plain text), it could be caught by the enemies or the messenger might read the secret messages. To avoid such a situation, Julius Caesar introduced a new method. Before going to the war front Julius Caesar and his generals agreed upon a secret number, say 3, which is the key of the cryptosystem. When Julius Caesar needed to send messages to the generals at the war front, he just cyclically shifted every letter of the command or instruction by 3 positions to the right. If we take the example of the English alphabets, without differentiating the lower and the upper case, we have altogether 26 letters. So if we shift cyclically each letter of the English alphabets 3 times to the right, A will be shifted to D, B will be shifted to $E, \ldots, X$ will be shifted to A, Y will be shifted to B and finally, Z will be shifted to C. If a message said "ROME", it would look like "URPH" (as $R \to U, O \to R, M \to P$ and $E \to H$). Now those who know the value of the shift to be 3 (in cryptography we call this value as "key"), they will shift cyclically each letter of the cypher text 3 times to the left and can easily recover the plain text.

Now think about the Caesarian shift cipher in today's view point. The question is: can we associate the Caesarian shift cipher with mathematics? Today, we have an important tool of mathematics, known as Cyclic Group. We can number each alphabet $A, B, \ldots, Z$ of English literature (not distinguishing the lower and upper cases) as $0, 1, \ldots, 25$, respectively. Since, each letter is shifted cyclically to the right (in the above example, it is shifted 3 times to the right), we can represent the set of English alphabets as the set $\mathbf{Z}_{26}$, the set of all integers modulo 26. We know that $(\mathbf{Z}_{26}, +)$ forms a cyclic group with respect to addition modulo 26 defined on $\mathbf{Z}_{26}$ properly. Now as each letter is shifted cyclically k times to the right, we can define the encryption function, e_k, from $\mathbf{Z}_{26}$ onto $\mathbf{Z}_{26}$ for the fixed $k \in \mathbf{Z}_{26}$, defined by $e_k(x) = x + k, \forall x \in \mathbf{Z}_{26}$. So the decryption function can be written as $d_k(x) = x - k$, $\forall x \in \mathbf{Z}_{26}$.

Now the question is: In today's computer age, is the shift cipher secure? Assume that an adversary only has a piece of cipher text along with the knowledge that this cipher text is obtained by using shift cipher. The adversary does not have the knowledge of the secret key k. However, it is very easy for the adversary to have a "cipher text only attack" as there are only 26 possible keys. So the adversary will try every key and will see which key decrypts the cipher text into a plain text that "makes sense". As mentioned earlier, such an attack on an encryption scheme is called a brute-force attack or exhaustive search attack. Clearly, any secure encryption scheme must not be vulnerable to such a brute-force attack; otherwise, it can be completely broken, irrespective of how sophisticated the encryption algorithm is. This leads to a trivial but important principle, called, the "sufficient key space principle". It states that

Any secure encryption scheme must have a key space that is not vulnerable to exhaustive search.

So we try to increase the size of the key space in the next cryptographic scheme.

12.4.2 Affine Cipher

Let us take $\mathcal{P} = \mathcal{C} = \mathbf{Z}_{26}$. Let $\mathcal{K} = \{(a, b) \in \mathbf{Z}_{26} \times \mathbf{Z}_{26} : \gcd(a, 26) = 1\}$. For $K = (a, b) \in \mathcal{K}$, let us define $e_k : \mathcal{P} \to \mathcal{C}$ and $d_k : \mathcal{C} \to \mathcal{P}$ by $e_K(x) = (ax + b) \mod 26$ and $d_K(y) = a^{-1}(y - b) \mod 26$, respectively, where $(x, y) \in \mathbf{Z}_{26}$. In this case, the size of the key space is $26 \cdot 12 = 312$ which is more than 26, the key space size of Caesarian shift cipher.

Remark In today's computer age, for an exhaustive search, a very powerful computer or many thousands of PC's that are distributed around the world may be used. Thus, the number of possible keys must be very large, must be an order of at least 2^{60} or 2^{70}. But we must emphasize that the "sufficient key space principle" gives a necessary condition for security, not a sufficient one. The next encryption scheme has a very large key space. However, we shall show that it is still insecure.

12.4.3 Substitution Cipher

Let us take $\mathcal{P} = \mathcal{C} = \mathbf{Z}_{26}$. Let $\mathcal{K}$ be the set of all possible permutations on the set $\{0, 1, 2, \ldots, 25\}$. For each permutation π, let us define $e_\pi : \mathcal{P} \to \mathcal{C}$ and $d_\pi : \mathcal{C} \to \mathcal{P}$ by $e_\pi(x) = \pi(x)$ and $d_\pi(y) = \pi^{-1}(y)$ respectively, where π^{-1} is the inverse permutation of π. Then one can verify that $(\mathcal{P}, \mathcal{C}, \mathcal{K}, \mathcal{E}, \mathcal{D})$ is a cryptosystem. The total number of possible keys is $26! = 403291461126605635584000000$ which is more than 4.0×10^{26}, a very large number.

Let us consider the plain text written in English language. In that case, the $\mathcal{P} = \mathcal{C} = \mathbf{Z}_{26}$. Let us take a secret key π as defined in Table 12.1. Then the plain text

Table 12.1 A permutation rule π in which A is replaced by d, i.e., $\pi(A) = d$, B is replaced by f, i.e., $\pi(B) = f$ and so on

A	B	C	D	E	F	G	H	I	J	K	L	M
d	f	h	l	n	a	c	e	g	i	k	m	o
N	O	P	Q	R	S	T	U	V	W	X	Y	Z
q	s	u	w	y	z	b	j	p	r	t	v	x

Table 12.2 Percentage of frequency of letters in English text

A	B	C	D	E	F	G	H	I	J	K	L	M
8.15	1.44	2.76	3.79	13.11	2.92	1.99	5.26	6.35	0.13	0.42	3.39	2.54
N	O	P	Q	R	S	T	U	V	W	X	Y	Z
7.10	8.00	1.98	0.12	6.83	6.10	10.47	2.46	0.92	1.54	0.17	1.98	0.08

"CRYPTOGRAPHYISFUN" becomes "hyvubscyduevgzajq". A brute-force attack on the key space for this cipher takes much longer than a lifetime, even using the most powerful computer known today. However, this does not necessarily mean that the cipher is secure. In fact, as we will show now, it is easy to break this scheme even though it has a very large key space. For example, an attacker can easily attack the substitution cipher, written in English language, because the letters in the English language (or any other human language) are not random. For instance, the letter q in English is always followed by the letter u in any word. Moreover, certain letters, such as e and t appear far much more frequently than other letters, such as j and q. Table 12.2 lists the letters with their typical frequencies in English text. So we see that the most frequent letter is e, followed by t, a, o, and n. It may also be useful to consider sequences of two or three consecutive letters called *digrams* and *trigrams*, respectively. The 30 most common digrams are (in decreasing order) TH, HE, IN, ER, AN, RE, ED, ON, ES, ST, EN, AT, TO, NT, HA, ND, OU, EA, NG, AS, OR, TI, IS, ET, IT, AR, TE, SE, HI, OF. The 12 most common trigrams are (in decreasing order) THE, ING, AND, HER, ERE, ENT, THA, NTH, WAS, ETH, FOR, DTH.

Based on these features of English language and the fact that in a substitution cipher a particular letter is always replaced by some other fixed letter, the attack, known as, *frequency analysis*, on the substitution may be done as follows:

- The number of different cipher text characters or combination of characters are counted to determine the frequencies of usages.
- The cipher text is examined for patterns, repeated series and common combinations.
- The cipher text characters are replaced with possible plain text equivalents using the known language characteristics.

Remark Note that Caesar Cipher is a special case of Substitution Cipher which includes only one of the 26! possible permutations on 26 elements.

All the examples discussed till now have a common feature, that is, all the plain text alphabets are going to a fixed cipher text alphabet. This feature leads to the following important definition.

Definition 12.4.1   A cryptosystem is said to be a mono-alphabetic cryptosystem if and only if each alphabetic character is mapped into a unique alphabetic character once the key of the cryptosystem is fixed.

As we have described, the frequency attack on the mono-alphabetic substitution cipher could be carried out, because the mapping of each alphabet was fixed. Thus such an attack may be avoided by mapping different instances of the same plain text alphabet to different cipher text alphabets. In that case, counting the character frequencies will not offer much information about the mapping. This feature leads to the following definition.

Definition 12.4.2   A cryptosystem is said to be a poly-alphabetic cryptosystem if and only if different instances of the same plain text alphabet are mapped into different cipher text alphabets.

In the next section we are going to introduce an example of a poly-alphabetic cryptosystem.

12.4.4 Vigenère Cipher

Let n be a positive integer. Define $C = P = K = (\mathbf{Z}_{26})^n$. For a key $K = (k_1, k_2, \ldots, k_n)$, we define

$$e_k(x_1, x_2, \ldots, x_n) = (x_1 + k_1, x_2 + k_2, \ldots, x_n + k_n) \bmod 26$$

$$d_k(y_1, y_2, \ldots, y_n) = (y_1 - k_1, y_2 - k_2, \ldots, y_n - k_n) \bmod 26$$

Intuitively, the Vigenère cipher works by applying multiple shift ciphers in sequence. Let us consider the following example.

Example 12.4.1   Let the plain text message be CRYPTOLOGY and the key be CIP. Let us represent the alphabets as elements of $\mathbf{Z}_{26}$ by identifying A by 0, B by 1 and so on.
Then the scheme is as follows:

C	R	Y	P	T	O	L	O	G	Y		2	17	24	15	19	14	11	14	6	24
C	I	P	C	I	P	C	I	P	C		2	8	15	2	8	15	2	8	15	2
e	z	n	r	b	d	n	w	v	a		4	25	13	17	1	3	13	22	21	0

Remark Note that the two different instances of the same plain text alphabet O in the word CRYPTOLOGY are mapped into different cipher text alphabets, namely d and w. If the key is a sufficiently long word (chosen at random), then breaking this cipher seems to be a very difficult task. Indeed, it was considered by many to be an unbreakable cipher and although it was invented in the 16th century a systematic attack on the scheme was only devised hundreds of years later. For more details, we refer the readers to the text books (Menezes et al. 1996; Stinson 2002; Katz and Lindell 2007).

12.4.5 Hill Cipher

Hill cipher was first introduced in 1929 by the mathematician Lester S. Hill (1929). This cipher may be thought of as the first systematic and simple poly-alphabetic cipher, in which Algebra, specially Linear Algebra was used. Intuitively, in the Hill cipher, the plain text is divided into groups of adjacent letters of the some fixed length, say n and then each such group is transformed into a different group of n letters. An arbitrary n-Hill cipher has as its key a given $n \times n$ non-singular matrix whose entries are integers $0, 1, 2, \ldots, t - 1$, where t is the size of the set of alphabets i.e., if the alphabets are the English letters only, then $t = 26$. The Hill cipher for the English alphabets may be explained as follows: For $n \geq 2$, let $\mathcal{P} = \mathcal{C} = \mathbf{Z}_{26}^n$ and $\mathcal{K} = \{K : K$ is a non-singular matrix of order n with entities from $\mathbf{Z}_{26}\}$. Let us take an $n \times n$ non-singular matrix $K = (k_{ij})_{n \times n}$ as our key. For $x = (x_1, x_2, \ldots, x_n) \in \mathcal{P}$ and $K \in \mathcal{K}$, the encryption rule $e_K(x) = y = (y_1, y_2, \ldots, y_n)$ is defined as follows:

$$(y_1, y_2, \ldots, y_n) = (x_1, x_2, \ldots, x_n) \begin{bmatrix} k_{11} & k_{12} & \cdots & k_{1n} \\ k_{21} & k_{22} & \cdots & k_{2n} \\ \cdots & \cdots & \cdots & \cdots \\ k_{n1} & k_{n2} & \cdots & k_{nn} \end{bmatrix}.$$

So idea is to take n linear combinations of the n alphabetic characters in one cipher text element. Now we shall show how to decrypt the cipher text. The above equation can be written as $y = xK$. As K is a non-singular matrix, for decryption, we can use, $d_K(y) = yK^{-1} = x$, where all the operations are performed in $\mathbf{Z}_{26}$.

Example 12.4.2 Let us consider a simple example of a Hill cipher with the set of alphabets as the English letters only, i.e., "a" to "z" only. While encrypting and decrypting, we shall represent the 26 characters in our alphabet in order by the non-negative integers $1, 2, \ldots, 26 \ (\equiv 0)$. Let $m = 2$, i.e., we are considering only the Hill 2-cipher with the key $K = \begin{bmatrix} 7 & 3 \\ 2 & 5 \end{bmatrix}$. Then the plain text element x could be written as $x = (x_1, x_2)$ and the cipher text element as $y = (y_1, y_2) = (x_1, x_2)\begin{bmatrix} 7 & 3 \\ 2 & 5 \end{bmatrix}$. Suppose Alice wants to send some secret message say "LOVE" to Bob over an insecure channel using the above mentioned Hill 2-cipher. Since, $m = 2$, Alice will break "LOVE" into two parts, each containing two characters i.e., "LO" and "VE". The corresponding integer values are $(12, 15)$ and $(22, 5)$, respectively, since L, O, V, E corresponds to the integers 12, 15, 22, and 5, respectively. Note that the matrix K is known to both Alice and Bob. For encryption Alice will compute

$$\left((12, 15) \begin{bmatrix} 7 & 3 \\ 2 & 5 \end{bmatrix} \right) \bmod 26 = (10, 7) = (j, g) \quad \text{and}$$

$$\left((22, 5) \begin{bmatrix} 7 & 3 \\ 2 & 5 \end{bmatrix} \right) \bmod 26 = (8, 13) = (h, m).$$

So Alice will send to Bob the cipher text "jghm" over an insecure channel. After receiving it, Bob will break "jghm" into "jg" and "hm" and convert it into $(10, 7)$ and $(8, 13)$ and will compute:

$$\left((10, 7) \begin{bmatrix} 7 & 3 \\ 2 & 5 \end{bmatrix}^{-1} \right) \bmod 26 = \left((10, 7) \begin{bmatrix} 19 & 25 \\ 8 & 11 \end{bmatrix} \right) \bmod 26 = (12, 15) = (l, o)$$

and

$$\left((8, 13) \begin{bmatrix} 7 & 3 \\ 2 & 5 \end{bmatrix}^{-1} \right) \bmod 26 = \left((8, 13) \begin{bmatrix} 19 & 25 \\ 8 & 11 \end{bmatrix} \right) \bmod 26 = (22, 5) = (v, e).$$

So Bob gets back the message "love" sent from Alice.

Remark For the cryptanalysis of the Hill cipher, we refer the readers to the text book (Stinson 2002).

In all of the above ciphers, if we observe carefully, we note that before sending the secret message, the sender and the receiver have to agree upon a common key secretly. The same key is used for both encryption and decryption. This type of cryptosystem is known as *private-key* or *symmetric key* cryptosystem. One drawback of a private-key cryptosystem is that it requires a prior secure communication of the common key k between sender and receiver, using a secure channel, before any cipher text is transmitted. In practice, this may be very difficult to achieve as the sender (say Alice) and the receiver (say Bob) may not have the luxury of meeting before hand or they may not have an access to a reasonably secure channel. Since Alice and Bob are not meeting in private, we have to assume that the adversary or the attacker is able to hear everything that Alice and Bob pass back and forth. Under this situation, is it possible for Alice and Bob to exchange a secret key for future secure communication? This leads to an interesting concept known as *public-key* or *asymmetric* cryptosystem.

12.5 Introduction to Public-Key Cryptography

Intuitively, in a public-key cryptosystem, there are two keys instead of one secret key. One of these keys is an encryption key, used by sender to encrypt the message and the other is a decryption key, used by the receiver to decrypt the cipher text. The most important thing is that the secrecy of encrypted messages should be preserved even against an adversary who knows the encryption key (but not the decryption key). Cryptosystems with this property are called asymmetric or public-key cryptosystems, in contrast to the symmetric or private-key encryption schemes. In a public-key cryptosystem the encryption key is called the public key, since it is kept open or published by the receiver so that anyone who wishes to send an encrypted message may do so and the decryption key is called the private key since it is kept completely private by the receiver.

There are certain advantages of public-key cryptography over the secret key cryptography. As the public-key encryption allows key distribution to be done over public channels, it simplifies the initial deployment of the system. As a result, the maintenance of the system becomes quite easy when parties join or leave the cryptosystem. Further, if we are going to use a public-key cryptosystem, then we do not need to store many secret keys for further secure communication. Even if all pairs of parties want the ability to communicate securely, each party needs only to store his/her own private key in a secure fashion. The public keys of the other parties can either be obtained when needed, or stored in a non-secure fashion. Finally, public-key cryptography plays an important role in the scenario where parties, who have never previously interacted, want the ability to communicate securely. For example, a merchant may post his public key on-line; any buyer making a purchase can obtain the merchant's public key when they need to encrypt their credit card information.

Asymmetric cryptography is a relatively new field. Whiteld Diffie and Martin Hellman, in 1976, published a paper entitled "New Directions in Cryptography" (Diffie and Hellman 1976) in which they formulated the concept of a public-key encryption system. A short time earlier, Ralph Merkle had independently invented a public-key construction for an undergraduate project at Berkeley, but this was little understood at the time. Merkle's paper entitled "Secure communication over insecure channels" appeared in Merkle (1982). However, it turns out that the concept of public-key encryption was originally discovered by James Ellis while working at the British Government Communications Headquarters (GCHQ). Ellis's discoveries in 1969 were classified as secret material by the British government and were not declassified and released until 1997, after his death (cf. Hoffstein et al. 2008).

The Diffie–Hellman key agreement was invented in 1976 during a collaboration between Whitfield Diffie and Martin Hellman and was the first practical method for establishing a shared secret over an insecure communication channel. The most important contribution of the paper by Diffie and Hellman (1976) was the definition of a Public Key Cryptosystem (PKC) and its associated components namely one-way functions and trapdoor information as defined below.

Definition 12.5.1 A one-way function is an invertible function that is easy to compute, but whose inverse is difficult to compute i.e., if any algorithm that attempts to compute the inverse in a reasonable amount of time will almost certainly fail, where the phrase almost certainly must be defined probabilistically.

A secure public-key cryptosystem is built using one-way functions that have a trapdoor which is defined below.

Definition 12.5.2   The trapdoor of a function is a piece of auxiliary information that allows the inverse of the function to be computed easily but without that auxiliary information it is computationally very hard to compute the inverse function.

Informally, a public-key (or asymmetric) cryptosystem consists of two types of keys, namely, a private key k_{priv} and a public key k_{pub}. Corresponding to each public/private-key pair (k_{priv}, k_{pub}), there is an encryption algorithm $e_{k_{pub}}$ and the corresponding decryption algorithm $d_{k_{priv}}$. The encryption algorithm $e_{k_{pub}}$ corresponding to k_{pub} is public of knowledge and easy to compute. Similarly, the decryption algorithm $d_{k_{priv}}$ must be easily computable by someone who knows the private key k_{priv}, but it should be very difficult to compute for someone who knows only the public key k_{pub}. The idea behind a public-key cryptosystem is that it might be possible to find a cryptosystem where it is computationally infeasible to determine the decryption function $d_{k_{priv}}$, given the encryption function $e_{k_{pub}}$ and the public key k_{pub} but with the knowledge of secret key k_{priv} it is easy to compute the inverse of $e_{k_{pub}}$, i.e., $d_{k_{priv}}$. In other words, the private key k_{priv} may be thought of as a trapdoor information for the function $e_{k_{pub}}$. Under this situation, say Alice wants to send a secret message to Bob. Then Alice will use the encryption rule $e_{k_{pub}}$ that is made public by Bob, to encrypt the secret message and will send the encrypted message to Bob through an insecure channel. A public-key cryptosystem is said to be secure if it is computationally infeasible to determine $d_{k_{priv}}$ given $e_{k_{pub}}$ and k_{pub}. In that case, only Bob can decrypt the cipher text to get the plain text. The main advantage of a public-key cryptosystem is that Alice can send an encrypted message to Bob, without any prior communication of a secret key, by using the public encryption rule $e_{k_{pub}}$ (hence the name public-key cryptosystem). The formal definition of public-key cryptosystem is as follows.

Definition 12.5.3   A public-key (or asymmetric) cryptosystem is a five-tuple $(\mathcal{P}, \mathcal{C}, \mathcal{K}, \mathcal{E}, \mathcal{D})$, where the following conditions are satisfied:

- $\mathcal{P}$, the plain text space, is a finite set of possible plain texts,
- $\mathcal{C}$, the cipher text space, is a finite set of possible cipher texts,
- $\mathcal{K}$, the key space, is a finite set of possible keys,
- For each $k = (k_{priv}, k_{pub}) \in \mathcal{K}$, there is an encryption rule $e_{k_{pub}} \in \mathcal{E}$ and the corresponding decryption rule $d_{k_{priv}} \in \mathcal{D}$. Each $e_{k_{pub}} : \mathcal{P} \to \mathcal{C}$ and $d_{k_{priv}} : \mathcal{C} \to \mathcal{P}$ are functions such that $d_{k_{priv}}(e_{k_{pub}}(x)) = x$ for every plain text element $x \in \mathcal{P}$.

It is quite interesting to note that Diffie and Hellman formulated this concept without finding a specified encryption/decryption pair of functions, although they did propose a similar method through which Alice and Bob can securely exchange a random piece of data whose value is not known initially to either one. Before describing the method, let us first give a brief introduction about the discrete logarithm problem.

12.5.1 Discrete Logarithm Problem (DLP)

The discrete logarithm problem is a well-known mathematical problem that plays an important role in public-key cryptography. The Diffie–Hellman Key Exchange protocol (Diffie and Hellman 1976) and ElGamal public-key cryptosystems are based on the discrete logarithm problem in a finite field $\mathbf{Z}_p$, where p is a large prime. Let us first explain the problem.

Let p be a large prime. Recall that $\mathbf{Z}_p^* = \mathbf{Z}_p \setminus \{(0)\}$ is a cyclic group of order $p - 1$ (see Chap. 10). Thus there exists a primitive element, say g of $\mathbf{Z}_p^*$. As a result, every non-zero element of $\mathbf{Z}_p$ is equal to some power of g. Further recall that, by Fermat's Little Theorem, $g^{p-1} = (1)$ and no smaller power of g is equal to (1). Hence all the elements of $\mathbf{Z}_p^*$ may be represented as $\mathbf{Z}_p^* = \{g, g^2, \ldots, g^{p-1}\}$. Now we are going to define the discrete logarithm problem in $\mathbf{Z}_p^*$.

Definition 12.5.4 The Discrete Logarithm Problem (DLP) in $\mathbf{Z}_p^*$ states that given a primitive element g of $\mathbf{Z}_p^*$ and an element $h \in \mathbf{Z}_p^*$, find x such that $g^x = h$. The number x is called the discrete logarithm of h modulo p to the base g and is denoted by $\log_g(p)$.

Remark Till date, in general, there is no efficient algorithm known to solve the DLP.

Though here, we define the discrete logarithm problem in terms of the primitive base g, in general this is not mandatory. We can take elements of any group and use the group law instead of multiplication. This leads to the most general form of the discrete logarithm problem as follows.

Definition 12.5.5 The Discrete Logarithm Problem (DLP) in a group $(G, \circ)$ states that, given an element g and h in G, find x such that

$$\underbrace{g \circ g \circ \cdots \circ g}_{x \text{ times}} = h.$$

Remark Discrete logarithm problem is not always hard. This hardness depends on the groups. For example, a popular choice of groups for discrete logarithm-based crypto-systems is Z_p^* where p is a prime number. However, if $p - 1$ is a product of small primes, then the Pohlig–Hellman algorithm can solve the discrete logarithm problem in this group very efficiently. That is why we always need a special type of prime p, known as safe prime which is of the form $2q + 1$, where q is a large prime, when using Z_p^* as the basis of discrete logarithm-based crypto-systems.

12.5.2 Diffie–Hellman Key Exchange

Until 1976, there was a belief that in cryptography, encryption could not be done without sharing a secret key among the sender and the receiver. However, Diffie and

Hellman noticed that there is a natural asymmetry in the world such that there are certain actions that can be easily performed but not easily reversed. For example, it is easy to multiply two large primes but difficult to recover these primes from their product. The existence of such phenomena motivates to construct an encryption scheme that does not rely on shared secrets, but rather one for which encrypting is "easy" but reversing this operation (i.e., decrypting) is infeasible for anyone other than the designated receiver. Using this observation, Diffie and Hellman solved the following interesting problem which looks quite infeasible. The problem may be stated as follows: Suppose Alice wants to share some secret information with Bob. This information may be a common key for future secure communication between Alice and Bob using symmetric key encryption. The only problem is that whatever communication will be made between Alice and Bob will be observed by the adversary. In this situation, the question is: how is it possible for Alice and Bob to share a key without making it available to the adversary? As mentioned earlier, at a first glance, it seems that it is quite impossible for Alice and Bob to solve the problem. But Diffie and Hellman first came up with a brilliant solution to this problem taking into account the difficulty of solving the discrete logarithm problem in $\mathbf{Z}_p^*$, where p is a large prime.

The steps for Diffie–Hellman key exchange protocol are as follows:

- At the first step, Alice and Bob agree on a large prime p and a generator, say g, of the cyclic group $\mathbf{Z}_p^*$.
- Alice and Bob make the values of p and g public, i.e., these values are known to all, in particular known to the adversary.
- Alice chooses a random integer a that she keeps secret, while at the same time Bob chooses randomly an integer b that he keeps secret.
- Alice and Bob use the secret integers a and b to compute $X \equiv g^a \pmod{p}$ and $Y \equiv g^b \pmod{p}$, respectively.
- Alice sends through an insecure channel the value X to Bob while Bob sends through an insecure channel the value Y to Alice. (Note that the adversary can see the values of X and Y as they are sent over insecure channels).
- After getting the value Y from Bob, Alice computes with the help of the secret value a, the value $X' \equiv Y^a \pmod{p}$, while Bob computes $Y' \equiv X^b \pmod{p}$ with the help of the secret value b that he has. Note that as $\mathbf{Z}_p^*$ is a commutative group, $X' \equiv Y^a \pmod{p} \equiv (g^b)^a \pmod{p} \equiv (g)^{ab} \pmod{p} \equiv (X^a)^b \pmod{p} \equiv Y' \pmod{p}$.
- Thus the common key for Alice and Bob is $g^{ab} \pmod{p}$.

Note that the adversary knows the values of $X \equiv g^a \pmod{p}$ and $Y \equiv g^b \pmod{p}$. The adversary also knows the values of g and p. To the adversary, only a and b are unknown. So if the adversary can solve the DLP, then he can find both a and b and thus can compute easily the shared secret value g^{ab} of Alice and Bob. At this point of time, though it seems that Alice and Bob are safe provided that the adversary is unable to solve the DLP, this is not quite correct. There is no doubt that one method of finding the shared value of Alice and Bob is to solve the DLP, but that is not the precise problem that the adversary is intended to solve. The adversary is actually intended to solve the following problem, known as Diffie–Hellman Problem (DHP).

Definition 12.5.6 Let p be a prime and g be a primitive element, i.e., the generator of the cyclic group $\mathbf{Z}_p^*$. The Diffie–Hellman Problem (DHP) is the problem of computing the value of g^{ab} (mod p) from the known values of g, p, g^a (mod p) and g^b (mod p).

Remark There is no doubt that the DHP is no harder than the DLP i.e., if some one can solve the DLP, then he can compute the secret values a and b of Alice and Bob, respectively, to compute their common key g^{ab}. However, the converse is less clear in the sense that if the adversary has an algorithm which can efficiently solve the DHP, then with that algorithm, can the adversary efficiently solve the DLP? The answer to this question is still not known.

In the next section we are going to discuss about the celebrated public-key encryption scheme known as RSA, based on factorization of integers and simple group theoretic techniques.

12.6 RSA Public Key Cryptosystem

In 1977, a year after the publication of the paper by Diffie and Hellman, three researchers at MIT developed a practical method to construct a public-key cryptosystem. This became known as RSA, after the initials of the three developers: Ron Rivest, Adi Shamir and Leonard Adelman. RSA is probably the most widely used public-key cryptosystem. It was patented in USA in 1983.

We now explain the RSA cryptosystem (Rivest et al. 1978). But before that we need a small introduction. Many of us have an impression that regarding computation, computer has the greatest power and computer can compute any thing very quickly given to it. But unfortunately, there are many computational works that even computers cannot do quickly. Let us consider two three digit prime numbers say $p = 101$ and $q = 113$. Let $n = pq$. Now if we give n to the computer and ask computer, by writing a suitable program, to factorize n, the computer will work readily. Now we consider p and q, each being a prime of 512 bits, for example, we may consider $p = 12735982667070981883239844093975974547292895820146911015162$ $603493560951020761841391134902665367989016813709401453182925 84413693$ $0973759846652210022179676 6571$ and $q = 98156335285410483440369801 9941$ $100653177550902406149945593433397033362991690486898903293727 2369426$ $1120467535604944588515163877922683718645770658848406 87021$, then $n = $ $12501173848581957366071203221142768531666009713556690031834829 12941$ $686579528012586798580489169947843472678336229158650831 5065615625486$ $63603060018793553021671872404768381128567128134 16806772774907378919$

Table 12.3 Algorithm for RSA public key cryptosystem

- **Key Generation Algorithm by Bob:**

  1. Generate two distinct large prime numbers p and q, each roughly of same size.
  2. Compute $n = pq$ and $\phi(n) = (p-1)(q-1)$.
  3. Select a random integer e with $1 < e < \phi(n)$, such that $\gcd(e, \phi(n)) = 1$.
  4. Use the extended Euclidean algorithm to find the integer d, $1 < d < \phi(n)$, such that $ed \equiv 1 \pmod{\phi(n)}$.
  5. Make public the public keys n and e (which are known to every body) and keep secret the private keys p, q, and d (which are known only to Bob).

- **Encryption Algorithm for Alice:**

  1. Obtain Bob's public key (n, e).
  2. Represent the message m as an integer from the set $\{0, 1, 2, \ldots, n - 1\}$.
  3. Compute $c \equiv m^e \pmod{n}$.
  4. Send the cipher text c to Bob.

- **Decryption Algorithm for Bob:**

  1. To obtain the plain text message m, Bob uses his private key d to get $m \equiv c^d \pmod{n}$.

535173884158630044034390278552191020034924578825282807318048667680131867741261022374538179761526720006374991 is an 1024 bit number having 309 many decimal digits. Now if we input n to a computer and ask it to factorize, then it will not be possible, even for a super computer, to factorize the number n in a reasonable time. So for a computer also the factorization of large number which is a product of two distinct large primes is considered to be a difficult or hard problem. This is because not yet any efficient algorithm for factorization has been discovered. This inability of computers added a strength to the public-key cryptographic scheme RSA which is described in the next section.

12.6.1 Algorithm for RSA Cryptosystem (Rivest et al. 1978)

Suppose Alice wants to send a secret message to Bob using the RSA public-key cryptosystem. Note that as Alice wants to send the message to Bob, Bob has to take the initiative to construct his public- and private-key pairs. This step is known as the key generation step. Alice then collects the public key of Bob and uses the publicly known encryption algorithm to get the cipher text. Alice then sends the cipher text to Bob through an insecure channel. After receiving the cipher text from Alice, using his private key, Bob decrypts the cipher text to get the plain text. The actual steps for RSA algorithm are described in Table 12.3.

Thus formally, the RSA cryptosystem is a five-tuple $(\mathcal{P}, \mathcal{C}, \mathcal{K}, \mathcal{E}, \mathcal{D})$, where $\mathcal{P} = \mathcal{C} = \mathbf{Z}_n$, n is the product of two distinct large prime numbers p and q, $\mathcal{K} = \{(n, p, q, e, d) : ed \equiv 1 \pmod{\phi(n)}\}$, $\phi(n)$, known as Euler ϕ function, is the number of positive integers not exceeding n and relatively prime to n. For each $K = (n, p, q, e, d) \in \mathcal{K}$, the encryption function e_K is defined by $e_K(x) =$

x^e (mod n) and the decryption function d_K is defined by $d_K(y) = y^d$ (mod n), where $x, y \in \mathbf{Z}_n$. The values n and e are considered to be the public keys while the values p, q, and d are considered to be the private keys.

12.6.2 Sketch of the Proof of the Decryption

A natural question that comes to our mind: Is really $m \equiv c^d$ (mod n)? The sketch of the proof is as follows.

It is given that $ed \equiv 1$ (mod $\phi(n)$). So there must exist some integer t such that $ed = 1 + t\phi(n)$. Since $m \in \{0, 1, 2, \ldots, n-1\}$, we consider the following four cases. Case 1: $\gcd(m, p) = p$ and $\gcd(m, q) = q$, (this case happens only when $m = 0$) Case 2: $\gcd(m, p) = 1$ and $\gcd(m, q) = 1$, Case 3: $\gcd(m, p) = 1$ and $\gcd(m, q) = q$ and finally, Case 4: $\gcd(m, p) = p$ and $\gcd(m, q) = 1$. If $\gcd(m, p) = 1$, then by Fermat's Theorem, $m^{p-1} \equiv 1$ (mod p) $\Rightarrow m^{t(p-1)(q-1)} \equiv 1$ (mod p) $\Rightarrow m^{1+t(p-1)(q-1)} \equiv m$ (mod p). Now if $\gcd(m, p) = p$, then also the above equality holds as both sides are equal to 0 modulo p. Hence in both the cases, $m^{ed} \equiv m$ (mod p). By the same argument it follows that $m^{ed} \equiv m$ (mod q). Finally, since p and q are distinct prime numbers, it follows that $m^{ed} \equiv m$ (mod n) and hence $c^d \equiv (m^e)^d \equiv m$ (mod n). Hence the result follows.

Note It is currently difficult to obtain the private key d from the public key (n, e). However, if one could factor n into p and q, then one could obtain the private key d. Thus the security of the RSA system is based on the assumption that factoring is difficult. The discovery of an easy method of factoring would "break" RSA cryptosystem.

Remark The encryption function $e_K(m)$ defined by $e_K(m) = m^e$ (mod n), for all $m \in \mathcal{P}$ and $K \in \mathcal{K}$, is a group homomorphism, as $e_K(m_1 m_2) = (m_1 m_2)^e$ (mod n) $= e_K(m_1)e_K(m_2)$. For this reason, RSA is known as homomorphic encryption scheme.

12.6.3 Some Attacks on RSA Cryptosystem

To understand some of the attacks on RSA cryptosystem, let us first review some of the properties of this cryptosystem. First of all this cryptosystem is deterministic or mono-alphabetic, i.e., the same plain text message will always be encrypted or mapped to the same cipher text. Further, the RSA encryption function is homomorphic, i.e., $e_K(x \cdot y) = e_K(x)e_K(y)$, for all $x \in \mathcal{P}$ and for all $K \in \mathcal{K}$. Based on these properties, we provide some of the following attacks on RSA cryptosystem.

- *Encrypting short messages using small encryption exponent e*: Suppose for the RSA encryption, we are going to encrypt a short message m using a small encryption exponent e. Let us illustrate this example for a particular small e assume e to be 3 and the short message m is such that $m < n^{1/3}$. The only information, regarding the plain text m, that is known to the adversary is that $m < n^{1/3}$. In that case, a very practical attack can be made by the adversary. Note that major advantage for the adversary is that the encryption of m does not involve any modular reduction since the integer m^3 is less than n. As a result, given the cipher text $c \equiv m^3 \pmod{n}$, the adversary can determine m by simply computing $m = c^{1/3}$ over the integers.

- *Encrypting same messages using same small encryption exponent e*: The above attack illustrates that short messages can be recovered easily from their small encryption exponent. Now we extend the attack to the case of arbitrary length messages as long as the same message m which is sent to multiple receivers with the same small encryption exponent e but with different values of n. Let us illustrate the attack with small exponent $e = 3$. Let the same message m be sent to three different receivers using the three public keys $pk_1 = (n_1, 3)$, $pk_2 = (n_2, 3)$ and $pk_3 = (n_3, 3)$, respectively, where $n_i = p_i q_i$, p_i and q_i are primes, $i = 1, 2, 3$. Then the cipher texts are $c_1 \equiv m^e \pmod{n_1}$, $c_2 \equiv m^e \pmod{n_2}$ and $c_3 \equiv m^e \pmod{n_3}$. As the cipher texts reach the receiver through insecure channels, the adversary can get hold of these cipher texts. Note that we may assume that $\gcd(n_i, n_j) = 1$, for $i \neq j$, otherwise, the adversary can decrypt the cipher text c_i to get the plain text m by factorizing n_i to get p_i and q_i. Let $n = n_1 n_2 n_3$. Then by the Chinese Remainder Theorem, the adversary can find a unique non-negative value $c < n$ such that $c \equiv c_1 \pmod{n_1}$, $c \equiv c_2 \pmod{n_2}$ and $c \equiv c_3 \pmod{n_3}$, i.e., $c \equiv m^3 \pmod{n_1}$, $c \equiv m^3 \pmod{n_2}$ and $c \equiv m^3 \pmod{n_3}$. As n_1, n_2 and n_3 are pairwise relatively prime, we have $c \equiv m^3 \pmod{n}$. Observe that $m < n_i$ for all $i = 1, 2, 3$. Hence $m^3 < n$. Thus using a similar technique as described in the last paragraph, the adversary can attack the scheme.

- *Encrypting same message using two relatively prime exponents of same modulus n*: Suppose Alice wants to send the same message m to two of her employees, say Bob1 and Bob2, using the public keys (n, e_1) and (n, e_2) of Bob1 and Bob2, respectively, where $\gcd(e_1, e_2) = 1$. Then the adversary can get hold of the cipher texts $c_1 \equiv m^{e_1} \pmod{n}$ and $c_2 \equiv m^{e_2} \pmod{n}$ of Bob1 and Bob2, respectively. As $\gcd(e_1, e_2) = 1$, by using Extended Euclidean algorithm, the adversary can find integers u and v such that $ue_1 + ve_2 = 1$. Then the adversary will compute $c_1^u \cdot c_2^v \pmod{n}$. Note that $c_1^u \cdot c_2^v \pmod{n} \equiv (m)^{ue_1 + ve_2} \pmod{n} \equiv m \pmod{n}$. Thus from c_1 and c_2, the adversary will get the message m.

- *Bidding attack*: Suppose one organization has decided to buy some software. Many software companies produce that software product. But the organization wants to select the software company through global tendering. That is the organization will publish an advertisement in which they will specify the requirement of the software and will ask for the software companies to submit quotations against that product. The organization will give the project to that software company having minimum quotation. As the organization wants open tendering for transparency, every body will see what is being sent to the organization. As a result, the software companies will not send the quotation (only the amount, e.g.,

20000000) as plain text form. So they must encrypt the quotation and will send the cipher text of that quotation. To get uniformity as well as high security, the organization decides to use RSA encryption scheme with large primes p and q. Each software company will get the public key (n, e) to encrypt their quotations. Until now it looks like the system is quite safe and useful. However, we shall show that due to the homomorphic property of the encryption, the RSA encryption is very much unsafe for the bidding or auction purpose. We shall show how an adversary can submit a quotation which is a modification of some valid quotation of some software company. Suppose initially there are k software companies who submitted their cipher text of the quotations, say $c_1, c_2, \ldots, c_k$ corresponding to the quotations $m_1, m_2, \ldots, m_k$, respectively. As the cipher texts will come to the organization through an insecure channel or the organization may keep the cipher texts public, the adversary will always be able to get hold of these cipher texts. The aim of the adversary is to submit quotations c_i' which is, say 10 % less than the original quotation m_i, for all $i = 1, 2, \ldots, k$. One thing we assume that m_i's are all multiples of 10. This assumption is quite justified as usually the amount in the quotations are multiples of 10. Thus to attack the scheme, the adversary will do the following:

- collect all the cipher texts c_i corresponding to the plain text m_i, $i = 1, 2, \ldots, k$.
- construct $c_i' = c_i \cdot (90)^e \cdot (100)^{-e}$, $i = 1, 2, \ldots, k$.
- send c_i' to the organization, $i = 1, 2, \ldots, k$.

Now let us see what is actually happening. Note that $c_i' = c_i \cdot (90)^e \cdot (100)^{-e} = (m \cdot 90 \cdot (100)^{-1})^e$. As p and q are large primes, $\gcd(100, n) = 1$. Thus the inverse of 100 in the ring Z_n exists and c_i' will be simply the cipher text of the plain text m_i' which is the 10 % decrease of the value of m_i. Thus $\min\{m_1', m_2', \ldots, m_k'\} < \min\{m_1, m_2, \ldots, m_k\}$. As a result, the adversary will always get the project. The similar attack may also be done in case of any on-line auction using RSA encryption scheme in which the adversary can always submit a price which is higher than all the actually submitted prices.

In the next section, we are going to describe another public-key cryptosystem based on discrete logarithm problem.

12.7 ElGamal Cryptosystem

Though Diffie–Hellman key exchange algorithm provided the first direction towards the invention of public-key cryptosystem, it did not achieve the full goal of being a public-key cryptosystem. The first public-key cryptosystem, as mentioned in the last section, was the RSA system. However, though RSA was historically the first one, the most natural development of a public-key cryptosystem based on the discrete log problem and Diffie–Hellman key exchange was developed by ElGamal (1985). In this section we are going to describe the version of the ElGamal cryptosystem which is based on the discrete logarithm problem over the finite field $\mathbf{Z}_p$. Note that the similar construction can also be made generally using the DLP in any group.

Table 12.4 Algorithm for ElGamal public key cryptosystem

- **Key Generation Algorithm by Bob:**

  1. Generate a large prime number p.
  2. Find a primitive element say g of $\mathbf{Z}_p^*$. Bob may take help of some trusted third party in generating the prime p and the primitive element.
  3. Choose $a, 1 \leq a \leq p-1$ and compute $X \equiv g^a \pmod{p}$.
  4. Keep a as a secret key and make public the entities g, p, and X as public keys.

- **Encryption Algorithm for Alice:**

  1. Obtain Bob's public key.
  2. Represent the message as an integer m in the interval $[1, p-1]$.
  3. Choose a random ephemeral key $k \in \mathbf{Z}_{p-1}$.
  4. Compute $c_1 \equiv g^k \pmod{p}$ and $c_2 \equiv mX^k \pmod{p}$.
  5. Send the cipher text (c_1, c_2) to Bob.

- **Decryption Algorithm for Bob:**

  1. To obtain the plain text message m, Bob uses his private key a to compute $m \equiv (c_1^a)^{-1} \cdot c_2 \pmod{p}$.

12.7.1 Algorithm for ElGamal Cryptosystem

Suppose Alice wants to send a secret message to Bob using the ElGamal public-key cryptosystem. The steps for ElGamal algorithm are described in Table 12.4.

Informally, the plain text m is "masked" by multiplying it by X^k to get c_2. The value $c_1 = g^k$ is also transmitted as a cipher text c_1. The receiver Bob who knows the secret key a is able to compute X^k. Thus Bob can "remove the mask" by computing $c_2(c_1^a)^{-1} \pmod{p}$. Thus formally, the ElGamal cryptosystem is a five-tuple $(\mathcal{P}, \mathcal{C}, \mathcal{K}, \mathcal{E}, \mathcal{D})$, where $\mathcal{P} = \mathbf{Z}_p^*$, $\mathcal{C} = \mathbf{Z}_p^* \times \mathbf{Z}_p^*$, $\mathcal{K} = \{(p, g, a, X) : X \equiv g^a \pmod{p}\}$ and p is a large prime. The value a is considered as a secret key and the entities g, p, and X are considered as the public keys. For each $K = (p, g, a, X) \in \mathcal{K}$ and for a secret random ephemeral key $k \in \mathbf{Z}_{p-1}$, the encryption function e_K is defined by $e_K(m, k) = (c_1, c_2)$, where $c_1 \equiv g^k \pmod{p}$ and $c_2 \equiv mX^k \pmod{p}$. For $c_1, c_2 \in \mathbf{Z}_p^*$, the decryption function is defined as $d_K(c_1, c_2) = c_2(c_1^a)^{-1} \pmod{p}$.

Remark 1 The ElGamal cryptosystem will be insecure if the adversary can compute the value of a, i.e., if the adversary can compute $\log_g X$, as in that case, the adversary will decrypt the cipher just like Bob does. Thus a necessary condition that the ElGamal cryptosystem to be secure is that the Discrete Logarithm Problem in $\mathbf{Z}_p^*$ is infeasible.

Remark 2 Due to the use of random ephemeral key $k \in \mathbf{Z}_{p-1}$, the encryption scheme becomes randomized. More specifically, there will be $p-1$ many options of the cipher texts that are the encryption of the same plain text, with the result that the encryption scheme as poly-alphabetic.

In the next section we are going to introduce another public-key cryptosystem which is a variant of the Rabin encryption scheme. The actual Rabin scheme is very similar to the RSA encryption scheme, but with one crucial difference: it is possible to prove that the Rabin encryption scheme is CPA-secure under the assumption that factoring is hard. This equivalence makes the scheme attractive.

12.8 Rabin Cryptosystem

The Rabin cryptosystem was published in 1979 by Michael O. Rabin. The Rabin cryptosystem was the first asymmetric cryptosystem for which the security is based on the fact that it is easy to compute square roots modulo a composite number N if the factorization of N is known, yet it appears difficult to compute square roots modulo N, when the factorization of N is unknown. In other words, the factoring assumption implies the difficulty of computing square roots modulo a composite. However, due to the deterministic encryption nature (i.e., mono-alphabetic nature) of the schemes, the actual Rabin encryption scheme was not chosen plaintext attack (CPA) secure. But a simple variant of original Rabin encryption provides a CPA security. In this section, we shall discuss about a variant of the original Rabin encryption scheme which is CPA-secure. For detailed security notions, the reader may refer the books (Adhikari et al. 2013; Katz and Lindell 2007).

The preliminaries required to understand the scheme is explained in the next section.

12.8.1 Square Roots Modulo N

Finding square roots modulo $N = pq$, where p and q are distinct primes plays an important role in constructing the Rabin public-key cryptosystem. The following proposition provides a characterization of an element y to be a quadratic residue modulo a composite integer N, where N is a product of two distinct odd primes.

Proposition 12.8.1 *Let $N = pq$ where p and q are distinct odd primes and $y \in \mathbf{Z}_N^*$. Let (y_p, y_q) denote the corresponding element in $\mathbf{Z}_p^* \times \mathbf{Z}_q^*$, where $y_p \equiv y \pmod{p}$ and $y_q \equiv y \pmod{q}$. Then y is a quadratic residue modulo N if and only if y_p is a quadratic residue modulo p and y_q is a quadratic residue modulo q.*

Proof Let y be a quadratic residue modulo N. Then there exists $x \in \mathbf{Z}_N^*$ such that $x^2 \equiv y \pmod{N}$. Let for $x \in \mathbf{Z}_N^*$, (x_p, x_q) be the corresponding element in $\mathbf{Z}_p^* \times \mathbf{Z}_q^*$, where $x_p \equiv x \pmod{p}$ and $x_q \equiv x \pmod{q}$. Thus $(x_p, x_q)^2 = (x_p^2, x_q^2)$ is the corresponding element of x^2 in $\mathbf{Z}_p^* \times \mathbf{Z}_q^*$, with the result that $(x_p^2, x_q^2) = (y_p, y_q)$, which implies that $x_p^2 \equiv y_p$ (mod p) and $x_q^2 \equiv y_q$ (mod q). Thus y_p is a quadratic residue modulo p and y_q is a quadratic residue modulo q.

Conversely, suppose y_p and y_q are quadratic residues modulo p and q, respectively. Then there exist x_p and x_q in $\mathbf{Z}_p^*$ and $\mathbf{Z}_q^*$, respectively, such that $x_p^2 \equiv y_p$ (mod p) and $x_q^2 \equiv y_q$ (mod q). Then arguing as above, we can say that there exist $x, y \in \mathbf{Z}_N^*$ such that $x^2 \equiv y$ (mod N), with the result that y is a quadratic residue modulo N. $\qquad\square$

The security of the Rabin cryptosystem is based on the fact that it is easy to compute square roots modulo an odd composite number $N = pq$ if the factorization of N is known, where p and q are distinct odd primes such that $p \equiv q \equiv 3$ (mod 4). However, there is no efficient method yet known to compute square roots modulo N if the factorization of N is not known. In fact, we shall show that computing square roots modulo N is equivalent to (in the sense that it is equally hard as) factoring N. For that let us prove the following proposition which shows that if we can compute the square roots modulo $N = pq$, where p and q are distinct odd primes, then we can find an efficient algorithm for factoring.

Proposition 12.8.2 *Let $N = pq$, where p and q are distinct odd primes. Let a_1 and a_2 be such that $a_1^2 \equiv a_2^2 \equiv b$ (mod N), but $a_1 \not\equiv \pm a_2$ (mod N). Then there exists an efficient algorithm to factor N.*

Proof As $a_1^2 \equiv a_2^2$ (mod N), N divides $(a_1 + a_2)(a_1 - a_2)$. Thus p divides $(a_1 + a_2)(a_1 - a_2)$ and hence p divides one of $(a_1 + a_2)$ or $(a_1 - a_2)$. Let p divide $a_1 + a_2$. The proof is similar if p divides $a_1 - a_2$. We claim that q does not divide $a_1 + a_2$, as otherwise, N will divide $a_1 + a_2$, contradicting the fact that $a_1 \not\equiv \pm a_2$ (mod N). Thus q does not divide $a_1 + a_2$ and hence $\gcd(N, a_1 + a_2) = p$. As the Euclidean algorithm to compute gcd of two integers is an efficient one, we can find p efficiently, thereby factor N efficiently. $\qquad\square$

Proposition 12.8.3 *Let $N = pq$, where p and q are distinct odd primes of the form $p \equiv q \equiv 3$ (mod 4). Then every quadratic residue modulo N has exactly one square root which is also a quadratic residue modulo N.*

Proof As $p \equiv q \equiv 3$ (mod 4), $\left(\frac{-1}{p}\right) \equiv (-1)^{\frac{p-1}{2}} \equiv (-1)$ (mod p), $\left(\frac{-1}{q}\right) \equiv (-1)^{\frac{q-1}{2}} \equiv (-1)$ (mod q), where $\left(\frac{-1}{p}\right)$ and $\left(\frac{-1}{q}\right)$ denote respectively the Legendre symbols of -1 modulo p and q, respectively. Let y be an arbitrary quadratic residue modulo N. As N is a product of two distinct primes, y will have four distinct square roots modulo N, namely (x_p, x_q), $(-x_p, x_q)$, $(x_p, -x_q)$ and $(-x_p, -x_q)$. We shall show that out of these four square roots modulo N, exactly one of them is a quadratic residue modulo N. Let $\left(\frac{x_p}{p}\right) = 1$ and $\left(\frac{x_q}{q}\right) = -1$. Similar arguments are applicable for other cases. Then $\left(\frac{-x_q}{q}\right) = \left(\frac{-1}{q}\right)\left(\frac{x_q}{q}\right) = (-1)(-1) = 1$, with the result

Table 12.5 A variant of
Rabin public key
cryptosystem

- **Key Generation Algorithm by Bob:**

  1. Choose two k-bit primes p and q such that $p \equiv q \equiv 3 \pmod 4$.
  2. Compute $N = pq$.
  3. Publish N as public key and p, q are kept secret as private keys.

- **Encryption Algorithm for Alice:**

  1. Collect the public key N.
  2. Let $m \in \{0, 1\}$ be the plain text.
  3. Choose $x \in_R \mathcal{QR}_N$ and compute $C = (c, c')$, where $c \equiv x^2 \pmod N$ and $c' = \mathrm{lsb}(x) \oplus m$, $\mathcal{QR}_N$ denotes the set of quadratic residues modulo N and lsb denotes the least significant bit of the binary representation of x.
  4. Send C to Bob.

- **Decryption Algorithm for Bob:**

  1. Collect $C = (c, c')$, sent by Alice.
  2. Compute unique $x \in_R \mathcal{QR}_N$ such that $x^2 \equiv c \pmod N$. [This is possible due to Proposition 12.8.3]
  3. Compute $m = \mathrm{lsb}(x) \oplus c'$.

that $-x_q$ is a quadratic residue modulo q. Thus by Proposition 12.8.1, $(x_p, -x_q)$ is a quadratic residue modulo N. By similar argument it can be shown that each of (x_p, x_q), $(-x_p, x_q)$ and $(-x_p, -x_q)$ cannot be a quadratic residue modulo N. □

Based on the above propositions, we are going to discuss a variant of Rabin Cryptosystem in Table 12.5 which can be shown to be chosen plain text attack (CPA) secure based solely on the assumption that factorization is hard. As the security notions are out of scope for this book, we are not discussing the proof of the CPA security.

12.8.2 Algorithm for a variant of Rabin Cryptosystem

Suppose Alice wants to send message to Bob. Then a variant of Rabin cryptosystem scheme is explained in Table 12.5.

In the Rabin encryption scheme, receiver is required to compute the modular square roots. So in the scheme, we first find an algorithm for computing square roots modulo a prime $p \equiv 3 \pmod 4$ and then extend that to compute square roots modulo a composite $N = pq$ with known factorization of N, where $p \equiv q \equiv 3 \pmod 4$. Before doing that let us start with the case for finding square roots modulo a prime $p \equiv 3 \pmod 4$.

Let $p = 4k + 3$, $k = 0, 1, 2, \ldots$. Then $\frac{p+1}{4}$ is an integer. Let a be a quadratic residue modulo p. Then $\left(\frac{a}{p}\right) = 1$. This implies that $a^{\frac{p-1}{2}} \equiv 1 \pmod p$. Multiplying

Table 12.6 Algorithm to compute four square roots of $a \in \mathbf{Z}_N^*$ modulo $N = pq$ having the knowledge of factorization of N, where $p \equiv q \equiv 3 \pmod 4$

- **Input:** Two primes p and q such that $N = pq$ and a quadratic residue $a \in \mathbf{Z}_N^*$, where $p \equiv q \equiv 3 \pmod 4$.
- **Method:**

  1. Compute $a_p \equiv a \pmod p$ and $a_q \equiv a \pmod q$.
  2. Using Lemma 12.8.1, compute two square roots x_p and $-x_p$ of a_p modulo p. Similarly compute x_q and $-x_q$ of a_q modulo q.
  3. Using the Chinese Remainder Theorem, compute the four elements x_1, x_2, x_3 and x_4 in $\mathbf{Z}_N^*$ corresponding to the elements (x_p, x_q), $(-x_p, x_q)$, $(x_p, -x_q)$ and $(-x_p, -x_q)$, respectively, in $\mathbf{Z}_p^* \times \mathbf{Z}_q^*$.

- **Output:** Four square roots of a modulo N.

both sides by a, we get $a \equiv a^{\frac{p-1}{2}+1} \equiv a^{2k+2} \equiv (a^{k+1})^2 \pmod p$. Thus $a^{k+1} \equiv a^{\frac{p+1}{4}} \pmod p$ is a square root of a modulo p. The other square root being $-a^{k+1} \pmod p$. This leads to the following lemma.

Lemma 12.8.1 *Let a be a quadratic residue modulo a prime p, where p is a prime of the form $4k + 3$, $k = 0, 1, 2, \ldots$. Then $a^{k+1} \pmod p$ and $-a^{k+1} \pmod p$ are two square roots of a modulo p.*

Proof As p is a prime, a has exactly two distinct square roots modulo p and if x is a square root modulo p, then $-x$ is also a square root modulo p.   □

Using Lemma 12.8.1 and the Chinese Remainder Theorem, we can describe the algorithm as in Table 12.6 to compute square roots of a modulo $N = pq$ provided the factorization of N is known, where p and q are primes of the form $p \equiv q \equiv 3 \pmod 4$.

12.9 Digital Signature

In general, a "conventional" handwritten signature of a person attached to a document is used to specify that the person is responsible for the document. In our daily life, we use signature while writing a letter, withdrawing money from bank through cheque or signing a contract, etc. We all have heard about signature, but what is then "digital signature"? Before we come to the concept of digital signature, let us first explore some of the examples from real life to review the concept of signature. Suppose Alice wants to sell her house for $125,000 and Bob wants to buy the house. Bob receives a signed letter from Alice mentioning that the price of the house is $125,000. When Bob reads the signed letter, he knows that this letter is indeed from Alice, that is, as long as it is not somebody else forging Alice's signature. Now the question is, how can Bob be absolutely sure that the signed letter comes from Alice and it is not a forgery? It may happen that somebody else, say an adversary, sends the letter to Bob in the name of Alice by copying her signature while actually Alice wanted to sell her house for \$225,000. It may further happen that Alice may change her mind to sell her house for \$325,000 and when Bob comes with the letter from Alice, she may claim that it must be a forgery: she never signed such a letter.

These lead to the following important features of signature:

- *Authenticity*: It convinces Bob that the signed letter indeed comes from Alice.
- *Unforgeability*: Nobody else but Alice could have signed the message.
- *Not reusability*: The same signature cannot be attached to another message.
- *Unalterability*: Once a message has been signed, it cannot be altered i.e., if Alice signs on a message m and then claims the same signature s for different message M, then there should be some mechanism through which Bob can check the correctness. As a result, no one will be able to use the same signature s for different messages.
- *Non-repudiation*: After signing a message, Alice cannot deny that she signed the message.

Now we are going to define a digital signature scheme. Public-key cryptosystem plays an important role in constructing and defining digital signature. Informally, any signature scheme consists of two components: one is a signing algorithm and the other is a verification algorithm. Suppose Alice wants to send a digitally signed message to Bob. Alice can sign a digital message m by using a signing algorithm sig_{sk} which is based on a secret key sk, known only to Alice, to produce a digital signature $\text{sig}_{\text{sk}}(m)$ for the message m. The resulting signature $\text{sig}_{\text{sk}}(m)$ can subsequently be verified by Bob using a publicly known verification algorithm ver_{pk} based on the public key pk. Thus given a message m and a purported signature s on m, the verification algorithm returns "true" if s is a valid signature on m, else it returns "no". The formal definition of a digital signature scheme is defined as follows.

Definition 12.9.1 A digital signature scheme is a five-tuple $(\mathcal{P}, \mathcal{A}, \mathcal{K}, \mathcal{S}, \mathcal{V})$, where the following conditions are satisfied:

- $\mathcal{P}$ is a finite set of possible messages.
- $\mathcal{A}$ is a finite set of possible signatures.
- $\mathcal{K}$ is a finite set of possible keys. $\mathcal{K}$ is known as keyspace.
- For each key $K = (\text{pk}, \text{sk}) \in \mathcal{K}$, there is a signing algorithm $\text{sig}_{\text{sk}} \in \mathcal{S}$ based on the private key sk and a corresponding verification algorithm $\text{ver}_{pk} \in \mathcal{V}$ based on the public key pk. Each $\text{sig}_{\text{sk}} : \mathcal{P} \to \mathcal{A}$ and $\text{ver}_{\text{pk}} : \mathcal{P} \times \mathcal{A} \to \{true, false\}$ are functions such that for all $(m, s) \in \mathcal{P} \times \mathcal{A}$, the following conditions are satisfied:

$$\text{ver}_{\text{pk}}(m, s) = \begin{cases} \text{true} & \text{if } s = \text{sig}_{\text{sk}}(m) \\ \text{false} & \text{if } s \neq \text{sig}_{\text{sk}}(m). \end{cases}$$

The pair (m, s) is called a digitally signed message.

Remark Let us now discuss some of the differences between the conventional pen-and-ink signature and the digital signature. In case of a traditional pen-and-ink signature, the signature is an integral part of some document, while in case of digital signature, the digital signature is not attached physically to the message that is signed. In that case, some algorithm is required to somehow "bind" the digital signature with the message. Another difference between these two is from the viewpoint of verification. In case of traditional signature, the validity of the signature is verified by comparing it with other stored authentic signatures. Clearly, this method is not secure as someone may forge someone else's signature. On the other hand, digital signatures can be verified by any one using the publicly verifiable algorithm. One of the major characteristics of digital signature is that the digital copy of signed digital message is exactly identical with the original one, while a photocopy of a pen-and-ink signature on some document can usually be distinguished from the original document. As a result, in case of digital signature, a certain care must be taken so that a same digital signature on some document should not be used more than once. For example, suppose Alice signs digitally a cheque of some bank to issue \$2000 to Bob. So Alice must take some measure while digitally signing the message, i.e., must put some date or some thing else so that Bob will not be able to reuse the same digitally signed cheque more than once.

Now let us explore a digital signature scheme which is based on RSA public-key cryptosystem.

12.9.1 RSA Based Digital Signature Scheme

In Sect. 12.6, we have already explored about the celebrated RSA algorithm for public-key cryptography. This RSA scheme may also be used to produce a very simple signature scheme, known as RSA-based digital signature scheme which is almost the reverse of the RSA public-key cryptosystem. Let us explain the algorithm.

Suppose Alice wants to sign digitally a message m and wants to send it to Bob. The algorithm is given in Table 12.7.

Remark 1 Any one can forge the RSA digital signature of Alice by choosing a random $s \in \mathbf{Z}_n$ and then computing $m \equiv s^e \pmod{n}$. Then clearly, $s = \text{sig}_d(m)$ is a valid signature on the random message m. However, not yet, any efficient method is found through which we can first choose a meaningful message m and then compute the corresponding signature y. In fact, it can be shown that if this could be done, then the RSA cryptosystem is insecure. Further, due to the homomorphic nature of the RSA cryptosystem, it is vulnerable to existential forgery in which it is possible to generate legitimate message and signature without knowing the private key d. For example, if (m_1, s_1) and (m_2, s_2) are two pairs of signed messages, then $(m_1 \cdot m_2, s_1 \cdot s_2)$ is also a legitimate signed message.

Table 12.7 Algorithm for RSA based signature scheme

- **Key Generation Algorithm by Alice:**

  1. Generate two distinct large prime numbers p and q, each roughly of same size.
  2. Compute $n = pq$ and $\phi(n) = (p-1)(q-1)$.
  3. Select a random integer e with $1 < e < \phi(n)$, such that $\gcd(e, \phi(n)) = 1$.
  4. Use the extended Euclidean algorithm to find the integer d, $1 < d < \phi(n)$, such that $ed \equiv 1 \pmod{\phi(n)}$.
  5. Make public the values of n and e (which are known to every body and hence considered as public keys) and keep secret the private keys p, q, and d (which are known only to Alice).

- **Signing Algorithm for Alice:**

  1. Represent the message as an integer m in the interval $[0, n-1]$.
  2. Compute the signature $\text{sig}_d(m) = s \equiv m^d \pmod{n}$.
  3. Send the pair (m, s) to Bob.

- **The Verification Algorithm for Bob:**

  1. Obtain the pair (m, s), the public keys n and e from Alice.
  2. Compute $\text{ver}_{e,n}(m, s)$. If $m \equiv s^e \pmod{n}$ i.e., if $\text{ver}_{e,n}(m, s) = true$, Bob accepts the signature, else rejects. This works because $s^e \equiv m^{ed} \pmod{n} \equiv m \pmod{n}$.

Remark 2 There has been a rich literature on various security notions of different types of signature schemes which are out of scope for this book. The reader may read the book (Katz 2010) for detailed and systematic analysis of signature schemes.

In the next section, we are going to discuss about a very important primitive known as oblivious transfer that plays an important role in modern cryptography.

12.10 Oblivious Transfer

Suppose Alice and Bob are very shy. However, they would like to know whether they are interested in each other or not. So one simple way could be a direct approach to each other. If both of them will show their interest towards each others, there will not be a problem. But if one of them will refuse, it could be a problem for the other due to face saving. So they need some protocol in which they will be able to figure out if they both agree but in such a way that if they do not then any of them who has rejected the matter has no clue about the other's opinion. Moreover they may want to be able to carry out this protocol publicly over a distance. This problem is known as dating or matching problem. At this point of time, it looks quite complicated. Well, both Alice and Bob may choose some trusted third party to whom both of them may confer their decisions and the trusted third party will declare whether both of them agree or not by keeping the decision of the individual secret for ever. But finding such a trusted third party is very difficult. However, for this kind of situation, an important primitive of cryptography, known as oblivious transfer, could be a useful tool.

Informally, an oblivious transfer is a protocol between two players, a sender S and a receiver R, such that it achieves the following: the sender S has two secret input bits b_0 and b_1 while the receiver R has a secret selection bit s. The protocol consists of a number of exchanges of information between the sender S and the receiver R. At the end of the protocol, the receiver R obtains the bit b_s but without having obtained any information about the other bit b_{1-s}. This is known as sender's security. On the other hand, the sender S does not get any information about the selection bit s chosen by the receiver R. This is known as the receiver's security.

Let the output of an oblivious transfer protocol be denoted by $OT(b_0, b_1, s) = b_s$ for the inputs of the secret bits b_0 and b_1 of the sender S and the secret selection bit s of the receiver R. Observe that b_s is actually equal to $(1 \oplus s) \cdot b_0 \oplus s \cdot b_1$, where "$\oplus$" denotes the addition of two bits modulo 2 and "$\cdot$" denotes the binary "AND" of two bits. Note that if $s = 0$, then $OT(b_0, b_1, 0) = (1 \oplus 0) \cdot b_0 \oplus 0 \cdot b_1 = b_0$ and if $s = 1$, then $OT(b_0, b_1, 1) = (1 \oplus 1) \cdot b_0 \oplus 1 \cdot b_1 = b_1$.

Now we again come to the problem of dating as discussed at the beginning of this section. To apply the oblivious transfer, let us first model this problem mathematically. Suppose, there are two players, say a Sender S and a Receiver R. In the context of the dating problem, we may assume Alice to be Sender and Bob to be Receiver. As the secret decisions of each of these players are either yes or no, we may assume that each of them has a secret bit, where 1 stands for "yes" and 0 stands for "no". Suppose the Sender S has the secret bit a and the Receiver R has the secret bit b. They actually want to compute $a \cdot b$, i.e., the logical "AND" of the bits a and b such that the following conditions are satisfied:

- None of the sender or receiver is led to accept a false result (this property is known as correctness).
- Both the sender and the receiver learn the result $a \cdot b$ (this property is known as fairness).
- Each learns nothing more than what is implied by the result and the own input (this property is known as privacy). For example, at the end of the oblivious transfer protocol for the dating problem, both Sender and the Receiver will learn $a \cdot b$. So for the sender (Alice), if $a = 1$ and she comes to know that $a \cdot b = 1$, then she gets the information that the secret bit b for the Receiver (Bob) must be 1, while if the secret bit a of the Sender (Alice) is 0 then $a \cdot b$ is always 0, irrespective of the input bit b of the receiver (Bob). Therefore the Receiver's (Bob's) choice for b remains unknown to the Sender's. A similar argument is also applicable when $b = 0$.

Now we are in a position to solve the dating problem by using an Oblivious Transfer protocol. Note that we have not yet said any thing about the existence of such Oblivious Transfer protocol. In the next section, we are going to provide an Oblivious Transfer protocol. Before that, assuming the existence of such Oblivious Transfer protocol, we are going to provide the following protocol that will solve the dating problem: Consider Alice to be the Sender and Bob to be the Receiver of an Oblivious Transfer protocol with the inputs for Alice to be the secret bits $b_0 = 0$ and $b_1 = a$ while the secret selection bit for Bob to be $s = b$. Then we have

$OT(0, a, b) = (b \oplus 1) \cdot 0 \oplus b \cdot a = b \cdot a = a \cdot b$. As an output of the Oblivious Transfer protocol, Bob receives $a \cdot b$. Finally, Bob sends $a \cdot b$ to Alice so that both of them learn the result $a \cdot b$.

Now let us see what are the conditions required to make the protocol secure. Note that to achieve the correctness and fairness, we have to assume that Bob indeed sends the correct value $a \cdot b$ to Alice. So we have to assume that both Alice and Bob follow the rules of the game but may try to learn as much as possible about the inputs of the other players. We call such players as semi-honest players. Further, note that if $b = 0$, then $a \cdot b$ is always 0, no matter what the value of the secret input a of Alice is. Thus the player Bob learns nothing more about a by the properties of the Oblivious Transfer protocol. On the other hand, if $a = 0$, then again from the fact that the Oblivious Transfer protocol does not leak information about b and the fact that in this case Bob returns 0 to Alice in the final step no matter what b is, Alice learns nothing more about b as a result of the complete protocol.

Assuming the existence of an oblivious transfer protocol, we have solved the dating problem. Now we are going to provide an oblivious transfer protocol based on RSA cryptosystem, as described in Sect. 12.6.

12.10.1 OT Based on RSA

To construct the RSA based oblivious transfer protocol, we assume that both the players, i.e., the Sender and the Receiver are semi-honest. It has been proved by Alexi et al. (1988) that in case of RSA cryptosystem, if a plain text m is chosen at random, then given just the cipher text $c \equiv m^e \pmod{n}$ along with the values of n and e, guessing the least significant bit of m i.e., $\text{lsb}(m)$ significantly better than at random is as hard as finding all bits of m. This is called a "hard-core" bit for the RSA encryption function. The hardness related to the hard-core bit is actually exploited to obtain an oblivious transfer by masking the two secret bits b_0 and b_1 of the Sender. Here the Sender has two secret bits b_0 and b_1 while the Receiver has one selection bit s. The Sender first starts the key set up steps for the RSA encryption by choosing two distinct primes p and q and compute $n = p \cdot q$. For the public key, the Sender chooses a random e such that $\gcd(e, \phi(n)) = 1$, where ϕ denotes the Euler phi function. Then for the decryption, the Sender finds d such that $ed \equiv 1 \pmod{n}$. Finally, the sender makes public the values of e, n, and keeps secret the values of d, p, and q. After the key setup step, the Sender sends the public values to the Receiver. The Receiver, having selection bit s, chooses a random plain text $m_s \pmod{n}$ and computes the cipher text $c_s \equiv m_s^e \pmod{n}$. The Receiver further selects a random integer c_{1-s} modulo n and consider it as another cipher text. Note that the Receiver can easily find the $\text{lsb}(m_s)$ while due to the hardness of hard-core bit for RSA encryption function, determining the $\text{lsb}(m_{1-s})$ from c_{1-s} is infeasible. The Receiver sends the cipher text pair (c_0, c_1) to the Sender. As the Sender has the secret key for decryption, she can decrypt the cipher text pair (c_0, c_1) to get

Table 12.8 Algorithm for OT based on RSA

Algorithm for Sender:

- Choose two distinct large primes p and q.
- Compute $n = pq$.
- Select a random integer e, $1 < e < \phi(n)$, such that $\gcd(e, \phi(n)) = 1$.
- Find the integer d, $1 < d < \phi(n)$, such that $ed \equiv 1 \pmod{\phi(n)}$.
- Send the public values n and e to the receiver.
- Keep secret the values p, q, and d.
- The Sender has two secret bits b_0 and b_1.

Algorithm for Receiver:

- Collect the public values n and e.
- The Receiver has a secret selection bit s.
- Choose a random plain text $m \in \mathbf{Z}_n$ and compute the cipher text $c_s \equiv m^e \pmod{n}$.
- Select c_{1-s} randomly from $\mathbf{Z}_n$ as another cipher text.
- Send the ordered pair (c_s, c_{1-s}) to the Sender.

Algorithm for Sender:

- Decrypt (c_0, c_1) to get (m_0, m_1).
- Compute $r_0 = \mathrm{lsb}(m_0)$ and $r_1 = \mathrm{lsb}(m_1)$.
- Mask the bits b_0 and b_1 by computing $b'_0 = b_0 \oplus r_0$ and $b'_1 = b_1 \oplus r_1$ and send (b'_0, b'_1) to receiver.

Algorithm for Receiver:

- Recover b_s by computing $b'_s \oplus r_s$.
- The bit b_{1-s} remains concealed since he cannot guess r_{1-s} with high enough probability.

the plain text (m_0, m_1). Let $r_0 = \mathrm{lsb}(m_0)$ and $r_1 = \mathrm{lsb}(m_1)$. Then the Sender masks the bits b_0 and b_1 by computing $b'_0 = b_0 \oplus r_0$ and $b'_1 = b_1 \oplus r_1$ and sends (b'_0, b'_1) to the Receiver. The Receiver recovers b_s by computing $b'_s \oplus r_s$. Note that the bit b_{1-s} remains concealed since the Receiver cannot guess r_{1-s} with high enough probability. Note that the selection bit s is unconditionally hidden from the Sender and security of the Sender is guaranteed due to the assumption that the receiver is semi-honest.

The actual algorithm is given in Table 12.8.

In the next section we are going to introduce a very important concept of cryptography, known as secret sharing.

12.11 Secret Sharing

Due to the recent development of computers and computer networks, huge amount of digital data can easily be transmitted or stored. But the transmitted data in networks or stored data in computers may easily be destroyed or substituted by enemies if the data are not enciphered by some cryptographic tools. So it is very important to restrict access of confidential information stored in a computer or in a certain nodes of a system. Access should be gained through a secret key, password or token. Again storing the secret key or password securely could be a problem. The best solution could be to memorize the secret key. But for large and complicated secret key, it is almost impossible to memorize the key. As a result, it should be stored safely. While storing data in a hard disk, the threats such as troubles of storage devices or attacks of destruction make the situation even worse. In order to prevent such attacks, we may make as many copies of the secret data as possible. But if we have many copies of the secret data, the secret may be leaked out and hence the number of the copies should be as small as possible. Under this circumstances, it is desirable that the secret key should be governed by a secure key management scheme. If the key or the secret data is shared among several participants in such a way that the secret data can only be reconstructed by a significantly large and responsible group acting in agreement, then a high degree of security is attained.

Shamir (1979) and Blakley (1979), independently, addressed this problem in 1979 when they introduced the concept of a threshold secret sharing scheme. A (t, n) *threshold scheme* is a method whereby n pieces of information, called *shares*, corresponding to the secret data or key K, are distributed to n participants so that the secret key can be reconstructed from the knowledge of any t or more shares and the secret key cannot be reconstructed from the knowledge of fewer than t shares.

But in reality, there are many situations in which it is desirable to have a more flexible arrangement for reconstructing the secret key. Given some n participants, one may want to designate certain authorized groups of participants who can use their shares to recover the key. This kind of scheme is called general secret sharing scheme.

Formally, a *general secret sharing scheme* is a method of sharing a secret K among a finite set of participants $\mathcal{P} = \{P_1, P_2, \ldots, P_n\}$ in such a way that

1. If the participants in $\mathcal{A} \subseteq \mathcal{P}$ are qualified to know the secret, then by pooling together their partial information, they can reconstruct the secret K,
2. Any set $\mathcal{B} \subset \mathcal{P}$ which is not qualified to know K, cannot reconstruct the secret K.

The key is chosen by a special participant $\mathcal{D}$, called the *dealer* and it is usually assumed that $\mathcal{D} \notin \mathcal{P}$. The dealer gives partial information, called *share* or *shadow*, to each participant to share the secret key K. The collection of subsets of participants that can reconstruct the secret in this way is called *access structure* Γ. Γ is usually monotone, that is, if $X \in \Gamma$ and $X \subseteq X' \subseteq \mathcal{P}$, then $X' \in \Gamma$. A *minimal qualified subset* $Y \in \Gamma$ is a subset of participants such that $Y' \notin \Gamma$ for all $Y' \subset Y$. The *basis* of Γ, denoted by Γ_0, is the family of all minimal qualified subsets. A secret sharing scheme is said to be *perfect* if the condition 2 of the above definition is strengthened as follows: Any unauthorized group of shares cannot be used to gain any information about the secret key that is if an unauthorized subset of participants $\mathcal{B} \subset \mathcal{P}$ pool their shares, then they can determine nothing more than any outsider about the value of the secret K.

12.11.1 Shamir's Threshold Secret Sharing Scheme

Shamir's original (t, n)-threshold scheme for $n \geq t \geq 2$, given in Shamir (1979), is based on polynomial interpolation of t points in a two dimensional plane. The basic intuition behind the scheme is that it requires two points to define a unique straight line that passes through these two pints, it requires three points to fully define a unique quadratic passing through these three points, it requires four points to fully define a unique cubic and so on. Thus in general, one can fit a unique polynomial of degree $k - 1$ to any set of k points that lie on the polynomial. To construct a (t, n)-threshold scheme, the dealer is going to choose a polynomial $f(x)$ of degree $t - 1$ in a two dimensional plane. Note that any t points on the curve $f(x)$ can determine the curve $f(x)$ uniquely, i.e., if we are given t or more points on the curve $f(x)$, we can uniquely reconstruct $f(x)$, while there will be infinitely many curves which pass through any $t - 1$ or less points on the curve $f(x)$. So $t - 1$ or less number of points on the curve $f(x)$ cannot reconstruct the curve $f(x)$ uniquely. Thus n points on the $t - 1$-degree curve $f(x) = a_0 + a_1 x + a_2 x^2 + \cdots + a_{t-1} x^{t-1}$, with $a_0 = $ *secret*, are chosen by the dealer for the n participants. To each participant, a point is given. Thus when t or more participants come together, they can reconstruct the polynomial $f(x)$ to get the constant coefficient $a_0 = secret$. But for any set of $t - 1$ or less participants come together, they cannot reconstruct a_0 uniquely. In fact they will have infinitely many choices for a_0. This intuitive idea helps us in building Shamir's (t, n)-threshold secret sharing scheme. Instead of taking a two dimensional Euclidean plane, Shamir proposed the scheme over a finite field $GF[q]$ having q elements such that the secret k and the number of participants n is less than q. Let $f(x)$ be a polynomial of degree at most $t - 1$ over the finite field $GF[q]$ having q elements. Assume that, for j $(1 \leq j \leq t)$ distinct elements x_j of $GF[q]$, the values of $f(x_j)$ are known. Hence the system of t linearly independent equations

$$f(x_j) = \sum_{i=0}^{t-1} a_i x_j^i,$$

in t unknowns $a_0, a_1, \ldots, a_{t-1}$, can be obtained. Lagrange's interpolation can now be used to determine uniquely the t unknowns. So the polynomial can be recovered from the t points. Shamir (1979) used this property of polynomial interpolation to construct a t-out-of-n threshold scheme.

For simplicity, take $GF[q]$ to be $\mathbf{Z}_q$, for some large prime q such that $n \leq q$. This set $\mathbf{Z}_q$ is the key space, i.e., the secret key K is to be chosen from $\mathbf{Z}_q$. In Shamir's scheme a secret K from $\mathbf{Z}_q$ and then a polynomial $f(x)$ of degree $t - 1$ are chosen by the Dealer in such a way that the constant term of the polynomial is K, i.e., $f(0) = K$. The participants are labeled $P_1, P_2, \ldots, P_n$. The dealer chooses n distinct non-zero elements from $\mathbf{Z}_q$, say $x_1, x_2, \ldots, x_n$ (that is why $q \geq n + 1$). These x_i's are made public. For $i = 1, 2, \ldots, n$, participant P_i is given the values $f(x_i)$ and x_i i.e., a point $(x_i, f(x_i))$ on the curve $f(x)$, as a share. When any t participants come together, they can use their shares to recover $f(x)$ and hence reconstruct the secret K as follows: Suppose the t participants $P_{i_1}, P_{i_2}, \ldots, P_{i_t}$, want to find the unknown value K. Each participants P_{i_k} has the knowledge of x_{i_k} and $f(x_{i_k})$. Since the secret polynomial is of degree at most $t - 1$, the participants may assume the form of the polynomial as

$$A(x) = a_0 + a_1 x + a_2 x^2 + \cdots + a_{t-1} x^{t-1},$$

where a_i's are unknown elements of $\mathbf{Z}_q$ with $a_0 = K$, which is actually retrieved as a key by the qualified set participants. The aim of the participants is to find a_0 from their shares. Now $f(x_{i_k}) = A(x_{i_k})$, $1 \le k \le t$. When the t participants come together, they can obtain t linear equations in the t unknowns $a_0, a_1, \ldots, a_{t-1}$, where all the operations are done in $\mathbf{Z}_q$. Now if the equations are linearly independent, the participants will get a unique solution to the system of equations and $a_0 = K$ will be revealed as the key. We show using Vandermonde matrix that these system of equations has always a unique solution. Note that this system of equations can be written as

$$a_0 + a_1 x_{i_1} + a_2 x_{i_1}^2 + \cdots + a_{t-1} x_{i_1}^{t-1} = f(x_{i_1})$$

$$a_0 + a_1 x_{i_2} + a_2 x_{i_2}^2 + \cdots + a_{t-1} x_{i_2}^{t-1} = f(x_{i_2})$$

$$\cdots$$

$$a_0 + a_1 x_{i_t} + a_2 x_{i_t}^2 + \cdots + a_{t-1} x_{i_t}^{t-1} = f(x_{i_t}).$$

This can be written in a matrix form as follows:

$$\begin{bmatrix} 1 & x_{i_1} & x_{i_1}^2 & \ldots & x_{i_1}^{t-1} \\ 1 & x_{i_2} & x_{i_2}^2 & \ldots & x_{i_2}^{t-1} \\ \cdots & \cdots & \cdots & \cdots & \cdots \\ 1 & x_{i_t} & x_{i_t}^2 & \ldots & x_{i_t}^{t-1} \end{bmatrix} \begin{bmatrix} a_0 \\ a_1 \\ \cdots \\ a_{t-1} \end{bmatrix} = \begin{bmatrix} f(x_{i_1}) \\ f(x_{i_2}) \\ \cdots \\ f(x_{i_t}) \end{bmatrix}.$$

The coefficient matrix is known as Vandermonde matrix and its determinant is of the form

$$\prod_{1 \le j < k \le t} (x_{i_k} - x_{i_j}) \ (\mathrm{mod} \ q).$$

Since $\mathbf{Z}_q$ is a field and x_i's are all distinct, the determinant of the coefficient matrix is always non-zero and hence the system of equations has a unique solution. This shows that any group of t participants can recover the key in this threshold scheme.

Now we will show what will happen if a group of $t - 1$ or less participants try to compute the secret key K? Assume that $t - 1$ participants wish to collaborate and try to guess the secret. The $t - 1$ participants generate a set of $t - 1$ equations in t unknowns. These equations have as their solution a set of q polynomials

$$f(x) = a_0 + \sum_{i=1}^{t-1} a_i x^i,$$

where a_0 ranges over all the elements of $\mathbf{Z}_q$. Hence each of the possible keys is equally likely. This proves that the scheme is perfectly secure.

Remark Note that Shamir's secret sharing scheme is a single use secret sharing scheme for single secret. For multi-use, multi-secret sharing schemes, the reader may refer to Das and Adhikari (2010).

In the next section, we are going to discuss about a special kind of secret sharing scheme, known as visual cryptography.

12.12 Visual Cryptography

Most of the secret sharing schemes are based on algebraic calculations in their realizations. But there are some different realizations from ordinal secret sharing schemes. Visual cryptography is one such secret sharing scheme. In visual cryptography, the problem is to encrypt some written material (handwritten notes, printed text, pictures, etc.) in a perfectly secure way in such a manner that the decoding may be done visually, without any cryptographic computations. The concept of visual cryptography was first proposed by Naor and Shamir (1994). Visual cryptographic scheme for a set $\mathcal{P}$ of n participants is a cryptographic paradigm that enables a secret image to be split into n shadow images called *shares*, where each *participant* in $\mathcal{P}$ receives one share. Certain qualified subsets of participants can "visually" recover the secret image with some loss of contrast, but other forbidden sets of participants have no information about the secret image. The collection of all qualified subsets is denoted by Γ_{Qual} and the collection of all forbidden subsets is denoted by Γ_{Forb}. The pair $(\Gamma_{\text{Qual}}, \Gamma_{\text{Forb}})$ is called the *access structure* of the scheme. A participant $P \in \mathcal{P}$ is an *essential* participant if there exists a set $X \subseteq \mathcal{P}$ such that $X \cup \{P\} \in \Gamma_{\text{Qual}}$ but $X \notin \Gamma_{\text{Qual}}$. Typically, in a (k, n)-Visual Cryptographic Scheme $((k, n)$-VCS), a page of secret image/text is encrypted to generate n pages of cipher text which may be printed on n transparency sheets. If $k - 1$ or less number of transparency sheets are superimposed, it will give no information on the secret image and will be indistinguishable from random noise. However, if any k of the transparencies are stacked together, the secret image will be revealed. So in a (k, n)-VCS, there is a secret image and a set of n persons, called *participants*. The secret image is split into n shadow images, called *shares* and each participant receives one share. $k - 1$ or less many participants cannot decipher the secret image from their shares but any k or more participants may together recover the secret image by photocopying the shares given to the participants onto transparencies and then stacking them. Since the reconstruction is done by human visual system, no computations are involved during decoding unlike traditional cryptographic schemes where a fair amount of computations is needed to reconstruct the plain text.

In this section, we shall discuss different techniques for black and white visual cryptography in which the secret is a black and white image made up of black and

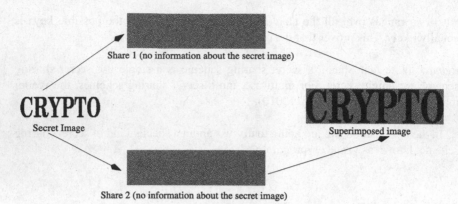

Fig. 12.1   (2, 2)-VCS for black and white image

white pixels. Two parameters are very important in visual cryptography, namely the pixel expansion and the contrast. Pixel expansion is the number of pixels, on the transparencies corresponding to the shares (each such pixel is called subpixel), needed to encode one pixel of the original secret image. On the other hand, the contrast is the clarity with which the reconstructed image is visible.

Now let us explain how a visual cryptographic scheme may be constructed.

12.12.1 (2, 2)-Visual Cryptographic Scheme

Suppose we have a secret black and white image i.e., the image is made up of only black and white pixels. Now suppose we want to distribute the secret image as shares among a set of two participants in such a way that from a single share it is not possible to decode the secret image, but if both of the participants come together and superimpose their shares they will be able to decode the secret image. This scheme is known as the (2, 2)-Visual Cryptographic Scheme (in short (2, 2)-VCS), illustrated in Fig. 12.1.

Naor and Shamir (1994) devised the following scheme for the (2, 2)-VCS. The algorithm specifies how to encode a single pixel and it would be applied for every pixel in the image to be shared. A pixel P is split into two pixels in each of the two shares (each such pixel in the shares is called subpixel). If P is white, then a coin toss is used to randomly choose one of the first two rows in Fig. 12.2.

If P is black, then a coin toss is used to randomly choose one of the last two rows in Fig. 12.2. Then the pixel is encrypted as two subpixels in each of the two shares, as determined by the chosen row in Fig. 12.2. Every pixels is encrypted using a new coin toss.

Suppose we look at the two subpixels, in the first share, corresponding to the pixel P in the secret image. One of these two subpixels is black and the other is white. Moreover, each of the two possibilities "black-white" and "white-black" is

Fig. 12.2 (2, 2)-VCS for
black and white image

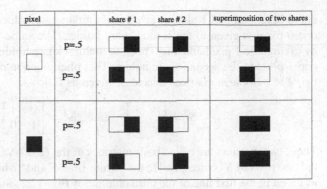

equally likely to occur, independent of whether the corresponding pixel in the secret image is black or white. Thus just by looking at the first share it is not possible to predict whether the subpixels in the share correspond to a black or white pixel in the secret image. The same argument is applicable for the second share also. Since all the pixels in the secret image were encrypted using independent random coin flips, there is no information to be gained by looking at any group of pixels on a share, either. This demonstrates the security of the scheme.

Now we consider the situation when two shares are superimposed. In Fig. 12.2 it is shown in the last column. If the pixel P in the original image is black then after superimposition of two shares, we get two black subpixels in the superimposed image, whereas if P is white then after superimposition of two shares we get one white and one black subpixels in the superimposed image. Thus we can say that in this particular case the reconstructed pixel has grey level of 1 if P is black and a grey level of 1/2 if P is white. So we see that in case of a white pixel in the secret image, there will be 50 % loss of contrast in the reconstructed image, but it should still be visible.

Now we shall explain mathematically the (2, 2)-VCS as described in Fig. 12.2. First let us try to explain the phenomenon of superimposition of transparencies mathematically. Note that when a black pixel, printed on a transparency, is superimposed over a white pixel, printed on a transparency (actually left blank), as a visual effect, we get a black pixel. Similarly, if both the pixels are black, the superimposed pixel will look black. The superimposed pixel will look white only when both the pixels are white. Now let us denote the superimposition operation by "$*$", a white pixel by 0 and a black pixel by 1. Then, as mentioned above, we have $1 * 1 = 1$, $1 * 0 = 1$, $0 * 1 = 1$ and $0 * 0 = 0$. If we look at the operation $*$ little carefully, it reveals that operation $*$ is noting but the Boolean "*or*" operation. Thus the superimposition of two pixels is simply the Boolean "*or*" operation.

Now let us explain the share distribution algorithm, as descried in Fig. 12.2, mathematically. As mentioned earlier, if the pixel of the secret image is 0, then one 0 pixel and one 1 pixel is given to the first share holder, while one 0 pixel and one 1 pixel is given to the second share holder. The same thing may also be done by giving one 1 pixel and one 0 pixel to the first share holder, while one 1 pixel and one 0 pixel to the second share holder. On the other hand, for a black pixel in the secret image, one 0 pixel and one 1 pixel is given to the first share holder, while one 1 pixel and one 0 pixel is given to the second share holder. The same thing may also be done by giving one 1 pixel and one 0 pixel to the first share holder, while one 0 pixel and one 1 pixel to the second share holder. This phenomenon may be explained by the two 2×2 Boolean matrices S^0 and S^1 given as follows:

$$S^0 = \begin{bmatrix} 0 & 1 \\ 0 & 1 \end{bmatrix} \quad \text{and} \quad S^1 = \begin{bmatrix} 0 & 1 \\ 1 & 0 \end{bmatrix}.$$

These two matrices are called basis matrices of the $(2, 2)$-VCS. First we observe that the two rows of S^0 correspond to the "white-black" and "white-black" combinations as given in the first line of the third column of Fig. 12.2. Similarly, the two rows of S^1 correspond to the "white-black" and "black-white" combinations as given in the third line of the third column of Fig. 12.2. Now if we interchange the columns of S^0, we see that the two rows of S^0 correspond to the "black-white" and "black-white" combinations as given in the second line of the third column of Fig. 12.2. Similarly, if we interchange the columns of S^1, we see that the two rows of S^1 correspond to the "black-white" and "white-black" combinations as given in the fourth line of the third column of Fig. 12.2. That is why, S^0 and S^1 are called the basis matrices because these matrices can construct the $(2, 2)$-VCS as follows.

Let the pixel P in the secret image be black. Since P is black, we use S^1, the basis matrix corresponding to the black pixel. We apply a random permutation on the columns of S^1 and let the resulting matrix be T^1. We give the first row of T^1 to the first participant as share and second row to the second participant as share. Similar procedure holds if P is a white pixel. Now for the security analysis, if we look at any individual share, it may be either "0 1" or "1 0". But both these patterns are present for both the black and white pixels. So just by looking at an individual share, it is not possible to predict correctly from which matrix it has come, i.e., with probability $\frac{1}{2}$, we can guess from which matrix it has come. Note that any person having no share can also guess with probability $\frac{1}{2}$. Thus with only one share, any participant will not have any extra privilege over any person having no share. But if we superimpose two shares, for black pixel we get "1 1", while for white pixel either we get "1 0" or "0 1". So we can distinguish the black and white pixels in the superimposed image. Thus to construct a VCS, we only need to construct the basis matrices S^0 and S^1.

One thing we note that in the above $(2, 2)$-VCS, for each pixel we are giving two subpixels to each share. So the size of the share will increase. The number of subpixels required to encode one pixel of the secret image (i.e., the number of columns of the basis matrices S^0 or S^1) is known as pixel expansion. In the above $(2, 2)$-VCS the pixel expansion is 2. Pixel expansion determines the size of the share. In visual cryptography we want the pixel expansion to be as small as possible.

Another parameter is very important in visual cryptography. That parameter is known as relative contrast. In order that the recovered image is clearly discernible, it is important that the grey level of a black pixel be darker than that of a white pixel. Informally, the difference in the grey levels of the two pixels is called *contrast*. We want the contrast to be as large as possible. Here the relative contrast is $\frac{2-1}{2} = 1/2$.

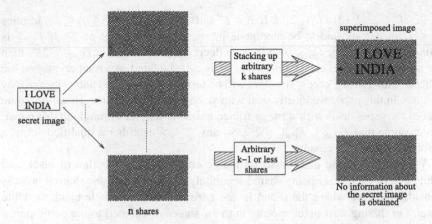

Fig. 12.3 (k, n)-VCS for black and white image

12.12.2 Visual Threshold Schemes

A (k, n)-threshold structure is any access structure $(\Gamma_{\text{Qual}}, \Gamma_{\text{Forb}})$ in which

$$\Gamma_0 = \big\{ B \subseteq \mathcal{P} : |B| = k \big\}$$

and

$$\Gamma_{\text{Forb}} = \big\{ B \subseteq \mathcal{P} : |B| \leq k - 1 \big\}.$$

In any (k, n)-threshold VCS, the image is visible if any k or more participants stack their transparencies (as shown in Fig. 12.3), but totally invisible if fewer than k transparencies are stacked together or analyzed by any other method. In a strong (k, n)-threshold VCS, the image remains visible if more than k participants stack their transparencies.

12.12.3 The Model for Black and White VCS

Let $\mathcal{P} = \{1, \ldots, n\}$ be a set of elements called *participants* and let $2^{\mathcal{P}}$ denote the set of all subsets of $\mathcal{P}$. Let Γ_{Qual} and Γ_{Forb} be subsets of $2^{\mathcal{P}}$, where $\Gamma_{\text{Qual}} \cap \Gamma_{\text{Forb}} = \emptyset$. We will refer to members of Γ_{Qual} as *qualified sets* and the members of Γ_{Forb} as *forbidden sets*. The pair $(\Gamma_{Qual}, \Gamma_{Forb})$ is called the *access structure* of the scheme. In general, $\Gamma_{Qual} \cup \Gamma_{Forb}$ need not be $2^{\mathcal{P}}$.

Let $\Gamma_0 = \{A \mid A \in \Gamma_{\text{Qual}} \wedge \forall A' \subset A, A' \notin \Gamma_{\text{Qual}}\}$ be the collection of all minimal qualified sets.

A participant $P \in \mathcal{P}$ is an *essential* participant if there exists a set $X \subseteq \mathcal{P}$ such that $X \cup \{P\} \in \Gamma_{\text{Qual}}$ but $X \notin \Gamma_{\text{Qual}}$. If a participant $P \in \mathcal{P}$ is not an essential participant, then we call the participant as a non-essential participant.

Let $\Gamma \subseteq 2^{\mathcal{P}} \setminus \{\emptyset\}$ $(\Gamma \subseteq 2^{\mathcal{P}})$. If $A \in \Gamma$ and $A \subseteq A' \subseteq \mathcal{P}$ $(A' \subseteq A \subseteq \mathcal{P})$ implies $A' \in \Gamma$ then Γ is said to be monotone increasing (decreasing) on $\mathcal{P}$. If Γ_{Qual} is monotone increasing, Γ_{Forb} is monotone decreasing and $\Gamma_{\text{Qual}} \cup \Gamma_{\text{Forb}} = 2^{\mathcal{P}}$, then the access structure is called *strong* and Γ_0 is called a *basis*. A visual cryptographic scheme with a strong access structure will be termed as a *strong visual cryptography scheme*. In this paper we mostly deal with strong access structures. However, some part of the paper deals with some restricted access structure. Throughout this paper, we presume that $\Gamma_{\text{Qual}} \cup \Gamma_{\text{Forb}} = 2^{\mathcal{P}}$. So any $X \subseteq \mathcal{P}$ is either a qualified set or a forbidden set of participants.

We further assume that the secret image consists of a collection of black and white pixels, each pixel being shared separately. To understand the sharing process consider the case where the secret image consists of just a single black or white pixel. On sharing, this pixel appears in the n shares distributed to the participants. However, in each share the pixel is subdivided into m subpixels. This m is called the pixel expansion i.e., the number of pixels, on the transparencies corresponding to the shares (each such pixel is called subpixel), needed to represent one pixel of the original image. The shares are printed on transparencies. So a "white" subpixel is actually an area where nothing is printed and left transparent. We assume that the subpixels are sufficiently small and close enough so that the eye averages them to some shade of grey.

In order that the recovered image is clearly discernible, it is important that the grey level of a black pixel be darker than that of a white pixel. Informally, the difference in the grey levels of the two pixel types is called *contrast*. We want the contrast to be as large as possible. Three variables control the perception of black and white regions in the recovered image: a threshold value (t), a relative contrast ($\alpha(m)$) and the pixel expansion (m). The *threshold value* is a numeric value that represents a grey level that is perceived by the human eye as the color black. The value $\alpha(m) \cdot m$ is the contrast, which we want to be as large as possible. We require that $\alpha(m) \cdot m \geq 1$ to ensure that black and white areas will be distinguishable.

Notations Consider an $n \times m$ Boolean matrix M and let $X \subseteq \{1, 2, \ldots, n\}$. By $M[X]$ we will denote the $|X| \times m$ submatrix obtained from M by retaining only the rows indexed by the elements of X. M_X will denote the Boolean "or" of the rows of $M[X]$. The *Hamming weight* $w(V)$ is the number of 1's in a Boolean vector V.

Definition 12.12.1 Let $(\Gamma_{\text{Qual}}, \Gamma_{\text{Forb}})$ be an access structure on a set of n participants. Two collections (multisets) of $n \times m$ Boolean matrices C_0 and C_1 constitute a visual cryptography scheme $(\Gamma_{\text{Qual}}, \Gamma_{\text{Forb}}, m)$-VCS if there exist values $\alpha(m)$ and $\{t_X\}_{X \in \Gamma_{\text{Qual}}}$ satisfying:

1. Any (qualified) set $X = \{i_1, i_2, \ldots, i_p\} \in \Gamma_{\text{Qual}}$ can recover the shared image by stacking their transparencies.

    (Formally, for any $M \in C_0$, M_X the "*or*" of the rows $i_1, i_2, \ldots, i_p$ satisfies $w(M_X) \leq t_X - \alpha(m) \cdot m$; whereas, for any $M \in C_1$ it results in $w(M_X) \geq t_X$.)

2. Any (forbidden) set $X = \{i_1, i_2, \ldots, i_p\} \in \Gamma_{\text{Forb}}$ has no information on the shared image.

   (Formally, the two collections of $p \times m$ matrices D_t, with $t \in \{0, 1\}$, obtained by restricting each $n \times m$ matrix in C_t to rows $i_1, i_2, \ldots, i_p$ are indistinguishable in the sense that they contain the same matrices with the same frequencies.)

12.12.4 Basis Matrices

To construct a visual cryptographic scheme, it is sufficient to construct the basis matrices corresponding to the black and white pixel. In the following, we formally define what is meant by basis matrices.

Definition 12.12.2 (Adapted from Blundo et al. (1999)) Let $(\Gamma_{\text{Qual}}, \Gamma_{\text{Forb}})$ be an access structure on a set $\mathcal{P}$ of n participants. A $(\Gamma_{\text{Qual}}, \Gamma_{\text{Forb}}, m)$-VCS with relative difference $\alpha(m)$ and a set of thresholds $\{t_X\}_{X \in \Gamma_{\text{Qual}}}$ is realized using the $n \times m$ basis matrices S^0 and S^1 if the following two conditions hold:

1. If $X = \{i_1, i_2, \ldots, i_p\} \in \Gamma_{\text{Qual}}$, then S_X^0, the "*or*" of the rows $i_1, i_2, \ldots, i_p$ of S^0, satisfies $w(S_X^0) \leq t_X - \alpha(m) \cdot m$; whereas, for S^1 it results in $w(S_X^1) \geq t_X$.
2. If $X = \{i_1, i_2, \ldots, i_p\} \in \Gamma_{\text{Forb}}$, the two $p \times m$ matrices obtained by restricting S^0 and S^1 to rows $i_1, i_2, \ldots, i_p$ are equal up to a column permutation.

12.12.5 Share Distribution Algorithm

We use the following algorithm to encode the secret image. For each pixel P in the secret image, do the following:

1. Generate a random permutation π of the set $\{1, 2, \ldots, m\}$.
2. If P is a black pixel, then apply π to the columns of S^1; else apply π to the columns of S^0. Call the resulting matrix T.
3. For $1 \leq i \leq n$, row i of T comprises the m subpixels of P in the ith share.

12.12.6 (2, n)-Threshold VCS

In this section we consider only $(2, n)$-VCS for black and white images. In Naor and Shamir (1994), Naor and Shamir first proposed a $(2, n)$-VCS for black and white images. They constructed the 2-out-of-n visual secret sharing scheme by considering the two $n \times n$ basis matrices S^0 and S^1 given as follows.

$$S^0 = \begin{bmatrix} 1 & 0 & 0 & \dots & 0 \\ 1 & 0 & 0 & \dots & 0 \\ \vdots & \vdots & \vdots & \ddots & \vdots \\ 1 & 0 & 0 & \dots & 0 \end{bmatrix}$$

$$S^1 = \begin{bmatrix} 1 & 0 & 0 & \dots & 0 \\ 0 & 1 & 0 & \dots & 0 \\ \vdots & \vdots & \vdots & \ddots & \vdots \\ 0 & 0 & 0 & \dots & 1 \end{bmatrix}$$

S^0 is a Boolean matrix whose first column comprises 1's and whose remaining entries are 0's. S^1 is simply the identity matrix of dimension n.

When we encrypt a white pixel, we apply a random permutation to the columns of S^0 to obtain matrix T. We then distribute row i of T to participant i. To encrypt a black pixel, we apply the permutation to S^1. A single share of a black or white pixel consists of a randomly placed black subpixel and $(n-1)$ white subpixels. Two shares of a white pixel have a combined Hamming weight of 1, whereas any two shares of a black pixel have a combined Hamming weight of 2, which looks darker. The visual difference between the two cases becomes clearer as we stack additional transparencies.

To exemplify this discussion, let us take a concrete example of a $(2, 4)$ VCS. The basis matrices S^0 and S^1 in this case are given by

$$S^0 = \begin{bmatrix} 1 & 0 & 0 & 0 \\ 1 & 0 & 0 & 0 \\ 1 & 0 & 0 & 0 \\ 1 & 0 & 0 & 0 \end{bmatrix} \quad \text{and} \quad S^1 = \begin{bmatrix} 1 & 0 & 0 & 0 \\ 0 & 1 & 0 & 0 \\ 0 & 0 & 1 & 0 \\ 0 & 0 & 0 & 1 \end{bmatrix}.$$

If one examines just a single share then it is impossible to determine whether it represents a share of a black or a white pixel since single shares, whether black or white, look alike. If two shares of a black pixel are superimposed together, we obtain two black and two white subpixels. Combining the shares of a white pixel yields only one black and three white subpixels. Therefore, on stacking two shares, a black pixel will look darker than a white pixel. In the above example the pixel expansion is 4 and the relative contrast for any two participants is 1/4.

12.12.7 Applications of Linear Algebra to Visual Cryptographic Schemes for (n, n)-VCS

Let us construct an (n, n)-VCS, i.e., the secret image is distributed among a set $\mathcal{P} = \{1, 2, \dots, n\}$ of n participants in such a way that if all the n participants superimpose their shares, they get back the secret image as a superimposed image, but a set of $n - 1$ or less number of participants will get no information about the secret image. So here $\Gamma_0 = \{\{1, 2, \ldots, n\}\}$. As we mentioned earlier, to construct the scheme, it is sufficient to construct the basis matrices. To generate the basis matrices, the dealer first associates with each participant i a variable x_i, $i = 1, 2, \ldots, n$. Then the dealer considers the two following systems of linear equations over $\mathbb{Z}_2$:

$$f_{\mathcal{P}} = 0 \tag{12.1}$$

$$f_{\mathcal{P}} = 1, \tag{12.2}$$

where $f_{\mathcal{P}} = x_1 + x_2 + \cdots + x_n$. Clearly, the set of all solutions of (12.1) over $\mathbb{Z}_2$ forms a vector space over $\mathbb{Z}_2$. Let S^0 and S^1 be the Boolean matrices whose columns are just all possible solutions of (12.1) and (12.2) respectively over the binary field. Then it can be shown that S^0 and S^1 satisfy all the two properties of the Definition 12.12.1 and thereby form basis matrices of the (n, n)-VCS having pixel expansion $m = 2^{n-1}$. For more details, the reader may refer to the following theorem from Adhikari (2013).

Theorem 12.12.1 *Let $\mathcal{P} = \{1, 2, \ldots, n\}$ be a set of n participants. Then there exists an (n, n)-VCS with black and white images having optimal pixel expansion $m = 2^{n-1}$ and relative contrast $\alpha(m) = 1/2^{n-1}$.*

To understand the scheme, let us consider the following example.

Example 12.12.1 Let us construct a $(4, 4)$-VCS scheme. As there are four participants, namely 1, 2, 3, and 4, we associate a variable x_i to the ith participant, for $i = 1, 2, 3, 4$. Now we consider the following two systems of linear equations over $\mathbb{Z}_2$:

$$x_1 + x_2 + x_3 + x_4 = 0 \tag{12.3}$$

$$x_1 + x_2 + x_3 + x_4 = 1. \tag{12.4}$$

Let S^0 and S^1 be the Boolean matrices whose columns are just all possible solutions of (12.3) and (12.4), respectively, over the binary field. Then S^0 and S^1 are given by

$$S^0 = \begin{bmatrix} 0 & 0 & 0 & 0 & 1 & 1 & 1 & 1 \\ 0 & 0 & 1 & 1 & 0 & 0 & 1 & 1 \\ 0 & 1 & 0 & 1 & 0 & 1 & 0 & 1 \\ 0 & 1 & 1 & 0 & 1 & 0 & 0 & 1 \end{bmatrix} \text{ and } S^0 = \begin{bmatrix} 0 & 0 & 0 & 0 & 1 & 1 & 1 & 1 \\ 0 & 0 & 1 & 1 & 0 & 0 & 1 & 1 \\ 0 & 1 & 0 & 1 & 0 & 1 & 0 & 1 \\ 1 & 0 & 0 & 1 & 0 & 1 & 1 & 0 \end{bmatrix}.$$

In $(4, 4)$-VCS, if all the four participants come together to superimpose their shares, then only the secret image should be visible. To ensure that we need to check that these two matrices really satisfy the conditions of the basis matrices as in Definition 12.12.1. Note that if $X = \{1, 2, 3, 4\}$, i.e., if the set X of four participants come together and if we assume $t_X = 8$, $m = 8$ and $\alpha(m) = \frac{1}{8}$, then S_X^0, the "or" of all the four rows of S^0, satisfies $w(S_X^0) \le t_X - \alpha(m) \cdot m = 8 - \frac{1}{8} \cdot 8 = 7$ whereas, for S^1, $w(S_X^1) \ge t_X = 8$. To check the second property of the Definition 12.12.1, we see that for any $X \subset \{1, 2, 3, 4\}$ with $|X| \le 3$, $S^0[X]$ and $S^1[X]$ are identical up to column permutation. For example, if we take $X = \{2, 3, 4\}$ and consider $S^0[X]$ and $S^1[X]$, then we see that both the matrices contain identical patters, i.e., exactly one $(0\ 0\ 1)^t$ column, exactly one $(0\ 1\ 0)^t$ column, exactly one $(1\ 0\ 0)^t$ column, exactly one $(1\ 1\ 1)^t$ column, exactly one $(0\ 0\ 0)^t$ column, exactly one $(0\ 1\ 1)^t$ column, exactly one $(1\ 0\ 1)^t$ column and exactly one $(1\ 1\ 0)^t$ column. So the restricted matrices $S^0[X]$ and $S^1[X]$ are identical up to column permutation.

12.12.8 Applications of Linear Algebra to Visual Cryptographic Schemes for General Access Structure

In this section we are going to show how liner algebra may play an important role in constructing visual cryptographic schemes for general access structure. Until now we have discussed about the schemes for $(2, n)$-VCS for (n, n)-VCS. But for more practical applications, we need more flexibility on the access structure. Let us start with an example.

Example 12.12.2 Let us consider a strong access structure on a set of 5 participants having the access structure $(\Gamma_{\text{Qual}}, \Gamma_{\text{Forb}})$ with $\Gamma_0 = \{\{1, 2\}, \{3, 4\}, \{2, 3\}, \{3, 5\}\}$. To construct a VCS for the above mentioned access structure, let us first divide the set Γ_0 into two parts namely $\Gamma_{01} = \{\{1, 2\}, \{3, 4\}\}$ and $\Gamma_{02} = \{\{2, 3\}, \{3, 5\}\}$. For Γ_{01}, let us consider the two following systems of linear equations over the binary field as follows:

$$\left. \begin{array}{r} x_1 + x_2 = 0 \\ x_3 + x_4 = 0 \\ x_5 = 0 \end{array} \right\} \tag{12.5}$$

$$\left. \begin{array}{r} x_1 + x_2 = 1 \\ x_3 + x_4 = 1 \\ x_5 = 0 \end{array} \right\} \tag{12.6}$$

Let S_1^0 and S_1^1 be the Boolean matrices whose columns are just all possible solutions of the above two systems of (12.5) and (12.6), respectively, over the binary field.

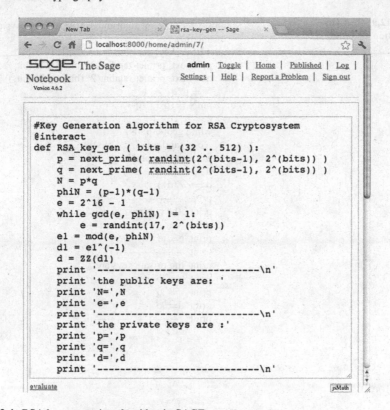

Fig. 12.4 RSA key generation algorithm in SAGE

Thus

$$S_1^0 = \begin{bmatrix} 0 & 0 & 1 & 1 \\ 0 & 0 & 1 & 1 \\ 0 & 1 & 0 & 1 \\ 0 & 1 & 0 & 1 \\ 0 & 0 & 0 & 0 \end{bmatrix} \quad \text{and} \quad S_1^1 = \begin{bmatrix} 0 & 0 & 1 & 1 \\ 1 & 1 & 0 & 0 \\ 0 & 1 & 0 & 1 \\ 1 & 0 & 1 & 0 \\ 0 & 0 & 0 & 0 \end{bmatrix}.$$

We further consider the two following systems of linear equations over the binary field:

$$\left. \begin{array}{r} x_2 + x_3 = 0 \\ x_3 + x_5 = 0 \\ x_1 = 0 \\ x_4 = 0 \end{array} \right\} \tag{12.7}$$

$$\left. \begin{array}{r} x_2 + x_3 = 1 \\ x_3 + x_5 = 1 \\ x_1 = 0 \\ x_4 = 0 \end{array} \right\} \tag{12.8}$$

Table 12.9 RSA key generation algorithm

```
#Key Generation algorithm for RSA Cryptosystem
@interact
def RSA_key_gen( bits = (32 .. 512) ):
  p = next_prime( randint(2^(bits-1), 2^(bits)) )
  q = next_prime( randint(2^(bits-1), 2^(bits)) )
  N = p*q
  phiN = (p-1)*(q-1)
  e = 2^16 - 1
  while gcd(e, phiN) != 1:
    e = randint(17, 2^(bits))
  e1 = mod(e, phiN)
  d1 = e1^(-1)
  d = ZZ(d1)
  print '——————————— \n '
  print ' the public keys are: '
  print 'N=',N
  print 'e=',e
  print '——————————— \n'
  print 'the private keys are:'
  print 'p=',p
  print 'q=',q
  print 'd=',d
  print '———————————\n'
```

As before,

$$S_2^0 = \begin{bmatrix} 0 & 0 \\ 0 & 1 \\ 0 & 1 \\ 0 & 0 \\ 0 & 1 \end{bmatrix} \quad \text{and} \quad S_2^1 = \begin{bmatrix} 0 & 0 \\ 1 & 0 \\ 0 & 1 \\ 0 & 0 \\ 1 & 0 \end{bmatrix}.$$

Finally,

$$S^0 = S_1^0 || S_2^0 = \begin{bmatrix} 0 & 0 & 1 & 1 & 0 & 0 \\ 0 & 0 & 1 & 1 & 0 & 1 \\ 0 & 1 & 0 & 1 & 0 & 1 \\ 0 & 1 & 0 & 1 & 0 & 0 \\ 0 & 0 & 0 & 0 & 0 & 1 \end{bmatrix} \quad \text{and}$$

$$S^1 = S_1^1 || S_2^1 = \begin{bmatrix} 0 & 0 & 1 & 1 & 0 & 0 \\ 1 & 1 & 0 & 0 & 1 & 0 \\ 0 & 1 & 0 & 1 & 0 & 1 \\ 1 & 0 & 1 & 0 & 0 & 0 \\ 0 & 0 & 0 & 0 & 1 & 0 \end{bmatrix}$$

form the basis matrices for the given access structure having pixel expansion 6.

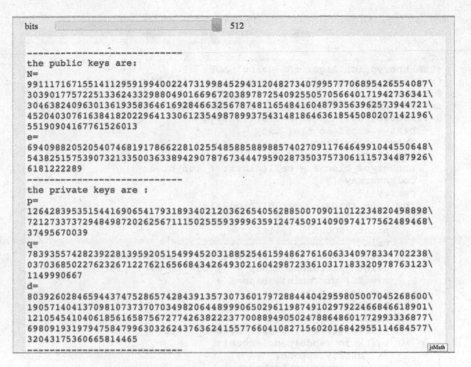

Fig. 12.5 Output of RSA key generation algorithm in SAGE

The above example actually follows from the following theorem proved in Adhikari (2013).

Theorem 12.12.2 *For any given strong access structure* $(\Gamma_{\mathrm{Qual}}, \Gamma_{\mathrm{Forb}})$ *on a set* $\mathcal{P} = \{1, 2, \ldots, n\}$ *of n participants with* $\Gamma_0 = \{B_1, B_2, \ldots, B_k\}$ *where* $B_i \subseteq \mathcal{P}, \forall i = 1, 2, \ldots, k$ *and for any permutation* $\sigma \in S_k$, *the symmetric group of degree k, there exists a strong visual cryptographic scheme* $(\Gamma_{\mathrm{Qual}}, \Gamma_{\mathrm{Forb}}, m)$ *on* $\mathcal{P}$ *with* $m = m_\sigma$, *where* m_σ *is given as follows:*

$$
m_\sigma = \begin{cases} \sum_{i=1}^{l} 2^{|B_{\sigma(2i-1)} \cup B_{\sigma(2i)}| - 2} & \text{if } k = 2l, l \geq 1 \\ \sum_{i=1}^{l} 2^{|B_{\sigma(2i-1)} \cup B_{\sigma(2i)}| - 2} + 2^{|B_{\sigma(2l+1)}| - 1} & \text{if } k = 2l+1, l \geq 0. \end{cases}
$$

Remark As a particular case of a general access structure, we can get a (k, n)-threshold scheme. For example, if we want to construct a $(3, 5)$-threshold VCS, we can consider the $\Gamma_0 = \{\{1, 2, 3\}, \{1, 2, 4\}, \{1, 2, 5\}, \{1, 3, 4\}, \{1, 3, 5\}, \{1, 4, 5\}, \{2, 3, 4\}, \{2, 3, 5\}, \{2, 4, 5\}, \{3, 4, 5\}\}$ and apply Theorem 12.12.2 to get the $(3, 5)$-VCS.

So, in general, we may have the following theorem for (k, n)-VCS as proved in Adhikari (2013).

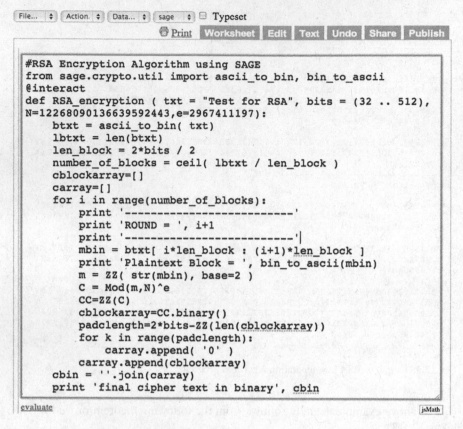

```
#RSA Encryption Algorithm using SAGE
from sage.crypto.util import ascii_to_bin, bin_to_ascii
@interact
def RSA_encryption ( txt = "Test for RSA", bits = (32 .. 512),
N=122680901366639592443,e=2967411197):
    btxt = ascii_to_bin( txt)
    lbtxt = len(btxt)
    len_block = 2*bits / 2
    number_of_blocks = ceil( lbtxt / len_block )
    cblockarray=[]
    carray=[]
    for i in range(number_of_blocks):
        print '---------------------------'
        print 'ROUND = ', i+1
        print '---------------------------'
        mbin = btxt[ i*len_block : (i+1)*len_block ]
        print 'Plaintext Block = ', bin_to_ascii(mbin)
        m = ZZ( str(mbin), base=2 )
        C = Mod(m,N)^e
        CC=ZZ(C)
        cblockarray=CC.binary()
        padclength=2*bits-ZZ(len(cblockarray))
        for k in range(padclength):
            carray.append( '0' )
        carray.append(cblockarray)
    cbin = ''.join(carray)
    print 'final cipher text in binary', cbin
```

Fig. 12.6 RSA encryption algorithm in SAGE

Theorem 12.12.3 *Let* $(\Gamma_{\text{Qual}}, \Gamma_{\text{Forb}})$ *be an access structure on a set* $\mathcal{P} = \{1, 2, \ldots, n\}$ *of n participants with* $\Gamma_0 = \{A \subseteq \mathcal{P} : |A| = k\}$, $2 \leq k \leq n$. *Then there exists a strong* (k, n)-*VCS with*

$$m_{\text{our}} = \begin{cases} l \cdot 2^{k-2}, & \text{if } l = \binom{n}{k} \text{ is even} \\ (l+1) \cdot 2^{k-2}, & \text{if } l = \binom{n}{k} \text{ is odd} \end{cases}$$

and relative contrast $\alpha(m) = \frac{1}{m_{\text{our}}}$.

Remark For further studies, the readers may refer (Adhikari and Sikdar 2003; Adhikari and Bose 2004; Adhikari et al. 2004, 2007; Adhikari and Adhikari 2007; Adhikari 2006, 2013; Ateniese et al. 1996a,b; Blundo et al. 1999, 2003; Naor and Shamir 1994).

In the next section, we are going to discuss about a nice application of open-source software, known as, SAGE for the implementation of cryptographic schemes with very large integers having many digits, such as, integers having more than 300

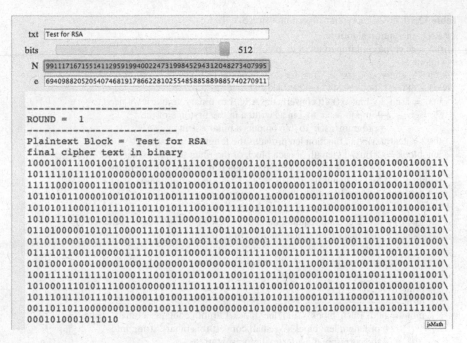

Fig. 12.7   Output of RSA encryption algorithm in SAGE

digits. For actual implementation of public-key cryptographic schemes, such big numbers are essential. In the next section, we show how to deal with these numbers.

12.13 Open-Source Software: SAGE

Let us start our discussion with open-source software. Free and open-source software (FOSS) or free/libre/open-source software (FLOSS) are a class of softwares that are not only free softwares but also open source in the sense that these are liberally licensed to grant users the right to use, copy, study, change, and improve their designs through the availability of their source codes. This novel approach has gained both momentum and acceptance, as the potential benefits have been increasingly recognized by both individuals and corporations. SAGE is one of such useful outcomes of such approach. SAGE (System for Algebra and Geometry Experimentation) is free as well as open-source mathematics software that is very much useful for research and teaching in algebra, geometry, number theory, cryptography, numerical computation and related areas. The overall goal of SAGE is to create a viable, free, open-source alternative to the costly mathematical computational softwares, such as, Maple, Mathematica, Magma, and MATLAB. SAGE is sometimes called SAGEMATH to distinguish it from other uses of the word. Historically, the first version of SAGE was released on 24 February 2005 as free and open-source software under the terms of the GNU General Public License. The originator and

Table 12.10 RSA encryption algorithm using SAGE

```
# RSA encryption algorithm
from sage.crypto.util import ascii_to_bin, bin_to_ascii
@interact
def RSA_encryption (txt = "Test for RSA", bits = (32 .. 512),
N=122680901366639592443,e=2967411197):
  btxt = ascii_to_bin( txt) #it converts the ASCII to binary, to use it we need to
                # import ascii_to_bin as written in the first line of the
                #program. ascii_to_bin outputs a binary string.
  lbtxt = len(btxt) #the function len provides the length of any string
  len_block = 2*bits/2 #length of each block of the plain text, note that if we take
                # len_block=2*bits, then it might exceed the range of N
  number_of_blocks = ceil( lbtxt / len_block ) #number of plain text blocks
  cblockarray=[ ] #it will contain the binary representation of c=m^e(mod n),
                #corresponding to each block
  carray=[ ] # it will contain required number of zeros to be appended at the left of
                # the binary representation of c=m^e(mod n) to make it a 2*bits
                # representation of c, corresponding to each block
  for i in range(number_of_blocks): #it runs ROUND=number_of_ blocks times.
      print '————————————'
      print 'ROUND = ', i+1
      print '————————————'
      mbin = btxt[ i*len_block: (i+1)*len_block ] #mbin contains a binary string
                # of length len_block. We shall convert this binary string into
                # the corresponding decimal representation
      print 'Plaintext Block = ', bin_to_ ascii(mbin)
      m = ZZ( str(mbin), base=2 ) #To compute m^e(mod N), we need to convert
                # the mbin into the corresponding decimal representation
      C = Mod(m,N)^e
      CC=ZZ(C)
      cblockarray=CC.binary() #To convert the CC into a binary of length 2*bits
          # and to store in cblockarray, we need to pad
          # (2*bits−ZZ(len(cblockarray)) zeros to the left of cblockarray
      padclength=2*bits-ZZ(len(cblockarray)) # number of zeros to be padded
      for k in range(padclength):
          carray.append( '0' ) # at the end of the for loop,
                        #carray contains padclength many zeros
      carray.append(cblockarray) # appends the contents of carray to the contents
                        # of cblockarray and stores it in carray
  # End of the main for loop
  cbin = ''.join(carray) # at the end of the original for loop, join the contents
          # of "carry" to get the final binary string representing the whole
          # cipher text stored in cbin.
  print 'final cipher text in binary', cbin #print cbin as the cipher text in binary
```

the leader of the SAGE project was William Stein, a mathematician at the University of Washington. SAGE uses the programming language Python. SAGE development uses both students and professionals for development. The development of SAGE is supported by both volunteer work and grants. The website for SAGE is http://www.sagemath.org. The software may be downloaded from the links available in that website. There is on-line computation facility available at that web-

Table 12.11   RSA decryption algorithm using SAGE

```
# RSA decryption algorithm
from sage.crypto.util import ascii_to_bin, bin_to_ascii
@interact
def RSA_decryption (ctext = '' 0011100011010000011100101100100l
1010111111101001000111110010100l'',
bits = (32 .. 512), d=3319087257124301681,N=6915083087689750517):
    f=[ ] # this is an array to be used latter on
    lctext = len(ctext) #it provides the length of the cipher text
    nb=lctext / (2*bits) # it provides the number of blocks
    cblockarray = [ ]
    for i in range(nb):
        print '_____'
        print 'ROUND = ', i+1
        print '_____'
        cblockarray = ctext[ i*2*bits: (i+1)*2*bits ] # it actually crops a block
                        # of binary string of length 2*bits from the array
                        # ctext to store it in the array cblockarray
        c = ZZ( str(cblockarray), base=2 ) # converts binary string into decimal
        P=Mod(c,N)^d # it calculates c^d (mod N), as a result, P becomes
                        # an element of the ring of integers modulo N
        M = ZZ(P) # converts element of the ring Z_n into an integer
        Mbin = M.binary() # Mbin contains the binary representation of M
        padlen = ZZ( 8 − ZZ(len(Mbin)).mod(8) ) #Finally we need to convert the
                        # binary to ASCII. So the length of Mbib must be a
                        # multiple of 8. padlen actually contains the number of zeros to
                        # be padded to the left of Mbin to make it a multiple of 8
        for k in range(padlen):
            f.append( '0' ) # the array f contains the number of zeros to be padded
        f.append( Mbin ) # it appends the content of Mbin to the content of f
    fbin = ' '.join(f) # it actually contains the plain text in the binary form
    ftxt = bin_to_ascii(fbin) # it converts the binary plain text into actual plain text
    print ' Final recovered message = ', ftxt # final recovered plain text
```

site. We can simply create an on-line account at either http://www.sagenb.org or
http://www.sagenb.kaist.ac.kr and use SAGE with the browser present at our computer.

In the next section, we are going to show how SAGE may be used to implement
three public-key cryptosystems, e.g., RSA, ElGamal, and Rabin cryptosystems with
large numbers.

12.13.1   Sage Implementation of RSA Cryptosystem

We have already discussed about the RSA cryptosystem in Sect. 12.6. Now we are
going to explain how RSA may be implemented using SAGE with large numbers.
Recall that, when Alice wants to send secret message to Bob using RSA cryptosystem, Alice and Bob have to run three algorithms, namely, key generation algorithm
(done by Bob), encryption algorithm (done by Alice) and decryption algorithm

```
#SAGE implementation of RSA Decryption
from sage.crypto.util import ascii_to_bin, bin_to_ascii
@interact
def RSA_decryption ( ctext =
"0011100011010000011100101100100110101111111101001000111110010
1001",bits =
(32 .. 512),d=3319087257124301681,N=6915083087689750517 ):
    f=[]
    lctext = len(ctext)
    nb=lctext/(2*bits)
    cblockarray = []
    for i in range(nb):
        print '---------------------------'
        print 'ROUND = ', i+1
        print '---------------------------'
        cblockarray = ctext[ i*2*bits : (i+1)*2*bits ]
        c = ZZ( str(cblockarray), base=2 )
        P=Mod(c,N)^d
        M = ZZ(P)
        Mbin = M.binary()
        padlen = ZZ( 8 - ZZ(len(Mbin)).mod(8) )
        for k in range(padlen):
            f.append( '0' )
        f.append( Mbin )
    fbin = ''.join(f)
    ftxt = bin_to_ascii(fbin)
    print 'Final recovered message = ', ftxt
```

ctext	1000100111001001010101100111110100000011011100000001110
bits	512
d	803926028465944374752865742843913573073601797288440429
N	991117167155141129591994002247319984529431204827340799 5

```
---------------------------------
ROUND =  1
---------------------------------
Final recovered message =  Test for RSA
```

Fig. 12.8 RSA decryption algorithm with output in SAGE

(done by Bob). First we shall explain the key generation algorithm to be implemented by Bob. Here we are going to define a function known as "RSA_key_gen" which takes as input the security parameter, i.e., the length of each of the primes (i.e., number of bits to represent p or q) and produces the public and private keys. Here we are using the concept "@interact" so that the function "RSA_key_gen" can take input i.e., the bit length of p or q, in an interactive way. Due to the use of "@interact", we can change the bit length (here 32 bits to 512 bits) interactively while running the program without changing the main body of the program as shown in Fig. 12.4. If we write "@interact def RSA_key_gen (bits = (16 .. 1024)):", then we can vary the length of the primes from 16 bits to 1024 bits interactively. In this

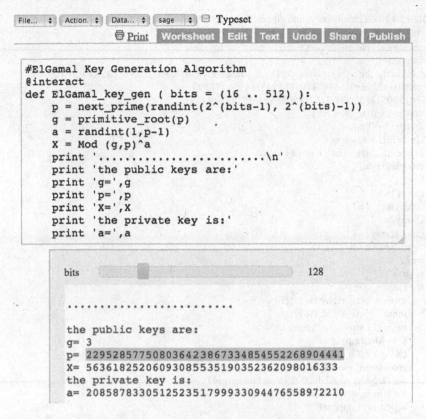

Fig. 12.9 ElGamal key generation algorithm with output in SAGE

Table 12.12 ElGamal key generation algorithm

```
#Key Generation algorithm for ElGamal Cryptosystem
@interact
def ELGAMAL_key_gen ( bits = (16 .. 512) ):
    p = next_prime( randint(2^(bits-1), 2^(bits)-1))
    g = primitive_root(p)
    a = randint(1,p-1)
    X = Mod (g,p)^a
    print '——————— \n'
    print ' the public keys are:'
    print ' g=' ,g
    print ' p=' ,p
    print ' X=' ,X
    print ' the private key is: '
    print ' a=',a
```

key generation program, we are going to use the following functions whose utilities are described below:

- **randint(a, b):** This function outputs a random integer in the interval $[a, b]$.

Table 12.13 ElGamal encryption algorithm using SAGE

```
# ElGamal encryption algorithm
from sage.crypto.util import ascii_to_bin, bin_to_ascii
@interact
def ElGamal_encryption ( txt = "Dr. Avishek Adhikari", bits = (16 .. 128), g = 10,
p = 2304302923946618747967135057445564451393,
X = 199001885002992147485058529117175239418):
    btxt = ascii_to_bin( txt )
    lbtxt = len(btxt)
    chr = bits # block size
    lb = ceil( lbtxt / chr ) # no of blocks
    cblockarray=[ ]
    f = [ ]
    carray1=[ ]
    cblockarray1=[ ]
    carray2=[ ]
    cblockarray2=[ ]
    r1 = randint(1,bits-1)
    r=ZZ(r1)
    print 'k=', r
    for i in range(lb):
        mbin = btxt[ i*chr: (i+1)*chr ]
        mbin = btxt[ i*chr: (i+1)*chr ]
        m = ZZ( str(mbin), base=2 )
        C = Mod(g,p)^r
        CC = ZZ(C)
        cblockarray1 = CC.binary()
        padclength1 = ZZ(bits-ZZ(len(cblockarray1)))
        for k in range(padclength1):
            carray1.append( '0' )
        carray1.append(cblockarray1)
        D = Mod(m*X^r,p)
        DD = ZZ(D)
        cblockarray2 = DD.binary()
        padclength2 = ZZ(bits-ZZ(len(cblockarray2)))
        for k in range(padclength2):
            carray2.append( '0' )
        carray2.append(cblockarray2)
    dbin = ''.join(carray2)
    cbin = ''.join(carray1)
    print ' final cipher text in binary c=', cbin
    print 'd=', dbin
```

- **next_prime(a):** This function outputs the next prime after a.
- **gcd(a, b):** This function outputs the gcd of a and b.
- **mod(a, n):** This function outputs an element of the ring $\mathbf{Z}_n$, i.e., mod(a, n) returns the equivalence class (a) in $\mathbf{Z}_n$.
- **e^(−1):** If e is an element of $\mathbf{Z}_n^*$, then e^(−1) represents the multiplicative inverse of e in $\mathbf{Z}_n^*$.

```
from sage.crypto.util import ascii_to_bin, bin_to_ascii
@interact
def ELGAMAL_test ( txt = "Dr. Avishek Adhikari", bits = (16 ..
128),g=3,p=2295285775080364238673348545522268904441,X=5636182520609308
5535190352362098016333):
    btxt = ascii_to_bin( txt )
    lbtxt = len(btxt)
    chr = bits # block size
    lb = ceil( lbtxt / chr ) # no of blocks
    f = []
    carray1=[]
    cblockarray1=[]
    carray2=[]
    cblockarray2=[]
    r1 = randint(1,bits-1)
    r=ZZ(r1)
    print 'k=', r
    for i in range(lb):
        mbin = btxt[ i*chr : (i+1)*chr ]
        m = ZZ( str(mbin), base=2 )
        C = Mod(g,p)^r
        CC=ZZ(C)
        cblockarray1=CC.binary()
        padclength1=ZZ(bits-ZZ(len(cblockarray1)))
        for k in range(padclength1):
            carray1.append( '0' )
        carray1.append(cblockarray1)
        D = Mod(m*X^r,p)
        DD=ZZ(D)
        cblockarray2=DD.binary()
        padclength2=ZZ(bits-ZZ(len(cblockarray2)))
        for k in range(padclength2):
            carray2.append( '0' )
        carray2.append(cblockarray2)
    dbin = ''.join(carray2)
    cbin = ''.join(carray1)
    print 'final cipher text in binary c=', cbin
```

Fig. 12.10 ElGamal encryption algorithm in SAGE

- **ZZ(d):** This function converts the element d from the ring $\mathbf{Z}_n$ to the element of the ring of integers $\mathbf{Z}$.
- **#:** "#" is used to comment a line.

Now the code for key generation algorithm in SAGE with documentation is written in Table 12.9. As screen short of the actual SAGE code with output for 512 bits are shown in Figs. 12.4 and 12.5, respectively.

Now we are going to explain the encryption algorithm. The idea of the encryption algorithm is as follows: Suppose Alice is going to encrypt a plain text message. Alice will use the function "RSA_encryption()" for encryption. This function takes as input the plain text stored in the array "txt", the security parameter (i.e., the number of bits required to represent p or q in binary) stored in the variable "bits", the public keys stored in the variables N and e. The program first converts the plain text (ASCII characters) into binary and stores it in the array "btxt". Suppose the

```
txt   Dr. Avishek Adhikari

bits  [                                            ]   128

  g   3
  p   2295285775080364238673348545522268904441
  X   5636182520609308553519035236209801633

k= 70
final cipher text in binary c=
0000000000000000000111101101101010010000111010011111101111100100000011111\
1101001010011111000001011111100111101000001101111101100100000000000000000\
0111101101101001000011101001111110111110010000000011111110100101001011111\
0000010111111001111010000011011111011001
d=
0110000010010110111110111010100100001110111100010101001010000010110100\
1111010011110001100110000010111101101101000101111010011011110100110011111\
1110100001001100000101100110100000011000000010010100000011110010101010110111\
01100100101111010111001100010000111011111
```

Fig. 12.11 Output of the ElGamal encryption algorithm in SAGE

length (calculated using "len(btxt)") of the binary string of the plain text is 3104 (note that it has to be a multiple of 8, as each ASCII character is represented by 8 bits). We subdivide the whole plain text string into certain blocks. So we need to calculate the block size. For example, if the length of the binary representation of p (or q) is 512 bits, then we choose the block size to be $512 * 2/2 = 512$ bits. So the number of blocks will be the ceiling of $(3104/512) = 7$, i.e., ceil(3104/512). For each such binary block, we convert it into the corresponding decimal representation, say m. Note that the value of m is always less than N = pq. Then we calculate c = m^e (mod N). Next we convert the decimal representation of c into binary. Note that we need to pad certain number of zeros at the left of this binary representation to make it a binary string of length 2*bits, where the variable "bits" represents the number of bits to represent p or q. Finally, we concatenate the content of all the sub-blocks to get the final binary representation of the cipher text. The program will output the binary representation of the corresponding cipher text.

In this program, we are going to use the following functions whose utilities are explained below.

- **from sage.crypto.util import ascii_to_bin:** The function "ascii_to_bin" is defined in sage.crypto.util. So to use "ascii_ to_bin", we need to import it from " sage.crypto.util".
- **from sage.crypto.util import bin_to_ascii:** The function "bin_to_ascii" is defined in sage.crypto.util. So to use "bin_to_ ascii", we need to import it from " sage.crypto.util".
- **RSA_encryption(txt, bits, N, e):** This function takes as input the plain text, the bit length, the public keys N and the encryption exponent e. This function outputs the binary representation of the cipher text.

Table 12.14 ElGamal decryption algorithm using SAGE

```
# ElGamal decryption algorithm
from sage.crypto.util import ascii_to_bin, bin_to_ascii
@interact
def ElGamal_decryption (c= ''00000000000000000000000000000000000000000000000
0000000000001101100011010110010011010110111000101110111101010000000000
000000000000000000000000000000000000000000000000000000000000000000000011
01100011010111001001101011011100010111011110101000000000000000000000'',
d= '' 010110010110001000110100110011000000100110101000011101111010010110
11001100001101101111100000101000111111010001000011010101000001001001010
1110111100011001001001100000100010101011111101101001011110101010000110100
1101111011110011101010001101001011001101011011101100'',bits =(16 .. 128),
a= 195316965957439562480200545011166321389,
p=230430292394661874796713505744556451393):
    f=[ ]
    lctext1 = len(c)
    lctext2 = len(d)
    nb=lctext1 / (bits)
    cblockarray1 = [ ]
    cblockarray2 = [ ]
    carray1=[ ]
    for i in range(nb):
        cblockarray1 = c[ i*bits: (i+1)*bits ]
        c1 = ZZ( str(cblockarray1), base=2 )
        cblockarray2 = d[ i*bits: (i+1)*bits ]
        d1 = ZZ( str(cblockarray2), base=2 )
        x=Mod(c1,p)^(-a)
        P=Mod(x*d1,p)
        M = ZZ(P)
        Mbin = M.binary()
        padlen = ZZ( 8 - ZZ(len(Mbin)).mod(8) )
        for k in range(padlen):
            f.append( '0' )      f.append( Mbin )
        print 'f=',f
    fbin = ''.join(f)    ftxt = bin_to_ascii(fbin)    print ' Final recovered message = ' , ftxt
```

- **ascii_to_bin( ):** It converts the ASCII to binary. To use it we need to import ascii_to_bin as written in the first line of the program. "ascii_to_bin" outputs a binary string.
- **bin_to_ascii():** It converts the binary to ASCII, to use it, we need to import ascii_to_bin as written in the first line of the program.
- **len(btxt):** "len" provides the length of a string stored in btxt.
- **ceil(z):** It gives the ceiling value of the rational number z.
- **Mod(a, N)$^\wedge$e:** It computes the value a^e (mod N).
- **CC.binary():** It converts the decimal value stored in CC into binary.
- **carray.append(cblockarray):** The function "append()" appends the contents of the array "carray" to the contents of the array "cblockarray" and stores it in the array "carray".
- **''.join(carray):** The function "join()" joins the contents of the array "carray" to get a string.

```
from sage.crypto.util import ascii_to_bin, bin_to_ascii
@interact
def Elgamal_decryption (c=
"00000000000000000011110110110101001000011010011111101111100100000001
11111010010100111110000010111111001111010000011011111011001000000000
00000000111011011010010000011101001111110111100100000001111110100
10100111110000010111111100111101000001101111011001", d=
"01100000100101101111101110101001000011110111100010101001010000001011
01001110100111100011001100000101111011101101000010111010011011101001
10011111110100010011000001011001101000001100000001001010000011110010
101011011011001001011110101110011000100001111101111" ,bits =(16 ..
128),a=208587833051252351799933094476558972210,p=229528577508036423867
334854552268904441):
    f=[]
    lctext1 = len(c)
    lctext2 = len(d)
    nb=lctext1/(bits)
    cblockarray1 = []
    cblockarray2 = []
    for i in range(nb):
        cblockarray1 = c[ i*bits : (i+1)*bits ]
        c1 = ZZ( str(cblockarray1), base=2 )
        cblockarray2 = d[ i*bits : (i+1)*bits ]
        d1 = ZZ( str(cblockarray2), base=2 )
        x=Mod(c1,p)^(-a)
        P=Mod(x*d1,p)
        M = ZZ(P)
        Mbin = M.binary()
        padlen = ZZ( 8 - ZZ(len(Mbin)).mod(8) )
        for k in range(padlen):
            f.append( '0' )

        f.append( Mbin )
        print 'f=',f
    fbin = ''.join(f)
    ftxt = bin_to_ascii(fbin)
    print 'Final recovered message = ', ftxt
```

evaluate

Fig. 12.12 ElGamal decryption algorithm in SAGE

The SAGE code with documentation is explained in Table 12.10.

The actual SAGE code for RSA encryption program with output are shown in Figs. 12.6 and 12.7, respectively.

We are now going to explain the decryption algorithm done by Bob. The RSA_decryption() function takes as input the cipher text in binary stored in the array "ctext", the security parameter stored in "bits", the decryption exponent d stored in "d" and the public value stored in "N". This function outputs the plain text back. The program works as follows: First the binary cipher text is divided into blocks, each of size 2*bits. Then each binary block is converted into decimal number, say c. For each of the decimal representation c, we need to compute c^d (mod N) to get M. Then we convert M into binary string of length a multiple of 8 by suitably padding zeros to the left. Append each such binary string obtained from each block one after

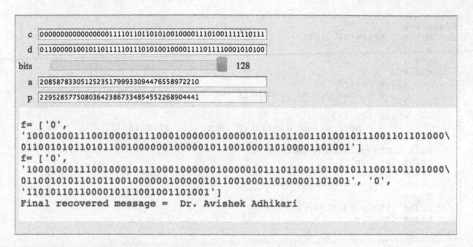

```
c  00000000000000000011110110110101001000011101001111110111
d  01100000100101101111101110101001000011110111100010101000
```
bits ██ ▌ 128
```
a  2085878330512523517999330944765589722210
p  2295285775080364238673348545522268904441
```

```
f= ['0',
'1000100011100100010111000100000010000010111011001101001011100110110\
0110010101011010110010000001000001011001000110100001101001']
f= ['0',
'1000100011100100010111000100000010000010111011001101001011100110110\
0110010101011010110010000001000001011001000110100001101001', '0',
'11010110110000101110010011101001']
Final recovered message =  Dr. Avishek Adhikari
```

Fig. 12.13 Output of the ElGamal decryption algorithm in SAGE

Table 12.15 Rabin key generation algorithm

```
#Key Generation algorithm for Rabin Cryptosystem
@interact
def Rabin_key_gen( bits = (16 .. 128) ):
    p=1
    q=1
    while(p==q):
        while p%4!=3:
            p = next_prime( randint(2^(bits-1), 2^(bits)) )
        while q%4!=3:
            q = next_prime( randint(2^(bits-1), 2^(bits)) )
    print '——————— \n '
    print 'The private keys are: '
    print 'p= ',p
    print 'q= ',q
    print '——————— \n '
    N=p*q
    print 'The public key is: '
    print 'N= ',N
```

another. Finally convert this binary string into ASCII characters to get back the original plain text. The SAGE code with documentation for RSA decryption is presented in Table 12.11.

A screen short of the actual SAGE code for RSA decryption algorithm and its output are shown in Fig. 12.8.

```
@interact
def Rabin ( bits=(16..128)):
    p=1
    q=1
    while(p==q):
        while p%4!=3:
            p = next_prime( randint(2^(bits-1), 2^(bits)) )

        while (q%4!=3):
            q = next_prime( randint(2^(bits-1), 2^(bits)) )|
    print '===================== '
    print 'The private keys are: '
    print 'p= ',p
    print 'q= ',q
    print '===================== '
    N=p*q
    print 'The public key is: '
    print 'N= ',N
```

```
bits                                                    16

=====================
The private keys are:
p=  58679
q=  38903
=====================
The public key is:
N=  2282789137
```

Fig. 12.14 Rabin key generation algorithm with output in SAGE

12.13.2 SAGE Implementation of ElGamal Plulic Key Cryptosystem

The EnGamal public-key cryptosystem is already explained in Sect. 12.7. The implementation of ElGamal is similar to that of RSA. Here also we need to implement three algorithms, namely, key generation algorithm (run by Bob), encryption algorithm (run by Alice) and the decryption (run by Bob). For the key generation, the function "ELGAMAL_key_gen()" takes as input the security parameter stored in the variable "bits" and out puts the public and the public values. In this algorithm, we use the function "primitive_root(n)" which returns a generator for the multiplicative group of integers modulo n, if one exists. The key generation algorithm for the ElGamal cryptosystem is in Table 12.12.

A screen short of the actual SAGE code for ElGamal key generation algorithm along with the output is shown in Fig. 12.9.

The SAGE code for ElGamal encryption is presented in Table 12.13.

A screen short of the actual SAGE code for ElGamal encryption algorithm along with the output are shown in Figs. 12.10 and 12.11, respectively.

The SAGE code for ElGamal decryption is presented in Table 12.14.

A screen short of the actual SAGE code for ElGamal decryption algorithm along with the output are shown in Fig. 12.12 and 12.13 respectively.

Table 12.16   Rabin encryption algorithm using SAGE

```
# Rabin encryption algorithm
from sage.crypto.util import least_significant_bits
from sage.crypto.util import ascii_to_bin
@interact
def Rabin_encryption (txt = ″A″,N=2245559221,bits=16):
  btxt = ascii_to_bin( txt ) # binary representation of the input in a multiple of 8
  C1_array=[ ] # the array will help to store the cipher text C1 in binary
  C2_array=[ ] # the array will help to store the cipher text C2 in binary
  C1_blockarray=[ ]
  lbtxt=len(btxt) # length of the input binary string
  R=IntegerModRing(N) # we shall work in the Ring Z_N, that is, the ring of
                      #integers modulo N
  for i in range(lbtxt):
    print '——————— \n '
    print 'ROUND = ', i+1
    print '——————— \n '
    m=btxt[i] #since btxt contains binary string, we need to convert it to
              # integer in the next step
    m = ZZ( str(m), base=2 ) # integer representation of m from string
    y=R.random_element()
    while (GCD(y,N)!=1):
        y=R.random_element() #y is a random element from Z_N*
    r=Mod(y,N)^2 # random element from QR_N and we need the integer value
    x=ZZ(r) # integer value of r, so x is a random element from QR_N
    c1=ZZ(Mod(x,N)^) # 1st pair of the cipher text in integer form,
                     # we need to put it in binary, so ZZ is required
    lsb=least_significant_bits(x, 1) # lsb is an array contains the lsb of x
                                     # in the zeroth location of lsb[ ]
    c2=(lsb[0]+m)%2 # 2nd pair of cipher text
    C1_blockarray=c1.binary() #c1 has to represent in binary representation of
                              # length of N, so we need to first make it binary and pad suitably
    padclength=ZZ(2*bits-ZZ(len(C1_blockarray)))
    for k in range(padclength):
        C1_array.append( '0' )
    C1_array.append(C1_blockarray) # the step in which the consecutive
                                   # binary representations of C1's are appended
    C2_blockarray=c2.binary() # though c2 is in binary, to append, we need to
                              # make it a string
    C2_array.append(C2_blockarray) # the step in which the consecutive binary
                                   # string representations of C2's are appended
  C1_bin = ''.join(C1_array)
  C2_bin = ''.join(C2_array)
  print 'The final Cipher for C1 is: ', C1_bin
  print 'The final Cipher for C2 is: ', C2_bin
```

12.13.3 SAGE Implementation of Rabin Plulic Key Cryptosystem

We have already explained the Rabin public-key cryptosystem in Sect. 12.8. In this section, we are going to implement this cryptosystem in SAGE. Like RSA and El-Gamal, Rabin cryptosystem also has three major parts, namely key generation, en-

Table 12.17   Rabin decryption algorithm using SAGE

```
# Rabin decryption algorithm
from sage.crypto.util import least_significant_bits, ascii_to_bin, bin_to_ascii
@interact
def Rabin_decryption(c1=" 01101111000111110000011101100100011001101011010
110011001000111100111110001110001111000011011110011001100110",
c2="00010011",p=34351,q=65371,bits=16):
  N=p*q
  R=IntegerModRing(N)
  f=[ ]
  len_c1=len(c1)
  blocks=len_c1/(bits*2) # no of blocks each of length of N i.e., 2*bits
  for i in range(blocks):
    cblockarray = c1[ i*2*bits: (i+1)*2*bits ]
    c = ZZ( str(cblockarray), base=2 ) # decimal representation of C1 in ith block
    cp=c%p
    cq=c%q
    ap=Mod(cp,p)^((p+1)/4) # to calculate the 4 values of sq roots
    aq=Mod(cq,q)^((q+1)/4) # to calculate the 4 values of sq roots
    ap1=ZZ(ap) # to calculate the Ligendre Symbol, int value is required. That is
                      # why we need to convert the ring element into integer value
    aq1=ZZ(aq)
    ap2=ZZ(-ap)
    aq2=ZZ(-aq)
    ls_ap1=legendre_symbol(ap1,p) # Ligendre symbol for ap1 modulo p
    ls_ap2=legendre_symbol(ap2,p)
    ls_aq1=legendre_symbol(aq1,q)
    ls_aq2=legendre_symbol(aq2,q)
    if ((ls_ap1==1) & (ls_aq1==1)): # If both the values are 1, then we can find
           # the required element x by using the Chinese Remainder Theorem (CRT)
        x=crt(ap1,aq1,p,q)
    if ((ls_ap1==1) & (ls_aq2==1)):
        x=crt(ap1,aq2,p,q)
    if ((ls_ap2==1) & (ls_aq1==1)):
        x=crt(ap2,aq1,p,q)
    if ((ls_ap2==1) & (ls_aq2==1)):
        x=crt(ap2,aq2,p,q)
    x1=ZZ(x) # As x is in the ring of integers modulo pq, to use the function
           # least_significant_bits(), we need to convert it in integer
    lsb_x1=least_significant_bits(x1, 1) # least significant bit of the element
    c_2_blockarray = c2[ i: (i+1)] # As c2 is a string of binary, we need to
           # convert it from string to integer in the next line
    cc2 = ZZ( str(c_2_blockarray), base=2 )
    bin_plain_text=(lsb_x1[0]+cc2)%2
    string_bin_plain_text=bin_plain_text.binary()
    f.append(string_bin_plain_text) # f is the final binary
       # array in string format to contain the bin plain text
# out of the main for loop
  fbin = "".join(f) # it joins the elements of f to use bin_to_ascii
  ftxt = bin_to_ascii(fbin)
  print 'Final recovered message = ', ftxt
```

```
from sage.crypto.util import least_significant_bits
from sage.crypto.util import ascii_to_bin
@interact
def Rabin_encryption (txt = "Bob",N=2282789137,bits=16):
    btxt = ascii_to_bin( txt )
    C1_array=[]
    C2_array=[]
    C1_blockarray=[]
    lbtxt=len(btxt)
    R=IntegerModRing(N)
    for i in range(lbtxt):
        m=btxt[i]
        m = ZZ( str(m), base=2 )
        y=R.random_element()
        while (GCD(y,N)!=1):
            y=R.random_element()
        r=Mod(y,N)^2
        x=ZZ(r)
        c1=ZZ(Mod(x,N)^2)
        lsb=least_significant_bits(x, 1)
        c2=(lsb[0]+m)%2
        C1_blockarray=c1.binary()
        padclength=ZZ(2*bits-ZZ(len(C1_blockarray)))
        for k in range(padclength):
            C1_array.append( '0' )
        C1_array.append(C1_blockarray)
        C2_blockarray=c2.binary()
        C2_array.append(C2_blockarray)
    C1_bin = ''.join(C1_array)
    C2_bin = ''.join(C2_array)
    print 'The final Cipher for C1 is : ', C1_bin
    print 'The final Cipher for C2 is : ', C2_bin
```

Fig. 12.15 Rabin encryption algorithm in SAGE

cryption, and decryption. In the key generation step, for a given security parameter "bits", we need two random primes p and q of length "bits" such that $p \equiv q \equiv 3 \pmod 4$.

The SAGE code for Rabin key generation is presented in Table 12.15.

A screen short of the actual SAGE code for Rabin key generation algorithm along with the output are shown in Fig. 12.14.

For encryption, we have to import the functions "least_significant_ bits" and "ascii_to_bin" by writing "from sage.crypto.util import least_significant_bits" and "from sage.crypto.util import ascii_to_ bin" at the very beginning of the encryption program. Here we shall deal with the Ring $\mathbf{Z}_N$, that is, the ring of integers modulo N. After the use of the command "R=IntegerModRing(N)", R becomes the ring of integers modulo N. We further require a random element from R. The command "R.random_ element()" returns a random element from the ring of integers modulo N. Further, the command "least_significant_bits(x, 1)" returns the least significant bit of x. As mentioned earlier, to use this function, we need to import it from " sage.crypto.util".

```
txt    Bob

N      2282789137

bits   16


The final Cipher for C1 is :
0011011001100011100101000000011100000010101111101111111111101010000001011\
0110111001011111010110000010111100000000101001100111010101011011011110011\
0001100101011000010001001110101000011110110101000000001111100111110111101\
1000001100101000101101001011110011110001001101101001101011110100001110\
1000000111100001001010111010000001111111100111000001000000001000000110111\
0001111011110001111100100111011001000000010011110111001001100111101001111\
0110100000101010000010011110100110001111101010010011111110011110001011110\
0011000010101000111000010110011001011010110100000100100011001100000011110001\
0100111111101001100011111000000000010110110101110100000111100110000010001\
0001001101001011001100001111001110110110011010010100101010010011000001011011\
1110100111011001011101011000001100011101101101111111
The final Cipher for C2 is :    1011100101101011110000010
```

Fig. 12.16   Output of the Rabin encryption algorithm in SAGE

For the encryption algorithm we use the function "Rabin_encryption()". As an input, this encryption function takes the plain text stored in "txt", the values N and the security parameter "bits". This function returns a pair of binary string of cipher texts stored in C1_bin and C2_bin.

For decryption algorithm, we use the function "Rabin_decryption()" which takes as input the two binary strings of cipher texts stored in the variables c1 and c2, the values of p, q, and the security parameter "bits". It outputs the original plain text. In this encryption algorithm, we have used all the previously explained functions except the two new functions, namely "legendre_symbol(a,p)" and the function "crt(a,b,p,q)". The function "legendre_symbol(a,p)" computes the Legendre symbol $binomap$, where p is a prime integer. On the other, the function "crt(a,b,p,q)" returns a solution to the congruences $x \equiv a \pmod{p}$ and $x \equiv b \pmod{q}$ using the Chinese Remainder Theorem.

The SAGE code for Rabin encryption scheme with documentation is presented in Table 12.16.

Screen shorts of the actual SAGE code for Rabin encryption algorithm along with the output are shown in Fig. 12.15 and Fig. 12.16, respectively.

The SAGE code for Rabin decryption scheme with documentation is presented in Table 12.17.

Screen shorts of the actual SAGE code for Rabin decryption algorithm along with the output are shown in Fig. 12.17 and Fig. 12.18, respectively.

12.14 Exercises

1. Find the primes p and q if $n = pq = 4386607$ and $\phi(n) = 4382136$.

```
from sage.crypto.util import  bin_to_ascii
@interact
def Rabin_decryption
(c1="00110110011000111001010000000111000000010101111101111111111010100000010110
1101110010111110101100000101111000000001010011001110101011101101111001100011000
1010110000100010011101010000111011010100000001111100111110111101000001100101
0001011010010111100111100010011101101001101011101000011110100000011110000100
0101110100000001111111001110001000000000000011011100011110111110001111100100
1101101000000001001110110011001001100111010011101000010101000010000100111101001
100011111010100100111111100111100010111100011000001010100011100001011001100101
1010110000010010001011000000011000101001111111010011000111110000000000010110
1011110100000011100100000100100100110100011101011001000011110011011100110100101
0010101001100000010110111101010011011001011101011000001100011101101011111
01110010110101110000010",c2="10111001011010111000010",p=58679,q=38903,bits=16):
    N=p*q
    R=IntegerModRing(N)
    f=[]
    len_c1=len(c1)
    blocks=len_c1/(bits*2)
    for i in range(blocks):
        cblockarray = c1[ i*2*bits : (i+1)*2*bits ]
        c = ZZ( str(cblockarray), base=2 )
        cp=c%p
        cq=c%q
        ap=Mod(cp,p)^((p+1)/4)
        aq=Mod(cq,q)^((q+1)/4)
        ap1=ZZ(ap)
        aq1=ZZ(aq)
        ap2=ZZ(-ap)
        aq2=ZZ(-aq)
        ls_ap1=legendre_symbol(ap1,p)
        ls_ap2=legendre_symbol(ap2,p)
        ls_aq1=legendre_symbol(aq1,q)
        ls_aq2=legendre_symbol(aq2,q)
        if ((ls_ap1==1) & (ls_aq1==1)):
            x=crt(ap1,aq1,p,q)
        if ((ls_ap1==1) & (ls_aq2==1)):
            x=crt(ap1,aq2,p,q)
        if ((ls_ap2==1) & (ls_aq1==1)):
            x=crt(ap2,aq1,p,q)
        if ((ls_ap2==1) & (ls_aq2==1)):
            x=crt(ap2,aq2,p,q)
        x1=ZZ(x)
        lsb_x1=least_significant_bits(x1, 1)
        c_2_blockarray = c2[ i: (i+1)]
        cc2 = ZZ( str(c_2_blockarray), base=2 )
        bin_plain_text=(lsb_x1[0]+cc2)%2
        string_bin_plain_text=bin_plain_text.binary()
        f.append(string_bin_plain_text)
    fbin = ''.join(f)
    ftxt = bin_to_ascii(fbin)
    print 'Final recovered message = ', ftxt
```

Fig. 12.17 Rabin decryption algorithm in SAGE

2. Decrypt each of the following Caesar encryptions by trying the various possible shifts until you obtain a meaningful text.

   (a) LWKLQNWKDWLVKDOOQHYHUVHHDELOOERDUGORYHOBDV
       DWUHH

   (b) UXENRBWXCUXENFQRLQJUCNABFQNWRCJUCNAJCRXWORW
       MB

evaluate

c1	0011011001100011100101000000011100000010101111101111111
c2	10111001011010111000010
p	58679
q	38903
bits	16

Final recovered message = Bob

Fig. 12.18 Output of the Rabin decryption algorithm in SAGE

    (c) BGUTBMBGZTFHNLXMKTIPBMAVAXXLXTEPTRLEXTOXKHHFY
        HKMAXFHNLX

3. Use the key $K = (5, 11) \in \mathcal{K}$ to encrypt the message "ILOVECRYPTOGRA PHY" using Affine cipher, assuming that $\mathcal{P} = \mathbf{Z}_{26}$. From the cipher text, describe an attack, assuming that the key K is unknown to the adversary. Calculate maximum how many trials the adversary has to make before breaking the scheme.

4. Give an example of a cryptographic scheme which is poly-alphabetic.

5. With the key *AVI* encrypt the plain text "ABSTRACTALGEBRA" using Vigenère Cipher.

6. With the key $k = \begin{bmatrix} 7 & 3 \\ 2 & 5 \end{bmatrix}$, encrypt the plain text "CRYPTOGRAPHYISFUN" using Hill 2-cipher.

7. What are the advantages of using public-key cryptosystem over private key cryptosystem.

8. Let g be a primitive root for $\mathbf{Z}_p$, for some prime p. Suppose that $x = a$ and $x = b$ are both integer solutions to the congruence $g^x \equiv y \pmod{p}$. Then

    (a) prove that $a \equiv b \pmod{p-1}$.
    (b) Prove that $\log_g(h_1 \cdot h_2) = \log_g(h_1) + \log_g(h_2)$ for all $h_1, h_2 \in \mathbf{Z}_p^*$.

9. Mount a man-in-middle attack for RSA cryptosystem.

10. Mount a bidding attack on RSA cryptosystem.

11. Construct basis matrices for a black and white visual cryptographic scheme with $\Gamma_0 = \{\{1, 2, 3\}, \{2, 3, 4\}\}$. What could be the pixel expansion and relative contrast of the scheme.

12. Using SAGE, implement the Diffie–Hellman key agreement protocol for a 512 bit prime.

13. Using SAGE, implement the bidding attack on RSA with 512 bit primes.

14. Using SAGE, implement the RSA signature scheme with 512 bit primes.

15. Using SAGE, implement Shamir's secret sharing scheme in the field $\mathbf{Z}_p$ with 512 bit prime p.

12.15 Additional Reading

We refer the reader to the books (Adhikari and Adhikari 2007; Menezes et al. 1996; Stinson 2002; Katz and Lindell 2007) for further details.

References

Adhikari, A.: An overview of black and white visual cryptography using mathematics. J. Calcutta Math. Soc **2**, 21–52 (2006)

Adhikari, A.: Linear algebraic techniques to construct black and white visual cryptographic schemes for general access structure and its applications to color images. Des. Codes Cryptogr. (2013). doi:10.1007/s10623-013-9832-5

Adhikari, A., Adhikari, M.R.: Introduction to Linear Algebra with Application to Basic Cryptography. Asian Books, New Delhi (2007)

Adhikari, A., Bose, M.: A new visual cryptographic scheme using Latin squares. IEICE Trans. Fundam. E **87-A** (5), 1998–2002 (2004)

Adhikari, A., Sikdar, S.: A new $(2, n)$-color visual threshold scheme for color images. In: Indocrypt'03. Lecture Notes in Computer Science, vol. 2904, pp. 148–161. Springer, Berlin (2003)

Adhikari, A., Dutta, T.K., Roy, B.: A new black and white visual cryptographic scheme for general access structures. In: Indocrypt'04. Lecture Notes in Computer Science, vol. 3348, pp. 399–413. Springer, Berlin (2004)

Adhikari, A., Kumar, D., Bose, M., Roy, B.: Applications of partially balanced and balanced incomplete block designs in developing visual cryptographic schemes. IEICE Trans. Fundam. E **90A** (5), 949–951 (2007)

Adhikari, A., Adhikari, M.R., Chaubey, Y.P.: Contemporary Topics in Mathematics and Statistics with Applications. Asian Books, New Delhi (2013)

Alexi, W., Chor, B., Goldreich, O., Schnorr, C.P.: RSA and Rabin functions: certain parts are as hard as the whole. SIAM J. Comput. **17**(2), 194–209 (1988)

Ateniese, G., Blundo, C., De Santis, A., Stinson, D.R.: Visual cryptography for general access structures. Inf. Comput. **129**, 86–106 (1996a)

Ateniese, G., Blundo, C., De Santis, A., Stinson, D.R.: Constructions and bounds for visual cryptography. In: auf der Heide, F.M., Monien, B. (eds.) 23rd International Colloquim on Automata, Languages and Programming (ICALP'96). Lecture Notes in Computer Science, vol. 1099, pp. 416–428. Springer, Berlin (1996b)

Blakley, G.R.: Safeguarding cryptographic keys. In: AFIPS 1979. National Computer Conference, vol. 48, pp. 313–317 (1979)

Blundo, C., De Santis, A., Stinson, D.R.: On the contrast in visual cryptography schemes. J. Cryptol. **12**(4), 261–289 (1999)

Blundo, C., D'arco, P., De Santis, A., Stinson, D.R.: Contrast optimal threshold visual cryptography. SIAM J. Discrete Math. **16**(2), 224–261 (2003)

Das, A., Adhikari, A.: An efficient multi-use multi-secret sharing scheme based on hash function. Appl. Math. Lett. **23**(9), 993–996 (2010)

Diffie, W., Hellman, M.: New directions in cryptography. IEEE Trans. Inf. Theory **IT-22**(6), 644–654 (1976)

ElGamal, T.: A public-key cryptosystem and a signature scheme based on discrete logarithms. IEEE Trans. Inf. Theory **31**(4), 469–472 (1985)

Hill, L.S.: Cryptography in an algebraic alphabet. Am. Math. Mon. **36**, 306–312 (1929)

Hoffstein, J., Pipher, J., Silverman, J.H.: An Introduction to Mathematical Cryptography. Springer, Berlin (2008)

Kahn, D.: The Codebreakers. Macmillan, New York (1967)

Katz, J.: Digital Signatures. Springer, New York (2010)

Katz, J., Lindell, Y.: Introduction to Modern Cryptography. Chapman & Hall/CRC, London/Boca Raton (2007)

Menezes, A., van Orschot, P.C., Vanstone, S.A.: Handbook of Applied Cryptography. CRC Press, Boca Raton (1996)

Merkle, R.C.: Secure communications over insecure channels. In: Secure Communications and Asymmetric Cryptosystems. AAAS Sel. Sympos. Ser., vol. 69, pp. 181–196. Westview, Boulder (1982)

Naor, M., Shamir, A.: Visual cryptography. In: Advance in Cryptography, Eurocrypt'94. Lecture Notes in Computer Science, vol. 950, pp. 1–12. Springer, Berlin (1994)

Rivest, R., Shamir, A., Adleman, L.: A method for obtaining digital signatures and public-key cryptosystems. Commun. ACM **21**(2), 120–126 (1978)

Shamir, A.: How to share a secret. Commun. ACM **22**(11), 612–613 (1979)

Stinson, D.R.: Cryptography Theory and Practice, 2nd edn. CRC Press, Boca Raton (2002)

Appendix A
Some Aspects of Semirings

Semirings considered as a common generalization of associative rings and distributive lattices provide important tools in different branches of computer science. Hence structural results on semirings are interesting and are a basic concept. Semirings appear in different mathematical areas, such as ideals of a ring, as positive cones of partially ordered rings and fields, vector bundles, in the context of topological considerations and in the foundation of arithmetic etc. In this appendix some algebraic concepts are introduced in order to generalize the corresponding concepts of semirings **N** of non-negative integers and their algebraic theory is discussed.

A.1 Introductory Concepts

H.S. Vandiver gave the first formal definition of a semiring and developed the theory of a special class of semirings in 1934. A semiring S is defined as an algebra $(S, +, \cdot)$ such that $(S, +)$ and $(S, \cdot)$ are semigroups connected by $a(b+c) = ab+ac$ and $(b+c)a = ba+ca$ for all $a, b, c \in S$. The set **N** of all non-negative integers with usual addition and multiplication of integers is an example of a semiring, called the semiring of non-negative integers. A semiring S may have an additive zero $\circ$ defined by $\circ + a = a + \circ = a$ for all $a \in S$ or a multiplicative zero 0 defined by $0a = a0 = 0$ for all $a \in S$. S may contain both $\circ$ and 0 but they may not coincide. Consider the semiring $(\mathbf{N}, +, \cdot)$, where **N** is the set of all non-negative integers;

$$a + b = \{\text{lcm of } a \text{ and } b, \text{ when } a \neq 0, b \neq 0\};$$

$$= 0, \quad \text{otherwise};$$

and

$$a \cdot b = \text{usual product of } a \text{ and } b.$$

Then the integer 1 is the additive zero and integer 0 is the multiplicative zero of $(\mathbf{N}, +, \cdot)$.

M.R. Adhikari, A. Adhikari, *Basic Modern Algebra with Applications*,
DOI 10.1007/978-81-322-1599-8, © Springer India 2014

In general, the concept of *multiplicative zero* 0 and *additive zero* ∘ may not coincide in a semiring. H.J. Weinert proved that both these concepts coincide if $(S, +)$ is a cancellative semigroup. Clearly, S has an absorbing zero (or a zero element) iff it has elements 0 and ∘ which coincide. Thus an absorbing zero of a semiring S is an element 0, such that $0a = a0 = 0$ and $0 + a = a + 0 = a$ for all $a \in S$.

A semiring S may have an identity 1 defined by $1a = a1 = a$, for all $a \in S$.

A semiring S is said to be additively commutative, iff $a + b = b + a$ for all $a, b \in S$. An additively commutative semiring with an absorbing zero 0 is called a hemiring. An additively cancellative hemiring is called a *halfring*. A multiplicatively cancellative commutative semiring with identity 1 is called a *semidomain*.

We mainly consider semirings S for which $(S, +)$ is commutative. If $(S, \cdot)$ is also commutative, S is called a commutative semiring. Moreover, to avoid trivial exceptions, each semiring S is assumed to have at least two elements.

A subset $A \neq \emptyset$ of a semiring S is called an ideal (left, right) of S iff $a + b \in A, sa \in A$ and $as \in A(sa \in A, as \in A)$ hold, for all $a, b \in A$ and all $s \in S$. An ideal A of S is called proper, iff $A \subset S$ holds, where $\subset$ denotes proper inclusion, and a proper ideal A is called maximal, iff there is no ideal B of S satisfying $A \subset B \subset S$. An ideal A of S is called trivial iff $A = S$ holds or $A = \{0\}$ (the latter clearly holds if S has an absorbing zero 0).

Many results in rings related to ideals have no analogues in semirings. As noted by Henriksen, it is not the case that any ideal A of a semiring S is the kernel of a homomorphism. To get rid of these difficulties, he defined in 1958, a more restricted class of ideals in a semiring which he called k-ideals. A k-ideal A of a semiring S is an ideal of S such that whenever $x + a \in A$, where $a \in A$ and $x \in S$, then $x \in A$. Iizuka defined a still more restricted class of ideals in semirings, which he called h-ideals. An h-ideal A of a semiring S is an ideal of S such that if $x + a + u = b + u$, where $x, u \in S$ and $a, b \in A$, then $x \in A$. It is clear that every h-ideal is a k-ideal. But examples disapprove the converse. Using only commutativity of addition, the following concepts and statements essentially due to Bourne, Zassenhaus, and Henriksen, are well known. For each ideal A of a semiring S, the k-closure $\bar{A} = \{a \in S : a + a_1 = a_2$ for some $a_i \in A\}$ is a k-ideal of S satisfying $A \subseteq \bar{A}$ and $\bar{\bar{A}} = \bar{A}$. Clearly, an ideal A of S is a k-ideal of S iff $A = \bar{A}$ holds. A proper k-ideal A of S is called a maximal k-ideal of S iff there is no k-ideal B of S satisfying $A \subset B \subset S$. An ideal A of a semiring S is called completely prime or prime iff $ab \in A$ implies $a \in A$ or $b \in A$, for all $a, b \in S$.

Let S and T be hemirings. A map $f : S \to T$ is said to be a hemiring homomorphism iff $f(s_1 + s_2) = f(s_1) + f(s_2); f(s_1 s_2) = f(s_2)f(s_2)$ and $f(0) = f(0)$.

A hemiring homomorphism $f : S \to T$ is said to be an N-homomorphism, iff whenever $f(x) = f(y)$ for some $x, y \in S$, there exist $r_i \in \ker f$ such that $x + r_1 = y + r_2$ holds.

An equivalence relation ρ on a semiring S is called a congruence relation iff $(a, b) \in \rho$ implies $(a + c, b + c) \in \rho, (ca, cb) \in \rho$ and $(ac, bc) \in \rho$ for all $c \in S$. A congruence relation ρ on a semiring S is said to be additively cancellative (AC) iff $(a + c, \ b + d) \in \rho$, and $(c, d) \in \rho$ imply $(a, b) \in \rho$. Each ideal A of S defines a congruence relation ρ_A on $(S, +, \cdot)$ given by $\rho_A = \{(x, y) \in S \times S : x + a_1 = $

$y + a_2$ for some $a_i \in A$}; this is known as *Bourne congruence*. The corresponding class semiring S/ρ_A, consisting of the classes $x\rho_A$, is also denoted by S/A. Each ideal A of S defines another type of congruence relation known as *Iizuka congruence* defined by $\sigma_A = \{(x, y) \in S \times S : x + a + u = y + b + u$ for some $a, b \in A$ and $u \in S\}$. The corresponding class semiring S/σ_A consists of the classes $x\sigma_A, x \in S$.

Bourne and Zassenhaus defined the zeroid $Z(S)$ of a semiring S as $Z(S) = \{z \in S : z + x = x$ for some $x \in S\}$; it is an h-ideal of S and is contained in every h-ideal of S. $Z(S)$ and S are called trivial h-ideals and other h-ideals (if they exist) are called proper h-ideals of S. A proper h-ideal A of S is called a maximal h-ideal iff there is no h-ideal B of S satisfying $A \subset B \subset S$. The h-closure $\bar{A}^h$ of each ideal A of a semiring S defined by $\bar{A}^h = \{x \in S : x + a_1 + u = a_2 + u$ for some $a_i \in A$ and $u \in S\}$ is the smallest h-ideal of S containing A.

A semiring S is said to be additively regular iff for each $a \in S$, there exists an element $b \in S$ such that $a = a + b + a$. If in addition, the element b is unique and satisfies $b = b + a + b$, then S is called an additively inverse semiring.

A semiring S is said to be semisubtractive iff for each pair a, b in S at least one of the equations $a + x = b$ or $b + x = a$ is solvable in S.

Let S be a semiring with absorbing zero 0. A left semimodule over S is a commutative additive semigroup M with a zero element $\mathbf{0}$ together with an operation

$$S \times M \to M; \qquad (a, x) \mapsto ax$$

called the *scalar multiplication* such that

$$\text{for all } a, b \in S, x, y \in M, \quad a(x + y) = ax + ay,$$

$$(a + b)x = ax + bx, \quad (ab)x = a(bx), \quad 0x = \mathbf{0}.$$

A right S-semimodule is defined in an analogous manner. Let H and T be hemirings. An (H, T) bi-semimodule M is both a left semimodule over H and a right semimodule over T such that $(bx)a = b(xa)$ for all $x \in M$, $b \in H$ and $a \in T$.

Let R be an arbitrary hemiring. Let S be also a hemiring. R is said to be an S-semialgebra iff R is a bi-semimodule over S such that $(ax)b = a(xb)$, for all $a, b \in R$ and $x \in S$.

An h-ideal A of the S-semialgebra R is said to be a left modular h-ideal iff there exists an element $e \in R$ such that

(i) each $x \in R$ satisfies $ex + a + z = x + b + z$ for some $z \in R$ and some $a, b \in A$;
(ii) each $s \in S$ satisfies $se + c + h = es + d + h$ for some $h \in R$ and some $c, d \in A$.

In this case e is called a left unit modulo A.

A pair (e_1, e_2) of elements of a semiring S is called an identity pair iff $a + e_1a = e_2a$ and $a + ae_1 = ae_2$, for all $a \in S$.

We define a right modular h-ideal in a similar manner. A congruence ρ on a semiring S is called a ring congruence iff the quotient semiring S/ρ is a ring. A semiring S who o and 1 is said to be h-Noetherian (k-Noetherian) iff it satisfies any one of the following three equivalent conditions:

(i) S satisfies the ascending chain condition on h-ideals (k-ideals);
(ii) The maximal condition for h-ideals (k-ideals) holds in S;
(iii) Every h-ideal (k-ideal) in S is finitely generated.

A.2 More Results on Semirings

In this section we present more results of algebraic theory of semirings which are closely related to the corresponding results of ring theory.

Theorem A.2.1 *Let S be a semiring such that $S = \langle a_1, a_2, \ldots, a_n \rangle$ is finitely generated. Then each proper k-ideal A of S is contained in a maximal k-ideal of S.*

Proof Let $k(A)$ be the set of all k-ideals B of S satisfying $A \subseteq B \subsetneqq S$, partially ordered by inclusion. Then $k(A) \neq \emptyset$, for A itself belongs to $k(A)$. Consider a chain $\{B_i, i \in I\}$ in $k(A)$. We claim that $B = \bigcup_{\langle i \in I \rangle} B_i$ is a proper k-ideal of S. Let $a, b \in B$ and $s \in S$. Then there exist $i, j \in I$ such that $a \in B_i$ and $b \in B_j$. As $\{B_i, i \in I\}$ forms a chain, either $B_i \subseteq B_j$ or $B_j \subseteq B_i$. For definiteness, we suppose $B_i \subseteq B_j$, so that both $a, b \in B_j$. Since B_j is a k-ideal of S, $a + b \in B_j \subseteq B$, $as, sa \in B_j \subseteq B$ for all $s \in S$. Again, if $a + x \in B$, $a \in B$, $x \in S$, then proceeding as above, $a + x, a \in B_j$ for some B_j of the above chain. Since B_j is a k-ideal, it follows that $x \in B_j \subseteq B$. As a result B is a k-ideal of S. Next we verify that B is a proper k-ideal of S. Suppose to the contrary $B = S = \langle a_1, a_2, \ldots, a_n \rangle$. Then each a_r would belong to some k-ideal B_{i_r} of the chain $\{B_i\}$, where $r = 1, 2, \ldots, n$. There being only finitely many B_{i_r}'s, one contains all others, let us call it B_t, while $t \in I$. Thus $a_1, a_2, \ldots, a_n$ all lie in this B_t. Consequently, $B_t = S$, which is clearly impossible. As a result $B \in k(A)$. Thus by Zorn's Lemma $k(A)$ has a maximal element as we were to prove. $\square$

Corollary *Let S be a semiring with 1. Then each proper k-ideal of S is contained in a maximal k-ideal of S.*

Proof The proof is immediate by $S = \langle 1 \rangle$. $\square$

We now consider conditions on a semiring S such that S has non-trivial k-ideals or maximal k-ideals, among others by the help of the congruence class semiring S/A defined by an ideal A of S. Let S' denote $S' = S \setminus \{0\}$, if S has 0, and $S' = S$, otherwise.

Definition A.2.1 A semiring S is said to satisfy the condition (C') iff for all $a \in S'$ and all $s \in S$, there are $s_1, s_2 \in S$ such that $s + s_1 a = s_2 a$ holds.

Clearly, if S has an identity 1, then (C') is equivalent to the condition (C), which states that $1 + s_1 a = s_2 a$ holds, for each $a \in S'$ and suitable $s_1, s_2 \in S$.

Example A.2.1 Let $\mathbf{Q}^+$ be the set of all non-negative rational numbers. Then $(\mathbf{Q}^+, +, \cdot)$ with usual operations is a semiring with 1 as identity satisfying condition (C). The same is true, more generally, for each positive cone P of a totally ordered skew fields (Fuchs 1963).

Example A.2.2 Let $\mathbf{N}$ be the set of all non-negative integers. Define $a + b = \max\{a, b\}$ and denote by ab the usual multiplication. Then $(\mathbf{N}, +, \cdot)$ is a semiring with 1 as identity which satisfies (C), since $1 + a = a$ holds for all $a \in \mathbf{N}^+$.

Lemma A.2.1 *If a semiring S with an absorbing zero 0 satisfies condition (C'), then $ab = 0$ for $a, b \in S$ implies $a = 0$ or $b = 0$.*

Proof By way of contradiction, assume $ab = 0$ and $a \neq 0 \neq b$. Then $s + s_1 a = s_2 a$ according to (C') yields $sb + s_1 ab = s_2 ab$, i.e., $sb = 0$ for all $s \in S$. Consequently $x + s_3 b = s_4 b$ implies $x = 0$, for all $s_3, s_4 \in S$, which contradicts (C') applied to the element $b \in S'$.   □

Theorem A.2.2 *Let S be a semiring. Then condition (C') implies that S contains only trivial k-ideals. The converse is true if $(S, \cdot)$ is commutative and provided that S has an element 0, then $Sa = \{sa : s \in S\} \neq \{0\}$ holds for all $a \in S'$.*

Proof Assume that S satisfies (C'). Let A be a k-ideal of S which contains at least one element $a \in S'$. Then $s + s_1 a = s_2 a$ according to (C') implies $s \in A$, for each $s \in S$, i.e., $A = S$. For the converse, our supplementary assumption on S shows that Sa is an ideal of S and that $Sa \neq \{0\}$ holds for each $a \in S'$ if S has an element 0. We assume that S has only trivial k-ideals. Then the k-ideal $\overline{Sa}$ coincides with S for each $a \in S'$, regardless whether S has an element 0 or not. Now $\overline{Sa} = \{s \in S : s + s_1 a = s_2 a$ for some $s_i \in S\} = S$ states that S satisfies condition (C').   □

Theorem A.2.3 *Let S be a commutative semiring with identity 1 and A a proper k-ideal of S. Then A is maximal iff the semiring $S/A = S/\rho_A$ satisfies condition (C), where ρ_A is the Bourne congruence on S.*

Proof Suppose A is a maximal k-ideal of S. Then A is the absorbing zero of S/A and $1\rho_A$ its identity. Consider any $c\rho_A \in (S/A)'$. Then $c \notin A$ holds and the smallest ideal B of S containing c and A consists of all elements sc, a and $sc + a$ for $s \in S$ and $a \in A$. From $A \subsetneq B$ it follows $\overline{B} = S$ and hence $1 + b_1 = b_2$ for suitable elements $b_1, b_2 \in B$. If $b_1 = s_1 c + a_1$ and $b_2 = s_2 c + a_2$ for suitable $s_i \in S$ and $a_i \in A$, then we obtain $1 + s_1 c + a_1 = s_2 c + a_2$. This implies $1\rho_A + (s_1\rho_A)(c\rho_A) = (s_2\rho_A)(c\rho_A)$. Discussions of other cases are similar. As a result S/ρ_A satisfies condition (C).

Conversely, assume (C) for S/A and let B be a k-ideal of S satisfying $A \subsetneq B$. Then there is an element $c \in B/A$, and $c\rho_A \in (S\backslash A)'$ yields $1\rho_A + (s_1\rho_A)(c\rho_A) = (s_2\rho_A)(c\rho_A)$ for suitable $s_i \in S$ by (C). Hence $(1 + s_1 c)\rho_A = (s_2 c)\rho_A$. Consequently, $1 + s_1 c + a_1 = s_2 c + a_2$ holds for some $a_i \in A$. Hence $1 + b_1 = b_2$ holds for some $b_1, b_2 \in B$. This implies $1 \in B$. Consequently, $B = S$. This shows that A is a maximal k-ideal of S. $\square$

[For more results, see Weinert et al. 1996].

In the rest of this section we only consider semirings S in which addition is commutative.

Definition A.2.2 An ideal A of a semirings S is said to be completely prime iff $ab \in A$ implies $a \in A$ or $b \in B$, for any $a, b \in S$.

Proposition A.2.1 *Let S be a commutative semiring with identity. Then each maximal k-ideal A of S is completely prime.*

Proof By Theorem A.2.3, the semiring S/A satisfies the conditions (C) and hence condition (C') (see Definition A.2.1). Since S/A has A as its absorbing zero, we can apply Lemma A.2.1 and find that S/A has no zero divisors. Hence $a\rho_A \neq A$ and $b\rho_A \neq A$ imply $(ab)\rho_A \neq A$, i.e., $a \notin A$ and $b \notin A$ imply $ab \notin A$, where ρ_A is the Bourne Congruence defined in Sect. A.1. $\square$

Concerning the converse of Proposition B.2.1, we show that a completely prime ideal A of a commutative semiring S with identity needs not be a k-ideal, and if it is one, A needs not be a maximal k-ideal of S.

Example A.2.3 (a) Let S be the set of all real numbers a satisfying $0 < a \leq 1$ and define $a + b = a \cdot b = \min\{a, b\}, \forall a, b \in S$. Then $(S, +, \cdot)$ is a commutative semiring with 1 as identity. Each real number r such that $0 < r < 1$ defines an ideal $A = \{a \in S : a \leq r\}$ of S which is completely prime. However, $r + 1 = r$ together with $r \in A$ and $1 \notin A$ show that A is not a k-ideal of S. The same is true if one includes 0 in these considerations (in this case 0 is an absorbing element but not a zero of $S \cup \{0\}$), but also if one adjoins 0 as an absorbing zero to S.

(b) The polynomial ring $\mathbf{Z}[x]$ over the ring $\mathbf{Z}$ of integers contains the subsemiring

$$S = \mathbf{N}[x] = \left\{ f(x) = \sum_{i=1}^{n} a_i x^i : a_i \in \mathbf{N} \right\},$$

which is commutative and has $1 \in \mathbf{N}$ as its identity. The ideal $A = \langle x \rangle$ of S consists of all $f(x) \in S$ such that $a_0 = 0$ holds. Then A is completely prime and a k-ideal of S. Now consider the set $B = \{f(x) \in S : a_0 \text{ is divisible by } 2\}$. Then B is a k-ideal of S, and $A \subsetneq B \subsetneq S \Rightarrow A$ is not a maximal k-ideal.

All maximal k-ideals of the semiring $\mathbf{N}$ are described below:

Proposition A.2.2 *Let $(\mathbf{N}, +, \cdot)$ be the semiring of non-negative integers under usual operations. Then $\mathbf{N}$ has exactly the k-ideals $\langle a \rangle = \{na : n \in \mathbf{N}\}$ for each $a \in \mathbf{N}$. Consequently, the maximal k-ideals of $\mathbf{N}$ are given by $\langle p \rangle$ for each positive prime p.*

Proof Clearly, each ideal $\langle a \rangle$ of **N** is a k-ideal. Now assume that $A \neq \{0\}$ is a k-ideal of **N**. Let a be the smallest positive integer contained in A, and b any element of A. Then $b = qa + r$ holds for some positive q and $r \in$ **N** satisfying $0 \leq r < a$. Since r belongs to the k-ideal A, it follows that $r = 0$, and hence $A = \langle a \rangle$. The last statement follows, since $\langle a \rangle \subseteq \langle b \rangle \Leftrightarrow b | a$. $\square$

Remark 1   None of the maximal k-ideals $\langle p \rangle$ of **N** is a maximal ideal of **N**. This follows since each ideal $A = \langle p \rangle$ is properly contained in the proper ideal $B = \{b \in$ **N** $: b \geq p\}$ of **N**.

Remark 2   Let $A = \langle p \rangle$, p is prime in Remark 1 and τ_A be the hull-kernel topology defined in A. Then

(a) (A, τ_A) is a connected space,
(b) (A, τ_A) is compact.

   (See Proposition A.6.1.)

Recall that a semiring S is said to be additively regular iff for each $a \in S$, $\exists$ an element $b \in S$ such that $a = a + b + a$. If in addition, the element b is unique and satisfies $b = b + a + b$, then S is called an additively inverse semiring. In an additively inverse semiring, the unique inverse b of an element a is usually denoted by a'.

Exercise (Karvellas 1974) Let S be an additively inverse semiring. Then

 (i) $x = (x')'$, $(x + y)' = y' + x'$, $(xy)' = x'y = xy'$ and $xy = x'y' \; \forall x, y \in S$.
(ii) $E^+ = \{x \in S : x + x = x\}$ is an additively commutative semilattice and an ideal of S.

Definition A.2.3 Let S be an additively inverse semiring in which addition is commutative and E^+ denote the set of all additive idempotents of S. A left k-ideal A of S is said to be full iff $E^+ \subseteq A$. A right k-ideal of S is defined dually. A non-empty subset I of S is called a full k-ideal iff I is both a left and a right full k-ideal.

Example A.2.4   (a) In a ring every ring ideal is a full k-ideal.
   (b) In a distributive lattice with more than two elements, a proper ideal is a k-ideal but not a full k-ideal.
   (c) $\mathbf{Z} \times \mathbf{N}^+ = \{(a, b) : a, b$ are integers and $b > 0\}$. Define $(a, b) + (c, d) = (a + c, \text{lcm}(b, d))$ and $(a, b)(c, d) = (ac, \gcd(b, d))$. Then $\mathbf{Z} \times \mathbf{N}^+$ becomes an additively inverse semiring in which addition is commutative. Let $A = \{(a, b) \in \mathbf{Z} \times \mathbf{N}^+ : a = 0\}$. Then A is a full k-ideal of $\mathbf{Z} \times \mathbf{N}^+$.

A.3   Ring Congruences and Their Characterization

We now define a ring congruence and characterize those ring congruences ρ on additively inverse semirings S in which addition is commutative and is such that

$-(a\rho) = a'\rho$, where a' denotes the inverse of a in S and $-(a\rho)$ denotes the additive inverse of $a\rho$ in the ring S/ρ.

Definition A.3.1 A congruence ρ on a semiring S is called a ring congruence iff the quotient semiring S/ρ is a ring.

Theorem A.3.1 *Let A be a full k-ideal of S. Then the relation $\rho_A = \{(a, b) \in S \times S : a + b' \in A\}$ is a ring congruence of S such that $-(a\rho_A) = a'\rho_A$.*

Proof Since $a + a' \in E^+ \subseteq A$ for all $a \in S$, it follows that ρ_A is reflexive. Let $a + b' \in A$. Now from W.O. Ex. 1(a), we find that $(a + b')' \in A$. Then $b + a' = (b')' + a' = (a + b')' \in A$. Hence ρ_A is symmetric. Let $a + b' \in A$ and $b + c' \in A$. Then $\bar{a} + b + b' + c' \in A$. Also $b + b' \in E^+ \subseteq A$. Since A is a k-ideal, we find that $a + c' \in A$. Hence ρ_A is an equivalence relation. Let $(a, b) \in \rho_A$ and $c \in S$. Then $a + b' \in A$. Since

$$(c + a) + (c + b)' = c + a + b' + c' = (a + b') + (c + c') \in A,$$

$$ca + (cb)' = ca + cb' = c(a + b') \in A \quad \text{and}$$

$$ac + (bc)' = ac + b'c = (a + b')c \in A,$$

it follows that ρ_A is a congruence on S. So we obtain the quotient semiring where addition and multiplication are defined by

$$a\rho_A + b\rho_A = (a + b)\rho_A \quad \text{and} \quad (a\rho_A)(b\rho_A) = (ab)\rho_A.$$

Now

$$a\rho_A + b\rho_A = (a + b)\rho_A = (b + a)\rho_A = b\rho_A + a\rho_A.$$

Let $e \in E^+$ and $a \in S$. Now $(e + a) + a' = e + (a + a') \in E^+$.
We find that $(e + a)\rho_A = a\rho_A$. Then $e\rho_A + a\rho_A = a\rho_A$.
Also

$$a\rho_A + a'\rho_A = (a + a')\rho_A = e\rho_A.$$

Hence $e\rho_A$ is the zero element and $a'\rho_A$ is the negative element of $a\rho_A$ in the ring S/ρ_A. $\square$

Theorem A.3.2 *Let ρ be a congruence of S such that S/ρ is a ring and $-(a\rho) = a'\rho$. Then there exists a full k-ideal A of S such that $\rho_A = \rho$.*

Proof Let $A = \{a \in S : (a, e) \in \rho \text{ for some } e \in E^+\}$. Since ρ is reflexive, it follows that $E^+ \subseteq A$. Then $A \neq \emptyset$, since $E^+ \neq \emptyset$. Let $a, b \in A$. Then there exist $e, f \in E^+$ such that $(a, e) \in \rho$ and $(b, f) \in \rho$. Then $(a + b, e + f) \in \rho$. But $e + f \in E^+$. Hence $a + b \in A$. Again for any $r \in S$, $(ra, re) \in \rho$ and $(ar, er) \in \rho$. But re and $er \in E^+$. Hence A is an ideal of S. Let $a + b \in A$ and $b \in A$. Then there exist $e, f \in E^+$ such that $(a + b, f) \in \rho$ and $(b, e) \in \rho$. Hence $f\rho = (a + b)\rho = a\rho + b\rho = a\rho + e\rho$.

But $f\rho$ and $e\rho$ are additive idempotents in the ring S/ρ. Hence $e\rho = f\rho$ is the zero element of S/ρ. As a result $a\rho$ is the zero element of S/ρ. Then $a\rho = e\rho$. This implies $a \in A$. So we find that A is a full k-ideal of S. Consider now the congruences ρ_A and ρ. Let $(a, b) \in \rho$. Then $(a + b', b + b') \in \rho$. But $b + b' \in E^+$. Hence $a + b' \in A$ and $(a, b) \in \rho_A$. Conversely, suppose that $(a, b) \in \rho_A$. Then $a + b' \in A$. Hence $(a + b', e) \in \rho$ for some $e \in E^+$. As a result, $e\rho = a\rho + b'\rho = a\rho - b\rho$ holds in the ring S/ρ. But $e\rho$ is the zero element of S/ρ. Consequently $a\rho = b\rho$. This show that $(a, b) \in \rho$ and hence $\rho_A = \rho$.   □

Remark (a) Certain types of ring congruences on an additive inverse semiring are characterized with the help of full k-ideals (see Theorems A.3.1 and A.3.2).

(b) It is shown that the set of all full k-ideals of an additively inverse semiring in which addition is commutative forms a complete lattice which is also modular (see W.O. Ex. 1(d)).

A.4   *k*-Regular Semirings

Definition A.4.1 Let S be an additively commutative semiring and $a \in S$. Then a is called *regular* (in the sense of Von Neuman) if $a = axa$ holds for some $x \in S$ and k-regular if there are $x, y \in S$ satisfying $a + aya = axa$. If all the elements of S have the corresponding property, S is called *a regular* or a k-regular semiring. Since aya can be added to both sides of $a = axa$, each regular semiring is also k-regular. If S happens to be a ring, both concepts coincide and each (additively commutative) semifield S is regular.

Example A.4.1 For an additively commutative semiring S with an identity 1, a condition C' is introduced in Definition A.2.1 by the property that each element $a \in S$ which is not multiplicatively absorbing zero satisfies $1 + s_1 a = s_2 a$ for suitable elements $s_i \in S$. This condition implies that S has only trivial k-ideals, and the converse holds if $(S, \cdot)$ is commutative (see Theorem A.2.2). Clearly, each semiring of this kind is k-regular, but not necessarily regular.

Lemma A.4.1 *Let S be an additively commutative and cancellative semiring and $R = D(S)$ its difference ring. If R is regular, then S is k-regular. The converse holds if S is semisubtractive, but not in general.*

Proof Let R be regular and $a \in S$. Then $a = ara$ holds for a suitable element $r \in R$, say $r = x - y$ for $x, y \in S$, which yields $a + aya = axa$. Now assume that S is k-regular and semisubtractive. Then each $r \neq 0$ of R satisfies $r = a$ or $r = -a$ for some $a \in S$, and $a + aya = axa$ for some $x, y \in S$ implies $a = a(x - y)a$ or $-a = a(y - x)a$ in R. For the last statement, let $\mathbf{Q}[z]$ be the polynomial ring over the field of rational numbers and $R = \mathbf{Q}[z]/(z^2)$. For brevity, we may write $a + bz$ for all elements of R. It was shown in Weinert (1963) that $S = \{a + bz \in R : a > 0\}$

is a subsemifield of R such that $R = D(S)$ holds. Hence S is even regular, whereas $zrz = 0$ for all $r \in R$ shows that R is not regular. □

Note that S can be replaced by $S \cup \{0\}$ in this example.

Example A.4.2 Let $M_{n,n}(H_0)$ be the semiring of all $n \times n$ matrices over the semifield H_0 of all non-negative rational numbers. Then $M_{n,n}(H_0)$ is k-regular, since the matrix ring $M_{n,n}(\mathbf{Q}) = D(M_{n,n}(H_0))$ is well known to be regular. Clearly, H_0 may be replaced in this example by any semifield S such that $D(S)$ is a field [see Weinert (1963)].

Lemma A.4.2 *Let S be a k-regular semiring. Then $A \cap B = \overline{BA}$ holds for each left k-ideal A and each right k-ideal B of S.*

Proof From $BA \subseteq A \cap B$, we get $\overline{BA} \subseteq \overline{A} \subseteq \overline{A}$ and $\overline{BA} \subseteq \overline{B} = B$. Hence $\overline{BA} \subseteq A \cap B$. To show the reverse inclusion, assume $a \in A \cap B$. Then $aza = a(za) \in BA$ for each $z \in S$ and $a + aya = axa$ imply $a \in \overline{BA}$. □

Theorem A.4.1 *For such a hemiring S, the set $K(S)$ of all k-ideals of S forms a complete lattice $(K(S), \subseteq)$. This lattice is distributive if*

(2.1) $A \cap B = \overline{AB}$ *holds for all k-ideals $A = \overline{A}$ and $B = \overline{B}$ of S, hence in particular if the hemiring is k-regular.*

Proof The intersection of any set of k-ideals of S is known to be a k-ideal of S, and S is a k-ideal of S. This yields the result that $(K(S) \subseteq)$ is a complete lattice, for which $A \wedge B = A \cap B$ is the infimum and $A \vee B = \overline{A + B}$ is the supremum of $A, B \in K(S)$. Moreover, as in each lattice, $A \wedge (B \vee C) \supseteq (A \wedge B) \vee (A \wedge C)$ holds for all $A, B, C \in K(S)$. So it remains to show the reverse inclusion. Now $A \wedge (B \vee C) = A \cap (\overline{B + C}) = \overline{A(\overline{B + C})}$ holds by (2.1). Suppose $x \in A(\overline{B + C})$. Then $x + u = v$ for some $u, v \in A(\overline{B + C})$. Hence $u = a_1 y_1$ and $v = a_2 v_2$ for suitable $a_i \in A$ and $y_i \in \overline{B + C}$. Then

$$y_1 + b_1 + c_1 = b_2 + c_2 \quad \text{and} \quad y_2 + b_3 + c_3 = b_4 + c_4$$

for suitable $b_i \in B$ and $c_i \in C$. Thus

$$a_1 y_1 + a_1 b_1 + a_1 c_1 = a_1 b_2 + a_1 c_2 \quad \text{yields} \quad u + a_1 b_1 + a_1 c_1 = a_1 b_2 + a_1 c_2,$$

which states $u \in \overline{AB + AC}$. Likewise we obtain $v \in \overline{AB + AC}$ and thus $x \in \overline{AB + AC}$ for $x + u = v$, since $\overline{AB + AC}$ is k-closed. Now $A \wedge (B \vee C) \subseteq \overline{AB + AC}$ and $\overline{AB + AC} \subseteq \overline{\overline{AB} + \overline{AC}} = \overline{(A \cap B) + (A \cap C)} = (A \wedge B) \vee (A \wedge C)$ complete the proof that the lattice $(K(S), \subseteq)$ is distributive.

We show in this context that, for each hemiring S, the condition (2.1) is equivalent to

(2.2) $A = \overline{AA}$ *holds for each k-ideal $A = \overline{A}$ of S.*

Clearly, (2.1) for $A = B$ yields (2.2), and $AB \subseteq A \cap B$ implies for k-ideals always $\overline{AB} \subseteq \overline{A \cap B} = A \cap B$. Applying (2.2) to the k-ideal $A \cap B$, we get the other inclusion of (2.1) by $A \cap B = \overline{(A \cap B)(A \cap B)} \subseteq \overline{AB}$.   □

Remark (2.2) follows from the (obviously stronger) condition that $A = A^2$ holds for all ideals of a hemiring S. The latter, for instance is satisfied if S is regular, implied by a result of Ashan (1993) that the complete lattice of all ideals of S is Brouwerian and thus distributive. However, none of the sufficient conditions for distributivity mentioned so far is necessary.

Example A.4.3 One easily checks that the set $S = \{0, a, b\}$ becomes a hemiring according to the following tables.

+	0	a	b		•	0	a	b
0	0	a	b		0	0	0	0
a	a	a	b		a	0	0	0
b	b	b	b		b	0	0	b

This hemiring has the k-ideals $\{0\} \subset \{0, a\} \subset S$ and one more ideal $\{0, b\}$ whose k-closure is S. Hence $(K(S), \subseteq)$ and obviously also the lattice of all ideals of S are distributive. However, $\{0, a\}^2 = \{0\}$ disproves (2.2) and thus k-regularity and clearly $A^2 = A$ for all ideals A of S.

Finally considering any ring as a hemiring, the k-ideals of R are just the ring-theoretical ideals of R, well known to form a modular lattice. So one could suspect that the latter holds also for the lattice $K(S)$ for each hemiring S. But the following example shows that the complete lattice $(K(S), \subseteq)$ of all k-ideals of a hemiring S does not need to be modular.

Remark For each k-regular hemiring S, the set $K(S)$ of all k-ideals of S forms a complete distributive lattice. In general, however, the lattice of all k-ideals of a hemiring does not need to be even modular.

Example A.4.4 Consider the set $S = \{0, a, b, s, c, d\}$. We define $(S, +)$ by the table

+	0	a	b	s	c	d
0	0	a	b	s	c	d
a	a	a	s	s	d	d
b	b	s	b	s	d	d
s	s	s	s	s	d	d
c	c	d	d	d	c	d
d	d	d	d	d	d	d

and $xy = 0$ for all $x, y \in S$. Instead of dealing with associativity of $(S, +)$ directly, note that $(S \backslash \{0\}, +)$ is the commutative and idempotent semigroup generated by $\{a, b, c, d\}$ subject to relations $a + c = d$ and $b + c = d$. Clearly, s is shorting for $a + b$. Now one checks that the subsets of S as shown in Fig. A.1 are k-ideals of this

Fig. A.1 k-ideals of S

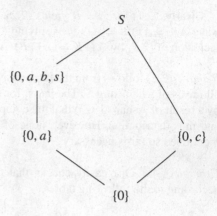

Fig. A.2 Commutativity of
the ring completion diagram

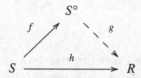

semiring. By a well known criterion, $\{0, a, b, s\} \cap \{0, c\} = \{0\}$ and $\{0, a\} \vee \{0, c\} = \overline{\{0, a\} + \{0, c\}} = \overline{\{0, a, c, d\}} = S$ show that $(K(S) \subseteq)$ is not modular.

Remark For one sided h-ideals and k-ideals in semirings, see Weinert et al. (1996).

A.5 The Ring Completion of a Semiring

The concept of a ring completion of a semiring arises through the process of passing from the semiring of non-negative integers to yield the ring of integers. This concept is used to study the stability properties of vector bundles (see Husemoller 1966, Chap. 8).

Definition A.5.1 Let S be a semiring with 0 (absorbing). The ring completion of S is a pair (S°, f), where S° is a ring and $f : S \to S^\circ$ is a semiring homomorphism such that for any semiring homomorphism $h : S \to R$, where R is an arbitrary ring, there exists a unique ring homomorphism $g : S^\circ \to R$ such that $g \circ f = h$, i.e., the diagram in Fig. A.2 is commutative.

We now prescribe a construction of S°.

Construction of S°: Consider the pairs $(a, b) \in S \times S$ and define an equivalence relation ρ on these pairs: (a, b) and (c, d) are said to be ρ-equivalent iff there exists

Fig. A.3 Commutative
diagram for ring completion

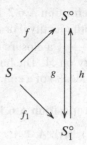

some $l \in S$ such that $a+d+l = c+b+l$. Denote the ρ-equivalent class of (a, b) by
$\langle a, b \rangle$. We may consider $\langle a, b \rangle = a - b$. Let $S°$ be the set of all ρ-equivalence class
$\langle a, b \rangle$. Define $\langle a, b \rangle + \langle c, d \rangle = \langle a+c, b+d \rangle$ and $\langle a, b \rangle \cdot \langle c, d \rangle = \langle ac+bd, bc+ad \rangle$.
These compositions are independent of the choice of the representatives of classes
and hence are well defined. Clearly, $\langle S°, +, \cdot \rangle$ is a ring with its zero element $0 =$
$\langle 0, 0 \rangle$ and negative of $\langle a, b \rangle$ being $\langle b, a \rangle$. The homomorphism $f : S \to S°$ is defined
by $f(x) = \langle x, 0 \rangle$.

Uniqueness of $(S°, f)$:Let $(S_1°, f_1)$ be another ring completion of $(S°, f)$. Then
there exist ring homomorphisms $g : S° \to S_1°$ and $h : S_1° \to S°$ making the diagram
in Fig. A.3 commutative.

It follows from the commutativity of the diagram that both $g \circ h$ and $h \circ g$ are
identity homomorphisms of rings. Hence the rings $S°$ and $S_1°$ are isomorphic.

Remark 1 $S°$ may be viewed as the free abelian group generated by the set S modulo the subgroup generated by $(a + b) + (-1)a + (-1)b, a, b \in S$.

Remark 2   The process of passing from a semigroup to a group is analogous and
yields its group completion.

A.6   Structure Spaces of Semirings

The structure spaces of semirings, formed by the class of prime k-ideals and prime
full k-ideals are considered in this section. More precisely, the properties such as
separation axioms, compactness and connectedness in these structure spaces are
studied. Finally, these properties for the semiring of non-negative integers are examined.

In this section we only consider semirings S for which $(S, +)$ is commutative.

Definition A.6.1 Let $A \neq \emptyset$ be any subset of a semiring S. Then the ideal generated
by A is the intersection of all ideals I of S containing A.

We recall the definition of hull-kernel topology and some of its properties (Gillman 1957; Kohis 1957; Slowikowski and Zawadowski 1955).

Definition A.6.2 Let $\mathcal{A}$ be the set of all proper prime ideals of a commutative ring with identity. For any subset A of $\mathcal{A}$, $\bar{A}$ given by $\bar{A} = \{I \in \mathcal{A} : \bigcap_{I_\alpha \in A} I_\alpha \subseteq I\}$ gives $A \mapsto \bar{A}$ a closure operator defining some topology τ_A called the hull-kernel topology on $\mathcal{A}$. The hull-kernel topology on the set $\mathcal{A}$ of all proper prime k-ideals (full k-ideals) of a commutative semiring with identity is defined in a similar way.

Proposition A.6.1

(a)   (i) $A \subseteq \bar{A}$,

     (ii) $\bar{\bar{A}} = \bar{A}$,

     (iii) $A \subseteq B \Rightarrow \bar{A} \subseteq \bar{B}$, *and*

     (iv) $\overline{A \cup B} = \bar{A} \cup \bar{B}$ *for all subsets* A, B *of* $\mathcal{A}$

(b)   (i) *For any commutative semiring* S *with identity* 1, *let* $\mathcal{A}$ *be the set of all proper prime k-ideals of S and for each* $a \in S$, $\Delta(a) = \{I \in \mathcal{A} : a \in I\}$. *If* $c\Delta(a) = \mathcal{A} \setminus \Delta(a)$ *then* $\{c\Delta(a) : a \in S\}$ *forms an open base for the hull-kernel topology on* $\mathcal{A}$.

     (ii) $(\mathcal{A}, \tau_A)$ *is a T_1-space iff no element of $\mathcal{A}$ is contained in any other element of* $\mathcal{A}$.

(c) *Let* $\mathcal{M}$ *be the set of all maximal k-ideals of S. Then* $(\mathcal{M}, \tau_\mathcal{M})$ *is a T_1-space, where $\tau_\mathcal{M}$ is the induced topology on $\mathcal{M}$ from* $(\mathcal{A}, \tau_A)$.

(d) *For the semiring* **N** *of non-negative integers, let* $A = \langle p \rangle$, p *is prime and $\mathcal{A}$ be the set of all prime ideals A of* **N**. *Then*

     (i) $(\mathcal{A}, \tau_A)$ *is a connected space;*

     (ii) $(\mathcal{A}, \tau_A)$ *is compact;*

where τ_A is the hull-kernel topology on $\mathcal{A}$.

Proof Left as an exercise.   □

We now study further properties of the structure spaces. Let I be any k-ideal of S. Define $\Delta(I) = \{I' \in \mathcal{A} : I \subseteq I'\}$.

Proposition A.6.2 *Any closed set in $\mathcal{A}$ is of the form $\Delta(I)$, where I is a k-ideal of S.*

Proof Let $\bar{A}$ be any closed set in $\mathcal{A}$, where $A \subseteq \mathcal{A}$. Let $A = \{I_\alpha : \alpha \in \Lambda\}$ and $I = \bigcap_{\alpha \in \Lambda} I_\alpha$. Then $\bar{A} = \Delta(I)$.   □

Theorem A.6.1 $(\mathcal{A}, \tau_A)$ *is a Hausdorff space if and only if for any distinct pair of elements I, J of $\mathcal{A}$, there exist $a, b \in S$ such that $a \notin J$, $b \notin I$ and there does not exist any element K of $\mathcal{A}$ such that $a \notin K$ and $b \notin K$.*

Proof Let $(\mathcal{A}, \tau_A)$ be Hausdorff. Then for any pair of distinct elements I, J of $\mathcal{A}$ there exist basic open sets $c\Delta(a)$ and $c\Delta(b)$ such that $I \in c\Delta(a)$, $J \in c\Delta(b)$ and $c\Delta(a) \cap c\Delta(b) = \emptyset$. It follows that $a \notin I$, $b \notin J$ and for any prime k-ideal K of S

for which $a, b \notin K$ implies $K \in c\Delta(a) \cap c\Delta(b)$, a contradiction, since $c\Delta(a) \cap c\Delta(b) = \emptyset$.

Conversely, let the given condition hold and $I, J \in \mathcal{A}$, $I \neq J$. Let $a, b \in S$ be such that $a \notin I, b \notin J$ and there does not exist any K of $\mathcal{A}$ such that $a \notin K, b \notin K$. Then $I \in c\Delta(a)$, $J \in c\Delta(b)$ and $c\Delta(a) \cap c\Delta(b) = \emptyset$, which proves that $(\mathcal{A}, \tau_{\mathcal{A}})$ is Hausdorff. $\square$

Corollary *If $(\mathcal{A}, \tau_{\mathcal{A}})$ is a T_2-space, then no proper prime k-ideal contains any other proper prime k-ideal. If $(\mathcal{A}, \tau_{\mathcal{A}})$ contains more than one element, then there exist $a, b \in S$ such that $\mathcal{A} = c\Delta(a) \cup c\Delta(b) \cup \Delta(I)$, where I is the k-ideal generated by a, b.*

Proof Suppose $(\mathcal{A}, \tau_{\mathcal{A}})$ is a T_2-space. Since every T_2-space is a T_1-space, $(\mathcal{A}, \tau_{\mathcal{A}})$ is a T_1-space. Hence by Proposition A.6.1 no proper prime k-ideal contains any other proper prime k-ideal. Now let $I, J \in \mathcal{A}$, where $I \neq J$. Then by Theorem A.6.1, there exist a, b in S such that $a \neq b, a \notin I, b \notin J$ and $c\Delta(a) \cap c\Delta(b) = \emptyset$. Let I be the k-ideal generated by a, b. Then since $c\Delta(a) \cap \Delta(b) = \emptyset$. Any element of $\mathcal{A}$, belongs to $c\Delta(a)$ or $c\Delta(b)$ or $\Delta(I)$, and therefore $c\Delta(a) \cup c\Delta(b) \cup \Delta(I) = \mathcal{A}$. $\square$

Theorem A.6.2 *$(\mathcal{A}, \tau_{\mathcal{A}})$ is a regular space if and only if for any $I \in \mathcal{A}$ and $a \notin I$, $a \in S$, there exist a k-ideal J of S and $b \in S$ such that $I \in c\Delta(b) \subseteq \Delta(J) \subseteq c\Delta(a)$.*

Proof Let $(\mathcal{A}, \tau_{\mathcal{A}})$ be a regular space. Then for any $I \in \mathcal{A}$ and any closed set $\Delta(J)$ not containing I, there exist disjoint open sets U, V such that $I \in U$ and $\Delta(J) \subseteq V$. If $a \notin I$, then $I \in c\Delta(a)$ and $\mathcal{A} \setminus c\Delta(a)$ is a closed set not containing I. Hence U, V are disjoint open sets containing I and $\mathcal{A} \setminus c\Delta(a)$, respectively. Then there exists $b \in S$ such that $I \in c\Delta(b) \subseteq \Delta(J) \subseteq c\Delta(a)$, where J is a k-ideal of S.

Conversely, let the given condition hold. Let $I \in \mathcal{A}$ and $\Delta(K)$ be any closed set not containing I. Let $a \notin I, a \in K$. Then by the given condition, there exist a k-ideal J and an element $b \in S$ such that $I \in c\Delta(b) \subseteq \Delta(J) \subseteq c\Delta(a)$. Obviously, $c\Delta(a) \cap \Delta(K) = \emptyset$. So, $c\Delta(b)$ and $\mathcal{A} \setminus \Delta(J)$ are two disjoint open sets containing I and $\Delta(K)$, respectively. Consequently, $(\mathcal{A}, \tau_{\mathcal{A}})$ is a regular space. $\square$

Theorem A.6.3 *$(\mathcal{A}, \tau_{\mathcal{A}})$ is a compact space if and only if for any collection $\{a_i\}$ of elements of S there exists a finite number of elements $a_1, a_2, \ldots, a_r$ in S such that for any $I \in \mathcal{A}$, there exists a_i such that $a_i \notin I$.*

Proof Let $(\mathcal{A}, \tau_{\mathcal{A}})$ be a compact space. Then the open cover $\{c\Delta(a) : a \in S\}$ has a finite subcover $\{c\Delta(a_i) : i = 1, \ldots, r\}$. Thus if $I \in \mathcal{A}$, then $I \in c\Delta(a_i)$ for some i which implies that $a_i \notin I$. Hence $a_i, \ldots, a_r$ are the required finite number of elements.

The converse follows from Kohis (1957, Lemma 3.1). $\square$

Corollary *If S is finitely generated, then $(\mathcal{A}, \tau_{\mathcal{A}})$ is compact.*

Proof Let $\{a_i : i = 1, \ldots, r\}$ be a set of generators of S. Then for any $I \in \mathcal{A}$, there exists a_i such that $a_i \notin I$, since I is a proper k-ideal. Hence by Theorem A.6.3. $(\mathcal{A}, \tau_{\mathcal{A}})$ is compact. $\qquad\square$

Proposition A.6.3 *Let $(\mathcal{S}, \tau_S)$ be the space of all proper prime full k-ideals of S. Then $(\mathcal{S}, \tau_S)$ is compact if and only if $E^+ = \{x \in S : x + x = x\} \neq \{0\}$.*

Proof Let $\{\Delta(I_\alpha) : \alpha \in \Lambda\}$ be any collection of closed sets in $\mathcal{S}$ with finite intersection property. Let I be the proper prime k-ideal which is also the full k-ideal generated by E^+. Since any prime full k-ideal J contains E^+, J contains I. Hence $I \in \bigcap_{\alpha \in \Lambda} \Delta(I_\alpha) \neq \emptyset$. Consequently, $(\mathcal{S}, \tau_S)$ is compact. $\qquad\square$

Definition A.6.3 A semiring S is said to be k-Noetherian if and only if it satisfies the ascending chain condition on k-ideals on S i.e., if and only if for the ascending chain $I_1 \subseteq I_2 \subseteq \cdots \subseteq I_n \cdots$ of k-ideals of S, there exists a positive integer m such that $I_n = I_m$ for all $n \geq m$.

Theorem A.6.4 *If S is a k-Noetherian semiring, then $(\mathcal{A}, \tau_{\mathcal{A}})$ is countably compact.*

Proof Let $\{\Delta(I_n)\}_{n=1}^\infty$ be a countable collection of closed sets in $\mathcal{A}$, with finite intersection property. Let $\langle A \rangle$ denote the prime k-ideal generated by A, where A ($\neq \emptyset$) is any subset of $\mathcal{A}$. Let us consider the following ascending chain of prime k-ideals: $I_1 \subseteq \langle I_1 \cup I_2 \rangle \subseteq \langle I_1 \cup I_2 \cup I_3 \rangle \subseteq \cdots$. Since S is k-Noetherian, there exists a positive integer m such that $\langle I_1 \cup I_2 \cup \cdots \cup I_n \rangle = \langle I_1 \cup I_2 \cup \cdots \cup I_{m+1} \rangle = \cdots$, which shows that $\langle I_1 \cup \cdots \cup I_m \rangle \in \bigcap_{n=1}^\infty \Delta(I_n) \neq \emptyset$. Hence $(\mathcal{A}, \tau_{\mathcal{A}})$ is countably compact. $\qquad\square$

Corollary *If S is a k-Noetherian semiring and $(\mathcal{A}, \tau_{\mathcal{A}})$ is second countable, then $(\mathcal{A}, \tau_{\mathcal{A}})$ is compact.*

Proof It follows from Theorem A.6.4 and the fact any open cover of a second countable space has a countable subcover. $\qquad\square$

Theorem A.6.5 *$(\mathcal{A}, \tau_{\mathcal{A}})$ is disconnected if and only if there exist a k-ideal I of S and a collection of points $\{a_\alpha\}_{\alpha \in \Lambda}$ of S not belonging to I such that if $I' \in \mathcal{A}$ and $a_\alpha \in I'$, $\forall \alpha \in \Lambda$, then $I \setminus I' \neq \emptyset$.*

Proof Let $(\mathcal{A}, \tau_{\mathcal{A}})$ be not connected. Then there exists a non-trivial open and closed subset of $\mathcal{A}$. Let I be the k-ideal of S for which $\Delta(I)$ is closed as well as open. Then $\Delta(I) = \bigcup_{\alpha \in \Lambda} c\Delta(a_\alpha)$, where $\{a_\alpha\}_{\alpha \in \lambda}$ is a collection of points of S. Now since $c\Delta(a_\alpha) \subseteq \Delta(I)$, $\forall \alpha \in \Lambda$ for any $I_\alpha \in c\Delta(a_\alpha)$ we have $I \subseteq I_\alpha$, and therefore $a_\alpha \notin I$ as $a_\alpha \notin I_\alpha$, $\forall \alpha \in \Lambda$. Now for any $I' \in \mathcal{A}$ and $a_\alpha \in I'$, $\forall \alpha \in \Lambda$ we have $I' \notin \Delta(I)$. Consequently, $I \nsubseteq I$ i.e., $I \setminus I' \neq \emptyset$.

Conversely, let the given condition hold. Then $\Delta(I) = \bigcup_{\alpha \in \Lambda} c\Delta(a_\alpha)$ is an open and closed non-trivial subset of $\mathcal{A}$ and hence $(\mathcal{A}, \tau_{\mathcal{A}})$ is disconnected. $\qquad\square$

Proposition A.6.4 (S, τ_S) *is connected space if* $E^+ \neq \{0\}$.

Proof Let I be the proper k-ideal generated by E^+. Then I belongs to any closed set $\Delta(I')$ of $\mathcal{A}$, since any full k-ideal of S contains E^+. Consequently any two closed sets of $\mathcal{A}$ are not disjoint. Hence (S, τ_S) is connected. $\qquad\square$

Example A.6.1 Let $S = \mathbf{N}$ be the set semiring of non-negative integers with respect to usual addition and multiplication. Then the prime proper k-ideals of S are $(p) = \{np : n \in \mathbf{N}\}$ where p is a prime number (Sen and Adhikari 1993). Let $\mathcal{A} = \{(p) : p \text{ is prime}\}$ and τ_A be the hull-kernel topology defined in $\mathcal{A}$. Since any prime proper k-ideal of $\mathbf{N}$ is not contained in other prime proper k-ideal, $(\mathcal{A}, \tau_A)$ is a T_1-space. Now let (p_1) and (p_2) be two distinct elements of $\mathcal{A}$ and let n_1, n_2 be two elements of S such that $n_1 \notin (p_1)$ and $n_2 \notin (p_2)$, then we can always find an element (p) of $\mathcal{A}$, such that $n_1 \notin (p)$ and $n_2 \notin (p)$. In fact one can take $p > n_1, n_2, p_1, p_2$ and p is prime. Hence by Theorem A.6.1, $(\mathcal{A}, \tau_A)$ is not a Hausdorff space. $c\Delta(n)$ is infinite and its complement, which is closed in $\mathcal{A}$, is finite. So any two non-trivial open sets intersect. Consequently, $(\mathcal{A}, \tau_A)$ is neither a T_2-space nor a regular space. For this reason $(\mathcal{A}, \tau_A)$ is a connected space. Clearly, $(\mathcal{A}, \tau_A)$ is compact by Theorem A.6.3.

A.7 Worked-Out Exercises and Exercises

Worked-Out Exercises (W.O. Ex)

1. Let S be an additively inverse semiring in which addition is commutative and E^+ denote the set of all additive idempotents of S.

   (a) Every k-ideal of S is an additively inverse subsemiring of S.
   (b) Let A be an ideal of S. Then

      (i) $\bar{A} = \{a \in S : a + x \in A \text{ for some } x \in A\}$ is a k-ideal of S such that $A \subseteq \bar{A}$.
      (ii) $\bar{A} = A \Leftrightarrow A$ is a k-ideal.

   (c) Let A and B be full k-ideals of S, then $\overline{A + B}$ is a full k-ideal of S such that $A \subseteq \overline{A + B}$ and $B \subseteq \overline{A + B}$.
   (d) If $I(S)$ denotes the set of full k-ideals of S, then $I(S)$ is a complete lattice which is also modular.

Solution (a) Let I be a k-ideal of S. Clearly I is a subsemiring of S. Let $a \in I$. Then

$$a + (a' + a) = a \in I.$$

Since I is a k-ideal, if follows $a' + a \in I$. Again this implies that $a' \in I$. Hence (a) follows.

(b) Let $a, b \in \bar{A}$. Then $a + x, b + y \in A$ for some $x, y \in A$. Now $a + x + b + y = (a + b) + (x + y) \in A$.

As $x + y \in A$, $a + b \in \bar{A}$. Next let $r \in S$, $ra + rx = r(a + x) \in A$.

As $rx \in A$, $ra \in \bar{A}$. Similarly, $ar \in \bar{A}$. As a result $\bar{A}$ is an ideal of S. Next, let c and $c + d \in \bar{A}$. Then there exist x and y in A such that $c + x \in A$ and $c + d + y \in A$.

Now $d + (c + x + y) = (c + d + y) + x \in A$ and $c + x + y \in A$.

Hence $d \in \bar{A}$ and $\bar{A}$ is a k-ideal of S. Since $a + a' \in A$ for all $a \in A$, it follows that $A \subseteq \bar{A}$.

(c) It can be shown that $A + B$ is an ideal of S. Then from (b), we find $\overline{A + B}$ is a k-ideal and $A + B \subseteq \overline{A + B}$. Now $E^+ \subseteq A, B$. Hence $E^+ \subseteq A + B \subseteq \overline{A + B}$. This implies that $\overline{A + B}$ is a full k-ideal. Let $a \in A$. Then

$$a = a + a' + a = a + (a' + a) \in A + B \quad \text{as } a' + a \in E^+ \subseteq B.$$

Hence $A \subseteq \overline{A + B}$ and similarly $B \subseteq \overline{A + B}$.

(d) We first note that $I(S)$ is a partially ordered set with respect to usual set inclusion. Let $A, B \in I(S)$. Then $A \cap B \in I(S)$ and from (c), $\overline{A + B} \in I(S)$. Define $A \wedge B = A \cap B$ and $A \vee B = \overline{A + B}$. Let $C \in I(S)$ be such that $A, B \subseteq C$. Then $A + B \subseteq C$ and $\overline{A + B} \subseteq \bar{C}$. But $\bar{C} = C$. Hence $\overline{A + B} \subseteq C$. As a result $\overline{A + B}$ is the lub of A, B. Thus we find that $I(S)$ is a lattice. Now E^+ is an ideal of S. Hence $E^+ \in I(S)$ and also $S \in I(S)$; consequently $I(S)$ is a complete lattice. Next suppose that $A, B, C \in I(S)$ such that

$$A \wedge B = A \wedge C \quad \text{and} \quad A \vee B = A \vee C \quad \text{and} \quad B \subseteq C.$$

Let $x \in C$. Then $x \in A \vee C = A \vee B = \overline{A + B}$. Hence there exists $a + b \in A + B$ such that $x + a + b = a_1 + b_1$ for some $a_1 \in A$, $b_1 \in B$.

Then $x + a + a' + b = a_1 + b_1 + a'$.

Now $x \in C$, $a + a' \in C$ and $b \in B \subseteq C$. Hence $a_1 + b_1 + a' \in C$. But $b_1 \in C$. Consequently, $a_1 + a' \in C \cap A = C \cap B$. Hence $a_1 + a' \in B$. So from $x + a + b = a_1 + b_1$ we find that $x + a + a' + b = a_1 + a' + b \in B$. But $(a + a') + b \in B$ is a k-ideal. Hence $x \in B$ and $B = C$. This proves that $I(S)$ is a modular lattice.

Exercises

1. Define homomorphism, monomorphism, epimorphism, isomorphism for semirings with the help of the corresponding definitions for semigroups and rings.

2. A mapping $f : (S, +, \cdot) \to (T, +, \cdot)$ is a semiring homomorphism iff both $f : (S, +) \to (T, +)$ and $f : (S, \cdot) \to (T, \cdot)$ are semigroup homomorphisms.

3. Let $f : (S, +, \cdot) \to (T, +, \cdot)$ be a semiring homomorphism. Then the homomorphic image $(f(S), +, \cdot)$ of $(S, +, \cdot)$ is also a semiring. If $(S, +, \cdot)$ is commutative, then $(f(S), +, \cdot)$ is also commutative. If f is an isomorphism then f^{-1} is also so.

4. Let H be a hemiring and $0 \neq a \in H$ and $(e_1, e_2) \neq (0, 0)$ be an identity pair of H. A pair $(c, d) \in H \times H$ is called an inverse relative of the identity pair (e_1, e_2) iff $e_1 + ca = e_2 + da$ and $e_1 + ac = e_2 + ad$. An additively cancellative hemiring H is called a division hemiring iff H contains an identity pair and every non-zero element of H is invertible in H. A hemiring H is said to be a left k-Artinian (left Artinian) iff H satisfies the d.c.c for k-ideals (ideals) of H.

(a) Let $H = \left\{ \begin{pmatrix} a & b \\ 0 & d \end{pmatrix} : a, b, d \in \mathbf{R} \text{ and } a, d \geq 5 \right\} \cup \left\{ \begin{pmatrix} 0 & 0 \\ 0 & 0 \end{pmatrix} \right\}$ is a division hemiring with $\left(\begin{pmatrix} 5 & 0 \\ 0 & 5 \end{pmatrix}, \begin{pmatrix} 6 & 0 \\ 0 & 6 \end{pmatrix} \right)$ as a left identity pair.

(b) A division hemiring H has no zero divisors.

(c) A division hemiring is multiplicatively cancellative.

(d) Let H be an additively cancellative hemiring with more than one element. Then H is multiplicatively cancellative k-Artinian iff H is a division hemiring.

5. A semiring $(S, +, \cdot)$ satisfying $|S| \geq 2$ is called a semifield iff the set of its non-zero elements S' forms a multiplicative group.

(a) Let $(S, +, \cdot)$ be a semifield satisfying $|S'| \geq 2$. Then the identity e of the group $(S', \cdot)$ is also the identity of $(S, +, \cdot)$ and $(S, +, \cdot)$ is multiplicatively cancellative.

(b) Let $(S, +, \cdot)$ be a semiring satisfying $|S'| \geq 2$. Then $(S, +, \cdot)$ is a semifield iff it satisfies one of the following statements:

   (i) $(S, +, \cdot)$ has a left identity e and each $a \in S'$ is invertible in $(S, \cdot)$;

   (ii) For arbitrary elements $a \in S', b \in S$, $\exists$ elements $x, y \in S$ such that $ax = b$ and $ya = b$.

(c) Let $(S, +, \cdot)$ be a semiring satisfying $|S| \geq 2$. If $(S, +, \cdot)$ has an identity and each $a \in S'$ has a (left) inverse $a' \in S$, then $(S, +, \cdot)$ is semifield.

6. A semiring S satisfying the condition (C') of Definition A.2.1 contains at most one left ideal consisting of a single element.

7. Let $S = (S, +)$ be a semimodule and $A \subset S$ a maximal k-closed subsemimodule of S. Then A is also k-closed in S i.e., $A \subset B = \overline{B} \subseteq S \Rightarrow B = S$ for each k-closed subsemimodule B of S.

8. If a semiring S satisfies the condition (C') of Definition A.2.1, it contains at most two left k-ideals, which are in fact two-sided ones, namely S and possibly, the ideal $\{0\}$ consisting of the multiplicatively absorbing element 0 of S.

9. Let S be a semiring. The each proper k-ideals (h-ideal) of S is contained in a maximal k-ideal (h-ideal) of S if $\exists$ a finitely generated ideal I of S such that $\overline{I} = S$.

10. Give an example of multiplicatively commutative semiring S which has only two h-ideals (viz. $Z(S)$ and S), but an infinite chain of k-ideal B_i satisfying

$$A = Z(S) \subset B_1 \subset B_2 \subset \cdots \subset S.$$

(In particular, $A = Z(S)$ is a maximal h-ideal of S, but not a maximal k-ideal.)

A.8 Additional Reading

We refer the reader to the books (Adhikari and Adhikari 2003; Adhikari and Das 1994; Adhikari et al. 1996; Ahsan 1993; Fuchs 1963; Gillman 1957; Glazek

1985; Golan 1992; Hazewinkel 1996; Hebich and Weinert 1993; Henriksen 1958; Husemoller 1966; Iizuka 1959; Sen and Adhikari 1992, 1993; Vandiver 1934; Weinert 1963; Weinert et al. 1996) for further details.

References

Adhikari, M.R., Adhikari, A.: Groups, Rings and Modules with Applications, 2nd edn. Universities Press, Hyderabad (2003)

Adhikari, M.R., Das, M.K.: Structure spaces of semirings. Bull. Calcutta Math. Soc. **86**, 313–317 (1994)

Adhikari, M.R., Sen, M.K., Weinert, H.J.: On k-regular semirings. Bull. Calcutta Math. Soc. **88**, 141–144 (1996)

Ahsan, J.: Fully idempotent semirings. Proc. Jpn. Acad., Ser. A **69**, 185–188 (1993)

Fuchs, L.: Partially Ordered Algebraic Systems. Addison-Wesley, Reading (1963)

Gillman, L.: Rings with Hausdorff structure space. Fundam. Math. **45**, 1–16 (1957)

Glazek, K.: A Short Guide Through the Literature on Semirings. Math. Inst. Univ. Wroclaw, Poland (1985)

Golan, J.S.: The Theory of Semirings with Applications in Mathematics and Theoretical Computer Science. Longman, Harlow (1992)

Hazewinkel, M. (ed.): Handbook of Algebra, vol. I. Elsevier, Amsterdam (1996)

Hebich, U., Weinert, H.J.: Algebraische Theorie und Anwendungen in der Informatik. Tember, Stuttgart (1993)

Henriksen, M.: Ideals in semirings with commutative addition. Not. Am. Math. Soc. **5**, 321 (1958)

Husemoller, D.: Fibre Bundles, 2nd edn. Springer, New York (1966)

Iizuka, K.: On the Jacobson radical of a semiring. Tohoku Math. J. **11**(2), 409–421 (1959)

Karvellas, P.H.: Inverse semirings. J. Aust. Math. Soc. **18**, 277–288 (1974)

Kohis, C.W.: The space of pair ideals of a ring. Fundam. Math. **45**, 1–27 (1957)

Sen, M.K., Adhikari, M.R.: On k-ideals of semirings. Int. J. Math. Math. Sci. **15**(2), 347–350 (1992)

Sen, M.K., Adhikari, M.R.: On maximal k-ideals of semirings. Proc. Am. Math. Soc. **113**(3), 699–703 (1993)

Slowikowski, W., Zawadowski, W.: A generalization of maximal ideals method of Stone and Gelfaand. Fundam. Math. **42**, 215–231 (1955)

Vandiver, H.S.: Note on a simply type of algebra in which cancellation law of addition does not hold. Bull. Am. Math. Soc. **40**, 914–920 (1934)

Weinert, H.J.: Über Halbringe und Halbkorper II. Acta Math. Acad. Sci. Hung. **14**, 209–227 (1963)

Weinert, H.J., Sen, M.K., Adhikari, M.R.: One sided k-ideals and h-ideals in semirings. Pannon. Hungary **7**(1), 147–162 (1996)

Appendix B
Category Theory

Category theory is a very important branch of modern mathematics. It has been growing quite rapidly both in contents and applicability to other branches of the subject. The concepts of categories, functors, natural transformations and duality form the foundation of category theory. These concepts were introduced during 1942–1945 by S. Eilenberg and S. Mac Lane.[1] Originally, the purpose of these notions was to provide a technique for classifying certain concepts such as that of natural isomorphism. In pedagogical methods in mathematics we compartmentalize mathematics into its different branches without emphasizing their interrelationships. But with the help of category theory one can move from one branch of mathematics to another. It provides a convenient language to tie together several notions and existing results of different branches of mathematics in a unified way. This language is conveyed in this chapter through modern algebra, algebraic topology, topological algebra, sheaf theory etc.

B.1 Categories

A category may be thought roughly as consisting of sets, possibly with additional structures, and functions, possibly preserving additional structures. More precisely, a category can be defined by the following characteristics.

Definition B.1.1 A category $\mathcal{C}$ consists of

(a) a class of objects $X, Y, Z, \ldots$ denoted by $\mathrm{ob}(\mathcal{C})$;
(b) for each ordered pair of objects X, Y a set of morphisms with domain X and range Y denoted by $\mathcal{C}(X, Y)$ or simply (X, Y); i.e., if $f \in (X, Y)$, then X is

---

[1] (i) Natural isomorphism in group theory. Proc. Natl. Acad. Sci. USA **28**, 537–544 (1942)
(ii) General theory of natural equivalence. Trans. Am. Math. Soc. **58**, 231–294 (1945)

M.R. Adhikari, A. Adhikari, *Basic Modern Algebra with Applications*,
DOI 10.1007/978-81-322-1599-8, © Springer India 2014

called the domain of f and Y is called the codomain (or range) of f: one also writes $f : X \to Y$ or $X \xrightarrow{f} Y$ to denote the morphism from X to Y;

(c) for each ordered triple of objects X, Y and Z and a pair of morphisms $f : X \to Y$ and $g : Y \to Z$, their composite denoted by $gf : X \to Z$ i.e., if $f \in (X, Y)$ and $g \in (Y, Z)$, then their composite gf satisfies the following two axioms:

(i) *associativity*: if $f \in (X, Y), g \in (Y, Z)$ and $h \in (Z, W)$, then $h(gf) = (hg)f \in (X, W)$;

(ii) *identity*: for each object Y in $\mathcal{C}$, there is a morphism $1_Y \in (Y, Y)$ such that if $f \in (X, Y)$, then $1_Y f = f$ and if $h \in (Y, Z)$, then $h 1_Y = h$. Clearly, 1_Y is unique.

If the class of objects is a set, the category is said to be *small*.

Example B.1.1   (i) Sets and functions form a category denoted by **S**.

(Here the class of objects is the class of all sets and for sets X and $Y, (X, Y)$ equals the set of functions from X to Y and the composition has the usual meaning, i.e., usual composition of functions.)

(ii) Sets and injections (or surjections or bijections) form a category.

(iii) Groups and homomorphisms form a category denoted by **Grp**.

(Here the class of objects is the class of all groups and for groups X and $Y, (X, Y)$ equals the set of homomorphisms from X to Y and the composition has the usual meaning.)

(iv) Rings and homomorphisms form a category denoted by **Ring**.

(v) Commutative rings and homomorphisms form a category.

(vi) R-modules and R-homomorphisms form a category denoted by $\mathbf{Mod}_R$.

(vii) Topological spaces and continuous maps form a category denoted by **Top**.

(viii) Finite sets and functions form a category.

(ix) Given a partial ordered set $(X, \leq)$, there is a category $\mathcal{C}$ whose objects are the elements of X and such that $\mathcal{C}(x, x')$ is either the singleton consisting of the ordered pair (x, x') or empty, according to whether $x \leq x'$ or $x \not\leq x'$, $1_x = (x, x)$ and $(q, r)(x, q) = (x, r)$ when $x \leq q \leq r$.

(Note that $\mathcal{C}(x, x')$ is not a set of functions.)

(x) Given a category $\mathcal{C}$, there is an opposite (dual) category $\mathcal{C}^0$ whose objects Y^0 are in one-to-one correspondence with the objects Y of $\mathcal{C}$ and whose morphisms $f^0 : Y^0 \to X^0$ are in one-to-one correspondence with the morphisms $f : X \to Y$ and for $g : Y \to Z$ in $\mathcal{C}$, $f^0 g^0$ is defined by $f^0 g^0 = (gf)^0$.

(xi) If $\mathcal{C}_1$ and $\mathcal{C}_2$ are categories, their product $\mathcal{C}_1 \times \mathcal{C}_2$ is the category whose objects are ordered pairs (Y_1, Y_2) of objects Y_1 in $\mathcal{C}_1$ and Y_2 in $\mathcal{C}_2$ and whose morphisms $(X_1, X_2) \to (Y_1, Y_2)$ are ordered pairs of morphisms (f_1, f_2), where $f_1 : X_1 \to Y_1$ in $\mathcal{C}_1$ and $f_2 : X_2 \to Y_2$ in $\mathcal{C}_2$. Similarly, there is a product of an arbitrary indexed family of categories.

(xii) Let $\mathcal{C}$ be a category. We define a category $\mathcal{C}^2$ in the following way: objects of $\mathcal{C}^2$ are the morphisms of $\mathcal{C}$; morphisms of $\mathcal{C}^2$ are certain pairs of morphisms of $\mathcal{C}$, for $f \in \mathcal{C}(A, B)$ and $g \in \mathcal{C}(C, D)$, the pair (α, β) is a morphism from f to g in $\mathcal{C}^2$ iff $\alpha : A \to C$ and $\beta : B \to D$ satisfy the commutativity relation: $\beta f = g \alpha$.

Note that as objects of $\mathcal{C}^2$ and $\mathcal{C} \times \mathcal{C}$ are different, $\mathcal{C}^2 \neq \mathcal{C} \times \mathcal{C}$.

(xiii) Exact sequences of R-modules and R-homomorphisms form a category.

Definition B.1.2 A subcategory $\mathcal{C}' \subset \mathcal{C}$ is a category such that

(a) the objects of $\mathcal{C}'$ are also objects of $\mathcal{C}$, i.e., $\text{ob}(\mathcal{C}') \subseteq \text{ob}(\mathcal{C})$;

(b) for objects X' and Y' of $\mathcal{C}'$, $\mathcal{C}'(X', Y') \subseteq \mathcal{C}(X', Y')$;

(c) if $f' : X' \to Y'$ and $g' : Y' \to Z'$ are morphisms of $\mathcal{C}'$, their composite in $\mathcal{C}'$ equals their composite in $\mathcal{C}$.

Definition B.1.3 A subcategory $\mathcal{C}'$ of $\mathcal{C}$ is said to be a full subcategory of $\mathcal{C}$ iff for objects X' and Y' in $\mathcal{C}'$, $\mathcal{C}(X', Y') = \mathcal{C}'(X', Y')$.

The category in Example B.1.1(ii) is a subcategory of the category in Example B.1.1(i) and the category in Example B.1.1(viii) is a full subcategory of the category in Example B.1.1(i).

Remark The categories in Examples B.1.1(iii)–(vii) are not subcategories of the category in Example B.1.1(i), because each object of one of the former categories consists of a set, endowed with an additional structure (hence, different objects in these categories may have the same underlying sets).

In category Example B.1.1(ix), the morphisms are not functions and so this category is not a subcategory of the category in Example B.1.1(i).

B.2 Special Morphisms

Let $\mathcal{C}$ be a category and $A, B, C, \ldots$ objects of $\mathcal{C}$.

Definition B.2.1 A morphism $f : A \to B$ in $\mathcal{C}$ is called a *coretraction* iff there is a morphism $g : B \to A$ in $\mathcal{C}$ such that $gf = 1_A$. In this case g is called a left inverse of f and f is called a right inverse of g and A is called a *retract* of B.

Dually we say that f is a retraction iff there is a morphism $g' : B \to A$ such that $fg' = 1_B$ in $\mathcal{C}$. In this case g' is called a right inverse of f.

Definition B.2.2 A two-sided *inverse* (or simply an inverse) of f is a morphism which is both a left inverse of f and a right inverse of f.

Lemma B.1 *If $f : A \to B$ in $\mathcal{C}$ has a left inverse and a right inverse, they are equal.*

Proof Let $g' : B \to A$ be a left inverse of f and $g'' : B \to A$ a right inverse of f, then $g'f = 1_A$ and $fg'' = 1_B$. Now $g' = g'1_B = g'(fg'') = (g'f)g'' = 1_Ag'' = g''$. $\qquad\square$

Definition B.2.3 A morphism $f : A \to B$ is called an *equivalence* (or an *isomorphism*) in a category $\mathcal{C}$ denoted by $f : A \approx B$ iff there is a morphism $g : B \to A$ which is a two-sided inverse of f.

Proposition B.2.1 *If $f : A \to B$ is a morphism in $\mathcal{C}$ such that f is both a retraction and a coretraction, then f is an equivalence.*

Proof It follows from Definition B.2.1 and Lemma B.1.   □

Remark An equivalence $f : A \approx B$ has a unique inverse denoted by $f^{-1} : B \to A$ and f^{-1} is also an equivalence.

Definition B.2.4 Two objects A and B in $\mathcal{C}$ are said to be equivalent iff there is an equivalence $f : A \approx B$ in $\mathcal{C}$.

Remark As the composite of equivalences is an equivalence, the relation of being equivalent is an equivalence relation in any set of objects of a category $\mathcal{C}$.

Definition B.2.5 A morphism $\alpha \in \mathcal{C}(A, B)$ is called a monomorphism (or monic) iff $\alpha f = ag \Rightarrow f = g$ for all pairs of morphisms f, g with codomain A and same domain in $\mathcal{C}$.

Definition B.2.6 A morphism $\alpha \in \mathcal{C}(A, B)$ is called an epimorphism (or epic) iff $f\alpha = g\alpha \Rightarrow f = g$ for all pairs of morphism f, g with domain B and same codomain in $\mathcal{C}$.

Remark The notion of an epimorphism is dual to that of a monomorphism in the sense that α is an epimorphism in $\mathcal{C}$ iff it is a monomorphism in its dual category $\mathcal{C}^0$. A coretraction is necessarily a monomorphism and a retraction is an epimorphism. Thus an isomorphism is both a monomorphism and an epimorphism in $\mathcal{C}$.

Proposition B.2.2 *If $\alpha : A \to B$ is a coretraction and also an epimorphism in $\mathcal{C}$, then it is an isomorphism in $\mathcal{C}$.*

Proof As α is a coretraction, there exists a morphism $\beta : B \to A$ such that $\beta\alpha = 1_A$. Then

$$(\alpha\beta)\alpha = \alpha(\beta\alpha) = \alpha 1_A = \alpha = 1_B\alpha. \tag{B.1}$$

Since α is an epimorphism, (B.1) shows that $\alpha\beta = 1_B$. Consequently α is both a retraction and a coretraction. Hence α is an isomorphism.   □

B.3 Functors

Our main interest in categories is in the maps from one category to another. Those maps which have the natural properties of preserving identities and composites are called *functors*. An algebraic representation of topology is a mapping from topology to algebra. Such a representation, formally called a functor, converts a topological problem into an algebraic one.

Definition B.3.1 Let $\mathcal{C}$ and $\mathcal{D}$ be categories. A *covariant functor* (or *contravariant functor*) T from $\mathcal{C}$ to $\mathcal{D}$ consists of

(i) an object function which assigns to every object X of $\mathcal{C}$ an object $T(X)$ of $\mathcal{D}$; and

(ii) a morphism function which assigns to every morphism $f : X \to Y$ in $\mathcal{C}$, a morphism $T(f) : T(X) \to T(Y)$ (or $T(f) : T(Y) \to T(X)$) in $\mathcal{D}$ such that

(a) $T(1_X) = 1_{T(X)}$;

(b) $T(gf) = T(g)T(f)$ (or $T(gf) = T(f)T(g)$) for $g : Y \to W$ in $\mathcal{C}$.

Example B.3.1 (i) There is a covariant functor from the category of groups and homomorphisms to the category of sets and functions which assigns to every group its underlying set. This functor is called a *forgetful functor* because it forgets the structure of a group.

(ii) Let R be a commutative ring. Given a fixed R-module M_0, there is a covariant functor π_{M_0} (or *contravariant functor* π^{M_0}) from the category of R-modules and R-homomorphisms to itself which assigns to an R-module M the R-module $\operatorname{Hom}_R(M_0, M)$ (or $\operatorname{Hom}_R(M, M_0)$ and if $\alpha : M \to N$ is an R-module homomorphism, then $\pi_{M_0}(\alpha) : \operatorname{Hom}_R(M_0, M) \to \operatorname{Hom}_R(M_0, N)$ is defined by $\pi_{M_0}(\alpha)(f) = \alpha f$ $\forall f \in \operatorname{Hom}_R(M_0, M)$ ($\pi^{M_0}(\alpha) : \operatorname{Hom}_R(N, M_0) \to \operatorname{Hom}_R(M, M_0)$ is defined by $\pi^{M_0}(\alpha)(f) = f\alpha$ $\forall f \in \operatorname{Hom}_R(N, M_0)$).

(iii) Let $\mathcal{C}$ be any category and $C \in \operatorname{ob}(\mathcal{C})$. Then there is a *covariant functor* $h_C : \mathcal{C} \to \mathcal{S}$ (category of sets and functions) defined by $h_C(A) = \mathcal{C}(C, A)$ (set of all morphisms from the object C to the object A in $\mathcal{C}$) $\forall$ objects $A \in \operatorname{ob}(\mathcal{C})$ and for $f : A \to B$ in $\mathcal{C}$, $h_C(f) : h_C(A) \to h_C(B)$ is defined by $h_C(f)(g) = fg$ $\forall g \in h_C(A)$ (the right hand side is the composite of morphisms in $\mathcal{C}$).

Its dual functor h^C defined in an usual manner is a contravariant functor.

(iv) For any category $\mathcal{C}$ there is a contravariant functor to its opposite category $\mathcal{C}^0$ which assigns to an object X of $\mathcal{C}$ the object X^0 of C^0 and to a morphism $f : X \to Y$ in $\mathcal{C}$ the morphism $f^0 : Y^0 \to X^0$ in $\mathcal{C}^0$.

Remark A functor from a category $\mathcal{C}$ to itself is sometimes called a functor on $\mathcal{C}$. Any contravariant functor on $\mathcal{C}$ corresponds to a covariant functor on $\mathcal{C}^0$ and vice versa. Thus any functor can be regarded as a covariant (or contravariant) functor on a suitable category. In spite of this, we consider covariant as well as contravariant functors on $\mathcal{C}$.

Definition B.3.2 A functor $T : \mathcal{C} \to \mathcal{D}$ is called

(i) faithful iff the mapping $T : \mathcal{C}(A, B) \to \mathcal{D}(T(A), T(B))$ is injective;

(ii) full iff the mapping $T : \mathcal{C}(A, B) \to \mathcal{D}(T(A), T(B))$ is surjective; and

(iii) an embedding iff T is faithful and $T(A) = T(B) \Rightarrow A = B$.

Definition B.3.3 A category C is called concrete iff there is a faithful functor T : $C \to S$.

B.4 Natural Transformations

In some occasions we have to compare functors with each other. We do this by means of suitable maps between functors.

Definition B.4.1 Let C and D be categories. Suppose T_1 and T_2 are functors of the same variance (either both covariant or both contravariant) from C to D. A *natural transformation* ϕ from T_1 to T_2 is a function from the objects of C to morphisms of D such that for every morphism $f : X \to Y$ in C the appropriate one of the following conditions hold:

$$\phi(Y)T_1(f) = T_2(f)\phi(X) \quad \text{(when } T_1 \text{ and } T_2 \text{ are both covariant functors)}$$

or

$$\phi(X)T_1(f) = T_2(f)\phi(Y) \quad \text{(when } T_1 \text{ and } T_2 \text{ are both contravariant functors).}$$

Definition B.4.2 Let C and D be categories and T_1, T_2 be functors of the same variance from C to D. If ϕ is a natural transformation from T_1 to T_2 such that $\phi(X)$ is an equivalence in D for each object X in C, then ϕ is called a *natural equivalence*.

Example B.4.1 Let R be a commutative ring and **Mod** be the category of R-modules and R-homomorphism, M and N be objects in **Mod**. Suppose $g : M \to N$ is a morphism in **Mod**. So by Example B.3.1(ii), π_M, π_N are both covariant functors and π^M, π^N are both contravariant functors from **Mod** to itself. Then there exists a natural transformation $g^* : \pi_N \to \pi_M$, where $g^*(X) : \pi_N(X) \to \pi_M(X)$ is defined by $g^*(X)(h) = hg$ for every object X in **Mod** and for all $h \in \pi_N(X)$; and a natural transformation

$$g_* : \pi_M \to \pi_N, \text{ where } g_*(X) \text{ is defined in an analogous manner.}$$

If g is an equivalence in **Mod**, then both the natural transformations g_* and g^* are natural equivalences.

Theorem B.4.1 (Yoneda's Lemma) *Let C be any category and T a covariant functor from C to S (category of sets and functions). Then for any object C in C, there is an equivalence $\theta = \theta_{C,T} : (h_C, T) \to T(C)$, where (h_C, T) is the class of natural transformations from the set valued functor h_C to the set valued functor T such that θ is natural in C and T.*

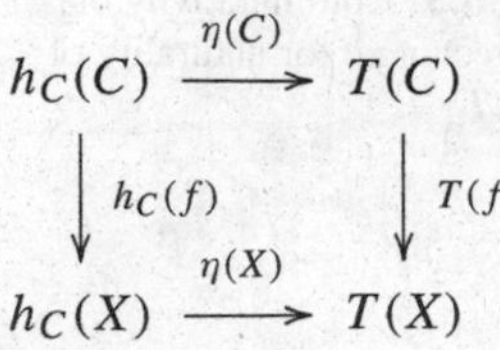

Fig. B.1 Commutativity of
the rectangle for natural
transformation η

$$
\begin{array}{ccc}
h_C(C) & \xrightarrow{\ \eta(C)\ } & T(C) \\
\downarrow{\scriptstyle h_C(f)} & & \downarrow{\scriptstyle T(f)} \\
h_C(X) & \xrightarrow{\ \eta(X)\ } & T(X)
\end{array}
$$

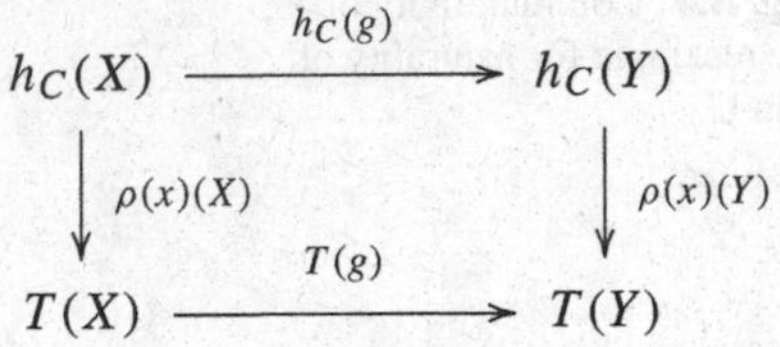

Fig. B.2 Commutativity of
the rectangle for natural
transformation ρ

$$
\begin{array}{ccc}
h_C(X) & \xrightarrow{\ h_C(g)\ } & h_C(Y) \\
\downarrow{\scriptstyle \rho(x)(X)} & & \downarrow{\scriptstyle \rho(x)(Y)} \\
T(X) & \xrightarrow{\ T(g)\ } & T(Y)
\end{array}
$$

Proof We claim that every object C in $\mathcal{C}$, (h_C, T) is a set. Let $\eta : h_C \to T$ be a natural transformation.

Then for each $f : C \to X$ in $\mathcal{C}$ the diagram in Fig. B.1 is commutative.

Hence

$$
T(f)\big(\eta(C)\big)(1_C) = \big(T(f)\eta(C)\big)(1_C) = \eta(X)h_C(f)(1_C) = \eta(X)(f) \qquad \text{(B.2)}
$$

since $h_C(1_C) = f1_C = f$ (see Example B.3.1(iii)).

This shows that for each X, the function $\eta(X)$ is completely determined by the element $\eta(C)(1_C) \in T(C)$. The latter being a set, so is (h_C, T).

Having dealt with this part we proceed with the main part.

We define

$$
\theta : (h_C, T) \to T(C) \quad \text{by } \theta(\eta) = \eta(C)(1_C) \in T(C) \quad \forall \eta \in (h_C, T). \qquad \text{(B.3)}
$$

We now define a function

$$
\rho : T(C) \to (h_C, T) \quad \text{by } \rho(x)(X)(f) = T(f)(x)
$$
$$
\forall x \in T(C),\ X \in \mathcal{C} \text{ and } f \in \mathcal{C}(C, X). \qquad \text{(B.4)}
$$

Then $\rho(x) \in (h_C, T)$. Now from the definition of ρ it follows that $\rho(x)(X) : \mathcal{C}(C, X) \to T(X)$ is a function in $\mathcal{S}$, because for $f \in \mathcal{C}(C, X)$, $T(f) \in \mathcal{S}(T(C), T(X)) \Rightarrow T(f)(x) \in T(X) \,\forall x \in T(C)$.

Now for $g : X \to Y$ in $\mathcal{C}$, the diagram in Fig. B.2 is commutative.

This is because

$$
\rho(x)(Y)h_C(g)(f) = \rho(x)(Y)(gf) = T(gf)(x) = T(g)\big(T(f)(x)\big)
$$
$$
= T(g)\rho(x)(X)(f) \quad \forall f \in h_C(X).
$$

Consequently, $\rho(x)$ is a natural transformation from h_C to T.

Fig. B.3 Commutativity of
the rectangle for naturality of
θ in T

$$
\begin{array}{ccc}
(h_C, T) & \xrightarrow{\;\theta=\theta_{C,T}\;} & T(C) \\[2pt]
\Big\downarrow{\scriptstyle N_*(\alpha)} & & \Big\downarrow{\scriptstyle \alpha(C)} \\[6pt]
(h_C, S) & \xrightarrow[\;\theta=\theta_{C,S}\;]{} & S(C)
\end{array}
$$

Fig. B.4 Commutativity of
the rectangle for naturality of
θ in C

$$
\begin{array}{ccc}
(h_C, T) & \xrightarrow{\;\theta_C=\theta_{C,T}\;} & T(C) \\[2pt]
\Big\downarrow{\scriptstyle N_*(f)} & & \Big\downarrow{\scriptstyle T(f)} \\[6pt]
(h_D, T) & \xrightarrow[\;\theta_D=\theta_{D,T}\;]{} & T(D)
\end{array}
$$

Now

$$(\rho\theta)(\eta) = \rho\big(\theta(\eta)\big) = \rho\big(\eta(C)(1_C)\big)$$

$$\Rightarrow\quad (\rho\theta)(\eta)(X)(f) = \rho\big(\eta(C)(1_C)\big)(X)(f) = Tf\big(\eta(C)(1_C)\big) \tag{B.5}$$

by (B.4) $= \eta(X)(f)$ by (B.2) $\forall X \in \mathcal{C}$ and $\forall f \in \mathcal{C}(C, X) \Rightarrow \rho\theta =$ identity.

Also,

$$\forall x \in T(C), \quad (\theta\rho)(x) = \theta\big(\rho(x)\big) = \rho(x)(C)(1_C)$$
$$\tag{B.6}$$
by (B.3) $= T(1_C)(X)$ by (B.4) $= 1_{T(C)}(x) = x \Rightarrow \theta\rho =$ identity.

Consequently, θ is an equivalence.

To show that θ is natural in T, we have to prove that the diagram in Fig. B.3
is commutative, where $\alpha : T \to S$ is any natural transformation from the set valued functor T to the set valued functor S and $N_*(\alpha) : (h_C, T) \to (h_C, S)$ is
defined by $N_*(\alpha)(\eta) = \alpha\eta$, the latter is given by $(\alpha\eta)(X) = \alpha(X)\eta(X)\ \forall X \in$
$\mathcal{C}$. Now $\alpha(C)\theta(\eta) = \alpha(C)\eta(C)(1_C) = (\alpha\eta)(C)(1_C)$ and $\theta N_*(\alpha)(\eta) = \theta(\alpha\eta) =$
$(\alpha\eta)(C)(1_C) \Rightarrow$ the above diagram commutes $\Rightarrow \theta$ is natural in T.

To show that θ is natural in C, we have to prove that the diagram in Fig. B.4 is
commutative.

For every morphism $f : C \to D$ in $\mathcal{C}$, where for each $X \in \mathcal{C}$ and $\eta \in (\eta_C, T)$,

$$N_*(f)(\eta)(X) : h_D(X) \to T(X) \quad \text{is defined by}$$

$$N_*(f)\big(\eta(X)\big)(g) = \eta(X)(gf) \quad \forall g \in h_D(X).$$

Chasing the element $\eta \in (h_C, T)$ anticlockwise, we have

$$\theta_D N_*(f)(\eta) = \big(N_*(f)(\eta)(D)\big)(1_D) = \eta(D)(f) \tag{B.7}$$

Fig. B.5 Commutativity of
the rectangle for naturality
of η

$$
\begin{array}{ccc}
h_C(C) & \xrightarrow{\;\;\eta(C)\;\;} & T(C) \\
\downarrow{\scriptstyle h_C(f)} & & \downarrow{\scriptstyle T(f)} \\
h_C(D) & \xrightarrow[\;\;\eta(D)\;\;]{} & T(D)
\end{array}
$$

and chasing η clockwise, we have

$$
T(f)\theta_C(\eta) = T(f)\eta(C)(1_C). \tag{B.8}
$$

Again by the naturality of η, the diagram in Fig. B.5 is commutative.
   Hence

$$
T(f)\eta(C)(1_C) = \eta(D)(h_C f)(1_C) = \eta(D)(f 1_C) = \eta(D)(f) \tag{B.9}
$$

Hence (B.7)–(B.9) show that θ is natural in T.   □

Remark For the dual result of the theorem, see Exercise 9.

Example B.4.2 Let Grp be the category of groups and homomorphisms $\mathbf{S}$ be the
category of sets and functions and $S : \text{Grp} \to \mathbf{S}$ be the forgetful functor which assigns to each group G its underlying set SG. Then

 (i) there is a natural equivalence from the covariant functor $h_\mathbf{Z}$ to the covariant
     functor $\mathbf{S}$; and
(ii) there is an equivalence $\theta : (S, S) \to S\mathbf{Z}$.

   [*Hint.* (i) Let G be an arbitrary object in Grp and $\eta : h_\mathbf{Z} \to S$ a natural transformation.
   Define $\eta(G) : h_\mathbf{Z}(G) \to SG$ by

$$
\eta(G)(f) = f(1) \quad \forall f \in \eta_\mathbf{Z}(G) = \text{Hom}(\mathbf{Z}, G),
$$

where 1 is the generator of the infinite cyclic group $\mathbf{Z}$.
   Again define $\rho(G) : SG \to h_\mathbf{Z}(G)$ by $\rho(G)(x) = f$, where f is the group homomorphism $: \mathbf{Z} \to G$ defined by $f(1) = x$.
   Then $\rho(G)\eta(G)(f) = \rho(G)(f(1)) = \rho(G)(x) = f$ and $\eta(G)\rho(G)(x) =
\eta(G)(f) = f(1) = x \Rightarrow \eta(G)$ is an equivalence $\forall G \in$ Grp $\Rightarrow \eta$ is a natural equiva-
lence.
   (ii) Take in particular $\mathcal{C} = $ Grp, $C = \mathbf{Z}$, $T = S$ in Yoneda's Lemma. Then it fol-
lows by (i) and Yoneda's Lemma that $(S, S) \cong (h_\mathbf{Z}, h_\mathbf{Z}) \cong h_\mathbf{Z}(\mathbf{Z}) = \text{Hom}(\mathbf{Z}, \mathbf{Z}) \cong
\mathbf{Z}$, the last equivalence is obtained by assigning to the group homomorphism
$f \in \text{Hom}(\mathbf{Z}, \mathbf{Z})$ the integer $f(1) \in \mathbf{Z}$.]

B.5 Presheaf and Sheaf

Let X be a topological space. A presheaf A (of abelian groups) on X is a contravariant functor from the category of open subsets of X and inclusions to the category of abelian groups and homomorphisms. In general, one may define presheaf with values in an arbitrary category. Thus, if each $A(U)$ is a ring for every open set $U \subset X$ and for each pair of open sets of U, V ($U \subset V$),

$\rho_{U,V} : A(V) \to A(U)$ is a ring homomorphism (called restriction) such that $\rho_{U,U} = $ identity and $\rho_{U,V} \rho_{V,W} = \rho_{U,W}$ when $U \subset V \subset W \subset X$, then A is called *a presheaf of rings*.

Similarly, let A be a presheaf of rings on X and suppose that B is a presheaf on X such that each $B(U)$ is an $A(U)$-module and $\rho_{U,V} : B(V) \to B(U)$ are module homomorphisms such that $\rho_{U,U} = $ identity and $\rho_{U,V} \rho_{V,W} = \rho_{U,W}$ for open sets $U \subset V \subset W \subset X$. Then B is said to be a presheaf of A-modules.

If M is an abelian group, then there is the 'constant presheaf' with $A(U) = M$ for all U and $\rho_{U,V} = $ identity $\forall U \subset V$. We also have the presheaf B assigning to U the group (under pointwise addition) $B(U)$ of all functions from U to M, where $\rho_{U,V}$ is the *canonical restriction*. If M is the group of all real numbers, we have the presheaf C with $C(U)$ being the group of all continuous real valued functions on U.

A *sheaf* (of abelian groups) on a topological space X is a pair $\mathcal{A} = (S, \pi)$, where

(i) S is a topological space;
(ii) $\pi : S \to X$ is a local homeomorphism, i.e., every point $\alpha \in S$ has an open neighborhood N in S such that $\pi | N$ is a homeomorphism between N and an open neighborhood of $\pi(\alpha)$ in X;
(iii) Each $\mathcal{A}_x = \pi^{-1}(x)$, for $x \in X$, is an abelian group (and is called the *Stalk* of S over x);
(iv) The group operations are continuous.

The meaning of (iv) is as follows: Let $S \oplus S = \{(\alpha, \beta) \in S \times S : \pi(\alpha) = \pi(\beta)\}$ be the subset of $S \times S$ with induced topology, the map $S \times S \to S$ defined by $(\alpha, \beta) \to (\alpha - \beta)$ is continuous (equivalently, the map $f : S \to S$, $\alpha \to -\alpha$ is continuous and the map $S \times S \to S$, $(\alpha, \beta) \to (\alpha + \beta)$ is continuous).

Similarly, we may, for example, define a sheaf of rings or a module (sheaf of modules) over a sheaf of rings. Thus for a sheaf of rings, each stalk is assumed to have the (given) structure of a ring and the map $S \oplus S \to S$, $(\alpha, \beta) \mapsto \alpha\beta$ is assumed to be continuous (in addition to (iv)). For a sheaf of R-module each stalk of S is an R-module and the module multiplication $S \to S$ defined by $\alpha \mapsto r\alpha$ is continuous for each $r \in R$.

The Canonical Presheaf of a Sheaf   Let $\mathcal{A} = (S, \pi)$ be a sheaf on X. A section of $\mathcal{A}$ over an open set $U \subset X$ is a continuous map $s : U \to S$ such that $\pi s : U \to U$ is the identity on U.

By (iii) the set of all sections of $\mathcal{A}$ over U is an abelian group denoted by $\mathcal{A}(U)$. Similarly, if $\mathcal{R}$ is a sheaf of rings, $\mathcal{R}((U))$ is a ring.

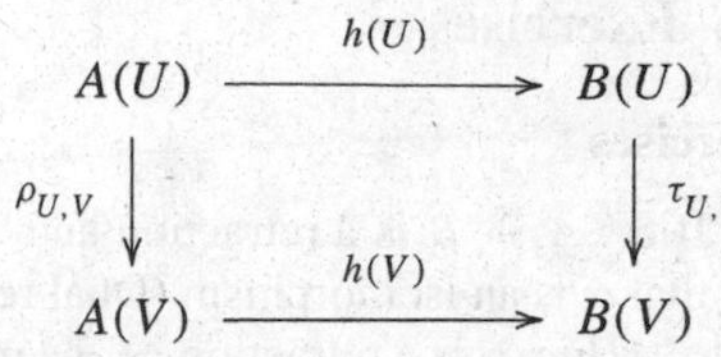

Fig. B.6 Commutativity of
the rectangle for a
homomorphism of presheaves

$$
\begin{array}{ccc}
A(U) & \xrightarrow{\;h(U)\;} & B(U) \\[2pt]
\rho_{U,V}\Big\downarrow & & \Big\downarrow \tau_{U,V} \\[2pt]
A(V) & \xrightarrow[\;h(V)\;]{} & B(V)
\end{array}
$$

Now we assign to each open set U of X the group $\mathcal{A}(U)$ of sections of the sheaf $\mathcal{A}$ over U, where $\mathcal{A}(U)$ is understood to be the zero group if $U = \emptyset$. If $V \subset U$, define

$$\rho_{U,V} : \mathcal{A}(U) \to \mathcal{A}(V) \tag{B.10}$$

to be the homomorphism which assigns to each section of $\mathcal{A}$ over U, its restriction to V (if $V = \emptyset$, we put $\rho_{U,V} = 0$). The assignment

$U \mapsto \mathcal{A}(U)$ for open sets $U \subset X$ defines a presheaf $\{\mathcal{A}(U), \rho_{U,V}\}$ on X. This presheaf is called the *canonical presheaf* of the sheaf $\mathcal{A} = (S, \pi)$ on the *presheaf of sections* of $\mathcal{A}$.

The Sheaf Generated by a Presheaf   Let A be a presheaf on X. For each open set $U \subset X$, consider the space $U \times A(U)$, where U has the subspace topology and $A(U)$ has the discrete topology. Form the disjoint union $E = \bigcup_{\langle V \subset X \rangle} (U \times A(U))$.

We consider the following equivalence relation ρ on E.

If $(x, s) \in U \times A(U)$ and $(y, t) \in V \times A(V)$, then $(x, s)\rho(y, t) \Leftrightarrow (x = y$ and $\exists$ an open neighborhood W of x with $W \subset U \cap V$ and $\rho_{U,W}(s) = \rho_{V,W}(t))$. Let $\mathcal{A}$ be the quotient space E/ρ and $\pi : \mathcal{A} \to X$ be the projection induced by the map $p : E \to X, (x, s) \mapsto x$. Then π is a local homeomorphism. Clearly, $\mathcal{A} = \pi^{-1}(x)$ is the direct limit of $A(U)$ for U ranging over the open neighborhoods of x. Thus the stalk $\mathcal{A}_x$ has a natural group structure. Clearly, the group operations in $\mathcal{A}$ are continuous (since they are in E). Thus $\mathcal{A}$ is a sheaf called the *sheaf generated* by the presheaf A.

Homomorphisms of Presheaves and Sheaves   We mainly consider presheaves or sheaves over a fixed base space X. A *homomorphism $h : A \to B$* of presheaves is a collection of homomorphisms $h_U : A(U) \to B(U)$ commuting the restrictions i.e., making the diagram in Fig. B.6 commutative for $V \subset U \subset X$. $\rho_{U,V}$ and $\tau_{U,V}$ are defined by (B.10) in Fig B.6. Then h is a natural transformation of functors.

A *homomorphism $f : \mathcal{A} \to \mathcal{A}'$* of sheaves $\mathcal{A} = (Y, \pi)$ and $\mathcal{A}' = (Y', \pi')$ on X is a continuous map $f : Y \to Y'$ such that $f(\mathcal{A}_x) \subset \mathcal{A}'_x \ \forall x \in X$ and the restriction $f_x : \mathcal{A}_x \to \mathcal{A}'_x$ of f to stalks is a homomorphism $\forall x \in X$.

A homomorphism of sheaves induces a homomorphism of their corresponding canonical presheaves.

Conversely, let $h : A \to B$ be a homomorphism of the presheaves. For each $x \in X$, h induces a homomorphism $h_x : \mathcal{A}_x = \lim_{\langle x \in U \rangle} A(U) \to \lim_{\langle x \in U \rangle} B(U) = \mathcal{B}_x$, and therefore, a map $\eta : \mathcal{A} \to \mathcal{B}$. If $s \in A(U)$, then h maps the section $\theta(s) \in \mathcal{A}(U)$ onto the section $\theta(h(s)) \in \mathcal{B}(U)$.

B.6 Exercises

Exercises

1. If $\alpha : A \to B$ is a retraction and also a monomorphism in a category $\mathcal{C}$, prove that α is an isomorphism (Dual result of Proposition B.2.2).

   [*Hint.* α is a retraction $\Rightarrow \exists$ a monomorphism $\beta : B \to A$ such that $\alpha\beta = 1_B$. Again $\alpha(\beta\alpha) = (\alpha\beta)\alpha = 1_B\alpha = \alpha 1_A \Rightarrow \beta\alpha = 1_A$ (as α is a monomorphism).]

2. Show that if $\alpha : A \to B$ is an epimorphism in $\mathcal{S}$, then $\alpha^0 \in \mathcal{S}^0$ is a monomorphism.

3. (a) Let T be a functor from a category $\mathcal{C}$ to a category $\mathcal{D}$. Show that T maps equivalences in $\mathcal{S}$ to equivalences in $\mathcal{D}$.

   (b) Show that the equivalences in the category

      (i) **S** are bijections;
      (ii) **Grp** are isomorphisms of groups;
      (iii) **Ring** are isomorphisms of rings;
      (iv) **Mod**$_R$ are isomorphisms of modules;
      (v) **Top** are homeomorphisms of topological spaces.

4. Let A be a subspace of a topological space X and $f : A \to Y$ continuous. Then f is said to have a continuous extension $F : X \to Y$ iff $F \circ i = f$, where $i : A \hookrightarrow X$ is the inclusion map.

   Let T be a covariant (or contravariant) functor from the category of topological spaces and continuous maps to a category $\mathcal{S}$. Show that a necessary condition that a map $f : A \to Y$ be extendable to X is that $\exists$ a morphism

$$\phi : T(X) \to T(Y) \quad \left(\text{or } \phi : T(Y) \to T(X)\right)$$

   in $\mathcal{S}$ such that

$$\phi \circ T(i) = T(f) \quad \left(\text{or } T(f) = T(i) \circ \phi\right).$$

5. Let $\mathbf{R}^n$ be the Euclidean n-space, with $\|x\| = \sqrt{\sum x_i^2}$, $E^n = n\text{-ball} = [x \in \mathbf{R}^n : \|x\| \leq 1\}$ and $S^{n-1} = (n-1)\text{-sphere} = \{x \in \mathbf{R}^n : \|x\| = 1\}$.

   Prove the following:

   (a) the identity map $1_{S^n} : S^n \to S^n$ cannot be extended to a continuous map $E^{n+1} \to S^n$;

   (b) **Brouwer Fixed Point Theorem** *Any continuous map* $f : E^{n+1} \to E^{n+1}$ *has a fixed point (i.e.,* $f(x) = x$ *for some* $x \in E^{n+1}$*).*

      [*Hint.* (a) If possible, let $f : E^{n+1} \to S^n$ be a continuous extension of 1_S^n. Then $f \circ i = 1_S^n$ (see Exercise 4). Assume that H_n is the homology functor such that $H_n(S^n) \cong \mathbf{Z}$ and $H_n(E^{n+1}) = 0$.

      Use H_n on $f \circ i = 1_S^n$ and obtain $H_n(S^n) \xrightarrow{H_n(i)} H_n(E^{n+1}) \xrightarrow{H_n(f)} H_n(S^n)$ such that the composite homomorphism is the identity homomorphism of $H_n(S^n)$ which is not possible as the composite homomorphism $\mathbf{Z} \to 0 \to \mathbf{Z}$ cannot be identity. For $n = 1$, the result (a) can be proved in a similar way by using the result that $\pi_1(S^1) \cong \mathbf{Z}$ and $\pi_1(E^2) = 0$ (see Sect. 2.10 of Chap. 2).

(b) Let $f : E^{n+1} \to E^{n+1}$ be a continuous map. If possible f has no fixed point (i.e. $f(x) \neq x$ for any $x \in E^{n+1}$). Then for any $x \in E^{n+1}$ join $f(x)$ to x by a line and move along the line in the direction from $f(x)$ to x until the unique point $r(x) \in S^n$ is reached. Then $r : E^{n+1} \to S^n$ is a continuous map extending 1_{S^n}, which contradicts the result of (a). The Brouwer fixed point theorem for dimension 2 as stated in (b) can be proved in a similar way by using the result that $\pi_1(S^1) \cong \mathbf{Z}$ and $\pi_1(E^2) = 0$ (see Sect. 2.10 of Chap. 2).]

6. Let S be a category and (X, Y) be the set of morphisms from X to Y in S. By keeping X fixed and varying Y, show that this set is an invariant of the equivalent sets in the sense that there is a bijective correspondence between the sets corresponding to the equivalent sets. Find the corresponding result when X is varied and Y is kept fixed.

   [*Hint.* Let $Y \approx Z$. So there exist $f : Y \to Z$ and $g : Z \to Y$ such that $g \circ f = 1_Y$ and $f \circ g = 1_Z$. Then $f_* : (X, Y) \to (X, Z)$ defined by $f_*(\alpha) = f \circ \alpha$ and $g_* : (X, Z) \to (X, Y)$ defined by $g_*(\beta) = g \circ \beta$ are such that $(g \circ f)_* = g_* \circ f_* = $ identity and $f_* \circ g_* = $ identity. Consequently f_* is a bijection.]

7. Let X and Y be objects of a category S and let $g : X \to Y$ be a morphism in S. Show that there is a natural transformation g^* from the covariant functor h_Y to the covariant functor h_X and a natural transformation g_* from the contravariant functor h^X to the contravariant functor h^Y. Further show that if g is an equivalence in S, both these natural transformations g_* and g^* are natural equivalences.

8. Let A and C be objects of a category $\mathcal{C}$. Using the Yoneda Lemma, show that $(h_C, h_A) \approx \mathcal{C}(A, C)$.

   [*Hint.* Take $T = h_A$. Then by Yoneda's Lemma $(h_C, h_A) \cong h_A(C) = \mathcal{C}(A, C)$.]

9. Prove the dual result of Theorem B.4.1:

   Let $\mathcal{C}$ be any category and T a contravariant functor from $\mathcal{C}$ to $\mathbf{S}$. Then for any object $C \in \mathcal{C}$, there is an equivalence $\theta : (h^C, T) \to T(C)$ such that θ is natural in C and T.

10. Let Htp denote the category of pointed topological spaces and homotopy classes of their base point preserving continuous maps and Grp be the category of groups and their homomorphisms.

    (a) If P is an H-group, show that there exists a contravariant function π^P from Htp to Grp. Further show that a homomorphism $\alpha : P \to P'$ between H-groups induces a natural transformation $\alpha_* : \pi^P \to \pi^{P'}$.

    (b) If P is a topological group, show that π^P is a contravariant functor from Htp to Grp.

    (c) π_1 is a covariant functor from Htp to Grp.

Proof (a) Using Theorems 2.9.1–2.9.3, Chap. 2, define $\pi^P : \text{Htp} \to \text{Grp}$, such that $\pi^P(X) = [X; P]$ and also for $f : X \to Y$ in Htp, $\pi^P(f) = f^* : [Y; P] \to [X; P]$ by $f^*[h] = [h \circ f]$, then π^P is a contravariant functor.

Again, for $[g] \in [Y; P]$, $f^*\alpha_*[g] = f^*[\alpha \circ g] = [(\alpha \circ g) \circ f]$ and $\alpha_* f^*[g] = \alpha_*[g \circ f] = [\alpha \circ (g \circ f)] \Rightarrow f^* \circ \alpha_* = \alpha_* \circ f^*$. Consequently, α_* is a natural transformation.

(b) If $g_1, g_2 : X \to P$ are base point preserving continuous maps, then $g_1 g_2 : X \to P$ is defined by $(g_1 g_2)(x) = g_1(x)g_2(x)$ $\forall x \in X$, where the right hand side is the group product in P. The law of composition carries over to give an operation on homotopy classes such that $[g_1][g_2] = [g_1 g_2]$. Then $[X; P]$ is a group. Hence (b) follows from (a).

(c) It follows from Sect. 2.10 of Chap. 2.   □

11. Let $C(X)$ be the ring of real valued continuous functions on a topological space X. Then C is a contravariant functor from the category **Top** of topological spaces and continuous functions to the category **Ring** of rings and homomorphisms.

12. Let Spec (R) be the spectrum space of a ring R endowed with Zariski topology (see Sect. 9.12). Then Spec is a contravariant functor from **Ring** to **Top**.

13. (a) Show that all chain complexes and chain maps (see Definition 9.11.4) form a category. We denote this category by **Comp**.

    (b) For each $n \in \mathbf{Z}$, show that $H_n :$ **Comp** $\to$ **Mod** is a covariant functor (see Definition 9.11.3).

    (c) For each $n \in \mathbf{Z}$, show that $H^n :$ **Comp** $\to$ **Mod** is a contravariant functor (see Definition 9.11.9).

    [*Hint.* For (a) and (b) use Proposition 9.11.4.]

14. Let **AB** denote the category of abelian groups and their homomorphisms.

    (a) For an abelian group G, let $T(G)$ denote its torsion group.
        Show that $T :$ **Ab** $\to$ **Ab** defines a functor if $T(f)$ is defined by $T(f) = f|T(G)$ for every homomorphism f in **Ab** such that

        (i) f is a monomorphism in **Ab** implies that $T(f)$ is also so;
        (ii) f is an epimorphism in **Ab** does not always imply that $T(f)$ is also so.

    (b) Let p be a fixed prime integer. Show that $T :$ **Ab** $\to$ **Ab** defines a functor, where the object function is defined by $T(G) = G/pG$ and the morphism function $T(f)$ is defined by $T(f) : G/pG \to H/pH$, $x + pG \mapsto f(x) + pH$ for every homomorphism $f : G \to H$ in **Ab** such that

        (i) f is an epimorphism in **Ab** implies that $T(f)$ is an epimorphism;
        (ii) f is a monomorphism in **Ab** does not always imply that $T(f)$ is a monomorphism.

B.7   Additional Reading

We refer the reader to the books (Adhikari and Adhikari 2003; Mac Lane 1997; Mitchell 1965; Spanier 1966) for further details.

References

Adhikari, M.R., Adhikari, A.: Groups, Rings and Modules with Applications, 2nd edn. Universities Press, Hyderabad (2003)

Mac Lane, S.: Categories for the Working Mathematician. Springer, Berlin (1997)

Eilenberg, S., Mac Lane, S.: Natural isomorphism in group theory. Proc. Natl. Acad. Sci. **28**, 537–545 (1942)

Eilenberg, S., Mac Lane, S.: Generalized theory of natural equivalences. Trans. Am. Math. Soc. **58**, 231–294 (1945)

Mitchell, B.: Theory of Categories. Academic Press, New York (1965)

Spanier, E.H.: Algebraic Topology. McGraw-Hill, New York (1966)

Appendix C
A Brief Historical Note

Modern algebra began with the work of French mathematician E. Galois (1811–1832) who died in a duel at a very young age. He created one of the most important theories in the history of algebra known as "The Theory of Galois". Before Galois, algebraists had mainly concentrated on the solutions of algebraic equations. Scipione dal Ferro, Tartaglia, and Cardano solved cubic equations, and Ferrari solved biquadratic equations completely by radicals. The most important predecessors of Galois are J.L. Lagrange (1736–1813), C.F. Gauss (1777–1855) and N.H. Abel (1802–1829).

Linear algebra arose through the study of the theory of solutions of linear equations and analytic geometry. The former contributed the algebraic formulas and the latter geometric images. The concept of vector spaces (linear spaces) became known around 1920. Hermann Weyl gave a formal definition. A study of matrices, determinants and their closely related topics is the main objective of linear algebra. Theories of determinants, quadratic forms, linear algebraic equations and linear differential equations were developed in the first third of the 19th century mainly by J.L. Lagrange and C.F. Gauss. Gauss studied in the fifth part of 'Disquisniones Arithmeticae' in 1801, the problem of reduction of the quadratic forms with integral coefficients to canonical forms by using invertible substitution of the variables with integral coefficients. A.L. Cauchy considered in his paper "Memoir on Functions" just two values equal in magnitude but opposite in sign under the permutations of the variables contained in them and extended the investigation of C.A. Vandermonde for determinants of small order. He viewed a determinant as a function of n^2 variables which he arranged in a square table. Some years later, C.G.J. Jacobi published a series of papers on theory of determinants and quadratic forms. The existing notation denoting a determinant by means of two vertical bars is introduced by Arthur Cayley, a British mathematician. His work on analytical geometry of dimension n and that of H. Grassmann in "The Science of Linear Extension" mark a turning point in the evolution of linear algebra. An important role in studying Linear Algebra is played by B. Riemann (1826–1866) in his famous paper "On the Hypothesis that lie at the Foundation of Geometry" and by F. Klein in his work on "n-Dimensional Space V and the Related Geometric Notion of Algebra" in 1870. On the other hand,

M.R. Adhikari, A. Adhikari, *Basic Modern Algebra with Applications*,
DOI 10.1007/978-81-322-1599-8, © Springer India 2014

Arthur Cayley introduced the concept of matrices in 1855 and published several notes in Crelle's Journal. The theory of determinants is much older than the theory of matrices. The German Mathematician Leibniz (1646–1716) introduced the concept of determinants in connection with the solvability of systems of linear equations communicated in a letter to L'Hospital dated April 28, 1693, but the term "determinant" was coined by C.F. Gauss in 1801. Cayley is the first mathematician to realize the importance of theory of matrices and laid the foundation of this theory. The progress in linear algebra during the last few decades focuses several techniques like rank-factorization, generalized inverses and singular value decomposition. Vector spaces over finite fields play a very important role in computer science, coding theory, design of experiments, combinatorics etc. Vector spaces over the rational field $\mathbf{Q}$ are very important in number theory and design of experiments and linear spaces over the complex field $\mathbf{C}$ are essential for the study of eigenvalues. The theory of linear spaces over an arbitrary field F is mainly developed with an eye to the model of linear spaces over $\mathbf{R}$.

Leonhard Euler (1707–1783)   He was born in Basel, Switzerland. He spent most of his time in St. Petersburg and Berlin. He joined the St. Petersburg Academy of Sciences in 1727. He went to Berlin in 1741. He returned to St. Petersburg in 1766, where he remained until his death. He is one of the greatest mathematician of the world. He published 886 papers and books. He is the inventor of the Euler "ϕ-function". His other important contributions are "Euler's formula", "Euler's theorem", "Eulerian angle", "Eulerian number", "Euler–Lagrange equation" etc. He represented the Königsberg bridge problem (concerning the seven bridges in Königsberg crossing the river Pregel) by a graph in which the areas are points and the bridges are edges. This study of the seven bridges of Königsberg is the beginning of combinatorial topology. His fundamental work in different areas makes his presence everywhere in mathematics. He died in 1783.

Joseph Louis Lagrange (1736–1813)   He was born on January 25, 1736 in Turin, Italy. His mathematical contribution to different branches of mathematics including number theory, theory of equations, differential equations, celestial mechanics, and fluid mechanics is of fundamental importance. In 1771, he presented an extremely valuable memoir to the Berlin Academy 'Réflections sur la théorie algébrique des équations'. In this paper he tried to prescribe a general method of solution for polynomials of degree greater than 4. He, however, failed to achieve his goal. But some new concepts introduced in this paper on permutations of roots stimulated his successors, like Abel and Galois to develop the necessary theory to find a general method of solution. This paper is considered to be one of the main sources from which modern group theory developed. In 1770, he proved his famous Lagrange's theorem in group theory. His presence is felt everywhere in mathematics. He died on April 10, 1813.

Carl Friedrich Gauss (1777–1855)   He was born on April 30, 1777 in a poor family, in Brunswick, Germany. At the age of 20, he published the first proof of the fundamental theorem of algebra. Lagrange considered special equations, such as the cyclotomic equation $x^n - 1 = 0$. But he did not go very far. The complete solution of this equation by means of radicals was given by Gauss in 1801 in his book on number theory 'Disquisitiones Arithmeticae', which laid the foundation of algebraic number theory. A second edition of his master piece was published in 1870 (15 years after his death). Gauss introduced the concept of congruence class $\mathbf{Z}_n$ of integers modulo n, notation i for $\sqrt{-1}$, the term complex numbers and Gaussian integers $\mathbf{Z}[i]$ and studied these extensively. He was the first mathematician to study the structures of fields and groups and established their close relationship. He referred to mathematics as 'Queen of Sciences'. Gauss did not have interest in teaching. He preferred his job as the Director of Observatory at Göttingen. But he accepted students like Dedekind, Dirichlet, Riemann, Eisenstein and Kummer. E.T. Bell remarked "Gauss lives everywhere in mathematics". He coined the term 'determinant' in 1801. He died on February 23, 1855.

Augustin Louis Cauchy (1789–1857)   He was born on August 21, 1789, in Paris, France. He came in touch with his neighbors Laplace and Bertholet in his childhood. He became an engineer in 1810 and started his mathematical research in 1811 with a problem from Lagrange on convex polyhedrons. He loved teaching. He solved the long standing Fermat's problem on polygon numbers in 1812. He published more than 800 papers and 8 books on mathematics, mathematical physics and celestial mechanics. His work on mathematics covered calculus, complex functions, algebra, differential equations, geometry and analysis. The notion of continuity is his contribution. His treatise on the definite integral submitted to the French Academy in 1814 forms a basis of the theory of complex functions. He made a distinction between permutations and substitutions. The n variables written in any order was called a permutation but a passage from one permutation to another written by 2-row notation called a substitution was introduced by him. We now call a substitution a permutation. He introduced the concept of 'group of substitutions' which was later called 'group of permutations'. Cauchy published a sequence of papers on substitutions during 1844–1846. The concepts of order of an element, a subgroup, and conjugates are found in his papers. He proved a theorem on a group of finite order, now called, Cauchy's Theorem in his honor. His work on determinants and matrices are also important. Cauchy extended the investigation of C.A. Vandermonde for determinants of small order. He viewed a determinant as a function of n^2 variables which he arranged in a square table. He died on May 22, 1857.

Niels Henrik Abel (1802–1829)   He was born on August 5, 1802, in Finnöy, Norway. While he was a school student, he was greatly influenced by his mathematics teacher Holmbëë to read the work of Euler, Lagrange, Laplace, and Cauchy. Inspired by their work he began to solve the then unsolved problem of solvability of the quintic equation. After a long effort, he proved in 1824 that a solution of this problem by radicals is impossible. He then published a leaflet in French entitled 'Mémorie sur les équations algébriques' in 1824. While proving this impossibility, he used some results obtained by Lagrange and Cauchy. The groups in which composition is commutative are now called Abelian in his honour. His work on elliptic functions revolutionized the theory of elliptic functions. He moved to Paris and Berlin in search of a teaching assignment. The information of his appointment as a Professor of Mathematics at the University of Berlin reached his home 2 days after his death from tuberculosis on April 6, 1829. An Abel Prize (equivalent to Nobel Prize) has been instituted in 2002 by Norway Government to mark his 200th Birth Anniversary.

C.G.J. Jacobi (1804–1851)   He published a number of papers on theory of determinants and quadratic forms. The identity introduced by him, called "The Jacobi Identity" is a relationship $[A, [B, C]] + [B, [C, A]] + [C, [A, B]] = 0$, between three elements A, B, and C, where $[A, B]$ is the commutator. The elements of a Lie algebra satisfy this identity. Jacobi's Identity has wide applications in science and engineering.

H. Grassmann (1809–1877)   His work on analytical geometry for dimension n appeared in "The Science of Linear Extension" and marked a turning point in the evolution of linear algebra.

Evariste Galois (1811–1832)   He was born on October 25, 1811 in Bour-la-Reine, near Paris, France. Galois twice failed at the entrance examination for the École Polytechnique. Galois presented his first paper on the solution of algebraic equations to the Académie des Sciences de Paris in May, 1829. He also presented his second paper on equations of prime degree to the Academy. Both the papers were sent to Cauchy but were lost for ever. He presented another paper on solution of algebraic equations to the Academy in February, 1830 which was sent to Fourier by the Academy. This paper was also lost for ever. Galois published in the "Bulletin des Sciences Mathématiques of Férussac" in April 1830 an article wherein he announced some of his main results of the lost papers. One theorem of the paper states "In order that an equation of prime degree is solvable by radicals, it is necessary and sufficient that, if two of its roots are known, the others can be expressed rationally. This shows that the general equation of degree 5 cannot be solved by radicals." He published two more papers in June, 1830 on resolutions of numerical equations and on the structure of finite field. Galois presented to the Academy a new version of his memoir entitled "Memoire sur les conditions de résolubilité des équations par radicaux" in January, 1831. This paper was published in 1846 (14 years after the death of Galois) by Liouville in the Journal "de mathématiques pures at appliqués II". An edition containing all preserved letters and manuscripts of Galois was published by Gauthier-Villars in 1962 under the title "Ecrits et mémoires mathématiques d'Evariste Galois".

Arthur Cayley (1821–1895)   He was born on August 16, 1821, in Cambridge, England. He became a lawyer in 1849. During his practice in the legal profession up to 1863, he wrote about 300 mathematical papers. He joined as Professor of Pure Mathematics at Cambridge in 1863 and worked there up to his death on January 26, 1895. He worked on mathematics, theoretical dynamics and mathematical astronomy. He wrote 966 papers and one book. In 1854, Cayley published two papers under the title "On the Theory of Groups Depending on the Symbolic Equation $\theta^n = 1$" in *Philosophical Magazine of Royal Society*, London, Vol. 7. In 1878, he gave the abstract definition of groups and formulated the problem: to find all finite groups of a given order n, and proved the famous Cayley Theorem for any finite group. He introduced multiplication table for a finite group known as Cayley Table. He introduced the concept of matrices in 1855 and published several notes in Crelle's Journal. He is the first mathematician who realized the importance of theory of matrices and laid the foundation work on this theory. He developed matrix theory and proved Cayley–Hamilton theorem. He is one of the earlier mathematicians who studied geometry of dimensions greater than 3. The existing notation denoting a determinant by means of two vertical bars was introduced by A. Cayley. He is considered as the founder of abstract group theory.

Leopold Kronecker (1823–1891)   He was born on December 7, 1823 in Germany. After 1870, the abstract notion of "groups" was developed in several stages, essentially due to Kronecker (1870) and Cayley (1878). The modern definition of a group by axioms was given for abelian groups by Kronecker in 1870. After the introduction of the concept of abstract group by Cayley and Kronecker, the theory of groups changed its character. Earlier to them, the main problem was to determine the structure of permutation groups under certain conditions as well as to determine the structure of finite dimensional continuous groups of transformations. But afterwards, the main problem became: to create a general theory of the structure of abstract groups, and to determine all finite groups of a given order. E.E. Kummer was his mathematics teacher. Kronecker was greatly inspired by Kummer for research. Kronecker's work on algebraic number theory placed him as one of the inventors of algebraic number theory along with Kummer and Dedekind. Kronecker is considered to be the first mathematician who clearly understood the wok of Galois. While Weierstrass and Cantor were creating modern analysis, Kronecker remarked "God made positive integers, all else is due to man". This remark badly affected Cantor. Kronecker died on December 29, 1891.

Bernhard Riemann (1826–1866)   His famous paper "On the Hypothesis that lie at the Foundation of Geometry" plays an important role in studying linear algebra. His idea on geometry of space has made a significant effect on the development of modern theoretical physics. He clarified the notion of integral by defining an integral called Riemann Integral.

Richard Dedekind (1831–1916)   He was born on October 6, in 1831 in Brunswick, Germany. He was in contact with Gauss, Dirichlet, Riemann. He completed his Ph.D. work under Gauss. He attended a series of lectures of Dirichlet on theory of numbers and lectures of Riemann on Abelian and elliptic functions. He became interested in analytic geometry and algebraic analysis, differential integral calculus and also in mechanics. He introduced the concept of "Dedekind cut" in 1872. He edited the work of Gauss, Dirichlet, Riemann. He coined the term "ideal" and developed ideal theory and unique factorization theory. While studying ideals in algebraic number fields, he introduced the concepts of ascending chain condition. He replaced the concept the permutation group by abstract group. He loved teaching on Galois theory. He died on February 12, 1916.

Peter Ludvig Mejdell Sylow (1832–1918)   He was born on December 12, 1832 in Oslo, Norway. In 1872, he published a paper of fundamental importance in *Math. Annalen* 5, 584–594. It contained eight theorems and extended Cauchy's result. The theorems in this paper are known as Sylow's Theorems. He is famous for his work on structural results in finite group theory. He died on September 7, 1918. Sylow and Lie prepared an edition on the complete work of Abel during 1883–1881 with Sylow as the main author according to Lie.

Camille Jordan (1838–1922)   He was born on January 5, 1838 in Lyons, France. He was an engineer but he took admission at École Polytechnique in 1855 to study mathematics. He became professor of analysis in École Polytechnique in 1876. Jordan published 120 papers in mathematics. He originated the notion of a bounded function. He proved his famous theorem: "A plane can be decomposed into two regions by a simple closed curve" (called Jordan curve theorem). He was greatly influenced by the work of Riemann. He used combinatorial methods to work in topology and introduced the concept of path homotopy. Jordan was basically an algebraist. He developed the theory of finite groups and its applications following Galois. He introduced the concept of composition series and proved the famous Jordan–Hölder Theorem and the concept of simple groups and epimorphisms. His mathematical work of 667 pages "Traité des substitutions et des équations algébriques" published in 1870, attracted many scholars like Sophus Lie from Norway and Felix Klein from Germany. In the preface to his "Traité", he acknowledged the contribution of his predecessors: Galois who invented the principles of Galois Theory, Betti who wrote a memoir, in which the complete sequence of Galois Theory had been rigorously established for the first time, Abel, Kronecker, and Cayley. He proved finiteness theorems and introduced the concept of simple groups. He studied the general linear groups and the fields of p elements (p is a prime) and applied his work to classical groups to determine the structure of Galois groups of equations. He died on January 22, 1922.

Sophus Lie (1842–1899)   He was born at Nordfjordeid, Norway, in December 1842. He made significant joint work with C.F. Klien. They went to Paris in 1870 and lived in adjacent rooms for 2 months. Their joint paper was published in 1871 in *Math. Annalen* 4, 424–429. They considered one dimensional continuous groups. In later years, they moved in different directions. S. Lie developed his theory of continuous maps and used it in investigating differential equations and C.F. Klein investigated discrete groups. S. Lie is the inventor of Lie Algebra. The fundamental idea of his Lie theory was published in his paper in *Math. Annalen* 16, in 1880. Lie's investigation on the integration of differential equations attracted himself to investigate groups of transformations transforming a differential equation into itself. He developed his theory of transformation groups to solve his integration problems. Continuous transformation groups (called Lie groups, named after Lie), commutator brackets and Lie algebras are essentially due to him. They have wide applications in quantum mechanics. He died in 1899.

Georg Ferdinand Ludwig Philipp Cantor (1845–1918)   He was born in March 1845. He was a German mathematician. He originated naive set theory. He is known as the father of set theory, which forms a strong foundation of different disciplines in modern science. There are two general approaches to set theory. The first one is called "naive set theory" and the second one is called "axiomatic set theory", also known as "Zermelo–Fraenkel set theory". These two approaches differ in a number of ways. While developing mathematical theory of sets to study the real numbers and Fourier series, Cantor established the importance of a bijective correspondence between two sets and introduced the concepts of infinite sets, well-ordered sets, cardinal and ordinal numbers of sets and their arithmetic. He proved that the set of real numbers is not countable by a process known as Cantor's diagonal process. This result shows the existence of transcendental numbers (which are not algebraic over the field $\mathbf{Q}$ of rational numbers) as the set A of algebraic numbers is countable. Transcendental numbers are precisely all the members of $\mathbf{R}$ of real numbers which are not algebraic over the field $\mathbf{Q}$. He died in January 6, 1918.

C. Felix Klein (1849–1925)   He did not consider groups in his first paper on Non-Euclidean geometry but he considered transformation groups of invertible transformations of a manifold in his second paper published in 1873. He defined group like Jordan. According to Klein, projective geometry and Euclidean geometry deal with properties of figures which are invariant under respective transformations. Klein-4 group is named in honor of Klein. His work on "n-dimensional vector spaces and the related geometric notion of algebra' in 1870 stimulated the study of linear algebra. Inspired by seminar lectures of Kronecker and Klein, the algebraist O.L. Hölder (1859–1937) completed the proof of so-called Jordan–Hölder theorem on composition series, which plays a fundamental role in group theory.

Jules Henri Poincaré (1854–1912)   He is a French mathematician and is known as the father of topology. He was born in 1854 in Nancy, Lorraine, France. He has published around 300 papers and several books spread over different areas of pure mathematics, celestial mechanics, fluid mechanics, optics, electricity, telegraphy, capillarity, elasticity, thermodynamics, potential theory, quantum theory, theory of relativity, physical cosmology and philosophy of science. The fundamental group defined in Chap. 3 was invented by Henri Poincaré in 1895 through his work Analysis Situs. This group is also called Poincaré's group in his honor. This group is associated to any given pointed topological space. It is a topological invariant in the sense that two homeomorphic pointed topological spaces have isomorphic fundamental groups. Intuitively, this group provides information about the basic shape, or holes, of the topological space. Poincaré's work in algebraic topology is mainly in geometric terms. The fundamental group is the first and simplest of the homotopy groups. Poincaré is the first mathematician who applied algebraic objects in homotopy theory. Historically, an idea of a fundamental group was in the study of Riemann surfaces by Bernhard Riemann, Henri Poincaré, and Felix Klein. This group has wide applications in different areas. For example, Brouwer fixed point theorem and fundamental theorem of algebra are proved in this book and an extension problem is solved by using homotopy theory and this group. His result on index of intersection of two subgroups of a finite group is known as Poincaré's theorem. He died in 1912 in Paris, France. His famous Poincaré conjecture, which was one of the most important long standing unsolved problems in mathematics till 2002–2003. The conjecture is "Every simply connected, closed 3-manifold is homeomorphic to the 3-sphere." Grigory Perelman, a Russian mathematician solved this problem. He was awarded a Fields Medal at the Madrid, Spain meeting of the International Congress of Mathematicians (ICM) for "his contributions to geometry and his revolutionary insights into the analytical and geometric structure of the Ricci flow". But he declined to accept it.

David Hilbert (1862–1943)   He was born on January 23, 1862 in Königsberg, Germany. The great mathematician Lindeman stimulated Hilbert to work on the theory of invariants. He proved his famous theorem known as Hilbert Basis theorem which made a revolution in algebraic geometry and ring theory. Greatly influenced by the axioms of Euclid, he proposed 21 axioms with their significance. While addressing the Second International Congress of Mathematicians (ICM) of 1900 at Paris, Hilbert posed 23 interesting problems for investigation including the continuum hypothesis, well ordering of reals, transcendence of powers of algebraic numbers, Riemann hypothesis, extension of the Principle of Dirichlet and others. He also worked on algebraic number theory, foundation of geometry, integral equation, calculus of variations, functional analysis and theoretical physics. He died on February 14, 1943.

Amalie Emmy Noether (1882–1935)   She was born on March 23, 1882 in Erlangen, Germany. Her father, Max Noether was a famous mathematician. She started her mathematics career at the University of Göttingen in 1903 as a non-regular student, as at that time girl students were not allowed to be admitted as regular students. However, she was permitted in 1904 to enroll at the University of Erlangen where her father taught. Hilbert invited her to Göttingen in 1915. After prolonged efforts of Hilbert, she was appointed an associate professor and taught there from 1922 to 1933. As she was a Jew she had to leave the university and went to USA in 1933 because of the rise of the Nazi regime. She gave lectures and did research work at Bryn Mawr College. In 1921 Noether extended the Dedekind theory of ideals and the representation theory of integral domains and rings of algebraic numbers for arbitrary commutative rings satisfying ascending chain condition. These rings are now called Noetherian ring. Motivated by Hilbert's axiomatization of Euclidean geometry, Noether became interested in an abstract axiomatic approach to ring theory. Noether developed a general representation theory of groups and algebras over arbitrary ground fields. Her 45 research papers are divided into four categories:

Category 1.   Group-Theoretic Foundations.
Category 2.   Non-commutative Ideal Theory.
Category 3.   Modules and Representations.
Category 4.   Representations of Groups and Algebras.

The basic concept of modern theory of rings came through the works of Noether and Artin during 1920s.

The ascending chain condition was introduced by Noether in her theory of ideals. The concept of Noetherian rings is her invention. She died on April 14, 1935.

Joseph Henry MecLagan Wedderburn (1882–1948)   He was born on February 26, 1882 in Forfar, Scotland. He published 38 papers from 1905 to 1928 and a book on matrices in 1934. He also worked on the structure of algebras over arbitrary fields, instead of algebras over the fields of complex numbers or real numbers. He proved the celebrated theorem in 1905 known as 'Wedderburn theorem' which asserts that a finite division ring is commutative. Its intrinsic beauty is for interlinking the number of elements in a certain algebraic system and the multiplication of that system. This result appears in many contexts and developed a large area of research. His work on finite algebra made a revolution in projective geometry with a finite number of points. For example, his proof of the geometrical result that Desargues configuration implies Pappus configuration is of immense intrinsic beauty. He died on October 9, 1948.

Emil Artin (1898–1962)   He was born on March 3, 1898 in Vienna, Austria. Artin generalized some results of Wedderburn on algebras over fields in 1927, by considering rings with descending chain condition. He worked on various areas of mathematics such as number theory, group theory, ring theory, field theory, geometric algebra, and algebraic topology. Artinian ring is his invention. He proved in 1927 the general laws of reciprocity, which cover all the previous laws of reciprocity up to the time of Gauss. He formulated Galois theory in an abstract setting, he published Galois theory, as used today by establishing a connection between field extension and the subgroups of automorphisms. He died on December 20, 1962.

Oscar Zariski (1899–1986)   He was born in the city of Kobrin, the then part of the Russian Empire. In 1920, the city Kobrin fell in independent Poland as per political agreement between Russia and Poland. He opted for the Polish nationality for his convenience to study mathematics. His work in the area of algebraic geometry is fundamental. The subject algebraic geometry arose through the study of the solution sets of polynomial equations and can be traced back to Descartes. But it becomes an important mathematical discipline in the 19th and 20th centuries. By the early twentieth century, the Italian school of algebraic geometers established many interesting results and investigated many basic problems of algebraic geometry. Zariski joined the strong group of the Italian school and by using modern algebra the school reformulated the subject. Zariski introduced a topology with the help of algebraic sets as closed sets. This topology is now known as Zariski topology. He was a professor of mathematics at Harvard University and made Harvard a world center for algebraic geometry and formed the basis for its twentieth century development.

C.1   Additional Reading

We refer the reader to the books (Adhikari and Adhikari 2003, 2004; Bell 1962; Birkoff and Mac Lane 2003; Hazewinkel et al. 2011; van der Waerden 1960) for further details.

References

Adhikari, M.R., Adhikari, A.: Groups, Rings and Modules with Applications, 2nd edn. Universities Press, Hyderabad (2003)

Adhikari, M.R., Adhikari, A.: Text Book of Linear Algebra: An Introduction to Modern Algebra. Allied Publishers, New Delhi (2004)

Bell, E.T.: Men of Mathematics. Simon and Schuster, New York (1962)

Birkhoff, G., Mac Lane, S.: A Survey of Modern Algebra. Universities Press, Andhra Pradesh (2003)

Hazewinkel, M., Gubareni, N., and Kirichenko, V.V.: Algebras, Rings and Modules, Vol. 1. Springer, New Delhi (2011)

van der Waerden, B.L.: A History of Algebra. Springer, Berlin (1960)

Index

Printed in the United States
By Bookmasters